THIRD EDITION

# VISUALIZING PSYCHOLOGY

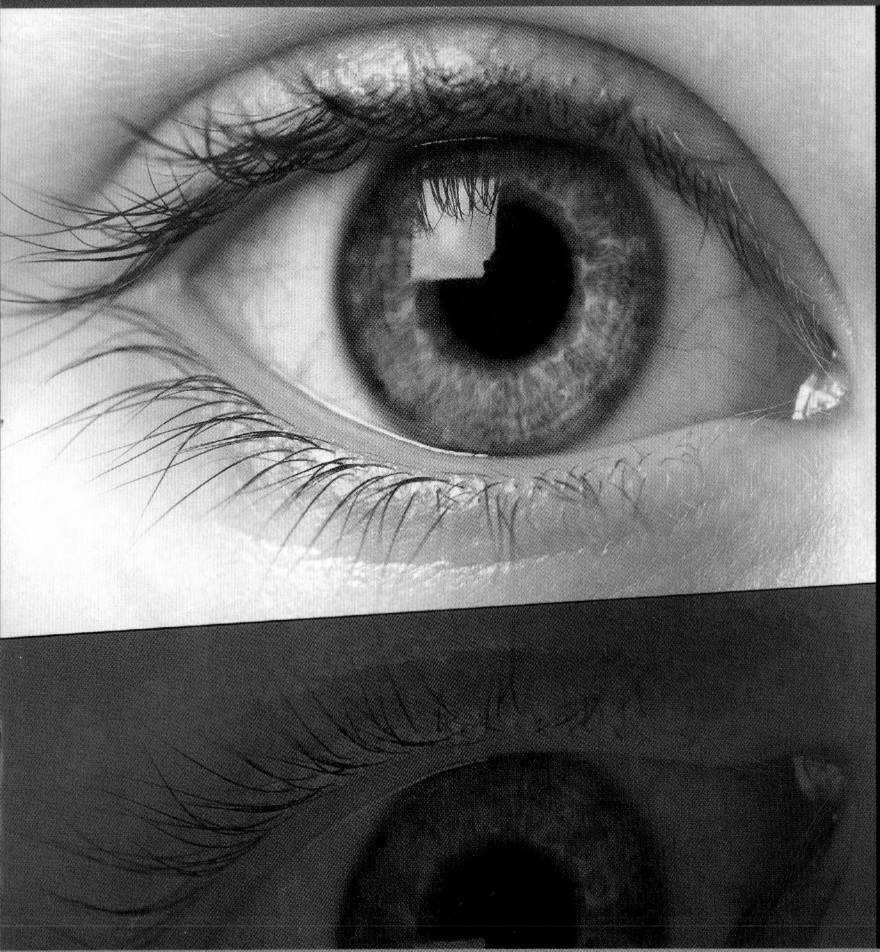
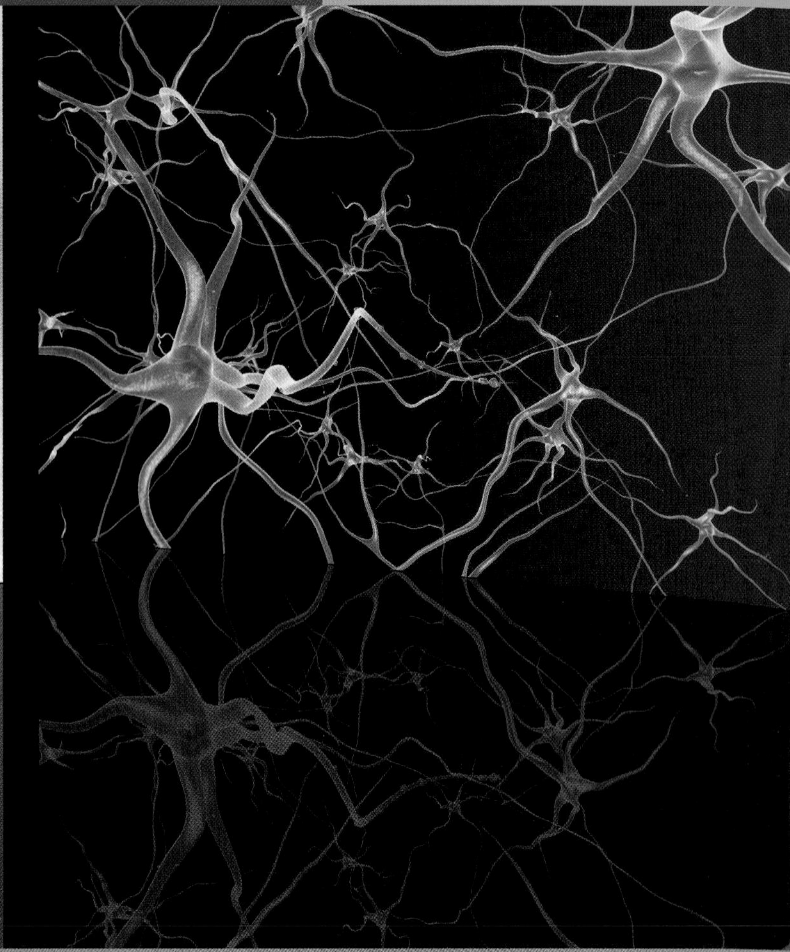

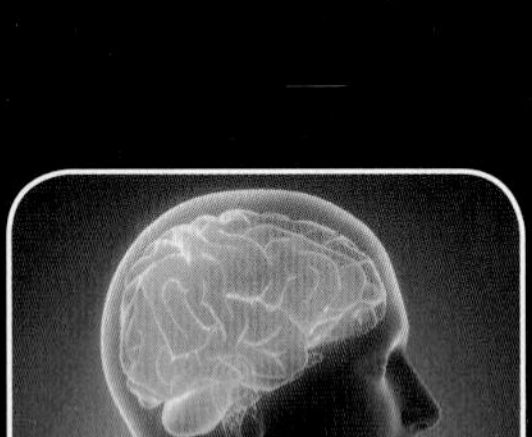

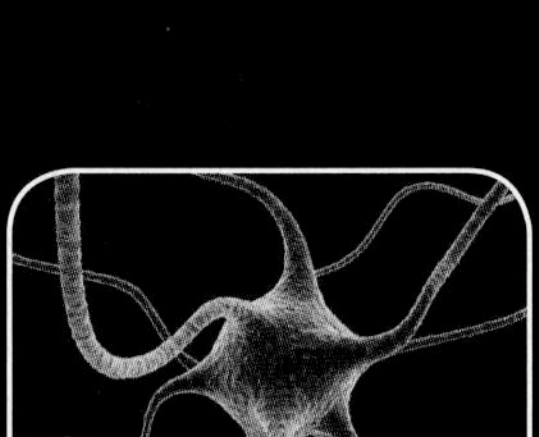

# VISUALIZING
# PSYCHOLOGY

**THIRD EDITION**


Siri Carpenter, Ph.D.

Karen Huffman
Palomar College



WILEY

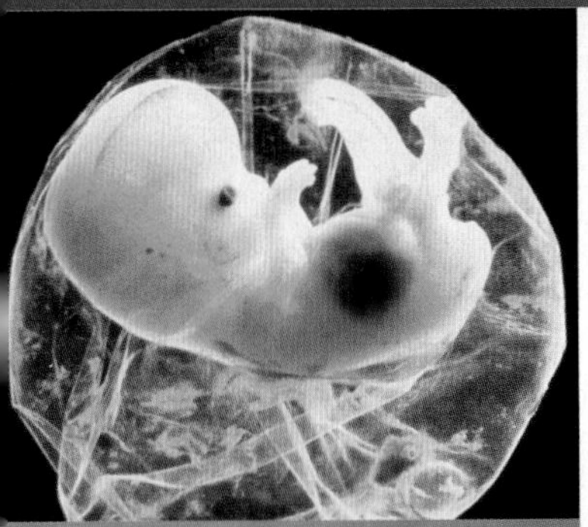

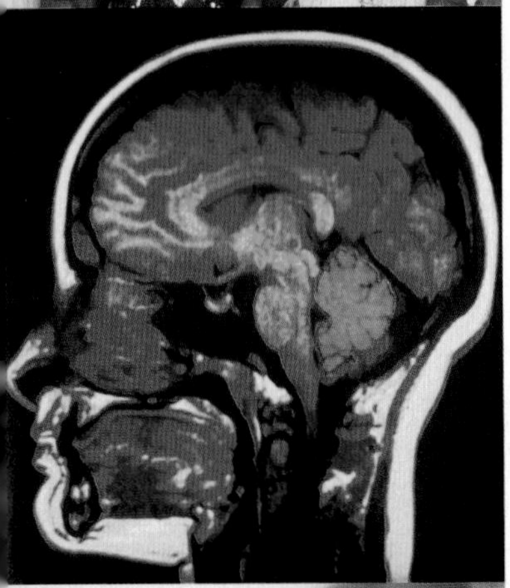

VICE PRESIDENT AND EXECUTIVE PUBLISHER Jay O'Callaghan
EXECUTIVE EDITOR Chris Johnson
DIRECTOR OF DEVELOPMENT Barbara Heaney
MANAGER, PRODUCT DEVELOPMENT Nancy Perry
PRODUCT DESIGNER Beth Tripmacher
EDITORIAL OPERATIONS MANAGER Lynn Cohen
EDITORIAL ASSISTANT Maura Gilligan
ASSISTANT CONTENT EDITOR Darnell Sessoms
ASSISTANT EDITOR Brittany Cheetham
MEDIA SPECIALIST Anita Castro
DIRECTOR, MARKETING COMMUNICATIONS Jeffrey Rucker
SENIOR MARKETING MANAGER Margaret Barrett
SENIOR CONTENT MANAGER Micheline Frederick
SENIOR PRODUCTION EDITOR Janet Foxman
CREATIVE DIRECTOR Harry Nolan
COVER DESIGN Harry Nolan
SENIOR DESIGNER Maureen Eide
SENIOR PHOTO EDITOR Mary Ann Price
PRODUCTION SERVICES Furino Production

COVER CREDITS Main Image: Pasieka/Science Source/Photo Researchers, Inc.
Filmstrip from top to bottom: © Christopher Futcher/iStockphoto, © Jana Blašková/iStockphoto, © Goldmund Lukic/iStockphoto, Laguna Design/Science Source/PhotoResearchers, Inc., © Djura Topalov/iStockphoto

This book was set in New Baskerville by cMPreparé printed and bound by Quebecor World. The cover was printed by Quebecor World.



This book is printed on acid-free paper. ∞

Founded in 1807, John Wiley & Sons, Inc. has been a valued source of knowledge and understanding for more than 200 years, helping people around the world meet their needs and fulfill their aspirations. Our company is built on a foundation of principles that include responsibility to the communities we serve and where we live and work. In 2008, we launched a Corporate Citizenship Initiative, a global effort to address the environmental, social, economic, and ethical challenges we face in our business. Among the issues we are addressing are carbon impact, paper specifications and procurement, ethical conduct within our business and among our vendors, and community and charitable support. For more information, please visit our website: www.wiley.com/go/citizenship.



**Library of Congress Cataloging-in-Publication Data**

Carpenter, Siri.
Visualizing psychology / Siri Carpenter, Karen Huffman. — 3rd ed.
p. cm.
Includes bibliographical references.
ISBN 978-1-118-38806-8 (pbk.)
1. Psychology—Textbooks. I. Huffman, Karen. II. Title.
BF121.C354 2013
150--dc23

2012045284

978-1-118-38806-8 (Main Book ISBN)
978-1-118-44978-3 (Binder-Ready Version ISBN)

Printed in the United States of America

10 9 8 7 6 5 4 3 2 1

# How Is Wiley Visualizing Different?

**Wiley Visualizing** is based on decades of research on the use of visuals in learning (Mayer, 2005).[1] The visuals teach key concepts and are pedagogically designed to **explain**, **present**, and **organize** new information. The figures are tightly integrated with accompanying text; the visuals are conceived with the text in ways that clarify and reinforce major concepts, while allowing students to understand the details. This commitment to distinctive and consistent visual pedagogy sets Wiley Visualizing apart from other textbooks.

The texts offer an array of remarkable photographs, maps, media, and film from photo collections around the world, including those of National Geographic. Wiley Visualizing's images are not decorative; such images can be distracting to students. Instead, they are purposeful and the primary driver of the content. These authentic materials immerse the student in real-life issues and experiences and support thinking, comprehension, and application.

Together these elements deliver a level of rigor in ways that maximize student learning and involvement. Wiley Visualizing has proven to increase student learning through its unique combination of text, photographs, and illustrations, with online video, animations, simulations and assessments.

**(1) Visual Pedagogy.** Using the Cognitive Theory of Multimedia Learning, which is backed up by hundreds of empirical research studies, Wiley's authors create visualizations for their texts that specifically support students' thinking and learning—for example, the selection of relevant materials, the organization of the new information, or the integration of the new knowledge with prior knowledge.

**(2) Authentic Situations and Problems.** *Visualizing Psychology 3e* benefits from National Geographic's more than century-long recording of the world and offers an array of remarkable photographs, maps, media, and film. These authentic materials immerse the student in real-life issues in psychology, thereby enhancing motivation, learning, and retention (Donovan & Bransford, 2005).[2]

**(3) Designed with Interactive Multimedia.** *Visualizing Psychology* is tightly integrated with *WileyPLUS*, our online learning environment that provides interactive multimedia activities in which learners can actively engage with the materials. The combination of textbook and *WileyPLUS* provides learners with multiple entry points to the content, giving them greater opportunity to explore concepts and assess their understanding as they progress through the course. *WileyPLUS* is a key component of the Wiley Visualizing learning and problem-solving experience, setting it apart from other textbooks whose online component is mere drill-and-practice.

## Wiley Visualizing and the *WileyPLUS* Learning Environment are designed as a natural extension of how we learn

To understand why the Visualizing approach is effective, it is first helpful to understand how we learn.

1. Our brain processes information using two main channels: visual and verbal. Our *working memory* holds information that our minds process as we learn. This "mental workbench" helps us with decisions, problem-solving, and making sense of words and pictures by building verbal and visual models of the information.
2. When the verbal and visual models of corresponding information are integrated in working memory, we form more comprehensive, lasting, mental models.
3. When we link these integrated mental models to our prior knowledge, stored in our *long-term memory*, we build even stronger mental models. When an integrated (visual plus verbal) mental model is formed and stored in long-term memory, real learning begins.

The effort our brains put forth to make sense of instructional information is called *cognitive load*. There are two kinds of cognitive load: productive cognitive load, such as when we're engaged in learning or exert positive effort to create mental models; and unproductive cognitive load, which occurs when the brain is trying to make sense of needlessly complex content or when information is not presented well. The learning process can be impaired when the information to be processed exceeds the capacity of working memory. Well-designed visuals and text with effective pedagogical guidance can reduce the unproductive cognitive load in our working memory.

[1] Mayer, R.E. (Ed) (2005). *The Cambridge Handbook of Multimedia Learning.* Cambridge University Press.
[2] Donovan, M.S., & Bransford, J. (Eds.) (2005). *How Students Learn: Science in the Classroom.* The National Academy Press. Available at http://www.nap.edu/openbook.php?record_id=11102&page=1

# Wiley Visualizing is designed for engaging and effective learning

The visuals and text in *Visualizing Psychology 3e* are specially integrated to present complex processes in clear steps and with clear representations, organize related pieces of information, and integrate related information with one another. This approach, along with the use of interactive multimedia, minimizes unproductive cognitive load and helps students engage with the content. When students are engaged, they're reading and learning, which can lead to greater knowledge and academic success.

Research shows that well-designed visuals, integrated with comprehensive text, can improve the efficiency with which a learner processes information. In this regard, SEG Research, an independent research firm, conducted a national, multisite study evaluating the effectiveness of Wiley Visualizing. Its findings indicate that students using Wiley Visualizing products (both print and multimedia) were more engaged in the course, exhibited greater retention throughout the course, and made significantly greater gains in content area knowledge and skills, as compared to students in similar classes that did not use Wiley Visualizing.[3]

The use of *WileyPLUS* can also increase learning. According to a white paper titled "Leveraging Blended Learning for More Effective Course Management and Enhanced Student Outcomes" by Peggy Wyllie of Evince Market Research & Communications, studies show that effective use of online resources can increase learning outcomes. Pairing supportive online resources with face-to-face instruction can help students to learn and reflect on material, and deploying multimodal learning methods can help students to engage with the material and retain their acquired knowledge.

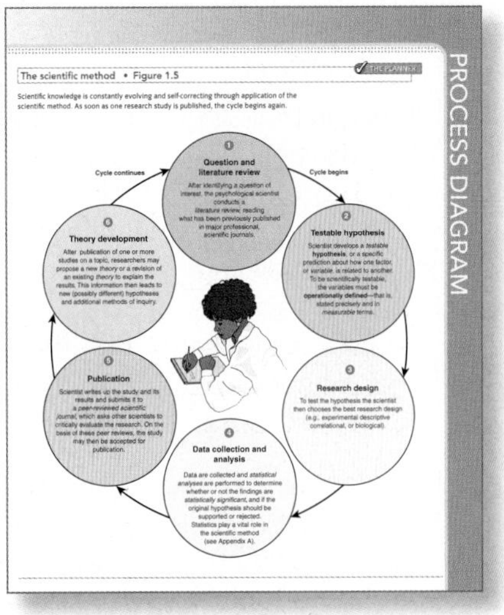


**The scientific method (Figure 1.5)** A logical progression of visuals and graphic features directs learners' attention to the underlying concept. The arrows visually display processes, helping students recognize relationships.

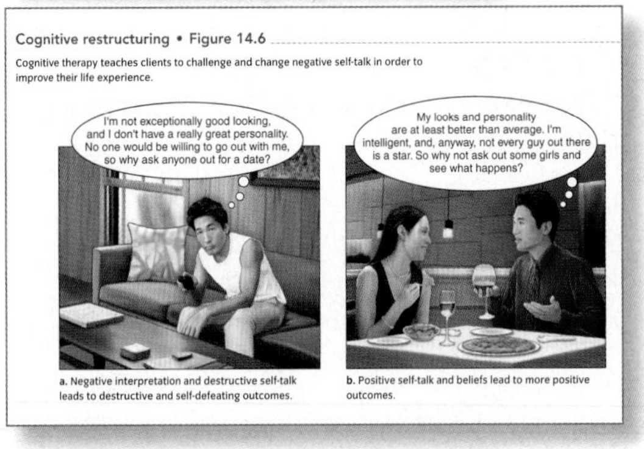


**Cognitive restructuring (Figure 14.6)** Images are paired so that students can compare and contrast them, thereby grasping the underlying concept. Adjacent captions eliminate split attention.

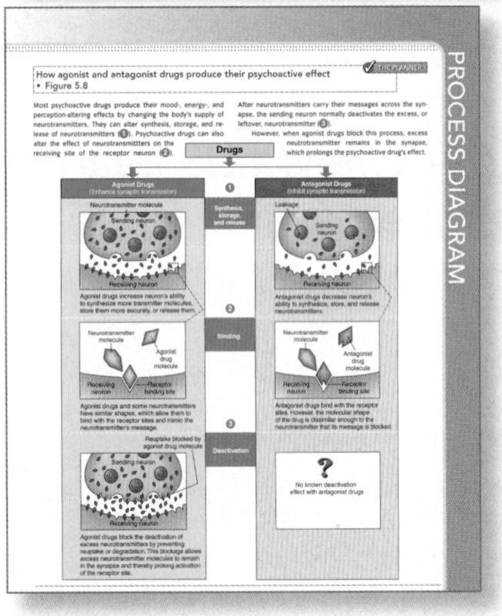


**How agonist and antagonist drugs produce their psychoactive effect (Figure 5.8)** Textual and visual elements are physically integrated. This eliminates split attention (when we must divide our attention between several sources of different information).

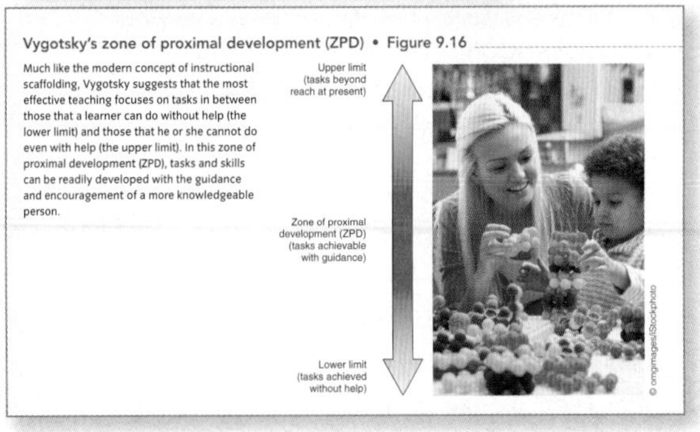


**Vygotsky's zone of proximal development (ZPD) (Figure 9.16)** This matrix visually organizes abstract information to reduce cognitive load.

[3]SEG Research (2009). Improving Student-Learning with Graphically-Enhanced Textbooks: A study of the Effectiveness of the Wiley Visualizing Series.

# How Are the Wiley Visualizing Chapters Organized?

Student engagement is more than just exciting videos or interesting animations—engagement means keeping students motivated to keep going. It is easy to get bored or lose focus when presented with large amounts of information, and it is easy to lose motivation when the relevance of the information is unclear. The design of *WileyPLUS* is based on cognitive science, instructional design, and extensive research into user experience. It transforms learning into an interactive, engaging, and outcomes-oriented experience for students.

## Each Wiley Visualizing chapter engages students from the start

Chapter opening text and visuals introduce the subject and connect the student with the material that follows.

**Chapter Introductions** illustrate key concepts in the chapter with intriguing stories and striking photographs.

**Chapter outlines** anticipate the content.

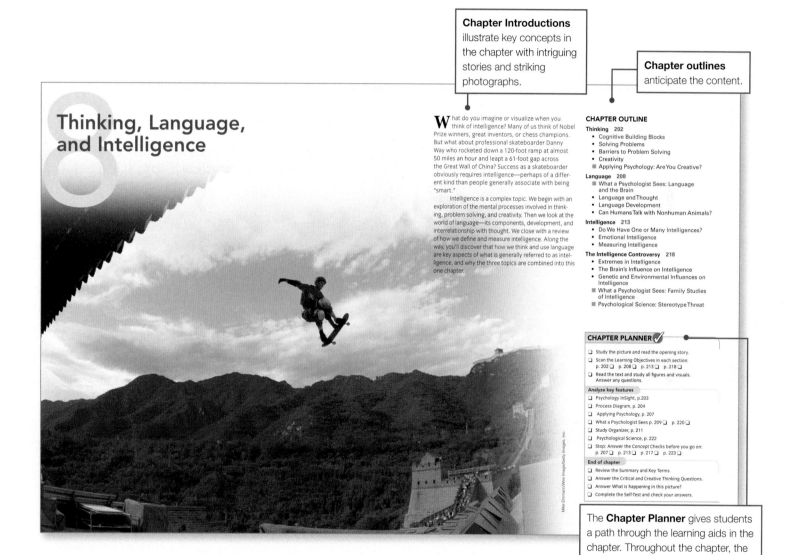

8 Thinking, Language, and Intelligence

What do you imagine or visualize when you think of intelligence? Many of us think of Nobel Prize winners, great inventors, or chess champions. But what about professional skateboarder Danny Way who rocketed down a 120-foot ramp at almost 50 miles an hour and leapt a 61-foot gap across the Great Wall of China? Success as a skateboarder obviously requires intelligence—perhaps of a different kind than people generally associate with being "smart."

Intelligence is a complex topic. We begin with an exploration of the mental processes involved in thinking, problem solving, and creativity. Then we look at the world of language—its components, development, and interrelationship with thought. We close with a review of how we define and measure intelligence. Along the way, you'll discover that how we think and use language are key aspects of what is generally referred to as intelligence, and why the three topics are combined into this one chapter.

Mike Ehrmann/Wire Image/Getty Images, Inc.

CHAPTER OUTLINE



CHAPTER PLANNER

- ❑ Study the picture and read the opening story.
- ❑ Scan the Learning Objectives in each section: p. 202 ❑ p. 208 ❑ p. 213 ❑ p. 218 ❑
- ❑ Read the text and study all figures and visuals. Answer any questions.

Analyze key features
- ❑ Psychology InSight, p. 203
- ❑ Process Diagram, p. 204
- ❑ Applying Psychology, p. 207
- ❑ What a Psychologist Sees p. 209 ❑ p. 220 ❑
- ❑ Study Organizer, p. 211
- ❑ Psychological Science, p. 222
- ❑ Stop: Answer the Concept Checks before you go on: p. 207 ❑ p. 213 ❑ p. 217 ❑ p. 223 ❑

End of chapter
- ❑ Review the Summary and Key Terms.
- ❑ Answer the Critical and Creative Thinking Questions.
- ❑ Answer What is happening in this picture?
- ❑ Complete the Self-Test and check your answers.

The **Chapter Planner** gives students a path through the learning aids in the chapter. Throughout the chapter, the Planner icon prompts students to use the learning aids and to set priorities as they study.

**WileyPLUS** Experience the chapter through a *WileyPLUS* course.

# Wiley Visualizing guides students through the chapter

The content of Wiley Visualizing gives students a variety of approaches—visuals, words, interactions, videos, and assessments—that work together to provide a guided path through the content.

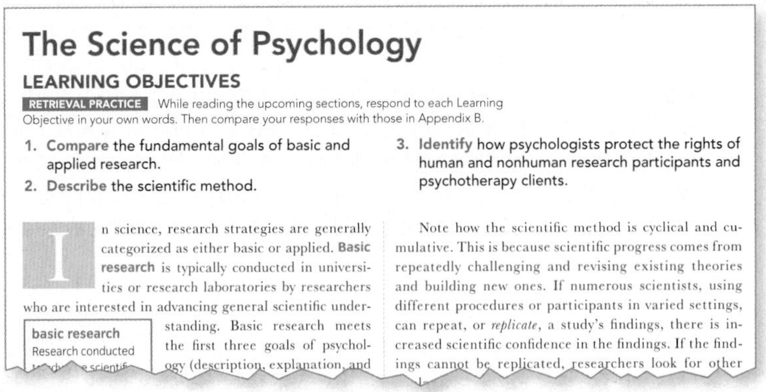

## The Science of Psychology

**LEARNING OBJECTIVES**

RETRIEVAL PRACTICE While reading the upcoming sections, respond to each Learning Objective in your own words. Then compare your responses with those in Appendix B.

1. **Compare** the fundamental goals of basic and applied research.
2. **Describe** the scientific method.
3. **Identify** how psychologists protect the rights of human and nonhuman research participants and psychotherapy clients.

In science, research strategies are generally categorized as either basic or applied. **Basic research** is typically conducted in universities or research laboratories by researchers who are interested in advancing general scientific understanding. Basic research meets the first three goals of psychology (description, explanation, and

**basic research** Research conducted

Note how the scientific method is cyclical and cumulative. This is because scientific progress comes from repeatedly challenging and revising existing theories and building new ones. If numerous scientists, using different procedures or participants in varied settings, can repeat, or *replicate*, a study's findings, there is increased scientific confidence in the findings. If the findings cannot be replicated, researchers look for other

**Learning Objectives** at the start of each section indicate in behavioral terms the concepts that students are expected to master while reading the section.

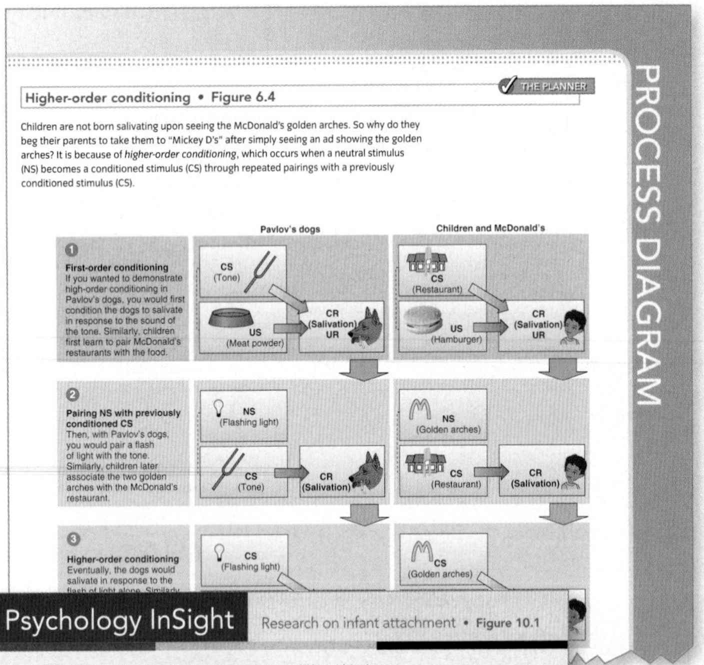

PROCESS DIAGRAM

THE PLANNER

**Higher-order conditioning • Figure 6.4**

Children are not born salivating upon seeing the McDonald's golden arches. So why do they beg their parents to take them to "Mickey D's" after simply seeing an ad showing the golden arches? It is because of *higher-order conditioning*, which occurs when a neutral stimulus (NS) becomes a conditioned stimulus (CS) through repeated pairings with a previously conditioned stimulus (CS).

Pavlov's dogs
Children and McDonald's
1 First-order conditioning
If you wanted to demonstrate high-order conditioning in Pavlov's dogs, you would first condition the dogs to salivate in response to the sound of the tone. Similarly, children first learn to pair McDonald's restaurants with the food.
CS (Tone)
US (Meat powder)
CR (Salivation) UR
CS (Restaurant)
US (Hamburger)
CR (Salivation) UR
2 Pairing NS with previously conditioned CS
Then, with Pavlov's dogs, you would pair a flash of light with the tone. Similarly, children later associate the two golden arches with the McDonald's restaurant.
NS (Flashing light)
CS (Tone)
CR (Salivation)
NS (Golden arches)
CS (Restaurant)
CR (Salivation)
3 Higher-order conditioning
Eventually, the dogs would salivate in response to the
CS (Flashing light)
CS (Golden arches)

**Process Diagrams** provide in-depth coverage of processes correlated with clear, step-by-step narrative, enabling students to grasp important topics with less effort.

Often in the form of a true/false quiz, **Myth Busters** challenges students to identify and correct common misconceptions about the topics in each chapter.

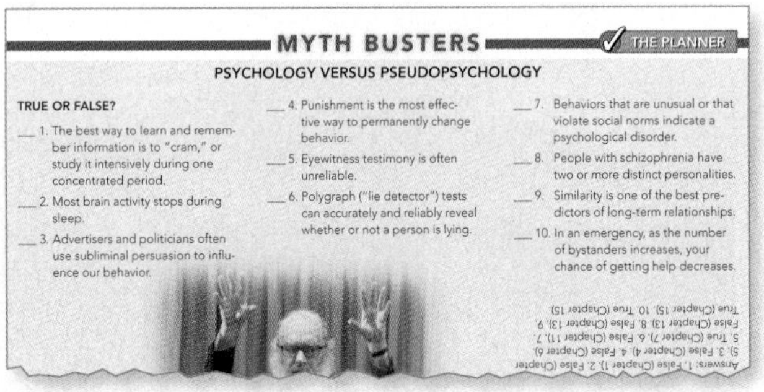

**MYTH BUSTERS** THE PLANNER

PSYCHOLOGY VERSUS PSEUDOPSYCHOLOGY

TRUE OR FALSE?

___ 1. The best way to learn and remember information is to "cram," or study it intensively during one concentrated period.

___ 2. Most brain activity stops during sleep.

___ 3. Advertisers and politicians often use subliminal persuasion to influence our behavior.

___ 4. Punishment is the most effective way to permanently change behavior.

___ 5. Eyewitness testimony is often unreliable.

___ 6. Polygraph ("lie detector") tests can accurately and reliably reveal whether or not a person is lying.

___ 7. Behaviors that are unusual or that violate social norms indicate a psychological disorder.

___ 8. People with schizophrenia have two or more distinct personalities.

___ 9. Similarity is one of the best predictors of long-term relationships.

___ 10. In an emergency, as the number of bystanders increases, your chance of getting help decreases.

Answers: 1. False (Chapter 1); 2. False (Chapter 5); 3. False (Chapter 6); 4. False (Chapter 6); 5. True (Chapter 7); 6. False (Chapter 11); 7. False (Chapter 13); 8. False (Chapter 13); 9. True (Chapter 15); 10. True (Chapter 15)

**Psychology InSight** Research on infant attachment • Figure 10.1

For most children, parents are the earliest and most important factor in a child's social development, and the attachment between parent and child is of particular interest to developmental psychologists.

**a. Strange situation procedure**
Mary Ainsworth and her colleagues (1967, 1978, 2010) found significant differences in the typical levels of attachment between infants and their mothers using a technique called the **strange situation procedure**, in which they observed how infants responded to the presence or absence of their mother and a stranger.

1. The baby plays while the mother is nearby.
2. A stranger enters the room, speaks to the mother, and approaches the child.
3. The mother leaves and the stranger stays in the room with an unhappy baby.
4. The mother returns and the stranger leaves.
5. The baby is reunited with the mother.

**b. Degrees of attachment**
Using the strange situation procedure, Ainsworth found that children could be divided into three groups: *Secure, anxious/avoidant*, and *anxious/ambivalent*. A later psychologist, Mary Main, added a fourth category, *disorganized/disoriented* (Main & Solomon, 1986, 1990).

Secure attachment
60%
Disorganized/disoriented attachment
15%
15%
10%
Anxious/ambivalent attachment
Anxious/avoidant attachment
Banana Stock/Age Fotostock

| |
|---|
| **Secure** Seeks closeness with mother when stranger enters. Uses her as a safe base from which to explore, shows moderate distress on separation from her, and is happy when she returns. |
| **Anxious/ambivalent** Infant becomes very upset when mother leaves the room and shows mixed emotions when she returns. |
| **Anxious/avoidant** Infant does not seek closeness or contact with the mother and shows little emotion when the mother departs or returns. |
| **Disorganized/disoriented** Infant exhibits avoidant or ambivalent attachment, often seeming either confused or apprehensive in the presence of the mother. |

**Psychology InSight** features are multipart visual sections that focus on a key concept or topic in the chapter, exploring it in detail or in broader context using a combination of photos, diagrams, and data.

**Applying Psychology** helps students relate psychological concepts to their own lives and understand how these concepts are applied in various sectors of society, such as the workplace.

## Applying Psychology

THE PLANNER

### Color Aftereffects

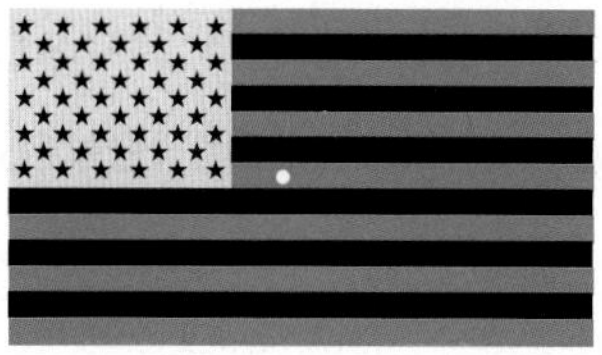

Try staring at the dot in the middle of this color-distorted American flag for 60 seconds. Then stare at a plain sheet of white paper. You should get interesting color aftereffects—red in place of green, blue in place of yellow, and white in place of black: a "genuine" American flag.

What happened? This is a good example of the opponent-process theory. As you stared at the green, black, and yellow colors, the neural systems that process those colors became fatigued. Then when you looked at the plain white paper, which reflects all wavelengths, a reverse opponent process occurred. Each fatigued receptor responded with its opposing red, white, and blue colors!

**Think Critically**

In what kinds of situations do you think color aftereffects are likely to occur?

**Think Critically** questions let students analyze the material and develop insights into essential concepts.

## Psychological Science

THE PLANNER

### Dream Variations and Similarities

What do people commonly dream about? Do men and women dream about different things? Are there differences between cultures in dream content? Scientists have discovered several interesting answers to these questions.

**a. Common dream themes**
A quick glance at this table shows that most people, at least in the Western world, dream a lot about being chased, sex, and misfortune.

| Rank | Dream content | Total prevalence |
|---|---|---|
| 1 | Chased or pursued, not physically injured | 81.5 |
| 2 | Sexual experiences | 76.5 |
| 3 | Falling | 73.8 |
| 4 | School, teachers, studying | 67.1 |
| 5 | Arriving too late, e.g., missing a train | 59.5 |
| 6 | Being on the verge of falling | 57.7 |
| 7 | Trying again and again to do something | 53.5 |
| 8 | A person now alive is dead | 54.1 |
| 9 | Flying or soaring through the air | 48.3 |
| 10 | Vividly sensing . . . a presence in the room | 48.3 |
| 11 | Failing an examination | 45.0 |
| 12 | Physically attacked (beaten, stabbed, raped) | 42.4 |

The data shown here are from a study of 1,181 Canadian college students (Nielsen et al., 2003). Total prevalence refers to the percentage of students reporting each dream.

*Source:* Nielsen, T. A., Zadra, A. L., Simard, V., Saucier, S., Stenstrom, P., Smith, C., & Kuiken, D. (2003). The typical dreams of Canadian university students. Dreaming, 13, 211–235. Copyright © 2003 Association for the Study of Dreams. [from Table 1, p. 217].

Best View Stock /Getty Images, Inc.

**b. Gender differences and similarities**
Men and women tend to share many common dream themes. But women are more likely to dream of children, family and familiar people, household objects, and indoor events, whereas men tend to dream more about strangers, violence, weapons, sexual activity, achievement, and outdoor events (Blume-Marcovici, 2010; Domhoff, 2003, 2007, 2010; Schredl, 2012).

**c. Culture and dreams**
Dreams involving basic human needs and fears (like sex, aggression, and death) seem to be found in all cultures. And children around the world often dream about large, threatening wild animals. In addition, dreams around the world typically include more misfortune than good fortune, and the dreamer is more often the victim of aggression than the cause of it (Domhoff, 2003, 2007, 2010; Domhoff & Schneider, 2008; Hall & Van de Castle, 1996).

iStockphoto

**Identify the Research Method**

1. What is the most likely research method used for the group of studies described above?
2. If you chose
   - the experimental method, label the IV, DV, experimental group, and control group.
   - the descriptive method, is this a naturalistic observation, survey, or case study?
   - the correlational method, is this a positive, negative, or zero correlation?
   - the biological method, identify the specific research tool (e.g., brain dissection, CT scan).

(Check your answers in Appendix C.)

Consciousness, Sleep, and Dreaming 127

**Identify the Research Method** helps reinforce the principles of the scientific method, while also building deeper appreciation and engagement with the latest research in psychology. Answers to these questions appear in Appendix C.

**Psychological Science** emphasizes the empirical, scientific nature of psychology by presenting expanded descriptions of current research findings, along with explanations of their significance and possible applications.

**What a Psychologist Sees** highlights a concept or phenomenon that would stand out to psychologists. Photos and figures are used to improve students' understanding of the usefulness of a psychological perspective and to develop their observational skills.

## WHAT A PSYCHOLOGIST SEES

THE PLANNER

### Enjoying Pizza—A Complex Experience

What happens when we eat pizza? Taste receptor cells, as well as sensory cells that respond to touch and temperature, are activated on our tongues. Information about taste, smell, texture, temperature, and appearance is sent to the brain, where it is integrated in various association regions of the cerebral cortex. These circuits, together with those that store memories related to our previous pizza experiences, work together to produce our perception of this particular slice.

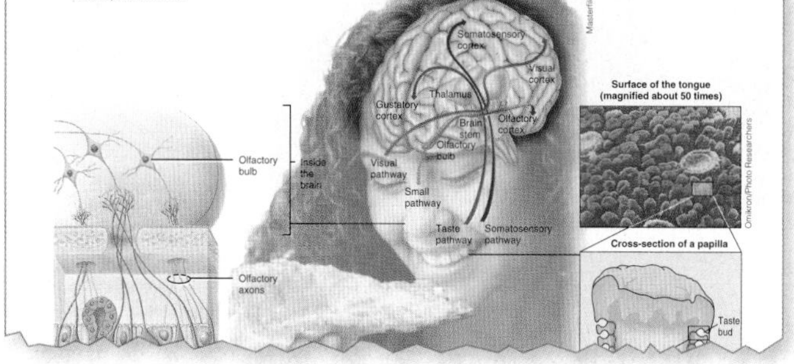


**Study Organizers** present material in a format that makes it easy to compare different aspects of a topic, thus providing students with a useful tool for enhancing their understanding of the topic and preparing for exams.

**Study Organizer 7.1** Types of long-term memory — THE PLANNER

LTM is divided and subdivided into various types of memory.

- **Varieties of long-term memory (LTM)**
  - **Explicit/declarative memory** — Memory with conscious recall
    - **Semantic memory** — Facts and general knowledge (e.g., bananas are yellow, 12 months in a year, spiders have eight legs)
    - **Episodic memory** — Personal experiences and events (e.g., your high school graduation, the birth of your first child)
  - **Implicit/nondeclarative memory** — Memory without conscious recall
    - **Procedural memory** — Motor skills and habits (e.g., how to drive a car, brush your teeth, ride a bike)
    - **Classically conditioned memory** — Conditioned responses to conditioned stimuli (e.g., phobias, some aspects of prejudice, and other attitudes)
    - **Priming** — Earlier exposure facilitates retrieval (e.g., heightened fears after reading a scary novel)

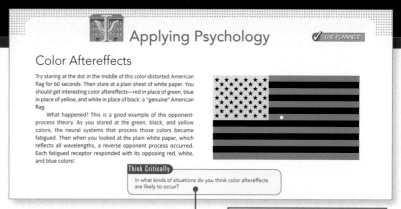

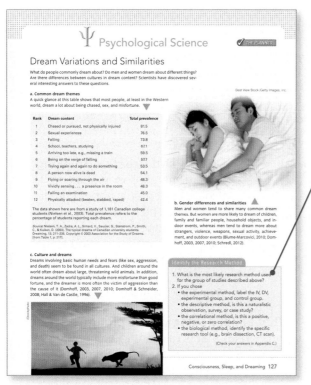

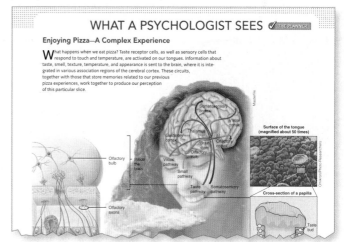

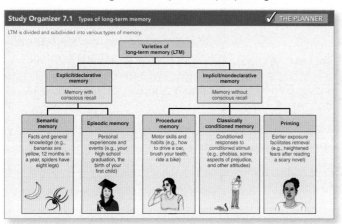

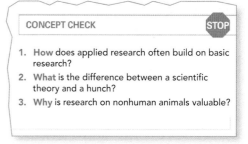

**CONCEPT CHECK** STOP

1. **How** does applied research often build on basic research?
2. **What** is the difference between a scientific theory and a hunch?
3. **Why** is research on nonhuman animals valuable?

Coordinated with the section-opening **Learning Objectives**, at the end of each section **Concept Check** questions allow students to test their comprehension of the learning objectives.

# Student understanding is assessed at different levels

Wiley Visualizing with *WileyPLUS* offers students lots of practice material for assessing their understanding of each study objective. Students know exactly what they are getting out of each study session through immediate feedback and coaching.

The **Summary** revisits each major section, with informative images taken from the chapter. These visuals reinforce important concepts.

## Summary

THE PLANNER

1 **Studying Development 230**

- **Developmental psychology** is the study of age-related changes in behavior and mental processes from conception to death. Development is an ongoing, lifelong process.
- The three most important debates or questions in human development are about *nature versus nurture* (including studies of **maturation** and **critical periods**), *stages versus continuity* (illustrated in the diagram), and *stability versus change*. For each question, most psychologists prefer an interactionist perspective.
- Developmental psychologists use two special techniques in their research: **cross-sectional design** and **longitudinal design**. Although both have valuable attributes, each also has disadvantages. Cross-sectional studies can confuse genuine age differences with **cohort effects**. On the other hand, ...nsive and time-consuming, ...d in generalizability.

**Stages versus continuity in development • Figure 9.2**

## Creative and Critical Thinking Questions

1. If you were forced to lose one type of memory—sensory, short-term, or long-term—which would you select? Why?
2. Why might students do better on a test if they take it in the same seat and classroom where they originally studied the material?
3. What might be the evolutionary benefit of heightened (but not excessive) arousal enhancing memory?
4. Why might advertisers of shoddy services or products benefit from "channel surfing" if the television viewer is skipping from news programs to cable talk shows to infomercials?
5. As an eyewitness to a crime, how could you use information in this chapter to improve your memory for specific details? If you were a juror, what would you say to the other jurors about the reliability of eyewitness testimony?

**Critical and Creative Thinking Questions** challenge students to think more broadly about chapter concepts. The level of these questions ranges from simple to advanced; they encourage students to think critically and develop an analytical understanding of the ideas discussed in the chapter.

## What is happening in this picture?

Most traits are polygenic, meaning they are controlled by more then one gene. These three brothers are demonstrating whether or not they can curl their tongues, one of the few traits that depend on only one dominant gene.

Courtesy Karen Huffman

**Think Critically**

1. Based on the photo, what can you predict about whether one or both of the boys' parents were "noncurlers"?
2. Can you imagine why humans and other animals have evolved to possess complex traits such as tongue curling?

**What is happening in this picture?** presents a photograph that is relevant to a chapter topic and illustrates a situation students are not likely to have encountered previously.

**Think Critically** questions ask students to apply what they have learned in order to interpret and explain what they observe in the image.

## Self-Test

RETRIEVAL PRACTICE Completing this self-test and comparing your answers with those in Appendix C provides immediate feedback and helpful practice for exams. Additional interactive, self-tests are available at www.wiley.com/college/carpenter.

1. The study of age-related changes in behavior and mental processes from conception to death is called ______.
   a. thanatology
   b. neo-gerontology
   c. developmental psychology
   d. longitudinal psychology
2. Development governed by automatic, genetically predetermined signals is called ______.
   a. growth
   b. natural progression
   c. maturation
   d. tabula rasa
3. Label the two basic types of research designs:

a. ____________

Group One
20-year-old participants
Group Two
40-year-old participants
Group Three
60-year-old participants

Research done in 2013

b. ____________

Study One
...

5. As shown in the diagram, at birth, an infant's head is ______ its body's size, whereas in adulthood, the head is ______ its body's size.
   a. 1/3; 1/4
   b. 1/3; 1/10
   c. 1/4; 1/10
   d. 1/4; 1/8

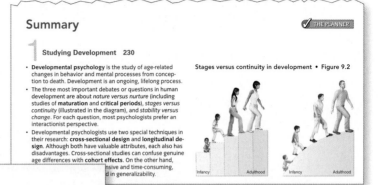

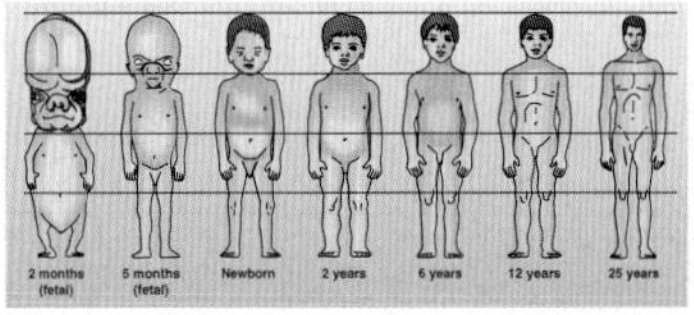


6. Which of the following is NOT true regarding infant sensory and perceptual development?
   a. Vision is almost 20/20 at birth.
   b. A newborn's sense of pain is highly developed at birth.
   c. An infant can recognize, and prefers, its own mother's breast milk by smell.
   d. An infant can recognize, and prefers, its own mother's breast milk by taste.

Visual end-of-chapter **Self-Tests** pose review questions that ask students to demonstrate their understanding of key concepts.

# Why *Visualizing Psychology*?

*The brain is wider than the sky.*
EMILY DICKINSON, 1830–1886

Welcome to the wonderful world of psychology! As poet Emily Dickinson suggests, each one of our human brains is wider than the sky—and so too is the field of psychology. Many students initially believe that psychologists only study and treat abnormal behavior, but as you'll discover throughout this text, the topics, research findings, and interests of psychological scientists are extraordinarily diverse. Neuroscience, stress, health, sensation, perception, states of consciousness, learning, memory, thinking, language, intelligence, lifespan development, motivation, emotion, personality, social psychology, and of course abnormality and therapy are just a few of the areas that we'll be exploring in *Visualizing Psychology, Third Edition*. We (your authors) are honored and pleased to invite you on a fascinating exploration of the complexities and nuances of behavior and mental processes—both human and nonhuman—that make the study of psychology so compelling.

As you might expect, the compelling (and rewarding) nature of psychology has attracted the attention and devotion of literally millions of readers, along with a multitude of psychology books. Why do we need another text? What makes *Visualizing Psychology* unique? Your two authors, and the editors and publisher of this text, all believe that active learning and critical thinking (two synonymous and inseparable terms) are key ingredients to true understanding and lifelong learning. Therefore, we have developed and incorporated a large set of active learning and critical thinking pedagogical tools that will help you, the reader, personally unlock the fascinating mysteries and excitement of psychology. These tools will also teach you how to apply the wealth of insights and knowledge from psychological science to your everyday life. Best of all, active learning and critical thinking can make your study and mastery of psychology easier and more rewarding.

As the name implies, *Visualizing Psychology* also is unique in its focus on visuals. Based in part on the old saying that a "picture is worth a thousand words," this text covers the basic content of a standard psychology text enhanced by an educationally sound and carefully designed visual art program.

Through this premier art program, combined with our strong emphasis on active learning and critical thinking, *Visualizing Psychology* provides readers with a new and innovative approach to the understanding of psychology's major issues, from stem cells to stereotyping. In the context of an engaging visual presentation, we offer solid discussions of critical psychological concepts, ranging from the impact of stress on health to the psychological foundations of prejudice.

This book is intended to serve as a broad overview of the entire field of psychology. Despite its shortened and condensed nature, *Visualizing Psychology*, like most other large survey and general education texts, contains a large number of unfamiliar terms and complex concepts. Do not be dismayed. The language of psychology is new to all but the most seasoned scholars. With a little hard work and concentrated study, you can master this material, and your work will pay off with immediate and unforeseen rewards that can last a lifetime.

As you can see, we feel passionate about psychology and believe that the study of psychology offers all of us an incomparable window into not only ourselves, but also to the world and the people who sustain us. We're eager to share our passion for psychology with you. We also welcome feedback from our readers. Please feel free to contact us at khuffman@palomar.edu.

## Organization

We noted earlier that psychology is a surprisingly diverse and complex field. To organize this diversity, our book is divided into 15 chapters that are arranged in a somewhat "microscopic/telescoping" fashion. We tend to move from the smallest element of behavior (the neuron and neuroscience) out to the largest (the group, culture, and social psychology). Here is a brief summary of the major topics explored in each chapter:

- **Chapter 1** describes psychology's history, its different theoretical perspectives and fundamental questions, and how psychologists go about answering those questions.
- **Chapter 2** explains the neural and other biological bases of behavior, and lays the groundwork for further discussions of biological foundations that appear in later chapters.
- **Chapter 3** examines critically important interactions among stress, health, and behavior.
- **Chapters 4 through 8** present aspects of cognition, including sensation, perception, consciousness, learning, memory, thinking, language, and intelligence. These chapters examine both cognition under healthy circumstances and cases where cognition goes awry. Throughout these discussions, we provide examples and exercises that connect basic research on cognition to real-world situations.
- **Chapters 9 and 10** explore human development across the lifespan, including physical, cognitive, social, moral, and personality development. We have organized these chapters topically, rather than chronologically, to help you appreciate the trajectory that each facet of development takes over the course of a lifetime.
- **Chapters 11 and 12** discuss processes and qualities that are integral to our most basic experiences and interactions with one another: motivation, emotion, and personality.

- **Chapter 13** addresses five major categories of psychological disorders. But first, we begin by discussing what constitutes "abnormal behavior" and how psychological disorders are identified and classified. We also explore how psychological disorders vary across cultures.
- **Chapter 14** describes and evaluates major forms of therapy, organizing the most widely used treatments into three groups: talk therapies, behavior therapies, and biomedical therapies.
- **Chapter 15**, covering social psychology, is in some ways the culmination of all the previous chapters, as there is no aspect of psychology that is irrelevant to how we think about, feel about, and act toward others. In this final chapter, we explore a range of social psychological phenomena, ranging from perceptions of others' intentions, to romantic attractions, to prejudice and discrimination.

## New to this edition

This Third Edition of *Visualizing Psychology* is dedicated to further enhancing your student learning experience through several new and unique features, including:

- **New *Psychological Science* features** *Visualizing Psychology* has always emphasized the empirical, scientific nature of psychology. This edition offers expanded descriptions of current research findings, explanations of their significance, and applications (e.g., Chapter 1, "Serious Problems with Multitasking," Chapter 10, "The Power of Resilience").
- **New *Identify the Research Method* questions** Given that the scientific method (and its various components) is one of the most common learning objectives in all of psychology, we believe students need repeated practice applying these concepts—beyond just the basic introduction traditionally provided in Chapter 1. To provide this practice, each of the *Psychological Science* features is followed by questions that prompt the reader to identify the research method, IV, DV, and so on. These interactive, self-testing activities help reinforce the core learning objective for the scientific method, while also building deeper appreciation and engagement with the latest research in psychology.
- **New *Myth Busters*** Why focus on myths? No one wants to be embarrassed by misinformation, and this natural desire to avoid being wrong not only increases student engagement, but it also has a significant educational side benefit. While studying the myths of psychology, students automatically and easily learn some of the most important terms and concepts—along with improving their critical thinking skills.
- **New increased emphasis on assessment** In order to meet the growing demand in higher education to show "results," new *Retrieval Practice* reminders appear with the Learning Objectives, Key Terms, and Self-Tests. Sample answers to the learning objectives now appear in Appendix B.
- **New *Psychology InSight* features** These features are specially designed multipart visual spreads that focus on a key concept or topic in the chapter, exploring it in detail or in broader context using a combination of photos and figures.
- **Enhanced visuals throughout the text** Photos, figures, diagrams, and other illustrations have been carefully examined and revised to increase their diversity and overall effectiveness as aids to learning.
- **Expanded coverage of important topics** This edition offers new or expanded discussions of topics such as sources of stress, positive reinforcement, mirror neurons, the misinformation effect, divergent thinking, the personal fable, parenting styles, nonverbal communication of emotion, the sharing of delusions on the Internet, and recent findings in support of psychoanalysis.
- **Updated *Applying Psychology* features** This feature helps students relate psychological concepts to their own lives and understand how these concepts are applied in various sectors of society, such as the workplace.
- **More opportunities for critical thinking** Each *Applying Psychology* box includes *Think Critically* questions designed to encourage students to critically evaluate the topic of the box. Many figure captions also include critical thinking questions to further enhance student comprehension and critical thinking skills.
- **Enhanced study aids** The carefully developed *Study Organizers* make it easy to compare different aspects of a topic, while also providing students with a useful tool for enhancing their understanding of the topic and preparing for exams. Among the topics treated in this way are the major psychological perspectives, properties of vision and hearing, schedules of reinforcement, stages of language development, parenting styles, and defense mechanisms.

# How Does Wiley Visualizing Support Instructors?

## Showcase Site

The Wiley Visualizing site hosts a wealth of information for instructors using Wiley Visualizing, including ways to maximize the visual approach in the classroom and a white paper titled "How Visuals Can Help Students Learn," by Matt Leavitt, instructional design consultant. Visit Wiley Visualizing at www.wiley.com/college/visualizing.

## WileyPLUS

This online teaching and learning environment integrates the entire digital textbook with the most effective instructor and student resources to fit every learning style. With *WileyPLUS*:

- Students achieve concept mastery in a rich, structured environment that's available 24/7.
- Instructors personalize and manage their course more effectively with assessment, assignments, grade tracking, and more.

*WileyPLUS* can be used with or in place of the textbook.

## Wiley Custom Select

Wiley Custom Select gives you the freedom to build your course materials exactly the way you want them. Offer your students a cost-efficient alternative to traditional texts. In a simple three-step process create a solution containing the content you want, in the sequence you want, delivered how you want. Visit Wiley Custom Select at http://customselect.wiley.com.

## Videos

More than 30 streaming videos from National Geographic's award-winning collection are available to students in the context of *WileyPLUS*. Below is a brief description of the videos available for each chapter.

**Chapter 1 Introduction and Research Methods**

1. *Among Wild Chimpanzees (3:48)* A young Jane Goodall speaks about her work in the wilds of Africa with primates.
2. *What Is Psychology? (0:54)* What makes us act the way we do? Psychology explores individual differences.

**Chapter 2 Neuroscience and Biological Foundations**

3. *Brain Surgery (4:33)* Brain surgery is performed on a young man's tumor while he is awake.
4. *Cool Quest (3:59)* MRIs map the activity of the brain, exposing "cool" and "uncool" images.
5. *Brain Bank (3:08)* The Harvard Brain Tissue Resource Center, known as the "Brain Bank," is the largest brain repository in the world.
6. *Brain Tumor Surgery (3:14)* A patient suffering from seizures discovers he has a massive brain tumor near the part of the brain that controls motor activity.
7. *MRI (0:38)* An actual patient undergoing an MRI, showing the various images that the MRI produces.

**Chapter 3 Stress and Health Psychology**

8. *Science of Stress (3:31)* How stress affects the body.

**Chapter 4 Sensation and Perception**

9. *Eye Trick Town (2:35)* In Italy, trompe l'oeil paintings that "trick" the eye into a perception of depth.
10. *Camels (1:25)* Photography of camels walking in the desert presents an interesting exploration into the relative nature of sensation and perception.

**Chapter 5 States of Consciousness**

11. *Sleep Walking (1:57)* The video suggests that during slow wave, non-REM sleep some people's lower part of the brain wakes up while the upper part of the brain responsible for awareness stays asleep.
12. *Bali: Trance (3:24)* Highlights of a festival in Bali where villagers come close to stabbing themselves while in a trance state.
13. *Peyote and the Huichol People (4:17)* The Huichol people ingest peyote, a mind-altering drug to enter the spirit world.

**Chapter 6 Learning**

14. *Animal Minds (1:15)* Rats are able to learn their way through a maze implying they may have a cognitive map.
15. *Thai Monkey (2:11)* Monkeys are taught how to retrieve coconuts through an official monkey training school, using both operative conditioning and modeling.

**Chapter 7 Memory**

16. *Taxi Drivers (4:06)* Video on the role that the hippocampus plays in consolidating memories, suggesting that there is a structural change to the brains of London taxi drivers.

**Chapter 8 Thinking, Language, Intelligence**

17. *Orangutan Language (3:23)* The orangutan language project at the National Zoo provides a stimulating environment where they learn a vocabulary of symbols and construct simple sentences.

**Chapter 9 Lifespan Development I**

18. *Feral Children (7:13)* A Western researcher studies a young boy found in the jungle of Uganda, Africa, living with wild monkeys, rescued, and raised by a couple.

**Chapter 10 Lifespan Development II**

**19.** *Taboo Childhood (2:04)* India allows imprisoned mothers to keep their children with them in jail—a very different value system than our Western view in the rearing of children.

**20.** *Taboo Sexuality: Eunuchs (3:53)* Highlights of a group of eunuchs in their struggle to make a life in the Indian culture where they are viewed as social outcasts.

**21.** *Coming of Age Rituals (4:38)* The coming-of-age rituals in the Fulani tribe where two young boys have a whipping match, while a girl is given a full face of tattoos.

**Chapter 11 Motivation and Emotion**

**22.** *Rodeo Clowns (2:28)* A window into the world of a dangerous profession where professional rodeo clown bullfighters risk life and limb to entertain.

**23.** *Heidi Howkins (3:22)* Howkins's climb of K2, the world's second tallest mountain, risking everything to become the first woman to do so.

**24.** *Fire Fighter Training (3:45)* Dedicated smokejumpers meeting the challenges season after season, demonstrating not only bravery but strong motivation.

**Chapter 12 Personality**

**25.** *Freud (2:31)* The life and times of Freud are highlighted, and some of the underlying tenets of psychoanalysis are explained.

**Chapter 13 Psychological Disorders**

**26.** *Philippines: Exorcism (5:03)* The exorcism of a teenage boy and a possible alternative explanation for his behavior are depicted.

**27.** *Mars Desert Research (3:08)* The Mars Desert Research Station program simulates conditions of a space mission to Mars, highlighting the psychological aspects of crew selection.

**28.** *Odd Festivals Around the World (3:04)* While these festivals in England, Thailand, and Italy may seem odd, they demonstrate the importance of cultural sensitivity.

**29.** *Hindu Festival (5:20)* Two brothers and other pilgrims proceed from Kuala Lumpur to the Batu Caves where they pierce their flesh with spikes and hooks, showing the importance of cultural sensitivity.

**Chapter 14 Therapy**

**30.** *What Is Psychology: Therapy? (2:28)* A clinical psychologist and a child psychiatrist are interviewed about the work they perform.

**31.** *Leeches for Curing Illness (2:46)* A look at the use of leeches in curing illness—how might this treatment be used in more mainstream medicine?

**Chapter 15 Social Psychology**

**32.** *Teeth Chiseling (3:48)* In Indonesia, a tribal chieftain's wife undergoes teeth chiseling to enhance her beauty—a matter of balance between the soul and body.

**33.** *Leg Stretching (3:31)* In China, a doctor performs leg-stretching operations to increase the height of patients, considering the cultural and psychological aspects, and the risks and benefits.

**34.** *The Maroons of Jamaica (1:56)* The story of the Maroons of Jamaica, winning independence from the British, and what their lives are like today.

## Book Companion Site

www.wiley.com/college/carpenter

All instructor resources (the Test Bank, Instructor's Manual, PowerPoint presentations, and all textbook illustrations and photos in jpeg format) are housed on the book companion site (www.wiley.com/college/carpenter). Student resources include self-quizzes and flashcards.

## PowerPoint Presentations

(available in *WileyPLUS* and on the book companion site)

A complete set of highly visual PowerPoint presentations—one per chapter—by Katie Townsend-Merino of Palomar College, is available online and in *WileyPLUS* to enhance classroom presentations. Tailored to the text's topical coverage and learning objectives, these presentations are designed to convey key text concepts, illustrated by embedded text art.

## Test Bank

(available in *WileyPLUS* and on the book companion site)

The visuals from the textbook are also included in the Test Bank by Melissa Acevedo of Westchester Community College. The Test Bank has a diverse selection of approximately 1200 test items including multiple-choice and essay questions, with at least 25% of them incorporating visuals from the book. The test bank is available online in MS Word files as a Computerized Test Bank, and within *WileyPLUS*. The easy-to-use test-generation program fully supports graphics, print tests, student answer sheets, and answer keys. The software's advanced features allow you to produce an exam to your exact specifications.

## Instructor's Manual

(available in *WileyPLUS* and on the book companion site)

The Instructor's Manual begins with the special introduction *Using Visuals in the Classroom*, prepared by Matthew Leavitt of the Arizona State University, in which he provides guidelines and suggestions on how to use the visuals in teaching the course. For each chapter, materials by Lynnel Kiely of the City Colleges of Chicago include suggestions and directions for using Web-based learning modules in the classroom and for homework assignments, as well as creative ideas for in-class activities.

Guidance is also provided on how to maximize the effectiveness of visuals in the classroom.

1. **Use visuals during class discussions or presentations.** Point out important information as the students look at the visuals, to help them integrate separate visual and verbal mental models.
2. **Use visuals for assignments and to assess learning.** For example, learners could be asked to identify samples of concepts portrayed in visuals.
3. **Use visuals to encourage group activities.** Students can study together, make sense of, discuss, hypothesize, or make decisions about the content. Students can work together to interpret and describe the diagram, or use the diagram to solve problems, conduct related research, or work through a case study activity.
4. **Use visuals during reviews.** Students can review key vocabulary, concepts, principles, processes, and relationships displayed visually. This recall helps link prior knowledge to new information in working memory, building integrated mental models.
5. **Use visuals for assignments and to assess learning.** For example, learners could be asked to identify samples of concepts portrayed in visuals.
6. **Use visuals to apply facts or concepts to realistic situations or examples.** For example, a familiar photograph, such as the Grand Canyon, can illustrate key information about the stratification of rock, linking this new concept to prior knowledge.

### Image Gallery

All photographs, figures, and other visuals from the text are online and in *WileyPLUS* and can be used as you wish in the classroom. These online electronic files allow you to easily incorporate images into your PowerPoint presentations as you choose, or to create your own handouts.

## Wiley Faculty Network

The Wiley Faculty Network (WFN) is a global community of faculty, connected by a passion for teaching and a drive to learn, share, and collaborate. Their mission is to promote the effective use of technology and enrich the teaching experience. Connect with the Wiley Faculty Network to collaborate with your colleagues, find a mentor, attend virtual and live events, and view a wealth of resources all designed to help you grow as an educator. Visit the Wiley Faculty Network at www.wherefacultyconnect.com.

# How Has Wiley Visualizing Been shaped by Contributors?

Wiley Visualizing and the *WileyPLUS* learning environment would not have come about without lots of people, each of whom played a part in sharing their research and contributing to this new approach.

### Academic Research Consultants

**Richard Mayer**, Professor of Psychology, UC Santa Barbara. His *Cognitive Theory of Multimedia Learning* provided the basis on which we designed our program. He continues to provide guidance to our author and editorial teams on how to develop and implement strong, pedagogically effective visuals and use them in the classroom.

**Jan L. Plass**, Professor of Educational Communication and Technology in the Steinhardt School of Culture, Education, and Human Development at New York University. He co-directs the NYU Games for Learning Institute and is the founding director of the CREATE Consortium for Research and Evaluation of Advanced Technology in Education.

**Matthew Leavitt**, Instructional Design Consultant. He advises the Visualizing team on the effective design and use of visuals in instruction and has made virtual and live presentations to university faculty around the country regarding effective design and use of instructional visuals.

### Independent Research Studies

SEG Research, an independent research and assessment firm, conducted a national, multisite effectiveness study of students enrolled in entry-level college Psychology and Geology courses. The study was designed to evaluate the effectiveness of Wiley Visualizing. You can view the full research paper at www.wiley.com/college/visualizing/huffman/efficacy.html.

### Instructor and Student Contributions

Throughout the process of developing the concept of guided visual pedagogy for Wiley Visualizing, we benefited from the comments and constructive criticism provided by the instructors and colleagues listed below. We offer our sincere appreciation to these individuals for their helpful reviews and general feedback.

## Visualizing Reviewers, Focus Group Participants, and Survey Respondents

James Abbott, *Temple University*
Melissa Acevedo, *Westchester Community College*
Shiva Achet, *Roosevelt University*
Denise Addorisio, *Westchester Community College*
Dave Alan, *University of Phoenix*
Sue Allen-Long, *Indiana University Purdue*
Robert Amey, *Bridgewater State College*
Nancy Bain, *Ohio University*
Corinne Balducci, *Westchester Community College*
Steve Barnhart, *Middlesex County Community College*
Stefan Becker, *University of Washington—Oshkosh*
Callan Bentley, *NVCC Annandale*
Valerie Bergeron, *Delaware Technical & Community College*
Andrew Berns, *Milwaukee Area Technical College*
Gregory Bishop, *Orange Coast College*
Rebecca Boger, *Brooklyn College*
Scott Brame, *Clemson University*
Joan Brandt, *Central Piedmont Community College*
Richard Brinn, *Florida International University*
Jim Bruno, *University of Phoenix*
William Chamberlin, *Fullerton College*
Oiyin Pauline Chow, *Harrisburg Area Community College*
Laurie Corey, *Westchester Community College*
Ozeas Costas, *Ohio State University at Mansfield*
Christopher Di Leonardo, *Foothill College*
Dani Ducharme, *Waubonsee Community College*
Mark Eastman, *Diablo Valley College*
Ben Elman, *Baruch College*
Staussa Ervin, *Tarrant County College*
Michael Farabee, *Estrella Mountain Community College*
Laurie Flaherty, *Eastern Washington University*
Susan Fuhr, *Maryville College*
Peter Galvin, *Indiana University at Southeast*
Andrew Getzfeld, *New Jersey City University*
Janet Gingold, *Prince George's Community College*
Donald Glassman, *Des Moines Area Community College*
Richard Goode, *Porterville College*
Peggy Green, *Broward Community College*
Stelian Grigoras, *Northwood University*
Paul Grogger, *University of Colorado*
Michael Hackett, *Westchester Community College*
Duane Hampton, *Western Michigan University*
Thomas Hancock, *Eastern Washington University*
Gregory Harris, *Polk State College*
John Haworth, *Chattanooga State Technical Community College*
James Hayes-Bohanan, *Bridgewater State College*
Peter Ingmire, *San Francisco State University*
Mark Jackson, *Central Connecticut State University*
Heather Jennings, *Mercer County Community College*
Eric Jerde, *Morehead State University*
Jennifer Johnson, *Ferris State University*
Richard Kandus, *Mt. San Jacinto College District*
Christopher Kent, *Spokane Community College*
Gerald Ketterling, *North Dakota State University*
Lynnel Kiely, *Harold Washington College*
Eryn Klosko, *Westchester Community College*
Cary T. Komoto, *University of Wisconsin—Barron County*
John Kupfer, *University of South Carolina*
Nicole Lafleur, *University of Phoenix*
Arthur Lee, *Roane State Community College*
Mary Lynam, *Margrove College*
Heidi Marcum, *Baylor University*
Beth Marshall, *Washington State University*
Dr. Theresa Martin, *Eastern Washington University*
Charles Mason, *Morehead State University*
Susan Massey, *Art Institute of Philadelphia*
Linda McCollum, *Eastern Washington University*
Mary L. Meiners, *San Diego Miramar College*
Shawn Mikulay, *Elgin Community College*
Cassandra Moe, *Century Community College*
Lynn Hanson Mooney, *Art Institute of Charlotte*
Kristy Moreno, *University of Phoenix*
Jacob Napieralski, *University of Michigan—Dearborn*
Gisele Nasar, *Brevard Community College, Cocoa Campus*
Daria Nikitina, *West Chester University*
Robin O'Quinn, *Eastern Washington University*
Richard Orndorff, *Eastern Washington University*
Sharen Orndorff, *Eastern Washington University*
Clair Ossian, *Tarrant County College*
Debra Parish, *North Harris Montgomery Community College District*
Linda Peters, *Holyoke Community College*
Robin Popp, *Chattanooga State Technical Community College*
Michael Priano, *Westchester Community College*
Alan "Paul" Price, *University of Wisconsin—Washington County*
Max Reams, *Olivet Nazarene University*
Mary Celeste Reese, *Mississippi State University*
Bruce Rengers, *Metropolitan State College of Denver*
Guillermo Rocha, *Brooklyn College*
Penny Sadler, *College of William and Mary*
Shamili Sandiford, *College of DuPage*
Thomas Sasek, *University of Louisiana at Monroe*
Donna Seagle, *Chattanooga State Technical Community College*
Diane Shakes, *College of William and Mary*
Jennie Silva, *Louisiana State University*
Michael Siola, *Chicago State University*
Morgan Slusher, *Community College of Baltimore County*
Julia Smith, *Eastern Washington University*
Darlene Smucny, *University of Maryland University College*
Jeff Snyder, *Bowling Green State University*
Alice Stefaniak, *St. Xavier University*
Alicia Steinhardt, *Hartnell Community College*
Kurt Stellwagen, *Eastern Washington University*
Charlotte Stromfors, *University of Phoenix*
Shane Strup, *University of Phoenix*
Donald Thieme, *Georgia Perimeter College*
Pamela Thinesen, *Century Community College*
Chad Thompson, *SUNY Westchester Community College*
Lensyl Urbano, *University of Memphis*
Gopal Venugopal, *Roosevelt University*
Daniel Vogt, *University of Washington – College of Forest Resources*
Dr. Laura J. Vosejpka, *Northwood University*
Brenda L. Walker, *Kirkwood Community College*

Stephen Wareham, *Cal State Fullerton*
Fred William Whitford, *Montana State University*
Katie Wiedman, *University of St. Francis*
Harry Williams, *University of North Texas*
Emily Williamson, *Mississippi State University*
Bridget Wyatt, *San Francisco State University*
Van Youngman, *Art Institute of Philadelphia*
Alexander Zemcov, *Westchester Community College*

### Student Participants

Karl Beall, *Eastern Washington University*
Jessica Bryant, *Eastern Washington University*
Pia Chawla, *Westchester Community College*
Channel DeWitt, *Eastern Washington University*
Lucy DiAroscia, *Westchester Community College*
Heather Gregg, *Eastern Washington University*
Lindsey Harris, *Eastern Washington University*
Brenden Hayden, *Eastern Washington University*
Patty Hosner, *Eastern Washington University*
Tonya Karunartue, *Eastern Washington University*
Sydney Lindgren, *Eastern Washington University*
Michael Maczuga, *Westchester Community College*
Melissa Michael, *Eastern Washington University*
Estelle Rizzin, *Westchester Community College*
Andrew Rowley, *Eastern Washington University*
Eric Torres, *Westchester Community College*
Joshua Watson, *Eastern Washington University*

## Reviewers of *Visualizing Psychology 3e*

Throughout the process of writing and developing this text and the visual pedagogy, we benefited from the comments and constructive criticism provided by the instructors listed below. We offer our sincere appreciation to these individuals for their helpful review:

### Revision Reviewers

Stacy Andersen, *Florida Gulf Coast University*
Michelle Bannoura, *Hudson Valley Community College*
Matthew Cole, *Lawrence Technological University*
Crystal Colter, *Maryville College*
Kristi Cordell-McNulty, *Angelo State University*
Julia Daniels, *Westchester Community College*
Stephanie Ding, *Del Mar College*
Judith Easton, *Austin Community College*
David Echevarria, *University of Southern Mississippi*
Lenore Frigo, *Shasta College*
Karen Giorgetti, *Youngstown State University*
Jerry Green, *Tarrant County College*
Nancy Gup, *Georgia Perimeter College*
Robert Guttentag, *The University of North Carolina, Greensboro*
Tracey Halverson, *Gwinnett Technical College*
Mark Hartlaub, *Texas A&M University, Corpus Christi*
Charles Huffman, *Georgia Southwestern State University*
Lynnel Kiely, *Harold Washington College—City Colleges of Chicago*
Nicole Korzetz, *Lee College*
Susan Kramer, *Bristol Community College*
Steven McCloud, *Borough of Manhattan Community College—CUNY*
Skip Mueller, *Oglethorpe University*
Paulina Multhaupt, *Macomb Community College*
Jilian Peterson, *Normandale Community College*
Michael Rader, *Northern Arizona University*
James Rodgers, *Hawkeye Community College*
Ariane Schratter, *Maryville College*
Allen Shoemaker, *Calvin College*
Elizabeth Tapp, *Galveston College*

### Virtual Focus Group Participants

Stacy Andersen, *Florida Golf Coast University*
Heidi Braunschweig, *Community College of Philadelphia*
Jarrod Calloway, *Northwest Mississippi Community College*
Lisa End-Berg, *Kennesaw State University*
Miranda Goodman-Wilson, *University of California, Davis*
Jerry Green, *Tarrant County College*
Jill Haasch, *Elizabeth City State University*
Laura Hebert, *Angelina College*
Zachary Hohman, *California State University, Fullerton*
Kathy Immel, *University of Wisconsin, Milwaukee*
Lisa Jackson, *Schoolcraft College*
Nicole Korzetz, *Lee College*
Paulina Multhaupt, *Macomb Community College*
KathryOleson, *Reed College*

### American Psychological Society Focus Group Participants

Michael Behen, *Wayne State University*
Todd Joseph, *Hillsborough Community College*
Catherine Matson, *Triton College*
Glenn Meyer, *Trinity University*
Allison O'Malley, *Butler University*
Robyn Oliver, *Roosevelt University*
Judith Wightman, *Kirkwood Community College*

## Reviewers of Past Editions of *Visualizing Psychology*

### Class Testing and Student Feedback

Sheree Barron, *Georgia College & State University*
Dale V. Doty, *Monroe Community College*
William Rick Fry, *Youngstown State University*
Andy Gauler, *Florida Community College at Jacksonville*
Bonnie A. Green, *East Stroudsburg University*
Janice A. Grskovic, *Indiana University Northwest*
Richard Keen, *Converse College*
Robin K. Morgan, *Indiana University Southwest*
Jack A. Palmer, *University of Louisiana at Monroe*
Marianna Rader, *Florida Community College at Jacksonville*
Melissa Terlecki, *Cabrini College*

### Manuscript Reviews

Marc W. Barnes, *Ivy Tech Community College*
Karen Bearce, *Mercer County Community College*
John Broida, *University of Southern Maine*
Tracie Burt, *Southeast Arkansas College*
Barbara Canaday, *Southwestern College*
Richard Cavasina, *California University of Pennsylvania*
Michelle Caya, *Community College of Rhode Island*
Diane Cook, *Gainesville State College*
Curt Dewey, *San Antonio College*
Dale V. Doty, *Monroe Community College*
Steve Ellyson, *Youngstown State University*
Nolen Embry-Bailey, *Bluegrass Community and Technical College*
Melanie Evans, *Eastern Connecticut State University*
William Rick Fry, *Youngstown State University*
Susan Fuhr, *Maryville College*
Matthew Tyler Giobbi, *Mercer County Community College*
Betsy Goldenberg, *University of Massachusetts, Lowell*
Jeffrey Henriques, *University of Wisconsin*
Kathryn Herbst, *Grossmont College*
Scott Husband, *University of Tampa*
Heather Jennings, *Mercer County Community College*
Richard Keen, *Converse College*
Dawn Mclin, *Jackson State University*
Jean Mandernach, *University of Nebraska at Kearney*
Jan Mendoza, *Brooks College/Golden West College*
Tara Mitchell, *Lock Haven University*
Ruby Montemayor, *San Antonio College*
Robin K. Morgan, *Indiana University Southeast*
Ronald Mulson, *Hudson Valley Community College*
Larry Peck, *Erie Community College–North*
Lori Perez, *Community College of Baltimore County*
Robin Popp, *Chattanooga State Technical Community College*
Marianna Rader, *Florida Community College at Jacksonville*
Christopher Smith, *Ivy Tech Community College*
Clayton Teem, *Gainesville State College*
Marianna Torres, *San Antonio College*
Karina Vargas, *San Antonio College*
Jameel Walji, *San Antonio College*
Colin William, *Ivy Tech Community College*

### American Psychological Association Focus Group Participants

Sheree Barron, *Georgia College & State University*
Joni Caldwell, *Union College*
Stephen Ray Flora, *Youngstown State University*
Regan Gurung, *University of Wisconsin, Green Bay*
Brett Heintz, *Delgado Community College*
J. Kris Leppien-Christensen, *Saddleback College*
Mike Majors, *Delgado Community College*
Debra Murray, *Viterbo University*
Jack A. Palmer, *University of Louisiana at Monroe*
Melissa S. Terlecki, *Cabrini College*

### American Psychological Society Focus Group Participants

Jonathan Bates, *Hunter College*
Michell E. Berman, *University of Southern Mississippi*
Will Canu, *University of Missouri-Rolla*
Patricia C. Ellerson, *Miami University of Ohio*
Renee Engeln-Maddox, *Loyola University Chicago*
Julie Evey, *University of Southern Indiana*
Bonnie A. Green, *East Stroudsburg University*
Janice A. Grskovic, *Indiana University Northwest*
Keith Happaney, *Lehman College*
Hector L. Torres, *Medial College of Wisconsin*

# Special Thanks

Our heartfelt thanks also go to the superb editorial and production teams at John Wiley and Sons who guided us through the challenging steps of developing this second edition. We thank in particular: Nancy Perry, Manager, Production Development; this edition would not exist were it not for Nancy's unflagging support, careful eye, and invaluable expertise.

We also owe an enormous debt of gratitude to Executive Editor Chris Johnson, who expertly launched and directed our process; Maura Gilligan, Editorial Assistant, who expertly juggled her multiple roles; Micheline Frederick, Senior Content Manager, who stepped in whenever we needed expert advice; Janet Foxman, Senior Production Editor, who guided the book through production; Jay O'Callaghan, Vice President and Executive Publisher, who oversaw the entire project; and Jeffrey Rucker, Marketing Director/Communications for Wiley Visualizing, and Margaret Barrett, Senior Marketing Manager, who adeptly represent the Visualizing imprint. In addition, we are deeply indebted to Rebecca Heider, our developmental editor, who contributed long hours of careful and patient editing. This type of "backstage" support requires a sharp, professionally trained mind and endless patience—two qualities that are seldom acknowledged (but deeply appreciated) by all authors.

Finally, we'd like to express our deepest gratitude to the entire team of ancillary authors: Test Bank, Melissa Acevedo, Westchester Community College; Instructor's Manual, Matthew Leavitt, Arizona State University, and Lynnel Kiely, City Colleges of Chicago; and the PowerPoint Presentations, Katie Townsend-Merino, Palomar College. Their shared dedication and professionalism will provide vital educational support to faculty and student alike.

We wish also to acknowledge the contributions of Vertigo Design for the interior design concept, and Harry Nolan, Wiley's Creative Director who gave art direction, refined the design and other elements and the cover. We appreciate the efforts of Mary Ann Price in researching and obtaining our text photos.

Our sincerest thanks are also offered to all who worked on the media and ancillary materials, including Lynn Cohen, Editorial Operations Manager, for her expert work in developing the video and electronic components, and a host of others who contributed to the wide assortment of ancillaries.

Next, we would like to offer our thanks to all the folks at Furino Production—particularly Jeanine Furino. Her incredible dedication, keen eye for detail, and desire for perfection can be seen throughout this book. The careful and professional approach of Jeanine and her staff was critical to the successful production of this edition.

All the writing, producing, and marketing of this book would be wasted without an energetic and dedicated sales staff. We wish to sincerely thank all the publishing representatives for their tireless efforts and good humor. It's a true pleasure to work with such a remarkable group of people.

From Siri Carpenter: thank you to my husband, Joe Carpenter, for his thoughtful advice and steadfast support throughout the production of this book. My appreciation also to colleagues who provided helpful feedback in one way or another: Tracy Banaszynski, Jennifer Randall Crosby, Brian Detweiler-Bedell, Jerusha Detweiler-Bedell, Meghan Dunn, and Kristi Lemm.

From Karen Huffman: continuing appreciation to my family and students who supported and inspired me. I also want to offer my heartfelt gratitude to two very special people, Richard Hosey and Rita Jeffries. Their careful editing, constructive feedback, professional research skills, and shared authorship were essential to this revision. Last, and definitely not least, I thank my beloved husband, Bill Barnard.

# Contents in Brief



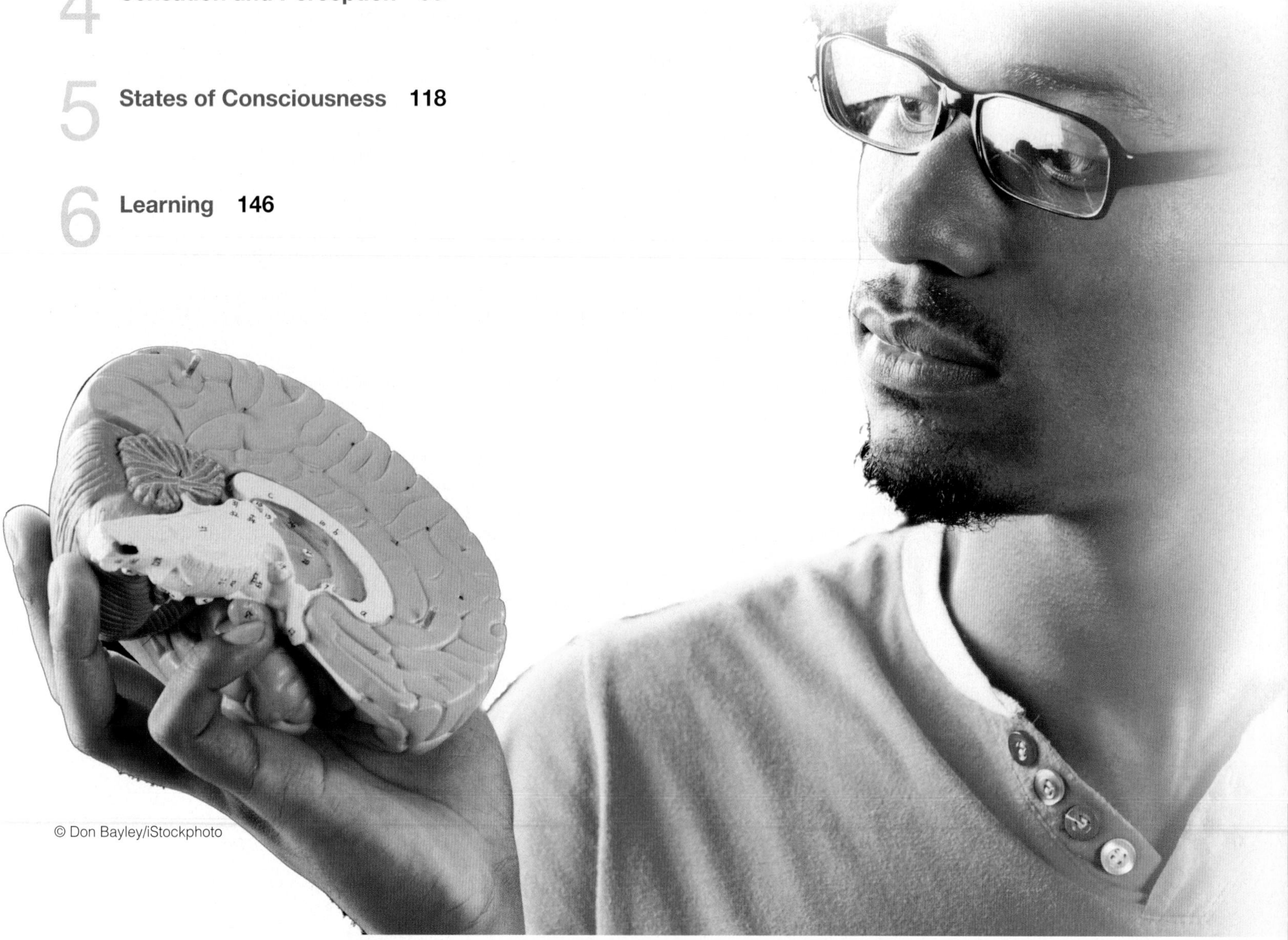

# Contents



Jeffrey Greenberg/Photo Researchers

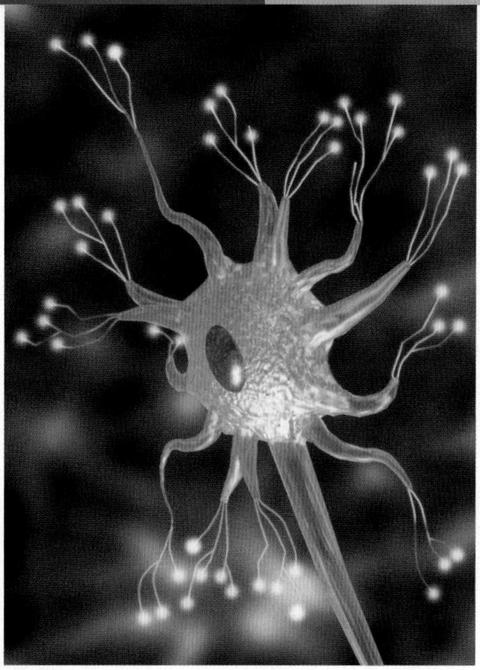

Alfred Pasieka/Photo Researchers

©AP/Wide World Photos

© RubberBall/Alamy Limited

Stockbyte/Getty Images, Inc.

PhotoDisc Green/Getty Images

James P. Blair/NG Image Collection

iStockphoto

Randy Olson/NG Image Collection

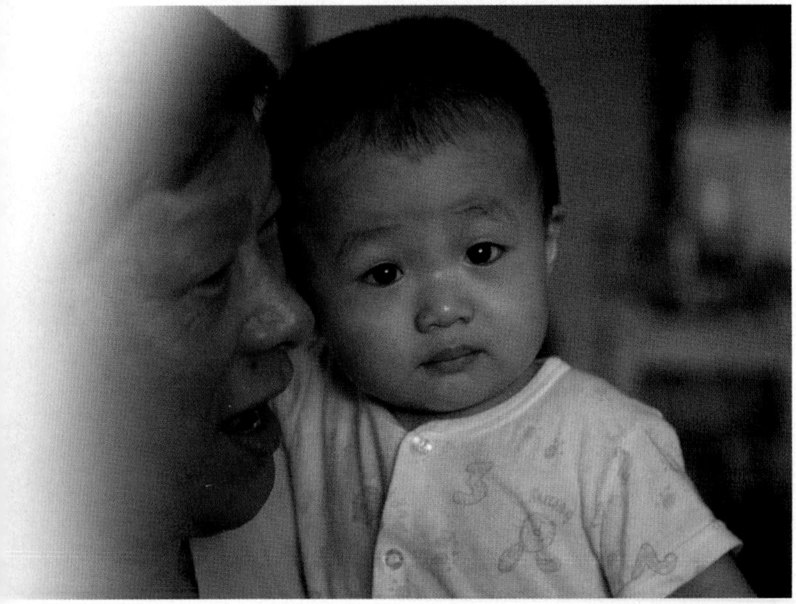

Elizabeth Crews/The Image Works

Marco Simoni/Getty Images

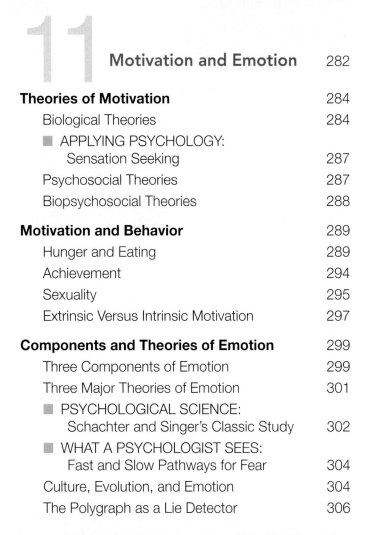

© Masterfile

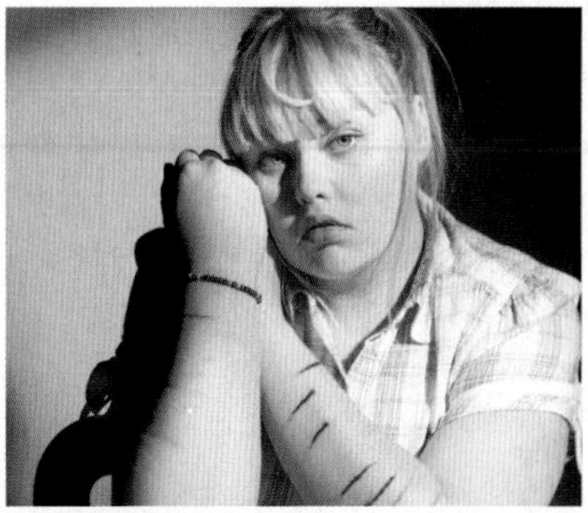

Peter Dazeley/Photographer's Choice/Getty Image, Inc.

© MarkBowden/iStockphoto

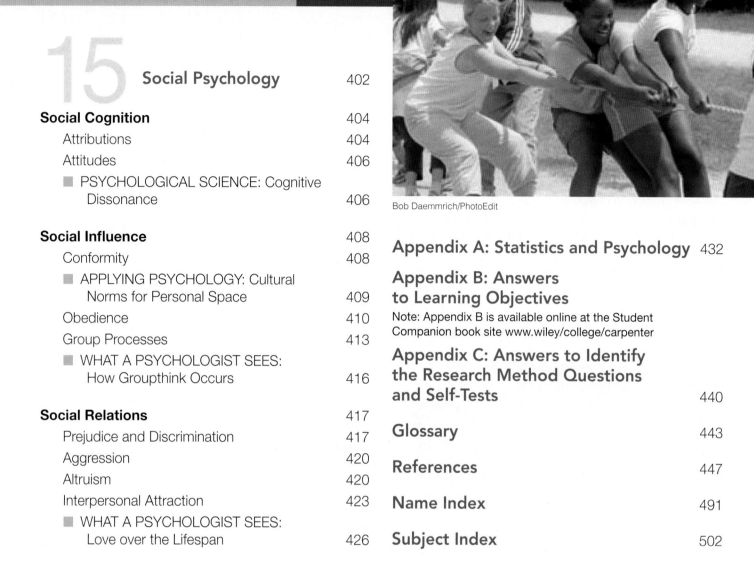

Bob Daemmrich/PhotoEdit

## Psychology InSight Features

**Multipart visual presentations that focus on a key concept or topic in the chapter**

**Chapter 1**
Three types of correlation

**Chapter 2**
Twin studies • How poisons and drugs affect our brain

**Chapter 3**
Cognitive appraisal and coping

**Chapter 4**
Measuring the senses • Selection • Four perceptual constancies

**Chapter 5**
Scientific study of sleep and dreaming • Your brain on "club" drugs

**Chapter 6**
Classical conditioning in everyday life • Operant conditioning in everyday life • Four key factors in observational learning

**Chapter 7**
Retrieval cues

**Chapter 8**
Concepts

**Chapter 9**
Brain development

**Chapter 10**
Research on infant attachment

**Chapter 11**
Extrinsic versus intrinsic motivation

**Chapter 12**
Personality congruence and mental health

**Chapter 13**
Four criteria for abnormal behavior

**Chapter 14**
How antidepressants affect the brain

**Chapter 15**
Helping

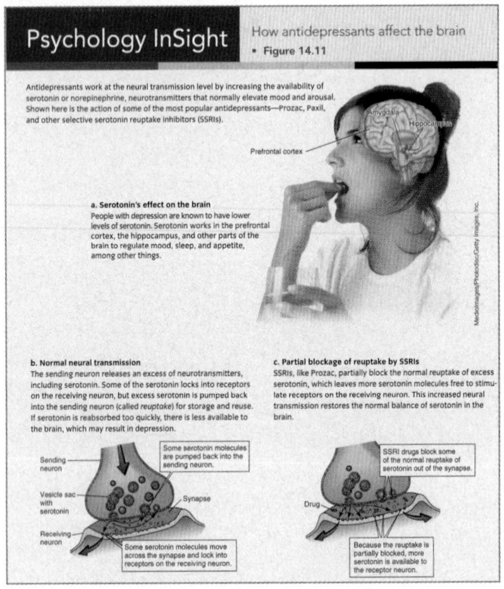


## Process Diagrams

**A series or combination of figures and photos that describe and depict a complex process**

**Chapter 1**
The scientific method • Experimental research design • Using the SQ4R method

**Chapter 2**
Communication *within* the neuron • Communication *between neurons*

**Chapter 3**
General adaptation syndrome (GAS) • Stress—An interrelated system

**Chapter 4**
How the eye sees • How the ear hears

**Chapter 5**
How agonistic and antagonistic drugs produce their psychoactive effect

**Chapter 6**
Pavlov's classical conditioning • Higher-order conditioning

**Chapter 7**
Encoding, storage, and retrieval

**Chapter 8**
Three steps to the goal

**Chapter 9**
Prenatal development

**Chapter 10**
Kohlberg's stages of moral development • Erikson's eight stages of psychosocial development

**Chapter 11**
Drive-reduction theory • Masters and Johnson's sexual response cycle

**Chapter 12**
Freud's five psychosexual stages of development

**Chapter 13**
Conditioning and phobias

**Chapter 14**
Aversion therapy

**Chapter 15**
Reducing cognitive dissonance

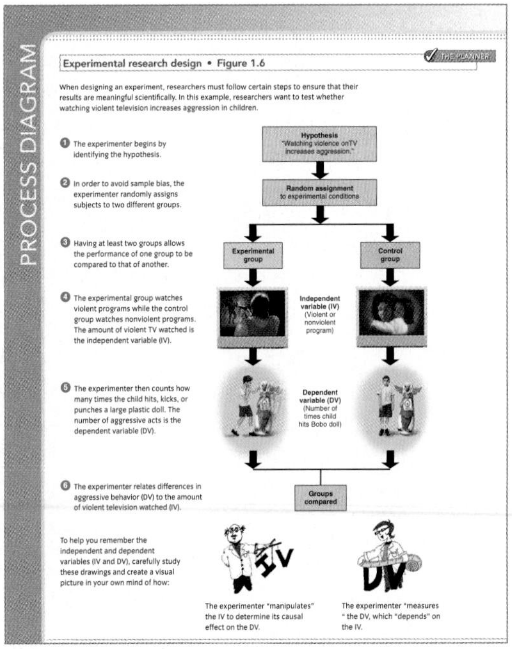

THIRD EDITION

# VISUALIZING PSYCHOLOGY

1

# Introduction and Research Methods

What do you *visualize* and picture in your mind's eye when you hear the words "psychology" and "psychologist"? When you flip through the pages of the text, does it match your expectations? Many students open this book and enter their first psychology class believing that psychology is mostly about abnormal behaviors, personal problems, and psychotherapy. Although psychologists do indeed study and treat psychological problems, we also investigate the brain and nervous system, stress, health, sensation, perception, states of consciousness, learning, memory, thinking, language, intelligence, lifespan development, motivation, emotion, personality, social psychology, and so much more.

As you'll soon discover, psychology is a living, dynamic field that affects virtually every aspect

of our lives. Our innermost thoughts, our relationships, our politics, our "gut" feelings, and our deliberate decisions are all shaped by a complex psychological system that affects us at every level, from the cellular to the cultural. Psychology encompasses not only humankind but our nonhuman compatriots as well—from rats and pigeons to cats and chimps.

We invite you to let us know how your study of psychology (and this text) affects you and your life. You can reach us at siri@nasw.org and khuffman@palomar.edu. We look forward to hearing from you.

Warmest regards,
Siri Carpenter and Karen Huffman

## CHAPTER OUTLINE

© Justin Horrocks/iStockphoto

## CHAPTER PLANNER

- ❑ Study the picture and read the opening story.
- ❑ Answer the Learning Objectives in each section:
  p. 4 ❑ p. 7 ❑ p. 12 ❑ p. 15 ❑ p. 23 ❑
- ❑ Read the text and study all figures and visuals. Answer any questions.

**Analyze key features**

- ❑ Myth Busters, p. 5
- ❑ Study Organizers p. 8 ❑ p. 15 ❑
- ❑ What a Psychologist Sees, p. 11
- ❑ Process Diagrams p. 13 ❑ p. 16 ❑ p. 24 ❑
- ❑ Psychology InSight, p. 20
- ❑ Applying Psychology, p. 25
- ❑ Psychological Science, p. 26
- ❑ Stop: Answer the Concept Checks before you go on.
  p. 6 ❑ p. 11 ❑ p. 14 ❑ p. 23 ❑ p. 26 ❑

**End of chapter**

- ❑ Review the Summary and Key Terms.
- ❑ Answer the Critical and Creative Thinking Questions.
- ❑ Answer What is happening in this picture?
- ❑ Complete the Self-Test and check your answers.

# Introducing Psychology

## LEARNING OBJECTIVES

**RETRIEVAL PRACTICE** While reading the upcoming sections, respond to each Learning Objective in your own words. Then compare your responses with those in Appendix B.

1. **Define** psychology.
2. **Identify** psychology's four main goals.
3. **Summarize** psychology's major career options.

The term **psychology** derives from the roots *psyche*, meaning "mind," and *logos*, meaning "word." Modern psychology is most commonly defined as the *scientific* study of **behavior** and **mental processes**. Behavior can be directly observed (crying, hitting, sleeping). Mental processes are private, internal experiences that cannot be directly observed (feelings, thoughts, memories).

For many psychologists, the most important part of the definition of psychology is the word *scientific*. Psychology places high value on **empirical evidence** that can be objectively tested and evaluated. Psychologists also emphasize **critical thinking** (Forshaw, 2013).

**psychology** The scientific study of behavior and mental processes.

**empirical evidence** Information acquired by direct observation and measurement using systematic scientific method.

**critical thinking** The process of objectively evaluating, comparing, analyzing, and synthesizing information.

Be careful not to confuse scientific psychology with **pseudopsychologies**, which only give the appearance of science. (*Pseudo* means "false.") *Pseudopsychology* is based on common beliefs, folk wisdom, or even superstitions, which are generally formed from unsupported information and do not follow the basics of empirical testing and research (i.e., the scientific method). Examples include believing in psychic powers, horoscopes, mediums, and many self-help, "pop psych" statements like: "I'm mostly left-brained or right-brained," or "We only use 10% of our brain." Given the popularity of these misleading beliefs, each chapter of this text offers special "Myth Busters" boxes, which will help you identify pseudopsychology and distinguish it from serious, scientific psychology (see *Myth Busters*).

## Psychology's Four Main Goals

In contrast to pseudopsychologies, which rely on unsubstantiated beliefs and opinions, psychology bases its findings on rigorous, scientific methods. When conducting their research, psychologists have four basic goals: to *describe*, *explain*, *predict*, and *change* behavior and mental processes.

1. **Description** Description tells "what" occurred. In some studies, psychologists attempt to *describe*, or name and classify, particular behaviors by making careful scientific observations. Description is usually the first step in understanding behavior. For example, if someone says, "Boys are more aggressive than girls," what does that mean? The speaker's definition of aggression may differ from yours. Science requires specificity.

2. **Explanation** An explanation tells "why" a behavior or mental process occurred. *Explaining* a behavior or mental process depends on discovering and understanding its causes. One of the most enduring debates in science has been the **nature–nurture controversy** (Tyson, Jones, & Elock, 2011). Are we controlled by biological and genetic factors (the nature side)? Or by environment and learning (the nurture side)? As you will see throughout the text, psychology (like all sciences) generally avoids "either-or" positions and focuses instead on *interactions*. Today, almost all scientists agree that most psychological, and even physical traits, reflect an interaction between nature and nurture. For example, research indicates that there are numerous interacting causes or explanations for aggression, including culture, learning, genes, brain damage, and high levels of testosterone (e.g., Bhanoo, 2011; Kelly et al., 2008; Pournaghash-Tehrani, 2011).

**nature–nurture controversy** The ongoing dispute over the relative contributors of nature (biological and genetic factors) and nurture (environment) to the development of behavior and mental processes.

3. **Prediction** Psychologists generally begin with description and explanation (answering the "whats" and

## MYTH BUSTERS

THE PLANNER

### PSYCHOLOGY VERSUS PSEUDOPSYCHOLOGY

**TRUE OR FALSE?**

___ 1. The best way to learn and remember information is to "cram," or study it intensively during one concentrated period.

___ 2. Most brain activity stops during sleep.

___ 3. Advertisers and politicians often use subliminal persuasion to influence our behavior.

___ 4. Punishment is the most effective way to permanently change behavior.

___ 5. Eyewitness testimony is often unreliable.

___ 6. Polygraph ("lie detector") tests can accurately and reliably reveal whether or not a person is lying.

___ 7. Behaviors that are unusual or that violate social norms indicate a psychological disorder.

___ 8. People with schizophrenia have two or more distinct personalities.

___ 9. Similarity is one of the best predictors of long-term relationships.

___ 10. In an emergency, as the number of bystanders increases, your chance of getting help decreases.

Answers: 1. False (Chapter 1). 2. False (Chapter 5). 3. False (Chapter 4). 4. False (Chapter 6). 5. True (Chapter 7). 6. False (Chapter 11). 7. False (Chapter 13). 8. False (Chapter 13). 9. True (Chapter 15). 10. True (Chapter 15).

Henry Groskinsky/Time Life Pictures/Getty Images

The magician James Randi has dedicated his life to educating the public about fraudulent pseudopsychologists. Along with the prestigious MacArthur Foundation, Randi has offered $1 million to "anyone who proves a genuine psychic power under proper observing conditions" (Randi, 1997; The Amazing Meeting, 2011). After many years, the money has never been collected.

"whys"). Then they move on to the higher-level goal of *prediction*, identifying the conditions under which a future behavior or mental process is likely to occur. For instance, knowing that alcohol leads to increased aggression (Ferguson, 2010; Mihic et al., 2009), we can predict that more fights will erupt in places where alcohol is consumed than in those where alcohol isn't consumed.

4. **Change** For some people, "change" as a goal of psychology brings to mind evil politicians or cult leaders "brainwashing" unknowing victims. However, to psychologists, *change* means applying psychological knowledge to prevent unwanted outcomes or bring about desired goals. In almost all cases, change as a goal of psychology is positive. Psychologists help people improve their work environment, stop addictive behaviors, become less depressed, improve their family relationships, and so on. Furthermore, as you may know from personal experience, it is very difficult (if not impossible) to change someone against her or his will. (*Joke question*: Do you know how many psychologists it takes to change a light bulb? *Answer*: None. The light bulb has to want to change.)

## Careers in Psychology

Many people think of psychologists only as therapists. While it's true that the fields of clinical and counseling psychology do make up the largest specialty area, there are many psychologists who have no connection with therapy. Instead,

| Sample careers and specialties in psychology Table 1.1 | |
|---|---|
| **Biopsychology/neuroscience** | Investigates the relationship between biology, behavior, and mental processes, including how physical and chemical processes affect the structure and function of the brain and nervous system |
| **Clinical psychology** | Specializes in the evaluation, diagnosis, and treatment of psychological disorders |
| **Cognitive psychology** | Examines "higher" mental processes, including thought, memory, intelligence, creativity, and language |
| **Comparative psychology** | Studies the behavior and mental processes of nonhuman animals; emphasizes evolution and cross-species comparisons |
| **Counseling psychology** | Overlaps with clinical psychology, but generally works with less seriously disordered individuals and focuses more on social, educational, and career adjustment |
| **Cross-cultural psychology** | Studies similarities and differences in and across various cultures and ethnic groups |
| **Developmental psychology** | Studies the course of human growth and development from conception to death |
| **Educational psychology** | Studies the processes of education and works to promote the intellectual, social, and emotional development of children in the school environment |
| **Environmental psychology** | Investigates how people affect and are affected by the physical environment |
| **Experimental psychology** | Examines processes such as learning, conditioning, motivation, emotion, sensation, and perception in humans and other animals (Note that psychologists working in almost all areas of specialization also conduct experiments) |
| **Forensic psychology** | Applies principles of psychology to the legal system, including jury selection, psychological profiling, assessment, and treatment of offenders |
| **Gender and/or cultural psychology** | Investigates how men and women and different cultures vary from one another and how they are similar |
| **Health psychology** | Studies how biological, psychological, and social factors affect health and illness |
| **Industrial/organizational psychology** | Applies principles of psychology to the workplace, including personnel selection and evolution, leadership, job satisfaction, employee motivation, and group processes within the organization |
| **Personality psychology** | Studies the unique and relatively stable patterns in a person's thoughts, feelings, and actions |
| **Positive psychology** | Examines factors related to optimal human functioning |
| **School psychology** | Collaborates with teachers, parents, and students within the educational system to help children with needs related to a disability and/or their academic and social progress; also provides evaluation and assessment of a student's functioning and eligibility for special services |
| **Social psychology** | Investigates the role of social forces in interpersonal behavior, including aggression, prejudice, love, helping, conformity, and attitudes |
| **Sports psychology** | Applies principles of psychology to enhance physical performance |

they work as researchers, teachers, or consultants in academic, business, industry, and government settings, or in a combination of settings (**Table 1.1**). For more information about what psychologists do—or how to pursue a career in psychology—check out the websites of the American Psychological Association and the Association for Psychological Science.

## CONCEPT CHECK STOP

1. **How** does psychology differ from pseudopsychology?
2. **What** are the two sides of the nature–nurture debate?
3. **What** is the largest specialty area of psychology?

# Origins of Psychology

## LEARNING OBJECTIVES

**RETRIEVAL PRACTICE** While reading the upcoming sections, respond to each Learning Objective in your own words. Then compare your responses with those in Appendix B.

1. **Describe** the different perspectives offered by early psychologists.
2. **Identify** the seven major perspectives of modern psychology.
3. **Explain** the central idea of the biopsychosocial model.

Although people have always been interested in human nature, it was not until the first psychological laboratory was founded in 1879 that psychology as a science officially began. As interest in the new field grew, psychologists adopted various perspectives on the "appropriate" topics for psychological research and the "proper" research methods. These diverse viewpoints and subsequent debates molded and shaped modern psychological science.

## Psychology's Past

Psychology's history as a science began in 1879 in Leipzig, Germany when Wilhelm Wundt (VILL-helm Voont), generally acknowledged as the "father of psychology," established the first psychological laboratory. Wundt and his followers were primarily interested in how we form sensations, images, and feelings. Their chief methodology was termed "introspection," which involved self-monitoring and reporting on conscious experiences (Baker, 2012; Goodwin, 2012).

A student of Wundt's, Edward Titchener, brought his ideas to the United States. Titchener's approach, now known as **structuralism**, sought to identify the basic building blocks, or structures, of mental life through introspection and then to determine how these elements combine to form the whole of experience. Because introspection could not be used to study animals, children, or more complex mental disorders, structuralism failed as a working psychological approach. Although short-lived, structuralism established a model for studying mental processes scientifically.

Structuralism's intellectual successor, **functionalism**, studied how the mind functions to enable humans and other animals to adapt to their environment. Functionalism was strongly influenced by Charles Darwin's theory of evolution (Green, 2009). William James was the leading force in the functionalist school (**Figure 1.1**). Although functionalism also eventually declined, it expanded the scope of psychology to include research on emotions and observable behaviors, initiated the psychological testing movement, and influenced modern education and industry.

During the late 1800s and early 1900s, while functionalism was prominent in the United States, the **psychoanalytic school** was forming in Europe. Its founder, Austrian physician Sigmund Freud, believed that a part of the human mind, the unconscious, contains thoughts, memories, and desires that lie outside personal awareness, yet still exert great influence. For example, according to Freud, a man who is cheating on his wife might slip up and say, "I wish you were her," when he consciously planned to say, "I wish

Bettmann/Corbis Images

**William James (1842–1910) • Figure 1.1**

William James broadened psychology to include animal behavior and biological processes. In the late 1870s, James established the first psychology laboratory in the United States, at Harvard University.

you were here." Such seemingly meaningless, so-called Freudian slips supposedly reveal a person's true unconscious desires and motives.

Freud also believed many psychological problems are caused by conflicts between "acceptable" behavior and "unacceptable," unconscious sexual or aggressive motives (Chapter 12). The theory provided a basis for a system of therapy known as *psychoanalysis* (Chapter 14).

## Modern Psychology

As summarized in **Study Organizer 1.1**, contemporary psychology reflects seven major perspectives: *psychodynamic*, *behavioral*, *humanistic*, *cognitive*, *biological*, *evolutionary*, and *sociocultural*. Although there are numerous differences between these seven perspectives, most psychologists recognize the value of each orientation and agree that no one view has all the answers.

Freud's nonscientific approach and emphasis on sexual and aggressive impulses have long been controversial, and today there are few strictly Freudian psychoanalysts left. However, the broad features of his theory remain in the modern **psychodynamic perspective**. The general goal of psychodynamic psychologists is to explore unconscious *dynamics*, internal motives, conflicts, and childhood experiences.

**Study Organizer 1.1** Modern psychology's seven major perspectives — THE PLANNER

| Perspectives | Major emphases | Sample research questions | |
|---|---|---|---|
| **Psychodynamic** | Unconscious drives, motives, conflicts, and childhood experiences | How do adult personality traits or psychological problems reflect unconscious processes and early childhood experiences? |  |
| **Behavioral** | Objective, observable, environmental influences on overt behavior; stimulus–response relationships and consequences for behavior | How do we learn both our good and bad habits? How can we increase desirable behaviors and decrease undesirable ones? |  |
| **Humanistic** | Free will, self-actualization, and human nature as naturally positive and growth-seeking | How can we promote a client's capacity for self-actualization and understanding of his or her own development? How can we promote international peace and reduce violence? |  |
| **Cognitive** | Thinking, perceiving, problem solving, memory, language, and information processing | How do our thoughts and interpretations affect how we respond in certain situations? How can we improve how we process, store, and retrieve information? | 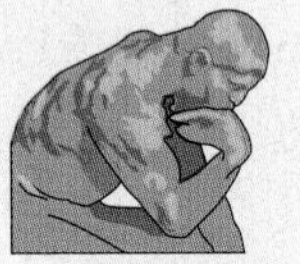 |
| **Biological** | Genetic and biological processes in the brain and other parts of the nervous system | How might changes in neurotransmitters or damage to parts of the brain lead to psychological problems and changes in behavior and mental processes? | 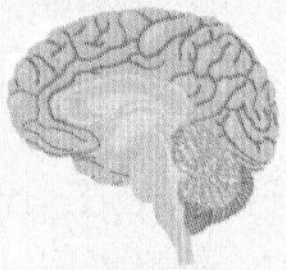 |
| **Evolutionary** | Natural selection, adaptation, and evolution of behavior and mental processes | How does natural selection help explain why we love certain people while we help or hurt others? Do we have specific genes for aggression and altruism? |  |
| **Sociocultural** | Social interaction and the cultural determinants of behavior and mental processes | How do the values and beliefs transmitted from our social and cultural environments affect our everyday psychological processes? |  |

In the early 1900s, another major perspective appeared that dramatically shaped the course of modern psychology. Unlike earlier approaches, the **behavioral perspective** emphasizes objective, observable environmental influences on overt behavior. Behaviorism's founder, John B. Watson (1913), rejected the practice of introspection and the influence of unconscious forces. Instead, Watson adopted Russian physiologist Ivan Pavlov's concept of *conditioning* (Chapter 6) to explain behavior as a result of observable stimuli (in the environment) and observable responses (behavioral actions).

Most early behaviorist research was focused on learning; nonhuman animals were ideal subjects for this research. One of the most well-known behaviorists, B. F. Skinner, was convinced that behaviorist approaches could be used to "shape" human behavior (**Figure 1.2**).

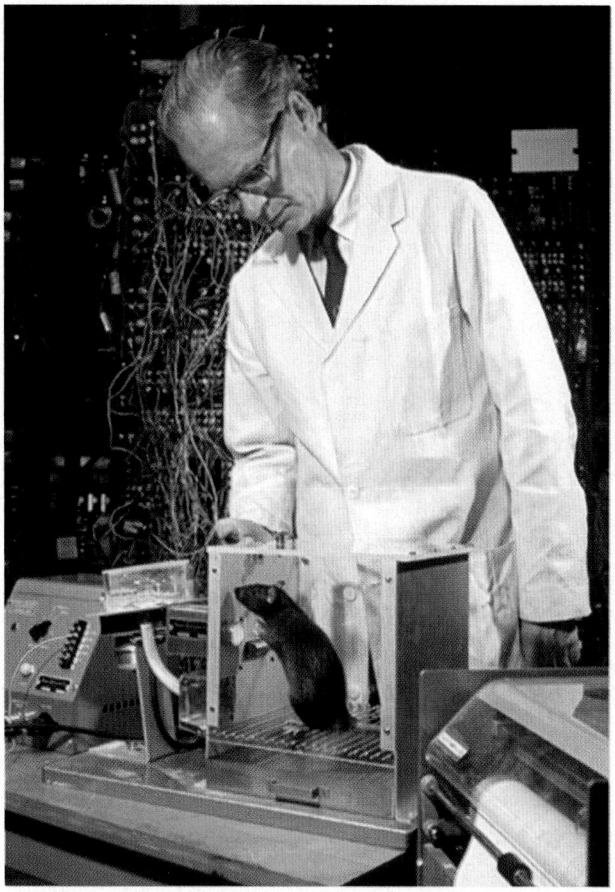

Nina Leen/Time Life Pictures/Getty Images

**B. F. Skinner (1904–1990) • Figure 1.2**

B. F. Skinner was one of the most influential psychologists of the twentieth century. Here he uses the so-called "Skinner box" to train a rat to press a lever for a reward.

Therapeutic techniques rooted in the behavioristic perspective have been most successful in treating observable behavioral problems, such as phobias and alcoholism (Miltenberger, 2011; Sarafino, 2012) (Chapters 6 and 14).

Although the psychoanalytic and behavioral perspectives dominated American psychology for some time, in the 1950s a new approach emerged—the **humanistic perspective**, which stressed *free will* (voluntarily chosen behavior) and *self-actualization* (a state of self-fulfillment). According to Carl Rogers and Abraham Maslow (two central humanist figures), all individuals naturally strive to develop and move toward self-actualization. Like psychoanalysis, humanist psychology developed an influential theory of personality and a form of psychotherapy (Chapters 12 and 14). The humanistic approach also led the way to a contemporary research specialty known as **positive psychology**—the scientific study of optimal human functioning (Cornum, Matthews, & Seligman, 2011; Diener, 2008; Seligman, 2003, 2011; Taylor & Sherman, 2008).

One of the most influential modern approaches, the **cognitive perspective**, recalls psychology's earliest days in that it emphasizes thoughts, perception, and information processing (Kellogg, 2011; Sternberg, 2012). Modern cognitive psychologists, however, study how we gather, encode, and store information using a vast array of mental processes. These include perception, memory, imagery, concept formation, problem solving, reasoning, decision making, and language. Many cognitive psychologists also use what is called an *information-processing approach*, likening the mind to a computer that sequentially takes in information, processes it, and then produces a response.

During the last few decades, scientists have explored the role of biological factors in almost every area of psychology. Using sophisticated tools and technologies, scientists who adopt this **biological perspective** examine behavior through the lens of genetics and biological processes in the brain and other parts of the nervous system.

The **evolutionary perspective** stresses natural selection, adaptation, and evolution of behavior and mental processes (Buss, 2011; Confer et al., 2010; Swami, 2011). Its proponents argue that natural selection favors behaviors that enhance an organism's reproductive success.

Finally, the **sociocultural perspective** emphasizes social interactions and cultural determinants of behavior and mental processes. Although we are often unaware of their influence, factors such as ethnicity, religion, occupation, and socioeconomic class have an enormous psychological impact on our mental processes and behavior (Berry et al., 2011; Gergen, 2010).

**Gender and cultural influences** During the late 1800s and early 1900s, most colleges and universities provided little opportunity for women and ethnic minorities, either as students or as faculty members. Nevertheless, one of the first women to be recognized in the field was Mary Calkins, who performed valuable research on memory, and in 1905 she served as the first female president of the APA. Her achievements are particularly noteworthy, considering the significant discrimination that she overcame. Even after she completed all the requirements for a Ph.D. at Harvard University and was described by William James as his brightest student, the university refused to grant the degree to a woman. The first woman to receive a Ph.D. in psychology was Margaret Floy Washburn (in 1894), who wrote several influential books and served as the second female president of the APA.

Francis Cecil Sumner became the first African American to earn a Ph.D. in psychology. He earned it from Clark University in 1920 and later chaired one of the country's leading psychology departments, at Howard University. In 1971, one of Sumner's students, Kenneth B. Clark, became the first African American to be elected APA president. Clark's research with his wife, Mamie Clark, documented the harmful effects of prejudice and directly influenced the Supreme Court's 1954 ruling against racial segregation in schools (**Figure 1.3**).

### Kenneth Clark (1914–2005) and Mamie Phipps Clark (1917–1985) • Figure 1.3

Kenneth Clark and his wife, Mamie Phipps Clark, conducted experiments with black and white dolls to study children's attitudes about race. This research and their expert testimony contributed to the U.S. Supreme Court's ruling that racial segregation in public schools was unconstitutional.

Library of Congress Prints and Photographs Division

### Ethnicities of recent doctorate recipients in psychology • Figure 1.4

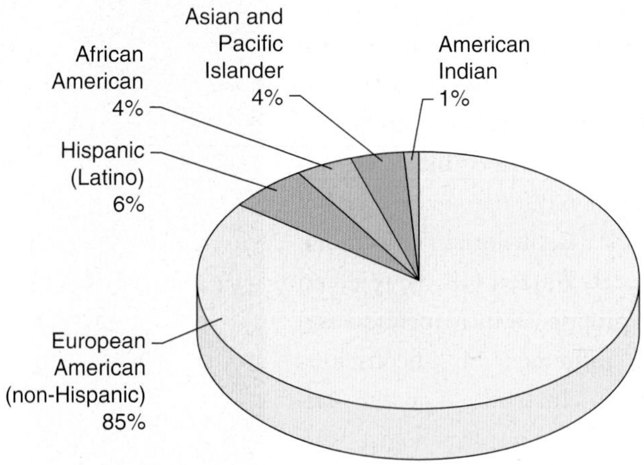


Sumner, Clark, Calkins, and Washburn, along with other important minorities and women, made significant and lasting contributions to the developing science of psychology. In recent years, people of color and women have been actively encouraged to pursue graduate degrees in psychology. However, European American (non-Hispanic) people still make up the majority of new doctorate recipients in psychology (**Figure 1.4**).

**Biopsychosocial model** The seven major perspectives, as well as gender and cultural influences, each make different contributions to modern psychology, which is why most contemporary psychologists do not adhere to one single intellectual perspective. Instead, a more integrative, unifying theme—the **biopsychosocial model**—has gained wide acceptance. This model views biological processes (e.g., genetics, brain functions, neurotransmitters, and evolution), psychological factors (e.g., learning, thinking, emotion, personality, and motivation), and social forces (e.g., family, school, culture, ethnicity, social class, and politics) as interrelated, inseparable influences that interact with the previously described seven major perspectives. (See *What a Psychologist Sees*.)

**biopsychosocial model** A unifying theme of modern psychology that considers biological, psychological, and social processes.

# WHAT A PSYCHOLOGIST SEES ✓ THE PLANNER

## The Biopsychosocial Model

When we look at an individual (**Figure a**), we don't always get a complete picture of her emotions and motivations. Stepping back to see the individual in a broader context (**Figure b**) can provide new insights. With this "bigger picture" (the child's immediate surroundings, her parents' guiding influence, and her group's enthusiasm for exciting sporting events) in mind, can you better understand why she might be feeling and behaving as she is? The biopsychosocial model recognizes that there is usually no single cause for our behavior or our mental states (**Figure c**). For example, our moods and feelings are often influenced by genetics and neurotransmitters (biological), our learned responses and patterns of thinking (psychological), and our socioeconomic status and cultural views of emotion (sociocultural).

a.

Tim Mantoani/Masterfile

b.

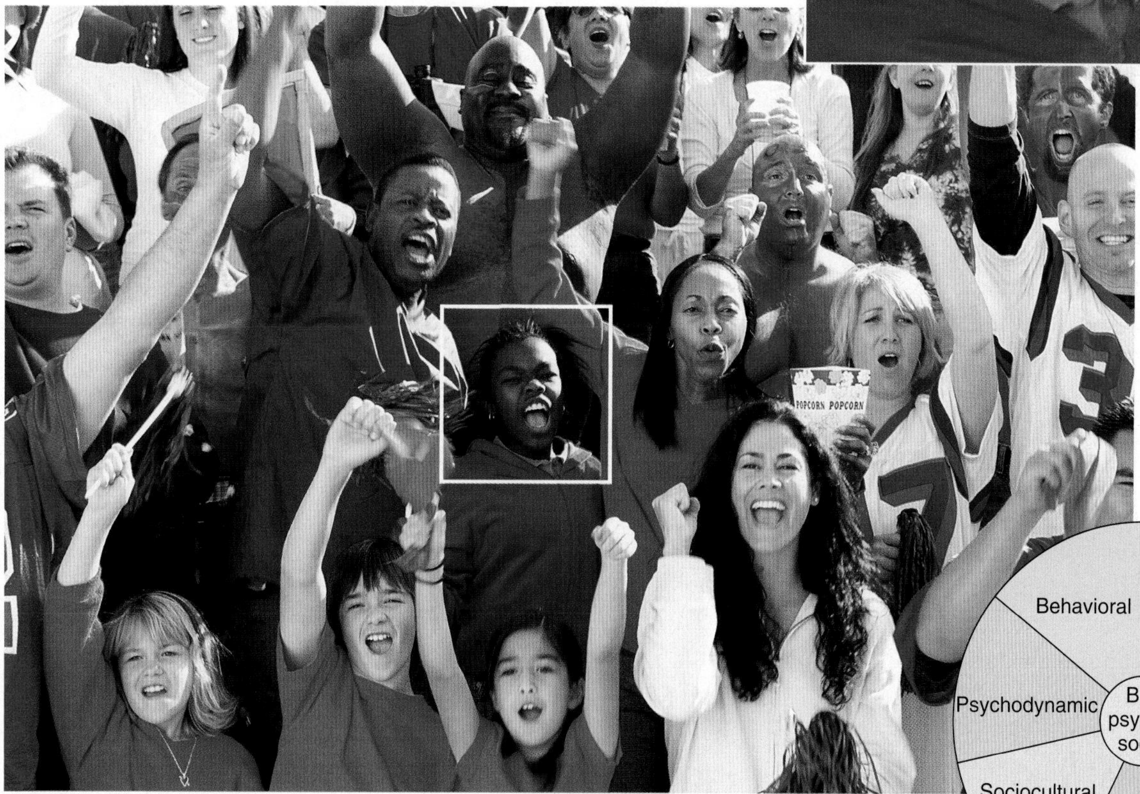

Tim Mantoani/Masterfile

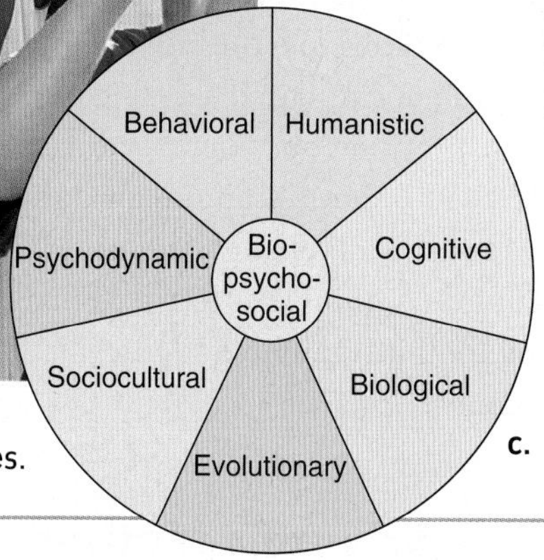


The biopsychosocial model combines and interacts with the seven major perspectives.

c.

**CONCEPT CHECK** STOP

1. **Why** did structuralism decline in popularity?
2. **How** does the biological perspective differ from the evolutionary perspective?
3. **Why** is the biopsychosocial model important to the field of psychology?

# The Science of Psychology

## LEARNING OBJECTIVES

**RETRIEVAL PRACTICE** While reading the upcoming sections, respond to each Learning Objective in your own words. Then compare your responses with those in Appendix B.

1. **Compare** the fundamental goals of basic and applied research.
2. **Describe** the scientific method.
3. **Identify** how psychologists protect the rights of human and nonhuman research participants and psychotherapy clients.

In science, research strategies are generally categorized as either basic or applied. **Basic research** is typically conducted in universities or research laboratories by researchers who are interested in advancing general scientific understanding. Basic research meets the first three goals of psychology (description, explanation, and prediction). For example, discoveries linking aggression to testosterone, genes, learning, and other factors came primarily from *basic research*.

**basic research** Research conducted to advance scientific knowledge rather than for practical application.

**applied research** Research designed to solve practical problems.

In contrast, **applied research** is generally conducted outside the laboratory, and it meets the fourth goal of psychology—to change behavior and mental processes as they relate to existing real-world problems. Applied research has designed programs for conflict resolution and counseling for perpetrators and victims of violence.

Basic and applied research frequently interact, with one building on the other. For example, after basic research documented a strong relationship between alcohol consumption and increased aggression, applied research led some sports stadium owners to limit or ban completely the sale of alcohol, and prohibit fans from bringing any liquid containers into the venue during the final quarter of football games and the last two innings of baseball games.

## The Scientific Method

Like scientists in any other field, psychologists follow strict, standardized procedures so that others can understand, interpret, and repeat or test their findings. Most scientific investigations involve six basic steps (**Figure 1.5**).

Note how the scientific method is cyclical and cumulative. This is because scientific progress comes from repeatedly challenging and revising existing theories and building new ones. If numerous scientists, using different procedures or participants in varied settings, can repeat, or *replicate*, a study's findings, there is increased scientific confidence in the findings. If the findings cannot be replicated, researchers look for other explanations and conduct further studies. When different studies report contradictory findings, researchers may average or combine the results of all such studies and reach conclusions about the overall weight of the evidence, a popular statistical technique called **meta-analysis**.

After many related findings have been collected and confirmed, scientists may generate a **theory** to explain the data through a systematic, interrelated set of concepts. In common usage, the term *theory* is often used to suggest that something is only a "mere hunch" or someone's personal opinion. In reality, scientific theories are evidence-based and subject to rigorous testing.

## Ethical Guidelines

The two largest professional organizations of psychologists, the American Psychological Association (APA) and the Association for Psychological Science (APS), both recognize the importance of maintaining high ethical standards in research, therapy, and all other areas of professional psychology. The preamble to the APA's publication *Ethical Principles of Psychologists and Code of Conduct* (2002, 2010) requires psychologists to maintain their competence, to retain objectivity in applying their skills, and to preserve the dignity and best interests of their clients, colleagues, students, research participants, and society.

THE PLANNER

## The scientific method • Figure 1.5

Scientific knowledge is constantly evolving and self-correcting through application of the scientific method. As soon as one research study is published, the cycle begins again.

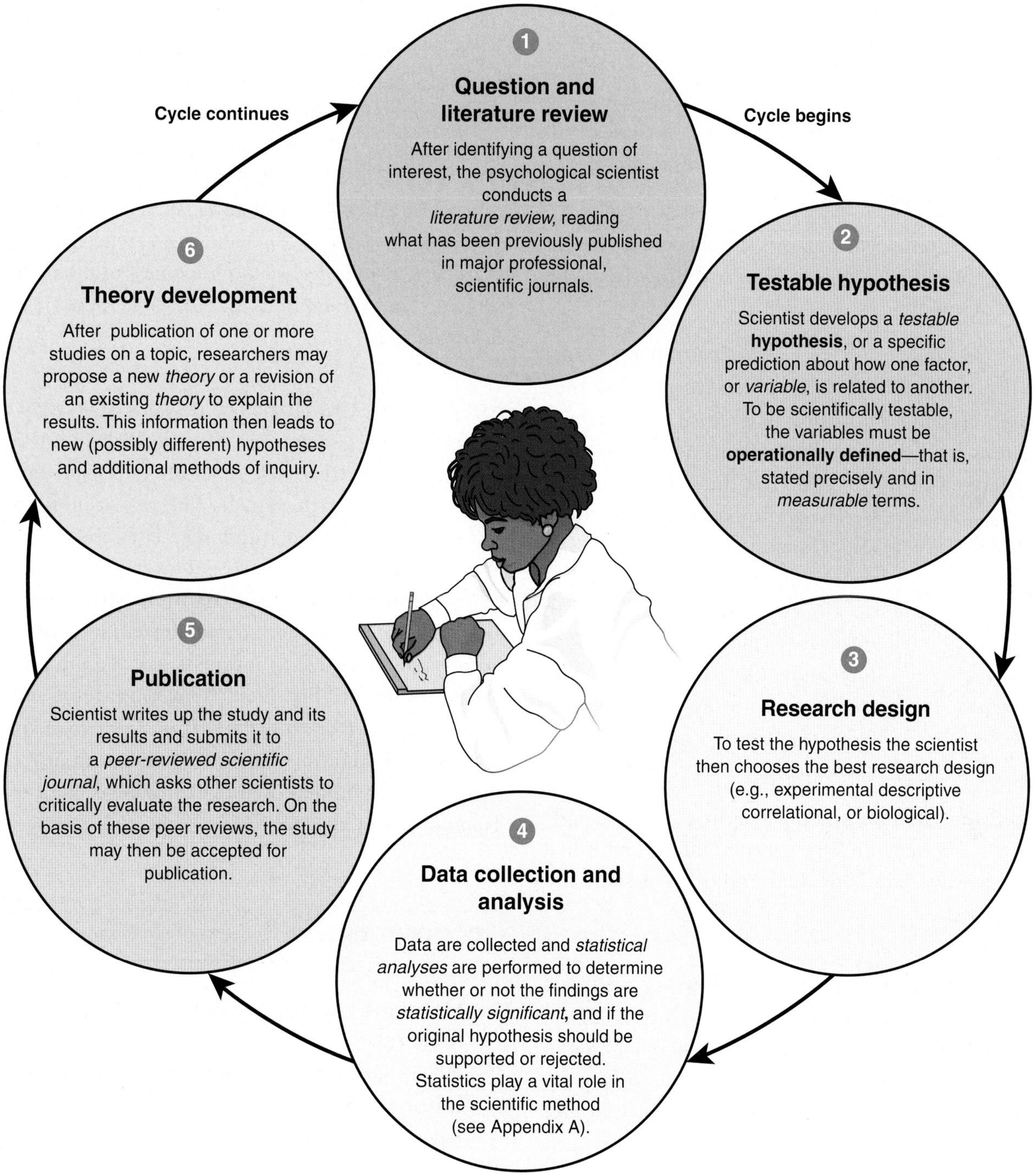

Cycle begins
1
Question and literature review
After identifying a question of interest, the psychological scientist conducts a *literature review,* reading what has been previously published in major professional, scientific journals.
2
Testable hypothesis
Scientist develops a *testable* **hypothesis**, or a specific prediction about how one factor, or *variable*, is related to another. To be scientifically testable, the variables must be **operationally defined**—that is, stated precisely and in *measurable* terms.
3
Research design
To test the hypothesis the scientist then chooses the best research design (e.g., experimental descriptive correlational, or biological).
4
Data collection and analysis
Data are collected and *statistical analyses* are performed to determine whether or not the findings are *statistically significant*, and if the original hypothesis should be supported or rejected. Statistics play a vital role in the scientific method (see Appendix A).
5
Publication
Scientist writes up the study and its results and submits it to a *peer-reviewed scientific journal*, which asks other scientists to critically evaluate the research. On the basis of these peer reviews, the study may then be accepted for publication.
6
Theory development
After publication of one or more studies on a topic, researchers may propose a new *theory* or a revision of an existing *theory* to explain the results. This information then leads to new (possibly different) hypotheses and additional methods of inquiry.
Cycle continues

## Respecting the rights of human participants

The APA has developed rigorous guidelines regulating research with human participants, including:

- *Informed consent* One of the first research principles is obtaining an **informed consent** from all participants before initiating an experiment. Participants should be aware of the nature of the study and significant factors that might influence their willingness to participate. This includes all physical risks, discomfort, or unpleasant emotional experiences.

**informed consent** A participant's agreement to take part in a study after being told what to expect.

- *Voluntary participation* Participants should be told they're free to decline to participate or to withdraw from the research at any time.

- *Restricted use of deception and debriefing* If participants know the true purpose behind certain studies, they may not respond naturally. Therefore, researchers sometimes need to *deceive* participants as to the actual design and reason for the research. But when deception is used, important guidelines and restrictions apply, including debriefing participants at the end of the experiment. **Debriefing** involves explaining the reasons for conducting the research and clearing up any participant's misconceptions, questions, or concerns.

**debriefing** Upon completion of the research, participants are informed of the study's design and purpose, and explanations are provided for any possible deception.

- *Confidentiality* All information acquired about people during a study must be kept private and not published in such a way that individual rights to privacy are compromised.

## Respecting the rights of nonhuman animals

Although they are involved in only 7 to 8% of psychological research (APA, 2009; ILAR, 2009; MORI, 2005) nonhuman animals—mostly rats and mice—have made significant contributions to almost every area of psychology. Without nonhuman animals in *medical research*, how would we test new drugs, surgical procedures, and methods for relieving pain? *Psychological research* with nonhuman animals has led to significant advances in virtually every area of psychology, including the brain and nervous system, health and stress, sensation and perception, sleep, learning, memory, stress, emotion, and so on.

Nonhuman animal research also has produced significant gains for animals themselves—for example, by suggesting more natural environments for zoo animals and more successful breeding techniques for endangered species.

Despite the advantages, using nonhuman animals in psychological research remains controversial (Guidelines for Ethical Conduct, 2008). While debate continues over ethical questions surrounding such research, psychologists take great care in handling research animals. Researchers also actively search for new and better ways to protect them (Guidelines for Ethical Conduct, 2008; Pope & Vasquez, 2011).

## Respecting the rights of psychotherapy clients

Professional organizations, like the APA and APS, as well as academic institutions, and state and local agencies all may require that therapists, like researchers, maintain the highest of ethical standards. They must also uphold their clients' trust. All personal information and therapy records must be kept confidential, with records being available only to authorized persons and with the client's permission. However, therapists are legally required to break confidentiality if a client threatens violence to him- or herself or to others, if a client is suspected of abusing a child or an elderly person, and in other limited situations (Campbell et al., 2010; Pope & Vasquez, 2011; Tyson & Jones, 2011).

### CONCEPT CHECK

STOP

1. How does applied research often build on basic research?
2. What is the difference between a scientific theory and a hunch?
3. Why is research on nonhuman animals valuable?

# Research Methods

## LEARNING OBJECTIVES

**RETRIEVAL PRACTICE** While reading the upcoming sections, respond to each Learning Objective in your own words. Then compare your responses with those in Appendix B.

1. **Explain** why only experiments can identify the cause and effect underlying particular patterns of behavior.
2. **Describe** the three key types of descriptive research.
3. **Define** positive and negative correlation.
4. **Summarize** important methods used in biological research.

Psychologists draw on four major types of psychological research—experimental, descriptive, correlational, and biological (**Study Organizer 1.2**). All have advantages and disadvantages, and most psychologists use several methods to study a single problem. In fact, when multiple methods lead to similar conclusions, scientists have an especially strong foundation for concluding that one variable does affect another in a particular way.

**experimental research** A carefully controlled scientific procedure that determines whether variables manipulated by the experimenter have a causal effect on other variables.

## Experimental Research

**Experimental research** is the most powerful research method because it allows the experimenter to manipulate and control the variables, and thereby determine cause and effect. Only through an experiment can researchers examine a single factor's effect on a particular behavior (Goodwin, 2011). That's because the only way to discover which of many factors has an effect is to experimentally isolate each one.

**Study Organizer 1.2 Psychology's four major research methods** THE PLANNER

| Method | Purpose | Advantages | Disadvantages | |
|---|---|---|---|---|
| **Experimental** (manipulation and control of variables) | Identify cause and effect (meets psychology's goal of *explanation*) | Allows researchers to have precise control over variables, helps identify cause and effect | Ethical concerns, practical limitations, artificiality of lab conditions, uncontrolled variables may confound results, researcher and participant biases | 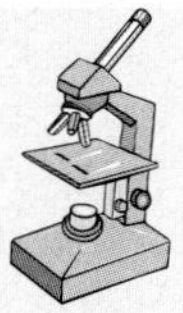 |
| **Descriptive** (naturalistic observation, surveys, case studies) | Observe, collect, and record data (meets psychology's goal of *description*) | Minimizes artificiality, makes data collection easier, allows description of behavior and mental processes as they occur | Little or no control over variables, researcher or participant biases, cannot identify cause and effect | 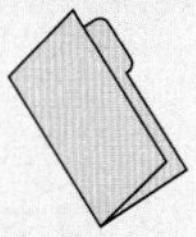 |
| **Correlational** (statistical analyses of relationships between variables) | Identify relationships and assess how well one variable predicts another (meets psychology's goal of *prediction*) | Helps clarify relationships between variables that cannot be examined by other methods and allows prediction | Little or no control over variables, cannot identify cause and effect | 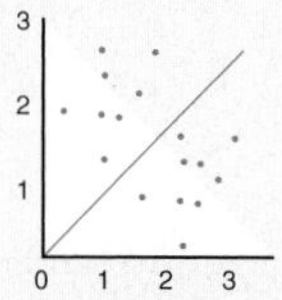  |
| **Biological** (studies the brain and other parts of the nervous system) | Identify contributing biological factors (meets one or more of psychology's goals) | Shares many or all of the advantages of experimental, descriptive, and correlational research | Shares many or all of the disadvantages of experimental, descriptive, and correlational research | 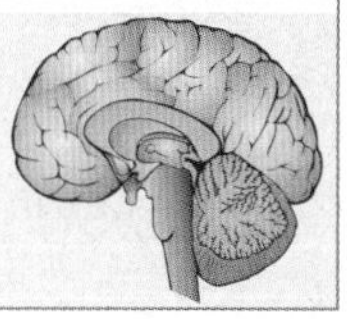 |

Note that the four methods are not mutually exclusive. Researchers may use two or more methods to explore the same topic.

PROCESS DIAGRAM

THE PLANNER

## Experimental research design • Figure 1.6

When designing an experiment, researchers must follow certain steps to ensure that their results are meaningful scientifically. In this example, researchers want to test whether watching violent television increases aggression in children.

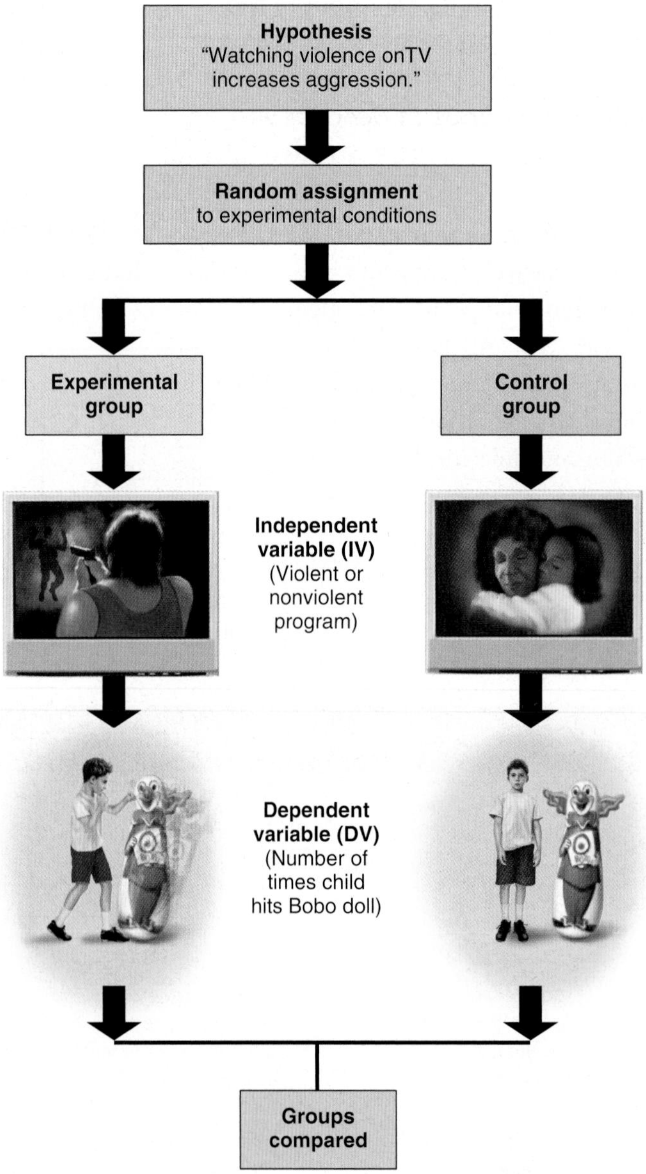


1. The experimenter begins by identifying the hypothesis.
2. In order to avoid sample bias, the experimenter randomly assigns subjects to two different groups.
3. Having at least two groups allows the performance of one group to be compared to that of another.
4. The experimental group watches violent programs while the control group watches nonviolent programs. The amount of violent TV watched is the independent variable (IV).
5. The experimenter then counts how many times the child hits, kicks, or punches a large plastic doll. The number of aggressive acts is the dependent variable (DV).
6. The experimenter relates differences in aggressive behavior (DV) to the amount of violent television watched (IV).

To help you remember the independent and dependent variables (IV and DV), carefully study these drawings and create a visual picture in your own mind of how:

The experimenter "manipulates" the IV to determine its causal effect on the DV.

The experimenter "measures " the DV, which "depends" on the IV.

As illustrated in **Figure 1.6**, an experiment has four critical components: an **independent variable (IV)**, a **dependent variable (DV)**, an **experimental group**, and a **control group**. All extraneous **confounding variables** (such as time of day, lighting conditions, and participants' age and gender) also must be held constant across experimental and control groups so that they do not affect the different groups' results.

**independent variable (IV)** A variable that is manipulated to determine its causal effect on the dependent variable.

**dependent variable (DV)** A variable that is measured; it is affected by (or dependent on) the independent variable.

In addition, a good scientific experiment also protects against potential sources of error from both the researcher and the participants (**Figure 1.7**).

If not controlled for, experimenters can unintentionally let their beliefs and expectations affect participants' responses, producing flawed results. Imagine what might happen if an experimenter breathed a sigh of relief when a participant gave a response that supported the researcher's hypothesis. One way to prevent such **experimenter bias** from destroying the validity of participants' responses is to establish objective methods for collecting and recording data. For example, an experimenter might use audiotape recordings to present stimuli and computers to record responses.

Experimenters also can skew their results when they assume that behaviors that are typical in their own culture are typical in all cultures—a bias known as **ethnocentrism**. One way to avoid this problem is to have researchers from two cultures conduct the same study twice, once with their own culture and once with at least one other culture. This kind of **cross-cultural sampling** isolates group differences in behavior that stem from researchers' ethnocentrism.

In addition to potential problems from the researcher, the participants themselves can also introduce error or bias into an experiment. First, **sample bias** can occur if a particular group of participants does not accurately reflect the composition of the larger population from which they are drawn. For example, critics suggest that much psychological literature is biased because it

**Potential research problems and solutions • Figure 1.7**

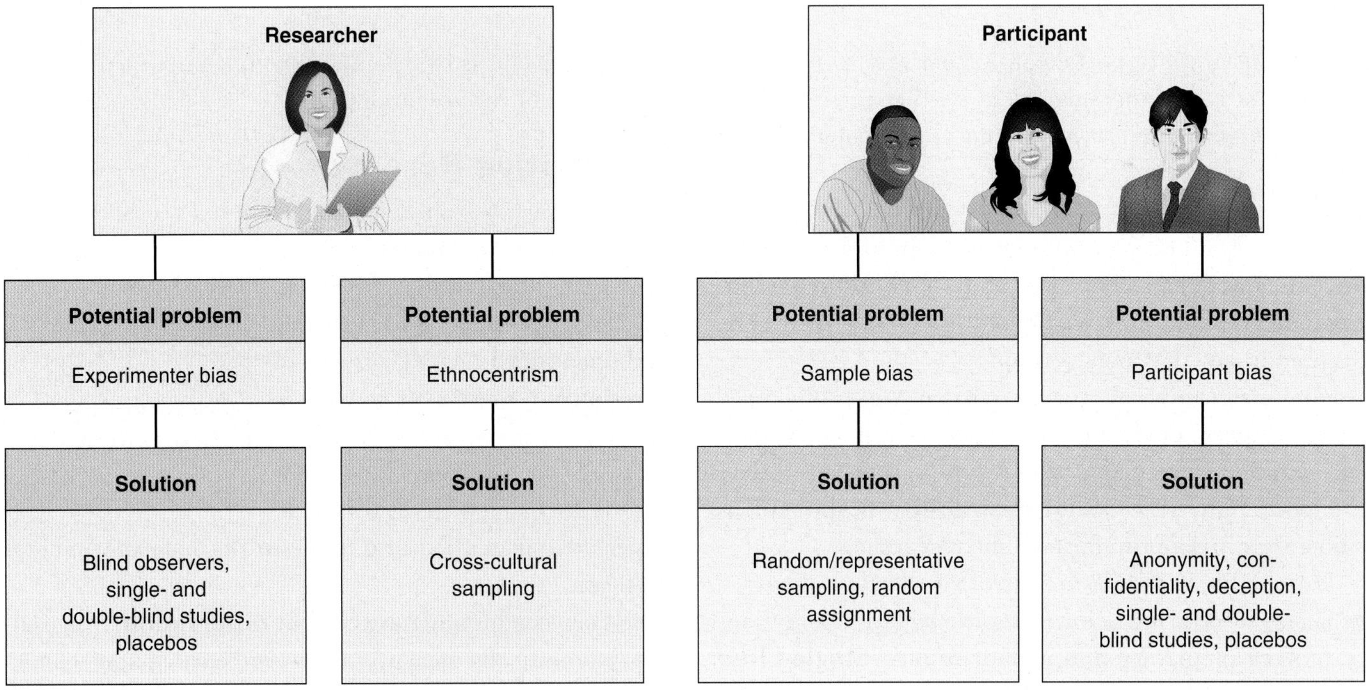

## A single- or double-blind experimental design • Figure 1.8

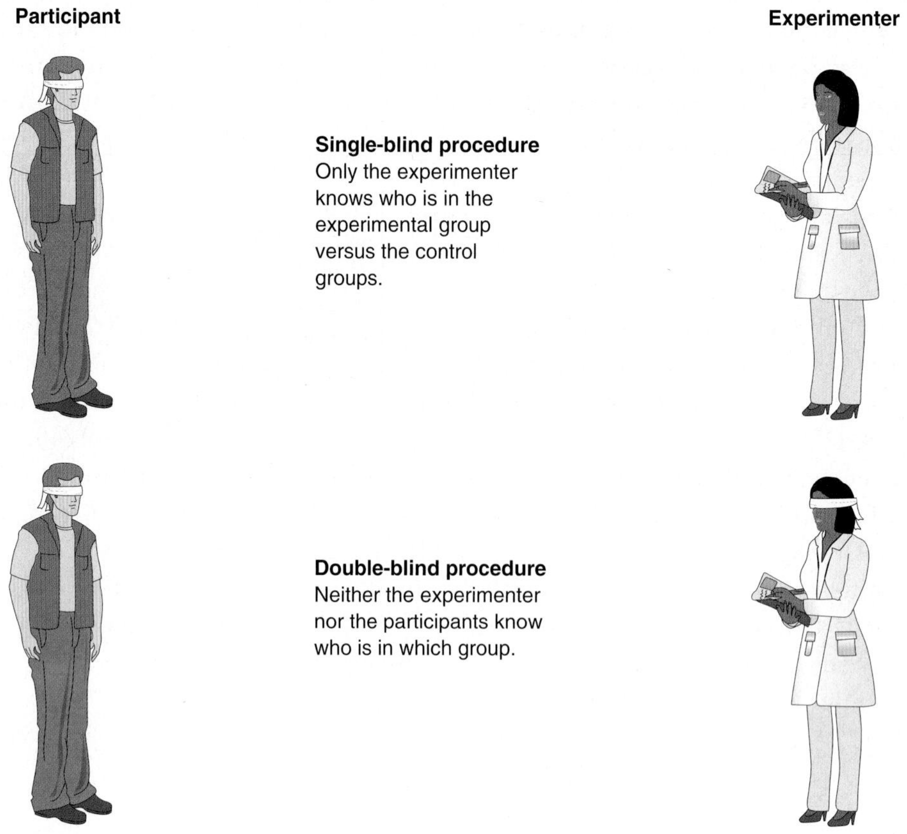

Participant
Experimenter
Single-blind procedure
Only the experimenter knows who is in the experimental group versus the control groups.
Double-blind procedure
Neither the experimenter nor the participants know who is in which group.


To test a new drug, experimenters administering the drug and/or the participants must be unaware (or "blind") as to who is receiving a *placebo* (a fake pill or injection) and who is receiving the drug itself. Researchers use placebos (thus treating experimental and control group participants exactly the same) because they know that the mere act of taking a pill or receiving an injection can change a participant's condition or responses, which is called "the placebo effect" (Barbui et al., 2011; Price, Finniss, & Benedetti, 2008).

primarily uses white participants. One way to minimize sample bias is to randomly select participants who constitute a *representative sample* of the entire population of interest.

Assigning participants to experimental groups using a chance or random system, such as a coin toss, also helps prevent sample bias. This procedure of **random assignment** ensures that each participant is equally likely to be assigned to any particular group.

Bias also can occur when participants are influenced by the experimenter or the experimental conditions. For example, participants may try to present themselves in a good light (the **social desirability response**) or deliberately attempt to mislead the researcher.

Researchers attempt to control for this type of participant bias by offering anonymous participation and other guarantees of privacy and confidentiality. **Single-blind studies**, **double-blind studies**, and **placebos** offer additional safeguards (**Figure 1.8**). Finally, one of the most effective (but controversial) ways of preventing participant bias is by deceiving them (temporarily) about the nature of the research project.

## Descriptive Research

Almost everyone observes and describes others in an attempt to understand them, but in conducting **descriptive research**, psychologists do it systematically and scientifically. The key types of descriptive research are *naturalistic observation*, *surveys*, and *case studies*. Most of the problems and safeguards discussed with regard to the experimental method also apply to the nonexperimental methods.

**descriptive research** Research method used to observe and describe behavior and mental processes without manipulating variables.

When conducting **naturalistic observation**, researchers systematically measure and record participants' behavior, without interfering. Many settings lend themselves to naturalistic observation, from supermarkets to airports to outdoor settings.

The chief advantage of naturalistic observation is that researchers can obtain data about natural behavior, rather than about behavior that is a reaction to an artificial experimental situation. But naturalistic observation can be difficult and time-consuming, and the lack of control by the researcher makes it difficult to conduct observations for behavior that occurs infrequently.

If a researcher wants to observe behavior in a more controlled setting, *laboratory observation* has many of the advantages of naturalistic observation, but with greater control over the variables (**Figure 1.9**).

Psychologists use **surveys** to measure a variety of psychological behaviors and attitudes (Goodwin, 2011). The survey technique includes tests, questionnaires, polls, and interviews. One key advantage of surveys is that researchers can gather data from many more people than is generally possible with other research methods. Unfortunately, most surveys rely on self-reported data, and not all participants are completely honest. Although they can help predict behavior, survey techniques cannot explain causes of behavior.

What if a researcher wants to investigate photophobia (fear of light)? In such a case, it would be difficult to find enough participants to conduct an experiment or to use surveys or naturalistic observation. For rare disorders or phenomena, researchers try to find someone who has the problem and study him or her intensively. Such in-depth studies of a single research participant are called **case studies** (Kazdin, 2011).

## Correlational Research

When nonexperimental researchers want to determine the degree of relationship (correlation) between two variables, they turn to **correlational research**. As the name implies, when any two variables are "co-related," a change in one is accompanied by a change in the other.

**correlational research** A method by which the researcher observes or measures (without directly manipulating) two or more variables to find relationships between them.

Using the *correlational method*, researchers first measure participants' responses or behaviors on variables of interest. Next,

**Laboratory observation • Figure 1.9**

In this type of observation, the researcher brings participants into a specially prepared room in the laboratory, with one-way mirrors, or an inconspicuous camera, and microphones. While hidden from view, the researcher can observe school children at work, families interacting, or the like.

Jeffrey Greenberg/Photo Researchers

# Psychology InSight

## Three types of correlation • Figure 1.10

Correlational research scores individuals on two factors, or variables. Each dot on these graphs (called *scatterplots*) represents one participant's score on both variables. When the dots are plotted on a graph, they demonstrate positive, negative, or zero correlation between the variables.

Chris Carroll/Corbis Images

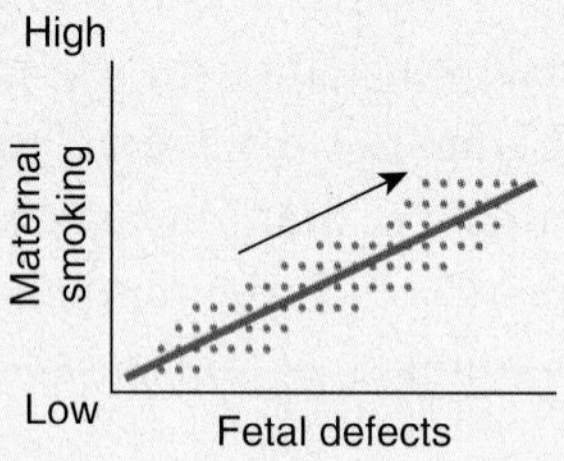


**a. Positive correlation**

In a positive correlation, the variables move (or vary) in the same direction—they increase or decrease together. Research shows there is a positive correlation between maternal smoking and fetal defects: the more a mother smokes, the greater the chances of fetal defects.

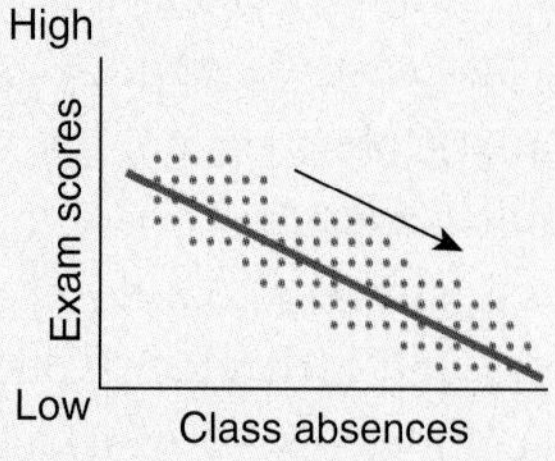


**b. Negative correlation**

In a negative correlation, the variables move in opposite directions—as one variable increases, the other decreases. For example, as the number of class absences increases, exam scores decrease.

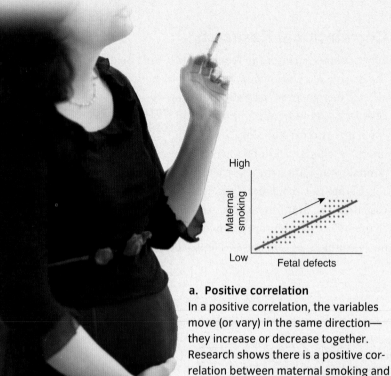

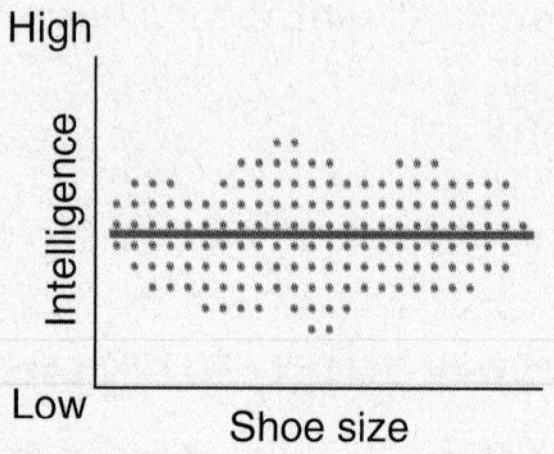


**c. Zero correlation**

In a zero correlation there is no relationship between two variables. For example, there is no correlation between intelligence and shoe size.

**Think Critically**

Can you identify which are the positive, negative, or zero correlations?

1. Health and exercise
2. Hours of TV viewing and student grades
3. Level of happiness and level of helpfulness.
4. Age of driver and weight of car.
5. Resale value and age of car

(Answers: positive, negative, positive, zero, negative)

researchers analyze these results using a statistical formula that gives a **correlation coefficient**. The sign (+ or −) indicates the *direction* of the correlation (positive or negative) (**Figure 1.10**).

> **correlation coefficient** Number indicating strength and direction of the relationship between two variables (from −1.00 to + 1.00).

Correlation coefficients are calculated by a formula (described in Appendix A) that produces a number ranging from −1.00 to +1.00, which indicates the *strength* of the relationship. Note that both a +1.00 and a −1.00 are the strongest possible relationship. As the number decreases and gets closer to 0.00, the relationship weakens. Thus, if you had a correlation of +.92 or −.92, you would have a *strong correlation*. By the same token, a correlation of +.15 or −.15 would represent a *weak correlation*. Correlations close to zero are often interpreted as representing no relationship between the variables—as is the relationship between broken mirrors and years of bad luck.

Correlational research is an important research method for psychologists because it often points to possible causation. It also can help us live safer and more productive lives. For example, correlational studies have repeatedly found high correlations between birth defects and a pregnant mother's use of tobacco and/or alcohol (e.g., Singer & Richardson, 2011). This kind of information enables us to reliably predict our relative risks and to make informed decisions. If you would like additional information about correlations and statistics, see Appendix A at the back of this text.

Unfortunately, people sometimes do not understand that a correlation between two variables does not mean that one variable causes another (**Figure 1.11**). They sometimes read media reports about correlations between stress and cancer, for example, or between family dynamics and homosexuality, and mistakenly infer that "stress causes cancer" or that "withdrawn fathers and overly protective mothers cause their sons to become homosexuals." It's important to realize that a third factor, perhaps genetics, may cause greater susceptibility to both of the correlated phenomena. Although correlational studies do sometimes point to possible causes, only the experimental method manipulates the independent variable under controlled conditions and, therefore, can support conclusions about cause and effect.

Before going on, it's important to recognize that although the experimental method is considered the "gold standard" in research, the other methods also offer unique information and significant value. For example, are you concerned about the increasing research on the dangers of cell phone use (e.g., Davis & Balzano, 2011; Park, 2011; Volkow et al., 2011)? If so, can you see how it might be impossible (and unethical) to do long-term experiments on humans? Instead, we may have to rely on the accumulating weight of the evidence from surveys, correlational studies, and nonhuman animal research. In the meantime, we can play it safe by investing in an inexpensive ear piece (Park, 2011).

## Correlation versus causation • Figure 1.11

Correlational, nonexperimental studies are important because they reveal associations between variables. To determine causation (what causes what), we would need to conduct an experiment, which controls the variables.

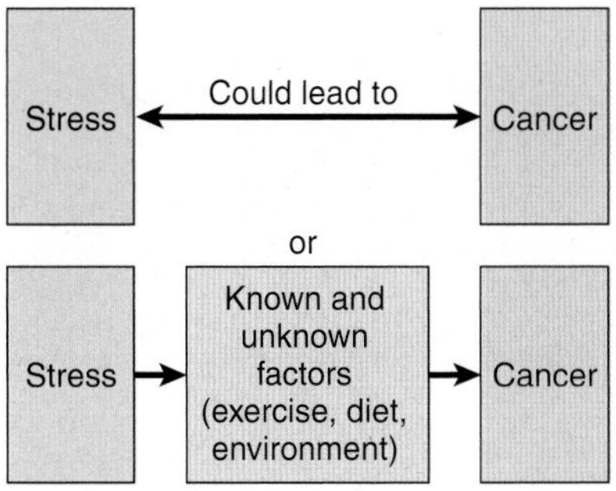


**a.** Research has found a strong correlation between stress and cancer (Chapter 3). However, this correlation does not tell us whether stress causes cancer, whether cancer causes stress, or whether other known and unknown factors, such as eating, drinking, and smoking, could contribute to both stress and cancer. Can you think of a way to study the effects of stress on cancer that is not correlational?

Norbert Schwerin/The Image Works

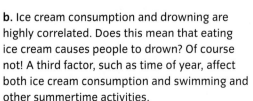

Chris Lowe/Photolibrary/Getty Images

**b.** Ice cream consumption and drowning are highly correlated. Does this mean that eating ice cream causes people to drown? Of course not! A third factor, such as time of year, affect both ice cream consumption and swimming and other summertime activities.

**Sample tools for biological research Table 1.2**

| | Method | Description | Purpose |
|---|---|---|---|
| 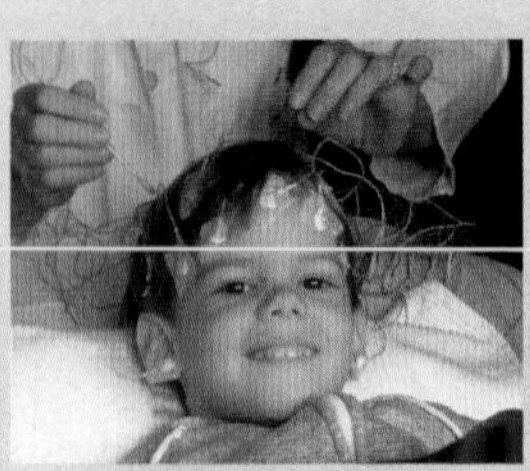 Larry Mulvehill/Photo Researchers, Inc. | **Electrical recordings** Electrical activity throughout the brain sweeps in regular waves across its surface, and the electroencephalogram (EEG) is a readout of this activity. | Using electrodes attached to the skin or scalp, brain activity is detected and recorded on an electroencephalogram (EEG). To explore deep brain activity, a thin wire probe is inserted inside the brain in or near a neuron. | Reveals areas of the brain most active during particular tasks or mental states, like reading or sleeping; also traces abnormal brain waves caused by brain malfunctions, like epilepsy or tumors. Intrabrain wire probes allow scientists to "see" individual neuron activity. |
| 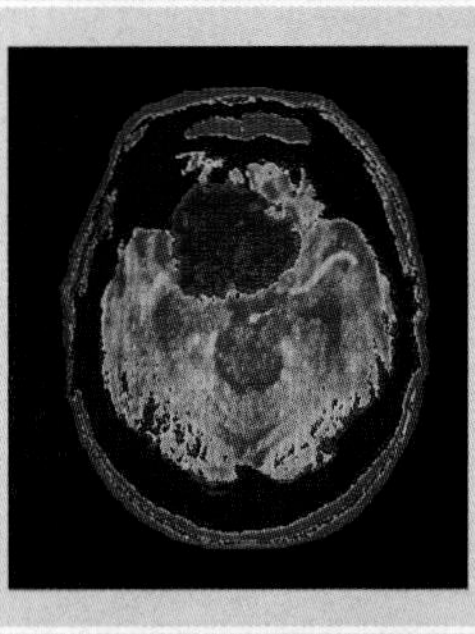 Mehau Kulyk/Photo Researchers, Inc. | **CT (computed tomography) scan** This CT scan used X-rays to locate a brain tumor, which is the deep purple mass at the top left. | Computer-created cross-sectional X-rays of the brain or other parts of the body, which produce 3-D images. Least expensive type of imaging and is widely used in research. | Reveals the effects of strokes, injuries, tumors, and other brain disorders. |
| 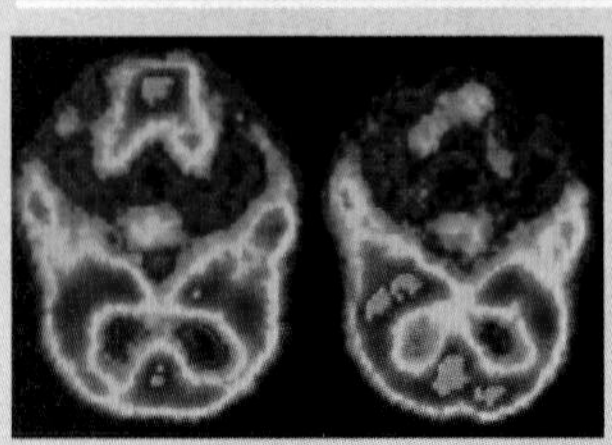 N.I. H/Photo Researchers, Inc. | **PET (positron emission tomography) scan** The left PET scan shows brain activity when the eyes are open, whereas the one on the right is with the eyes closed. Note the increased activity, red and yellow, in the left photo when the eyes are open. | Radioactive form of glucose injected into the bloodstream; scanner records amount of glucose used in particularly active areas of the brain, producing a computer-constructed picture of the brain. | Originally designed to detect abnormalities, now used to identify brain areas active during ordinary activities (reading, singing). |
| 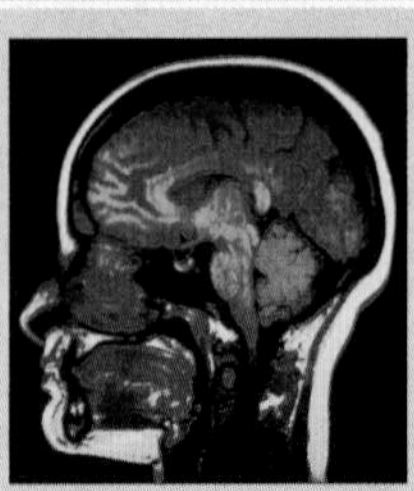 Scott Camazine/Photo Researchers, Inc. | **MRI (magnetic resonance imaging)** Note the fissures and internal structures of the brain. The throat, nasal airways, and fluid surrounding the brain are dark. | Powerful electromagnets produce high-frequency magnetic field that is passed through the brain. | Produces high-resolution three-dimensional pictures of the brain useful for identifying abnormalities and mapping brain structures and function. |
| 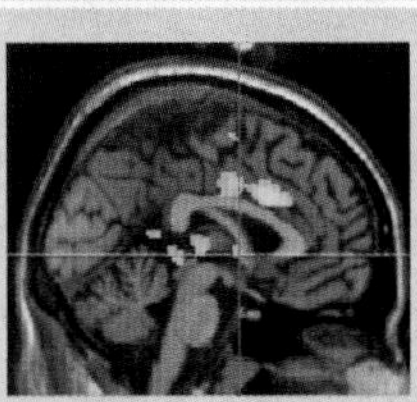 Science Photo Library/Photo Researchers, Inc. | **fMRI (functional magnetic resonance imaging)** | Newer, faster version of the MRI that detects blood flow by picking up magnetic signals from blood, which has given up its oxygen to activate brain cells. | Measures blood flow, which indicates areas of the brain that are active or inactive during ordinary activities or responses (like reading or talking); also shows changes associated with various disorders. |
| 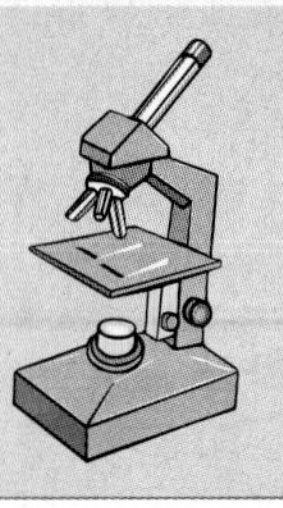 | **Other Methods:** (a) **Cell body or tract (myelin) staining** (b) **Microinjections** (c) **Intrabrain electrical recordings** | (a) Colors/stains selected neurons or nerve fibers. (b) Injects chemicals into specific areas of the brain. (c) Records activity of one or a group of neurons inside the brain. | Increases overall information of structure and function through direct observation and measurement. |

## Biological Research

**Biological research** examines the biological processes that are involved in our feelings, thoughts, and behavior.

**biological research** The scientific study of the brain and other parts of the nervous system

The earliest explorers of the brain *dissected* the brains of deceased humans and conducted experiments on other animals using *lesioning* techniques (systematically destroying brain tissue to study the effects on behavior and mental processes). By the mid-1800s, this research had produced a basic map of the nervous system, including some areas of the brain. Early researchers also relied on clinical observations and case studies of living people who had experienced injuries, diseases, and disorders that affected brain functioning.

Modern researchers still use such methods, but they also employ other techniques to examine biological processes that underlie our behavior (**Table 1.2**). For example, recent advances in brain science have led to various types of brain-imaging scans, which can be used in both clinical and laboratory settings. Most of these methods are relatively *noninvasive*—that is, their use does not involve breaking the skin or entering the body .

**CONCEPT CHECK** STOP

1. **How** do psychologists guard against bias?
2. **What** are the advantages and disadvantages of naturalistic research?
3. **What** is meant by the axiom "correlation is not causation"?
4. **What** does it mean to call a procedure "noninvasive"?

# Strategies for Student Success

## LEARNING OBJECTIVES

**RETRIEVAL PRACTICE** While reading the upcoming sections, respond to each Learning Objective in your own words. Then compare your responses to those in Appendix B.

1. **Describe** the steps you can take to improve your study habits.
2. **Explain** how visual features can enhance learning.
3. **Identify** how you might improve your current time management habits.

In this section, you will find several well-researched study strategies and techniques guaranteed to make you a more successful college student. They are grouped together under *study habits*, *visual learning*, and *time management.*

## Study Habits

Do you sometimes read a paragraph many times and still remember nothing from it? Here are four ways to successfully read (and remember) information in this and most texts.

**1. Familiarization** The first step is to familiarize yourself with the general text so that you can take full advantage of its contents. In *Visualizing Psychology*, the Preface, Table of Contents, Glossary (both running through the text of each chapter and at the end of each chapter), References, Name Index, and Subject Index will help give you a bird's-eye view of the rest of the text. In addition, as you scan the book to familiarize yourself with its contents, you should also take note of the many tables, figures, photographs, and special feature boxes, all of which will enhance your understanding of the subject.

**2. Active reading** The next step is to make a conscious decision to actively read and learn the material. Reading a text is NOT like reading a novel or fun articles on the Internet! You must tell your brain to slow down, focus on details, and save the material for future recall.

PROCESS DIAGRAM

## Using the SQ4R method • Figure 1.12

THE PLANNER

Follow these steps to improve the effectiveness of your reading.

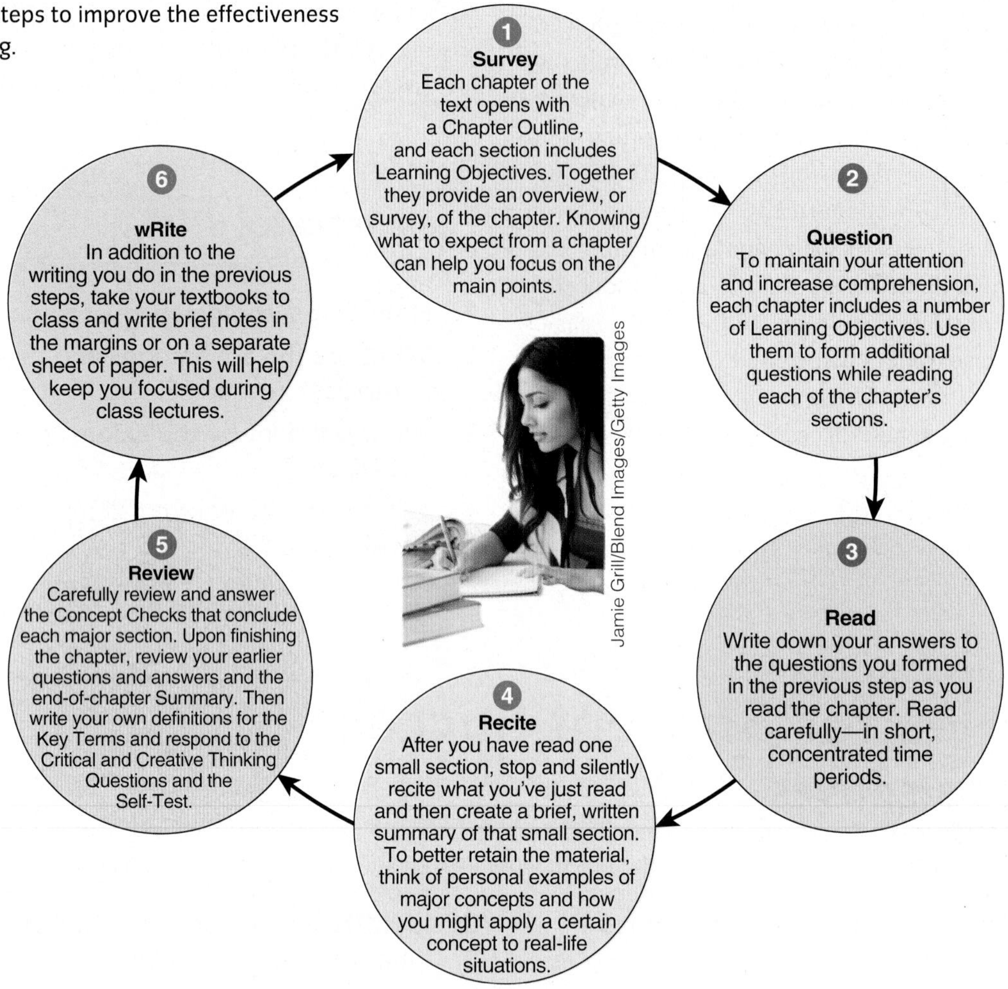

1
Survey
Each chapter of the text opens with a Chapter Outline, and each section includes Learning Objectives. Together they provide an overview, or survey, of the chapter. Knowing what to expect from a chapter can help you focus on the main points.
2
Question
To maintain your attention and increase comprehension, each chapter includes a number of Learning Objectives. Use them to form additional questions while reading each of the chapter's sections.
3
Read
Write down your answers to the questions you formed in the previous step as you read the chapter. Read carefully—in short, concentrated time periods.
4
Recite
After you have read one small section, stop and silently recite what you've just read and then create a brief, written summary of that small section. To better retain the material, think of personal examples of major concepts and how you might apply a certain concept to real-life situations.
5
Review
Carefully review and answer the Concept Checks that conclude each major section. Upon finishing the chapter, review your earlier questions and answers and the end-of-chapter Summary. Then write your own definitions for the Key Terms and respond to the Critical and Creative Thinking Questions and the Self-Test.
6
wRite
In addition to the writing you do in the previous steps, take your textbooks to class and write brief notes in the margins or on a separate sheet of paper. This will help keep you focused during class lectures.
Jamie Grill/Blend Images/Getty Images

One of the best ways to read actively is to use the **SQ4R method**, which was developed by Francis Robinson (1970). The initials stand for six steps in effective reading: **S**urvey, **Q**uestion, **R**ead, **R**ecite, **R**eview, and w**R**ite. As you might have guessed, *Visualizing Psychology* was designed to incorporate each of these steps (**Figure 1.12**).

**3. Distributed study** Distributed, spaced practice is a much more efficient way to study and learn than massed practice (Chapter 7). That is, you will learn material more thoroughly if you distribute your study over time, rather than trying to cram all the information in at once.

**4. Overlearning** Many people tend to study new material just to the point where they can recite the information, but they do not attempt to understand it more deeply. For best results, you should be sure to *overlearn*. In other words, be sure to fully understand how key terms and concepts are related to one another, and also be able to generate examples other than the ones in the text.

In addition, you should repeatedly review the material (by visualizing the phenomena that are described and explained in the text and by rehearsing what you have learned) until the information is firmly locked in place. Overlearning is particularly important if you suffer from test anxiety.

For additional strategies to improve your study habits see *Applying Psychology*.

## Visual Learning

Our brains are highly tuned to visual cues as well as verbal cues. Photographs, drawings, and other graphical information help us solidify our understanding, organize and internalize new material, recognize patterns and interrelationships, and think creatively.

In some books, photographs and illustrations merely repeat, visually, concepts that are also stated in words. *Visualizing Psychology* is different. The photographs, drawings, diagrams, and graphs in this book carry their own weight—that is, they serve a specific instructional purpose, above and beyond what is stated in words. They are as essential as the text itself; be sure to pay attention to them. (The Illustrated Book Tour in the Preface describes these features in detail.)

## Time Management

If you find that you can't always strike a good balance between work, study, and social activities or that you aren't always good at budgeting your time, here are five basic time-management strategies:

- *Establish a baseline.* Before attempting any changes, simply record your day-to-day activities for one to two weeks. You may be surprised by how you spend your time.
- *Set up a realistic schedule.* Make a daily and weekly "to do" list, including all required activities, basic maintenance tasks (like laundry, cooking, child care, and eating), and a reasonable amount of "down time." Then create a daily schedule of activities that includes time for each of these. To make permanent time-management changes, shape your behavior, starting with small changes and building on them.
- *Reward yourself.* Give yourself immediate, tangible rewards for sticking with your daily schedule.
- *Maximize your time.* Time-management experts, such as Alan Lakein (1998), suggest that you should try to minimize the amount of time you spend worrying and complaining and fiddling around getting ready to study ("fretting and prepping"). Also be alert for hidden "time opportunities"—spare moments that

# Applying Psychology

THE PLANNER

## Improving Your Grade

Here are several specific strategies for grade improvement and test taking that can further improve your student success:

- *Note taking* Effective note taking depends on active listening. Ask yourself, "What is the main idea?" Write down key ideas and supporting details and examples.
- *Understanding your professor* The amount of lecture time spent on various topics is generally a good indication of what the instructor considers important.
- *General test taking* On multiple-choice exams, carefully read each question and all of the alternative answers before responding. Be careful to respond to all questions, and make sure that you have recorded your answers correctly. Finally, practice your test taking by responding to the Concept Check questions, the Critical and Creative Thinking Questions, and the Self-Test in each chapter.
- *Skills courses* Improve your reading speed and comprehension and your word-processing/typing skills by taking additional courses designed to develop these specific abilities.
- *Additional resources* Don't overlook important human resources. Your instructors, roommates, classmates, friends, and family members often provide useful tips and encouragement.

© Chris Schmidt/iStockphoto

**Think Critically**

1. What factors might prevent you from reading test questions carefully and responding accurately?
2. How could a friend or roommate help you improve your grades on tests?

# Psychological Science

THE PLANNER

## Serious Problems with Multitasking

*By Thomas Frangicetto*
*(Northampton Community College, Bethlehem, PA)*

Do you believe you're more efficient and productive when you multitask (e.g., surfing the net while listening to class lectures, or texting your friends while driving a car)? If so, try this test.

**a.** Using a stopwatch, test to see how fast you can name the color of each rectangular box.

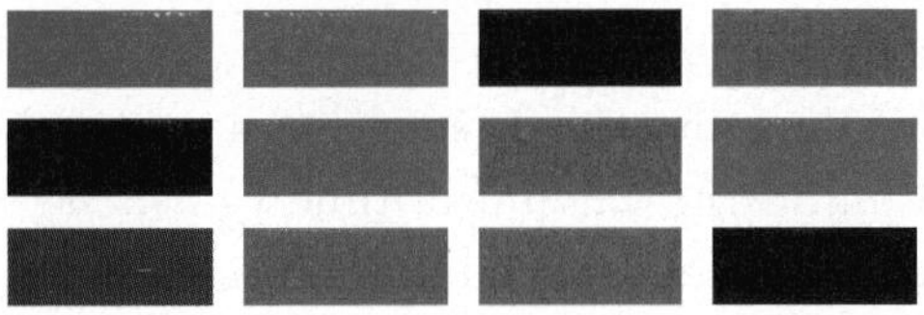

**b.** Now, time yourself to see how fast you can state the color of ink used to print each word, ignoring what each word says.

| | | | |
|---|---|---|---|
| GREEN | RED | BROWN | RED |
| BROWN | GREEN | GREEN | BLUE |
| GREEN | BROWN | RED | BLUE |

Most people find it takes more time and that they make more errors on **b.** than on **a.** Your tendency to read the word in **b.**, instead of saying the color of ink as instructed, is known as the *Stroop effect*, after the psychologist who discovered it.

Just as the distraction of trying to ignore the words in **b.** greatly increased your time and errors, trying to ignore or answer a cell-phone call while listening to a lecture or driving a car can lead to increased errors on exams and increased odds of a serious traffic accident. In short, it's difficult and sometimes dangerous to do two things at once—especially when one task is more highly practiced.

The problems with multitasking were documented in a study at Stanford University of "high-tech jugglers"—people who are consistently exposed to several sources of electronic information at the same time.

Using 100 college students as their population, the researchers ran a series of three tests on attention to detail, memory, and ability to switch from one task to another. In each test, the subjects were divided into two groups, those who regularly multitask and those who don't. The outcomes for all three tests were consistent: The heavy multitaskers were outperformed by the nonmultitaskers.

In short, the researchers found that the self-described multitaskers paid less attention to detail, displayed poorer memory, and had more trouble switching from one task to another compared to subjects who preferred doing only one task at a time (Ophir, Nass, & Wagner, 2009). Like the Stroop test, subjects failed to selectively attend to what was relevant versus irrelevant.

### Identify the Research Method

1. What is the most likely research method for the Stanford University (Ophir, Nass, & Wagner) study of multitasking?
2. If you chose
   - the experimental method, label the IV, DV, experimental group, and control group.
   - the descriptive method, is this a naturalistic observation, survey, or case study?
   - the correlational method, is this a positive, negative, or zero correlation?
   - the biological method, identify the specific research tool (e.g., brain dissection, CT scan).

(Check your answers in Appendix C.)

normally go to waste, which you might instead use productively.

- *Avoid multitasking*. You may feel like you are using your time efficiently when you do more than one thing at once, for example, texting your friends while listening to classroom lectures. To test this assumption, try the exercise in *Psychological Science*.

### CONCEPT CHECK

STOP

1. **What** are the elements of active reading?
2. **How** does this textbook take advantage of your visual learning abilities?
3. **What** can you do to maximize your time?

# Summary

THE PLANNER

## 1 Introducing Psychology 4

- **Psychology** is the scientific study of **behavior** and **mental processes**. The discipline places high value on **empirical evidence** and **critical thinking**. **Pseudopsychologies**, such as psychic powers, are not based on scientific evidence, as James Randi demonstrates in his performances, shown in the photo.

Myth Busters: Psychology Versus Pseudopsychology

- Psychology's four basic goals are to describe, explain, predict, and change behavior and mental processes through the use of scientific methods. One of the most enduring debates in science has been the **nature–nurture controversy**.
- Psychologists work as therapists, researchers, teachers, and consultants in a wide range of settings.

## 2 Origins of Psychology 7

- Wilhelm Wundt, considered the father of psychology, and his followers, including Edward Titchener, were interested in studying conscious experience. Their approach, **structuralism**, sought to identify the basic structures of mental life through introspection.
- The **functionalism** school, led by William James, studied how the mind functions to enable humans and other animals to adapt to their environment.
- The **psychoanalytic school**, founded by Sigmund Freud, emphasized the influence of the unconscious mind, which lies outside personal awareness.
- Contemporary psychology reflects seven major perspectives: **psychodynamic**, **behavioral**, **humanistic**, **cognitive**, **biological**, **evolutionary**, and **sociocultural**. Most contemporary psychologists embrace a unifying perspective known as the **biopsychosocial model**, illustrated in this diagram.

What a Psychologist Sees: The Biopsychosocial Model

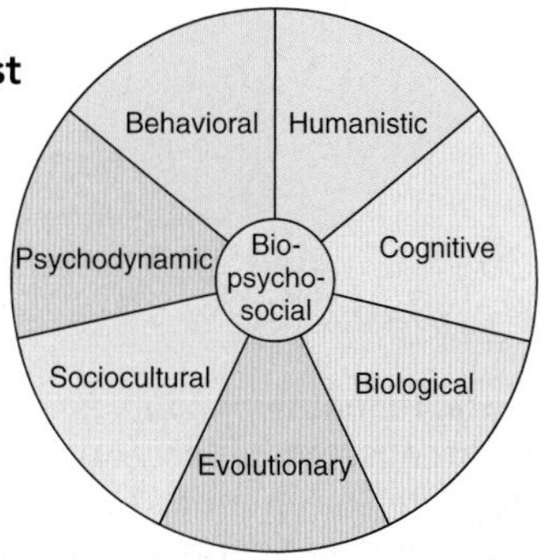


- Despite early societal and cultural limitations, women and minorities have made important contributions to psychology. Pioneers include Margaret Floy Washburn, Mary Calkins, Francis Cecil Sumner, and Kenneth and Mamie Clark.

## 3 The Science of Psychology 12

- **Basic research** is aimed at advancing general scientific understanding, whereas **applied research** works to address real-world problems.
- Most scientific investigations involve six basic steps, collectively known as the **scientific method**, shown in the diagram. Scientific progress comes from repeatedly challenging and revising existing **theories** and building new ones.

The scientific method • Figure 1.5

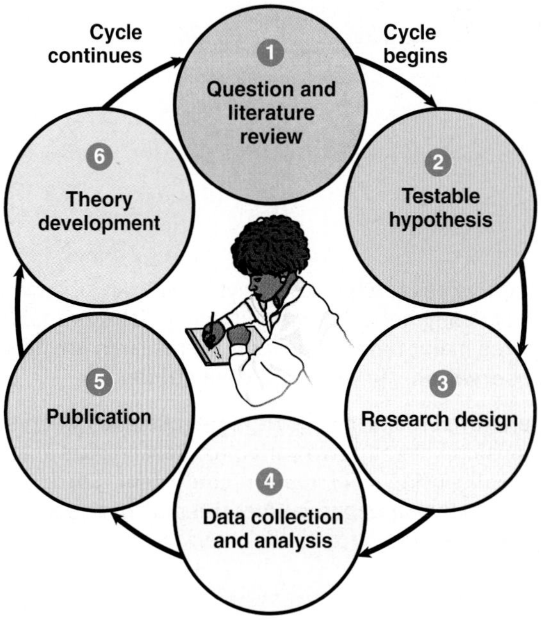

- Psychologists must maintain high ethical standards. This includes respecting the rights of both human and nonhuman research participants and of psychotherapy clients.

- **Informed consent** and **debriefing** are critical elements of any research that involves human participants. Researchers and clinicians are held professionally responsible for their actions by the APA, by their institutions, and by local and state agencies.

## 4 Research Methods 15

- **Experimental research** manipulates and controls variables to determine cause and effect. An experiment has four critical components: **independent** and **dependent variables**, and **experimental** and **control groups**. A good scientific experiment protects against potential sources of error from both the researcher and the participants through means such as **single-blind** and **double-blind studies**, shown in the diagram.

### A single- or double-blind experimental design • Figure 1.8

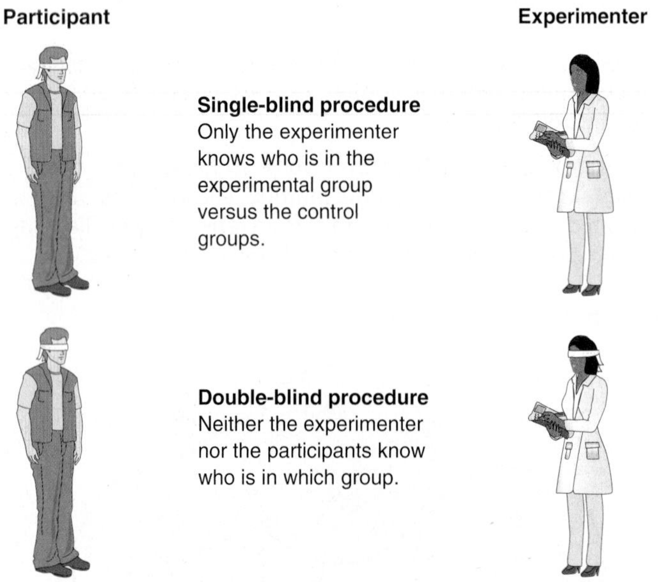

Participant
Experimenter
Single-blind procedure
Only the experimenter knows who is in the experimental group versus the control groups.
Double-blind procedure
Neither the experimenter nor the participants know who is in which group.


- **Descriptive research** involves systematically observing and describing behavior without manipulating variables. The three major types of descriptive research are **naturalistic observation**, **surveys**, and **case studies**.

- **Correlational research** allows researchers to observe the relationship between two variables. Researchers analyze their results using a **correlation coefficient**. Correlations can be positive or negative. A correlation between two variables does not necessarily mean that one causes the other.

- **Biological research** focuses on internal, biological processes that are involved in our feelings, thoughts, and behavior. Recent advances in brain imaging have improved scientists' ability to examine these processes and to do so noninvasively.

## 5 Strategies for Student Success 23

- Several well-documented techniques will help readers understand and absorb the material in this book most completely. Tips for improved study habits include familiarization, active reading through the **SQ4R method** (shown in the diagram), distributed study, and overlearning.

### Using the SQ4R method • Figure 1.12

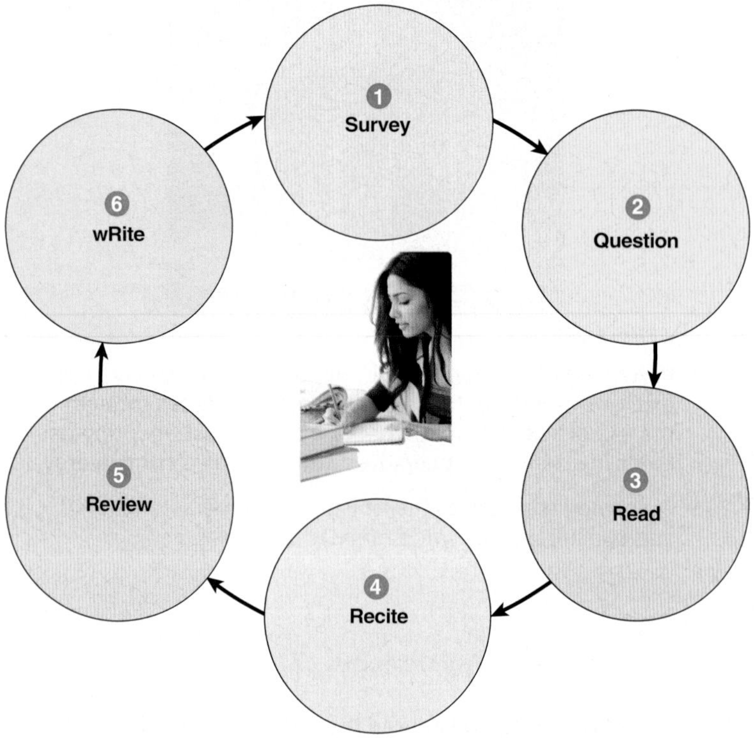


- In addition to paying attention to visual learning and time management, several other strategies for grade improvement and test taking can further improve course performance. These include more effective note taking, understanding the professor, better test-taking strategies, study skills courses, and other helpful resources.

# Key Terms

**RETRIEVAL PRACTICE** Write a definition for each term before turning back to the referenced page to check your answer.

- applied research 12
- basic research 12
- behavior 4
- behavioral perspective 9
- biological perspective 9
- biological research 23
- biopsychosocial model 10
- case study 19
- cognitive perspective 9
- confounding variable 17
- control group 17
- correlation coefficient 20
- correlational research 19
- critical thinking 4
- cross-cultural sampling 17
- debriefing 14
- dependent variable (DV) 17
- descriptive research 18
- double-blind study 18
- empirical evidence 4
- ethnocentrism 17
- evolutionary perspective 9
- experimental group 17
- experimental research 15
- experimenter bias 17
- functionalism 7
- humanistic perspective 9
- hypothesis 13
- independent variable (IV) 17
- informed consent 14
- mental process 4
- meta-analysis 12
- naturalistic observation 18
- nature–nurture controversy 4
- operational definition 13
- placebo 18
- positive psychology 9
- pseudopsychology 4
- psychology 4
- psychoanalytic school 7
- psychodynamic perspective 8
- random assignment 18
- sample bias 17
- scientific method 12
- single-blind study 18
- social desirability response 18
- sociocultural perspective 9
- SQ4R method 24
- structuralism 7
- survey 19
- theory 12

# Creative and Critical Thinking Questions

1. Psychologists are among the least likely to believe in psychics, palmistry, astrology, and other paranormal phenomena. Why might that be?
2. Which psychological perspective would most likely be used to study and explain why some animals, such as newly hatched ducks or geese, follow and become attached to (or imprinted on) the first large moving object they see or hear?
3. Why is the scientific method described as a cycle rather than as a simple six-step process?
4. Imagine that a researcher recruited research participants from among her friends and then assigned them to experimental or control groups based on their gender. Why might this be a problem?
5. Which modern methods of examining how the brain influences behavior are noninvasive?
6. What do you think keeps most people from fully employing the strategies for student success presented in this chapter?

# What is happening in this picture?

Nonhuman animals are sometimes used in psychological research when it would be impractical or unethical to use human participants. These mice are subjects in psychological research. Opinions are sharply divided on the question of whether nonhuman animal research is ethical.

**Think Critically**

1. What research questions might require the use of nonhuman animals?
2. What safeguards help ensure the proper treatment of these animals?

Jonathan Selig/Getty Images

# Self-Test

**RETRIEVAL PRACTICE** Completing this self-test and comparing your answers with those in Appendix C provides immediate feedback and helpful practice for exams. Additional interactive self-tests are available at www.wiley.com/college/carpenter.

1. In this text, psychology is defined as the _____.
   a. science of conscious and unconscious forces on behavior
   b. empirical study of the mind
   c. scientific study of the mind
   d. scientific study of behavior and mental processes

2. According to your textbook, the goals of psychology are to _____.
   a. explore the conscious and unconscious functions of the human mind
   b. understand, compare, and analyze human behavior
   c. improve psychological well-being in all individuals from conception to death
   d. describe, explain, predict, and change behavior and mental processes

3. _____ is generally acknowledged to be the father of psychology.
   a. Sigmund Freud
   b. B. F. Skinner
   c. Wilhelm Wundt
   d. William James

4. The biopsychosocial model is known as a(n) _____.
   a. integrative, unifying model
   b. concept formation
   c. consolidation model
   d. eclectic conceptualization

5. The term *basic research* is BEST defined as research that _____.
   a. is basic to one field only
   b. is intended to advance scientific knowledge rather than for practical application
   c. is done to get a grade or a tenured teaching position
   d. solves basic problems encountered by humans and animals in a complex world

6. Identify and label the six steps in the scientific method.

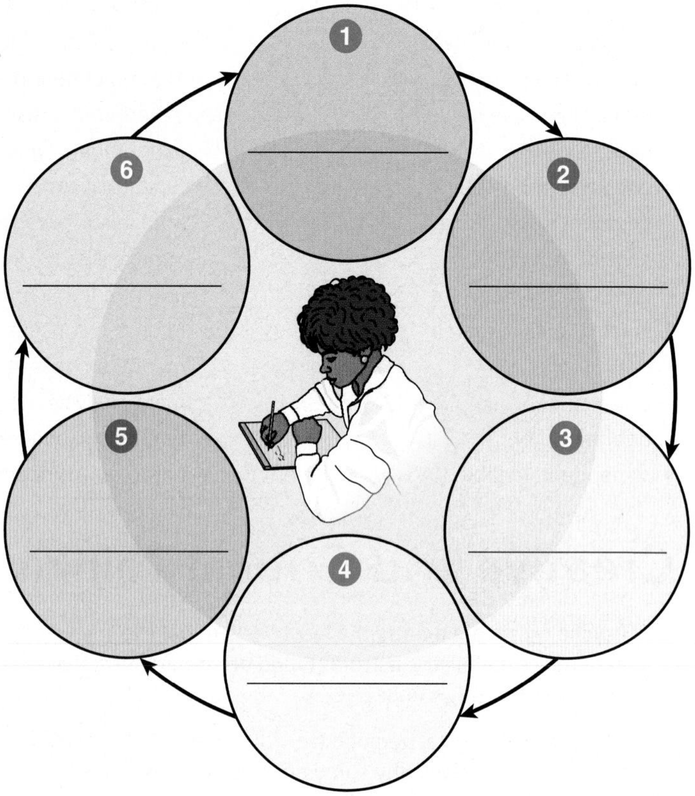


7. A(n) _____ allows variables to be stated precisely and in measurable terms.
   a. meta-analysis
   b. theory
   c. independent observation
   d. operational definition

**8.** According to your text, debriefing is _____.

a. interviewing subjects after a study to find out what they were thinking during their participation

b. explaining the design and purpose of the study, and any deception used when the study is over.

c. disclosing potential physical and emotional risks, and the nature of the study prior to its beginning

d. interviewing subjects after a study to determine whether any deception used was effective in preventing them from learning the true purpose of the study

**9.** _____ are manipulated; _____ are measured.

a. Dependent variables; independent variables

b. Surveys; experiments

c. Statistics; correlations

d. Independent variables; dependent variables

**10.** If researchers gave participants varying amounts of a new "memory" drug and then gave them a story to read and measured their scores on a quiz, the _____ would be the IV, and the _____ would be the DV.

a. response to the drug; amount of the drug

b. experimental group; control group

c. amount of the drug; quiz scores

d. researcher variables; extraneous variables

**11.** The **BEST** definition of a double-blind study is research in which _____.

a. nobody knows what they are doing

b. neither the participants in the treatment group nor the control group know which treatment is being given to which group

c. both the researcher and the participants are unaware of who is in the experimental and control groups

d. two control groups (or placebo conditions) must be used

**12.** In a case study, a researcher is most likely to _____.

a. interview many research subjects who have a single problem or disorder

b. conduct an in-depth study of a single research participant

c. choose and investigate a single topic

d. use any of these options, which describe different types of case studies

**13.** In _____ research, a researcher observes or measures (without directly manipulating) two or more variables to find relationships between them, without implying a causal relationship.

a. experimental

b. correlational

c. basic

d. applied

**14.** Which of the following correlation coefficients indicates the strongest relationship?

a. $+.43$

b. $-.64$

c. $-.72$

d. 0.00

**15.** Label the six steps of the SQ4R study method.

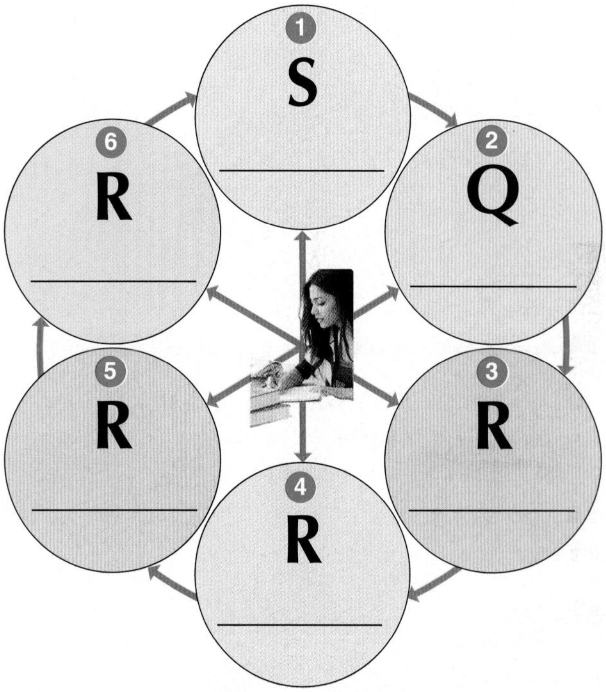


**THE PLANNER**

Review your Chapter Planner on the chapter opener and check off your completed work.

# 2 Neuroscience and Biological Foundations

*The brain is the last and grandest biological frontier, the most complex thing we have yet discovered in our universe. It contains hundreds of billions of cells interlinked through trillions of connections. The brain boggles the mind.*

*James Watson*
*(Nobel Prize Winner)*

What do you *visualize* and anticipate you'll learn while studying this particular chapter? When you flip through the pages and see so many figures and diagrams of the brain and nervous system, do you wonder why these topics are covered in psychology?

Ancient cultures, including the Egyptian, Indian, and Chinese, believed the heart was the center of all thoughts and emotions. Today we know that the brain and the rest of the nervous system are the power behind our psychological life and much of our physical being. This chapter introduces you to the important and exciting field of **neuroscience** and **biopsychology**, the scientific study of the *biology* of behavior and mental processes, while also providing a foundation for understanding several fascinating discoveries and essential biological processes discussed throughout the text.

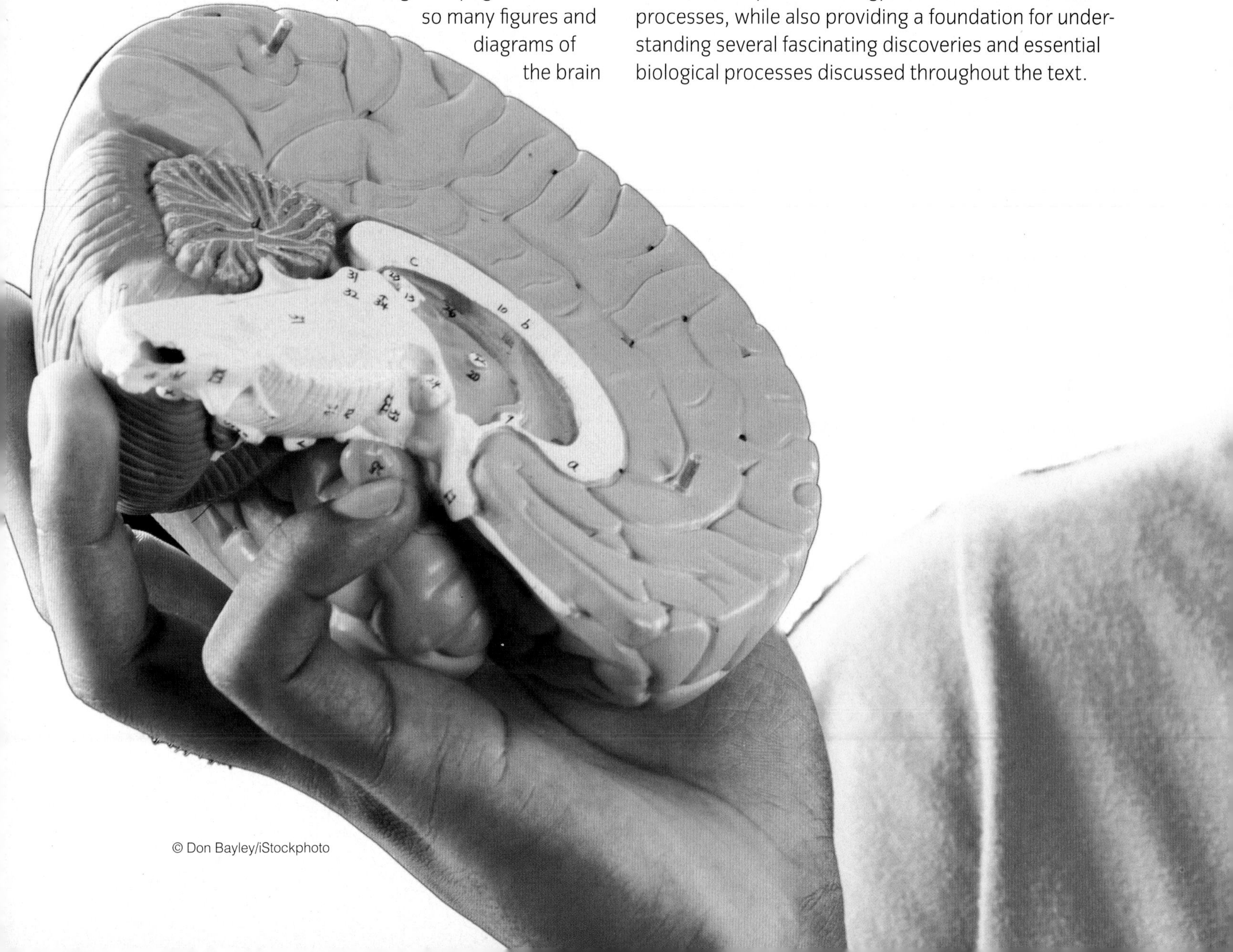

## CHAPTER OUTLINE



## CHAPTER PLANNER ✓

- ❑ Study the picture and read the opening story.
- ❑ Answer the Learning Objectives in each section:
  p. 34 ❑ p. 38 ❑ p. 44 ❑ p. 50 ❑
- ❑ Read the text and study all figures and visuals. Answer any questions.

**Analyze key features**

- ❑ Myth Busters p. 34 ❑ p. 59 ❑
- ❑ Psychology InSight p. 36 ❑ p. 41 ❑
- ❑ Process Diagrams p. 39 ❑ p. 40 ❑ p. 46 ❑
- ❑ Study Organizer, p. 45
- ❑ Applying Psychology, p. 47
- ❑ What a Psychologist Sees p. 49 ❑ p. 58 ❑
- ❑ Psychological Science, p. 55
- ❑ Stop: Answer the Concept Checks before you go on.
  p. 37 ❑ p. 44 ❑ p. 49 ❑ p. 59 ❑

**End of chapter**

- ❑ Review the Summary and Key Terms.
- ❑ Answer the Critical and Creative Thinking Questions.
- ❑ Answer What is happening in this picture?
- ❑ Complete the Self-Test and check your answers.

# Our Genetic Inheritance

## LEARNING OBJECTIVES

**RETRIEVAL PRACTICE** While reading the upcoming sections, respond to each Learning Objective in your own words. Then compare your answers with those in Appendix B.

1. **Describe** how genetic material passes from one generation to the next.
2. **Review** the approaches that scientists take to explore human inheritance.
3. **Explain** the process of natural selection.

Millions of years of evolution have contributed to what we are today. Our ancestors foraged for food, fought for survival, and passed on traits that were selected and transmitted down through the generations. How do these transmitted traits affect us today? For answers, psychologists often turn to **behavioral genetics** (how heredity and environment affect us) and **evolutionary psychology** (how the natural process of adapting to our environment affects us).

You may already have a lot of ideas about heredity and the biological bases of behavior. Before you read further, take the *Myth Busters* quiz to find out whether your ideas are fact or fiction.

**behavioral genetics** The study of the relative effects of genetic and environmental influences on behavior and mental processes.

**evolutionary psychology** The study of how natural selection and adaptation help explain behavior and mental processes.

## Behavioral Genetics

**Genes** are the most important and basic building blocks of our biological inheritance (Kalat, 2013) (**Figure 2.1**). Each of our human characteristics and behaviors is related to the presence or absence of particular genes, which control the transmission of traits. For some traits, such as blood type, a single pair of genes (one from each parent) determines what characteristics we will possess. But most traits are determined by a combination of many genes.

When the two genes for a given trait conflict, the outcome depends on whether the gene is *dominant* or *recessive*. A dominant gene reveals its trait whenever the gene is present. In contrast, the gene for a recessive trait will normally be expressed only if the other gene in the pair is also recessive.

It was once assumed that characteristics such as eye color, hair color, or height were the result of either one dominant gene or two paired recessive genes. But modern geneticists believe that each of these characteristics is *polygenic*, meaning they are controlled by multiple genes. Many polygenic traits like height or

## MYTH BUSTERS

THE PLANNER

### OUR AMAZING BRAINS

**TRUE OR FALSE?**

___ 1. Our brains are hard-wired and can't be "rewired."

___ 2. Most people are either left-brained or right-brained.

___ 3. When we learn something new, our brains become more wrinkled.

___ 4. We generally use only 10% of our brains.

___ 5. Brain damage is always permanent.

___ 6. The human brain is the largest in the animal kingdom.

___ 7. People remain conscious for several minutes after decapitation.

___ 8. You can get holes in your brain from drug use.

___ 9. Alcohol kills brain cells.

___ 10. If a trait is heritable, that means it can't be changed.

Answers: All of these statements are false. Details were provided in this chapter, and/or in the following sources: Lilienfeld, 2010; http://www.smithsonianmag.com/science-nature/Top-Ten-Myths-About-the-Brain.html; http://health.howstuffworks.com/human-body/systems/nervous-system/10-brain-myths1.htm

## Hereditary code • Figure 2.1

The genes we inherit from our biological parents determine many of our physical traits (like height and hair color) and also are possible determinants of numerous behaviors (such as aggression, sensation-seeking, and homosexuality).

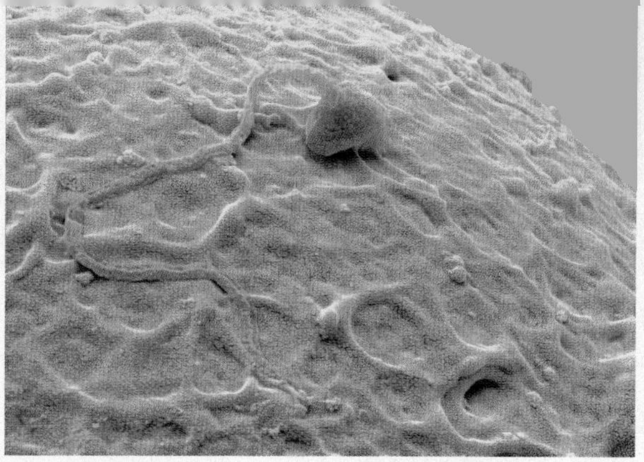

**a. Conception**
At the moment of conception, a father's sperm and a mother's egg each contribute 23 chromosomes, for a total of 46.

Thierry Berrod, Mona Lisa Production/Photo Researchers

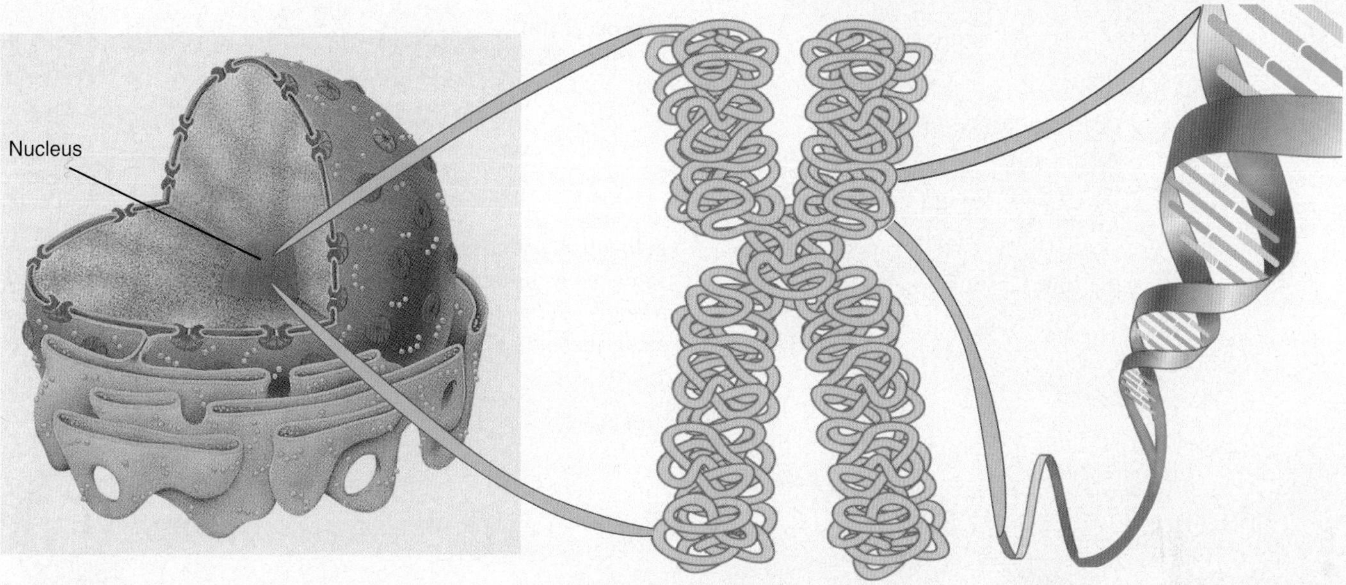


**b. Cell nucleus**
Each cell in the human body (except red blood cells) contains a nucleus.

**c. Chromosomes**
Each cell nucleus contains 46 chromosomes, which are threadlike molecules of DNA (deoxyribonucleic acid).

**d. Genes**
Each DNA molecule contains thousands of genes, which are the most basic units of heredity.

intelligence also are affected by environmental and social factors (**Figure 2.2**). Fortunately, most serious genetic disorders are not transmitted through a dominant gene. Can you guess why?

How do scientists research human inheritance? To determine the relative influences of heredity or environment on complex traits like aggressiveness, intelligence, or sociability, scientists rely on studies of twins, families, and genetic abnormalities.

## Gene–environment interaction • Figure 2.2

Children who are malnourished may not reach their full potential genetic height or maximum intelligence. Can you see how environmental factors interact with genetic factors to influence many traits?

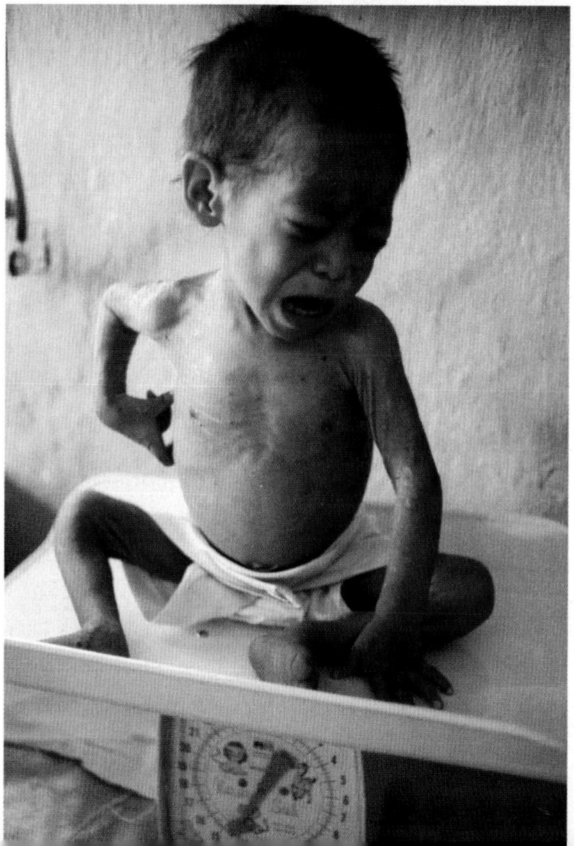

W.E. Garrett/NG Image Collection

# Psychology InSight

## Twin studies • Figure 2.3

Because both identical and fraternal twins share the same parents and develop in relatively the same environment, they provide a valuable "natural experiment." If heredity influences a trait or behavior to some degree, identical twins should be more alike than fraternal twins.

Ira Block/NG Image Collection

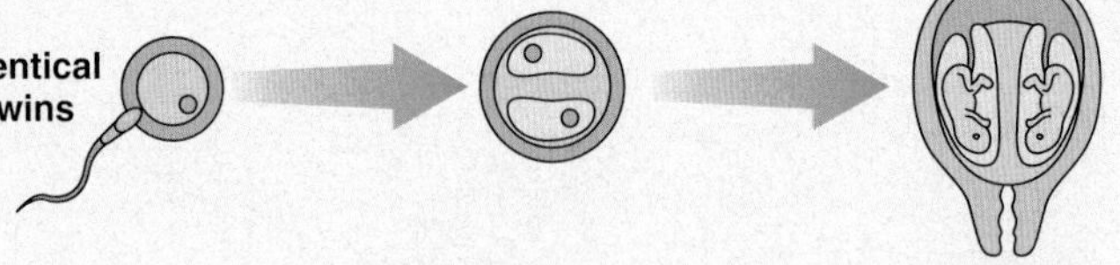

Identical twins
Same sex only


**a.** Identical (*monozygotic*—one egg) twins share 100% of the same genes because they develop from a single egg fertilized by a single sperm. They also share the same placenta and have the same sex and same genetic makeup.

Tony Freeman/PhotoEdit

Fraternal twins
Same or opposite sex


**b.** Fraternal (*dizygotic*—two eggs ) twins share, on average, 50% of their genes because they are formed when two separate sperm fertilize two separate eggs. Although they share the same general environment within the womb, they are no more genetically similar than nontwin siblings. They're simply nine-month "womb mates."

Twin studies have provided a wealth of information on the relative effects of heredity on behavior (**Figure 2.3**). For example, studies of intelligence show that identical twins have almost identical IQ scores, whereas fraternal twins are only slightly more similar in their IQ scores than are nontwin siblings (Bouchard, 2004; Plomin, 1999). The difference suggests a genetic influence on intelligence.

In addition to twin studies, psychologists interested in behavioral genetics also study entire families with biological and/or adopted children. If a specific trait is inherited, blood relatives should show increased trait similarity, compared with unrelated people, and closer relatives, like siblings, should be more similar than distant relatives. Studies of families with adopted children provide additional, unique advantages (**Figure 2.4**).

Family studies have shown that many traits and mental disorders, such as intelligence, sociability, and depression, do indeed run in families.

Finally, research in behavioral genetics explores disorders and diseases that result when genes malfunction. For example, researchers believe that genetic or chromosomal abnormalities are important factors in Alzheimer's disease and schizophrenia.

## Adoption studies • Figure 2.4

If adopted children are more like their biological family in some trait, then genetic factors probably had the greater influence. Conversely, if adopted children resemble their adopted family, even though they do not share similar genes, then environmental factors may predominate.

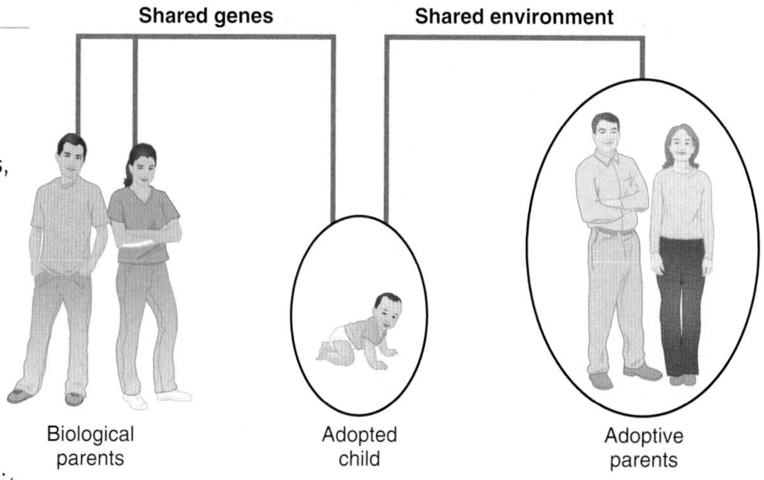


Findings from these three methods have allowed behavioral geneticists to estimate the *heritability* of various traits. That is, to what degree are individual differences a result of genetic, inherited factors rather than differences in the environment? If genetics contributed nothing to the trait, it would have a heritability estimate of 0%. If a trait was completely due to genetics, it would have a heritability estimate of 100%. Keep in mind, however, that heritability estimates apply to groups, not individuals (**Figure 2.5**).

As we've seen, behavioral genetics studies help explain the role of heredity (nature) and the environment (nurture) in our individual behavior. To increase our understanding of genetic dispositions, we also need to look at universal behaviors transmitted from our evolutionary past.

## Evolutionary Psychology

Evolutionary psychology suggests that many behavioral commonalities, from eating to fighting with our enemies, emerged and remain in human populations because they helped our ancestors (and ourselves) survive (Buss, 2011; Confer et al., 2010). This perspective stems from the writings of Charles Darwin (1859), who suggested that natural forces select traits that are adaptive to the organism's survival. This process of **natural selection** occurs when a particular genetic trait gives an organism a reproductive advantage over others. Some people mistakenly believe that natural selection means "survival of the fittest." But what really matters is *reproduction*—the survival of the genome. Because of natural selection, the fastest or otherwise most fit organisms will be most likely to live long enough to reproduce and thereby pass on their genes to the next generation.

*Genetic mutations* also help explain behavior. Everyone likely carries at least one gene that has mutated, or changed from the original. Very rarely, a mutated gene will be significant enough to change an individual's behavior. It might cause someone to be more social, more risk taking, more shy, more careful. If the gene then gives the person reproductive advantage, he or she will be more likely to pass on the gene to future generations. However, this mutation doesn't guarantee long-term survival. A well-adapted population can perish if its environment changes too rapidly.

## Height and heritability • Figure 2.5

Height has one of the highest heritability estimates—around 90% (Plomin, 1990). However, it's impossible to predict with certainty an individual's height from a heritability estimate. What other factors might have contributed to the difference in height between this mother and daughter?

Yellow Dog Productions/Taxi/Getty Images

### CONCEPT CHECK

STOP

1. **What** is the difference between a dominant and a recessive trait?
2. **Why** doesn't knowing the heritability of a trait predict what characteristics a given individual will have?
3. **How** are heredity and evolution linked to human behavior?

# Neural Bases of Behavior

## LEARNING OBJECTIVES

**RETRIEVAL PRACTICE** While reading the upcoming sections, respond to each Learning Objective in your own words. Then compare your responses with those in Appendix B.

1. **Identify** the key role and features of neurons.
2. **Explain** how neurons communicate throughout the body.
3. **Describe** the role of hormones in the endocrine system.

Your brain and the rest of your nervous system essentially consist of **neurons**. Each one is a tiny information-processing system with thousands of connections for receiving and sending electrochemical signals to other neurons. Each human body may have as many as one *trillion* neurons. (Be careful not to confuse the term *neuron* with the term *nerve*. Nerves are large bundles of axons outside the brain and spinal cord.)

**neuron** Nerve cell that processes and transmits information; basic building block of the nervous system responsible for receiving and transmitting electrochemical information.

Neurons are held in place and supported by **glial cells**, which make up about 90% of the brain's total cells. They also supply nutrients and oxygen, perform cleanup tasks, and insulate one neuron from another so that their neural messages are not scrambled. In addition, they play a direct role in nervous system communication (Arriagada et al., 2007; Wieseler-Frank, Maier, & Watkins, 2005; Zillmer, Spiers, & Culbertson, 2011). However, the "star" of the communication show is still the neuron.

No two neurons are alike, but most share three basic features: **dendrites**, the **cell body**, and an **axon** (**Figure 2.6**). To remember how information travels through the neuron, think of these three in reverse alphabetical order: *Dendrite* → *Cell Body* → *Axon*.

### The structure of a neuron • Figure 2.6

Arrows indicate direction of information flow: dendrites → cell body → axon → terminal buttons of axon.

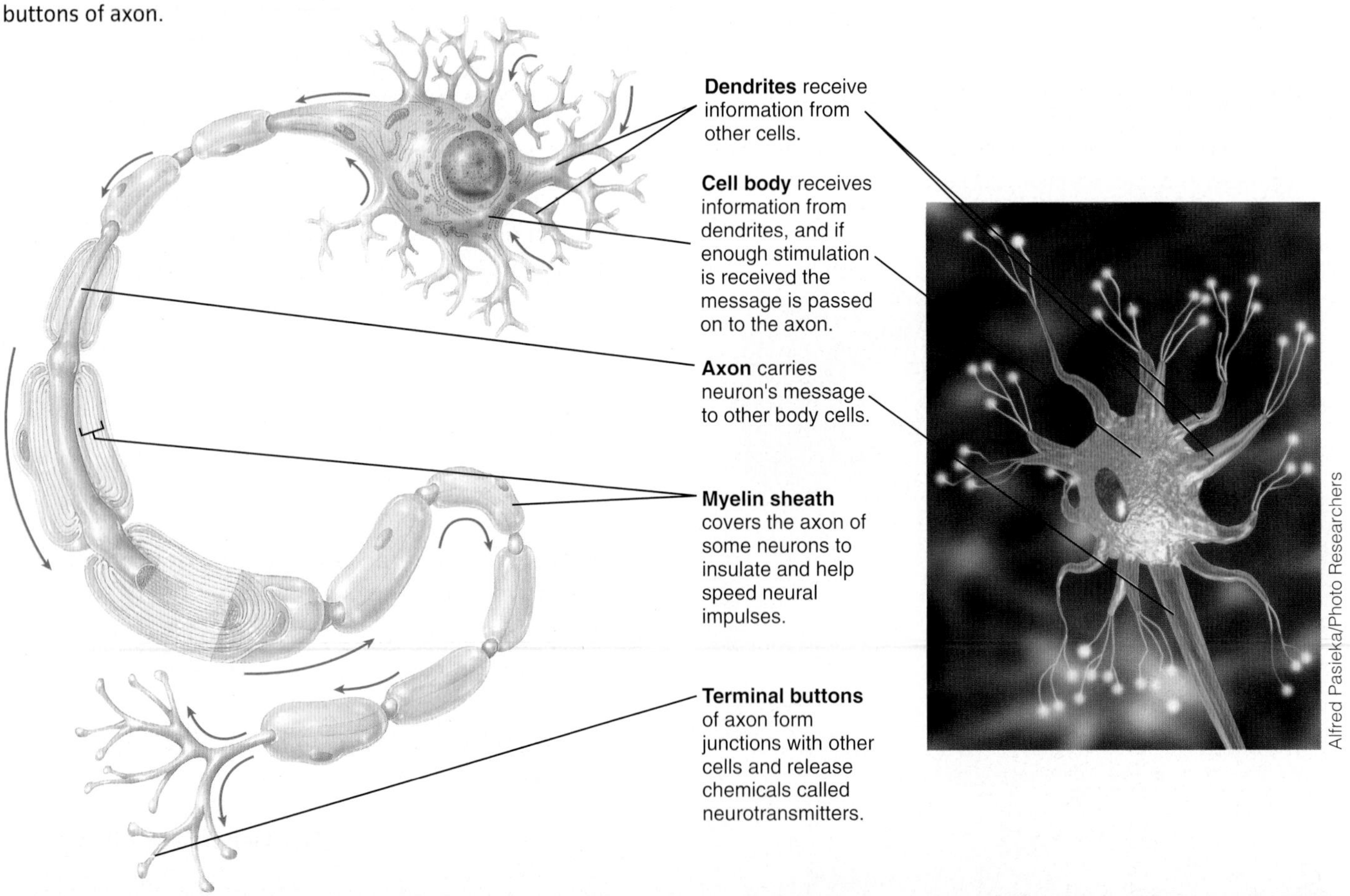

**Dendrites** receive information from other cells.
**Cell body** receives information from dendrites, and if enough stimulation is received the message is passed on to the axon.
**Axon** carries neuron's message to other body cells.
**Myelin sheath** covers the axon of some neurons to insulate and help speed neural impulses.
**Terminal buttons** of axon form junctions with other cells and release chemicals called neurotransmitters.
Alfred Pasieka/Photo Researchers

## Communication *within* the neuron • Figure 2.7

THE PLANNER

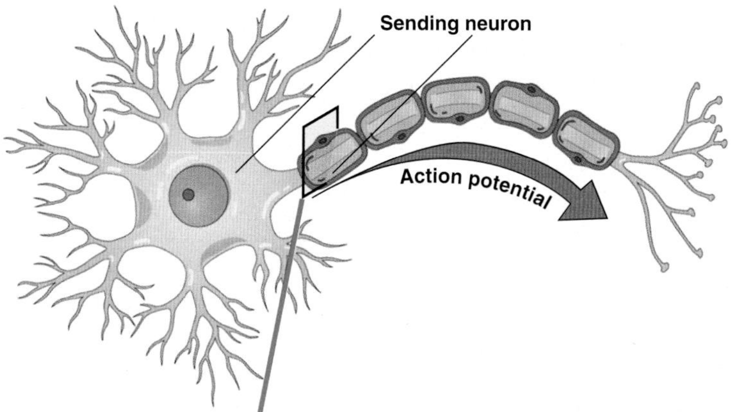


The process of neural communications begins within the neuron itself when the dendrites and cell body receive information and conduct it toward the axon. From there, the information travels down the entire length of the axon via a brief traveling electrical charge called an *action potential,* which can be described in three steps:

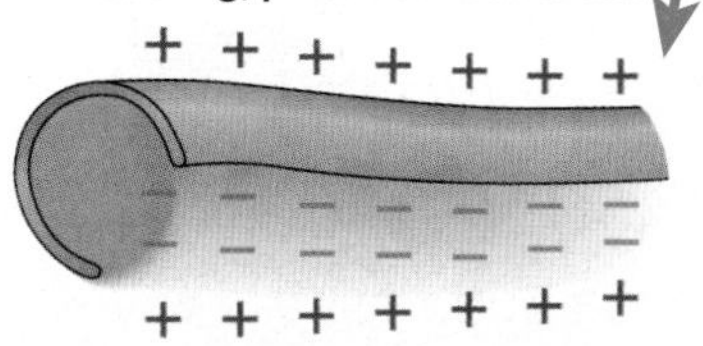


**1 Resting potential**
When an axon is not stimulated, it is in a polarized state, called the *resting potential.* "At rest," the fluid inside the axon has more negatively charged ions than the fluid outside. This results from the selective permeability of the axon membrane and a series of mechanisms, called *sodium-potassium pumps*, which pull potassium ions in and pump sodium ions out of the axon. The inside of the axon has a charge of about −70 millivolts relative to the outside.

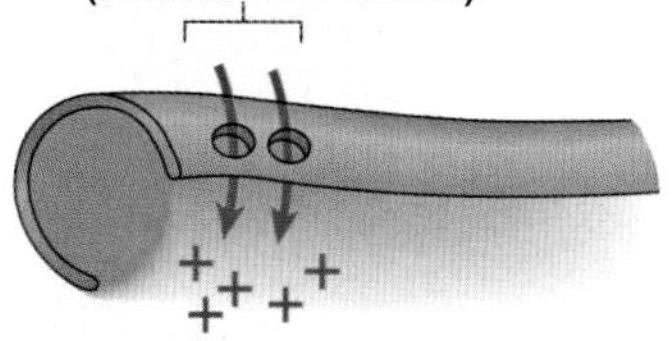


**2 Action potential initiation**
When an "at rest" axon membrane is stimulated by a sufficiently strong signal, it produces an action potential (or depolarization). This action potential begins when the first part of the axon opens its "gates" and positively charged sodium ions rush through. The additional sodium ions change the previously negative charge inside the axon to a positive charge—thus depolarizing the axon.

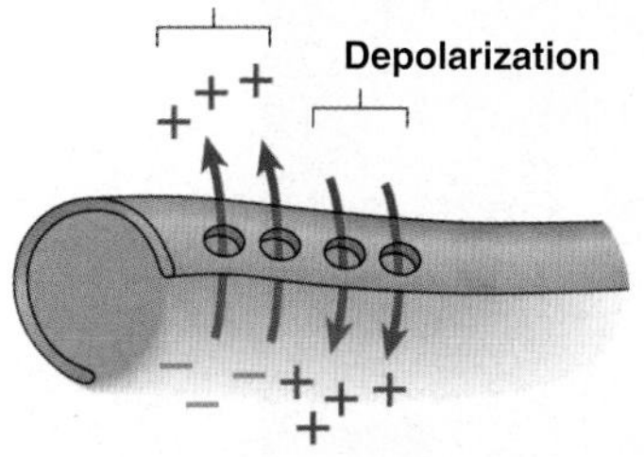


**3 Spreading of action potential and repolarization**
The initial depolarization (or action potential) of Step 2 produces a subsequent imbalance of ions in the adjacent axon membrane. This imbalance thus causes the action potential to spread to the next section. Meanwhile, "gates" in the axon membrane of the initially depolarized section open and potassium ions flow out, thus allowing the first section to repolarize and return to its resting potential.

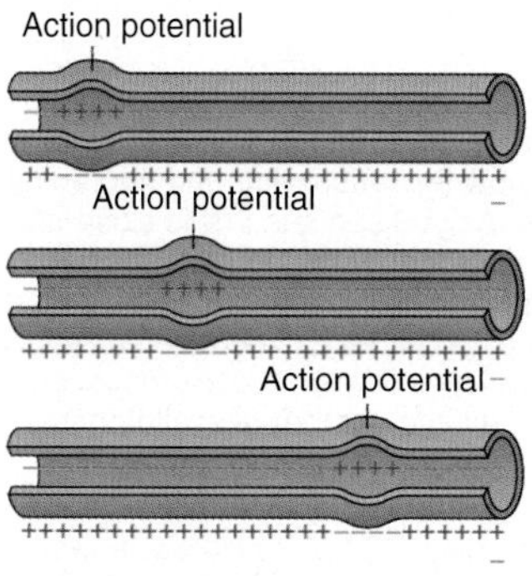


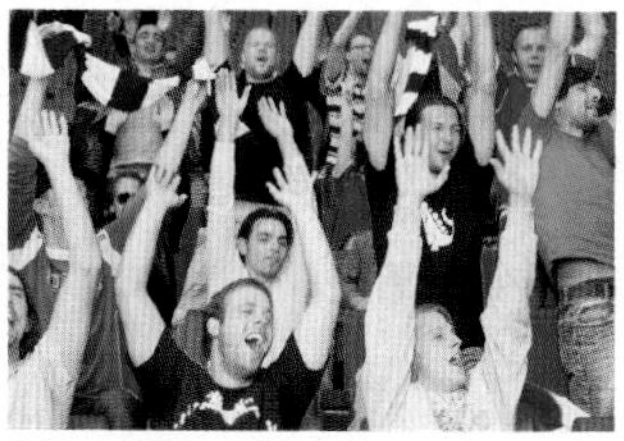

© ImageSource/SuperStock

**Overall summary**
The sequential process of depolarization and repolarization moving the action potential from the cell body to the terminal buttons is similar to fans at an athletic event doing the "wave." One section of fans first stands up and waves their arms (action potential initiation), then sits down (resting potential), and then the "wave" spreads on. Note that this is an "all-or-none" event.

## How Do Neurons Communicate?

A neuron's basic function is to transmit information throughout the nervous system. Neurons "speak" in a type of electrical and chemical language. The process of neural communication begins within the neuron itself, when the dendrites and cell body receive electrical "messages." These messages move along the axon in the form of a neural impulse, or **action potential** (**Figure 2.7**).

THE PLANNER

## Communication *between* neurons • Figure 2.8

Within the neuron (Figure 2.7), messages travel electrically. Between neurons, messages are transmitted chemically. The three steps below summarize this chemical transmission.

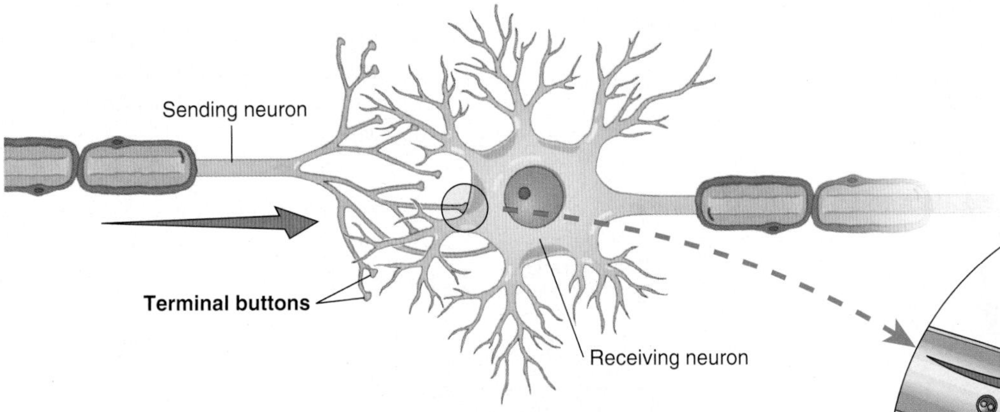


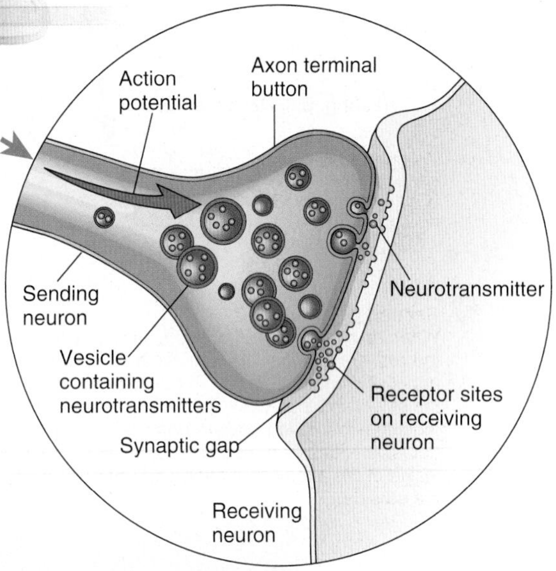


**1 Sending a chemical signal**
When an action potential reaches the branching axon terminals, it triggers the terminal buttons at the axon's end to open and release thousands of neurotransmitters into the **synapse**, the tiny opening between the sending and receiving neuron. These chemicals then move across the synaptic gap and attach to the membranes of the receiving neuron. In this way, they carry the message from the sending neuron to the receiving neuron.

**2 Receiving a chemical signal**
After a chemical message flows across the synaptic gap, it attaches to the receiving neuron. It's important to know that each receiving neuron gets multiple neurotransmitter messages. As you can see in this close-up photo, the axon terminals from thousands of other nearby neurons almost completely cover the cell body of the receiving neuron. It's also important to understand that neurotransmitters deliver either excitatory or inhibitory messages, and that the receiving neuron will only produce an action potential and pass along the message if the number of excitatory messages outweigh the inhibitory messages.

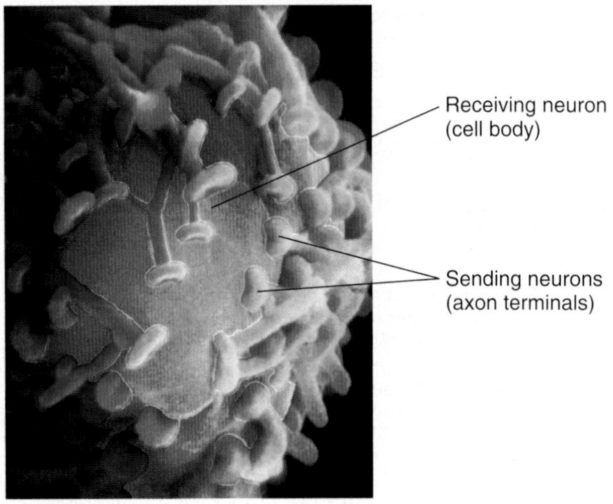


Photo Researchers, Inc.

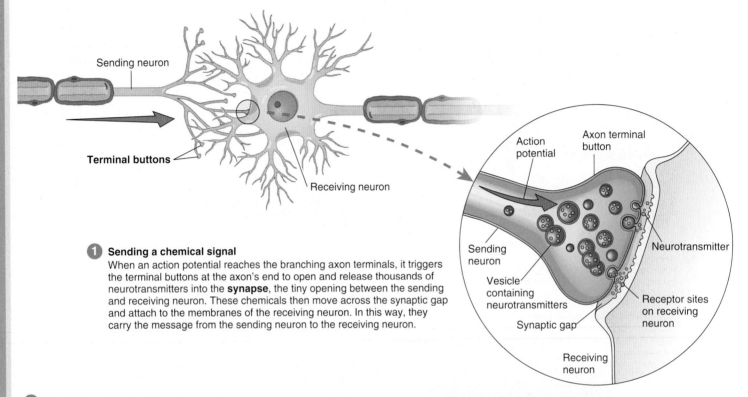

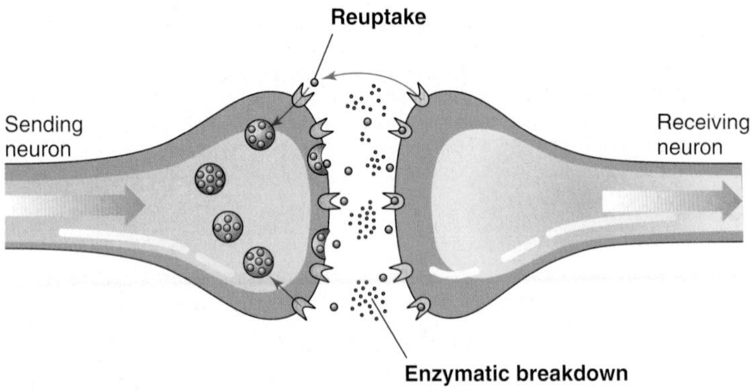


**3 Dealing with leftovers**
Given that some neurons have thousands of receptors, which are only responsive to specific neurotransmitters, what happens to excess neurotransmitters or to those that do not "fit" into the adjacent receptor sites? The sending neuron normally reabsorbs the excess (called "reuptake"), or they are broken down by special enzymes.

Nerve impulses move much more slowly than electricity through a wire. A neural impulse travels along a bare axon at only about 10 meters per second. (Electricity moves at 36 million meters per second.)

Some axons, however, are enveloped in fatty insulation, the **myelin sheath**. This sheath blankets the axon, with the exception of periodic *nodes,* points at which the myelin is very thin or absent. In a myelinated axon, the nerve impulse moves about 10 times faster than in a bare axon because the action potential jumps from node to node rather than traveling along the entire axon.

Communication *within* the neuron (Figure 2.7) is not the same as communication *between* neurons (**Figure 2.8**). Within the neuron, messages travel electrically, whereas between neurons the messages are carried across the *synaptic gap* via chemicals called **neurotransmitters**.

**neurotransmitters** Chemicals that neurons release, which affect other neurons.

Researchers have discovered hundreds of substances that function as neurotransmitters. For example, some **agonist drugs** enhance or "mimic" the action of particular neurotransmitters, whereas **antagonist drugs** block or inhibit the effects (**Figure 2.9**).

## Psychology InSight

### How poisons and drugs affect our brain • Figure 2.9

Foreign chemicals, like poisons and drugs, can mimic or block ongoing actions of neurotransmitters, thus interfering with normal functions.

**a. Normal neurotransmission**

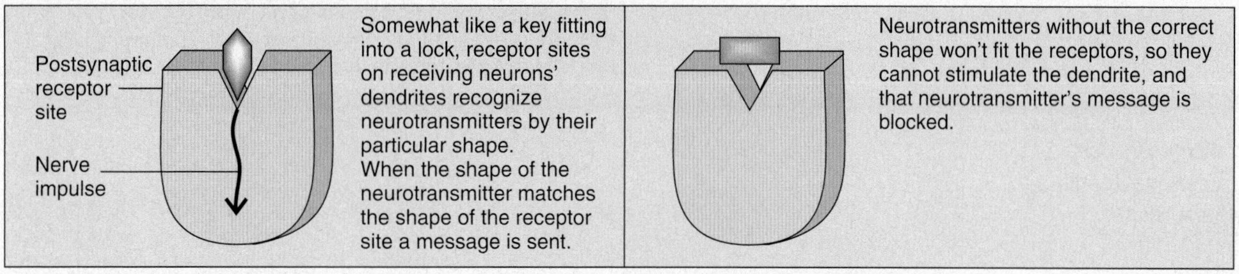

Postsynaptic receptor site
Nerve impulse
Somewhat like a key fitting into a lock, receptor sites on receiving neurons' dendrites recognize neurotransmitters by their particular shape. When the shape of the neurotransmitter matches the shape of the receptor site a message is sent.
Neurotransmitters without the correct shape won't fit the receptors, so they cannot stimulate the dendrite, and that neurotransmitter's message is blocked.

Normal neurotransmitter activation

Blocked neurotransmitter activation

**b. How poisons and drugs affect neurotransmission**

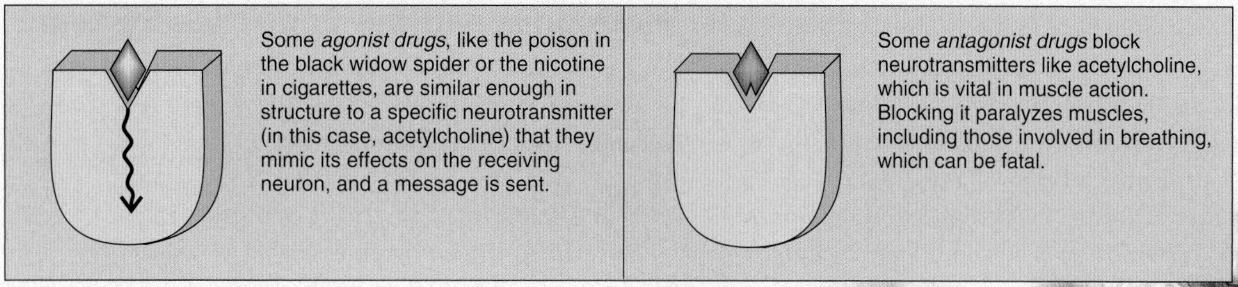

Some *agonist drugs*, like the poison in the black widow spider or the nicotine in cigarettes, are similar enough in structure to a specific neurotransmitter (in this case, acetylcholine) that they mimic its effects on the receiving neuron, and a message is sent.
Some *antagonist drugs* block neurotransmitters like acetylcholine, which is vital in muscle action. Blocking it paralyzes muscles, including those involved in breathing, which can be fatal.

Agonist drug "mimics" neurotransmitter

Antagonist drug fills receptor space and blocks neurotransmitter

**c.** Most snake venom and some other poisons, like *botulinum* toxin (Botox®), seriously affect normal muscle contraction. Ironically, these same poisons are sometimes used to treat certain medical conditions involving abnormal muscle contraction—as well as for some cosmetic purposes.

altrendo images/ Getty Images

**How neurotransmitters affect us Table 2.1**

| Neurotransmitter | Known or suspected effects |
|---|---|
| **Acetylcholine (ACh)** | Muscle action, learning, memory, REM (rapid-eye-movement) sleep, emotion; decreased ACh plays a suspected role in Alzheimer's disease |
| **Dopamine (DA)** | Movement, attention, memory, learning, and emotion; excess DA associated with schizophrenia; too little DA linked with Parkinson's disease; also plays a role in addiction and the reward system |
| **Endorphins** | Mood, pain, memory, learning, blood pressure, appetite, and sexual activity |
| **Epinephrine (or adrenaline)** | Emotional arousal, memory storage, and metabolism of glucose necessary for energy release |
| **GABA (gamma-aminobutyric acid)** | Neural inhibition in the central nervous system; tranquilizing drugs, like Valium, increase GABA's inhibitory effects and thereby decrease anxiety |
| **Norepinephrine (NE) [or noradrenaline (NA)]** | Learning, memory, dreaming, emotion, waking from sleep, eating, alertness, wakefulness, reactions to stress; low levels of NE associated with depression; high levels of NE linked with agitated, manic states |
| **Serotonin** | Mood, sleep, appetite, sensory perception, arousal, temperature regulation, pain suppression, and impulsivity; low levels of serotonin associated with depression |

AP/Wide World Photos

Which neurotransmitters best explain this professional tennis player's exceptional skill?

One benefit of studying your brain and its neurotransmitters is that it will help you understand some common medical problems. For example, we know that decreased levels of the neurotransmitter dopamine are associated with Parkinson's disease (PD), whereas excessively high levels of dopamine appear to contribute to some forms of schizophrenia. **Table 2.1** presents additional examples of the better understood neurotransmitters.

Perhaps the best-known neurotransmitters are the endogenous opioid peptides, commonly known as **endorphins** (a contraction of *endogenous* [self-produced] and *morphine*). These chemicals mimic the effects of opium-based drugs such as morphine—they elevate mood and reduce pain. They also affect memory, learning, blood pressure, appetite, and sexual activity.

## Hormones and the Endocrine System

We've just seen how the nervous system uses neurotransmitters to transmit messages throughout the body. A second type of communication system also exists. This second system is made up of a network of glands, called the **endocrine system** (**Figure 2.10**). Rather than neurotransmitters, this system uses **hormones** to carry its messages (**Figure 2.11**).

**hormones** Chemicals manufactured by endocrine glands and circulated in the bloodstream to produce bodily changes or maintain normal bodily function.

Your endocrine system has several important functions. It helps regulate long-term bodily processes, such as growth and sexual characteristics.

## The endocrine system • Figure 2.10

This figure shows the major endocrine glands, along with some internal organs to help you locate the glands.

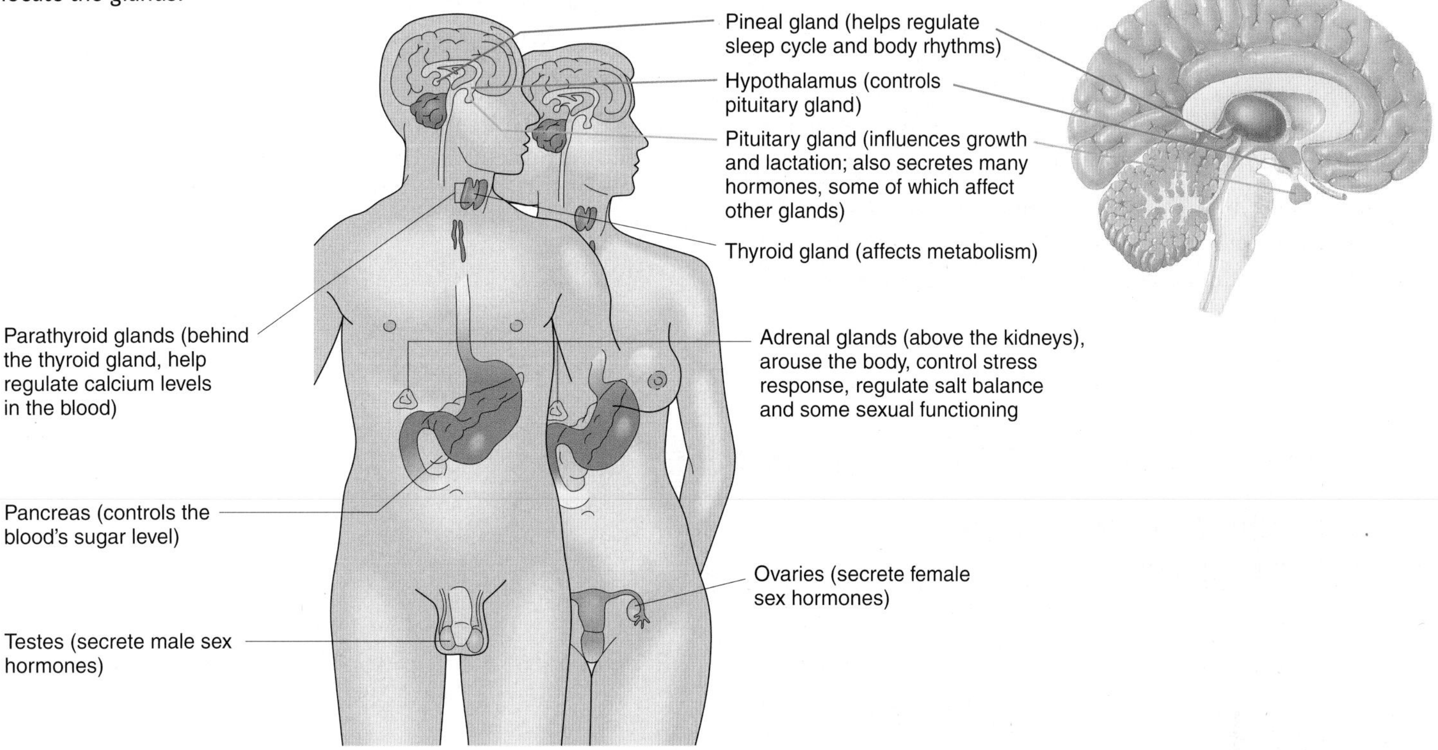


## Why do we need two communication systems? • Figure 2.11

You can think of neurotransmitters as individual e-mails that you send to particular people. Neurotransmitters deliver messages to specific receptors, which other neurons nearby probably don't "overhear." Hormones, in contrast, are like a global e-mail message that you send to everyone in your address book. Endocrine glands release hormones directly into the bloodstream, which travel throughout the body, carrying messages to any cell that will listen. Hormones also function like your global e-mail recipients forwarding your message to yet more people. For example, a small part of the brain called the hypothalamus releases hormones that signal the pituitary (another small brain structure), which stimulates or inhibits the release of other hormones.

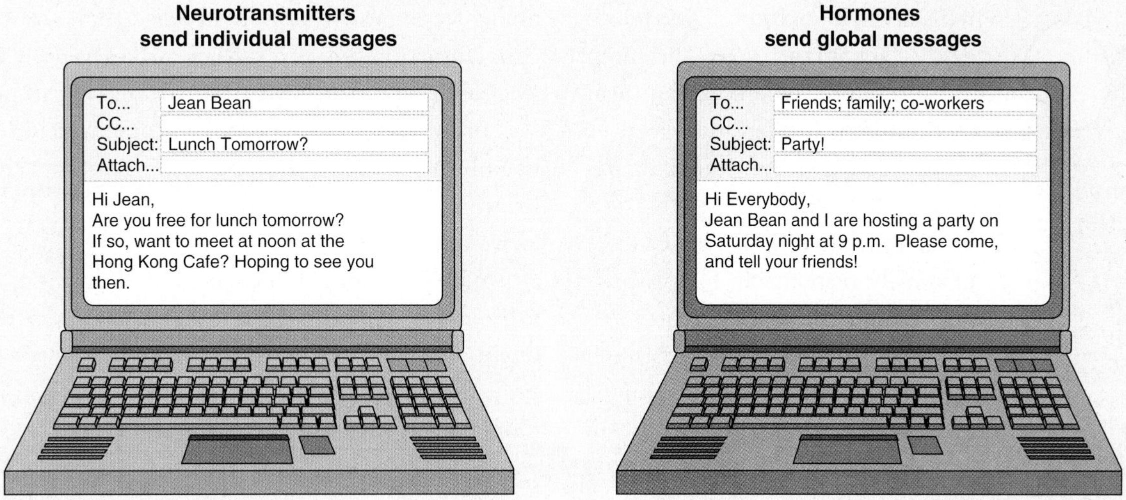

Neurotransmitters send individual messages
To... Jean Bean
CC...
Subject: Lunch Tomorrow?
Attach...
Hi Jean,
Are you free for lunch tomorrow?
If so, want to meet at noon at the Hong Kong Cafe? Hoping to see you then.
Hormones send global messages
To... Friends; family; co-workers
CC...
Subject: Party!
Attach...
Hi Everybody,
Jean Bean and I are hosting a party on Saturday night at 9 p.m. Please come, and tell your friends!

It also maintains ongoing bodily processes (such as digestion and elimination). Finally, hormones control the body's response to emergencies. In times of crisis, the hypothalamus sends messages through two pathways—the neural system and the endocrine system (primarily the pituitary). The pituitary sends hormonal messages to the adrenal glands (located right above the kidneys). The adrenal glands then release **cortisol**, a "stress hormone" that boosts energy and blood sugar levels, *epinephrine* (commonly called adrenaline), and *norepinephrine* (or nonadrenaline). (Remember that these same chemicals also can serve as neurotransmitters.)

**CONCEPT CHECK** STOP

1. **Why** is myelin important?
2. **How** does an action potential move along an axon?
3. **What** are the differences between communication through hormones and through neurotransmitters?

# Nervous System Organization

## LEARNING OBJECTIVES

**RETRIEVAL PRACTICE** While reading the upcoming sections, respond to each Learning Objective in your own words. Then compare your responses with those in Appendix B.

1. **Identify** the major elements of the nervous system.
2. **Explain** the role of the central nervous system (CNS).
3. **Describe** the components of the peripheral nervous system (PNS).

Have you heard the expression "Information is power?" Nowhere is this truer than in the human body. Without information, we could not survive. Neurons within our nervous system must take in sensory information from the outside world and then decide what to do with the information. Just as the circulatory system handles blood, which conveys chemicals and oxygen, our nervous system uses chemicals and electrical processes that convey information.

The nervous system is divided and subdivided into several branches (see **Study Organizer 2.1**). The main branch includes the brain and a bundle of nerves that form the *spinal cord*. Because this system is located in the center of your body (within your skull and spine), it is called the **central nervous system (CNS)**. The CNS is primarily responsible for processing and organizing information.

The second major branch of your nervous system includes all the nerves outside the brain and spinal cord. This **peripheral nervous system (PNS)** carries messages (action potentials) to and from the central nervous system to the periphery of the body. Now, let's take a closer look at the CNS and the PNS.

**central nervous system (CNS)** The brain and spinal cord.

**peripheral nervous system (PNS)** All the nerves and neurons outside the brain and spinal cord that connect the CNS to the rest of the body.

### Central Nervous System (CNS)

The central nervous system (CNS) is the branch of the nervous system that makes us unique. Most other animals can smell, run, see, and hear far better than we can. But thanks to our CNS, we can process information and adapt to our environment in ways that no other animal can. Unfortunately, our CNS is also incredibly fragile. Unlike neurons in the PNS that can regenerate and require less protection, serious damage to neurons in the CNS is usually permanent. However, the brain may not be as "hard wired" as we once believed.

Scientists long believed that after the first two or three years of life, humans and most animals are unable to repair or replace damaged neurons in the brain or spinal cord. We now know that the brain is capable of lifelong **neuroplasticity** and **neurogenesis**.

**neuroplasticity** The brain's ability to reorganize and change its structure and function throughout the lifespan.

**neurogenesis** The division and differentiation of nonneuronal cells to produce neurons.

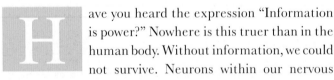

**Study Organizer 2.1** The nervous system

THE PLANNER

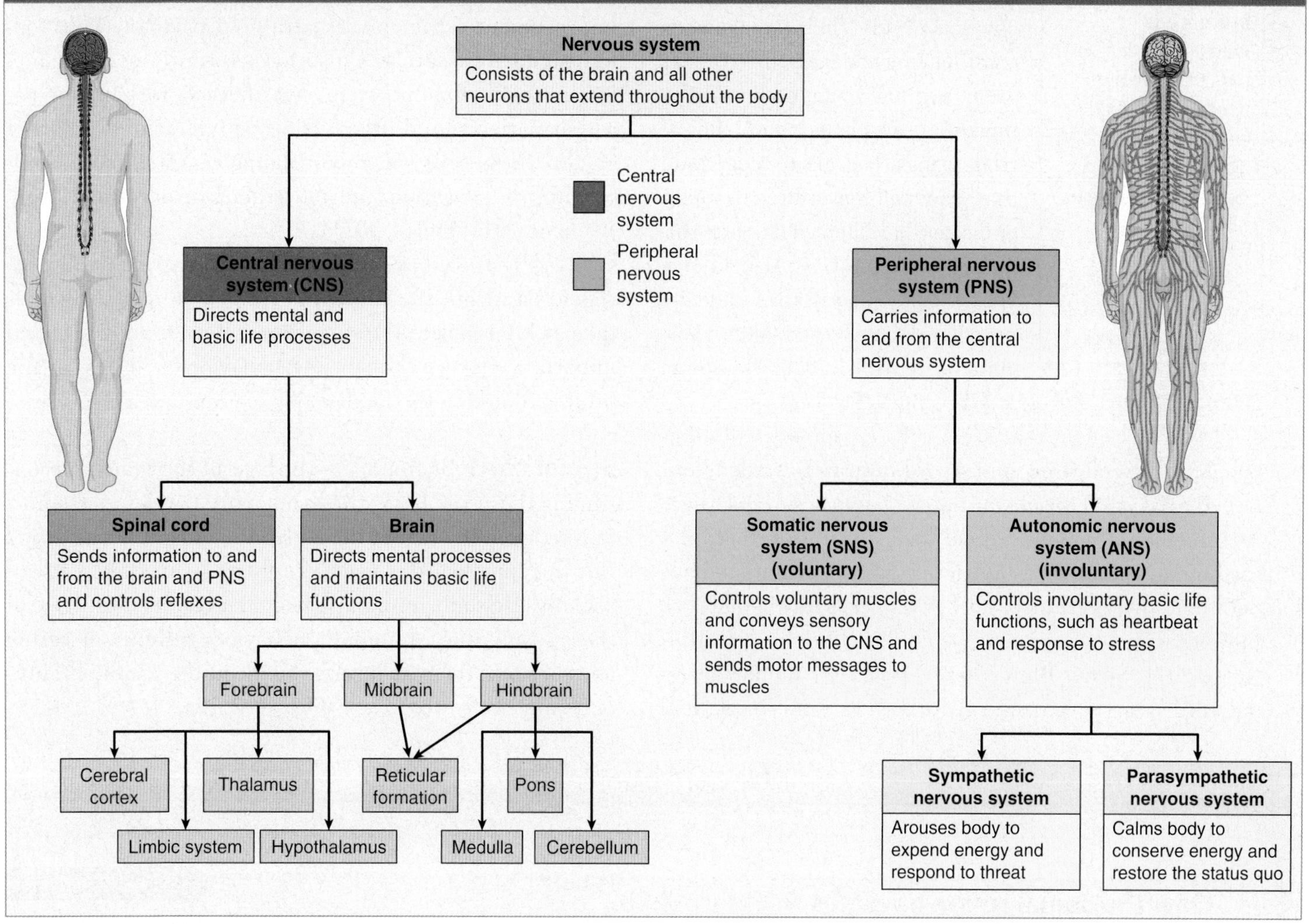

Nervous system
Consists of the brain and all other neurons that extend throughout the body
Central nervous system
Peripheral nervous system
Central nervous system (CNS)
Directs mental and basic life processes
Peripheral nervous system (PNS)
Carries information to and from the central nervous system
Spinal cord
Sends information to and from the brain and PNS and controls reflexes
Brain
Directs mental processes and maintains basic life functions
Somatic nervous system (SNS) (voluntary)
Controls voluntary muscles and conveys sensory information to the CNS and sends motor messages to muscles
Autonomic nervous system (ANS) (involuntary)
Controls involuntary basic life functions, such as heartbeat and response to stress
Forebrain
Midbrain
Hindbrain
Cerebral cortex
Thalamus
Reticular formation
Pons
Limbic system
Hypothalamus
Medulla
Cerebellum
Sympathetic nervous system
Arouses body to expend energy and respond to threat
Parasympathetic nervous system
Calms body to conserve energy and restore the status quo

**Neuroplasticity** Rather than being a fixed, solid organ, the brain is capable of changing its structure and function as a result of usage and experience (Park & Bischof, 2011; Romero et al., 2008; Rossignol et al., 2008). This "rewiring" is what makes our brains so wonderfully adaptive. For example, it makes it possible for us to learn a new sport or a foreign language.

Remarkably, this rewiring has even helped "remodel" the brain following strokes. For example, psychologist Edward Taub and his colleagues (2002, 2004, 2007, 2010) have had success working with stroke patients (**Figure 2.12**).

**Neurogenesis** Our brains continually replace lost cells with new cells that originate deep within the brain and migrate to become part of its circuitry. The source of these

## A breakthrough in neuroscience • Figure 2.12

By immobilizing the unaffected arm or leg and requiring rigorous and repetitive exercise of the affected limb, psychologist Edward Taub and colleagues "recruit" stroke patients' intact brain cells to take over for damaged cells. The therapy has restored function in some patients as long as 21 years after their strokes.

Courtesy Taub Therapy Clinc/UAB Media Relations

newly created cells is neural **stem cells**—rare, immature cells that can grow and develop into any type of cell. Their fate depends on the chemical signals they receive. Stem cells have been used for bone marrow transplants, and clinical trials using stem cells to repopulate or replace cells devastated by injury or disease have helped patients suffering from strokes, Alzheimer's, Parkinson's, epilepsy, stress, and depression (Chang et al., 2005; Fleischmann & Welz, 2008; Hampton, 2006, 2007; Leri, Anversa, & Frishman, 2007).

**stem cells**
Precursor (immature) cells that give birth to new specialized cells; a stem cell holds all the information it needs to make bone, blood, brain—any part of a human body—and can also copy itself to maintain a stock of stem cells.

Does this mean that stem cell transplants will allow people paralyzed from spinal cord injuries to walk again? Despite risky and unsubstantiated claims of success, there are no currently approved cell transplantation therapies for human spinal cord injuries. However, scientists have had some success transplanting stem cells into spinal cord injured *animals* (Cusimano et al., 2012; Lee et al., 2011; Rossi, 2011; Sieber-Blum, 2010). When the damaged spinal cord was viewed several weeks later, the implanted cells had survived and spread throughout the injured area. More important, the transplant animals also showed some improvement in previously paralyzed parts of their bodies. Medical researchers also have recently begun safety testing of embryonic stem cell therapy for human paralysis patients, and future trials will determine whether or not these cells will repair damaged spinal cords and/or improve sensation and movement in paralyzed areas (Conger, 2011; Fields, 2011).

Now that we have discussed neuroplasticity and neurogenesis within the central nervous system (CNS), let's take a closer look at the spinal cord. Because of its central importance for psychology and behavior, we'll discuss the brain in detail in the next major section.

### Spinal cord

Beginning at the base of the brain and continuing down the back, the spinal cord carries vital information from the rest of the body into and out of the brain. But the spinal cord doesn't simply relay messages. It can also initiate some automatic behaviors on its own. We call these involuntary, automatic behaviors **reflexes** or **reflex arcs** because the response to the incoming stimuli is automatically "reflected" back (**Figure 2.13**).

PROCESS DIAGRAM

## How the spinal reflex operates • Figure 2.13

THE PLANNER

In a simple reflex arc, a sensory receptor responds to stimulation and initiates a neural impulse that travels to the spinal cord. This signal then travels back to the appropriate muscle, which then contracts. The response is automatic and immediate in a reflex because the signal only travels as far as the spinal cord before action is initiated, not all the way to the brain. The brain is later "notified" of the action when the spinal cord sends along the message. What might be the evolutionary advantages of the reflex arc?

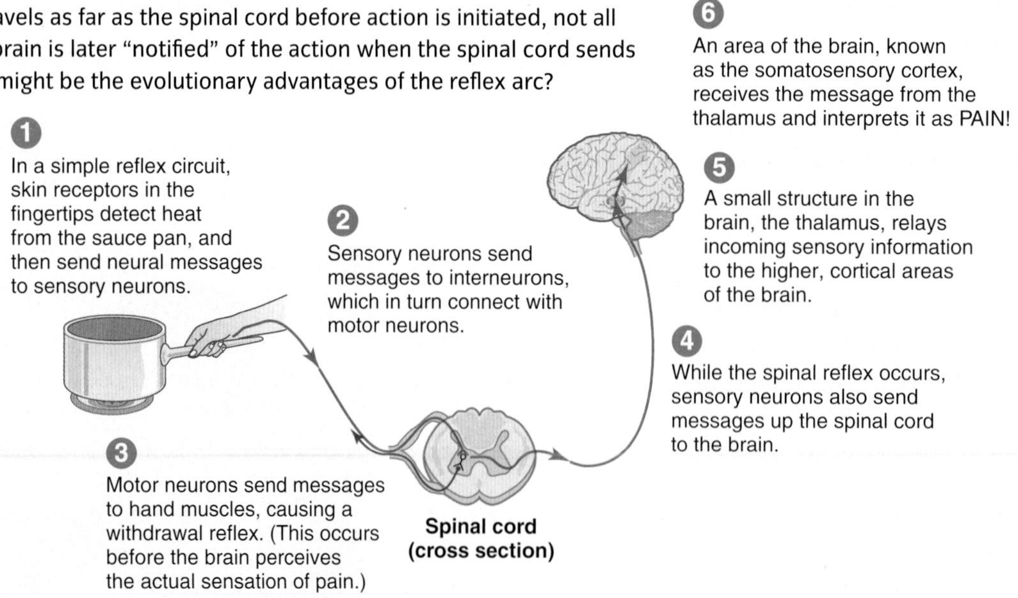

6
An area of the brain, known as the somatosensory cortex, receives the message from the thalamus and interprets it as PAIN!
1
In a simple reflex circuit, skin receptors in the fingertips detect heat from the sauce pan, and then send neural messages to sensory neurons.
2
Sensory neurons send messages to interneurons, which in turn connect with motor neurons.
5
A small structure in the brain, the thalamus, relays incoming sensory information to the higher, cortical areas of the brain.
4
While the spinal reflex occurs, sensory neurons also send messages up the spinal cord to the brain.
3
Motor neurons send messages to hand muscles, causing a withdrawal reflex. (This occurs before the brain perceives the actual sensation of pain.)
Spinal cord (cross section)

## Applying Psychology

THE PLANNER

# Testing for Reflexes

If you have a newborn or young infant in your home, you can easily (and safely) test for these simple reflexes. (Note: Most infant reflexes disappear within the first year of life. If they reappear in later life, it generally indicates damage to the central nervous system.)

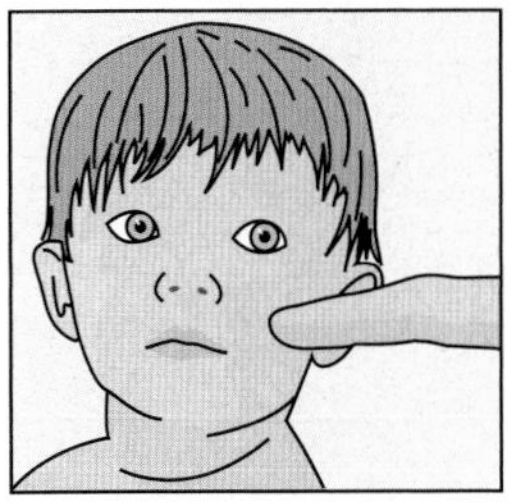

**a. Rooting reflex**
Lightly stroke the cheek or side of the mouth, and watch how the infant automatically (reflexively) turns toward the stimulation and attempts to suck.

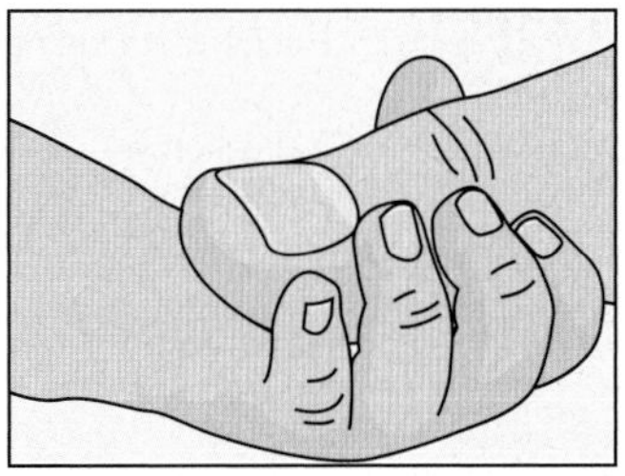

**b. Grasping reflex**
Place your finger in the infant's palm and note the automatic grasp.

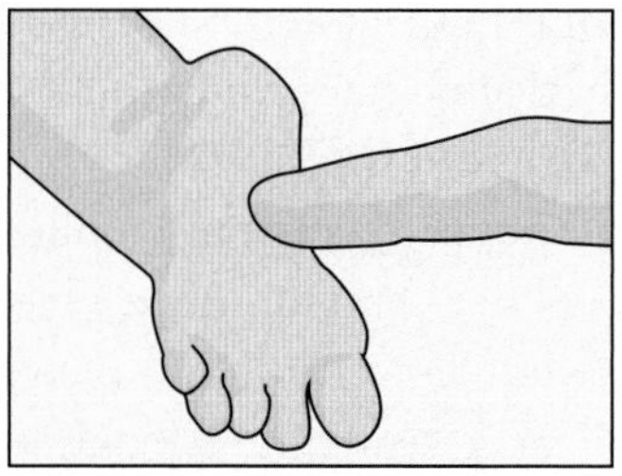

**c. Babinski reflex**
Lightly stroke the sole of the infant's foot, and the toes will fan out and the foot will twist inward.

**Think Critically**

1. What might happen if infants lacked these reflexes?
2. Can you imagine why most infant reflexes disappear within the first year?

We're all born with numerous reflexes, many of which fade over time (see *Applying Psychology*). But even as adults, we still blink in response to a puff of air in our eyes, gag when something touches the back of the throat, and urinate and defecate in response to pressure in the bladder and rectum. Reflexes even influence our sexual responses. Certain stimuli, such as the stroking of the genitals, can lead to arousal and the reflexive muscle contractions of orgasm in both men and women. However, in order to have the passion, thoughts, and emotion we normally associate with sex, the sensory information from the stroking and orgasm must be carried to the brain.

## Peripheral Nervous System (PNS)

The *peripheral nervous system (PNS)* is just what it sounds like—the part that involves nerves *peripheral* to (or outside of) the brain and spinal cord. The chief function of the peripheral nervous system (PNS) is to carry information to and from the central nervous system. It links the brain and spinal cord to the body's sense receptors, muscles, and glands.

The chief function of the peripheral nervous system (PNS) is to carry information to and from the central nervous system. It links the brain and spinal cord to the body's sense receptors, muscles, and glands.

The PNS is subdivided into the somatic nervous system and the autonomic nervous system. The **somatic nervous system (SNS)** consists of all the nerves that connect to sensory receptors and skeletal muscles. The name comes from the term soma, which means "body," and the somatic nervous system plays a key role in communication throughout the entire body. In a kind of "two-way street," the somatic nervous system (also called the skeletal nervous system) first carries sensory information to the brain and spinal cord (CNS), and then carries messages from the CNS to skeletal muscles (see again Study Organizer 2.1).

The other subdivision of the PNS is the **autonomic nervous system (ANS)**. The ANS is responsible for involuntary tasks, such as heart rate, digestion, pupil dilation, and breathing. Like an automatic pilot, the ANS can sometimes be consciously overridden. But as its name implies, the autonomic system normally operates on its own (autonomously).

**somatic nervous system (SNS)** Subdivision of the peripheral nervous system (PNS) that connects the sensory receptors and controls the skeletal muscles.

**autonomic nervous system (ANS)** Subdivision of the peripheral nervous system (PNS) that controls involuntary functions. It includes the *sympathetic* nervous system and the *parasympathetic* nervous system.

The autonomic nervous system is further divided into two branches, the sympathetic and parasympathetic, which tend to work in opposition to each other to regulate the functioning of such target organs as the heart, the intestines, and the lungs. Like two children on a teeter-totter, one will be up while the other is down, but they essentially balance each other out. *What a Psychologist Sees* describes a familiar example of the interaction between the sympathetic and parasympathetic nervous systems.

During stressful times, either mental or physical, the **sympathetic nervous system** mobilizes bodily resources to respond to the stressor. This emergency response is often called the "fight-or-flight" response. If you noticed a dangerous snake coiled and ready to strike, your sympathetic nervous system would increase your heart rate, respiration, and blood pressure, stop your digestive and eliminative processes, and release hormones, such as cortisol, into the bloodstream. The net result of sympathetic activation is to get more oxygenated blood and energy to the skeletal muscles, thus allowing you to cope with the stress—to "fight or flee."

In contrast to the sympathetic nervous system, the **parasympathetic nervous system** is responsible for returning your body to its normal functioning by slowing your heart rate, lowering your blood pressure, and increasing your digestive and eliminative processes.

The sympathetic nervous system provides an adaptive, evolutionary advantage. At the beginning of human evolution, when we faced a dangerous bear or an aggressive human intruder, there were only two reasonable responses—fight or flight. This automatic mobilization of bodily resources can still be critical even in modern times. However, less life-threatening events, such as traffic jams, also activate our sympathetic nervous system. Our bodies still respond to these sources of stress with sympathetic arousal. As the next chapter discusses, ongoing sympathetic system response to such chronic, daily stress can become detrimental to our health.

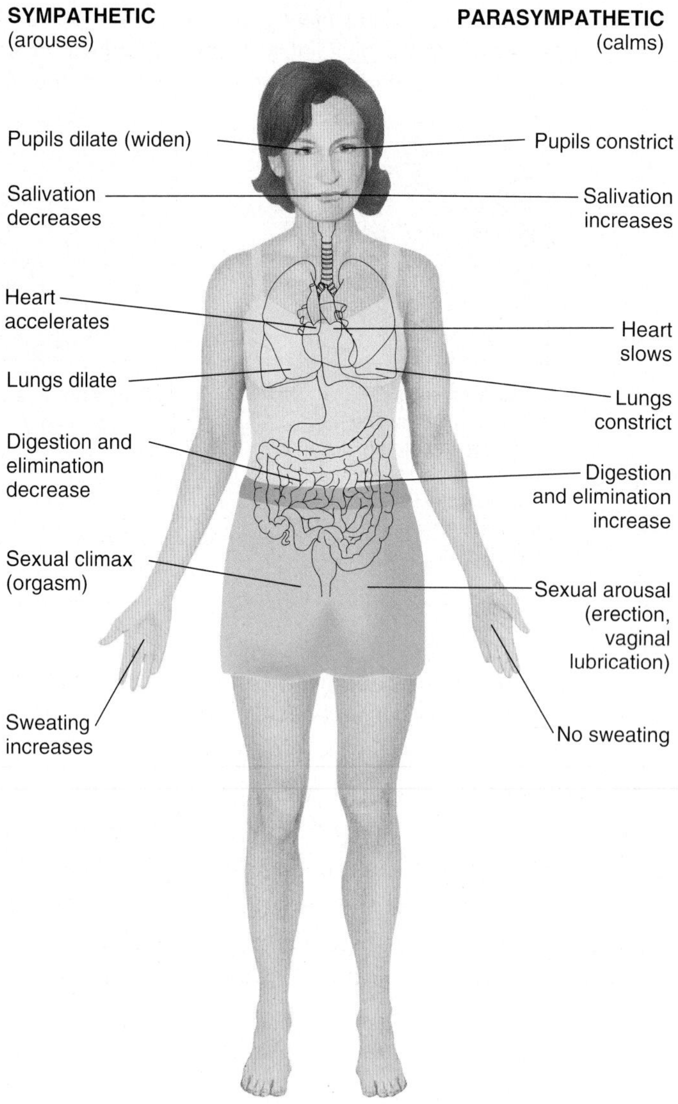

# WHAT A PSYCHOLOGIST SEES

THE PLANNER

## Autonomic Nervous System and Sexual Arousal

The complexities of sexual interaction—and in particular, the difficulties that couples sometimes have in achieving sexual arousal or orgasm—illustrate the balancing act between the sympathetic and parasympathetic nervous systems.

Simon Watson/Botanica/Getty Images

**Parasympathetic dominance**

Sexual arousal and excitement require that the body be relaxed enough to allow increased blood flow to the genitals—in other words, the nervous system must be in *parasympathetic dominance.* Parasympathetic nerves carry messages from the central nervous system directly to the sexual organs, allowing for a localized response (increased blood flow and genital arousal).

Latin Stock Collection/©Corbis

**Sympathetic dominance**

During strong emotions, such as anger, anxiety, or fear, the body shifts to *sympathetic dominance* and blood flow to the genitals and other organs decreases as the body readies for "fight or flight." As a result, the person is unable (or less likely) to become sexually aroused. Any number of circumstances—for example, performance anxiety, fear of unwanted pregnancy or disease, or tensions between partners—can trigger sympathetic dominance.

### CONCEPT CHECK

STOP

1. **What** are the different responsibilities of the central and peripheral nervous systems?
2. **How** is the spinal cord responsible for reflexes?
3. **What** are the opposing roles of the sympathetic and parasympathetic nervous systems?

# A Tour Through the Brain

## LEARNING OBJECTIVES

**RETRIEVAL PRACTICE** While reading the upcoming sections, respond to each Learning Objective in your own words. Then compare your responses with those in Appendix B.

1. **Identify** the major structures of the hindbrain, midbrain, and forebrain.
2. **Summarize** the major roles of the lobes of the cerebral cortex.
3. **Describe** how the brain is divided into two specialized hemispheres.

We begin our exploration of the brain at the lower end, where the spinal cord joins the base of the brain, and move upward through the skull. As we move from bottom to top, "lower," basic processes, like breathing, generally give way to more complex mental processes (**Figure 2.14** and **Figure 2.15**).

### Lower-Level Brain Structures

Brain size and complexity vary significantly from species to species. For example, fish and reptiles have smaller, less complex brains than do cats and dogs. The most complex brains belong to whales, dolphins, and higher primates such as chimps, gorillas, and humans. The billions of neurons that make up the human brain control much of what we think, feel, and do. Certain brain structures are specialized to perform certain tasks, a process known as **localization of function**. However, most parts of the brain perform integrating, overlapping functions.

**localization of function** Specialization of various parts of the brain for particular functions.

**Hindbrain** You are deep in REM sleep as you begin your last dream of the night. Your vital signs increase as the dream gets more exciting. But then your sleep—and your dream—are shattered by a buzzing alarm clock. All your automatic behaviors and survival responses in this scenario are controlled or influenced by parts of the hindbrain. The **hindbrain** includes the medulla, pons, and cerebellum.

**Damage to the brain • Figure 2.14**

On January 8, 2011, a lone gunman fired his gun into a crowd in Arizona, killing six people and seriously wounding 13 others, including Representative Gabby Giffords. Despite being critically wounded by a bullet that entered through the front of her head and exited through the back, Representative Giffords later made significant improvements in her ability to walk, speak, read, and write. You'll understand and appreciate the importance of these achievements as you study the upcoming coverage of the various parts of the brain.

Win McNamee/Getty Images

## The human brain • Figure 2.15

This profile drawing highlights key structures and functions of the right half of the brain. As you read about each of these structures, keep this drawing in mind and refer back to it as necessary. (The diagram shows the brain structures as if the brain were split vertically down the center and the left hemisphere were removed.)

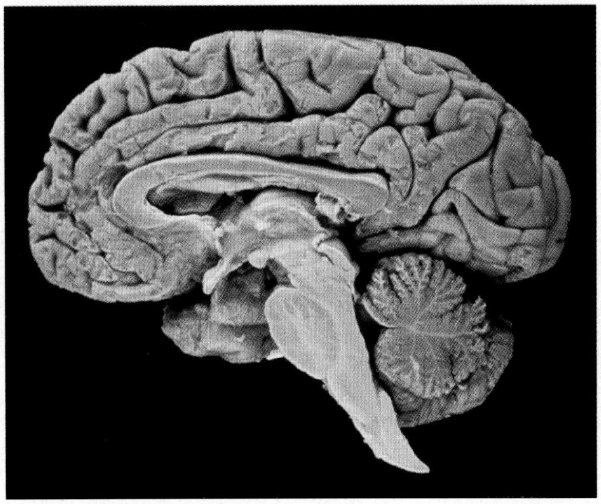

Science Pictures Limited/Photo Researchers

**Actual photo of the brain**
This deceased human's brain is sliced down the center, splitting it into right and left halves. The right half is shown in this photo.

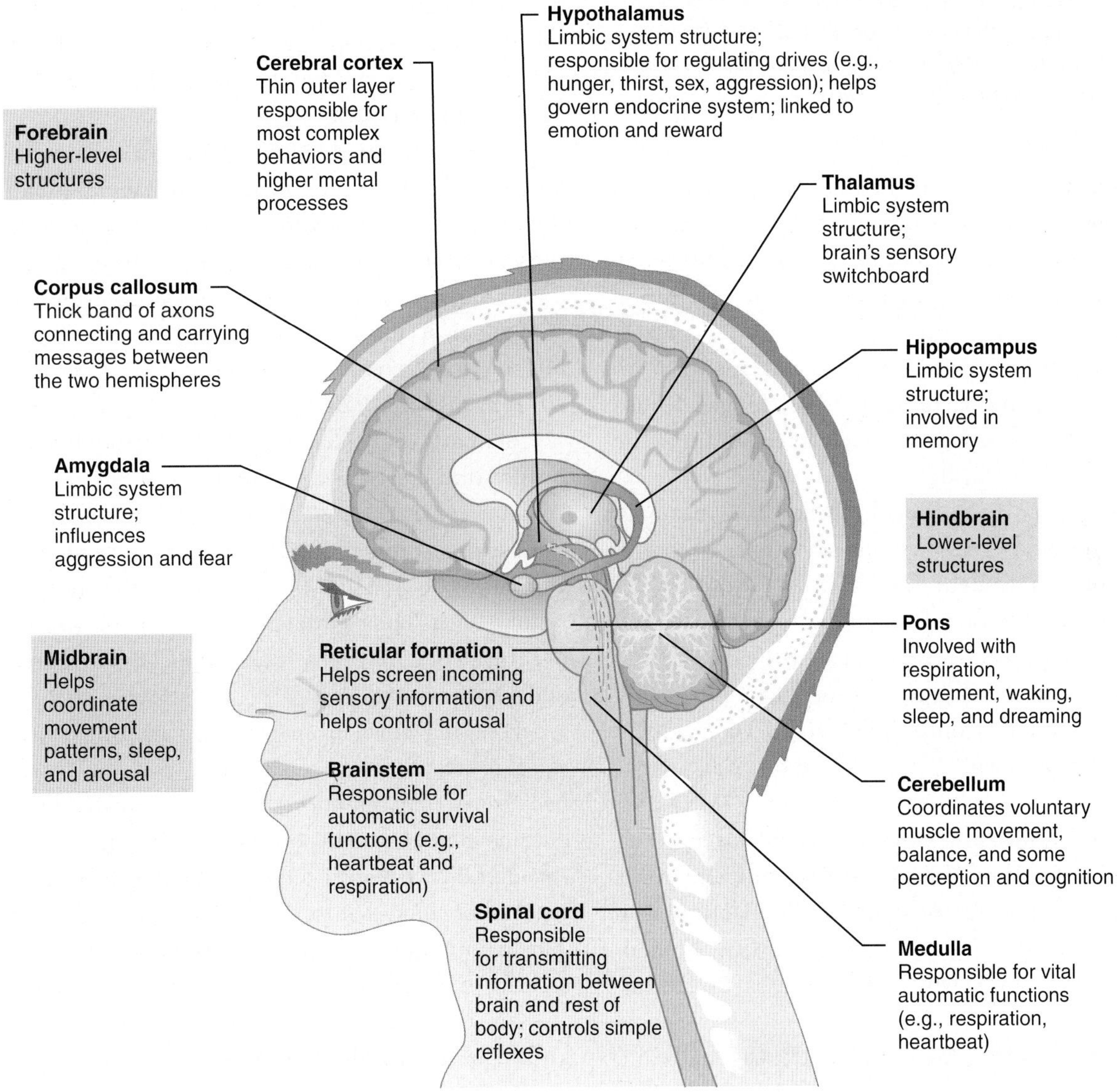

Forebrain Higher-level structures
Cerebral cortex Thin outer layer responsible for most complex behaviors and higher mental processes
Hypothalamus Limbic system structure; responsible for regulating drives (e.g., hunger, thirst, sex, aggression); helps govern endocrine system; linked to emotion and reward
Thalamus Limbic system structure; brain's sensory switchboard
Corpus callosum Thick band of axons connecting and carrying messages between the two hemispheres
Hippocampus Limbic system structure; involved in memory
Amygdala Limbic system structure; influences aggression and fear
Hindbrain Lower-level structures
Pons Involved with respiration, movement, waking, sleep, and dreaming
Midbrain Helps coordinate movement patterns, sleep, and arousal
Reticular formation Helps screen incoming sensory information and helps control arousal
Brainstem Responsible for automatic survival functions (e.g., heartbeat and respiration)
Cerebellum Coordinates voluntary muscle movement, balance, and some perception and cognition
Spinal cord Responsible for transmitting information between brain and rest of body; controls simple reflexes
Medulla Responsible for vital automatic functions (e.g., respiration, heartbeat)

Bill Fritsch/Getty Images, Inc.

**Walk the line • Figure 2.16**

The cerebellum, responsible for smooth and precise movements, is one of the first areas of the brain to be affected by alcohol. Asking drivers to perform tasks like walking the white line is a common measure of potential drunk driving.

The **medulla** is essentially an extension of the spinal cord, with many nerve fibers passing through it carrying information to and from the brain. It also controls many essential automatic bodily functions, such as respiration and heart rate.

The **pons** is involved in respiration, movement, sleeping, waking, and dreaming (among other things). It also contains axons that cross from one side of the brain to the other (*pons* is Latin for "bridge").

The cauliflower-shaped **cerebellum** (little brain in Latin) is, evolutionarily, a very old structure. It coordinates fine muscle movement and balance (**Figure 2.16**). Researchers using functional magnetic resonance imaging (fMRI) also have shown that parts of the cerebellum are important for some memory, sensory, perceptual, cognitive, language, and learning (Bellebaum & Daum, 2011; Thompson, 2005; Woodruff-Pak & Disterhoft, 2008).

**Midbrain** The **midbrain** helps us orient our eye and body movements to visual and auditory stimuli, and works with the pons to help control sleep and level of arousal. It also contains a small structure involved with the neurotransmitter dopamine, which deteriorates in Parkinson's disease.

Running through the core of the hindbrain, midbrain, and brainstem is the **reticular formation (RF)**. This diffuse, finger-shaped network of neurons helps screen incoming sensory information and alerts the higher brain centers to important events. Without your reticular formation, you would not be alert or perhaps even conscious.

**Forebrain** The **forebrain** is the largest and most prominent part of the human brain. It includes the cerebral cortex, hypothalamus, limbic system, and thalamus (**Figure 2.17**). The first three structures are located near the top of the brainstem. The cerebral cortex (discussed separately in the next section) is wrapped above and around them. (*Cerebrum* is Latin for "brain," and *cortex* is Latin for "covering" or "bark.")

The **thalamus** integrates input from the senses, and it may also be involved in learning and memory (Kalat, 2013). Think of the thalamus as an air traffic control center that receives information from all aircraft and directs them to landing or takeoff areas. The thalamus receives input from nearly all sensory systems, except smell, and directs the information to the appropriate cortical areas. The thalamus also transmits some higher brain information to the cerebellum and medulla.

Because the thalamus is the brain's major sensory relay center to the cerebral cortex, damage or abnormalities might cause the cortex to misinterpret or not receive vital sensory information. Interestingly, brain-imaging research links thalamus abnormalities to schizophrenia, a serious psychological disorder involving problems with sensory filtering and perception (Bor et al., 2011; De Witte et al., 2011; Preuss et al., 2005).

Beneath the thalamus lies the kidney bean-sized **hypothalamus** (*hypo-* means "under"). It has been called the "master control center" for emotions and many basic motives such as hunger, thirst, sex, and aggression (Banich & Compton, 2011; Hull, 2011; Lenz & McCarthy, 2010). It regulates the body's internal environment, including temperature control, which it accomplishes by regulating the endocrine system.

Hanging down from the hypothalamus, the *pituitary gland* is usually considered the master endocrine gland because it releases hormones that activate the other endocrine glands. The hypothalamus influences the pituitary through direct neural connections and through release of its own hormones into the blood supply of the pituitary. The hypothalamus also directly influences some important aspects of behavior, such as eating and drinking patterns.

An interconnected group of forebrain structures, known as the **limbic system**, is located roughly along the border between the cerebral cortex and the lower-level brain structures.

## Structures of the forebrain • Figure 2.17

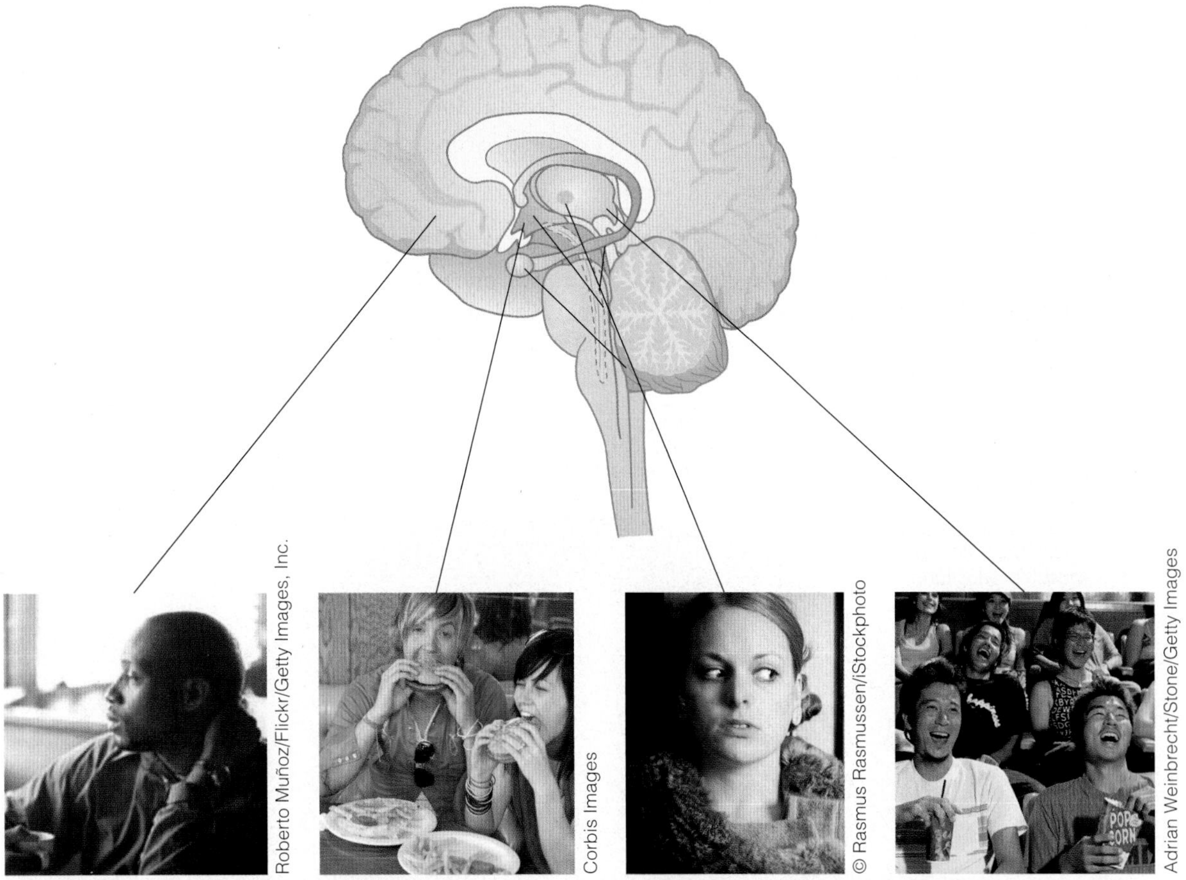


**Cerebral cortex** Governs higher mental processes

**Hypothalamus** Controls basic drives, such as hunger

**Limbic system** Involved in emotions, drives, and memory

**Thalamus** Brain's sensory switchboard

The limbic system is generally responsible for emotions, drives, and memory. However, the major focus of interest in the limbic system, and particularly the **amygdala**, has been its production and regulation of aggression and fear (LeDoux, 1998, 2002, 2007; Pessoa, 2010; Ritchey, LaBar & Cabeza, 2011). Another well-known function of the limbic system is its role in pleasure or reward (Dackis & O'Brien, 2001; Olds & Milner, 1954). Even though limbic system structures and neurotransmitters are instrumental in emotions, the frontal lobes of the cortex also play an important role.

## The Cerebral Cortex

The gray, wrinkled **cerebral cortex** is responsible for most complex behaviors and higher mental processes. It plays such a vital role that many consider it the essence of life. In fact, physicians may declare a person legally dead when the cortex dies, even when the lower-level brain structures and the rest of the body are fully functioning.

**cerebral cortex** The thin surface layer on the cerebral hemispheres that regulates most complex behavior, including receiving sensations, motor control, and higher mental processes.

Although the cerebral cortex is only about one-eighth of an inch thick, it's made up of approximately 30 billion neurons and nine times as many glial cells. It contains numerous "wrinkles" called *convolutions*, which increases the surface area while allowing it to fit in the restricted space of the skull.

The full cerebral cortex and the two cerebral hemispheres beneath it closely resemble an oversized walnut. The division, or *fissure*, down the center marks the left and right *hemispheres* of the brain, which make up about 80% of the brain's weight. They are mostly filled with axon connections between the cortex and the other brain structures. Each hemisphere controls the opposite side of the body.

The cerebral hemispheres are divided into eight distinct areas or lobes—four in each hemisphere (**Figure 2.18**). Like the lower-level brain structures, each lobe specializes in somewhat different tasks—another example of *localization of function*. However, some functions overlap between lobes.

**Frontal lobes** The large **frontal lobes** coordinate messages received from the other three lobes. An area at the very back of the frontal lobes, known as the *motor cortex*, instigates all voluntary movement. In the lower left frontal lobe lies *Broca's area*. In 1865, French physician Paul Broca discovered that damage to this area causes

## Lobes of the brain • Figure 2.18

This is a view of the brain's left hemisphere showing its four lobes—*frontal*, *parietal*, *temporal*, and *occipital*. The right hemisphere has the same four lobes. Divisions between the lobes are marked by visibly prominent folds. Keep in mind that Broca's and Wernicke's areas are only in the left hemisphere.

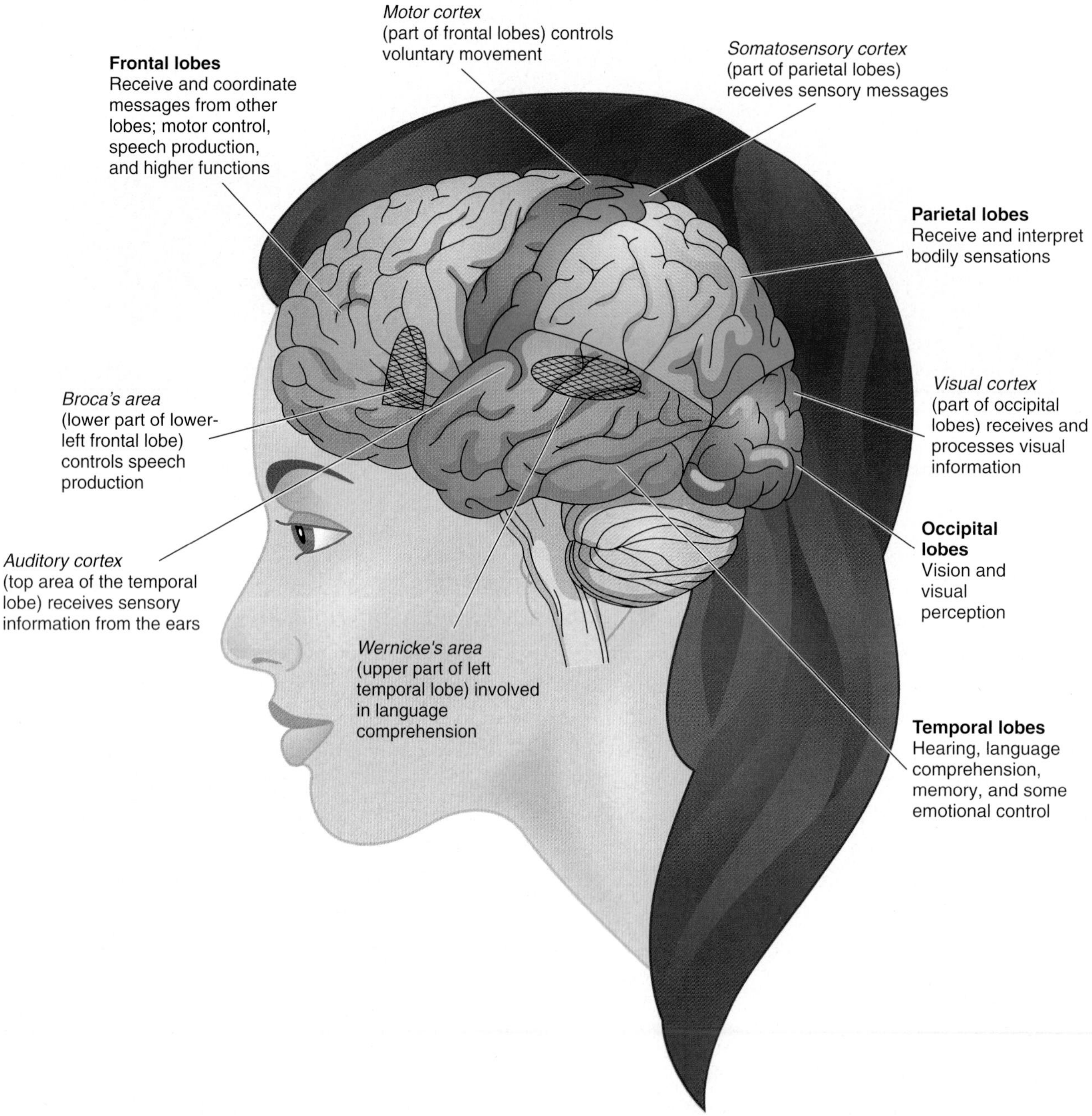

Motor cortex
(part of frontal lobes) controls voluntary movement
Frontal lobes
Receive and coordinate messages from other lobes; motor control, speech production, and higher functions
Somatosensory cortex
(part of parietal lobes) receives sensory messages
Parietal lobes
Receive and interpret bodily sensations
Broca's area
(lower part of lower-left frontal lobe) controls speech production
Visual cortex
(part of occipital lobes) receives and processes visual information
Occipital lobes
Vision and visual perception
Auditory cortex
(top area of the temporal lobe) receives sensory information from the ears
Wernicke's area
(upper part of left temporal lobe) involved in language comprehension
Temporal lobes
Hearing, language comprehension, memory, and some emotional control

THE PLANNER

## Phineas Gage—Myths Versus Facts

From the collection of Jack and Beverly Wilgus

In 1848, a 25-year-old railroad foreman named Phineas Gage had a metal rod (13 ½ pounds, 3 feet 7 inches long, and 1 ¼ inches in diameter) accidentally blown through the front of his face—destroying much of his brain's left frontal lobe. Amazingly, Gage was immediately able to sit up, speak, and move around, and only received medical treatment about 1 ½ hours later. After his wound healed, he tried to return to work, but was fired because the previously friendly, efficient, and capable foreman was now "fitful, impatient, and lacking in deference to his fellows" (Macmillan, 2000). In the words of his friends: "Gage was no longer Gage" (Harlow, 1868).

This so-called American Crowbar Case is often cited in current texts and academic papers as one of the earliest in-depth studies of an individual's survival after massive damage to the brain's frontal lobes. The evidence is clear that Gage did experience several dramatic changes in his behavior and personality after the accident, but the extent and permanence of these changes are in dispute. Most accounts of post-accident Gage report him as impulsive and unreliable until his death. However, the fact is that later, more reliable evidence shows a remarkably recovered Gage, who spent many years driving stagecoaches, a job that required demanding motor, cognitive, and interpersonal skills (Macmillan 2000, 2008; Macmillan & Lena, 2010).

So why bother reporting this controversy? As you'll note throughout this text, we include several "Myth Buster" boxes in order to clarify and correct popular misconceptions in psychology. We also include it because Phineas Gage's story highlights how a small set of reliable facts can be distorted and shaped to fit existing beliefs and scientific theories. For example, at the time of Gage's accident, little was known about how the brain functioned and damage to it was believed to be largely irreversible. Can you see how our current knowledge of *neurogenesis* and *neuroplasticity* (p. 44) now explains the previously ignored evidence of Gage's significant recovery in later life?

### Identify the Research Method

1. What is the most likely research method used to study the case of Phineas Gage?
2. If you chose
   - the experimental method, label the IV, DV, experimental group, and control group.
   - the descriptive method, is this a naturalistic observation, survey, or case study?
   - the correlational method, is this a positive, negative, or zero correlation?
   - the biological method, identify the specific research tool (e.g., brain dissection, CT scan).

(Check your answers in Appendix C.)

difficulty in speech, but not language comprehension. This type of impaired language ability is known as *Broca's aphasia*.

Finally, the frontal lobes control most higher functions that distinguish humans from other animals, such as thinking, personality, emotion, and memory. Abnormalities in the frontal lobes are often observed in patients with schizophrenia (Chapter 13). And as the case of Phineas Gage (see *Psychological Science*) and other research indicate, damage to the frontal lobe affects personality, motivation, drives, creativity, self-awareness, initiative, reasoning, and emotional behavior (see again Figure 2.18).

## Body representation of the motor cortex and somatosensory cortex • Figure 2.19

This drawing represents a vertical cross section taken from the left hemisphere's motor cortex and right hemisphere's somatosensory cortex. If body areas were truly proportional to the amount of tissue on the motor and somatosensory cortices, our bodies would look like the oddly shaped human figures draped around the outside edge of the cortex.

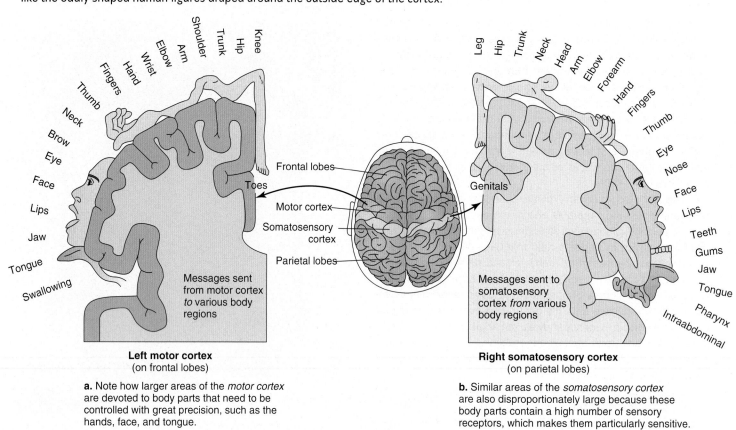

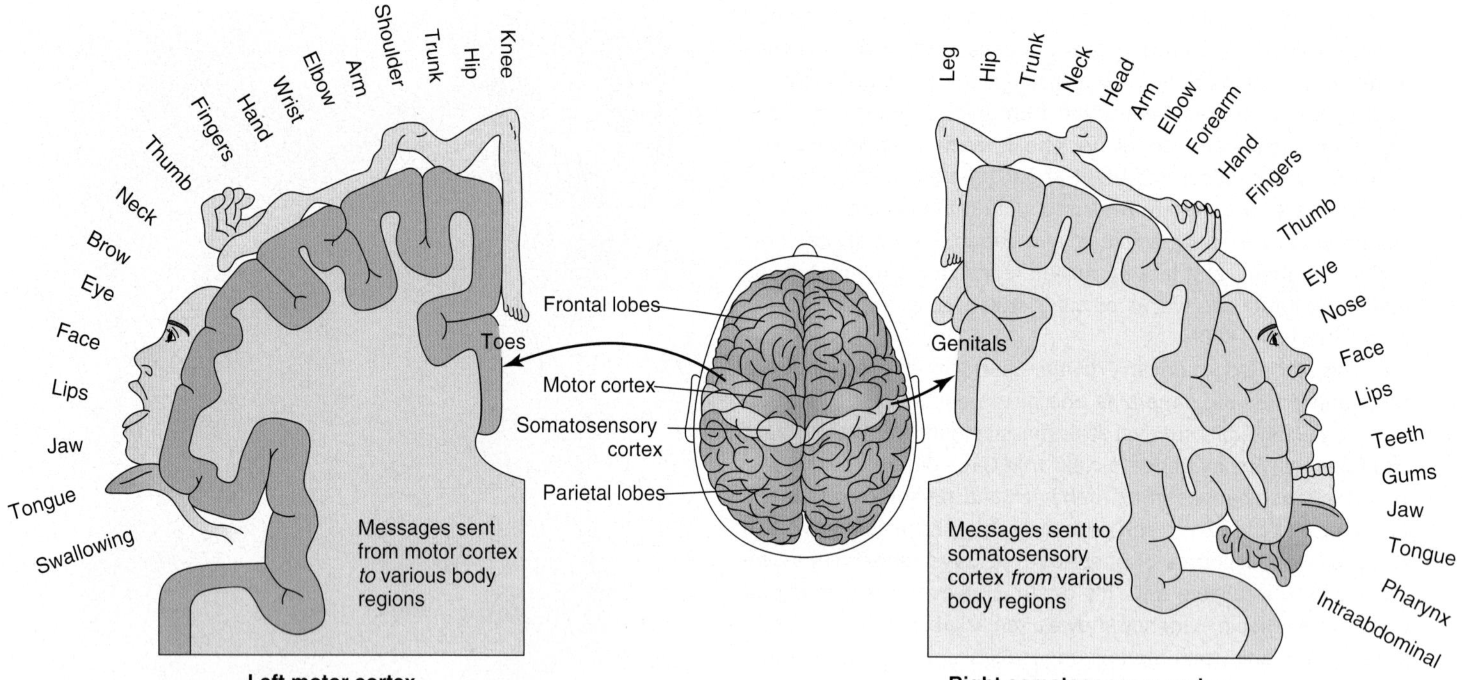


**Left motor cortex**
(on frontal lobes)

**a.** Note how larger areas of the *motor cortex* are devoted to body parts that need to be controlled with great precision, such as the hands, face, and tongue.

**Right somatosensory cortex**
(on parietal lobes)

**b.** Similar areas of the *somatosensory cortex* are also disproportionately large because these body parts contain a high number of sensory receptors, which makes them particularly sensitive.

**Parietal lobes, temporal lobes, and occipital lobes** The **parietal lobes** receive and interpret bodily sensations including pressure, pain, touch, temperature, and location of body parts. A band of tissue on the front of the parietal lobe, called the *somatosensory cortex*, receives information about touch in different body areas. Areas of the body with more somatosensory and motor cortex devoted to them (such as the hands and face) are most sensitive to touch and have the most precise motor control (**Figure 2.19**).

The **temporal lobes** are responsible for hearing, language comprehension, memory, and some emotional control. The *auditory cortex* (which processes sound) is located at the top front of each temporal lobe. This area processes incoming sensory information and sends it to the parietal lobes, where it is combined with other sensory information.

An area of the left temporal lobe, *Wernicke's area*, is involved in language comprehension. About a decade after Broca's discovery, German neurologist Carl Wernicke noted that patients with damage in this area could not understand what they read or heard, but they could speak quickly and easily. However, their speech was often unintelligible because it contained made-up words, sound substitutions, and word substitutions. This syndrome is now referred to as *Wernicke's aphasia*.

The **occipital lobes** are responsible, among other things, for vision and visual perception. Damage to the occipital lobe can produce blindness, even though the eyes and their neural connection to the brain are perfectly healthy.

**Association areas** One of the most popular myths in psychology is that we use only 10% of our brain. This myth might have begun with early research showing that approximately three-fourths of the cortex is "uncommitted" (with no precise, specific function responsive to electrical brain stimulation). These areas are not dormant, however. They are clearly involved in interpreting, integrating, and acting on information processed by other parts of the brain. They are called **association areas** because they associate, or connect, various areas and functions of the brain. The association areas in the frontal lobe, for example, help in decision making and planning. Similarly, the association area right in front of the motor cortex is involved in the planning of voluntary movement.

## Two Brains in One?

We mentioned earlier that the brain's left and right cerebral hemispheres control opposite sides of the body. Each hemisphere also has separate areas of specialization. (This is another example of *localization of function*, yet it is technically referred to as *lateralization*.)

Early researchers believed that the right hemisphere was "subordinate" or "nondominant" to the left, with few special functions or abilities. In the 1960s, landmark **split-brain research** began to change this view.

The primary connection between the two cerebral hemispheres is a thick, ribbon-like band of nerve fibers under the cortex called the **corpus callosum** (Figure 2.20). In some rare cases of severe epilepsy, when other forms of treatment have failed, surgeons cut the corpus callosum to stop the spread of epileptic seizures from one hemisphere to the other. Because this operation cuts the only direct communication link between the two hemispheres, it reveals what each half of the brain can do in isolation from the other. The resulting research has profoundly improved our understanding of how the two halves of the brain function.

### Views of the corpus callosum • Figure 2.20

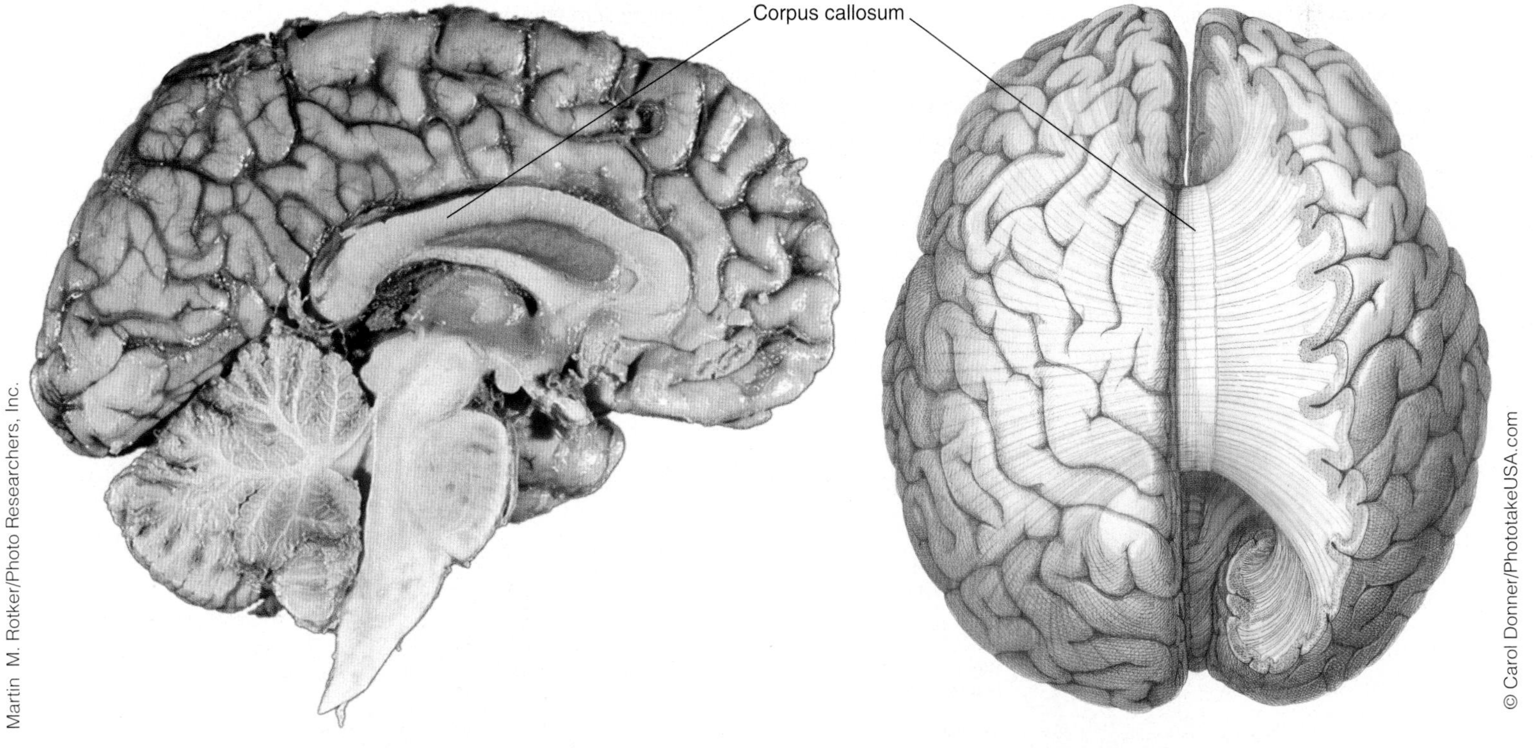


**a.** This is a photo of the left hemisphere of a deceased human brain, which has been sliced vertically from the top to the bottom to display the inside structures of the brain. Keep in mind that in actual split-brain surgery on live patients, only some fibers within the corpus callosum are cut (NOT the lower brain structures), and that this surgery is only performed in rare cases of intractable epilepsy.

**b.** In this illustration, the top layers of the cerebral cortex are removed to reveal a top-down view of the fibers, or *axons*, of the corpus callosum, which connect and transfer information between the two hemispheres.

# WHAT A PSYCHOLOGIST SEES

THE PLANNER

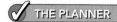

## Split-Brain Research

Experiments on split-brain patients often present visual information to only the patient's left or right hemisphere, which leads to some intriguing results. For example,

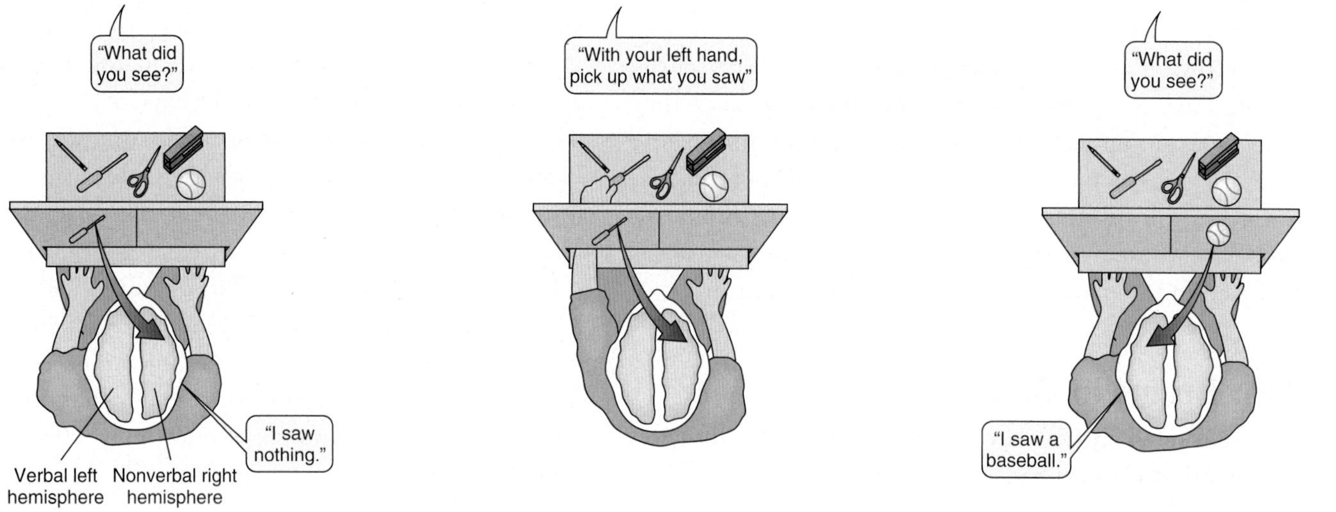


**a.** When a split-brain patient is asked to stare straight ahead while a photo of a screwdriver is flashed only to the right hemisphere, he will report that he "saw nothing."

**b.** However, when asked to pick up with his left hand what he saw, he can reach through and touch the items hidden behind the screen and easily pick up the screwdriver.

**c.** When the left hemisphere receives an image of a baseball, the split-brain patient can easily name it.

Assuming you have an intact, nonsevered corpus callosum, if the same photos were presented to you in the same way, you could easily name both the screwdriver and the baseball. Can you explain why? The answers lie in our somewhat confusing visual wiring system:

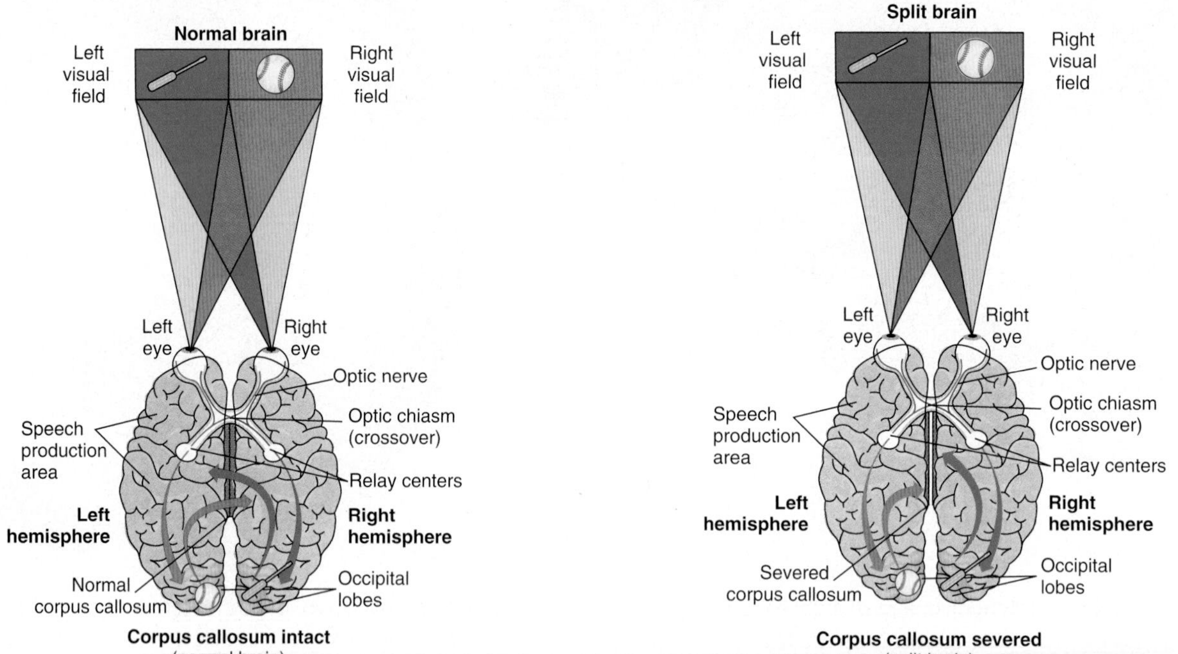


**Corpus callosum intact** (normal brain)

**Corpus callosum severed** (split brain)

**d.** As you can see, our eyes connect to our brains in such a way that, when we look straight ahead, information from the left visual field (the blue line) travels to our right hemisphere, and information from the right visual field (the red line) travels to our left hemisphere. The messages received by either hemisphere are then quickly sent to the other across the corpus callosum.

**e.** When the corpus callosum is severed, and information is presented only to the right hemisphere, a split-brain patient cannot verbalize what he sees because the information cannot travel to the opposite (verbal) hemisphere.

## MYTH BUSTERS

THE PLANNER

### MYTH OF THE "NEGLECTED RIGHT BRAIN"

Courses and books directed at "right-brain thinking" and "drawing on the right side of the brain" often promise to increase your intuition, creativity, and artistic abilities by waking up your neglected and underused right brain. Contrary to this myth, research has clearly shown that the two hemispheres work together in a coordinated, integrated way, with each making important contributions.

If you are a member of a sports team, you can easily understand this principle. Just as you and your teammates often "specialize" in different jobs, the hemispheres also somewhat divide their workload. However, team players and both hemispheres are generally aware of what the others are doing.

In our tour of the nervous system, the principles of localization of function, lateralization, and specialization are common—dendrites receive information, the occipital lobes specialize in vision, and so on. Keep in mind, however, that, like a good sports team, all parts of the brain and nervous system play overlapping and synchronized roles.

For example, when someone has a brain stroke and loses his or her ability to speak, we know that this generally points to damage on the left hemisphere because this is where Broca's area, which controls speech production, is located (see again Figure 2.18). However, when specific regions of the brain are injuried or destroyed, we now know that its functions can sometimes be picked up by a neighboring region—even the opposite hemisphere.

Although most split-brain surgery patients generally show very few outward changes in their behavior, other than fewer epileptic seizures, the surgery does create a few unusal responses. For example, one split-brain patient reported that when he dressed himself, he sometimes pulled his pants down with his left hand and up with his right (Gazzaniga, 2009). The subtle changes in split-brain patients normally appear only with specialized testing. See *What a Psychologist Sees* for a visual and description of this type of specialized test.

Popularized accounts of split-brain research have led to some exaggerated claims and unwarranted conclusions about differences between the left and right hemispheres. Read the *Myth Busters* to separate fact from fiction.

**CONCEPT CHECK** STOP

1. **What** are some examples of localization of function in the forebrain?
2. **How** do Broca's aphasia and Wernicke's aphasia differ?
3. **How** has split-brain research informed our understanding of the brain?

# Summary

THE PLANNER

## 1 Our Genetic Inheritance 34

- **Neuroscience/biopsychology** studies how biological processes relate to behavioral and mental processes.
- **Genes** (dominant or recessive) hold the code for inherited traits, as shown in the diagram. **Behavioral geneticists** use studies of twins, families, and genetic abnormalities to help determine the relative influences of heredity and environment on various traits.
- **Evolutionary psychology** suggests that many behavioral commonalities emerged and remain in human populations through **natural selection**, because they helped ensure our "genetic survival."

**Hereditary code: Genes**
**• Figure 2.1**

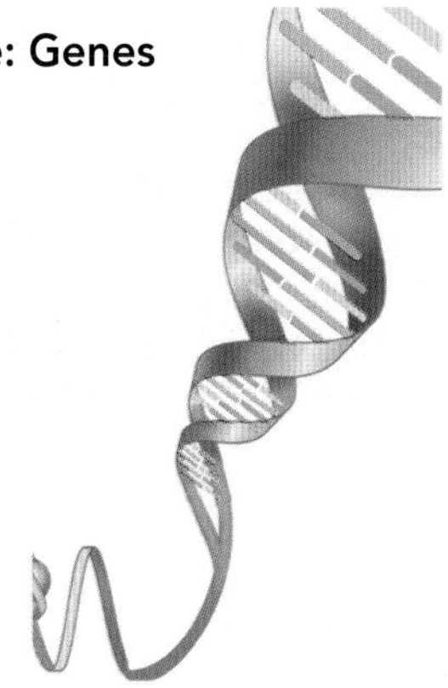

## 2 Neural Bases of Behavior 38

- **Neurons**, supported by **glial cells**, receive and send electrochemical signals to other neurons and to the rest of the body. Their major components are **dendrites**, a **cell body**, and an **axon**.
- Within a neuron, a neural impulse, or **action potential**, moves along the axon, as shown in the diagram. An action potential is an all-or-none event.

**Communication *within* the neuron • Figure 2.7**

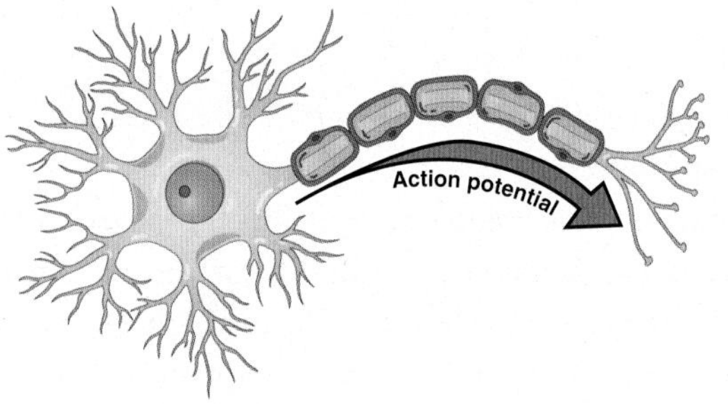


- Neurons communicate with each other using **neurotransmitters**, which are released at the synapse and attach to the receiving neuron. Neurons receive input from many synapses. Hundreds of different neurotransmitters regulate a wide variety of physiological processes. Many **agonist** and **antagonist drugs** and poisons act by mimicking or interfering with neurotransmitters.
- The **endocrine system** uses **hormones** to broadcast messages throughout the body. The system regulates long-term bodily processes, maintains ongoing bodily processes, and controls the body's response to emergencies.

## 3 Nervous System Organization 44

- The **central nervous system (CNS)** includes the brain and spinal cord, as shown in the diagram. The CNS allows us to process information and adapt to our environment in ways that no other animal can. The spinal cord transmits information between the brain and the rest of the body, and initiates involuntary reflexes. Although the CNS is very fragile, recent research shows that the brain is capable of lifelong **neuroplasticity** and **neurogenesis**. Neurogenesis is made possible by **stem cells**.

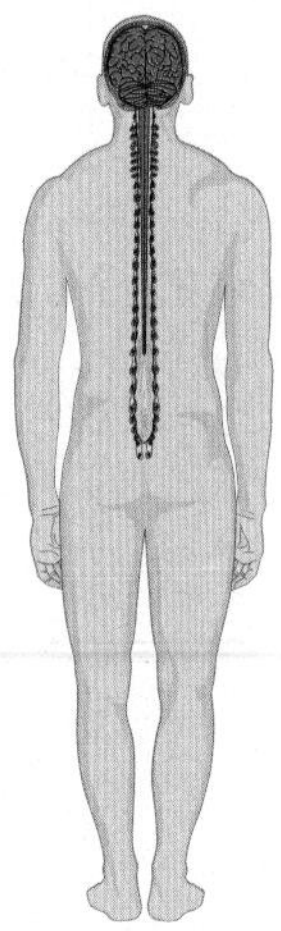

**Study Organizer 2.1 • The nervous system**

- The **peripheral nervous system (PNS)** includes all the nerves outside the brain and spinal cord. It links the brain and spinal cord to the body's sense receptors, muscles, and glands. The PNS is subdivided into the **somatic nervous system (SNS)** and the **autonomic nervous system (ANS)**.
- The ANS includes the **sympathetic nervous system** and the **parasympathetic nervous system**. The sympathetic nervous system mobilizes the body's "fight-or-flight" response. The parasympathetic nervous system returns the body to its normal functioning.

## 4 A Tour Through the Brain 50

- As shown in the diagram, the brain is divided into the **hindbrain**, the **midbrain**, and the **forebrain**. The brainstem includes parts of each of these. Certain brain structures are specialized to perform certain tasks (**localization of function**).

**The human brain • Figure 2.15**

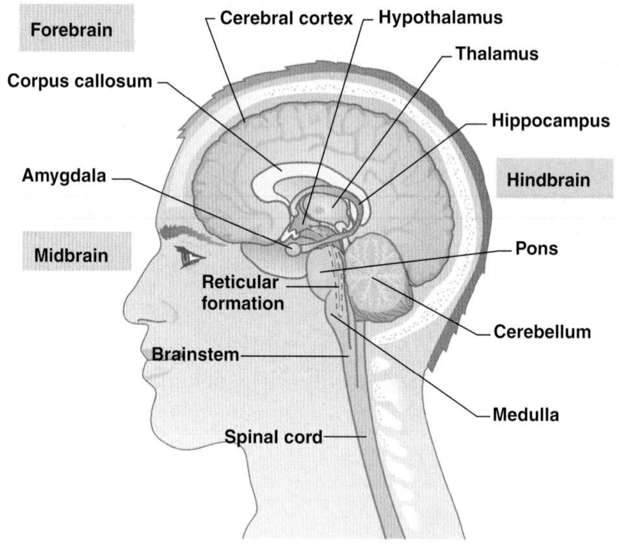


- The hindbrain (including the **medulla**, **pons**, and **cerebellum**) controls automatic behaviors and survival responses.
- The midbrain helps us orient our eye and body movements, helps control sleep and arousal, and is involved with the neurotransmitter dopamine. The **reticular formation (RF)** runs through the core of the hindbrain, midbrain, and brainstem.
- Forebrain structures (including the **cerebral cortex, hypothalamus, limbic system,** and **thalamus**) integrate input from the senses, control basic motives, regulate the body's internal environment, and regulate emotions, learning, and memory.
- The **cerebral cortex**, part of the forebrain, governs most higher processing and complex behaviors. It is divided into two hemispheres, each controlling the opposite side of the body. The **corpus callosum** links the hemispheres. Each

hemisphere is divided into **frontal**, **parietal**, **temporal**, and **occipital lobes**. Each lobe specializes in somewhat different tasks, but a large part of the cortex is devoted to integrating actions performed by different brain regions.

- Split-brain research shows that each hemisphere performs somewhat different functions, although they work in close communication.

# Key Terms

**RETRIEVAL PRACTICE** Write a definition for each term before turning back to the referenced page to check your answer.

- action potential 39
- agonist drugs 41
- antagonist drugs 41
- amygdala 53
- association areas 57
- autonomic nervous system (ANS) 47
- axon 38
- behavioral genetics 34
- biopsychology 32
- cell body 38
- central nervous system (CNS) 44
- cerebellum 52
- cerebral cortex 53
- corpus callosum 57
- cortisol 44
- dendrite 38
- endocrine system 42
- endorphin 42
- evolutionary psychology 34
- forebrain 52
- frontal lobe 54
- gene 34
- glial cell 38
- hindbrain 50
- hormone 42
- hypothalamus 52
- limbic system 52
- localization of function 50
- medulla 52
- midbrain 52
- myelin sheath 41
- natural selection 37
- neurogenesis 44
- neuron 38
- neuroplasticity 44
- neuroscience 32
- neurotransmitter 41
- occipital lobe 56
- parasympathetic nervous system 48
- parietal lobe 56
- peripheral nervous system (PNS) 44
- pons 52
- reflex/reflex arc 46
- reticular formation (RF) 52
- somatic nervous system (SNS) 47
- split-brain research 57
- stem cell 46
- sympathetic nervous system 48
- synapse 40
- temporal lobe 56
- thalamus 52

# Critical and Creative Thinking Questions

1. Imagine that scientists were able to identify specific genes linked to serious criminal behavior, and it was possible to remove or redesign these genes. Would you be in favor of this type of gene manipulation? Why or why not?
2. From an evolutionary perspective, can you explain why people are more likely to help family members than strangers?
3. Why is it valuable for scientists to understand how neurotransmitters work at a molecular level?
4. What are some everyday examples of neuroplasticity—that is, of how the brain is changed and shaped by experience?
5. Imagine you are giving a speech. Name the cortical lobes involved in the following behaviors:
   a. Seeing faces in the audience
   b. Hearing questions from the audience
   c. Remembering where your car is parked when you are ready to go home
   d. Noticing that your new shoes are too tight and hurting your feet

# What is happening in this picture?

Courtesy Karen Huffman

Most traits are polygenic, meaning they are controlled by more then one gene. These three brothers are demonstrating whether or not they can curl their tongues, one of the few traits that depend on only one dominant gene.

**Think Critically**

1. Based on the photo, what can you predict about whether one or both of the boys' parents were "noncurlers"?
2. Can you imagine why humans and other animals have evolved to possess complex traits such as tongue curling?

# Self-Test

**RETRIEVAL PRACTICE** Completing this self-test and comparing your answers with those in Appendix C provides immediate feedback and helpful practice for exams. Additional interactive, self-tests are available at www.wiley.com/college/carpenter.

1. Behavioral genetics is the study of ______.
   a. the relative effects of behavior and genetics on survival
   b. the relative effects of heredity and environment on behavior and mental processes
   c. the relative effects of genetics on natural selection
   d. how genetics affects correct behavior
2. Evolutionary psychology studies ______.
   a. the ways in which humans adapted their behavior to survive and evolve
   b. the ways in which humankind's behavior has changed over the millennia
   c. the ways in which humans can evolve to change behavior
   d. the ways in which natural selection and adaptation can explain behavior and mental processes
3. This is a measure of the degree to which a characteristic is related to genetic, inherited factors.
   a. heritability
   b. inheritance
   c. the biological ratio
   d. the genome statistic
4. The term ______ refers to a process that occurs when a genetic trait gives an organism a reproductive advantage over others.
   a. natural selection
   b. devolution
   c. survival of the fastest
   d. reproductive success
5. Label the following parts of a neuron, the cell of the nervous system responsible for receiving and transmitting electrochemical information:
   a. dendrites
   b. cell body
   c. axon
   d. myelin sheath
   e. terminal buttons of axon

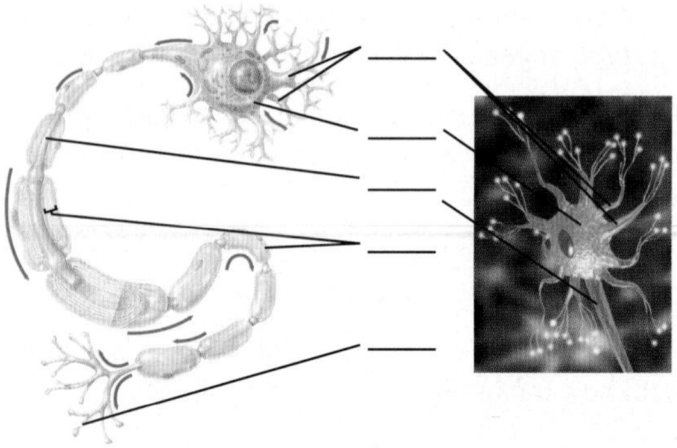

6. Your textbook's definition of an action potential is ______.
   a. the likelihood that a neuron will take action when stimulated
   b. the tendency for a neuron to be potentiated by neurotransmitters
   c. a neural impulse that carries information along the axon of a neuron
   d. the firing of a nerve, either toward or away from the brain

7. Too much of this neurotransmitter may be related to schizophrenia, whereas too little of this neurotransmitter may be related to Parkinson's disease.
   a. acetylcholine
   b. dopamine
   c. norepinephrine
   d. serotonin

8. Chemicals that are manufactured by endocrine glands and circulated in the bloodstream to change or maintain bodily functions are called ______.
   a. vasopressors
   b. gonadotropins
   c. hormones
   d. steroids

9. Label the main glands of the endocrine system (pineal, pituitary, adrenal, thyroid, hypothalamus).

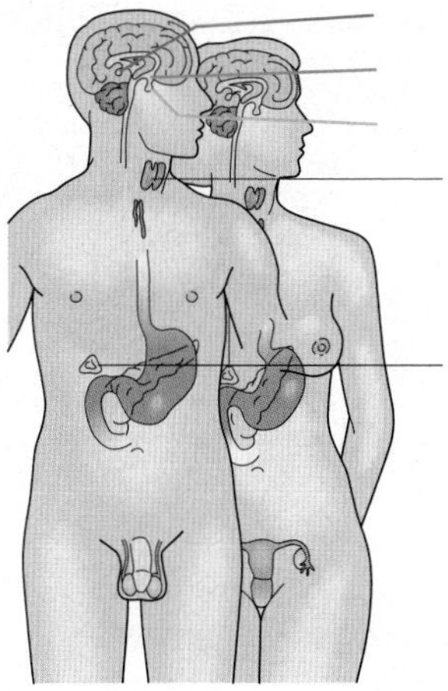

10. The central nervous system ______.
    a. consists of the brain and spinal cord
    b. is the most important nervous system
    c. includes the automatic and other nervous systems
    d. all of these options

11. The peripheral nervous system ______.
    a. is composed of the spinal cord and peripheral nerves
    b. is less important than the central nervous system
    c. is contained within the skull and spinal column
    d. includes all the nerves and neurons outside the brain and spinal cord

12. The ______ nervous system is responsible for fight or flight, whereas the ______ nervous system is responsible for maintaining calm.
    a. central; peripheral
    b. parasympathetic; sympathetic
    c. sympathetic; parasympathetic
    d. autonomic; somatic

13. Label the following structures/areas of the brain:

| | |
|---|---|
| a. corpus callosum | d. thalamus |
| b. amygdala | e. hippocampus |
| c. cerebellum | f. cerebral cortex |

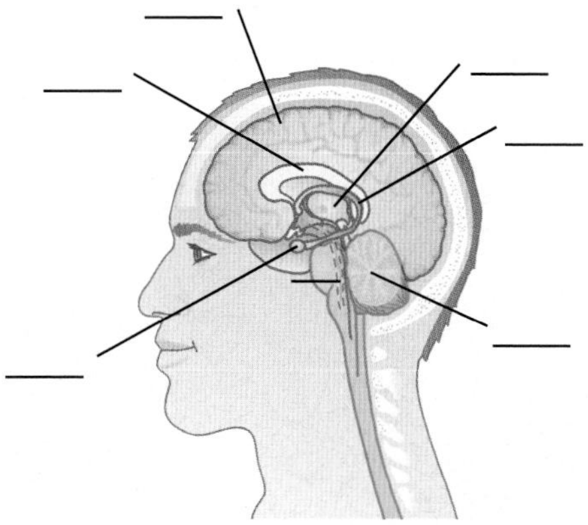

14. Label the four lobes of the brain:
    a. frontal lobe
    b. parietal lobe
    c. temporal lobe
    d. occipital lobe

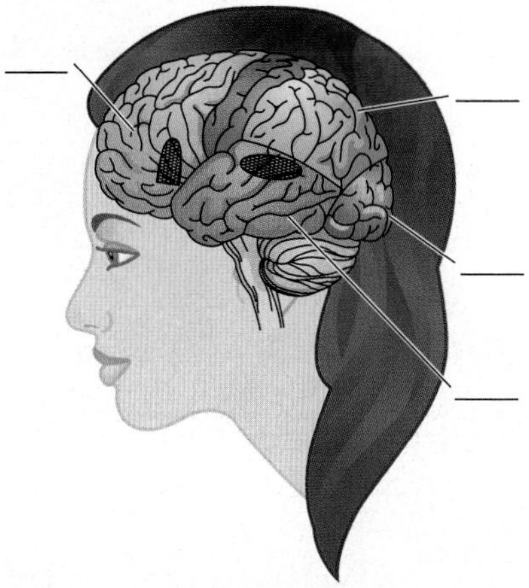

15. Although the left and right hemispheres sometimes perform differing, specialized functions, they are normally in close communication and share functions thanks to the ______.
    a. thalamus
    b. sympathetic nervous system
    c. cerebellum
    d. corpus callosum

**THE PLANNER** ✓

Review your Chapter Planner on the chapter opener and check off your completed work.

# 3 Stress and Health Psychology

What picture or visualization comes to mind when you think about stress? A high-powered executive facing a difficult decision is what many people think of as the epitome of stress. But, as this photo of an unhappy family picnic shows, stress is an everyday, inescapable part of all our lives. As a student you also experience chronically high levels of tension over the rising costs of tuition and fees combined with internal and external pressures to do well on your exams and assignments.

How do you manage these and other stressors? What do you do to protect your health? The good news is that in this chapter we'll explore the latest

information on how biological, psychological, and social factors (the *biopsychosocial model*) affect illness as well as health and well-being. You'll also discover ways to protect your health, how to manage your stress, and how to change your risky behaviors. Welcome to the practical and exciting world of *Stress and Health Psychology!*

## CHAPTER OUTLINE



## CHAPTER PLANNER ✓

- ❑ Study the picture and read the opening story.
- ❑ Answer the Learning Objectives in each section: p. 66 ❑ p. 72 ❑ p. 76 ❑
- ❑ Read the text and study all figures and visuals. Answer any questions.

**Analyze key features**

- ❑ Myth Busters, p. 66
- ❑ Applying Psychology p. 67 ❑ p. 81 ❑
- ❑ Study Organizers p. 68 ❑ p. 80 ❑
- ❑ Process Diagrams p. 69 ❑ p. 70 ❑
- ❑ Psychological Science, p. 72
- ❑ What a Psychologist Sees, p. 77
- ❑ Psychology InSight, p. 78
- ❑ Stop: Answer the Concept Checks before you go on. p. 71 ❑ p. 76 ❑ p. 81 ❑

**End of chapter**

- ❑ Review the Summary and Key Terms.
- ❑ Answer the Critical and Creative Thinking Questions.
- ❑ Answer What is happening in this picture?
- ❑ Complete the Self-Test and check your answers.

# Understanding Stress

## LEARNING OBJECTIVES

**RETRIEVAL PRACTICE** While reading the upcoming sections, respond to each Learning Objective in your own words. Then compare your responses with those in Appendix B.

1. **Explain** the sources and effects of stress.
2. **Review** the three phases of the general adaptation syndrome (GAS).
3. **Describe** the SAM system and the HPA axis.

Anything that places a demand on the body can cause **stress**. The trigger that prompts the stressful reaction is called a **stressor**. Stress reactions can occur in response to either internal cognitive stimuli, external stimuli, or environmental stimuli (Marks et al., 2011; Sanderson, 2013). There is a lot of discussion of stress in popular culture, but not all of it is supported by science. Check your knowledge of stress by taking the quiz in *Myth Busters.*

**stress** The body's nonspecific response to any demand made on it; physical and mental arousal to situations or events that we perceive as threatening or challenging.

Pleasant or beneficial stress, such as moderate exercise, is called **eustress**. Stress that is unpleasant or objectionable, as from chronic illness, is called **distress** (Selye, 1974). The total absence of stress would mean the total absence of stimulation, which would eventually lead to death. Because health psychology has been chiefly concerned with the negative effects of stress, we will adhere to convention and use the word "stress" to refer primarily to harmful or unpleasant stress.

### MYTH BUSTERS

#### WHAT DO YOU KNOW ABOUT STRESS?

**TRUE OR FALSE?**

___ 1. Even positive events, like graduating from college and getting married, are major sources of stress.
___ 2. Small, everyday hassles can impair your immune system functioning.
___ 3. People in chronically stressful professions are particularly prone to "burnout."
___ 4. Stress causes cancer.
___ 5. Having a positive attitude helps fight off cancer,
___ 6. Hardy personality types may cope better with stress.
___ 7. Having a cynical, hostile Type A personality contributes to heart disease.
___ 8. Ulcers are caused primarily or entirely by stress.
___ 9. Friends are one of your best health resources.
___ 10. You can control, or minimize, most of the negative effects of stress.

Answers: Three out of the 10 questions are false. Check the chapter for the correct answers.

## Sources of Stress

Although stress is pervasive in all our lives, psychogical science has focused on seven major sources (**Figure 3.1**). For example, early stress researchers Thomas Holmes and Richard Rahe (1967) believed that any **life change** that required some adjustment in behavior or lifestyle could cause some degree of stress. They also believed that exposure to numerous stressful events within a short period could have a direct detrimental effect on health.

To investigate the relationship between change and stress, Holmes and Rahe created a Social Readjustment Rating Scale (SRRS) that asked people to check off all the life events they had experienced in the previous year (see *Applying Psychology*).

**Seven major sources of stress • Figure 3.1**

Although stress can come from a variety of sources, researchers commonly focus on these seven major sources.

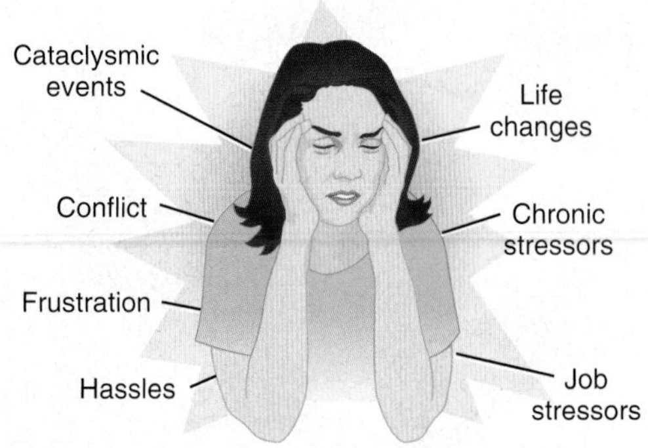

# Applying Psychology

THE PLANNER

## Measuring Life Changes

To score yourself on the Social Readjustment Rating Scale (SRRS), add up the "life change units" for all life events you have experienced during the last year and compare your score with the following standards: 0–150 = No significant problems; 150–199 = Mild life crisis (33% chance of illness); 200–299 = Moderate life crisis (50% chance of illness); 300 and above = Major life crisis (80% chance of illness).

| Life events | Life change units | Life events | Life change units |
|---|---|---|---|
| Death of spouse | 100 | Son or daughter leaving home | 29 |
| Divorce | 73 | Trouble with in-laws | 29 |
| Marital separation | 65 | Outstanding personal achievement | 28 |
| Jail term | 63 | Spouse begins or stops work | 26 |
| Death of a close family member | 63 | Begin or end school | 26 |
| Personal injury or illness | 53 | Change in living conditions | 25 |
| Marriage | 50 | Revision of personal habits | 24 |
| Fired at work | 47 | Trouble with boss | 23 |
| Marital reconciliation | 45 | Change in work hours or conditions | 20 |
| Retirement | 45 | Change in residence | 20 |
| Change in health of family member | 44 | Change in schools | 20 |
| Pregnancy | 40 | Change in recreation | 19 |
| Sex difficulties | 39 | Change in church activities | 19 |
| Gain of a new family member | 39 | Change in social activities | 18 |
| Business readjustment | 39 | Mortgage or loan for lesser purchase (car, major appliance) | 17 |
| Change in financial state | 38 | | |
| Death of a close friend | 37 | Change in sleeping habits | 16 |
| Change to different line of work | 36 | Change in number of family get-togethers | 15 |
| Change in number of arguments with spouse | 35 | Change in eating habits | 15 |
| Mortgage or loan for major purchase | 31 | Vacation | 13 |
| Foreclosure on mortgage or loan | 30 | Christmas | 12 |
| Change in responsibilities at work | 29 | Minor violations of the law | 11 |

*Source:* Reprinted from the *Journal of Psychosomatic Research*, Vol. III; Holmes and Rahe: "The Social Readjustment Rating Scale," 213–218, 1967, with permission from Elsevier.

The SRRS scale is an easy and popular way to measure stress. Cross-cultural studies have shown that most people rank the magnitude of stressful events in similar ways (De Coteau, Hope, & Anderson, 2003; Thoits, 2010). But the SRRS is not foolproof. For example, it only shows a correlation between stress and illness; it does not prove that stress actually causes illnesses. Moreover, not all stressful situations are **cataclysmic events**, such as a terrorist attack, or single events like a death or a birth. **Chronic stressors**, such as war, a bad marriage, poor working conditions, or a repressive political climate, can be significant too (Enoch, 2011; Torpy, Lynm, & Glass, 2007). Even the stress of low-frequency noise is associated with measurable hormonal and cardiac changes (Waye et al., 2002). Our social lives can also be chronically stressful because making and maintaining friendships involves considerable thought and energy (Sias et al., 2004).

Given the recent global economic meltdown, one of our most pressing concerns is **job stress**, which includes unemployment, keeping or changing jobs, and job performance (Hoppe, 2011; O'Neill & Davis, 2011). The most stressful jobs are those that make great demands on performance and concentration but allow little creativity or opportunity for advancement (Sanderson, 2013; Straub, 2011).

Stress at work can also cause serious stress at home, not only for the worker but for other family members as well. In our private lives, divorce, child and spousal abuse, alcoholism, and money problems can place severe stress on all members of a family (Aboa-Éboulé, 2008; Aboa-Éboulé et al. 2007; Rosenström et al., 2011).

In addition to job stressors, stress can arise when we experience **conflict**—that is, when we are forced to make a choice between at least two incompatible alternatives. There are three basic types of conflict as shown in **Study Organizer 3.1**.

Generally, approach–approach conflicts are the easiest to resolve and produce the least stress. Avoidance–avoidance conflicts, on the other hand, are usually the most difficult because all choices lead to unpleasant results. The longer any conflict exists or the more important the decision, the more stress a person will experience.

The minor **hassles** of daily living also can pile up and become a major source of stress. We all share many hassles, such as time pressures and financial concerns. But our reactions to these hassles vary. Persistent hassles can lead to a form of physical, mental, and emotional exhaustion known

**Study Organizer 3.1** Types of conflict — THE PLANNER

| Conflict | Description/resolution | Example/resolution | |
|---|---|---|---|
| **Approach–approach** | Forced choice between two options both of which have equally desirable characteristics | Two equally desirable job offers, but you must choose one of them because you're broke | 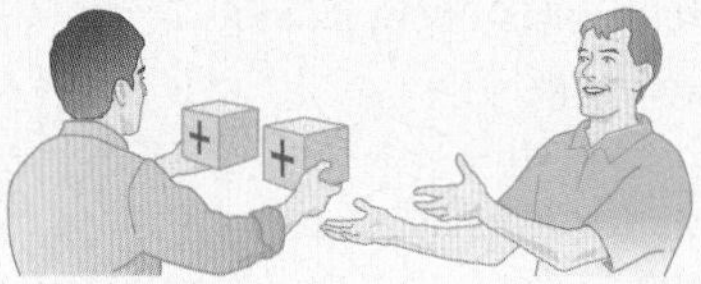 |
| + + | Generally easiest and least stressful conflict to resolve | You make a pro/con list and/or "flip a coin." | |
| **Avoidance–avoidance** | Forced choice between two options both of which have equally undesirable characteristics | Two equally undesirable job offers, but you must choose one of them because you're broke |  |
| − − | Difficult, stressful conflict, generally resolved with a long delay and considerable denial | You make a pro/con list and/or "flip a coin," and then delay the decision hoping for additional job offers. | |
| **Approach–avoidance** | Forced choice within one option, which has equally desirable and undesirable characteristics | One high-salary job offer requiring you to relocate to an undesirable location leaving all your friends and family |  |
| + − | Difficult, stressful conflict, generally resolved with delay and/or partial approach | You make a pro/con list, "flip a coin," delay the decision, and/or make a *partial approach* by taking the job until you get a better offer. | |

Lisa Peardon/Taxi/Getty

**Think Critically**

Judging by the expression on this man's face, he's obviously experiencing some form of conflict.

- Can you explain how this could be both an avoidance-avoidance and/or an approach-avoidance conflict?

## General adaptation syndrome (GAS) • Figure 3.2

THE PLANNER

The three phases of this syndrome (*alarm, resistance*, and *exhaustion*) focus on the biological response to stress—particularly the "wear and tear" on the body with prolonged stress.

1. **Alarm phase**
When surprised or threatened, your body enters an alarm phase during which your resistance to stress is temporarily suppressed, while your arousal is high (e.g., increased heart rate and blood pressure) and blood is diverted to your skeletal muscles to prepare for "fight-or-flight" (Chapter 2).

2. **Resistance phase**
If the stress continues, your body rebounds to a phase of increased resistance. Physiological arousal remains higher than normal, and there is an outpouring of stress hormones. During this resistance stage, people use a variety of coping methods. For example, if your job is threatened, you may work longer hours and give up your vacation days

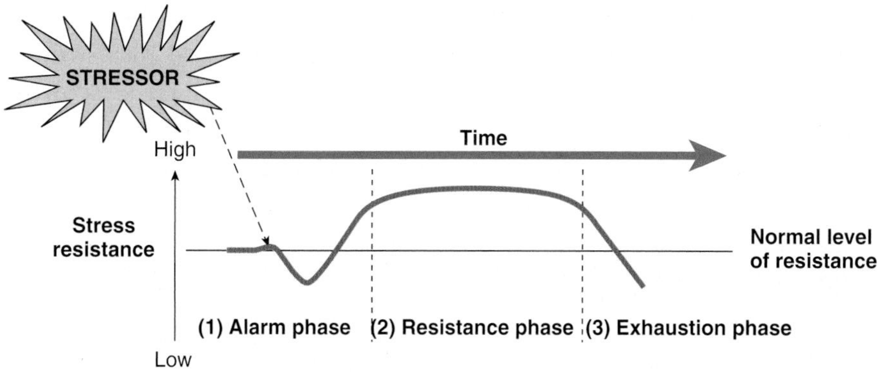


3. **Exhaustion phase**
Your body's resistance to stress can only last so long before exhaustion sets in. During this final phase, you become more susceptible to serious illnesses, and possibly irreversible damage to your body. Selye maintained that one outcome of this exhaustion phase for some people is the development of *diseases of adaptation*, including asthma, ulcers, and high blood pressure. Unless away of relieving stress is found, the eventual result may be complete collapse and death.

Art Resource

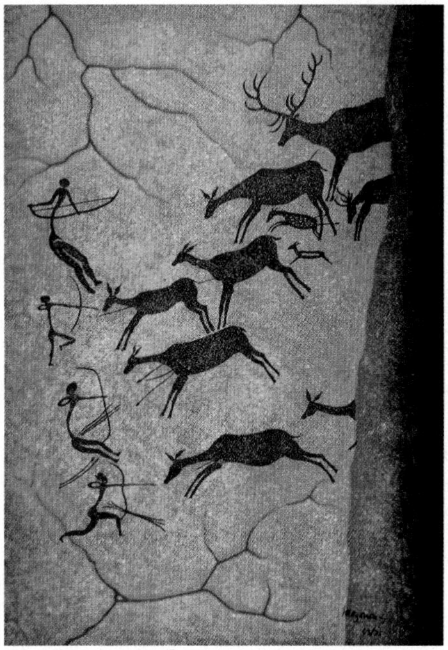

**Stress in ancient times**
As shown in these ancient cave drawings, the automatic fight-or-flight response was adaptive and necessary for early human survival. However, in modern society it occurs as a response to ongoing situations where we often cannot fight or flee. This repeated arousal can be detrimental to our health.

as **burnout** (Sarafino, 2011; Tümkaya, 2007). Over time, some people in chronically stressful professions become emotionally drained. Burnout can cause more work absences, less productivity, and increased risk for physical problems.

Some authorities believe that hassles can be more significant than major life events in creating stress (Stefanek et al., 2012; Kubiak et al., 2008). Divorce is extremely stressful, but it may be so because of the increased number of hassles—a change in finances, child-care arrangements, longer working hours, and so on.

Like hassles, **frustration**, a negative emotional state resulting from a blocked goal, can cause stress. And the more motivated we are, the more frustrated we are when our goals are blocked.

## Effects of Stress

When mentally or physically stressed, our bodies undergo several biological changes that can be detrimental to health. In 1936, Canadian physician Hans Selye (SELL-yay) described a generalized physiological reaction to stress, which he called the **general adaptation syndrome (GAS)**. The GAS occurs in three phases—*alarm*, *resistance*, and *exhaustion*—activated by efforts to adapt to any stressor, whether physical or psychological (**Figure 3.2**).

**general adaptation syndome (GAS)** Selye's three-stage (alarm, resistance, exhaustion) reaction to chronic stress.

Most of Selye's ideas about the GAS pattern of stress response have proven to be correct, but he did wrongly propose that all stressors had similar effects. Today, we know that different stressors evoke different responses and that people vary widely in their reactions. For example, one of the most interesting ways that people differ in their stress response has to do with gender. Men more often choose "fight or flight" whereas women "tend and befriend." They take care of themselves and their children (tending), and form strong social bonds with others (befriending) (Taylor, 2006, 2012).

Some researchers believe these differences are hormonal in nature. Although oxytocin is released during stress in both men and women, the female's higher level of estrogen tends to enhance oxytocin, which results in more calming and nurturing feelings. In contrast, the hormone testosterone, which men produce in higher levels during stress, reduces the effects of oxytocin.

What is Selye's most important take-home message? *Our bodies are relatively well designed for temporary stress but poorly equipped for prolonged stress*. The same biological processes that are adaptive in the short run, such as the fight-or-flight response, can become hazardous in the long run. A look at the diagram of the **SAM** (sympatho-adreno-medullary) **system** and **HPA** (hypothalamic-pituitary-adrenal) **axis** will help you understand how this happens (**Figure 3.3**).

*Cortisol*, a key element of the HPA axis, plays a critical role in the long-term effects of stress. For example, increased cortisol levels initially help us fight stressors, but if these levels stay high, the body's disease-fighting immune system is suppressed.

PROCESS DIAGRAM

THE PLANNER

## Stress—an interrelated system • Figure 3.3

Under stress, the sympathetic nervous system prepares us for immediate action—to "fight or flee." Parts of the brain and endocrine system then kick in to maintain our arousal. How does this happen?

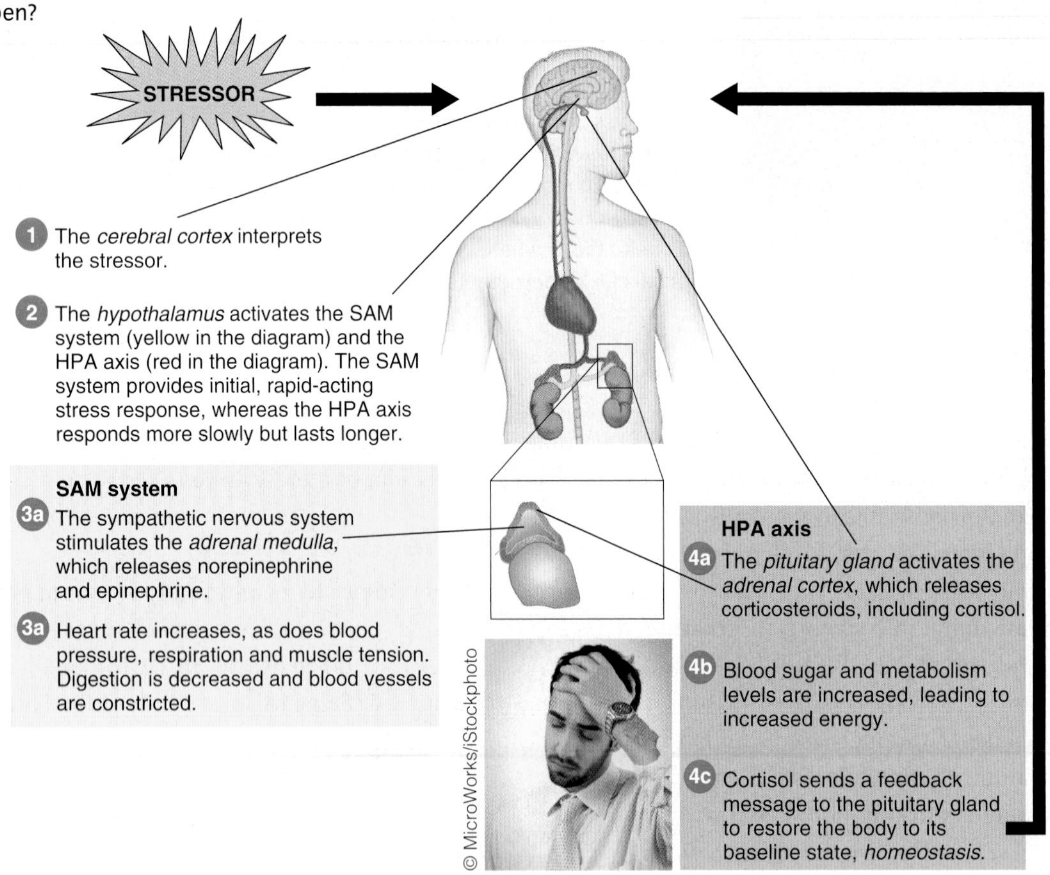

STRESSOR
1 The *cerebral cortex* interprets the stressor.
2 The *hypothalamus* activates the SAM system (yellow in the diagram) and the HPA axis (red in the diagram). The SAM system provides initial, rapid-acting stress response, whereas the HPA axis responds more slowly but lasts longer.
SAM system
3a The sympathetic nervous system stimulates the *adrenal medulla*, which releases norepinephrine and epinephrine.
3a Heart rate increases, as does blood pressure, respiration and muscle tension. Digestion is decreased and blood vessels are constricted.
HPA axis
4a The *pituitary gland* activates the *adrenal cortex*, which releases corticosteroids, including cortisol.
4b Blood sugar and metabolism levels are increased, leading to increased energy.
4c Cortisol sends a feedback message to the pituitary gland to restore the body to its baseline state, *homeostasis*.
© MicroWorks/iStockphoto

Prolonged elevation of cortisol also contributes to hypertension, depression, posttraumatic stress disorder (PTSD), drug and alcohol abuse, and even low-birth-weight infants (Bagley, Weaver, & Buchanan, 2011; Johnson, Delahanty, & Pinna, 2008; Stalder et al., 2010; Straub, 2011). As if this long list of ill-effects weren't enough, severe or prolonged stress can also produce overall physical deterioration, premature aging, and even death.

In sum, the SAM system and HPA axis are designed to increase our energy for dealing with emergencies. However, continued arousal, and its chemical onslaught, may deplete our body's energy reserves, and thereby contribute to various stress-related health problems (Lundberg, 2011; Pace & Heim. 2011).

**Stress and the immune system** The discovery of the relationship between stress and the immune system is very important. When the immune system is impaired, we are at greatly increased risk of suffering from a number of diseases, including cancer, bursitis, colitis, Alzheimer's disease, rheumatoid arthritis, periodontal disease, and even the common cold (Carroll et al., 2011; Cohen et al., 2002; Cohen & Lemay, 2007; Dantzer et al., 2008; Gasser & Raulet, 2006: Segerstrom & Miller, 2004).

Knowledge that psychological factors have considerable control over infectious diseases has upset the long-held assumption in biology and medicine that these diseases are "strictly physical." The clinical and theoretical implications are so important that a new field of biopsychology has emerged: **psychoneuroimmunology**.

> **psychoneuro-immunology** [sye-koh-NEW-roh-IM-you-NOLL-oh-gee] The interdisciplinary field that studies the effects of psychological factors on the immune system.

**Stress and cognitive functioning** Finally, what happens to our brain and thought processes when we're under stress? As we've just seen, cortisol helps us deal with immediate dangers by increasing our immunity and mobilizing our energy resources. However, it also can prevent the retrieval of existing memories, as well as the laying down of new memories and general information processing (Almela et al., 2011; Hurlemann et al., 2007; Mahoney et al., 2007; Pechtel & Pizzagalli, 2011). This interference with cognitive functioning helps explain why you may forget important information during an important exam and why people may become dangerously confused during a fire and can't find the fire exit. The good news is that once the cortisol "washes out," memory performance generally returns to normal levels.

In addition to the problems with cognitive functioning during acute stress, Robert Sapolsky (1992, 2003, 2007) has shown that prolonged stress can permanently damage the hippocampus, a key part of the brain involved in memory (Chapter 7). Furthermore, once the hippocampus is damaged, it cannot provide proper feedback to the hypothalamus, so cortisol continues to be secreted and a vicious cycle can develop (**Figure 3.4**).

**Stress and the brain • Figure 3.4**

Cortisol released in response to stress damages the brain, triggering a vicious cycle.

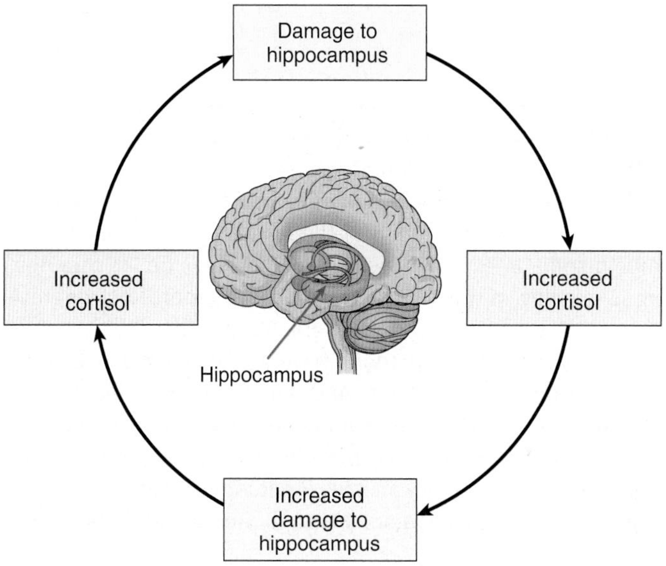


**CONCEPT CHECK** STOP

1. **What** are the three types of conflict that can cause stress?
2. **What** occurs during the exhaustion phase of the GAS?
3. **Why** does chronic stress threaten the immune system?

# Stress and Illness

## LEARNING OBJECTIVES

**RETRIEVAL PRACTICE** While reading the upcoming sections, respond to each Learning Objective in your own words. Then compare your responses with those in Appendix B.

1. **Explain** how biological and psychological factors can jointly influence the development of gastric ulcers.
2. **Explain** the relationship between stress and cancer.
3. **Define** the personality patterns that can influence how we respond to stress.
4. **Describe** the key symptoms of posttraumatic stress disorder (PTSD).

As we've just seen, stress has dramatic effects on our bodies. This section explores how stress is related to four serious illnesses—gastric ulcers (discussed in *Psychological Science*), cancer, cardiovascular disorders, and posttraumatic stress disorder (PTSD).

## Psychological Science

THE PLANNER

### Does Stress Cause Gastric Ulcers?

**Gastric ulcers** are lesions to the lining of the stomach (and duodenum—the upper section of the small intestine) that can be quite painful. In extreme cases, they may even be life threatening. Beginning in the 1950s, psychologists reported strong evidence that stress can lead to ulcers. Studies found that people who live in stressful situations have a higher incidence of ulcers than people who don't. And numerous experiments with laboratory animals have shown that stressors, such as shock or confinement to a very small space for a few hours, can produce ulcers in some laboratory animals (Andrade & Graeff, 2001; Bhattacharya & Muruganandam, 2003; Gabry et al., 2002; Takahashi et al., 2012).

The relationship between stress and ulcers seemed well established until researchers reported a bacterium (*Helicobacter pylori* or *H. pylori*), which appears to be associated with ulcers. Because many people prefer medical explanations, like bacteria or viruses, to psychological or *psychosomatic* ones, the idea of stress as a cause of ulcers was largely abandoned.

Modern research shows that ulcer patients do have the *H. pylori* bacterium in their stomachs. It clearly damages the stomach wall, and antibiotic treatment does help many patients. However, approximately 75% of normal control subjects' stomachs also have the bacterium. This suggests that the bacterium may cause the ulcer, but only in people who are compromised by stress. Furthermore, behavior modification and other psychological treatments, used alongside antibiotics, can help ease ulcers. In other words, although stress *by itself* does not cause ulcers, it is a contributing factor along with biological factors (Fink, 2011).

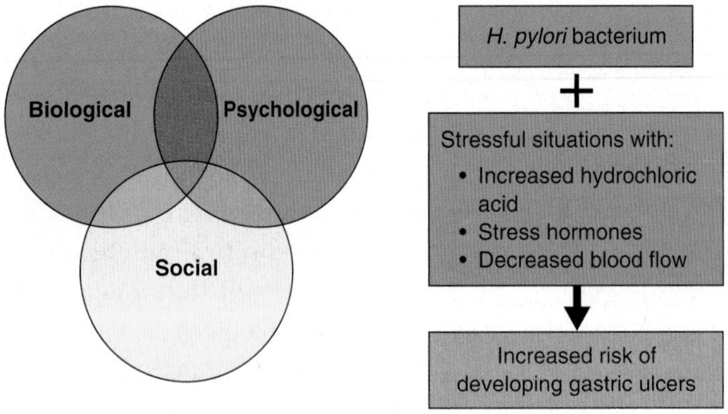


#### Identify the Research Method

1. What is the most likely research method used to study the relationship between stress and gastric ulcers in humans?
2. If you chose
   - the experimental method, label the IV, DV, experimental group, and control group.
   - the descriptive method, is this a naturalistic observation, survey, or case study?
   - the correlational method, is this a positive, negative, or zero correlation?
   - the biological method, identify the specific research tool (e.g., brain dissection, CT scan).

(Check your answers in Appendix C.)

## Cancer

**Cancer** is among the leading causes of death for adults in the United States. It occurs when a particular type of primitive body cell begins rapidly dividing and then forms a tumor that invades healthy tissue. In a healthy person, whenever cancer cells start to multiply, the immune system checks the uncontrolled growth by attacking the abnormal cells (**Figure 3.5**). Unless destroyed or removed, the tumor eventually damages organs and causes death. More than 100 types of cancer have been identified. They appear to be caused by an interaction between environmental factors (such as diet, smoking, and pollutants) and inherited predispositions.

It's important to note that research does *not* support the popular myths that stress *causes* cancer or that positive attitudes can fight off cancer (Coyne & Tennen, 2010; Lilienfeld et al., 2010; Surtees et al., 2009).

However, this is not to say that developing a positive attitude and reducing our stress levels aren't worthy health goals (Dempster et al., 2011; O'Brien & Moorey, 2010). As you read earlier, stress causes the adrenal glands to release hormones that suppress the immune system, and a compromised immune system is less able to resist infection or to fight off cancer cells (Ben-Eliyahu, Page, & Schleifer, 2007; Bernabé et al., 2011; Kemeny, 2007; Krukowski et al., 2011). For example, when researchers interrupted the sleep of 23 men and then measured their *natural killer cells* (a type of immune system cell), they found the number of killer cells was 28% below average (Irwin et al., 1994). Can you see how staying up late studying for an exam (or partying) can decrease the effectiveness of your immune system? Fortunately, these researchers found that a normal night's sleep after the deprivation returned the killer cells to their normal levels.

### The immune system in action • Figure 3.5

Stress can compromise the immune system, but the actions of a healthy immune system are shown here. The round red structures are leukemia cells. Note how the yellow killer cells are attacking and destroying the cancer cells.

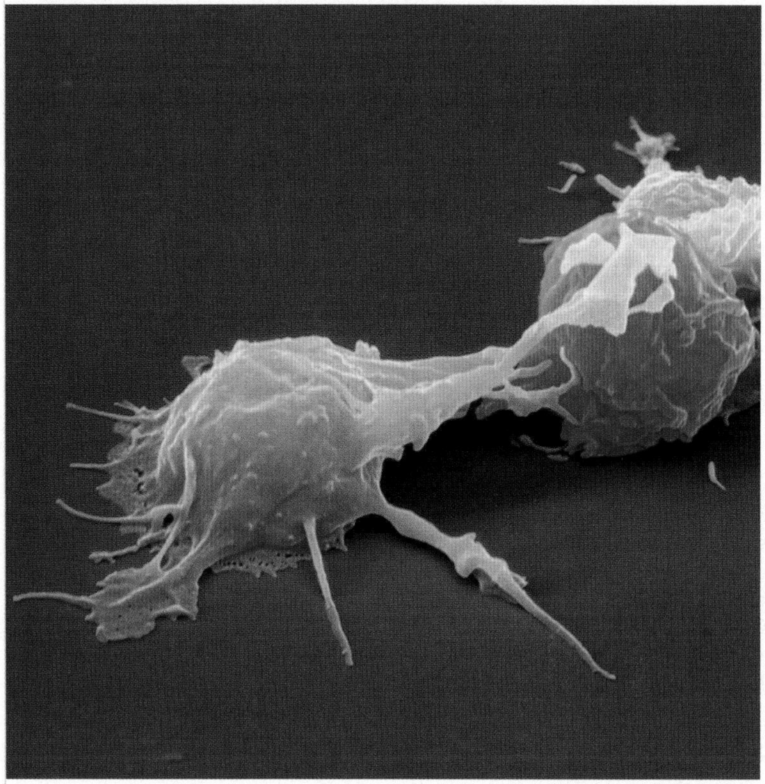

Eye of Science/Photo Researchers, Inc.

## Cardiovascular Disorders

Cardiovascular disorders contribute to over half of all deaths in the United States (American Heart Association, 2012). Understandably, health psychologists are concerned because stress is a major contributor to these deaths (Aboa-Éboulé et al., 2007; Landsbergis et al., 2011; Montoro-Garcia, Shantsila, & Lip, 2011; Phillips & Hughes, 2011).

**Heart disease** is a general term for all disorders that eventually affect the heart muscle and lead to heart failure. *Coronary heart disease* occurs when the walls of the coronary arteries thicken, reducing or blocking the blood supply to the heart. Symptoms of such disease include *angina* (chest pain due to insufficient blood supply to the heart) and *heart attack* (death of heart muscle tissue). Controllable factors that contribute to heart disease include stress, smoking, certain personality characteristics, obesity, a high-fat diet, and lack of exercise (Aboa-Éboulé, 2008; Scarborough et al., 2012; Straub, 2011).

When the body is stressed, the autonomic nervous system releases epinephrine (adrenaline) and cortisol into the bloodstream. These hormones increase heart rate and release fat and glucose from the body's stores to give muscles a readily available source of energy.

If no physical "fight-or-flight" action is taken (and this is most likely the case in our modern lives), the fat that was released into the bloodstream is not burned as fuel. Instead, it may adhere to the walls of blood vessels. These fatty deposits are a major cause of blood supply blockage, which causes heart attacks.

**Type A and Type B personalities** The effects of stress on heart disease may be amplified if a person tends to be hard driving, competitive, ambitious, impatient, and hostile—a **Type A personality**. The antithesis of the Type A personality is the **Type B personality**, which is characterized by a laid-back, relaxed attitude toward life.

> **Type A personality** Behavior characteristics including intense ambition, competition, exaggerated time urgency, and a cynical, hostile outlook.
>
> **Type B personality** Behavior characteristics consistent with a calm, patient, relaxed attitude.

Among Type A characteristics, hostility is the strongest predictor of heart disease (Allan, 2011; Mittag & Maurischat, 2004). In particular, the constant stress associated with cynical hostility—constantly being "on watch" for problems—translates physiologically into higher blood pressure, heart rates, and levels of stress-related hormones. In addition, people who are hostile, suspicious, argumentative, and competitive tend to have more interpersonal conflicts. This can heighten autonomic activation, leading to increased risk of cardiovascular disease (Boyle et al., 2004; Bunde & Suls, 2006; Haukkala et al., 2010; Williams, 2010).

Health psychologists have developed two types of behavior modification for people with Type A personalities. The shotgun approach aims to eliminate or modify all Type A behaviors. The major criticism of this approach is that it eliminates not only undesirable Type A behaviors but also desirable traits, such as ambition. In contrast, the target behavior approach focuses on only those Type A behaviors that are likely to cause heart disease—namely, cynical hostility.

**Hardiness** Have you ever wondered why some people survive in the face of great stress (personal tragedies, demanding jobs, and even a poor home life) while others do not? Suzanne Kobasa was among the first to study this question (Kobasa, 1979; Maddi et al., 2012; Vogt et al., 2008). Examining male executives with high levels of stress, she found that some people are more resistant to stress than others because of a personality factor called **hardiness**, a resilient type of optimism that comes from three distinctive attitudes: *commitment, control,* and *challenge*.

First, hardy people feel a strong sense of commitment to both their work and their personal life. They also make intentional commitments to purposeful activity and problem solving. Second, these people see themselves as being in control of their lives, rather than as victims of their circumstances. Finally, hardy people look at change as a challenge, an opportunity for growth and improvement—not as a threat (**Figure 3.6**).

The important lesson from this research is that hardiness is a learned behavior, not something based on luck or genetics. If you're not one of the hardy souls, you can develop the trait.

Of course, Type A personality and lack of hardiness are not the only controllable risk factors associated with heart disease. *Smoking, obesity*, and *lack of exercise* are also very important factors. Smoking restricts blood circulation, and obesity stresses the heart by causing it to pump more blood to the excess body tissue. A high-fat diet, especially one that is high in cholesterol, contributes to the fatty deposits that clog blood vessels. Lack of exercise contributes to weight gain and prevents the body from obtaining important exercise benefits, including strengthened heart muscle, increased heart efficiency, and the release of neurotransmitters, such as serotonin, that alleviate stress and promote well-being.

**Hardiness in action • Figure 3.6**

Based on his smile and cheery wave, it looks like this patient may be one of those lucky "hardy souls." Can you see how an optimistic attitude toward commitment, control, and challenge might help him cope and recuperate from his serious injuries?

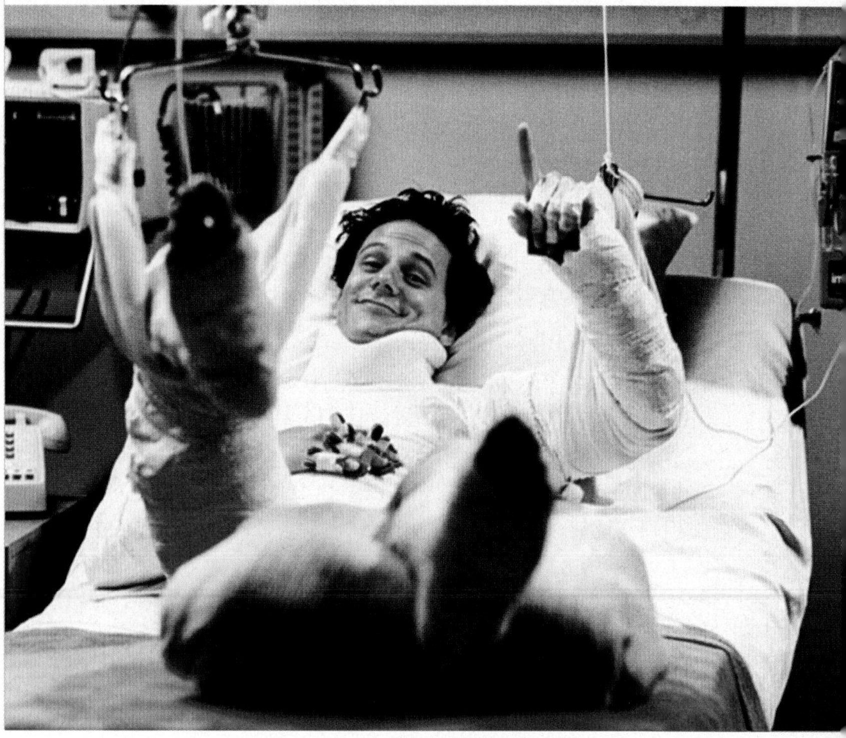

Stuart Hughes/Corbis

## Posttraumatic Stress Disorder (PTSD)

One of the most powerful examples of the effects of severe stress is **posttraumatic stress disorder (PTSD)** (Baker, Nievergelt, & O'Conner, 2011; Pace & Heim, 2011; Ruzek et al., 2011). Children as well as adults can experience the symptoms of PTSD, which include feelings of terror and helplessness during the trauma and recurrent flashbacks, nightmares, impaired concentration, and emotional numbing afterward. These symptoms may continue for months or years after the event. Some victims of PTSD turn to alcohol and other drugs, which often compound the problem (Cougle et al., 2011; Kaysen et al., 2008; McCart et al., 2011; Sullivan & Holt, 2008).

**posttraumatic stress disorder (PTSD)** An anxiety disorder following exposure to a life-threatening or other extreme event that evoked great horror or helplessness. It is characterized by flashbacks, nightmares, and impaired functioning.

During the Industrial Revolution, workers who survived horrific railroad accidents sometimes developed a condition very similar to PTSD. It was called "railway spine" because experts thought the problem resulted from a twisting or concussion of the spine. Later, doctors working with combat veterans referred to the disorder as "shell shock" because they believed it was a response to the physical concussion caused by exploding artillery. Today, we know that PTSD is caused by any exposure to extraordinary stress (**Figure 3.7**).

Joel Sartore/NG Image Collection

**Coping with extreme trauma • Figure 3.7**

Can you see how finding sources of help for those who have experienced trauma is important for both practical and psychological reasons?

**Table 3.1** summarizes the primary symptoms of PTSD and offers five important tips for coping with traumatic events. PTSD's essential feature is *severe anxiety* (a

**Identifying PTSD and coping with crisis Table 3.1**

| Primary symptoms of posttraumatic stress disorder (PTSD) | Five important tips for coping with crisis |
|---|---|
| • Reexperiencing the event through vivid memories or flashbacks<br>• Feeling "emotionally numb"<br>• Feeling overwhelmed by what would normally be considered everyday situations<br>• Diminished interest in performing normal tasks or pursuing usual interests<br>• Crying uncontrollably<br>• Isolating oneself from family and friends and avoiding social situations<br>• Relying increasingly on alcohol or drugs to get through the day<br>• Feeling extremely moody, irritable, angry, suspicious, or frightened<br>• Having difficulty falling or staying asleep, sleeping too much, or experiencing nightmares<br>• Feeling guilty about surviving the event or being unable to solve the problem, change the event, or prevent the disaster<br>• Feeling fear and sense of doom about the future | 1. If you have experienced a traumatic event, recognize your feelings about the situation and talk to others about your fears. Know that these feelings are a normal response to an abnormal situation.<br>2. If you know someone who has been traumatized, be willing to patiently listen to their account of the event, pay attention to their feelings, and encourage them to seek counseling if necessary.<br>3. Be patient with people. Tempers are short in times of crisis, and others may be feeling as much stress as you.<br>4. Recognize normal crisis reactions, such as sleep disturbances and nightmares, withdrawal, reverting to childhood behaviors, and trouble focusing on work or school.<br>5. Take time with your children, spouse, life partner, friends, and coworkers to do something you enjoy. |

*Source:* American Counseling Association, 2006, and adapted from Pomponio, 2002.

state of constant or recurring alarm and fearfulness) that develops after experiencing a traumatic event, learning about a violent or unexpected death of a family member, or even being a witness or bystander to violence (American Psychiatric Association, 2002). According to the American Psychological Association (APA) (2011) website, about 70% of U.S. adults have experienced a severe traumatic event, and one out of five go on to develop symptoms of PTSD.

**CONCEPT CHECK** STOP

1. **What** does it mean to say that stress is a contributing factor to gastric ulcers?
2. **How** does stress suppress the immune system?
3. **Which** aspect of the Type A personality is most strongly related to heart disease?
4. **What** is the primary feature of PTSD?

# Health Psychology and Stress Management

## LEARNING OBJECTIVES

**RETRIEVAL PRACTICE** While reading the upcoming sections, respond to each Learning Objective in your own words. Then compare your responses with those in Appendix B.

1. **Explain** what health psychologists do.
2. **Compare** emotion-focused and problem-focused forms of coping.
3. **Review** the eight major resources for combating stress.

The study of how biological, psychological, and social factors affect health and illness, **health psychology**, is a growing field in psychology. In this section, we will first discuss the work of health psychologists and then explore how we cognitively appraise potential stressors, cope with perceived threats, and make use of eight resources for stress management.

**health psychology** The study of how biological, psychological, and social factors affect health and illness.

### What Does a Health Psychologist Do?

Health psychologists are interested in how people's lifestyles and activities, emotional reactions, ways of interpreting events, and personality characteristics influence their physical health and well-being.

As researchers, health psychologists are particularly interested in how changes in behavior can improve health outcomes (Suls, Davidson, & Kaplan, 2010). They also emphasize the relationship between stress and the immune system. As we discovered earlier, a normally functioning immune system helps defend against disease. And a suppressed immune system leaves the body susceptible to a number of diseases.

As practitioners, health psychologists can work as independent clinicians or as consultants alongside physicians, physical and occupational therapists, and other health care workers. The goal of health psychologists is to reduce psychological distress or unhealthy behaviors. They also help patients and families make critical decisions and prepare psychologically for surgery or other treatment. Health psychologists have become so involved with health and illness that medical centers are one of their major employers (Considering a Career, 2011).

Health psychologists also educate the public about health *maintenance*. They provide information about the effects of stress, smoking, alcohol, lack of exercise, and other health issues (see *What a Psychologist Sees*). In addition, health psychologists help people cope with conditions such as chronic pain, diabetes, and high blood pressure, as well as unhealthful behaviors such as anger expression and lack of assertiveness. If you're interested in a career in this field, check with your college's counseling or career center, and the website included at www.wiley.com/college/carpenter.

# WHAT A PSYCHOLOGIST SEES

THE PLANNER

## Nicotine and the Biopsychosocial Model

Most people know that smoking is bad for their health and that the more they smoke, the more at risk they are. Tobacco use endangers both smokers and those who breathe secondhand smoke, so it's not surprising that health psychologists are concerned with preventing smoking and getting those who already smoke to stop.

The first puff on a cigarette is rarely pleasant. So why do people start smoking? The answer can be found in the biopsychosocial model (Figure a). From a biological perspective, nicotine is highly addictive. So once a person begins to smoke, there is a biological need to continue (Figure b). Nicotine addiction appears to be very similar to heroin, cocaine, and alcohol addiction (Abadinsky, 2011; Wang, Yuan, & Li, 2011).

From a psychological and social perspective, smokers learn to associate smoking with pleasant things, such as good food, friends, and sex (Figure c). Is it any wonder that many psychologists and medical personnel believe that the best way to reduce the number of smokers is to stop people from ever taking that first puff?

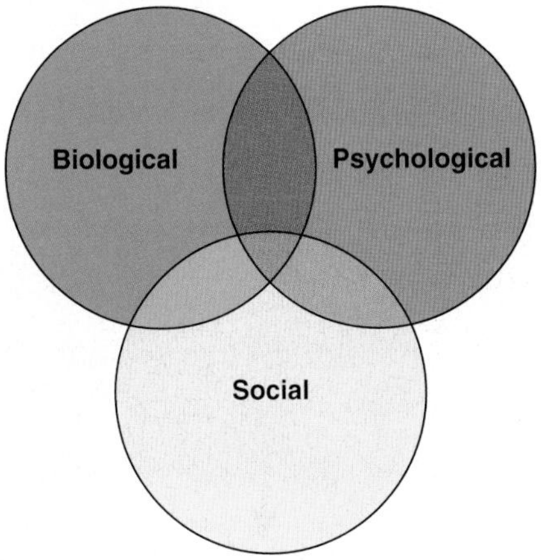


**a. The biopsychosocial model**
All three factors in the *biopsychosocial model* interact in the issue of smoking and addiction.

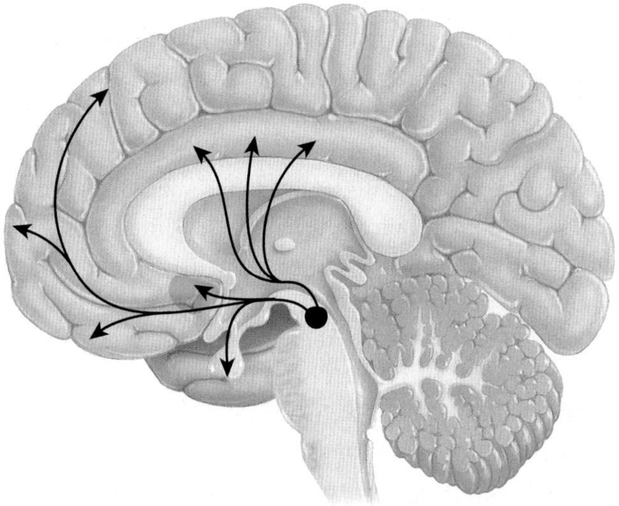

**b. The brain's reward center pathway**
Nicotine quickly increases the release of acetylcholine and norepinephrine in our brain, increasing alertness, concentration, memory, and feelings of pleasure. Nicotine also stimulates the release of dopamine, the neurotransmitter most closely related to reward centers in the brain (Lee & Messing, 2011; Marti et al., 2012).

Michael Siluk/Index Stock/Age Fotostock

**c. Smokers unite?**
Antismoking laws may have made quitting smoking even more difficult for some. Being forced to gather together outside to smoke may forge stronger social bonds among smokers and strengthen their resolve to "suffer the tyranny" of antismoking laws.

## Coping with Stress

Because we can't escape stress, we need to learn how to effectively cope with it. Simply defined, **coping** is an attempt to manage stress in some effective way. It is not one single act but a process that allows us to deal with various stressors (**Figure 3.8**).

Our level of stress generally depends on both our interpretation of and our reaction to stressors. **Emotion-focused forms of coping** are emotional or cognitive strategies that help us manage a stressful situation. For example, suppose you were refused a highly desirable job. You might reappraise the situation and decide that the job wasn't the right match for you or that you weren't really qualified or ready for it.

**emotion-focused forms of coping** Managing one's emotional reactions to a stressor.

When faced with unavoidable stress, people commonly use **defense mechanisms**. That is, they unconsciously distort reality to protect their egos and to avoid anxiety. For instance, fantasizing about what you will do on your next vacation can help relieve stress. Sometimes, however, defense mechanisms can be destructive. For example, people often fabricate excuses for not attaining particular goals. Such *rationalization* keeps us from seeing a situation more clearly and realistically and can prevent us from developing valuable skills or qualities.

Emotion-focused forms of coping that are accurate reappraisals of stressful situations and that do not distort reality may alleviate stress in some situations (Giacobbi, Foore, & Weinberg, 2004; Hoar, Evans, & Link, 2012). Many times, however, it is necessary and more effective to use **problem-focused forms of coping**, which deal directly with the situation or the stressor to eventually decrease or eliminate it (Jopp & Schmitt, 2010; Peng et al., 2012). These direct coping strategies include:

**problem-focused forms of coping** Dealing directly with a stressor to decrease or eliminate it.

- identifying the stressful problem,
- generating possible solutions,
- selecting the appropriate solution,
- applying the solution to the problem, thus eliminating the stress.



# Psychology InSight

Research suggests that our emotional response to an event depends largely on how we interpret the event. First we appraise events and then we choose either emotion- or problem-focused methods of coping, or a combination of the two.

**a.** According to some psychologists, we appraise events in two steps, primary appraisal and secondary appraisal.

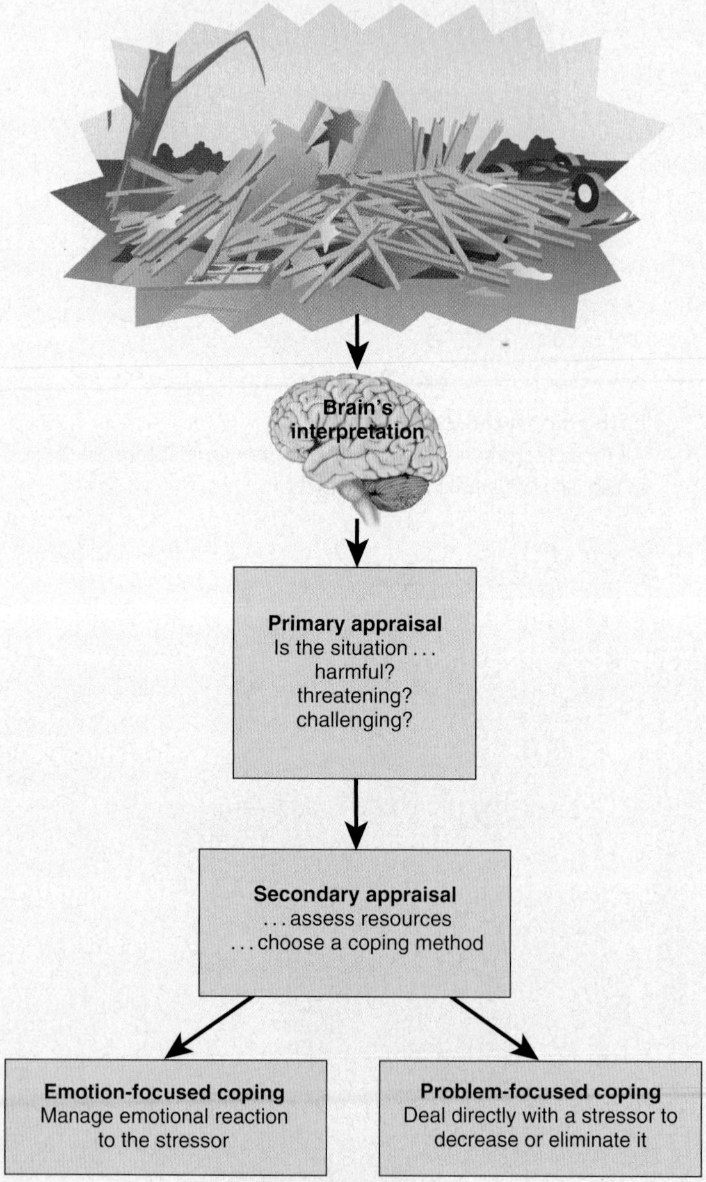

**b.** In some situations an emotion-focused strategy can allow people to step back from an especially overwhelming problem. These women share a tearful embrace in response to the destruction of their home.

©AP/Wide World Photos

©AP/Wide World Photos

**c.** Once the affected individuals are better able to handle their initially overwhelming emotions, they are in position to reappraise and use a problem-focused approach to look for solutions. Here, a community comes together to rebuild.

## Study Organizer 3.2 Eight ways to cope with stress

THE PLANNER

**Exercise**

Exercising and keeping fit helps minimize anxiety, depression, and tension, which are associated with stress. Exercise also helps relieve muscle tension; improves cardiovascular efficiency; and increases strength, flexibility, and stamina.

OJO Images/ SUPERSTOCK

**Positive beliefs**

A positive self-image and attitude can be especially significant coping resources. Even temporarily raising self-esteem reduces the amount of anxiety caused by stressful events. Also, hope can sustain a person in the face of severe odds, as is often documented in news reports of people who have triumphed over seemingly unbeatable circumstances.

© Wojciech Gajda/ iStockphoto

**Social skills**

People who acquire social skills (such as knowing appropriate behaviors for certain situations, having conversation starters up their sleeves, and expressing themselves well) suffer less anxiety than people who do not. In fact, people who lack social skills are more at risk for developing illness than those who have them. Social skills not only help us interact with others but also communicate our needs and desires, enlist help when we need it, and decrease hostility in tense situations.

© Henk Badenhorst/ iStockphoto

**Social support**

Having the support of others helps offset the stressful effects of divorce, the loss of a loved one, chronic illness, pregnancy, physical abuse, job loss, and work overload. When we are faced with stressful circumstances, our friends and family often help us take care of our health, listen, hold our hands, make us feel important, and provide stability to offset the changes in our lives.

NewsCom

**Control**

Believing that you are the "master of your own destiny" is an important resource for effective coping. People with an **external locus of control** feel powerless to change their circumstances and are less likely to make healthy changes, follow treatment programs, or positively cope with a situation. Conversely, people with an **internal locus of control** believe that they are in charge of their own destinies and are therefore able to adopt more positive coping strategies.

© REMI OCHLIK/IP3/ MAXPPP/NewsCom

**Material resources**

Money increases the number of options available for eliminating sources of stress or reducing the effects of stress. When faced with the minor hassles of everyday living, or when faced with chronic stressors or major catastrophes, people with money and the skills to effectively use it generally fare better and experience less stress than people without money.

© Juanmonino/ iStockphoto

**Sense of humor**

Research shows that humor is one of the best ways to reduce stress. The ability to laugh at oneself, and at life's inevitable ups and downs, allows us to relax and gain a broader perspective. In short: "Don't sweat the small stuff."

Corbis/ SUPERSTOCK

**Relaxation**

There are a variety of relaxation techniques. Biofeedback is often used in the treatment of chronic pain, but it is also useful in teaching people to relax and manage their stress. **Progressive relaxation** helps reduce or relieve the muscular tension commonly associated with stress (see *Applying Psychology)*. Using this technique, patients first tense and then relax specific muscles, such as those in the neck, shoulders, and arms. This technique teaches people to recognize the difference between tense and relaxed muscles.

Dynamic Graphics Value/SUPERSTOCK

# Applying Psychology

THE PLANNER

## Progressive Relaxation

You can use progressive relaxation techniques anytime and anywhere you feel stressed, such as before or during an exam. Here's how:

1. Sit in a comfortable position, with your head supported.
2. Start breathing slowly and deeply.
3. Let your entire body relax. Release all tension. Try to visualize your body getting progressively more relaxed with each breath.
4. Systematically tense and release each part of your body, beginning with your toes. Curl them tightly while counting to 10. Now, release them. Note the difference between the tense and relaxed state. Next, tense your feet to the count of 10. Then relax them and feel the difference. Continue upward with your calves, thighs, buttocks, abdomen, back muscles, shoulders, upper arms, forearms, hands and fingers, neck, jaw, facial muscles, and forehead.

Try practicing progressive relaxation twice a day for about 15 minutes. You will be surprised at how quickly you can learn to relax—even in the most stressful situations.

## Resources for Healthy Living

A person's ability to cope effectively depends on the stressor itself—its complexity, intensity, and duration—and on the type of coping strategy used. It also depends on available resources. Eight important resources for healthy living and stress management are exercise, positive beliefs, social skills, social support, control, material resources, sense of humor, and relaxation (Archer, 2011; Chou, Chiao, & Fu, 2011; Krypel & Henderson-King, 2010; Marks et al., 2011; McLoyd, 2011; Veselka et al., 2010). These resources are described in **Study Organizer 3.2** and *Applying Psychology*.

**CONCEPT CHECK** STOP

1. **Why** is it so difficult for people to quit smoking?
2. **What** is a defense mechanism?
3. **Why** is exercise important for counteracting stress?

# Summary

THE PLANNER

## 1 Understanding Stress 66

- **Stress** is the body's nonspecific response to any demand placed on it. Pleasant or beneficial stress is called **eustress**. Stress that is unpleasant or objectionable is called **distress**. There are seven major sources of (stress), as shown in the diagram. There also are three basic types of conflict: **approach–approach, avoidance–avoidance,** and **approach–avoidance.**

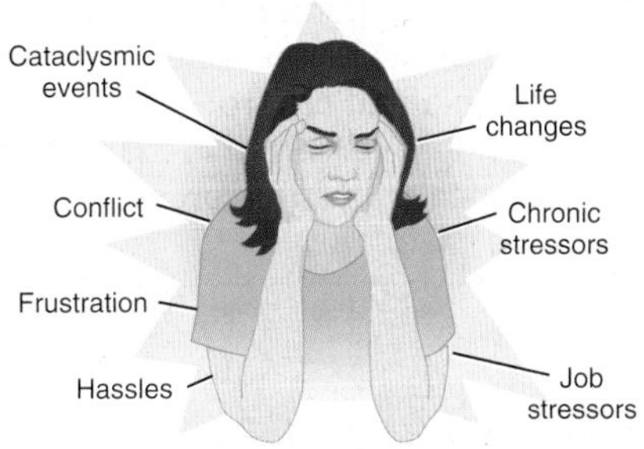


**Seven major sources of stress • Figure 3.1**

- The **general adaptation syndrome (GAS)** describes the body's three-stage reaction to stress. The phases are the initial alarm reaction, the resistance phase, and the exhaustion phase (if resistance to stress is not successful).
- The **SAM system** and the **HPA axis** control significant physiological responses to stress. The SAM system prepares us for immediate action; the HPA axis responds more slowly but lasts longer. Researchers' understanding that psychological factors influence many diseases has spawned a new field of biopsychology called **psychoneuroimmunology.**

## 2 Stress and Illness 72

- Stress increases the risk of developing **gastric ulcers** among people who have the *H. pylori* bacterium in their stomachs.
- Stress makes the immune system, as shown in the photo, less able to resist infection and cancer development.

**The immune system in action • Figure 3.5**

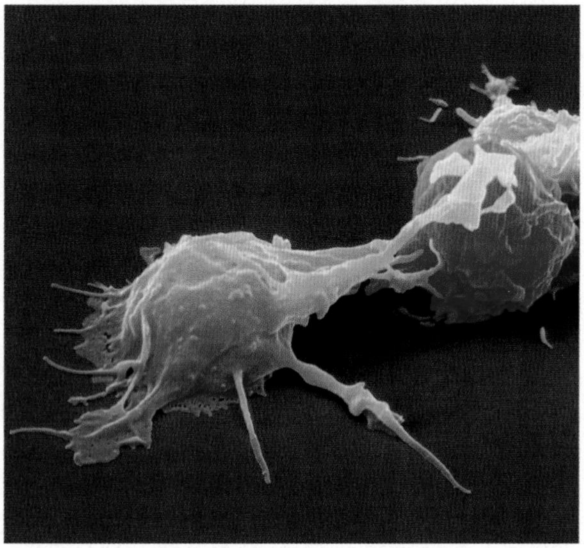

- Increased stress hormones can cause fat to adhere to blood vessel walls, increasing the risk of heart attack. Having a **Type A personality** amplifies the risk of stress-related heart disease. The **Type B personality** is its antithesis.
- Exposure to extraordinary stress can cause **posttraumatic stress disorder (PTSD)**, a type of anxiety disorder characterized by flashbacks, nightmares, and impaired functioning.

## 3 Health Psychology and Stress Management 76

- **Health psychology**, the study of how biological, psychological, and social factors affect health and illness, is a growing field in psychology.
- Our level of stress generally depends on both our interpretation of and our reaction to stressors, as shown in the diagram. **Emotion-focused forms of coping** are emotional or cognitive strategies that change how we view a stressful situation. Some emotion-focused **defense mechanisms**, such as rationalization, can be destructive because they prevent us from developing valuable skills or qualities. It is often more effective to use **problem-focused forms of coping**. People often combine problem-focused and emotion-focused coping strategies to resolve complex stressors or to respond to a stressful situation that is in flux.

**Cognitive appraisal and coping • Figure 3.8**

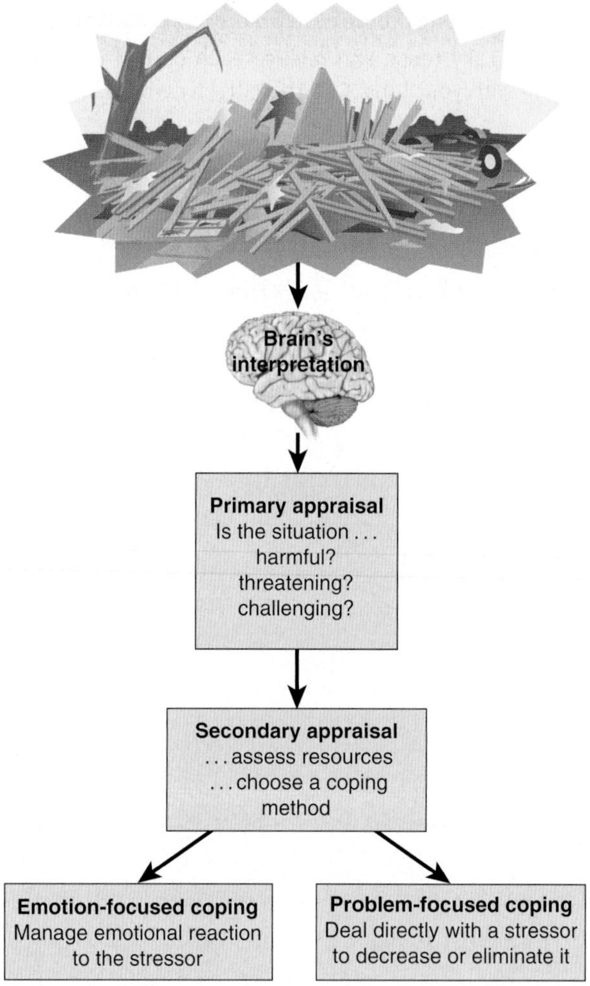


- Eight important resources for healthy living and stress management are exercise, positive beliefs, social skills, social support, control, material resources, relaxation, and a sense of humor. A more specific look at control tells us that believing that one controls one's own fate—an **internal locus of control**—is a healthy way to manage stress. Having an **external locus of control** means believing that chance or outside forces beyond one's control determine one's fate.

# Key Terms

**RETRIEVAL PRACTICE** Write a definition for each term before turning back to the referenced page to check your answer.

- approach–approach conflict 68
- approach–avoidance conflict 68
- avoidance–avoidance conflict 68
- burnout 69
- cancer 73
- cataclysmic event 67
- chronic stressor 67
- conflict 68
- coping 78
- defense mechanism 78
- distress 66
- emotion-focused forms of coping 78
- eustress 66
- external locus of control 80
- frustration 69
- gastric ulcer 72
- general adaptation syndrome (GAS) 69
- hardiness 74
- hassle 68
- heart disease 73
- health psychology 76
- HPA axis 70
- internal locus of control 80
- job stress 67
- life change 66
- posttraumatic stress disorder (PTSD) 75
- problem-focused forms of coping 78
- progressive relaxation 80
- psychoneuroimmunology 71
- SAM system 70
- stress 66
- stressor 66
- Type A personality 74
- Type B personality 74

# Critical and Creative Thinking Questions

1. What are the major sources of stress in your life?
2. If you were experiencing a period of high stress, what would you do to avoid illness?
3. Why are smoking-prevention efforts often aimed at adolescents?
4. What are the benefits of involving a person's family in behavior modification programs to treat chronic pain?
5. Which forms of coping do you most often draw on?
6. Which of the resources for stress management seem most important to you?

# What is happening in this picture?

Illene MacDonald/PhotoEdit

The American Medical Association considers alcohol to be the most dangerous and physically damaging of all drugs (American Medical Association, 2012). After tobacco, it is the leading cause of premature death in most countries, and it is a major contributor to divorce and domestic and child abuse (Abadinsky, 2011; Levinthal, 2011; Livingston, 2011).

## Think Critically

1. How might the child in this photo be affected biologically, psychologically, and socially (the biopsychosocial model) by her mother's drinking?
2. What could a health psychologist do to improve the well-being of this family?
3. Would you like to be a health psychologist? Why or why not?

# Self-Test

**RETRIEVAL PRACTICE** Completing this self-test and comparing your answers with those in Appendix C provide immediate feedback and helpful practice for exams. Additional interactive, self-tests are available at www.wiley.com/college/carpenter.

1. In your text, the physical and mental arousal to situations that we perceive as threatening or challenging is called ______.
   a. distress
   b. eustress
   c. stress
   d. none of these options

2. When John saw his girlfriend kissing another man at a party, he became very upset. In this situation, watching someone you love kiss a potential competitor is ______, and it illustrates ______.
   a. a stressor; distress
   b. eustress; a stressor
   c. distress; a stressor
   d. a stressor; eustress

3. Label the seven major sources of stress on the diagram.

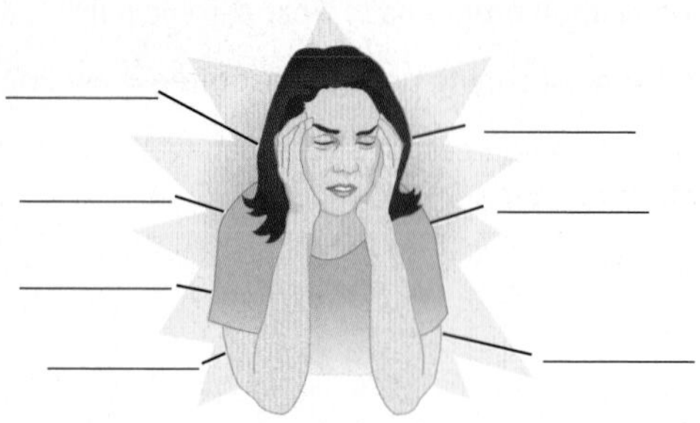

4. ______ is one of the largest sources of CHRONIC stress for adults.
   a. Birth
   b. Work
   c. Christmas
   d. Moving

5. A state of physical, emotional, and mental exhaustion attributable to long-term involvement in emotionally demanding situations is called ______.
   a. primary conflict
   b. technostress
   c. burnout
   d. secondary conflict

6. In an *approach–approach conflict,* we must choose between two or more goals that will lead to ______, whereas in an *avoidance–avoidance conflict,* we must choose between two or more goals that will lead to ______.
   a. less conflict; more conflict
   b. frustration; hostility
   c. a desirable result; an undesirable result
   d. effective coping; ineffective coping

7. As Michael watches his instructor pass out papers, he suddenly realizes this is the first major exam, and he is unprepared. Which phase of the GAS is he most likely experiencing?
   a. resistance
   b. alarm
   c. exhaustion
   d. phase out

8. Label the three key parts of the *HPA axis,* which enables us to deal with chronic stressors:
   a. hypothalamus
   b. pituitary
   c. adrenal cortex
   d. cerebral cortex
   e. adrenal medulla

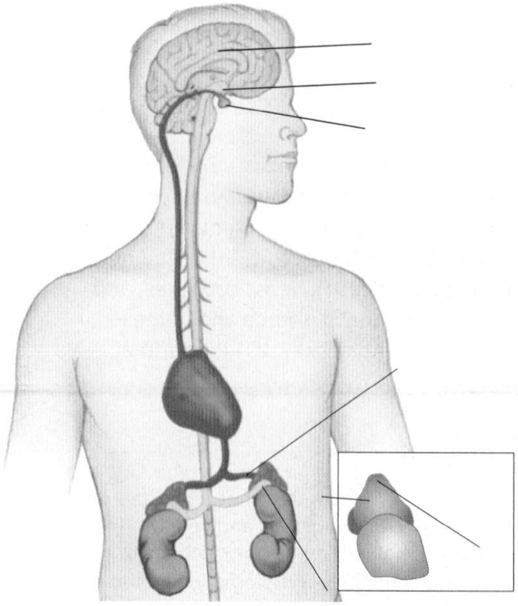

9. ______ is the field that studies the effects of psychological factors on the immune system.
   a. Psychosomatology
   b. Neurobiology
   c. Psychoneuroimmunology
   d. Biopsychology

10. Research suggests that one particular *Type A* characteristic is MOST associated with heart disease. What characteristic is this?
   a. intense ambition
   b. impatience
   c. cynical hostility
   d. time urgency

11. People who experience flashbacks, nightmares, and impaired functioning following a life-threatening or other horrifying event may be ______.
   a. suffering from a psychosomatic illness
   b. experiencing posttraumatic stress disorder
   c. having a nervous breakdown
   d. weaker than people who take such events in stride

12. Which of the following is TRUE of *health psychology*?
   a. It studies the relationship between psychological behavior and physical health.
   b. It studies the relationship between social factors and illness.
   c. It emphasizes wellness and the prevention of illness.
   d. All of these statements are true.

13. *Emotion-focused forms of coping* are based on changing your ______ when faced with stressful situations.
   a. feelings
   b. behavior
   c. strategies
   d. all of these options

14. Research suggests that people with a higher ______ have less psychological stress than those with a higher ______.
   a. external locus of control; internal locus of control
   b. internal locus of control; external locus of control
   c. emotion-focused coping style; problem-focused coping style
   d. none of these options

15. The women in this photo are demonstrating the power of ______, one of the eight resources for healthy living.

**THE PLANNER**

Review your Chapter Planner on the chapter opener and check off your completed work.

# 4 Sensation and Perception

What do you think about or *visualize* when you hear the words "sensation" or "perception"? Are you surprised to discover that the young woman in this photo sees numbers as colors? Similarly, when presented with a high-pitched tone, a musician reported, "It looks like fireworks tinged with a pink-red hue. The color feels rough and unpleasant, and it has an ugly taste—rather like that of a briny pickle" (Luria, 1968). This musician was describing a rare condition known as **synesthesia**, which means "mixing of the senses." People with synesthesia routinely blend their sensory experiences. They may "see" temperatures, "hear" colors, or "taste" shapes.

To appreciate how extraordinary synesthesia is, we first must understand the basic processes of normal, nonblended sensations. For example, how do we turn light and sound waves from the environment into something our brain can comprehend? To do this, we must have both a means of detecting stimuli and a means of converting them into a language the brain can understand. How we get the outside world inside to our brains, and what our brains do with this information, are the key topics of this chapter.

Jessica Scott/The Grand Rapids Press/Landov LLC

## CHAPTER OUTLINE



## CHAPTER PLANNER ✓

- ❑ Study the picture and read the opening story.
- ❑ Answer the Learning Objectives in each section:
  p. 88 ❑ p. 93 ❑ p. 98 ❑ p. 101 ❑
- ❑ Read the text and study all figures and visuals. Answer any questions.

**Analyze key features**

- ❑ Study Organizer, p. 88
- ❑ Myth Busters p. 89 ❑ p. 98 ❑ p. 101 ❑
- ❑ Psychology InSight p. 90 ❑ p. 103 ❑ p. 108 ❑
- ❑ Psychological Science, p. 91
- ❑ Process Diagrams p. 95 ❑ p. 96 ❑
- ❑ What a Psychologist Sees, p. 99
- ❑ Applying Psychology p. 102 ❑ p. 109 ❑
- ❑ Stop: Answer the Concept Checks before you go on.
  p. 93 ❑ p. 97 ❑ p. 101 ❑ p. 112 ❑

**End of chapter**

- ❑ Review the Summary and Key Terms.
- ❑ Answer the Critical and Creative Thinking Questions.
- ❑ Answer What is happening in this picture?
- ❑ Complete the Self-Test and check your answers.

# Understanding Sensation

## LEARNING OBJECTIVES

**RETRIEVAL PRACTICE** While reading the upcoming sections, respond to each Learning Objective in your own words. Then compare your responses with those in Appendix B.

1. **Describe** how raw sensory stimuli are converted to signals in the brain.
2. **Explain** how our experience of stimuli can be measured.
3. **Describe** why adapting to sensory stimuli provides an evolutionary advantage.

Psychologists are keenly interested in our senses—*sensation* is the mind's window to the outside world. We're equally interested in *perception*—how the brain gives meaning to sensory information. **Sensation** begins with specialized receptor cells located in our sense organs (eyes, ears, nose, tongue, skin, and internal body tissues). When sense organs detect an appropriate stimulus (light, mechanical pressure, chemical molecules), they convert it into neural impulses (action potentials) that are transmitted to our brain. The brain then assigns meaning to this sensory information—a process called **perception** (**Study Organizer 4.1**).

**sensation** The process of detecting, converting, and transmitting raw sensory information from the external and internal environments to the brain.

**perception** The process of selecting, organizing, and interpreting sensory information into meaningful patterns.

**Study Organizer 4.1** Sensation and perception ✓ THE PLANNER

SENSATION PERCEPTION

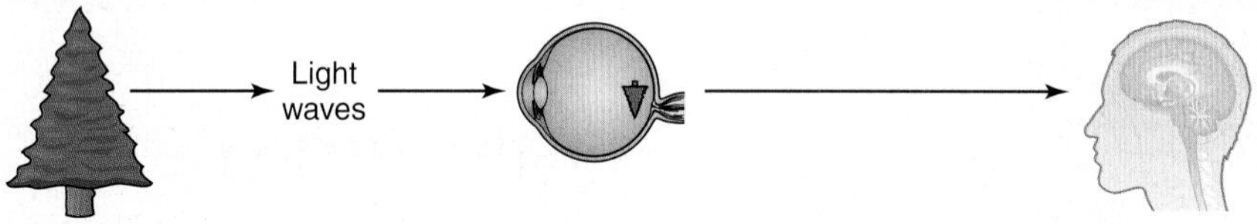


| Sense | Stimulus | Receptors | Brain |
|---|---|---|---|
| **Vision** | Light waves | Light-sensitive rods and cones in eye's retina 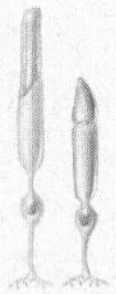 | Visual cortex in the occipital lobe |
| **Audition** (hearing) | Sound waves | Pressure-sensitive hair cells in ear's cochlea  | Auditory cortex in the temporal lobe |
| **Olfaction** (smell) | Molecules dissolved on nose's mucous membranes | Neurons in the nose's olfactory epithelium  | Temporal lobe and limbic system |
| **Gustation** (taste) | Molecules dissolved on tongue | Taste buds on tongue's surface  | Limbic system, somatosensory cortex, and frontal lobe |
| **Body Senses** | Variety of stimuli | Variety of receptors  | Motor cortex in the frontal lobe and the somatosensory cortex in the parietal lobe |

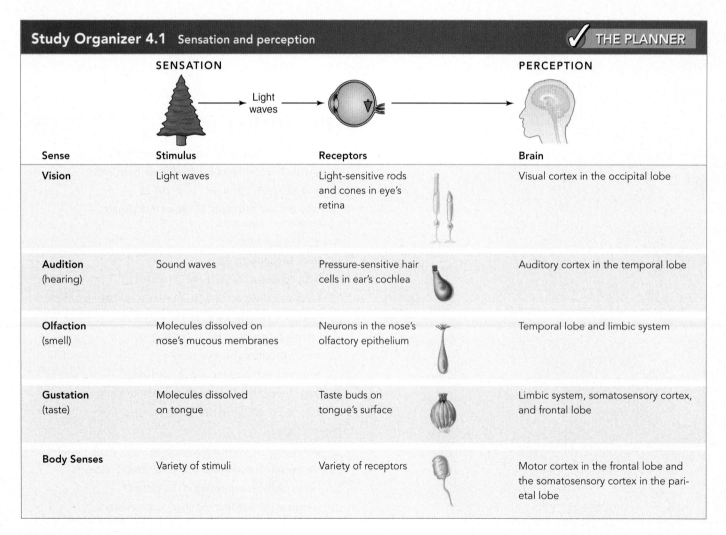

## MYTH BUSTERS

### SENSATIONAL FACT AND FICTION

**TRUE OR FALSE?**

___ 1. Subliminal advertising is very effective.

___ 2. Humans have one blind spot in each eye.

___ 3. Most color-blind people can only see the world in black and white.

___ 4. Loud music can lead to permanent hearing loss.

___ 5. There are discrete spots on our tongues for specific tastes.

___ 6. Humans have only five senses.

___ 7. Illusions are not the same as hallucinations.

___ 8. People tend to see what they expect to see.

___ 9. There is strong scientific evidence for ESP.

___ 10. We generally remember our "hits" and forget our "misses."

Answers: 1. False, 2. True, 3. False, 4. True, 5. False, 6. False, 7. True, 8. True, 9. False, 10. True. (Detailed answers appear in this chapter.)

THE PLANNER

Find out what you already know about sensation and perception by taking the quiz in *Myth Busters*.

## Processing

Our eyes, ears, skin, and other sense organs all contain special cells called receptors, which receive and process sensory information from the environment. For each sense, these specialized cells respond to a distinct stimulus, such as sound waves or odor molecules. During the process of **transduction**, the receptors convert the stimulus into neural impulses, which are sent to the brain. For example, in hearing, tiny receptor cells in the inner ear convert mechanical vibrations from sound waves into electrochemical signals. These signals are carried by neurons to the brain, where specific sensory receptors detect and interpret the information. How does our brain differentiate between sensations, such as sounds and smells? Through a process known as **coding**, different physical stimuli are interpreted as distinct sensations because their neural impulses travel by different routes and arrive at different parts of the brain (**Figure 4.1**).

We also have structures that purposefully reduce the amount of sensory information we receive. Without this natural filtering of stimuli we would constantly hear blood rushing through our veins and feel our clothes brushing against our skin. Some level of filtering is needed so the brain is not overwhelmed with unnecessary information.

All species have evolved selective receptors that suppress or amplify information for survival. In the process of **sensory reduction**, we filter incoming sensations and analyze the sensations that the body does register before a neural impulse is finally sent to the various parts of the brain. Humans, for example, cannot sense ultraviolet light, microwaves, the ultrasonic sound of a dog whistle, or infrared heat patterns from warm-blooded animals, as some other animals can.

**sensory reduction** Filtering and analyzing incoming sensations before sending a neural message to the cortex.

### Sensory processing within the brain • Figure 4.1

Neural messages from the various sense organs must travel to the brain in order for us to see, hear, smell, and so on. Shown here are the primary locations in the cerebral cortex for vision, hearing, taste, smell, and somatosensation (which includes touch, pain, and temperature sensitivity).

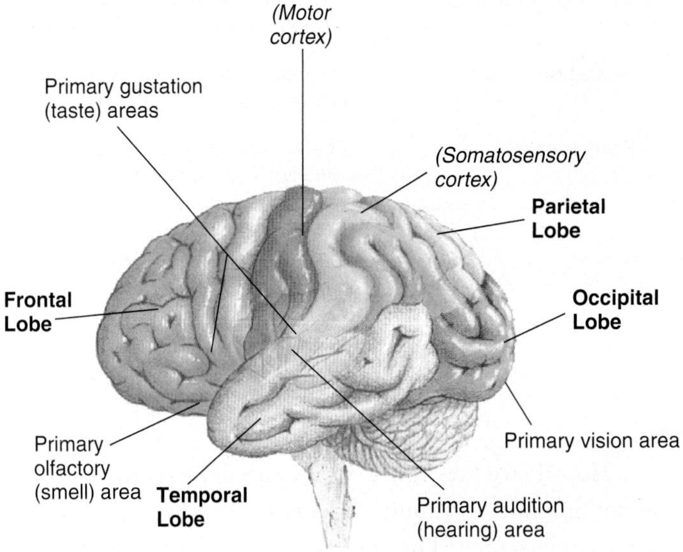

# Psychology InSight Measuring the senses • Figure 4.2

Psychophysics measures our experience of sensory stimuli. Your **absolute threshold** is the smallest amount of a stimulus you can reliably detect. Your **difference threshold**, or *just noticeable difference* (JND), is the minimum difference you can reliably detect.

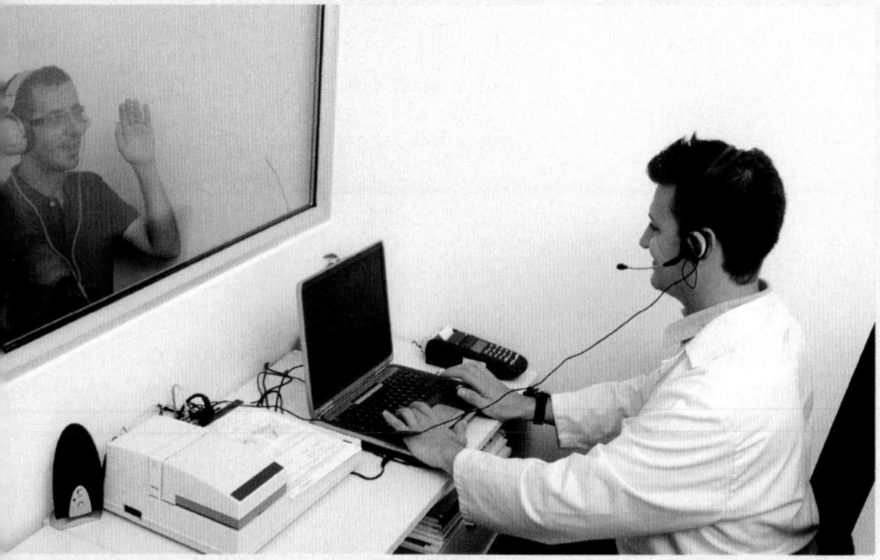

© Carmen Martínez Banús/iStockphoto

**a.** To measure your senses, an examiner presents a series of signals that vary in intensity, and asks you to report which signals you can detect. In a hearing test, the softest level at which you can consistently hear a tone is your absolute threshold. The examiner then compares your threshold with those of people with normal hearing to determine whether you have hearing loss.

**b.** Researchers have established absolute thresholds for various senses.

| Sense | Absolute threshold |
|---|---|
| **Vision** | A candle flame seen from 30 miles away on a clear, dark night |
| **Audition** (hearing) | The tick of an old-fashioned watch at 20 feet |
| **Olfaction** (smell) | One drop of perfume spread throughout a six-room apartment |
| **Gustation** (taste) | One teaspoon of sugar in two gallons of water |
| **Body senses** | A bee's wing falling on your cheek from a height of about half an inch |

**c.** A dog's absolute and difference thresholds for smell are far more sensitive than those of a human. For this reason, specially trained dogs provide invaluable help in sniffing out dangerous plants, animals, drugs, and explosives, tracking criminals, and assisting in search-and-rescue operations. Some researchers believe dogs can even detect chemical signs of certain illnesses, such as diabetes or cancer (Akers & Denbow, 2008).

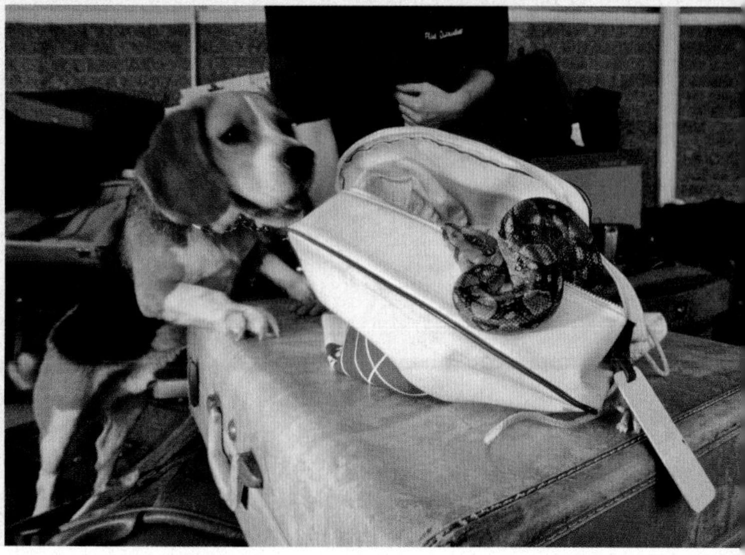

Karen Kasmauskil/NG Image Collection

How do we know this? How can scientists measure the exact amount of stimulus energy it takes to trigger a conscious experience? The answers come from the field of **psychophysics**, which studies and measures the link between the physical characteristics of stimuli and the sensory experience of them (see **Figure 4.2**).

There is an interesting and ongoing controversy about whether we can be affected by subliminal stimuli, sensations below our absolute threshold, even when we're not aware of them. Years ago many people believed movie theaters were manipulating consumers by flashing messages such as "Eat popcorn" and "Drink Cola-Cola" on the screen so briefly that viewers weren't aware of seeing them. Read about current research into this topic in *Psychological Science*.

## Psychological Science

THE PLANNER

# Subliminal Perception

HECTOR!
CHECK THIS OUT!
"SUBLIMINALLY SEDUCE WOMEN INSTANTLY!"
CAN YOU BELIEVE OUR LUCK?
"WIN OVER ANY WOMAN YOU WANT IN LESS THAN AN HOUR! INSTANTLY AND PERMANENTLY CHANGE THE WAY WOMEN TREAT YOU... GUARANTEED!"
SURELY YOU'RE NOT SUGGESTING THAT WE--
THEY COULDN'T PUT IT ON THE INTERNET IF IT WASN'T TRUE!

It is, in fact, possible to perceive something without conscious awareness, called **subliminal perception** (Aarts, 2007; Bermeitinger et al., 2009; Boccato et al., 2008; Cleeremans & Sarrazin, 2007; Martens, Ansorge, & Kiefer, 2011). Experimental studies commonly use an instrument called a *tachistoscope* to flash images too quickly for conscious recognition, but slowly enough to be registered.

In one study the experimenter flashed one of two pictures subliminally (either a happy face, like the one shown in the photo, or an angry face) followed by a neutral face. They found this subliminal presentation evoked matching unconscious facial expressions in the participant's own facial muscles (Dimberg, Thunberg, & Elmehed, 2000).

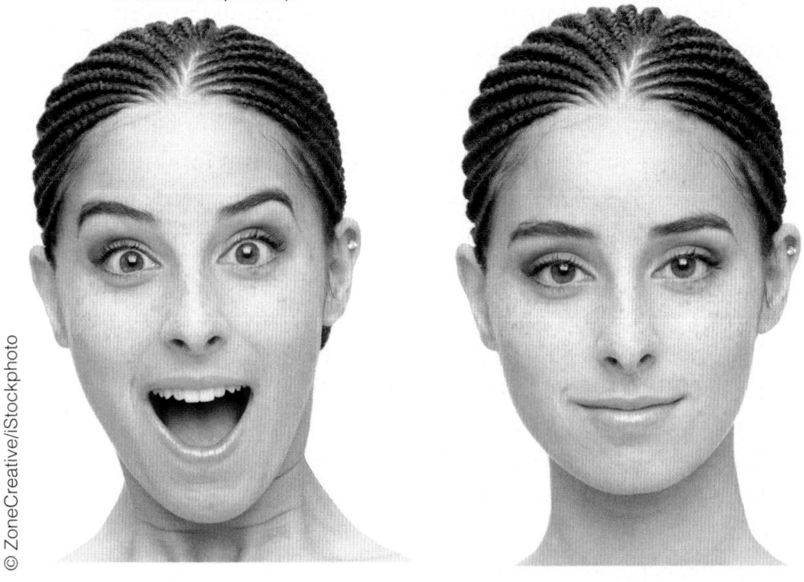



Despite this and much other evidence that *subliminal perception* occurs, that does not mean that such processes lead to *subliminal persuasion*. Subliminal stimuli are basically *weak* stimuli. At best, they have a modest (if any) effect on consumer behavior, and absolutely *no effect on* the minds of youth listening to rock music or citizens' voting behavior (Bermeitinger et al., 2009; Dijksterhuis, Aarts, & Smith, 2005; Fennis & Stroebe, 2010; Karremans, Stroebe, & Claus, 2006). As for subliminal tapes promising to help you lose weight or relieve stress, save your money. Blank "placebo" tapes appear to be just as effective!

### Identify the Research Method

1. What is the most likely research method for this study of subliminal perception by Dimberg, Thunberg, & Elmehed, 2000?
2. If you chose
   - the experimental method, label the IV, DV, experimental group, and control group.
   - the descriptive method, is this a naturalistic observation, survey, or case study?
   - the correlational method, is this a positive, negative, or zero correlation?
   - the biological method, identify the specific research tool (e.g., brain dissection, CT scan).

(Check your answers in Appendix C.)

Otto Greule Jr/Getty Images

**Sidelined? • Figure 4.3**

The body's abliity to inhibit pain perception sometimes makes it possible for athletes to "play through" painful injuries. Do you think the potential for lasting damage is too high a price to pay for a medal or trophy?

## Sensory Adaptation

Imagine that friends have invited you to come visit their beautiful new baby. As they greet you at the door, you are overwhelmed by the odor of a wet diaper. Why don't your friends do something about that smell? The answer lies in the previously mentioned sensory reduction and **sensory adaptation**. When a constant stimulus is presented for a length of time, sensation often fades or disappears. In sensory adaptation, receptors within our sensory system get "tired" and actually fire less frequently.

**sensory adaptation** Repeated or constant stimulation decreases the number of sensory messages sent to the brain, which causes decreased sensation.

Sensory adaptation makes sense from an evolutionary perspective. We can't afford to waste attention and time on unchanging, normally unimportant stimuli. "Turning down the volume" on repetitive information helps the brain cope with an overwhelming amount of sensory stimuli and allows time to pay attention to change. Sometimes, however, adaptation can be dangerous, as when people stop paying attention to a small gas leak in the kitchen.

Although some senses, like smell and touch, adapt quickly, we never completely adapt to visual stimuli or to extremely intense stimuli, such as the odor of ammonia or the pain of a bad burn. From an evolutionary perspective, these limitations on sensory adaptation aid survival, for example, by reminding us to avoid strong odors and heat or to take care of that burn.

If we don't adapt to pain, how do athletes keep playing despite painful injuries? In certain situations, the body releases natural painkillers called *endorphins* (see Chapter 2), which inhibit pain perception (**Figure 4.3**).

In addition to endorphin release, one of the most accepted explanations of pain perception is the **gate-control theory**, first proposed by Ronald Melzack and Patrick Wall (1965). According to this theory, the experience of pain depends partly on whether the neural message gets past a "gatekeeper" in the spinal cord. Normally, the gate is kept shut, either by impulses coming down from the brain or by messages coming from large-diameter nerve fibers that conduct most sensory signals, such as touch and pressure. However, when body tissue is damaged, impulses from smaller pain fibers open the gate.

**gate-control theory** The theory that pain sensations are processed and altered by mechanisms within the spinal cord.

According to the gate-control theory, massaging an injury or scratching an itch can temporarily relieve discomfort because pressure on large-diameter neurons interferes with pain signals. Messages from the brain can also control the pain gate, explaining how athletes and soldiers can carry on despite excruciating pain. When we are soothed by endorphins or distracted by competition or fear, our experience of pain can be greatly diminished. On the other hand, when we get anxious or dwell on our pain, we can intensify it (Roth et al., 2007; Sullivan, 2008; Sullivan, Tripp, & Santor, 1998). Ironically, well-meaning friends who ask chronic pain sufferers about their pain may unintentionally reinforce and increase it (Jolliffe & Nicholas, 2004).

Research also suggests that the pain gate may be chemically controlled, that a neurotransmitter called *substance P* opens the pain gate, and that endorphins close it (Bianchi et al., 2008; Cesaro & Ollat, 1997; Papathanassoglou, et al., 2010). Other research (Melzack, 1999; Snijders et al., 2010; Vertosick, 2000) finds that when normal sensory input is disrupted, the brain can generate pain and other sensations on its own. Amputees sometimes continue to feel pain (and itching or tickling) long after a limb has been amputated. This *phantom limb pain* occurs because nerve cells send conflicting messages to the brain. The brain interprets this "static" as pain because it arises in the area of the spinal cord responsible for pain signaling. When amputees are fitted with prosthetic limbs and begin using them, phantom pain often disappears because the brain is somehow tricked into believing that there is no longer a missing limb (Crawford, 2008; Gracely, Farrell, & Grant, 2002).

**CONCEPT CHECK** STOP

1. **How** do we convert sensory information into signals the brain can understand?
2. **What** is the distinction between the absolute threshold and the difference threshold?
3. **What** factors govern pain perception?

# How We See and Hear

## LEARNING OBJECTIVES

**RETRIEVAL PRACTICE** While reading the upcoming sections, respond to each Learning Objective in your own words. Then compare your responses with those in Appendix B.

1. **Identify** the three major characteristics of light and sound waves.
2. **Explain** how the eye captures and focuses light energy and how it converts it into neural signals.
3. **Describe** the path that sound waves take in the ear.
4. **Summarize** the two theories that explain how we distinguish among different pitches.

Many people mistakenly believe that what they see and hear is a copy of the outside world. In fact, vision and hearing are the result of what our brains create in response to light and sound waves. Three physical properties of light and sound (*wavelength*, *amplitude*, and *complexity*) help determine our sensory experience of them (**Figure 4.4**).

### Properties of vision and hearing • Figure 4.4

Vision and hearing are based on wave phenomena, similar to ocean waves. Waves have a certain distance between them (the *wavelength*), and they pass by you at intervals. If you counted the number of passing waves in a set amount of time (e.g., 5 waves in 60 seconds), you could calculate the *frequency* (the number of complete wavelengths that pass a point in a given time). Longer wavelength means lower frequency, and vice versa.

In addition, waves also have the characteristics of height (technically called *amplitude*)—some large, and others small. Finally, waves also vary in complexity, or range. Some have a very simple, uniform shape. Others are a combination of waves of different wavelengths and heights.

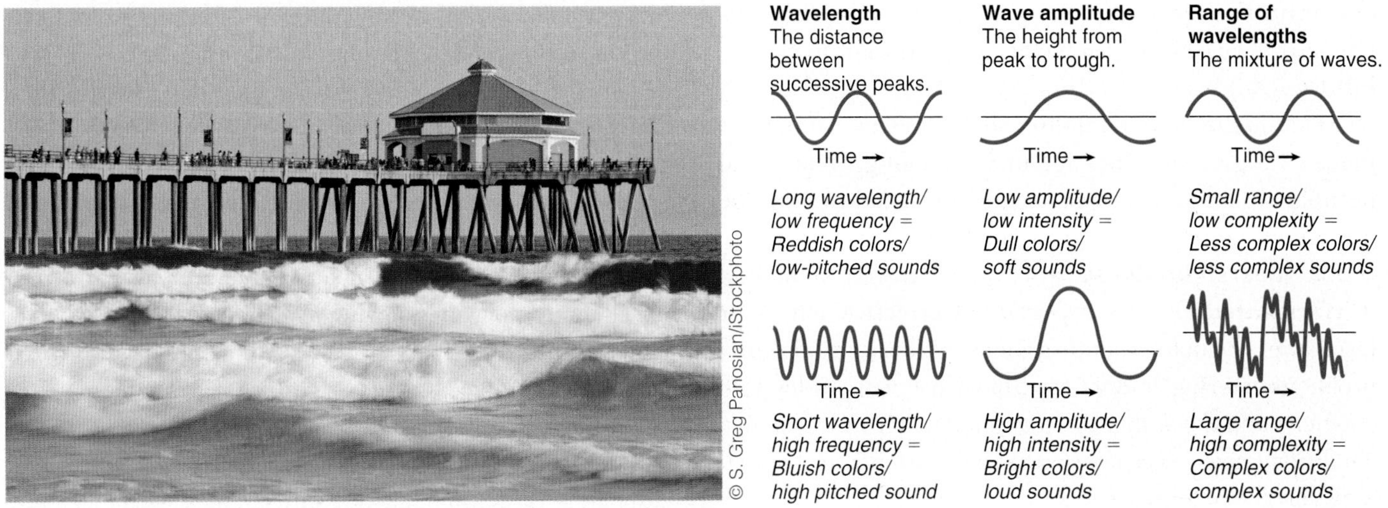

## The electromagnetic spectrum • Figure 4.5

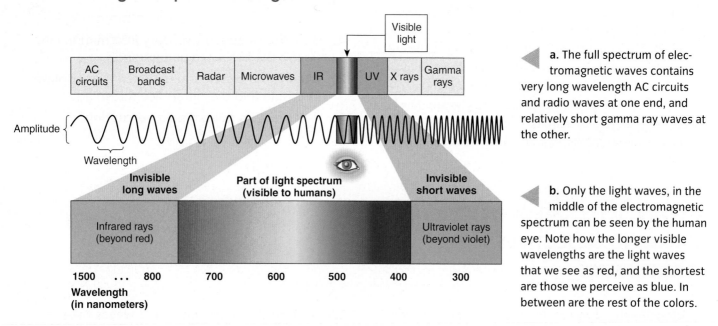

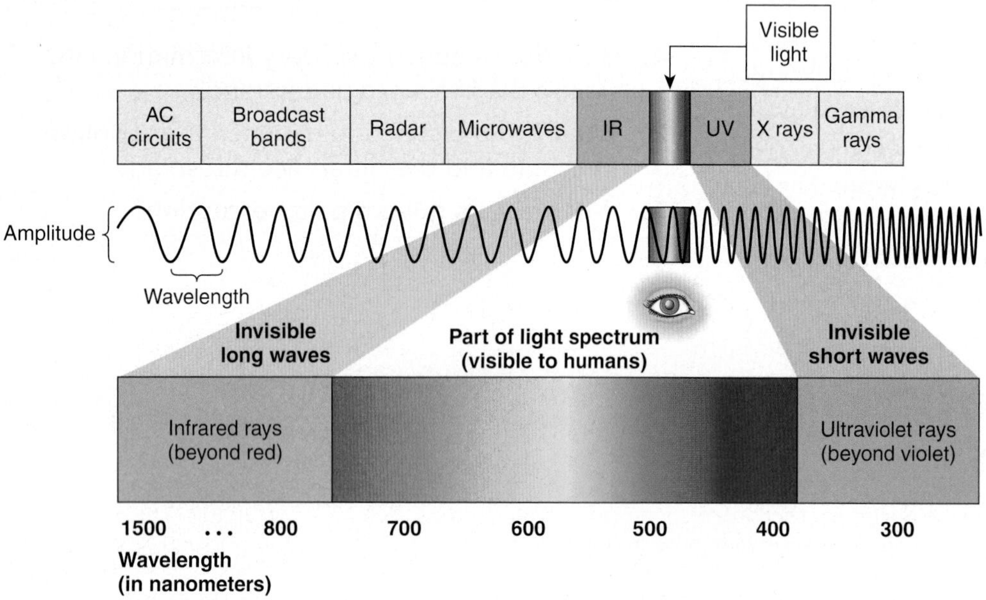


**a.** The full spectrum of electromagnetic waves contains very long wavelength AC circuits and radio waves at one end, and relatively short gamma ray waves at the other.

**b.** Only the light waves, in the middle of the electromagnetic spectrum can be seen by the human eye. Note how the longer visible wavelengths are the light waves that we see as red, and the shortest are those we perceive as blue. In between are the rest of the colors.

## Vision

Do you know that major league batters can routinely hit a 90-mile-per-hour fastball four-tenths of a second after it leaves the pitcher's hand? How can the human eye receive and process information that fast? To understand the marvels of vision, we need to start with the basics—how light waves are a form of electromagnetic energy and how they are only a small part of the full electromagnetic spectrum (**Figure 4.5**).

Now, to fully appreciate how our eyes turn these light waves into what we call *vision*, we need to examine the various structures in our eyes that first capture and focus the light waves, and then transform (transduce) them into neural messages (action potentials) that our brain then processes into what we consciously see. (Be sure to carefully study this step-by-step process in **Figure 4.6**.)

Thoroughly understanding the processes detailed in Figure 4.6 gives us clues for understanding some visual peculiarities. For example, small abnormalities in the eye sometimes cause images to be focused in front of the retina (**nearsightedness**, also called *myopia*) or behind it (**farsightedness**, or *hyperopia*). Corrective lenses or laser surgery can fix or improve most of these visual acuity problems. During middle age, most people's lenses lose elasticity and the ability to accommodate for near vision, a condition known as *presbyopia* that can normally be treated with corrective lenses.

If you walk into a dark movie theater on a sunny afternoon, you will at first be blinded. This is because the pigment inside the rods at the back of your eye has been bleached by the bright light, making the rods temporarily nonfunctional. It takes a second or two for the rods to become functional enough to see. This process of **dark adaptation** continues for 20 to 30 minutes. **Light adaptation**, the adjustment that takes place when you go from darkness to a bright setting, takes about 7 to 10 minutes and is the work of the cones.

## Hearing

The sense or act of hearing, officially known as **audition**, has a number of important functions, ranging from alerting us to dangers to helping us communicate with others. In this section we talk first about sound waves, then about the ear's anatomy and function, and, finally, about problems with hearing.

Like the visual process, which transforms (transduces) light waves into vision, the auditory system is designed to convert sound waves into hearing. Sound waves are produced by air molecules moving in a particular wave pattern. For example, an impact (a tree hitting the ground) or vibrating objects (vocal cords or guitar strings) create waves of compressed and expanded air (like ripples on a lake circling out from a tossed stone). When our ears detect and respond to these small air pressure changes, we hear!

## How the eye sees • Figure 4.6

THE PLANNER

Various structures of your eye work together to focus the light waves from the outside world. Receptor cells in your retina (rods and cones) then convert these waves into messages sent along the optic nerve to be interpreted by your brain.

3 The muscularly controlled lens then focuses incoming light into an image on the light-sensitive *retina*, located on the back surface of the fluid-filled eyeball. Note how the image of the flower is inverted when it is projected on to the retina. The brain later reverses the visual input into the final image that we perceive.

2 The light then passes through the *pupil*, a small adjustable opening. Muscles in the *iris* allow the *pupil* to dilate or constrict in response to light intensity or emotional factors.

1 Light first enters through the *cornea,* which helps protect the eye and focus incoming light rays.

4 In the **retina**, light waves are detected and transduced into neural signals by vision receptor cells (rods and cones).

5 The **fovea**, a tiny pit filled with cones, is responsible for our sharpest vision.

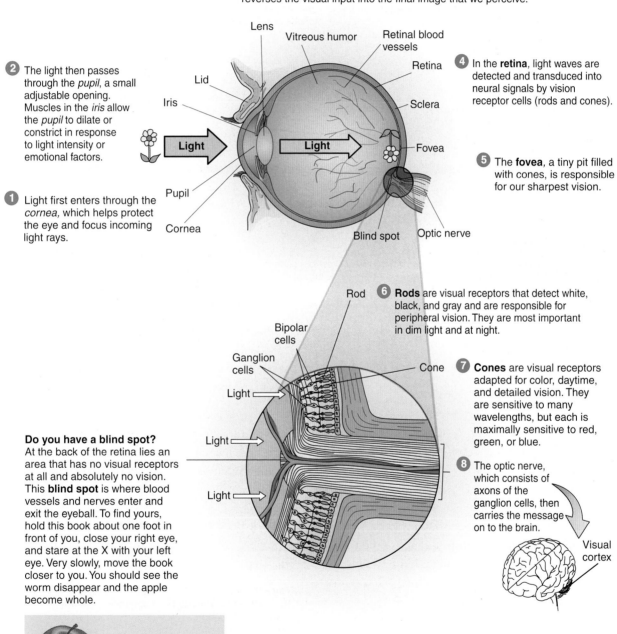

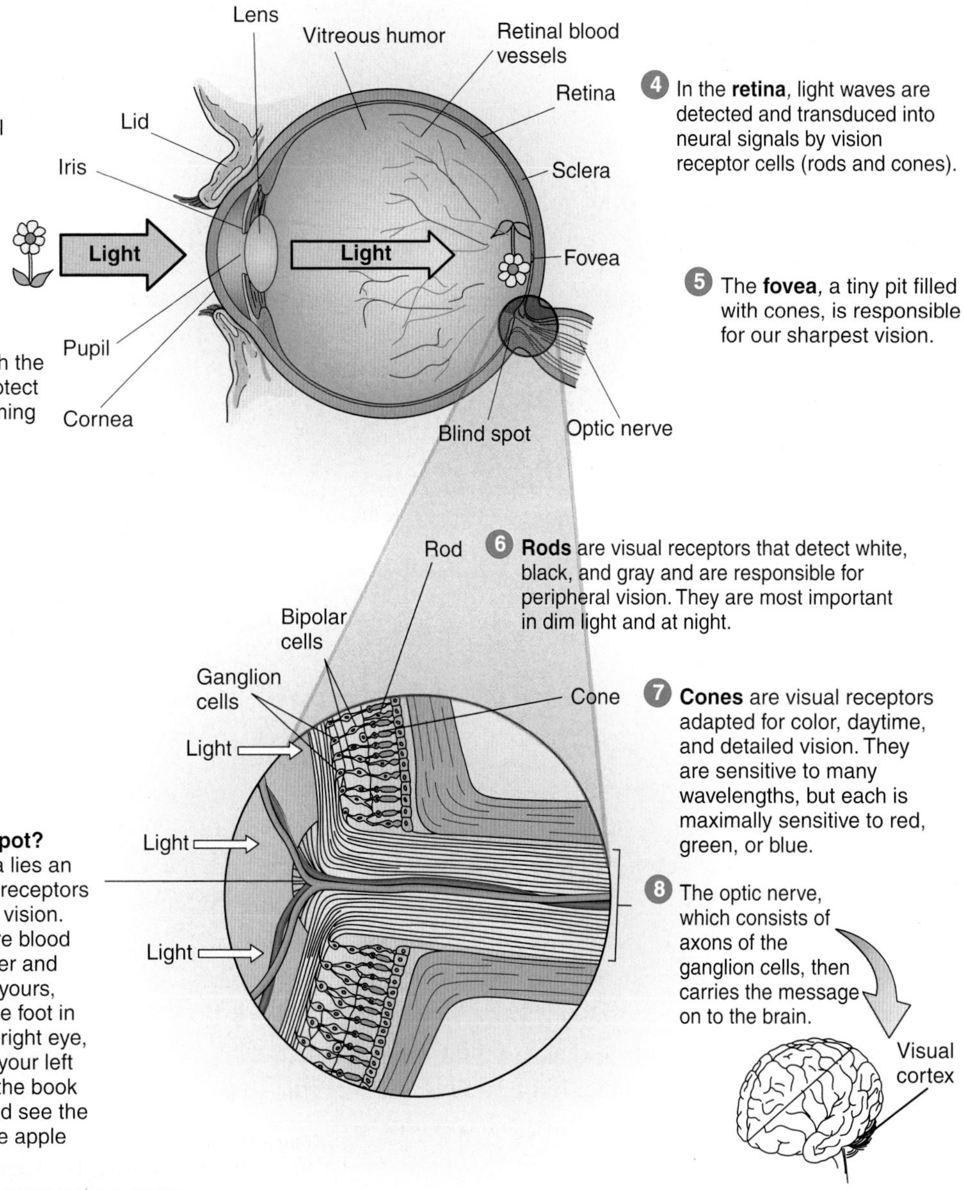


6 **Rods** are visual receptors that detect white, black, and gray and are responsible for peripheral vision. They are most important in dim light and at night.

7 **Cones** are visual receptors adapted for color, daytime, and detailed vision. They are sensitive to many wavelengths, but each is maximally sensitive to red, green, or blue.

8 The optic nerve, which consists of axons of the ganglion cells, then carries the message on to the brain.

**Do you have a blind spot?**
At the back of the retina lies an area that has no visual receptors at all and absolutely no vision. This **blind spot** is where blood vessels and nerves enter and exit the eyeball. To find yours, hold this book about one foot in front of you, close your right eye, and stare at the X with your left eye. Very slowly, move the book closer to you. You should see the worm disappear and the apple become whole.

PROCESS DIAGRAM

## How the ear hears • Figure 4.7

THE PLANNER

The **outer ear** captures and funnels sound waves into the eardrum. Then three tiny bones in the **middle ear** pick up the eardrum's vibrations and transmit them to the **inner ear's** cochlea. This snail-shaped **cochlea** then transforms (transduces) the sound waves into neural messages (action potentials) that our brain processes into what we consciously hear.

1. The **outer ear** captures and funnels sound waves onto the tympanic membrane (ear drum).
2. Vibrations of the tympanic membrane strike the **middle ear's** ossicles (hammer, anvil, and stirrup). Then the stirrup hits the oval window.
3. Vibrations of the oval window create waves in the **inner ear's** cochlear fluid which deflects the basilar membrane. This movement bends the hair cells.
4. The hair cells communicate with the auditory nerve, which sends neural impulses to the brain.

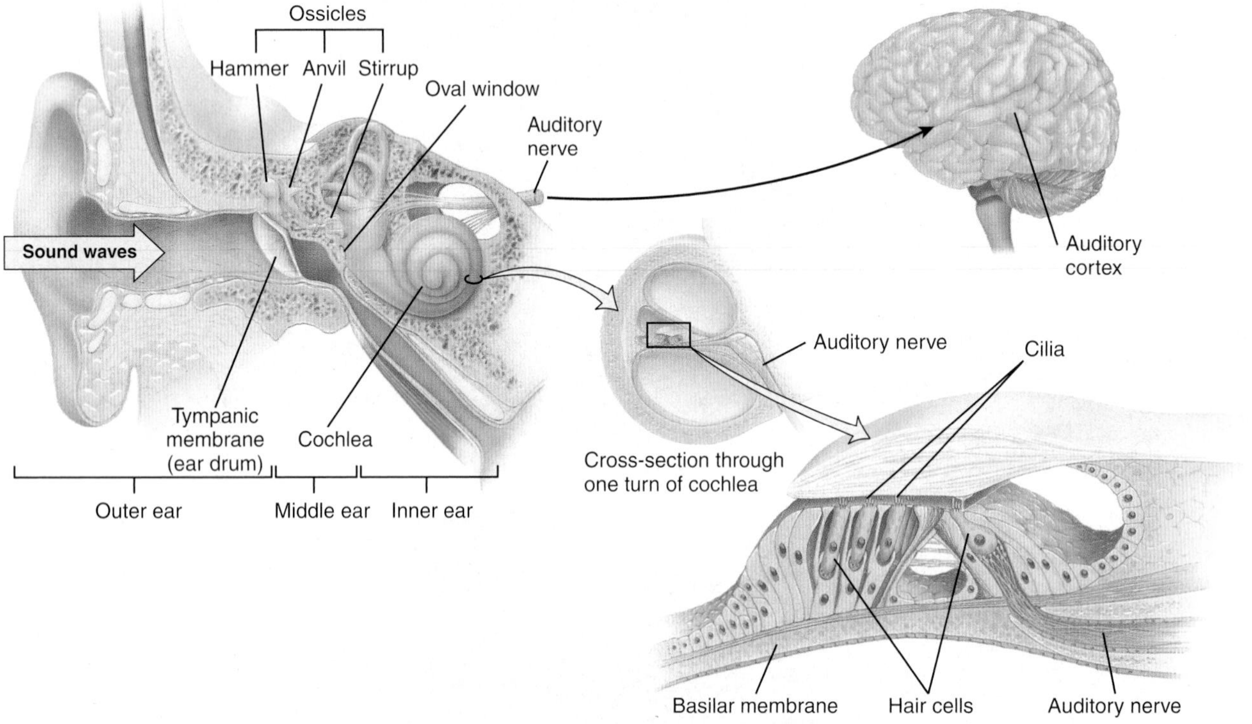


To understand how our ears turn sound waves into hearing, we need to study the step-by-step process depicted in **Figure 4.7**.

The mechanisms determining how we distinguish among sounds of different pitches (low to high) differ, depending on the **frequency** of the sound waves. According to **place theory**, different high-frequency sound waves (which produce high-pitched sounds) maximally stimulate the hair cells at different locations along the basilar membrane (Figure 4.7).

Hearing for low-pitched sounds works differently. According to **frequency theory,** low-pitched sounds cause hair cells along the basilar membrane to bend and fire neural messages (action potentials) at the same rate as the frequency of that sound. For example, a sound with a frequency of 90 hertz would produce 90 action potentials per second in the auditory nerve. In sum, both place and frequency theories are correct, but place theory best explains how we hear high-pitched sounds, whereas frequency theory best explains how we hear low-pitched sounds (**Table 4.1**).

**How the ear distinguishes among sounds of different pitches Table 4.1**

| Theory | Frequency | Pitch | Example |
|---|---|---|---|
| **Place theory** Hair cells are stimulated at different locations on basilar membrane. | High frequency | High-pitched sounds | A squeal or a child's voice |
| **Frequency theory** Hair cells fire at the same rate as the frequency for the sound. | Low frequency | Low-pitched sounds | A growl or a man's voice |

## For whom the bell tolls • Figure 4.8

Stealthy teenagers now have a biological advantage over their teachers: a cell phone ringtone that sounds at 17 kilohertz—too high for adult ears to detect. The ringtone is an ironic offshoot of another device using the same sound frequency. That invention, dubbed the Mosquito, was designed to help shopkeepers annoy and deter loitering teens.

Interestingly, as we age, we tend to lose our ability to hear high-pitched sounds, while still being able to hear low-pitched sounds (**Figure 4.8**).

Whether we detect a sound as soft or loud depends on its intensity. Waves with high peaks and low valleys produce loud sounds; those that have relatively low peaks and shallow valleys produce soft sounds. The relative loudness or softness of sounds is measured on a scale of *decibels* (**Figure 4.9**).

Hearing loss can stem from two major causes: (1) **conduction deafness**, or middle-ear deafness, which results from problems with the mechanical system that conducts sound waves to the inner ear, and (2) **nerve deafness**, or inner-ear deafness, which involves damage to the cochlea, hair cells, or auditory nerve.

Disease and biological changes associated with aging can cause nerve deafness. But the most common (and preventable) cause of nerve deafness is continuous exposure to loud noise, which can damage hair cells and lead to permanent hearing loss. Even brief exposure to really loud sounds, like a stereo or headphones at full blast, a jackhammer, or a jet airplane engine, can cause permanent nerve deafness (see again Figure 4.9).

Although most conduction deafness is temporary, damage to the auditory nerve or receptor cells is almost always irreversible. The only treatment for nerve deafness is a small electronic device called a *cochlear implant*. If the auditory nerve is intact, the implant bypasses hair cells to stimulate the nerve. Currently, cochlear implants produce only a crude approximation of hearing, but the technology is improving. It's best to protect your sense of hearing by avoiding exceptionally loud noises, wearing earplugs when such situations cannot be avoided, and paying attention to bodily warnings of possible hearing loss, including a change in your normal hearing threshold and *tinnitus*, a whistling or ringing sensation in your ears.

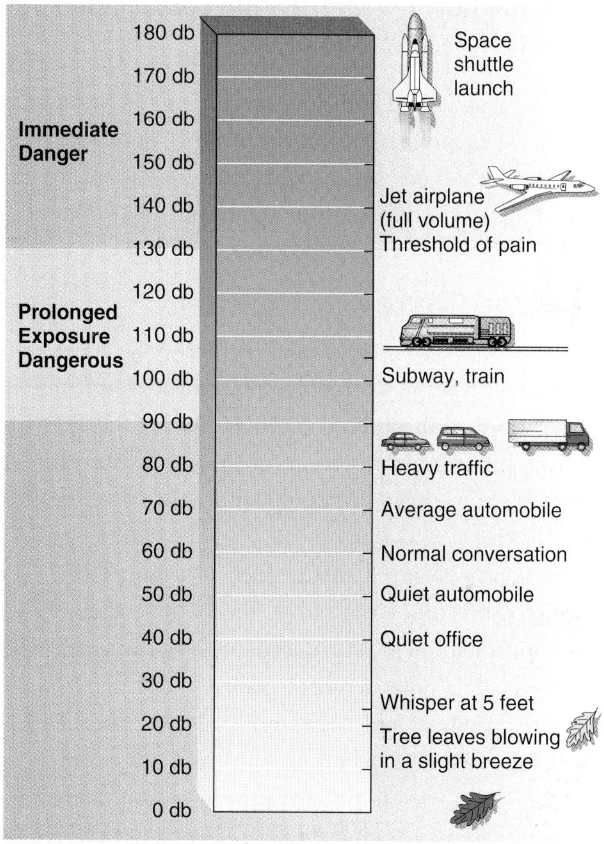


## How loud is too loud? • Figure 4.9

The loudness of a sound is measured in *decibels*, and the higher a sound's decibel reading, the more damaging it is to the ear. Chronic exposure to loud noise, such as loud music or heavy traffic—or brief exposure to really loud sounds, such as a stereo at full blast, a jackhammer, or a jet engine—can cause permanent nerve deafness. Disease and biological changes associated with aging can also cause nerve deafness.

### CONCEPT CHECK

STOP

1. **What** part of the electromagnetic spectrum is visible to human eyes?
2. **What** are the rods and cones responsible for?
3. **What** causes deafness?

# Our Other Important Senses

## LEARNING OBJECTIVES

**RETRIEVAL PRACTICE** While reading the upcoming sections, respond to each Learning Objective in your own words. Then compare your responses with those in Appendix B.

1. **Explain** how the information contained in food and liquid molecules reaches the brain.
2. **Describe** how the information contained in odor molecules reaches the brain.
3. **Identify** the body senses.

Vision and audition may be the most prominent of our senses, but the others—smell, taste, and the body senses—are also important for gathering information about our environment.

## Smell and Taste

Smell and taste are sometimes referred as the *chemical senses* because they both involve chemoreceptors that are sensitive to certain chemical molecules. Smell and taste receptors are located near each other and often interact so closely that we have difficulty separating the sensations (*What a Psychologist Sees*).

Our sense of smell, **olfaction,** is remarkably useful and sensitive. We possess more than 1,000 types of olfactory receptors, allowing us to detect more than 10,000 distinct smells. The nose is more sensitive to smoke than any electronic detector, and—through practice—blind people can quickly recognize others by their unique odors.

Some research on **pheromones**—compounds found in natural body scents that may affect various behaviors—supports the idea that these chemical odors increase sexual behaviors in humans (Marazziti et al., 2010; Savie, Berglund, & Lunstrom, 2007; Thornhill et al., 2003). However, other findings question the results (Hays, 2003), suggesting that human sexuality is far more complex than that of other animals—and more so than perfume advertisements would have you believe.

Today, the sense of taste, **gustation**, may be the least critical of our senses. In the past, however, it probably contributed to our survival. The major function of taste, aided by smell, is to help us avoid eating or drinking harmful substances. Because many plants that taste bitter contain toxic chemicals, an animal is more likely to survive if it avoids bitter-tasting plants (Cooper et al., 2002; Kardong, 2008; Skelhorn et al., 2008). Humans and other animals have a preference for sweet foods, which are generally nonpoisonous and are good sources of energy.

Many food and taste preferences also are learned from childhood experiences and cultural influences. For example, many Japanese children eat raw fish, and some Chinese children eat chicken feet as part of their normal diet. American children might consider these foods "yucky." However, most American children love cheese, which children in many other cultures find repulsive. Cultural and learning experiences also help explain why adults who are told that a bottle of wine is $90 versus its real $10 price report that it tastes better. Ironically, these false expectations actually trigger more brain activity in areas that respond to pleasant experiences (Plassmann et al., 2008).

When we take away the sense of smell, there are five distinct tastes: sweet, sour, salty, bitter, and umami. Umami means "delicious" or "savory" and refers to sensitivity to an amino acid called glutamate (Chandrashekar et al., 2006; Jinap & Hajeb, 2010; McCabe & Rolls, 2007). Glutamate is found in meats, meat broths, and monosodium glutamate (MSG).

Taste receptors respond differentially to food molecules of different shapes. The major taste receptors (*taste buds*) are clustered on our tongues within little bumps called *papillae* (see again *What a Psychologist Sees*). Read the *Myth Busters* to learn more about taste receptors.

### MYTH BUSTERS

#### THE TRUTH ABOUT TASTE

Scientists once believed there were specific areas of the tongue dedicated to detecting bitter, sweet, salty, and other tastes. However, we now know that taste receptors, like smell receptors, respond differentially to the varying shapes of food and liquid molecules. The major taste receptors (or taste buds) are clustered on our tongues within little bumps called papillae. But a small number of taste receptors are also found in the palate and the back of our mouths. Thus, even people without a tongue experience some taste sensations.

# WHAT A PSYCHOLOGIST SEES

THE PLANNER

## Enjoying Pizza—A Complex Experience

What happens when we eat pizza? Taste receptor cells, as well as sensory cells that respond to touch and temperature, are activated on our tongues. Information about taste, smell, texture, temperature, and appearance is sent to the brain, where it is integrated in various association regions of the cerebral cortex. These circuits, together with those that store memories related to our previous pizza experiences, work together to produce our perception of this particular slice.

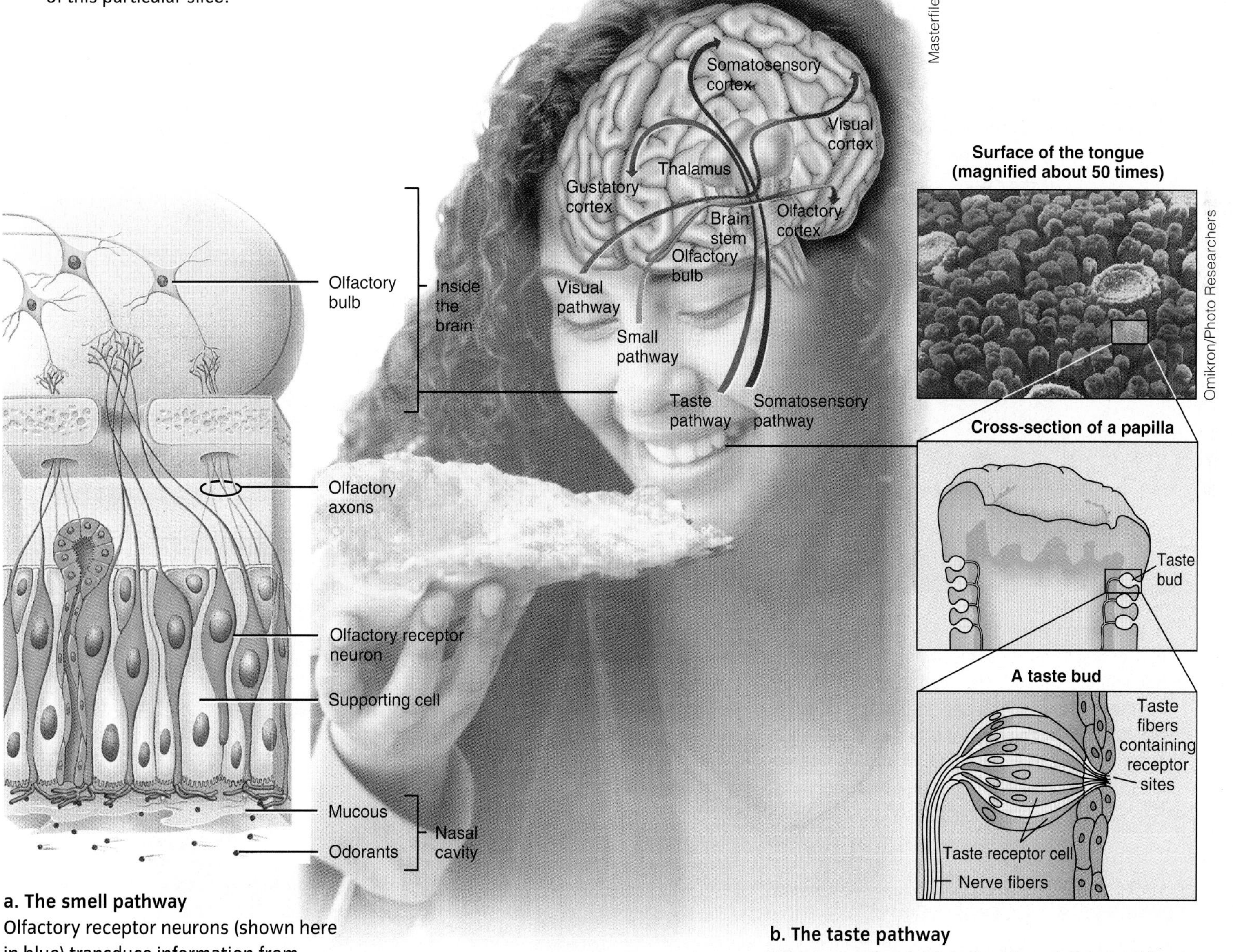


**a. The smell pathway**
Olfactory receptor neurons (shown here in blue) transduce information from odorant molecules that enter the nose. The olfactory nerve carries this information into the olfactory bulb, where most information related to smell is processed before being sent on to other parts of the brain.

**b. The taste pathway**
When we eat and drink, liquids and dissolved foods flow over papillae (the lavender circular areas) and into their pores to the taste buds, which contain the receptors for taste. These nerves carry information into the brain stem, thalamus, gustatory cortex, and somatosensory cortex.

## The Body Senses

The senses that tell the brain how the body is oriented, where and how the body is moving, and what it touches or is touched by are called the body senses. They include the **skin senses**, the **vestibular sense**, and **kinesthesis** (Figure 4.10).

The skin senses reflect the fact that our skin is sensitive to touch (pressure), temperature, and pain (Figure 4.10a). The concentration and depth of the receptors for each of these stimuli vary. For example, touch receptors are most concentrated on the face and fingers and least so in the back and legs. Some receptors respond to more than one type of stimulation—for example, itching, tickling, and vibrating sensations seem to be produced by light stimulation of both pressure and pain receptors.

### The body senses • Figure 4.10

The body senses tell the brain what it touches and is touched by, how the body is moving, and how the body is oriented.

**a. Skin senses**
The tactile senses rely on a variety of receptors located in different parts of the skin. Both humans and nonhuman animals are highly responsive to touch.

Roy Toft/NG Image Collection

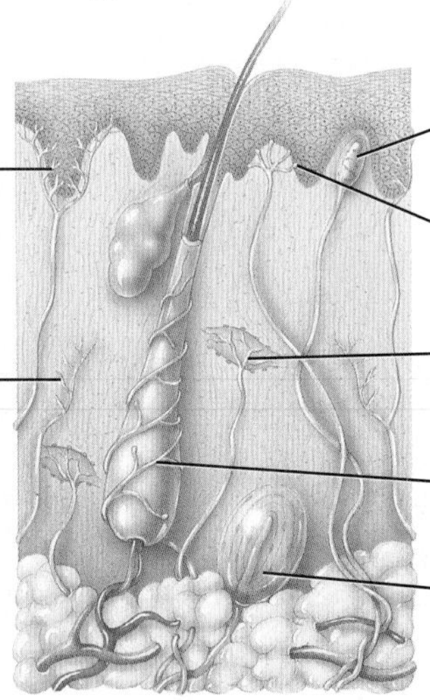


**b. Vestibular sense**
Part of the "thrill" of amusement park rides comes from our vestibular sense becoming confused. The vestibular sense is used by the eye muscles to maintain visual fixation and sometimes by the body to change body orientation. We can become dizzy or nauseated if the vestibular sense becomes "confused" by boat, airplane, or automobile motion. Children between ages 2 and 12 years have the greatest susceptibility to motion sickness.

© RubberBall/Alamy Limited

**c. Kinesthesis**
This performer's finely tuned kinesthetic sense provides information about her bodily posture, orientation, and movement.

© Janis Litavnieks/iStockphoto

We have more skin receptors for cold than for warmth, and we don't seem to have any "hot" receptors at all. Instead, our cold receptors detect not only coolness but also extreme temperatures—both hot and cold (Craig & Bushnell, 1994).

The vestibular sense is responsible for balance—it informs the brain of how the body, and particularly the head, is oriented with respect to gravity and three-dimensional space. When the head moves, liquid in the *semicircular canals*, located in the inner ear, moves and bends hair cell receptors. At the end of the semicircular canals are the *vestibular sacs*, which contain hair cells sensitive to the specific angle of the head—straight up and down or tilted. Information from the semicircular canals and the *vestibular sacs* is converted to neural impulses that are then carried to the appropriate section of the brain (**Figure 4.10b**).

Kinesthesis is the sense that provides the brain with information about bodily posture, orientation, and movement of individual body parts. Kinesthetic receptors are found throughout the muscles, joints, and tendons of the body. They tell the brain which muscles are being contracted or relaxed, how our body weight is distributed, where our arms and legs are in relation to the rest of our body, and so on (**Figure 4.10c**).

**CONCEPT CHECK** STOP

1. **Why** are our senses of smell and taste called the chemical senses?
2. **How** are taste and smell important to our survival?
3. **How** do the vestibular and kinesthetic senses differ?

# Understanding Perception

## LEARNING OBJECTIVES

**RETRIEVAL PRACTICE** While reading the upcoming sections, respond to each Learning Objective in your own words. Then compare your responses with those in Appendix B.

1. **Document** the relationship between selective attention, feature detectors, and habituation.
2. **Describe** the four ways we organize sensory data.
3. **Summarize** the factors involved in perceptual interpretation.

We are ready to move from sensation and the major senses to perception, the process of selecting, organizing, and interpreting incoming sensations into useful mental representations of the world.

Normally, our perceptions agree with our sensations. When they do not, the result is an **illusion**, a false or misleading impression produced by errors in the perceptual process or by actual physical distortions, as in desert mirages. Be careful not to confuse illusions with hallucinations or delusions. Read the *Myth Busters* to learn the difference.

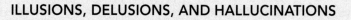

Cross Creek Pictures/Kobal Collection/The Picture Desk

## MYTH BUSTERS

THE PLANNER

### ILLUSIONS, DELUSIONS, AND HALLUCINATIONS

If you watched the 2011 movie, *Black Swan*, you may remember scenes showing Nina (Natalie Portman) bending her legs like a swan's, having webbed feet, and growing black feathers. Were these *illusions*, *delusions*, or *hallucinations*? For psychologists, the differences are important.

Unlike illusions, which are produced by *true* errors in the perceptual process, hallucinations are *false* sensory experiences that occur without an external stimulus, such as hearing voices during a psychotic episode or seeing "pulsating flowers" after using LSD [lysergic acid diethylamide] or other hallucinogenic drugs. Delusions refer to false beliefs, often of persecution or grandeur, which may accompany drug or psychotic experiences.

# Applying Psychology

THE PLANNER

## Optical Illusions

As you may have noticed, this text highlights numerous popular *myths* about psychology because it's important to understand and correct our misperceptions. For similar reasons, you need to know how illusions mislead our normal information processing and recognize that "seeing is believing, but seeing isn't always believing correctly" (Lilienfeld et al., 2010, p.7).

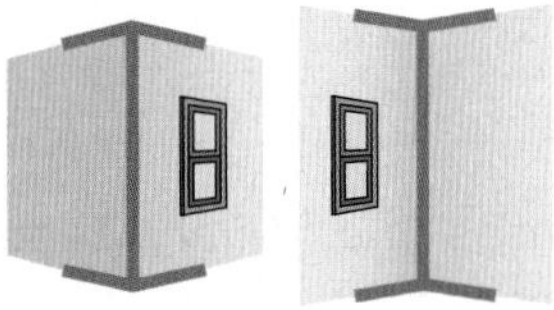

**a. Müller-Lyer illusion**
Which vertical line is longer? In fact, the two vertical lines are the same length, but psychologists have learned that people who live in urban environments normally see the one on the right as longer. This is because they have learned to make size and distance judgments from perspective cues created by right angles and horizontal and vertical lines of buildings and streets.

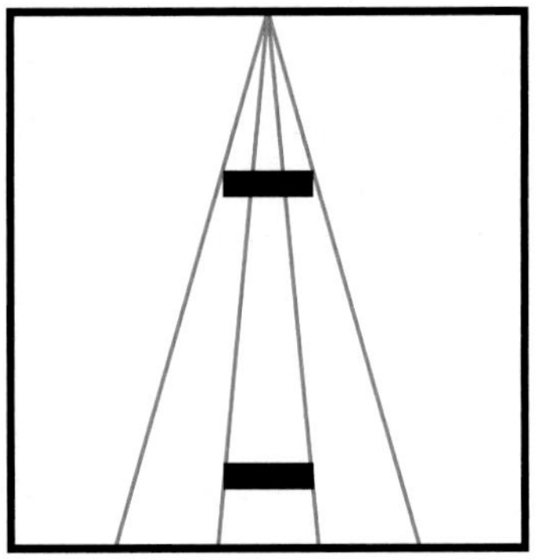

**b. Ponzo illusion**
Which of the two horizontal lines is longer? In fact, both lines are the exact same size, but the converging lines provide depth cues telling you that the top dark, horizontal, line is farther away than the bottom line and therefore much larger.

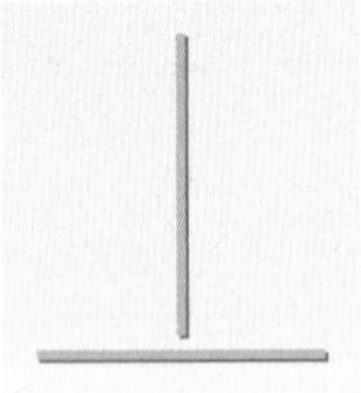

**c. The horizontal-vertical illusion**
Which is longer, the horizontal (flat) or the vertical (standing) line? People living in areas where they regularly see long straight lines, such as roads and train tracks, perceive the horizontal line as shorter because of their environmental experiences.

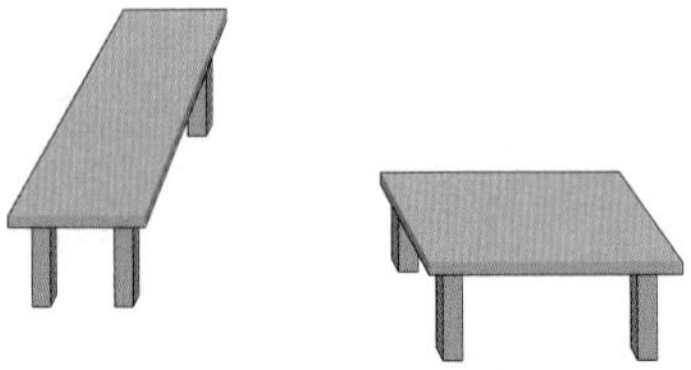

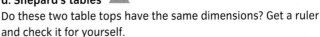

**d. Shepard's tables**
Do these two table tops have the same dimensions? Get a ruler and check it for yourself.

**Think Critically**

1. What do you think causes "errors" in the perceptual process such as the optical illusions described here?
2. When watching films of moving cars, the wheels appear to go backwards. Can you explain this common visual illusion?

See *Applying Psychology* for more information about how illusions provide psychologists with a tool for studying the normal process of perception.

### Selection

In almost every situation, we confront more sensory information than we can reasonably pay attention to. Three major factors help us focus on some stimuli and ignore others: **selective attention**, **feature detectors**, and **habituation** (**Figure 4.11**).

**selective attention** Filtering out and attending only to important sensory messages.

**feature detectors** Specialized brain cells that respond to specific features of the stimulus, like movement, shape, and angles.

Certain basic mechanisms for perceptual selection are built into the brain. For example, through the process of selective attention (see **Figure 4.11a**), the brain picks out the information that is important to us and discards the rest (Haab et al., 2011; Prado, Carp, & Weissman, 2011).

**habituation** The tendency of the brain to ignore environmental factors that remain constant.

In humans and other animals, the brain contains specialized cells, called feature detectors, that respond only to certain stimuli (see **Figure 4.11b**).

Other examples of the brain's ability to filter experience are evidenced by habituation. Apparently, the brain is "pre-wired" to pay more attention to changes in the environment than to stimuli that remain constant. For example, when orthodontic braces are first applied or when they are tightened, they can be very painful. After a while, however, awareness of the pain diminishes (see **Figure 4.11c**).

As advertisers and political operatives well know, people tend to automatically select stimuli that are intense, novel, moving, and contrasting. For sheer volume of sales (or votes), the question of whether you like the ad is irrelevant. If it gets your attention, that's all that matters.

## Psychology InSight

Selection • **Figure 4.11**

Ryan McVay/Photodisc Green/Getty

**a. Selective attention**
Have you noticed that when you're at a noisy party you can still select and attend to the voices of people you find interesting, or that you can suddenly pick up on another group's conversation if someone in that group mentions your name? These are prime examples of *selective attention*, or the so- called "cocktail party phenomenon."

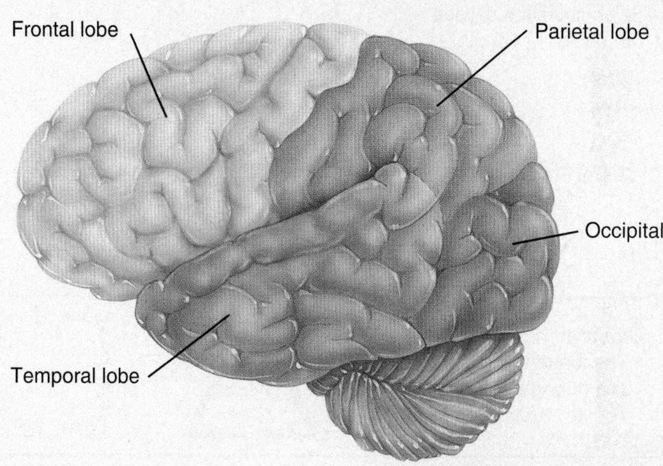


**b. Feature detectors**
Studies with humans have found *feature detectors* in the temporal and occipital lobes that respond maximally to faces. Problems in these areas can produce a condition called *prosopagnosia (prosopon means* "face" and *agnosia* means "failure to know").

Interestingly, people with prosopagnosia can recognize that they are looking at a face. But they cannot say whose face is reflected in a mirror, even if it is their own or that of a friend or relative (Barton, 2008; Palermo et al., 2011; Stollhoff et al., 2011).

© Brankica Tekic/iStockphoto

Cleo Photo/Alamy

**c. Habituation**
These two girls' brains may "choose to ignore" their painful braces. (Sensory adaptation may have also occurred. Over time the girls' pressure sensors send fewer messages to the brain.)

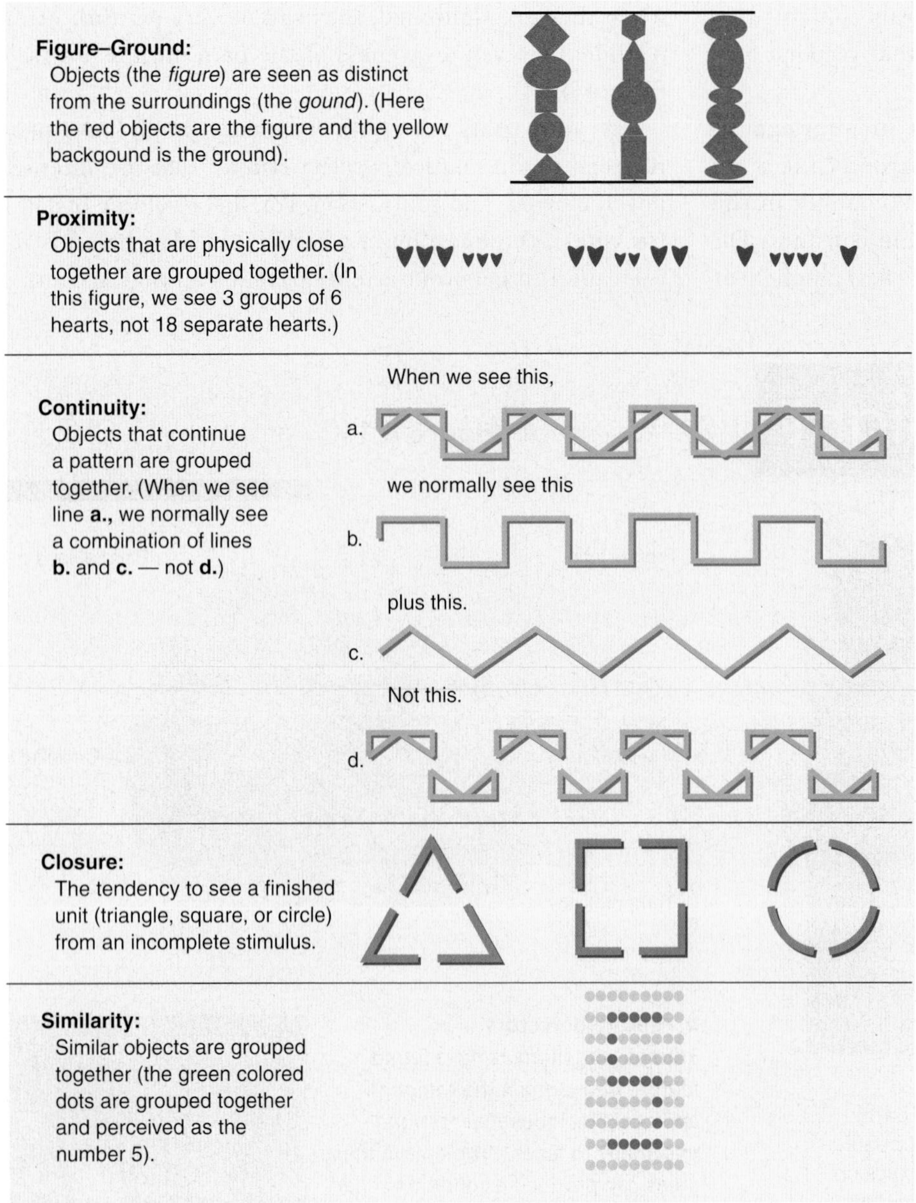

Figure–Ground:
Objects (the *figure*) are seen as distinct from the surroundings (the *gound*). (Here the red objects are the figure and the yellow backgound is the ground).
Proximity:
Objects that are physically close together are grouped together. (In this figure, we see 3 groups of 6 hearts, not 18 separate hearts.)
Continuity:
Objects that continue a pattern are grouped together. (When we see line **a.,** we normally see a combination of lines **b.** and **c.** — not **d.**)
When we see this,
a.
we normally see this
b.
plus this.
c.
Not this.
d.
Closure:
The tendency to see a finished unit (triangle, square, or circle) from an incomplete stimulus.
Similarity:
Similar objects are grouped together (the green colored dots are grouped together and perceived as the number 5).


## Gestalt principles of organization • Figure 4.12

Gestalt principles are based on the notion that we all share a natural tendency to force patterns onto whatever we see. Although the examples of the Gestalt principles in this figure are all visual, each principle applies to other modes of perception as well. For example, the Gestalt principle of *contiguity* cannot be shown because it involves nearness in time, not visual nearness. You also may have experienced *aural figure and ground effects* at a movie or a concert when there was a conversation going on close by and you couldn't sort out what sounds were the background and what you wanted to be your focus.

## Organization

Raw sensory data are like the parts of a watch—they must be assembled in a meaningful way before they are useful. We organize sensory data in terms of form, depth, constancy, and color.

**Form perception** Gestalt psychologists were among the first to study how the brain organizes sensory impressions into a **gestalt**—a German word meaning "form" or "whole." They emphasized the importance of organization and patterning in enabling us to perceive the whole stimulus rather than perceiving its discrete parts as separate entities. The Gestaltists proposed several laws of organization that specify how people perceive form (**Figure 4.12**).

The most fundamental Gestalt principle of organization is our tendency to distinguish between figure (our main focus of attention) and ground (the background or surroundings).

Your sense of figure and ground is at work in what you are doing right now—reading. Your brain is receiving sensations of black lines and white paper, but your brain is organizing these sensations into black letters and words on a white background. You perceive the letters as the figure and the white as the ground. If you make a great effort, you might be able to force yourself to see the

## Reversible and impossible figures • Figure 4.13

a. This so-called *reversible figure* demonstrates alternating figure–ground relations. It can be seen as a woman looking in a mirror or as a skull, depending on what you see as figure or ground.

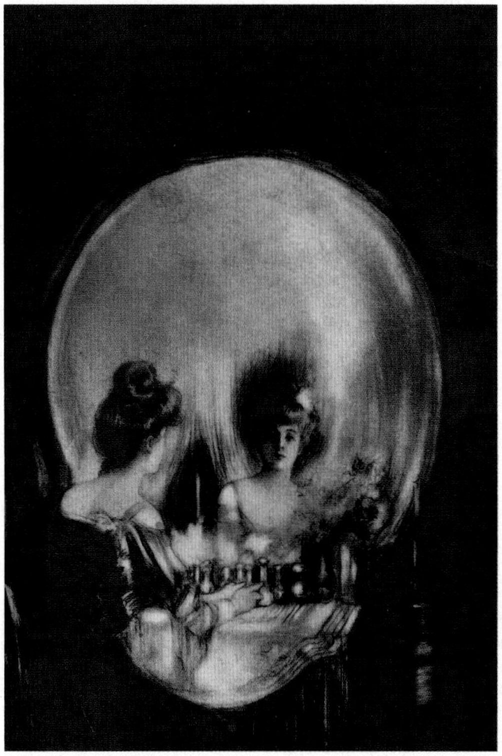

Rykoff Collection/Corbis

b. When you first glance at the famous painting by the Dutch artist M.C. Escher, you detect specific features of the stimuli and judge them as sensible figures. But as you try to sort and organize the different elements into a stable, well-organized whole, you realize they don't add up—they're impossible. There is no one-to-one correspondence between your actual sensory input and your final perception.

page reversed, as though a black background were showing through letter-shaped holes in a white foreground. There are times, however, when it is very hard to distinguish the figure from the ground, as can be seen in **Figure 4.13a**. This is known as a *reversible figure*. Your brain alternates between seeing the light areas as the figure and as the ground.

Like reversible figures, **impossible figures** help us understand perceptual principles—in this case, the principle of form organization (**Figure 4.13b**).

**Depth perception** In our three-dimensional world, the ability to perceive the depth and distance of objects—as well as their height and width—is essential. We usually rely most heavily on vision to perceive distance and depth.

**Depth perception** is learned primarily through experience. However, research using an apparatus called the *visual cliff* (**Figure 4.14**) suggests that some depth perception is inborn.

## Visual cliff • Figure 4.14

Crawling infants hesitate or refuse to move to the "deep end" of the visual cliff (Gibson & Walk, 1960), indicating that they perceive the difference in depth. (The same is true for baby animals that walk almost immediately after birth.) Even two-month-old infants show a change in heart rate when placed on the deep versus shallow side of the visual cliff (Banks & Salapatek, 1983).

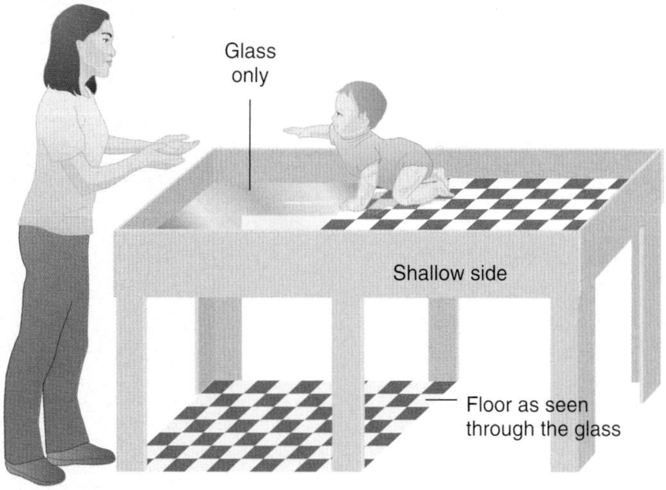

## Binocular depth cues • Figure 4.15

How do we perceive a three-dimensional world with a two-dimensional receptor system? One mechanism is the interaction of both eyes to produce binocular cues.

**a. Retinal disparity**

Stare at your two index fingers a few inches in front of your eyes with their tips half an inch apart. Do you see the "floating finger"? Move it farther away and the "finger" will shrink. Move it closer and it will enlarge. Because our eyes are about 2½ inches apart, objects at different distances (such as the "floating finger") project their images on different parts of the retina, an effect called **retinal disparity.** Far objects project on the retinal area near the nose, whereas near objects project farther out, closer to the ears.

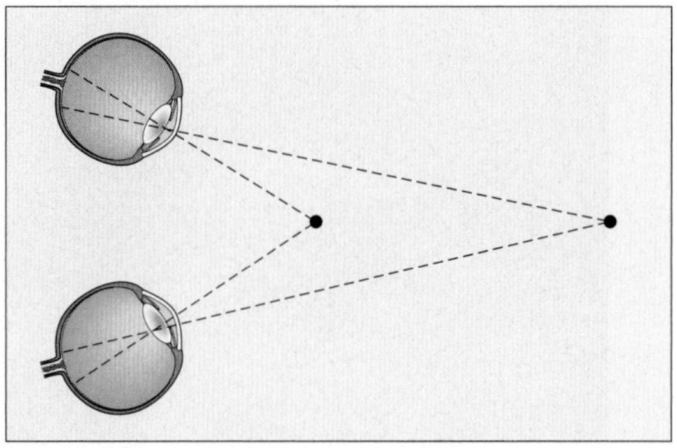

**b. Convergence**

Hold your index finger at arm's length in front of you and watch it as you bring it closer until it is right in front of your nose. The amount of strain in your eye muscles created by the **convergence**, or turning inward of the eyes, is used as a cue by your brain to interpret distance.

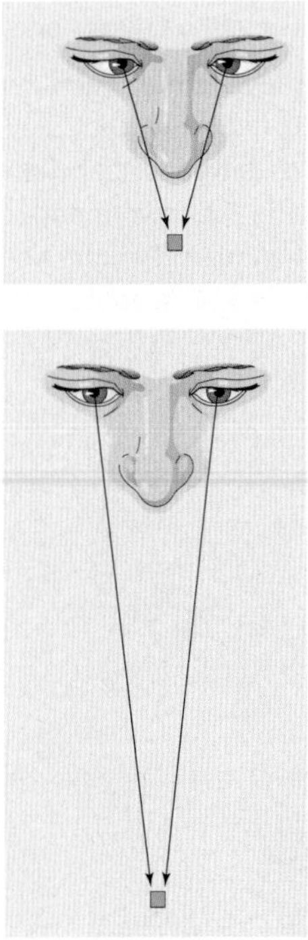

**c. Applying retinal disparity and convergence**

Why are coaches always reminding players to "keep their eyes on the ball"? Convergence and depth perception are better when you are looking directly at an object, rather than out of the corner of your eye. Thus, if you turn your body or your head so that you look straight at your tennis opponent or the pitcher, you will more accurately judge the distance of the ball and thereby be more likely to swing at the right time.

Edith Held/©Corbis

## Monocular depth cues • Figure 4.16

The binocular (two eyes) cues of retinal disparity and convergence are inadequate in judging distances longer than the length of a football field. Luckily, we have several monocular (one eye) cues available separately to each eye.

ZAH/NewsCom

**a.** Artists use monocular depth cues such as linear perspective, light and shadow, and interposition to give depth to a flat image, such as this sidewalk painting by artist Julian Beever.

Digital Vision/Getty Images, Inc.

**b. Motion parallax** (also known as *relative motion*) refers to the fact that when we are moving, close objects appear to whiz by, whereas farther objects seem to move more slowly or remain stationary. This effect can easily be seen when traveling by car or train.

One mechanism by which we perceive depth is the interaction of both eyes to produce **binocular cues** (**Figure 4.15**). The other involves **monocular cues**, which work with each eye separately (**Figure 4.16**).

Artists can use some of the following monocular cues to trick our eyes into seeing depth on a flat canvas (Figure 4.16a):

**Linear perspective** Parallel lines converge, or angle toward one another, as they recede into the distance.

**Interposition** Objects that obscure or overlap other objects are perceived as closer.

**Relative size** Close objects cast a larger retinal image than distant objects.

**Texture gradient** Nearby objects have a coarser and more distinct texture than distant ones.

**Aerial perspective** Distant objects appear hazy and blurred compared to close objects because of intervening atmospheric dust or haze.

**Light and shadow** Brighter objects are perceived as being closer than darker objects.

**Relative height** Objects positioned higher in our field of vision are perceived as farther away.

Two additional monocular cues for depth perception, *accommodation* of the lens of the eye and *motion parallax* (Figure 4.16b), cannot be used by artists. In **accommodation**, muscles that adjust the shape of the lens as it focuses on an object send neural messages to the brain, which interprets the signal to perceive distance. For near objects, the lens bulges; for far objects, it flattens.

# Psychology InSight

Four perceptual constancies • **Figure 4.17**

Without **perceptual constancy**, things would seem to grow as we got closer to them, change shape as our viewing angle changed, and change color as light levels changed.

**a. Size constancy**

Although the person in the foreground of the photo is much larger than the trees behind, we perceive her to be normal size because of size constancy. Interestingly, size constancy, like all constancies, appears to develop from learning and experience. Studies of people who have been blind since birth, and then have their sight restored, find they have little or no size constancy (Sacks, 1995).

RubberBall Productions/Getty Images, Inc.

**b. Shape constancy**

As the coin is rotated, it changes shape, but we still perceive it as the same coin because of shape constancy.

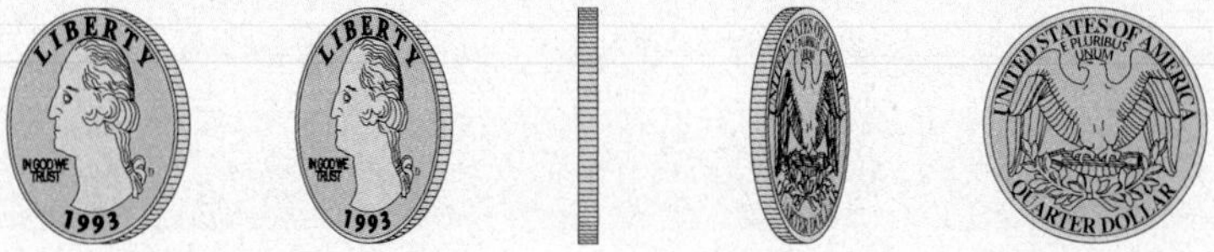


The so-called Ames room illusion (shown in the diagram) also illustrates size and shape constancy. To the viewer, peering through the peephole, the Ames room appears to be a normal cubic-shaped room. But the true shape is trapezoidal. In addition, the walls are slanted and the floor and ceiling are at an incline. Because our brains mistakenly assume the two people are the same distance away, we compensate for the apparent size difference by making the person on the left appear much smaller.

**c. Color constancy** and **brightness constancy**

We perceive the dog's fur in this photo as having a relatively constant hue (or color) and brightness despite the fact that the wavelength of light reaching our retinas may vary as the light changes.

GK Hart/Vikki Hart/Getty Images, Inc.

**Constancy perception** To organize our sensations into meaningful patterns, we develop **perceptual constancies**, the learned tendency to perceive the environment as stable despite changes in an object's size, shape, color and brightness (**Figure 4.17**).

**Color perception** Our color vision is as remarkable as our ability to perceive depth and distance. Humans may be able to discriminate among seven million different hues, and research conducted in many cultures suggests that we all seem to see essentially the same colored world (Davies, 1998). Furthermore, studies of infants old enough to focus and move their eyes show that they are able to see color nearly as well as adults (Knoblauch, Vital-Durand, & Barbur, 2000; Werner & Wooten, 1979).

Although we know color is produced by different wavelengths of light, the actual way in which we perceive color is a matter of scientific debate. Traditionally, there have been two theories of color vision, the trichromatic (three-color theory) and the opponent-process theory. The **trichromatic theory** (from the Greek word *tri*—meaning "three," and *chroma*—meaning "color") was first proposed by Thomas Young in the early nineteenth century and was later refined by Herman von Helmholtz and others. According to the trichromatic theory, we have three "color systems," as they called them—one system that is maximally sensitive to red, another maximally sensitive to green, and another maximally sensitive to blue (Young, 1802). The proponents of this theory demonstrated that mixing lights of these three colors could yield the full spectrum of colors we perceive. Unfortunately this theory has its flaws. One is that it doesn't explain *color aftereffects*, a phenomenon you can experience in the *Applying Psychology* box.

**trichromatic theory** The theory that color perception results from mixing three distinct color systems—red, green, and blue.

The **opponent-process theory**, proposed by Ewald Hering later in the nineteenth century, also suggested the three color systems, but he suggested that each system is sensitive to two opposing colors—blue and yellow, red and green, black and white—in an "on-off" fashion. In other words, each color receptor responds either to blue or yellow or to red or green, with the black-or-white systems responding to differences in brightness levels. This theory makes a lot of sense because when different-colored lights are combined, people are unable to see reddish greens and

**opponent-process theory** The theory that color perception is based on three systems of color receptors, each of which responds in an on-off fashion to opposite-color stimuli: blue-yellow, red-green, and black-white.

## Applying Psychology

THE PLANNER

### Color Aftereffects

Try staring at the dot in the middle of this color-distorted American flag for 60 seconds. Then stare at a plain sheet of white paper. You should get interesting color aftereffects—red in place of green, blue in place of yellow, and white in place of black: a "genuine" American flag.

What happened? This is a good example of the opponent-process theory. As you stared at the green, black, and yellow colors, the neural systems that process those colors became fatigued. Then when you looked at the plain white paper, which reflects all wavelengths, a reverse opponent process occurred. Each fatigued receptor responded with its opposing red, white, and blue colors!

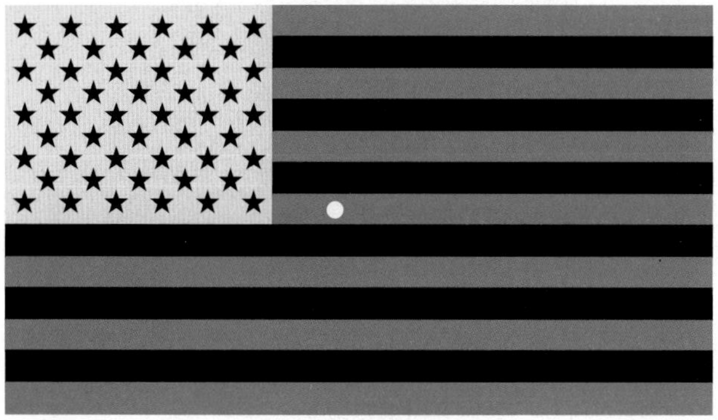

**Think Critically**

In what kinds of situations do you think color aftereffects are likely to occur?

bluish yellows. In fact, when red and green lights or blue and yellow lights are mixed in equal amounts, we see white.

In 1964, research by George Wald and colleague Paul Brown showed we do have three kinds of cones in the retina and confirmed the central proposition of the trichromatic theory. At nearly the same time, R. L. DeValois (1965) was studying electrophysiological recording of cells in the optic nerve and optic pathways to the brain. He discovered that cells respond to color in an opponent-fashion in the thalamus. The findings reconciled the once-competing trichromatic and opponent-process theories. Now, we know that color is processed in a trichromatic fashion at the level of the cones, in the retina, and in an opponent fashion at the level of the optic nerve and the thalamus, in the brain.

**Color-deficient vision** Most people perceive three different colors—red, green, and blue—and are called *trichromats*. However, a small percentage of the population has a genetic deficiency in the red-green system, the blue-yellow system, or both. Those who perceive only two colors are called *dichromats*. People who are sensitive to only the black-white system are called *monochromats*, and they are totally color blind. If you'd like to test yourself for red-green color blindness, see **Figure 4.18**.

### Color-deficient vision • Figure 4.18

Are you color blind? People who suffer red-green deficiency have trouble perceiving the number within this design. Although we commonly use the term *color blindness*, most problems are color confusion rather than color blindness. Furthermore, most people who have some color blindness are not even aware of it.

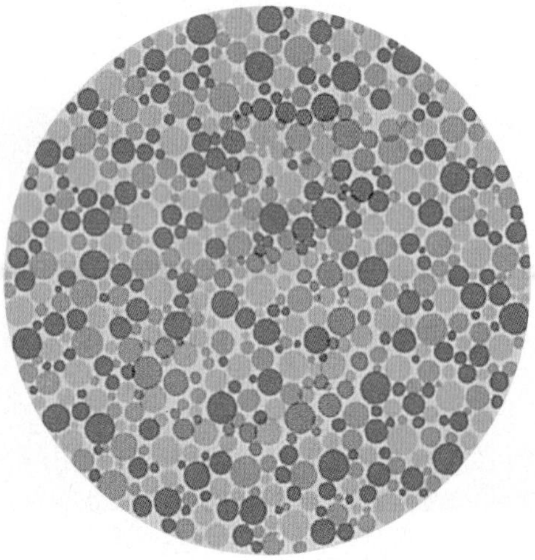

## Interpretation

After selectively sorting through incoming sensory information and organizing it, the brain uses this information to explain and make judgments about the external world. This final stage of perception—**interpretation**—is influenced by several factors, including perceptual adaptation, perceptual set, frame of reference, and bottom-up or top-down processing.

Imagine that your visual field is suddenly inverted and reversed. Things you expect to be on your right are now on your left, and things you expect to be above your head are now below your head. Do you think you could ever adapt to this upside-down world?

To answer that question, psychologist George Stratton (1896) invented special lenses and wore them for eight days. For the first few days, he had a great deal of difficulty navigating in this environment and coping with everyday tasks. But by the third day, he noted:

> Walking through the narrow spaces between pieces of furniture required much less care than hitherto. I could watch my hands as they wrote, without hesitating or becoming embarrassed thereby.

By the fifth day, Stratton had almost completely adjusted to his strange perceptual environment, and when he later removed the headgear, he quickly readjusted.

Stratton's experiment illustrates the critical role that **perceptual adaptation** plays in how we interpret the information that our brains gather. Without his ability to adapt his perceptions to a skewed environment, Stratton would not have been able to function. His brain's ability to "retrain" itself to his new surroundings allowed him to create coherence out of what would otherwise have been chaos.

Our previous experiences, assumptions, and expectations also affect how we interpret and perceive the world, by creating a **perceptual set**, or a readiness to perceive in a particular manner, based on expectations—in other words, we largely see what we expect to see (**Figure 4.19**). In one study involving participants who were members of a Jewish organization (Erdelyi & Applebaum, 1973), collections of symbols were briefly flashed on a screen. When a swastika was at the center, the Jewish participants were less likely to recognize and remember the other symbols.

## Perceptual set? • Figure 4.19

When you look at this drawing, do you see a young woman looking back over her shoulder, or an older woman with her chin buried in a fur collar? It may depend on your age. Younger students tend to first see a young woman, and older students first see an older woman. Although basic sensory input stays the same, your brain's attempt to interpret ambiguous stimuli creates a type of perceptual dance, shifting from one interpretation to another (Gaetz et al., 1998).

Can you see how the life experiences of the Jewish subjects led them to create a perceptual set for the swastika?

How we perceive people, objects, or situations is also affected by the **frame of reference**, or context. An elephant is perceived as much larger when it is next to a mouse than when it stands next to a giraffe.

Finally, recall that we began this chapter by discussing how we receive sensory information (sensation) and work our way upward to the top levels of perceptual processing (perception). Psychologists refer to this type of information processing as **bottom-up processing** (Mulckhuyse et al., 2008; Prouix, 2007). In contrast, **top-down processing** begins with "higher," "top"-level processing involving thoughts, previous experiences, expectations, language, and cultural background and works down to the sensory level (Freeman & Ambady, 2011; Latinus, VanRullen, & Taylor, 2010; Schuett et al., 2008; Zhaoping & Guyader, 2007) (**Figure 4.20**).

## Bottom-up, top-down • Figure 4.20

When first learning to read, you used bottom-up processing. You initially learned that certain arrangements of lines and "squiggles" represented specific letters. You later realized that these letters make up words.

*Now, yuor aiblity to raed uisng top-dwon prcessoing mkaes it psosible to unedrstnad thsi sntenece desipte its mnay mssipllengis.*

Jose Luis Pelaez/Iconica/Getty Images, Inc.

**Science and ESP** So far in this chapter, we have talked about sensations provided by our eyes, ears, nose, mouth, and skin. What about a so-called sixth sense? Can some people perceive things that cannot be perceived with the usual sensory channels, by using **extrasensory perception (ESP)**? People who claim to have ESP profess to be able to read other people's minds (telepathy), perceive objects or events that are inaccessible to their normal senses (clairvoyance), predict the future (precognition), or move or affect objects without touching them (psychokinesis).

As we discussed in Chapter 1, all of these claims fall under the name pseudopsychology and the claims of psychics such as these have been successfully debunked (Irwin, 2008; Lancaster, 2007; Nickell, 2001; Shaffer & Jadwiszczok, 2010). Perhaps the most serious weakness of parapsychology is the failure of replication by rivals in independent laboratories (Hyman, 1996; Nisbet, 2000). (Recall from Ch. 1, p. 12, that replicability is a core requirement for scientific acceptance.) Furthermore, findings in ESP are notoriously "fragile" in that they do not hold up to intense scrutiny (Alcock, 2011; Hyman, 1996; Nisbet, 2000).

So why do so many people believe in ESP? One reason is that, as mentioned earlier in the chapter, our motivations and interests often influence our perceptions, driving us to selectively attend to things we want to see or hear. In addition, the subject of extrasensory perception (ESP) often generates strong emotional responses. When individuals feel strongly about an issue, they sometimes fail to recognize the faulty reasoning underlying their beliefs.

Belief in ESP is particularly associated with illogical or noncritical thinking. For example, people often fall victim to the *fallacy of positive instances*, or the *confirmation bias*, noting and remembering events that confirm personal expectations and beliefs (the "hits"), and ignoring nonsupportive evidence (the "misses"). Other times, people fail to recognize chance occurrences for what they are. Finally, human information processing often biases us to notice and remember the most vivid information—such as a detailed (and spooky) anecdote or a heartfelt personal testimonial.

**CONCEPT CHECK** STOP

1. **Why** do we experience perceptual illusions?
2. **What** kinds of cues do we use to perceive depth and distance?
3. **What** factors entice some people to believe in ESP?

# Summary

THE PLANNER

## 1 Understanding Sensation 88

- **Sensation** is the process by which we detect stimuli and convert them into neural signals (**transduction**). During **coding**, the neural impulses generated by different physical stimuli travel by separate routes and arrive at different parts of the brain. In **sensory reduction**, we filter and analyze incoming sensations.
- **Psychophysics** measures our experience of sensory stimuli, as shown in the photo. The **absolute threshold** is the smallest amount of a stimulus needed to detect a stimulus, and the difference threshold, or just noticeable difference (JND), is the smallest change in stimulus intensity that a person can detect.
- In **sensory adaptation**, sensory receptors fire less frequently with repeated stimulation, so that over time, sensation decreases.
- According to the **gate-control theory**, our experience of pain depends partly on whether the neural message gets past a "gatekeeper" in the spinal cord, which researchers believe is chemically controlled.

**Measuring the senses • Figure 4.2**

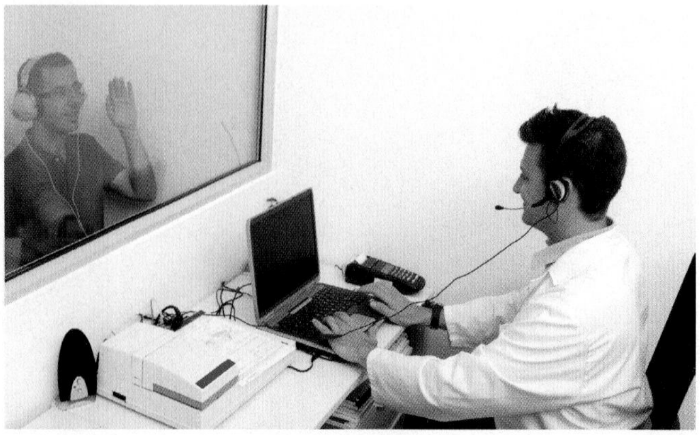

## 2 How We See and Hear 93

- Light and sound move in waves. Light waves are a form of electromagnetic energy, and sound waves are produced when air molecules move in a particular wave pattern. Both light waves and sound waves vary in length, height, and range.
- Light enters the eye at the front of the eyeball, as shown in the diagram. The cornea protects the eye and helps focus light rays. Muscles in the iris dilate or constrict the pupil. The lens further focuses light, adjusting to allow focusing on objects at different distances. At the back of the eye, incoming light waves reach the **retina**, which contains light-sensitive **rods** and **cones**. A network of neurons in the retina transmits neural information to the brain.

How the eye sees • Figure 4.6

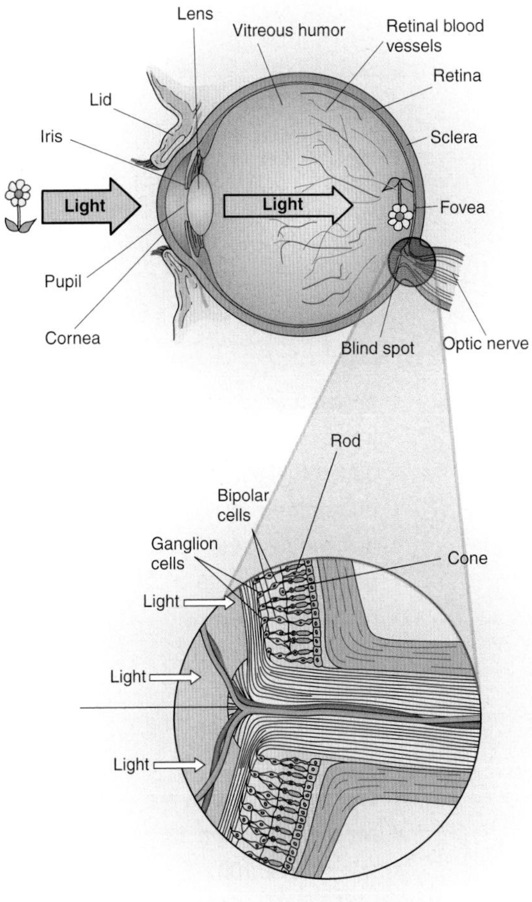


- The **outer ear** gathers sound waves; the **middle ear** amplifies and concentrates the sounds; and the **inner ear** changes the mechanical energy of sounds into neural impulses. The frequency and intensity of sounds determine how we distinguish among sounds of different pitches and loudness, respectively.

## 3 Our Other Important Senses 98

- Smell and taste, sometimes called the chemical senses, involve chemoreceptors that are sensitive to certain chemical molecules. In **olfaction**, odor molecules stimulate receptors in the olfactory epithelium, in the nose, as shown in the diagram. The resulting neural impulse travels to the olfactory bulb, where the information is processed before being sent elsewhere in the brain. Our sense of taste (**gustation**) involves five tastes: sweet, sour, salty, bitter, and umami (umami means "savory" or "delicious"). The taste buds are clustered on our tongues within the papillae.

What a Psychologist Sees: The smell pathway

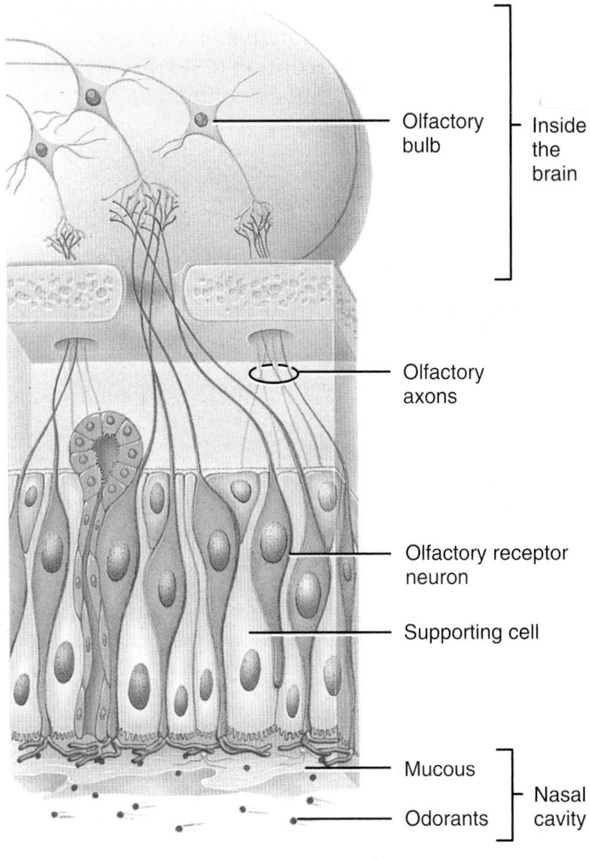


- The body senses tell the brain how the body is oriented, where and how it is moving, and what it touches, or is touched by. They include the **skin senses**, the **vestibular sense**, and **kinesthesis**.

## 4 Understanding Perception 101

- **Perception** is the process of selecting, organizing, and interpreting incoming sensations into useful mental representations of the world. **Selective attention** allows us to filter out unimportant sensory messages. **Feature detectors** are specialized cells that respond only to certain sensory information. **Habituation** is the tendency to ignore stimuli that remain constant. People tend to automatically select stimuli that are intense, novel, moving, and contrasting.
- To be useful, sensory data must be assembled in a meaningful way. We organize sensory data in terms of form, depth, constancy, and color. Size and shape constancy are demonstrated in the Ames room, shown in the diagram. Traditionally there have been two theories of color vision: the **trichromatic theory** and the **opponent-process theory**.
- **Perceptual adaptation**, **perceptual set**, **frame of reference**, and **bottom-up processing** versus **top-down processing** affect our interpretation of what we sense and perceive. **Extrasensory perception (ESP)** research has produced "fragile" results that do not hold up to scientific scrutiny.

**Four perceptual constancies • Figure 4.17**

# Key Terms

**RETRIEVAL PRACTICE** Write a definition for each term before turning back to the referenced page to check your answers.

# Critical and Creative Thinking Questions

1. Sensation and perception are closely linked. What is the central distinction between the two?
2. If we sensed and attended equally to each stimulus in the world, the amount of information would be overwhelming. What sensory and perceptual processes help us lessen the din?
3. If we don't adapt to pain, why is it that people can sometimes "tune out" painful injuries?
4. What senses, outside of hearing, would likely be impaired if a person were somehow missing all of the apparatus of the ear (including the outer, middle, and inner ear)?
5. Can you explain how your own perceptual sets might create prejudice or discrimination?

# What is happening in this picture?

This man willingly endures what would normally be excruciating pain.

Lindsey Hebberd/Corbis

**Think Critically**

1. What psychological and biological factors might make this possible for him?
2. Do you think this man would feel more pain, or less, if his friends and family members were frequently and solicitously asking how he was feeling?

# Self-Test

**RETRIEVAL PRACTICE** Completing this self-test and comparing your answers with those in Appendix C provide immediate feedback and helpful practice for exams. Additional interactive, self-tests are available at www.wiley.com/college/carpenter.

**1.** Coding is the process of converting ______.

a. neural impulses into mental representations of the world

b. receptors into transmitters

c. a particular sensory stimulus into a specific sensation

d. receptors into neural impulses

**2.** Sensory reduction refers to the process of ______.

a. reducing your dependence on a single sensory system

b. decreasing the number of sensory receptors that are stimulated

c. filtering and analyzing incoming sensations before sending a neural message to the cortex

d. reducing environmental sensations by physically preventing your sensory organs from seeing, hearing, etc.

3. Identify the parts of the eye, placing the appropriate label on the diagram:

| | |
|---|---|
| cornea | rod |
| iris | cone |
| pupil | fovea |
| lens | blind spot |
| retina | |

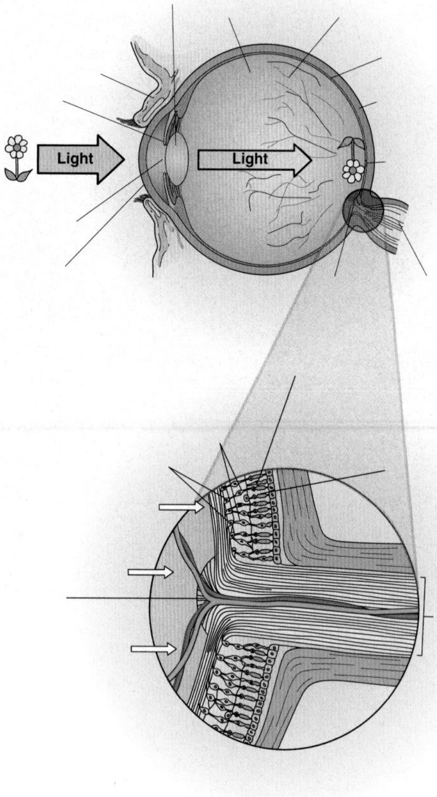


4. A visual acuity problem that occurs when the cornea and lens focus an image in front of the retina is called ______.

a. farsightedness

b. hyperopia

c. myopia

d. presbyopia

5. Identify the parts of the ear, placing the appropriate label on the diagram:

| | |
|---|---|
| auditory nerve | stirrup |
| tympanic membrane | oval window |
| hammer | cochlea |
| anvil | |

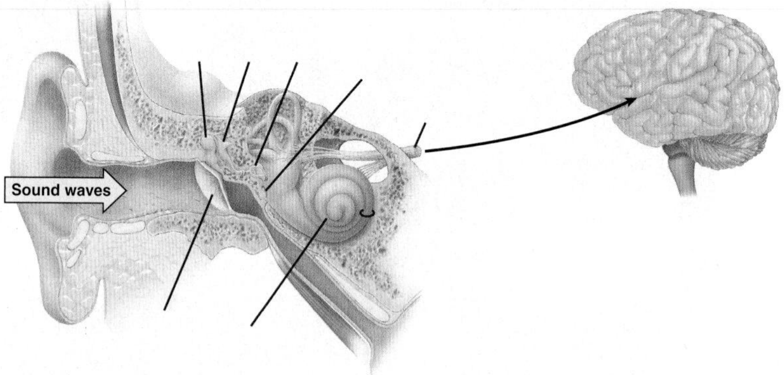


6. Chronic exposure to loud noise can cause permanent ______.

a. auditory illusions

b. auditory hallucinations

c. nerve deafness

d. conduction deafness

7. Most information related to smell is processed in the ______.

a. nasal cavity

b. temporal lobe

c. olfactory bulb

d. parietal lobe

8. Most of our taste receptors are found on the ______.

a. olfactory bulb

b. gustatory cells

c. taste buds

d. frenulum

9. Identify which of these photos, 1 or 2, illustrates:
   a. vestibular sense: photo ______
   b. kinesthetic sense: photo ______

1.

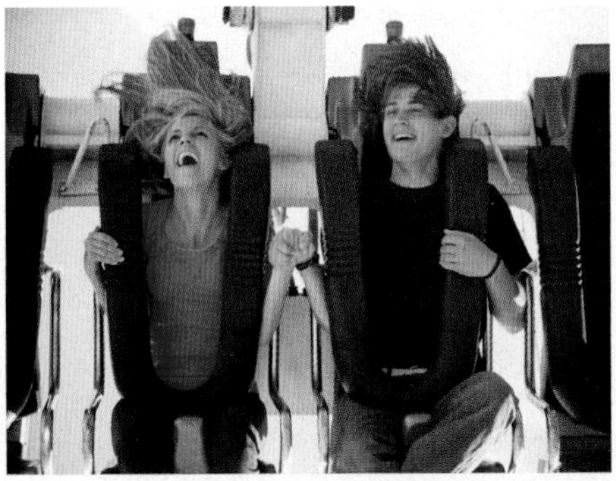

2.

10. When the brain is sorting out and attending only to the most important messages from the senses, it is engaged in the process of ______.
   a. sensory adaptation
   b. sensory habituation
   c. selective attention
   d. selective sorting

11. In the ______ shown here, the discrepancy between figure and ground is too vague, and you may have difficulty perceiving which is figure and which is ground.
   a. delusion
   b. vertical illusion
   c. reversible figure
   d. hallucination

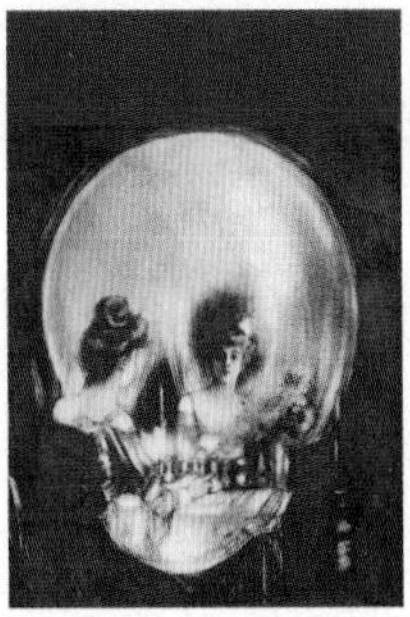

12. The tendency for the environment to be perceived as remaining the same even with changes in sensory input is called ______.
   a. perceptual constancy
   b. the constancy of expectation
   c. an illusory correlation
   d. Gestalt's primary principle

13. The ______ theory of color vision, proposed by Thomas Young, says that color perception results from mixing three distinct color systems.
   a. triocular
   b. trichromatic
   c. tripigment
   d. opponent-process

14. As shown in this figure, a readiness to perceive in a particular manner is known as ______.
   a. sensory adaptation
   b. perceptual set
   c. habituation
   d. frame of reference

15. People who supposedly have ESP claim they have which of the following abilities?
   a. telepathy
   b. precognition
   c. clairvoyance
   d. any of these options

**THE PLANNER**

Review your Chapter Planner on the chapter opener and check off your completed work.

# 5 States of Consciousness

Can you imagine what it would be like if your mind and body were fully functioning, but you could not move or speak? In 1995 Jean-Dominique Bauby suffered a stroke that left him with a condition known as "locked-in syndrome." Without a voice or the ability to move, he could communicate only by blinking his left eye. Yet Bauby remained alert—so much so that he was able to dictate a memoir, *The Diving Bell & the Butterfly*, by "winking" one letter at a time.

What if Bauby had not been able to open or move his eye? If he couldn't communicate, would he still be "conscious"? What exactly is consciousness? Is it simple awareness? How can we study our consciousness when we require our consciousness to study everything? Can we use our consciousness to study our consciousness?

In this chapter, we begin with the definition and description of consciousness. Then we examine how consciousness changes because of circadian rhythms, sleep, and dreams. We also look at how psychoactive drugs, meditation, and hypnosis affect consciousness.

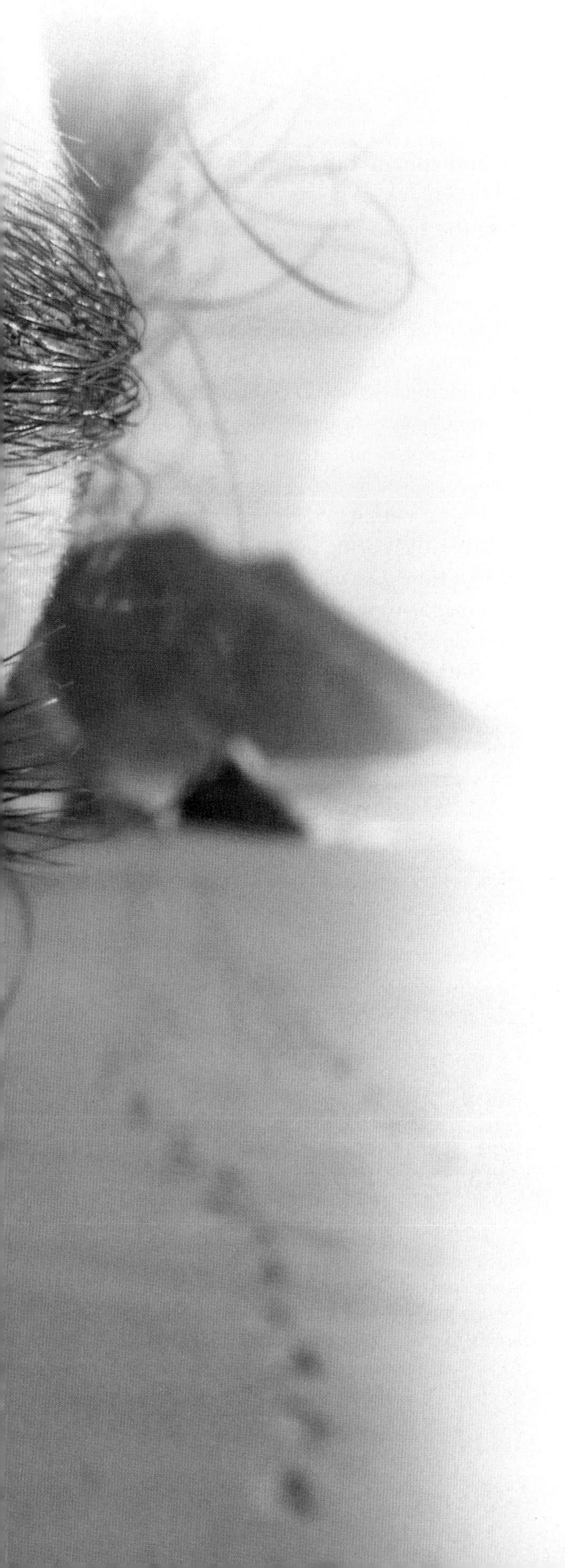

Rui Almeida Fotografia/Getty Images, Inc.

## CHAPTER OUTLINE



## CHAPTER PLANNER ✓

- ❑ Study the picture and read the opening story.
- ❑ Answer the Learning Objectives in each section:
  p. 120 ❑ p. 130 ❑ p. 137 ❑
- ❑ Read the text and study all figures and visuals. Answer any questions.

**Analyze key features**

- ❑ Applying Psychology p. 122 ❑ p. 128 ❑
- ❑ Psychology InSight p. 123 ❑ p. 136 ❑
- ❑ Myth Busters p. 124 ❑ p. 126 ❑ p. 133 ❑ p. 140 ❑
- ❑ Psychological Science, p. 127
- ❑ Process Diagram, p. 131
- ❑ What a Psychologist Sees, p. 138
- ❑ Stop: Answer the Concept Checks before you go on:
  p. 130 ❑ p. 137 ❑ p. 140 ❑

**End of chapter**

- ❑ Review the Summary and Key Terms.
- ❑ Answer the Critical and Creative Thinking Questions.
- ❑ Answer What is happening in this picture?
- ❑ Complete the Self-Test and check your answers.

# Consciousness, Sleep, and Dreaming

## LEARNING OBJECTIVES

**RETRIEVAL PRACTICE** While reading the upcoming sections, respond to each Learning Objective in your own words. Then compare your responses with those in Appendix B.

1. **Compare** the different forms and levels of consciousness.
2. **Explain** the elements of the circadian "clock."
3. **Review** the stages of sleep.
4. **Compare and contrast** the theories of why we sleep and dream.
5. **Summarize** the types of sleep disorders.

William James, the first American psychologist, likened **consciousness** to a stream that's constantly changing yet always the same. It meanders and flows, sometimes where the person wills and sometimes not. However, through the process of *selective attention* (Chapter 4), we can control our consciousness by deliberate concentration and full attention. For example, at the present moment you are, hopefully, fully awake and concentrating on the words on this page. At times, however, your control may weaken, and your stream of consciousness may drift to thoughts of a computer you want to buy, a job, or an attractive classmate.

**consciousness** An organism's awareness of its own self and surroundings (Damasio, 1999).

**alternate state of consciousness (ASC)** The mental states found during sleep, dreaming, psychoactive drug use, hypnosis, and so on.

In addition to meandering and flowing, your "stream of consciousness" also varies in depth. Consciousness is not an all-or-nothing phenomenon—conscious or unconscious. Instead, it exists along a continuum (**Figure 5.1**).

Other than being awake, two of the more common states of consciousness are sleep and dreaming. You may think of yourself as being unconscious while you sleep, but that's not true. Rather, you are in an **alternate state of consciousness (ASC)**. In this chapter, you will

**Levels of awareness • Figure 5.1**

Consciousness exists on a continuum from low awareness to high awareness.

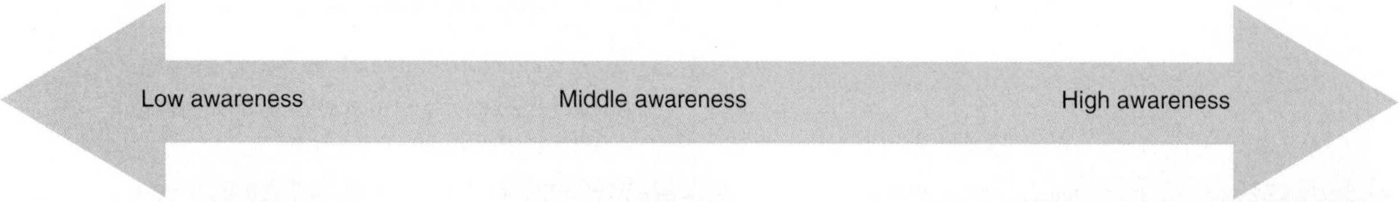

**a.** At the lowest levels of awareness are sleeping, dreaming, anesthesia, and coma.

**b.** You use **automatic processing** for activities that require minimal attention, such as walking while you talk on the phone.

**c.** You use **controlled processing** for activities that require focused, maximum attention, such as learning to drive a car or studying for an exam.

## What controls circadian rhythms? • Figure 5.2

Circadian rhythms are regulated by a part of the hypothalamus called the *suprachiasmatic nucleus* (SCN).

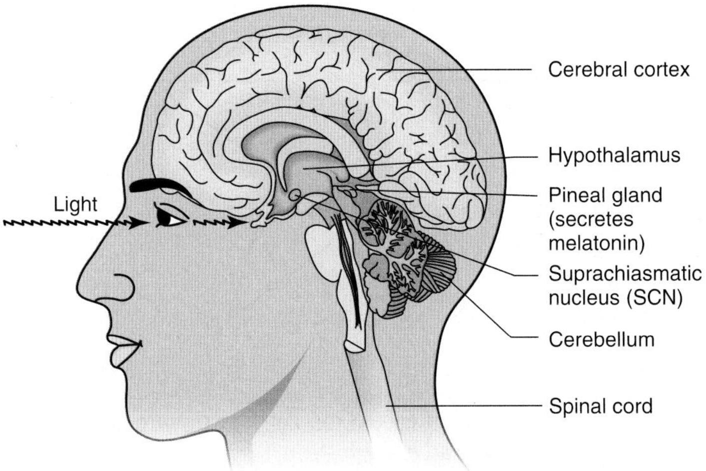


**a.** The SCN receives information about light and darkness from the eyes and then sends control messages to the *pineal gland*, which releases the hormone *melatonin*. Like other feedback loops in the body, the level of melatonin in the blood is sensed by the SCN, which then can modify the output of the pineal to maintain the "desired" level.

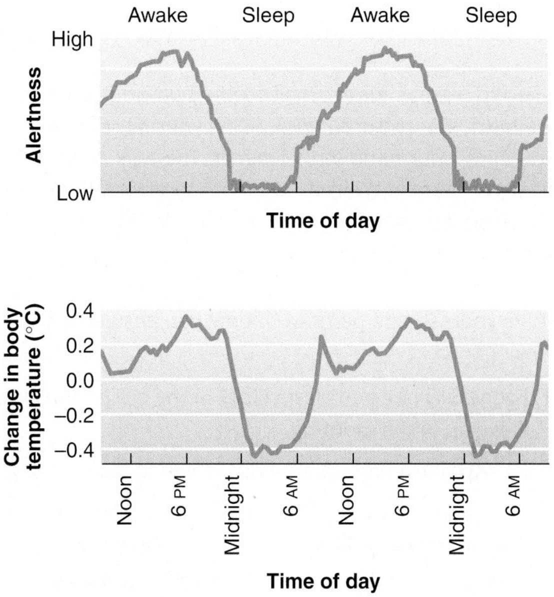


**b.** Melatonin is thought to influence sleep, alertness, and body temperature. Note how our degree of alertness and core body temperature rise and fall in similar ways.

learn about this state of consciousness and why we spend so much time in it.

## Circadian Rhythms

To understand sleep and dreaming, we need to first explore the topic of **circadian rhythms**. Most animals have adapted to our planet's cycle of days and nights by developing a pattern of bodily functions that wax and wane over each 24-hour period. Our alertness, core body temperature, moods, learning efficiency, blood pressure, metabolism, and pulse rate all follow these circadian rhythms (Daan, 2011; Karatsoreos et al., 2011; Kyriacou & Hastings, 2010; Sack et al., 2007). Usually, these activities reach their peak during the day and their low point at night (see **Figure 5.2**).

> **circadian** [ser-KAY-dee-an] **rhythms** The biological changes that occur on a 24-hour cycle (in Latin, *circa* means about, and *dies* means day).

Disruptions in circadian rhythms cause increased fatigue, decreased concentration, sleep disorders, and other health problems (Daan, 2011; Karatsoreos et al., 2011; Kyriacou & Hastings, 2010; Salvatore et al., 2008). Many physicians, nurses, police, and others have rotating work schedules (about 20% of employees in the United States) and do manage to function well. Unfortunately, studies do find that shift work and sleep deprivation lead to decreased concentration and productivity—and increased accidents (Dawson et al., 2011; Dorrian et al., 2011; Pruchnick, Wu, & Belenky, 2011; Williamson et al., 2011). Research suggests that shifting from days to evenings to nights makes it easier to adjust to rotating shift schedules—probably because it's easier to go to bed later than the reverse. Productivity and safety also increase when shifts are rotated every three weeks versus every week.

Like shift work, flying across several time zones can also cause fatigue and irritability—symptoms of jet lag,

# Applying Psychology

THE PLANNER

## Sleep Deprivation

Take the following test to deterine whether you are sleep deprived.

**Part 1** Set up a small mirror next to the text and see if you can trace the black star pictured here, using your nondominant hand while watching your hand in the mirror. The task is difficult, and sleep-deprived people typically make many errors. If you are not sleep deprived, it may be difficult to trace the star, but you'll probobly trace it accurately.

**Part 2** Give yourself one point each time you answer yes to the following:

_____ 1. I generally need an alarm clock or my cell phone alarm to wake up in the morning.

_____ 2. I have a hard time getting out of bed in the morning.

_____ 3. I try to only take late morning or early afternoon college classes because it's so hard to wake up early.

_____ 4. People often tell me that I look tired and sleepy.

_____ 5. I often struggle to stay awake during class, especially in warm rooms.

_____ 6. I find it hard to concentrate and often nod off while I'm studying.

_____ 7. I often feel sluggish and sleepy in the afternoon.

_____ 8. I need several cups of coffee or other energy drinks to make it through the day.

_____ 9. My friends often tell me I'm less moody and irritable when I've had enough sleep.

_____ 10. I tend to get colds and infections, especially around final exams.

_____ 11. When I get in bed at night, I generally fall asleep within four minutes.

_____ 12. I try to catch up on my sleep debt by sleeping as long as possible on the weekends.

*Sources*: Kaida et al., 2008; Mathis & Hess, 2009; National Sleep Foundation, 2012; Smith, Robinson, & Segal, 2012, Winerman, 2004.

---

which also correlates with decreased alertness, mental agility, and efficiency, as well as exacerbation of psychiatric disorders (Leglise, 2008; Paul et al., 2011; Sack, 2010). Jet lag tends to be worse when we fly eastward because our bodies adjust more easily to going to bed later versus trying to force ourselves to go to sleep earlier than normal.

What about long-term **sleep deprivation**? Exploring the scientific effects of severe sleep loss is limited by both ethical and practical concerns. For example, sleep deprivation increases stress, making it difficult to separate the effects of sleep deprivation from those of stress. Take the two-part test in *Applying Psychology* to find out whether you might be sleep deprived.

Nonetheless, researchers have learned that, like disrupted circadian cycles, sleep deprivation poses several hazards. These include reduced cognitive and motor performance, irritability and other mood alterations, decreased self-esteem, and increased *cortisol* levels (a sign of stress) (Doane et al., 2010; Martella, Casagrande, & Lupiáñez, 2011; Orzel-Gryglewska, 2010).

The consequences of such impairments are wide-ranging, from threatening students' school performance to endangering physical health. Perhaps the most frightening danger is that lapses in attention among sleep-deprived pilots, physicians, truck drivers, and other workers cause serious accidents and cost thousands of lives each year (Dorrian, Sweeney, & Dawson, 2011; Williamson et al., 2011; Yegneswaran & Shapiro, 2007).

### Stages of Sleep

Surveys and interviews can provide some information about the nature of sleep, but researchers in sleep laboratories use a number of sophisticated instruments to study physiological changes during sleep.

Imagine that you are a participant in a sleep experiment. When you arrive at the sleep lab, you are assigned one of several bedrooms. The researcher hooks you up to various physiological recording devices (**Figure 5.3a**). You will probably need a night or two to adapt to the equipment before the researchers can begin to monitor your typical night's sleep. At first, you enter a relaxed, *presleep* state. As you continue relaxing, your brain's electrical activity slows even further. In the course of about an hour, you progress through four distinct stages of sleep (Stages 1 through 4), each progressively deeper (**Figure 5.3b**). Then the sequence begins to reverse itself. Although we don't necessarily go through all sleep stages in this sequence, during the course of a night, people usually complete four to five cycles of light to deep sleep and back, each lasting about 90 minutes (**Figure 5.3c**).

# Psychology InSight

## Scientific study of sleep and dreaming • Figure 5.3

Data collected in sleep labs has helped scientists understand the stages of sleep.

**a.** Participants in sleep research labs wear electrodes on their heads and bodies to measure brain and bodily responses during the sleep cycle. An **electroencephalograph (EEG)** detects and records brain-wave changes by means of small electrodes on the scalp. Other electrodes measure muscle activity and eye movements.

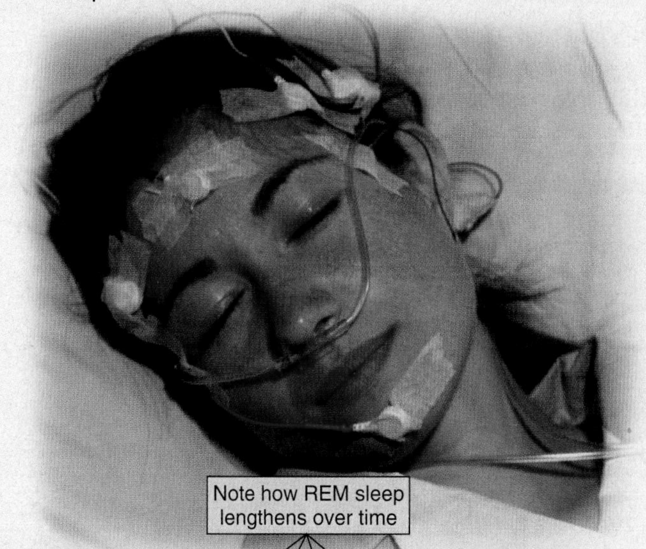



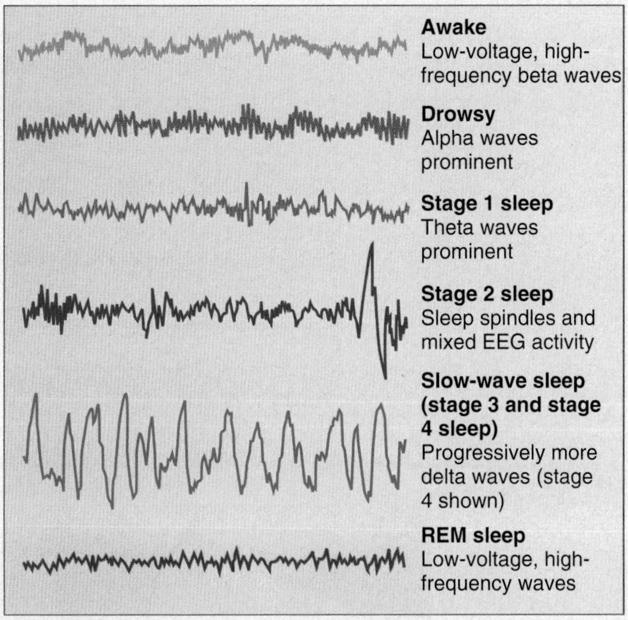


**b.** The stages of sleep, defined by telltale changes in brain waves, are indicated by the jagged lines. The compact brain waves of alertness gradually lengthen as we drift into Stages 1–4.

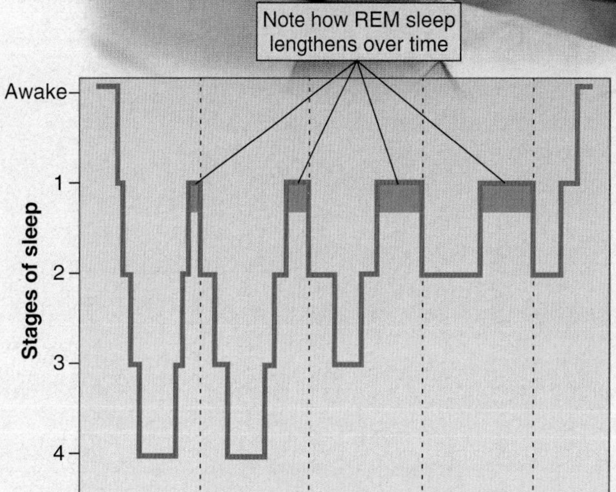


**c.** Your first sleep cycle generally lasts about 90 minutes from awake and alert, downward through Stages 1–4, and then back up through Stages 3, 2, and REM. If you sleep 8 hours, you'll typically go through approximately four or five sleep cycles (as shown by the vertical dotted lines). The overall amount of REM sleep increases as the night progresses, while the amount of deep sleep (Stages 3 and 4) decreases.

Figure 5.3b also shows an interesting phenomenon that occurs at the end of the first sleep cycle (and subsequent cycles). You reverse back through Stages 3 and 2. Then your scalp recordings abruptly display a pattern of small-amplitude, fast-wave activity, similar to an awake, vigilant person's brain waves. Your breathing and pulse rates become fast and irregular, and your genitals likely show signs of arousal. Yet your musculature is deeply relaxed and unresponsive. Because of these contradictory qualities, this stage is sometimes referred to as *paradoxical sleep*.

During this stage of paradoxical sleep, rapid eye movements occur under your closed eyelids. Researchers refer to this stage as **rapid-eye-movement (REM) sleep**. When awakened from REM sleep, people almost always report dreaming. Because REM sleep is so different from the other periods of sleep, Stages 1 through 4 are often collectively referred to as **non-rapid-eye-movement (NREM) sleep**. Dreaming also occurs during NREM sleep, but less frequently. Recent research also shows that dream reports from NREM sleep are very similar to those from REM sleep (Chellappa et al., 2011; Wamsley et al., 2007).

## Why Do We Sleep and Dream?

There are many myths and misconceptions about this topic (see *Myth Busters*). Fortunately scientists have carefully studied what sleep and dreaming do for us and why we spend approximately 25 years of our life in these alternate state of consciousness (ASCs).

**Four sleep theories** The four theories about the benefits of sleep are:

1. **Adaptation/protection theory** Sleep may have evolved because animals needed to protect themselves from predators that are more active at night (Acerbi & Nunn, 2011; Capellini et al., 2010; Siegel, 2008; Swami, 2011). As you can see in **Figure 5.4**, animals with the highest need to feed themselves and the lowest ability to hide tend to sleep the least.

## MYTH BUSTERS

THE PLANNER

### COMMON MYTHS ABOUT SLEEP AND DREAMS

Before reading on, test your personal knowledge of sleep and dreaming by reviewing these common myths.

- ***Myth: Everyone needs 8 hours of sleep a night to maintain sound mental and physical health.*** Although the average is 7.6 hours of sleep a night, some people get by on an incredible 15 to 30 minutes. Others may need as much as 11 hours (Colrain, 2011; Daan, 2011; Doghramji, 2000; Maas et al., 1999).
- ***Myth: Dreams have special or symbolic meaning.*** Many people mistakenly believe dreams can foretell the future, reflect unconscious desires, have secret meaning, can reveal the truth, or contain special messages, but scientific research finds little or no support for these beliefs (Blum, 2011; Carey, 2009; Domhoff, 2010; Hobson et al., 2011; Lilienfeld et al., 2010; Morewedge & Norton, 2009).
- ***Myth: Some people never dream.*** In rare cases, adults with certain brain injuries or disorders do not dream (Solms, 1997). But otherwise, virtually all adults regularly dream. Even people who firmly believe they never dream report dreams if they are repeatedly awakened during an overnight study in a sleep laboratory. Children also dream regularly. For example, between ages 3 and 8, they dream during approximately 20 to 28% of their sleep time (Foulkes, 1993, 1999). Apparently, almost everyone dreams, but some people don't remember their dreams.
- ***Myth: Dreams last only a few seconds and only occur in REM sleep.*** Research shows that most dreams occur in "real time." For example, a dream that seemed to last 20 minutes probably did last approximately 20 minutes (Dement & Wolpert, 1958). Dreams also occur in NREM sleep.
- ***Myth: When genital arousal occurs during sleep, it means the sleeper is having a sexual dream.*** When sleepers are awakened during this time, they are no more likely to report sexual dreams than at other times.
- ***Myth: Most people only dream in black and white, and blind people don't dream.*** People frequently report seeing color in their dreams. Those who are blind do dream, but only report visual images if they lost their sight after approximately age 7 (Lilienfeld et al., 2010).
- ***Myth: Dreaming of dying can be fatal.*** This is a good opportunity to exercise your critical thinking skills. Where did this myth come from? Although many people have personally experienced and recounted a fatal dream, how would we scientifically prove or disprove this belief?

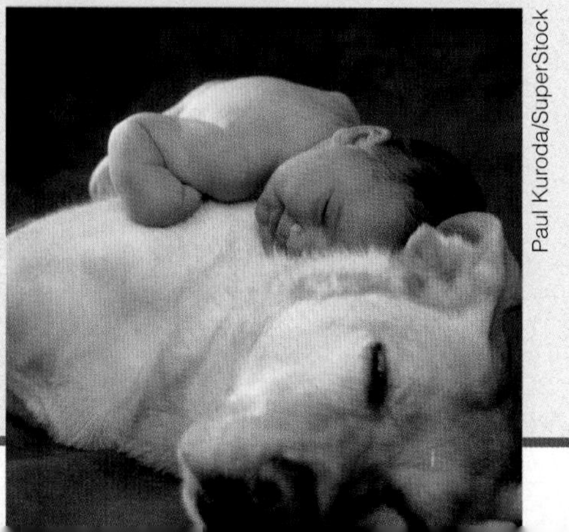

Paul Kuroda/SuperStock

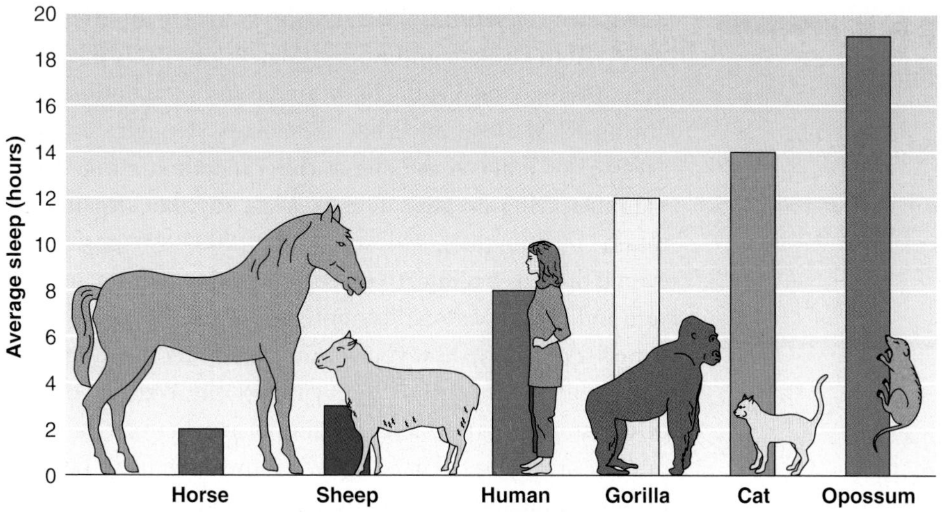


## Average daily hours of sleep for different mammals • Figure 5.4

According to the adaptation/protection theory, differences in diet and number of predators affect different species' sleep habits. For example, opossums sleep many hours each day because they are relatively safe in their environment and are able to easily find food and shelter. In comparison, sheep and horses sleep very little because their diets require constantly foraging for food.

2. **Repair/restoration theory** Sleep helps us recuperate from depleting daily activities. Essential factors in our brain or body are apparently repaired or replenished while we sleep. We recover not only from physical fatigue but also from emotional and intellectual demands (Colrain, 2011). When deprived of REM sleep, most people "catch up" later by spending more time than usual in this state (the so-called *REM rebound*), which supports the theory that sleep (particularly the REM stage) serves an important biological need.
3. **Growth/development theory** Deep sleep coincides with the release of growth hormones from the pituitary gland—particularly in children. As we age, we release fewer of these hormones, grow less, and sleep less.
4. **Learning/memory theory** Sleep is important for learning and the consolidation, storage, and maintenance of memories (Drosopoulos, Harrer, & Born, 2011; Fogel & Smith, 2011; Sara, 2010; Silvestri & Root, 2008). This is particularly true for REM sleep, which increases after periods of stress or intense learning. And fetuses, infants, and young children, who generally are learning more than adults, spend a large percentage of their sleep time in REM sleep (**Figure 5.5**). In addition, REM sleep occurs only in mammals of higher intelligence (Rechtschaffen & Siegel, 2000; Rial et al., 2010). This theory may have

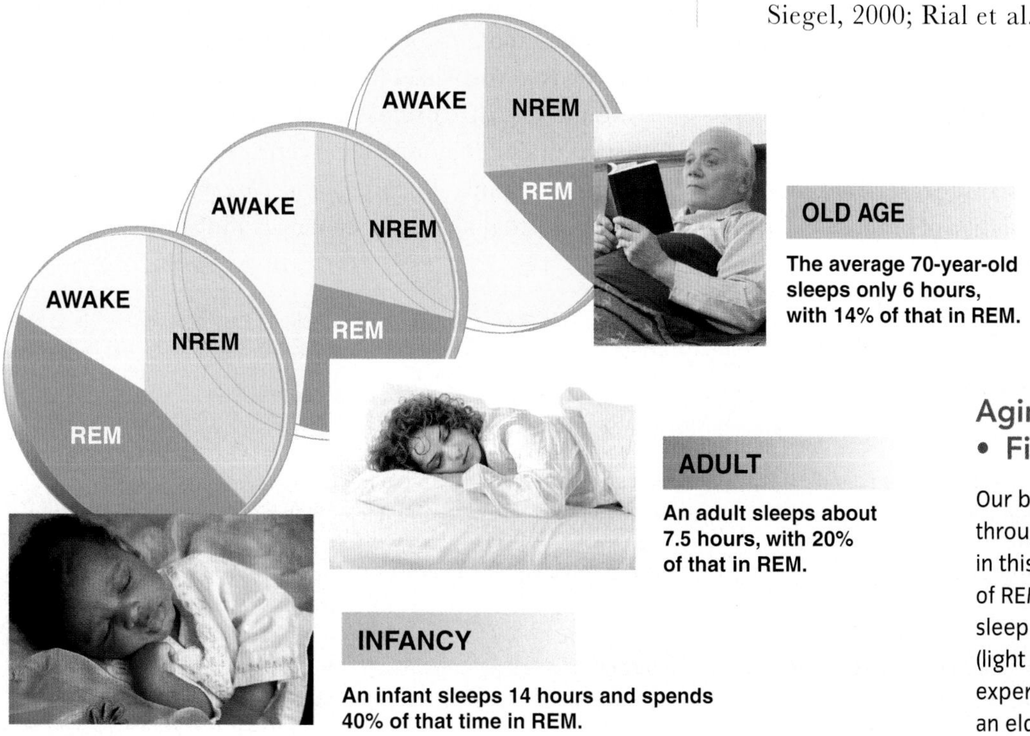




## Aging and the sleep cycle • Figure 5.5

Our biological need for sleep changes throughout our life span. The pie charts in this figure show the relative amounts of REM sleep (dark blue), non-REM sleep (medium blue), and awake time (light blue) that the average person experiences as an infant, an adult, and an elderly person.

## MYTH BUSTERS

### SLEEP-ASSISTED LEARNING

***Myth: It is easy to learn new, complicated things, like a foreign language, while asleep.*** Although some learning can occur doing the lighter stages (1 and 2) of sleep, the processing and retention of this material is minimal (Lilienfeld et al., 2010; Ogilvie, Wilkinson, & Alison, 1989). Wakeful learning is much more effective and efficient.

contributed to some misconceptions about learning new things during sleep (see *Myth Busters*).

**Three dream theories** Now let's look at three theories of why we dream—and whether dreams carry special meaning or information.

One of the oldest and most scientifically controversial explanations for why we dream is Freud's **wish-fulfillment view**. Freud proposed that unacceptable desires, which are normally repressed, rise to the surface of consciousness during dreaming. We avoid anxiety, Freud believed, by disguising our forbidden unconscious needs (what Freud called the dream's latent content) as symbols (manifest content). For example, a journey is supposed to symbolize death; horseback riding and dancing would symbolize sexual intercourse; and a gun might represent a penis.

Most modern scientific research does not support Freud's view (Domhoff, 2004; Dufresne, 2007; Siegel, 2010). Critics also say that Freud's theory is highly subjective and that the symbols can be interpreted according to the particular analyst's view or training.

In contrast to Freud, a biological view called the **activation–synthesis hypothesis** suggests that dreams are a by-product of random stimulation of brain cells during REM sleep (Hobson, 1999, 2005; Wamsley & Stickgold, 2010). Alan Hobson and Robert McCarley (1977) proposed that specific neurons in the brain stem fire spontaneously during REM sleep and that the cortex struggles to "synthesize" or make sense out of this random stimulation by manufacturing dreams. This is *not* to say that dreams are totally meaningless. Hobson (1999, 2005) suggests that even if dreams begin with essentially random brain activity, your individual personality, motivations, memories, and life experiences guide how your brain constructs the dream.

Finally, some researchers support the **cognitive view of dreaming** that dreams are simply another type of information processing. That is, our dreams help us to periodically sift and sort our everyday experiences and thoughts. For example, some research reports strong similarities between dream content and waking thoughts, fears, and concerns (Domhoff, 2005, 2007, 2010; Erlacher & Schredl, 2004). See *Psychological Science* to learn more about research findings about dreams.

## Sleep Disorders

An estimated two-thirds of American adults suffer from sleep problems, and about 25% of children under age 5 have a sleep disturbance (Lader, Cardinali, & Pandi-Perumal, 2006; National Sleep Foundation, 2007; Wilson & Nutt, 2008). Each year, Americans spend more than $98 million on nonprescription sleep medications, and about half that amount on caffeine tablets for daytime use. One in five adults is so sleepy during the day that it interferes with daily activities, and 20% of all drivers have fallen asleep for a few seconds at the wheel (a phenomenon called *microsleep*).

Mental health professionals divide sleep disorders into two major diagnostic categories: *dyssomnias*, which describe problems sleeping, and *parasomnias*, which describe abnormal sleep disturbances (**Table 5.1**).

**Summarizing sleep disorders Table 5.1**

| Sleep Disorders | Characteristics |
|---|---|
| **Dyssomnias** | |
| Insomnia | Persistent difficulty falling asleep or staying asleep, or waking up too early |
| Narcolepsy | Sudden, irresistible onset of sleep during waking hours, characterized by sudden sleep attacks while standing, talking, or even driving |
| Sleep apnea | Repeated interruption of breathing while asleep, causing loud snoring or poor-quality sleep |
| **Parasomnias** | |
| Nightmares | Bad dreams during REM sleep, which significantly disrupt sleep |
| Night terrors | Abrupt awakenings with feelings of panic during NREM sleep, which significantly disrupt sleep |

# Ψ Psychological Science

**✓ THE PLANNER**

# Dream Variations and Similarities

What do people commonly dream about? Do men and women dream about different things? Are there differences between cultures in dream content? Scientists have discovered several interesting answers to these questions.

### a. Common dream themes
A quick glance at this table shows that most people, at least in the Western world, dream a lot about being chased, sex, and misfortune. ▼

Best View Stock /Getty Images, Inc.

| Rank | Dream content | Total prevalence |
|------|---------------|------------------|
| 1 | Chased or pursued, not physically injured | 81.5 |
| 2 | Sexual experiences | 76.5 |
| 3 | Falling | 73.8 |
| 4 | School, teachers, studying | 67.1 |
| 5 | Arriving too late, e.g., missing a train | 59.5 |
| 6 | Being on the verge of falling | 57.7 |
| 7 | Trying again and again to do something | 53.5 |
| 8 | A person now alive is dead | 54.1 |
| 9 | Flying or soaring through the air | 48.3 |
| 10 | Vividly sensing . . . a presence in the room | 48.3 |
| 11 | Failing an examination | 45.0 |
| 12 | Physically attacked (beaten, stabbed, raped) | 42.4 |

The data shown here are from a study of 1,181 Canadian college students (Nielsen et al., 2003). Total prevalence refers to the percentage of students reporting each dream.

*Source*: Nielsen, T. A., Zadra, A. L., Simard, V., Saucier, S., Stenstrom, P., Smith, C., & Kuiken, D. (2003). The typical dreams of Canadian university students. Dreaming, 13, 211–235. Copyright © 2003 Association for the Study of Dreams. [from Table 1, p. 217].

### b. Gender differences and similarities ▲
Men and women tend to share many common dream themes. But women are more likely to dream of children, family and familiar people, household objects, and indoor events, whereas men tend to dream more about strangers, violence, weapons, sexual activity, achievement, and outdoor events (Blume-Marcovici, 2010; Domhoff, 2003, 2007, 2010; Schredl, 2012).

### c. Culture and dreams
Dreams involving basic human needs and fears (like sex, aggression, and death) seem to be found in all cultures. And children around the world often dream about large, threatening wild animals. In addition, dreams around the world typically include more misfortune than good fortune, and the dreamer is more often the victim of aggression than the cause of it (Domhoff, 2003, 2007, 2010; Domhoff & Schneider, 2008; Hall & Van de Castle, 1996). ▼

iStockphoto

### Identify the Research Method

1. What is the most likely research method used for the group of studies described above?
2. If you chose
   - the experimental method, label the IV, DV, experimental group, and control group.
   - the descriptive method, is this a naturalistic observation, survey, or case study?
   - the correlational method, is this a positive, negative, or zero correlation?
   - the biological method, identify the specific research tool (e.g., brain dissection, CT scan).

(Check your answers in Appendix C.)

**dyssomnia** A problem in the amount, timing, and quality of sleep, including insomnia, sleep apnea, and narcolepsy.

The most common **dyssomnia** is **insomnia**. Although it's normal to have trouble sleeping before an exciting event, as many as one in ten people has persistent difficulty falling asleep or staying asleep, or wakes up too early. Nearly everybody has insomnia at some time in their life (Bastien, 2011; Colrain, 2011; Wilson & Nutt, 2008). A telltale sign of insomnia is that the person feels poorly rested the next day. Most people with serious insomnia have other medical or psychological disorders as well (Riemann & Voderholzer, 2003; Taylor, Lichstein, & Durrence, 2003).

Unfortunately, nonprescription insomnia pills generally don't work. Prescription tranquilizers and barbiturates do help people sleep, but they decrease Stage 4 and REM sleep, seriously affecting sleep quality. In the short term, drugs such as Ambien, Dalmane, Xanax, Halcion, and Lunesta may be helpful in treating sleep problems related to anxiety and acute, stressful situations. However, chronic users run the risk of psychological and physical drug dependence (Leonard, 2003; McKim, 2002). *Applying Psychology* offers recommendations for alleviating sleep problems without medication.

**Narcolepsy** is another serious dyssomnia, characterized by sudden and irresistible onsets of sleep during normal waking hours. Narcolepsy afflicts about 1 person in 2,000 and generally runs in families (Billiard, 2007; Pedrazzoli et al., 2007; Raggi et al., 2011). During an attack, REM-like sleep suddenly intrudes into the waking state of consciousness. Victims may experience sudden, incapacitating attacks of muscle weakness or paralysis (known as cataplexy). Such people may fall asleep while walking, talking, or driving a car. Although long naps each day and stimulant or antidepressant drugs may help reduce the frequency of attacks, both the causes and cure of narcolepsy are still unknown (**Figure 5.6**).

# Applying Psychology

THE PLANNER

## Natural Sleep Aids

Are you wondering what is recommended for sleep problems? Research finds consistent benefits from behavior therapy (Constantino et al., 2007; Smith et al., 2005). You can use these same techniques in your own life. For example, when you're having a hard time going to sleep, don't keep checking the clock and worrying about your loss of sleep. Instead, remove all TVs, stereos, and books, and limit the use of the bedroom to sleep. If you need additional help, try some of the following:

**During the day**

*Exercise*. Daily physical activity works away tension. But don't exercise vigorously late in the day, or you'll get fired up instead.

*Keep regular hours*. An erratic schedule can disrupt biological rhythms. Get up at the same time each day.

*Avoid stimulants*. Coffee, tea, soft drinks, chocolate, and some medications contain caffeine. Nicotine may be an even more potent sleep disrupter.

*Avoid late meals and heavy drinking*. Overindulgence can interfere with your normal sleep pattern.

*Stop worrying*. Focus on your problems at a set time earlier in the day.

*Use presleep rituals*. Follow the same routine every evening: listen to music, write in a diary, meditate.

**In bed**

*Use progressive muscle relaxation*. Alternately tense and relax various muscle groups.

*Apply yoga*. These gentle exercises help you relax.

*Use fantasies*. Imagine yourself in a tranquil setting. Feel yourself relax.

*Use deep breathing*. Take deep breaths, telling yourself you're falling asleep.

*Try a warm bath*. This can induce drowsiness because it sends blood away from the brain to the skin surface.

## Narcolepsy • Figure 5.6

Research on specially bred narcoleptic dogs has found degenerated neurons in certain areas of the brain (Siegel, 2000). Whether human narcolepsy results from similar degeneration is a question for future research. Note how this hungry puppy has lapsed suddenly from alert wakefulness to deep sleep even when offered his preferred food.

Perhaps the most serious dyssomnia is **sleep apnea**. People with sleep apnea may fail to breathe for a minute or longer and then wake up gasping for breath. When they do breathe during their sleep, they often snore. Although people with sleep apnea are often unaware of it, repeated awakenings result in insomnia and leave the person feeling tired and sleepy during the day. Sleep apnea seems to result from blocked upper airway passages or from the brain's ceasing to send signals to the diaphragm, thus causing breathing to stop. This disorder can lead to high blood pressure, stroke, and heart attack (Billiard, 2007; Bourke et al., 2011; Furukawa et al., 2010; Nikolaou et al., 2011).

Treatment for sleep apnea depends partly on its severity. If the problem occurs only when you're sleeping on your back, sewing tennis balls to the back of your pajama top may help remind you to sleep on your side. Because obstruction of the breathing passages is related to obesity and heavy alcohol use (Christensen, 2000), dieting and alcohol restriction are often recommended. For others, surgery, dental appliances that reposition the tongue, or machines that provide a stream of air to keep the airway open may be the answer.

Recent findings suggest that even "simple" snoring (without the breathing stoppage characteristic of sleep apnea) can lead to heart disease and possible death (Stone & Redline, 2006). Although occasional mild snoring remains somewhat normal, chronic snoring is a possible "warning sign that should prompt people to seek help" (Christensen, 2000, p. 157).

The second major category of sleep disorders, **parasomnias**, includes abnormal sleep disturbances such as **nightmares** and **night terrors** (Figure 5.7).

> **parasomnias** The abnormal disturbances occurring during sleep, including nightmares, night terrors, sleepwalking, and sleeptalking.

**Sleepwalking**, which tends to accompany night terrors, usually occurs during NREM sleep. (Recall that large muscles are paralyzed during REM sleep, which explains why sleepwalking normally occurs during NREM sleep.) **Sleeptalking** can occur during any stage of sleep, but appears to arise most commonly during NREM sleep. It can consist of single indistinct words or long, articulate sentences. It is even possible to engage some sleep talkers in a limited conversation.

## Nightmare or night terrors? • Figure 5.7

Nightmares, or bad dreams, occur toward the end of the sleep cycle, during REM sleep. Less common but more frightening are night terrors, which occur early in the cycle, during Stage 3 or Stage 4 of NREM sleep. Like the child in this photo, the sleeper may sit bolt upright, screaming and sweating, walk around, and talk incoherently, and the person may be almost impossible to awaken.

Nightmares, night terrors, sleepwalking, and sleep talking are all more common among young children, but they can also occur in adults, usually during times of stress or major life events (Billiard, 2007; Hobson & Silvestri, 1999). Patience and soothing reassurance at the time of the sleep disruption are usually the only treatment recommended for both children and adults.

**CONCEPT CHECK** STOP

1. **What** is the difference between controlled and automatic processing?
2. **What** are the effects of sleep deprivation and disruption of circadian rhythms?
3. **Why** is REM sleep important?
4. **What** piece of evidence supports the adaptation/protection theory of sleep?
5. **How** do dyssomnias and parasomnias differ?

# Psychoactive Drugs

## LEARNING OBJECTIVES

**RETRIEVAL PRACTICE** While reading the upcoming sections, respond to each Learning Objective in your own words. Then compare your responses with those in Appendix B.

1. **Explain** how psychoactive drugs affect nervous system functioning.
2. **Compare** the four major categories of psychoactive drugs.
3. **Describe** the effects of club drugs on the nervous system.

Have you noticed how difficult it is to have a logical, nonemotional discussion about drugs? In our society, where the most popular **psychoactive drugs** are caffeine, tobacco, and ethyl alcohol, people often become defensive when these drugs are grouped with illicit drugs such as marijuana and cocaine. Similarly, marijuana users are disturbed that their drug of choice is grouped with "hard" drugs like heroin. Most scientists believe that there are good and bad uses of all drugs. How drug use differs from drug abuse and how chemical alterations in consciousness affect a person, psychologically and physically, are important topics in psychology.

**psychoactive drugs** Chemicals that change conscious awareness, mood, or perception.

Psychoactive drugs influence the nervous system in a variety of ways. Alcohol, for example, has a diffuse effect on neural membranes throughout the nervous system. Most psychoactive drugs, however, act in a more specific way: by either enhancing a particular neurotransmitter's effect (an **agonist drug**) or inhibiting it (an **antagonist drug**) (**Figure 5.8**).

Is drug abuse the same as drug addiction? The term **drug abuse** generally refers to drug taking that causes emotional or physical harm to oneself or others. The drug consumption is also typically compulsive, frequent, and intense. **Addiction** is a broad term referring to a condition in which a person feels compelled to use a specific drug. People now use the term to describe almost any type of compulsive activity, from working to surfing the Internet (Padwa & Cunningham, 2010; Ross et al., 2010). In fact, recent research has shown that risky trading in financial markets can create a high that is indistinguishable from that experienced by a drug addiction (Zweig, 2007).

For the sake of clarity, many researchers use the term **psychological dependence** to refer to the mental desire or craving to achieve a drug's effects. They use the term **physical dependence** to refer to changes in bodily processes that make a drug necessary for minimum daily functioning. Physical dependence appears most clearly when the drug is withheld and the user undergoes painful **withdrawal** reactions, including physical pain and intense cravings. After repeated use of a drug, many of the body's physiological processes adjust to higher and higher levels of the drug, producing a decreased sensitivity called **tolerance**.

THE PLANNER

## How agonist and antagonist drugs produce their psychoactive effect • Figure 5.8

Most psychoactive drugs produce their mood-, energy-, and perception-altering effects by changing the body's supply of neurotransmitters. They can alter synthesis, storage, and release of neurotransmitters (1). Psychoactive drugs can also alter the effect of neurotransmittters on the receiving site of the receptor neuron (2).

After neurotransmitters carry their messages across the synapse, the sending neuron normally deactivates the excess, or leftover, neurotransmitter (3).

However, when agonist drugs block this process, excess neutrotransmitter remains in the synapse, which prolongs the psychoactive drug's effect.

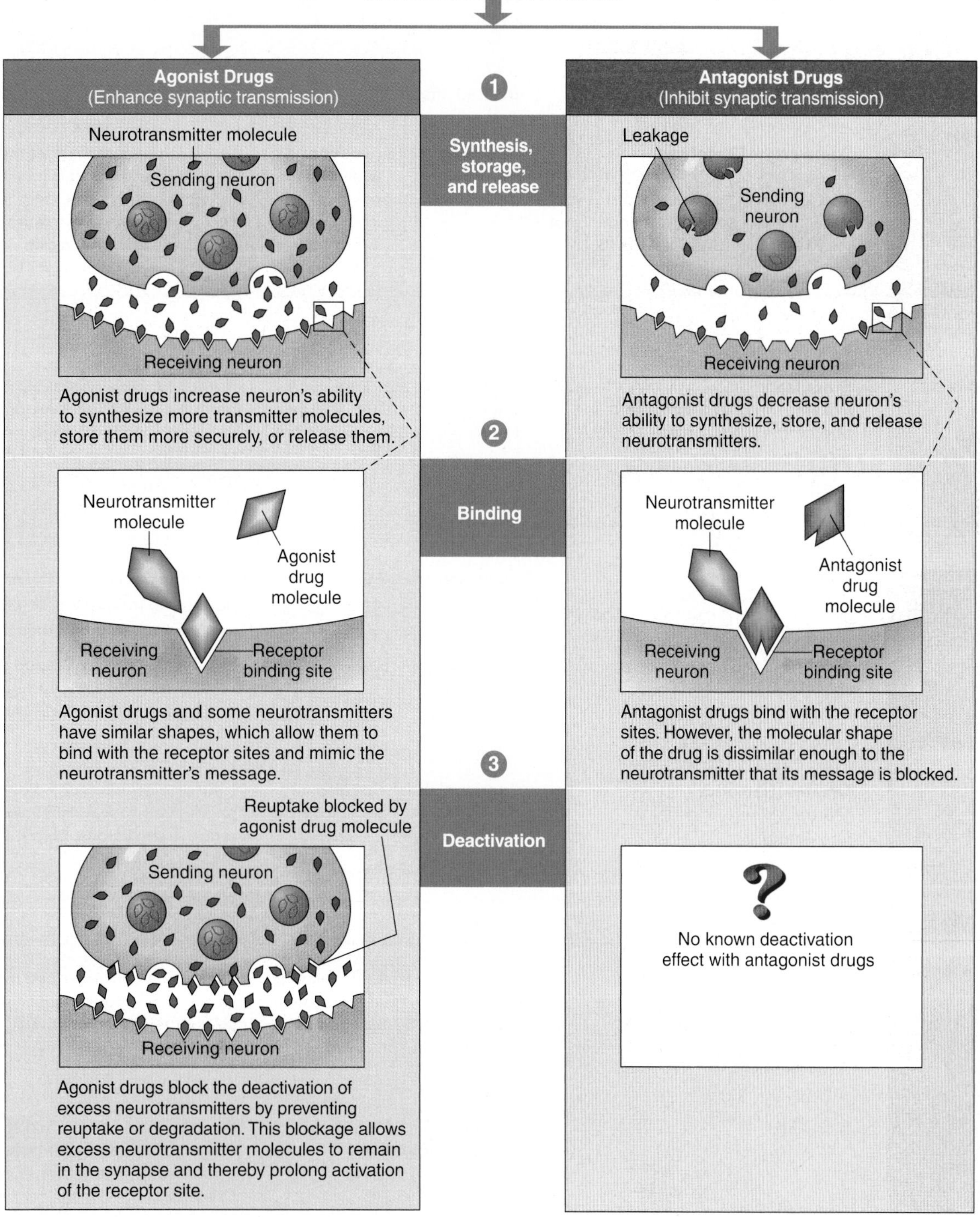
Drugs
Agonist Drugs
(Enhance synaptic transmission)
Antagonist Drugs
(Inhibit synaptic transmission)
1
Synthesis, storage, and release
Neurotransmitter molecule
Sending neuron
Receiving neuron
Agonist drugs increase neuron's ability to synthesize more transmitter molecules, store them more securely, or release them.
Leakage
Sending neuron
Receiving neuron
Antagonist drugs decrease neuron's ability to synthesize, store, and release neurotransmitters.
2
Binding
Neurotransmitter molecule
Agonist drug molecule
Receiving neuron
Receptor binding site
Agonist drugs and some neurotransmitters have similar shapes, which allow them to bind with the receptor sites and mimic the neurotransmitter's message.
Neurotransmitter molecule
Antagonist drug molecule
Receiving neuron
Receptor binding site
Antagonist drugs bind with the receptor sites. However, the molecular shape of the drug is dissimilar enough to the neurotransmitter that its message is blocked.
3
Deactivation
Reuptake blocked by agonist drug molecule
Sending neuron
Receiving neuron
Agonist drugs block the deactivation of excess neurotransmitters by preventing reuptake or degradation. This blockage allows excess neurotransmitter molecules to remain in the synapse and thereby prolong activation of the receptor site.
?
No known deactivation effect with antagonist drugs

Tolerance leads many users to escalate their drug use and to experiment with other drugs in an attempt to recreate the original pleasurable altered state. Sometimes, using one drug increases tolerance for another. This is known as **cross-tolerance**. Developing tolerance or cross-tolerance does not prevent drugs from seriously damaging the brain, heart, liver, and other organs.

Psychological dependence is no less damaging than physical dependence. The craving in psychological dependence can be strong enough to keep the user in a constant drug-induced state—and to lure an "addict" back to a drug habit long after he or she has overcome physical dependence.

## Four Drug Categories

Psychologists divide psychoactive drugs into four broad categories: depressants, stimulants, opiates, and hallucinogens. **Table 5.2** provides examples of each and their effects.

**Effects of the major psychoactive drugs Table 5.2**

| | Category | Desired effects | Undesirable effects |
|---|---|---|---|
|  Richard Nowitz/NG Image Collection | **Depressants (Sedatives)** Alcohol, barbiturates, anxiolytics, also known as antianxiety drugs or tranquilizers (Xanax), Rohypnol (roofies), Ketamine (Special K), 6HB | Tension reduction, euphoria, disinhibition, drowsiness, muscle relaxation | Anxiety, nausea, disorientation, impaired reflexes and motor functioning, amnesia, loss of consciousness, shallow respiration, convulsions, coma, death |
|  Taylor S. Kennedy/NG Image Collection | **Stimulants** Cocaine, amphetamine, methamphetamine (crystal meth), MDMA (Ecstasy) | Exhilaration, euphoria, high physical and mental energy, reduced appetite, perceptions of power, sociability | Irritability, anxiety, sleeplessness, paranoia, hallucinations, psychosis, elevated blood pressure and body temperature, convulsions, death |
|  Sam Abell/NG Image Collection | Caffeine | Increased alertness | Insomnia, restlessness, increased pulse rate, mild delirium, ringing in the ears, rapid heartbeat |
| | Nicotine | Relaxation, increased alertness, sociability | Irritability, increased blood pressure, stomach pains, vomiting, dizziness, cancer, heart disease, emphysema |
| 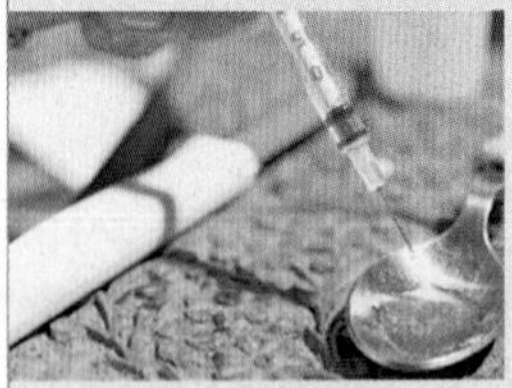 Uwe Schmid/Okapia/Photo Researchers | **Opiates (Narcotics)** Morphine, heroin, codeine, OxyContin | Euphoria, "rush" of pleasure, pain relief, prevention of withdrawal discomfort, sleep | Nausea, vomiting, constipation, painful withdrawal, shallow respiration, convulsions, coma, death |
|  Joel Sartore/NG Image Collection | **Hallucinogens (Psychedelics)** LSD (lysergic acid diethylamide), mescaline (extract from the peyote cactus), psilocybin (magic mushrooms). | Heightened aesthetic responses, euphoria, mild delusions, hallucinations, distorted perceptions and sensation | Panic, nausea, longer and more extreme delusions, hallucinations, perceptual distortions ("bad trips"), psychosis |
| | Marijuana | Relaxation, mild euphoria, increased appetite | Perceptual and sensory distortions, hallucinations, fatigue, lack of motivation, paranoia, possible psychosis |

## Alcohol's effect on the body and behavior • Figure 5.9

Alcohol's effects are determined primarily by the amount that reaches the brain. Because the liver breaks down alcohol at the rate of about one ounce per hour, the number of drinks and the speed of consumption are both very important. People can die after drinking large amounts of alcohol in a short period of time. In addition, men's bodies are more efficient at breaking down alcohol. Even after accounting for differences in size and muscle-to-fat ratio, women have a higher blood alcohol level than men following equal doses of alcohol.

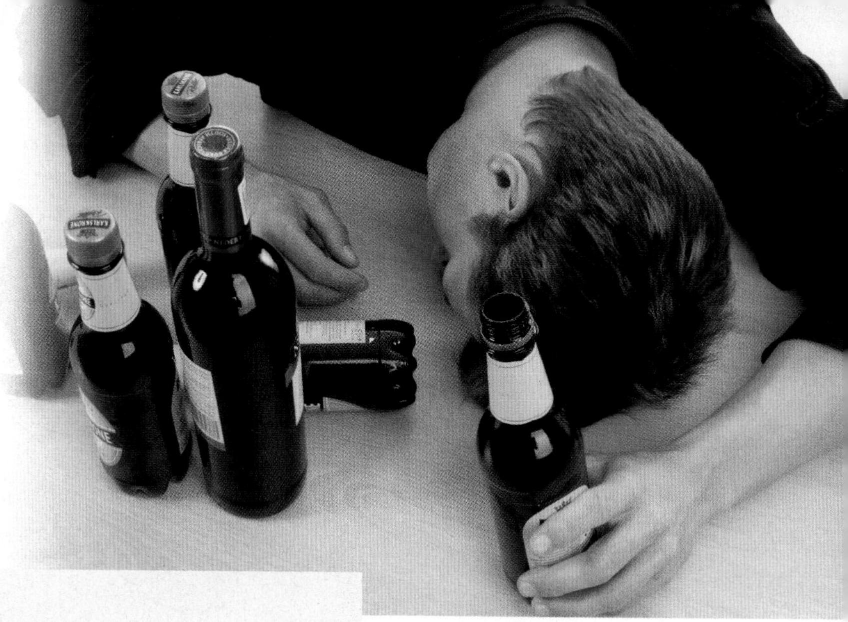

McPHOTO/Blickwinkel/Age Fotostock America, Inc.

| Number of drinks[a] in two hours | Blood alcohol (content (%)[b] | Effect |
|---|---|---|
| (2) | 0.05 | Relaxed state; increased sociability |
| (3) | 0.08 | Everyday stress lessened |
| (4) | 0.10 | Movements and speech become clumsy |
| (7) | 0.20 | Very drunk; loud and difficult to understand; emotions unstable |
| (12) | 0.40 | Difficult to wake up; incapable of voluntary action |
| (15) | 0.50 | Coma and/or death |

[a]A drink refers to one 12-ounce beer, a 4-ounce glass of wine, or a 1.25-ounce shot of hard liquor.

[b]In America, the legal blood alcohol level for "drunk driving" varies from 0.05 to 0.12.

**Depressants** (sometimes called "downers") act on the central nervous system to suppress or slow bodily processes and to reduce overall responsiveness. Because tolerance and dependence (both physical and psychological) are rapidly acquired with these drugs, there is strong potential for abuse.

Although alcohol is primarily a depressant, at low doses it has stimulating effects, thus explaining its reputation as a "party drug." As consumption increases, symptoms of drunkenness appear (**Figure 5.9**). Take the quiz in *Myth Busters* to discover if some of your ideas about alcohol are really misconceptions.

## MYTH BUSTERS

THE PLANNER

### WHAT'S YOUR ALCOHOL IQ?

**TRUE OR FALSE?**

___ 1. Alcohol increases sexual desire.
___ 2. Alcohol helps you sleep.
___ 3. Alcohol kills brain cells.
___ 4. It's easier to get drunk at high altitudes.
___ 5. Switching among different types of alcohol is more likely to lead to drunkenness.
___ 6. Drinking coffee or taking a cold shower are great ways to sober up after heavy drinking.
___ 7. Alcohol warms the body.
___ 8. You can't become an alcoholic if you only drink beer.
___ 9. Alcohol's primary effect is as a stimulant.
___ 10. People only experience impaired judgment after drinking if they show obvious signs of intoxication.

Answers: All of these are false. Detailed answers are provided in this chapter and in Lilienfeld et al., 2010.

## Cocaine: An agonist drug in action • Figure 5.10

Cocaine is an agonist drug that acts as a stimulant.

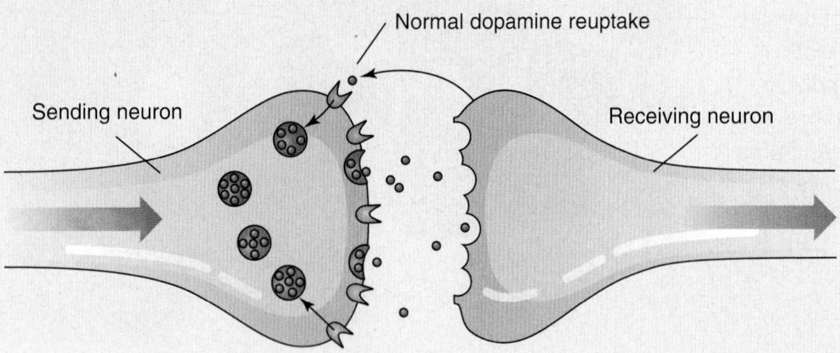


**a.** After releasing neurotransmitter into the synapse, the sending neuron normally reabsorbs (or reuptakes) excess neurotransmitter back into the terminal buttons.

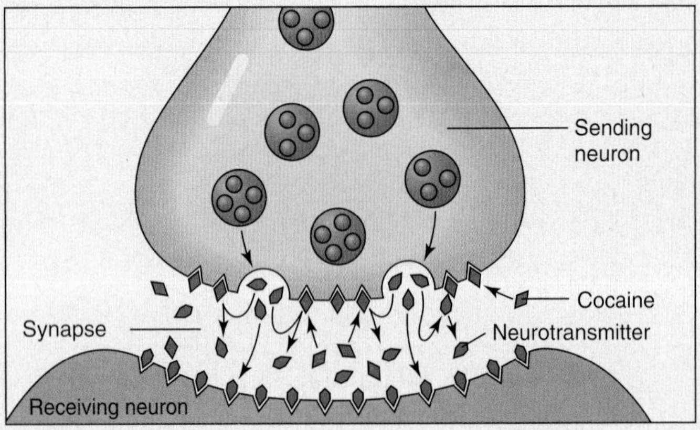


**b.** However, if cocaine is present in the synapse, it blocks the reuptake of dopamine, serotonin, and norepinephrine, which allows extra time for absorption by the receiving neuron. This extra processing time intensifies the normal mood-altering effect of these three mood- and energy-activating neurotransmitters.

Todd Gipstein/NG Image Collection

Alcohol should not be combined with any other drug; combining alcohol and barbiturates—both depressants—is particularly dangerous. Together, they can relax the diaphragm muscles to such a degree that the person literally suffocates.

Depressants suppress central nervous system activity, whereas **stimulants** (uppers) increase the overall activity and responsiveness of the central nervous system.

Cocaine is a powerful central nervous system stimulant extracted from the leaves of the coca plant. It produces feelings of alertness, euphoria, well-being, power, energy, and pleasure. But it also acts as an *agonist drug* to block the reuptake of our body's natural neurotransmitters that produce these same effects (**Figure 5.10**).

Although cocaine was once considered a relatively harmless "recreational drug," even small initial doses can be fatal because cocaine interferes with the electrical system of the heart, causing irregular heartbeats and, in some cases, heart failure. It also can produce heart attacks and strokes by temporarily constricting blood vessels (Abadinsky, 2011; NIDA, 2010; Westover, McBride, & Haley, 2007). The most dangerous form of cocaine is the smokeable, concentrated version known as crack or rock. Its lower price makes it affordable and attractive to a large audience. And its greater potency makes it more highly addictive.

Even legal stimulants can lead to serious problems. For example, the U.S. Public Health Service considers cigarette smoking the single most preventable cause of death and disease in the United States. Researchers have found that nicotine activates the same brain areas (nucleus accumbens) as cocaine—a dangerous stimulant well known for its addictive potential (Dandekar et al.,

## How opiates create physical dependence • Figure 5.11

Psychoactive drugs such as opiates affect the brain and body in a variety of ways.

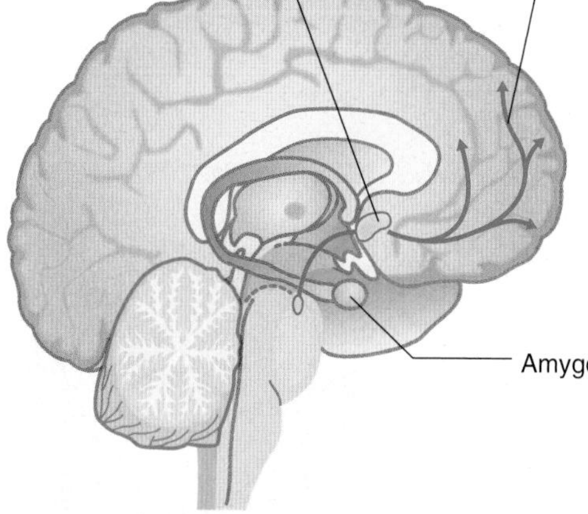


a. Most researchers believe that increased dopamine activity in this so-called *reward pathway* of the brain accounts for the reinforcing effects of most addictive drugs.

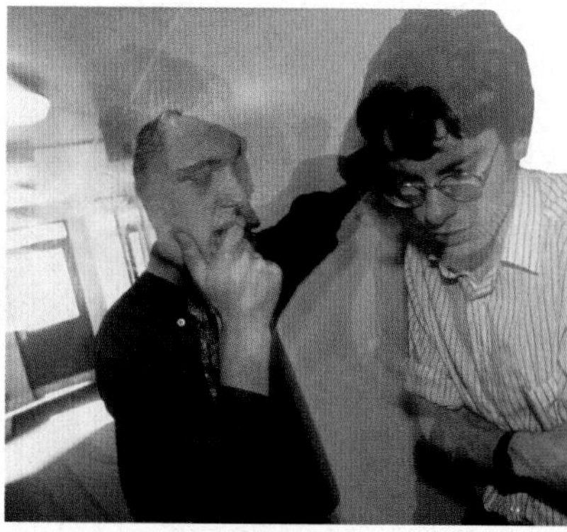

Jim Varney/Science PhotoLibrary/
Photo Researchers, Inc.

b. Absence of the drug triggers withdrawal symptoms (e.g., intense pain and cravings).

2011; Kuhn & Gallinat, 2011; Tronci & Balfour, 2011; Zanetti, Picciotto, & Zoli, 2007). Nicotine's effects (relaxation, increased alertness, diminished pain and appetite) are so powerfully reinforcing that some people continue to smoke even after having a cancerous lung removed.

**Opiates** (or narcotics), which are derived from the opium poppy, are used medically to relieve pain (Maisto, Galizio, & Connors, 2011). They mimic the brain's natural endorphins (Chapter 2), which numb pain and elevate mood. This creates a dangerous pathway to drug abuse. After repeated flooding with artificial opiates, the brain eventually reduces or stops the production of its own opiates. If the user later attempts to stop, the brain lacks both the artificial and normal level of painkilling chemicals, and withdrawal becomes excruciatingly painful (**Figure 5.11**).

So far, we have discussed three of the four types of psychoactive drugs: depressants, stimulants, and opiates. One of the most intriguing alterations of consciousness comes from **hallucinogens**, drugs that produce sensory or perceptual distortions, including visual, auditory, and kinesthetic hallucinations. Some cultures have used hallucinogens for religious purposes, as a way to experience "other realities" or to communicate with the supernatural. In Western societies, most people use hallucinogens for their reported "mind-expanding" potential.

Hallucinogens are commonly referred to as psychedelics (from the Greek for "mind manifesting"). They include mescaline (derived from the peyote cactus), psilocybin (derived from mushrooms), phencyclidine (chemically derived), and LSD (lysergic acid diethyl-amide, derived from ergot, a rye mold).

LSD, or acid, is a synthetic substance that produces dramatic alterations in sensation and perception. Perhaps because the LSD experience is so powerful, few people "drop acid" on a regular basis. Nevertheless, LSD can be an extremely dangerous drug. Bad LSD trips can be terrifying and may lead to accidents, deaths, or suicide. Dangerous flashbacks may unpredictably recur long after the initial ingestion. They can be brought on by stress, fatigue, marijuana use, illness, or occasionally by the individual's intentional effort (Abadinsky, 2011; Levinthal, 2011).

Marijuana is also classified as a hallucinogen even though it has some properties of a depressant (it induces drowsiness and lethargy) and some of a narcotic (it acts as a weak painkiller). In low doses, marijuana produces mild

# Psychology InSight Your brain on "club" drugs • Figure 5.12

Although club drugs can produce desirable effects (e.g., Ecstasy's feeling of great empathy and connectedness with others), almost all psychoactive drugs may cause serious health problems—in some cases, even death (Chamberlin & Saper, 2009; Jaehne et al., 2011; National Institute on Drug Abuse, 2012).

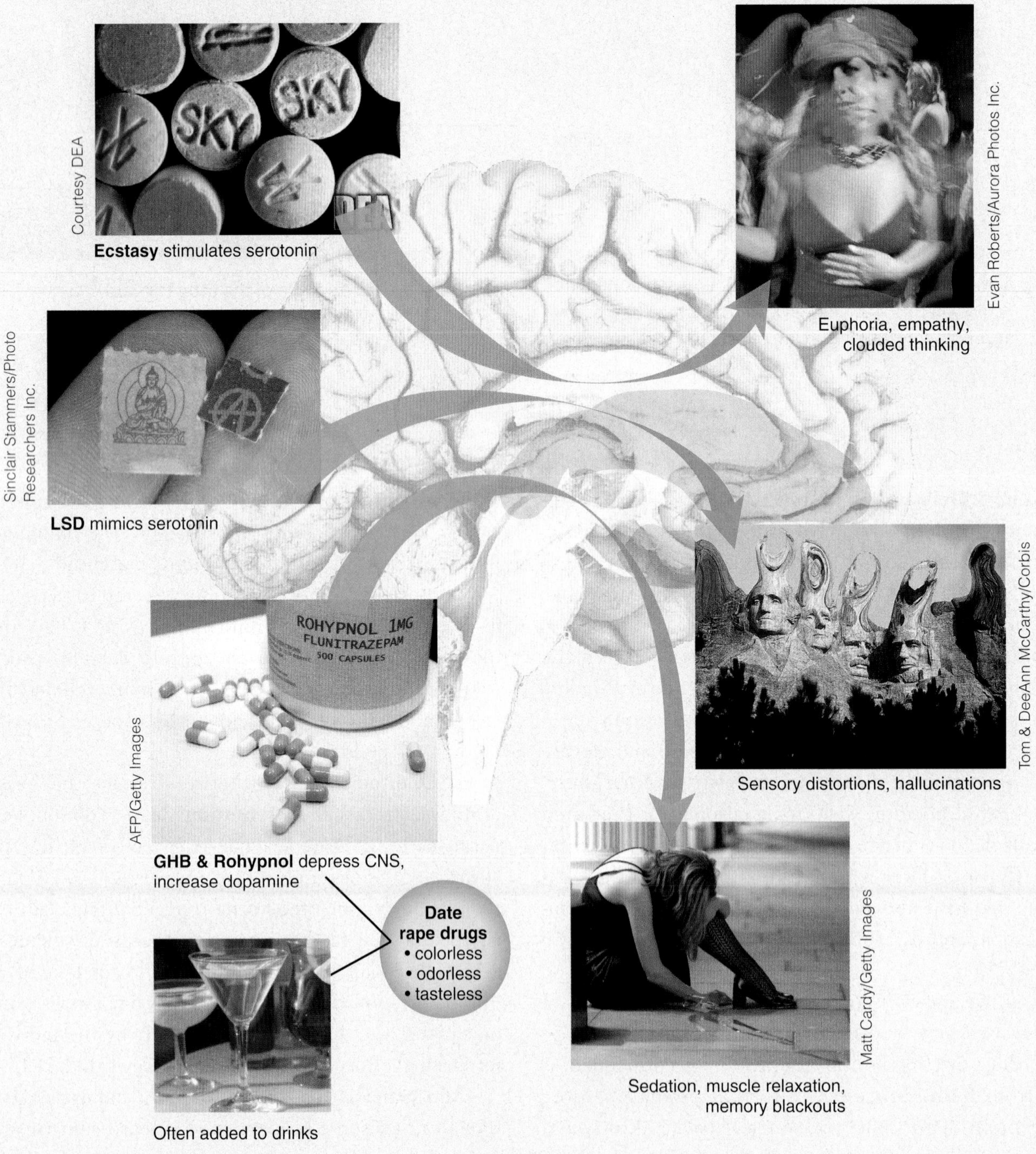

euphoria; moderate doses lead to an intensification of sensory experiences and the illusion that time is passing slowly. High doses may produce hallucinations, delusions, and distortions of body image (Köfalvi, 2008; Ksir, Hart, & Ray, 2008). The active ingredient in marijuana (cannabis) is THC, or tetrahydrocannabinol, which attaches to receptors that are abundant throughout the brain.

Some research has found marijuana to be therapeutic in treating glaucoma (an eye disease) and in alleviating the nausea and vomiting associated with chemotherapy, and in dealing with other health problems (Darmani & Crim, 2005; Fogarty et al., 2007; Green & De-Vries, 2010; Köfalvi, 2008).

Chronic marijuana use can lead to throat and respiratory disorders, impaired lung functioning and immune response, declines in testosterone levels, reduced sperm count, and disruption of the menstrual cycle and ovulation (Hall & Degenhardt, 2010; Levinthal, 2011; Skinner et al., 2011). While some research supports the popular belief that marijuana serves as a "gateway" to other illegal drugs, other studies find little or no connection (Jacquette, 2010; Ksir, Hart, & Ray, 2008; Sabet, 2007; Tarter et al., 2006).

Marijuana also can be habit-forming, but few users experience the intense cravings associated with cocaine or opiates. Withdrawal symptoms are mild because the drug dissolves in the body's fat and leaves the body very slowly, which explains why a marijuana user can test positive for days or weeks after the last use.

## Club Drugs

As you may know from television or newspapers, psychoactive drugs like Rohypnol (the "date rape drug") and MDMA (3-4 methylenedioxymethamphetamine, commonly known as Ecstasy) are fast becoming some of our nation's most popular drugs of abuse, especially at all-night dance parties. Other "club" drugs, like GHB (gamma-hydroxybutyrate), ketamine (Special K), methamphetamine (crystal meth), and LSD, are also gaining in popularity (Abadinsky, 2011; Hopfer, 2011; Weaver & Schnoll, 2008). (See **Figure 5.12**).

**CONCEPT CHECK** STOP

1. **How** do agonist drugs differ from antagonist drugs?
2. **Why** are opiates particularly habit-forming?
3. **What** dangerous situations can result from use of club drugs?

# Meditation and Hypnosis

## LEARNING OBJECTIVES

**RETRIEVAL PRACTICE** While reading the upcoming sections, respond to each Learning Objective in your own words. Then compare your responses with those in Appendix B.

1. **Describe** the effect of meditation on the nervous system.
2. **Explain** the features of the hypnotic state.

As we have seen, factors such as sleep, dreaming, and psychoactive drug use can create altered states of consciousness. Changes in consciousness also can be achieved by means of meditation and hypnosis.

## Meditation

"Suddenly, with a roar like that of a waterfall, I felt a stream of liquid light entering my brain through the spinal cord . . . I experienced a rocking sensation and then felt myself slipping out of my body, entirely enveloped in a halo

# WHAT A PSYCHOLOGIST SEES

## Benefits of Meditation

During meditation (**Figure a**), time pressures and worries decrease, resulting in a sensation of peace and timelessness. In addition, research has verified that meditation can produce dramatic changes in basic physiological processes, including heart rate, oxygen consumption, sweat gland activity, and brain activity. Meditation has also been somewhat successful in reducing pain, anxiety and stress, lowering blood pressure, and improving negative moods (Evans et al., 2008; Grant, Courtemanche, & Rainville, 2011; Yu et al., 2011). Studies have even implied that meditation can change the body's parasympathetic response (Sathyaprabha et al., 2008; Young & Taylor, 1998) and increase structural support for the sensory, decision-making, and attention-processing centers of the brain (Lazar et al., 2005) (**Figures b** and **c**).

**a.** Some meditation techniques, such as t'ai chi and hatha yoga, involve body movements and postures, while in other techniques the meditator remains motionless, chanting or focusing on a single point, like a candle flame.

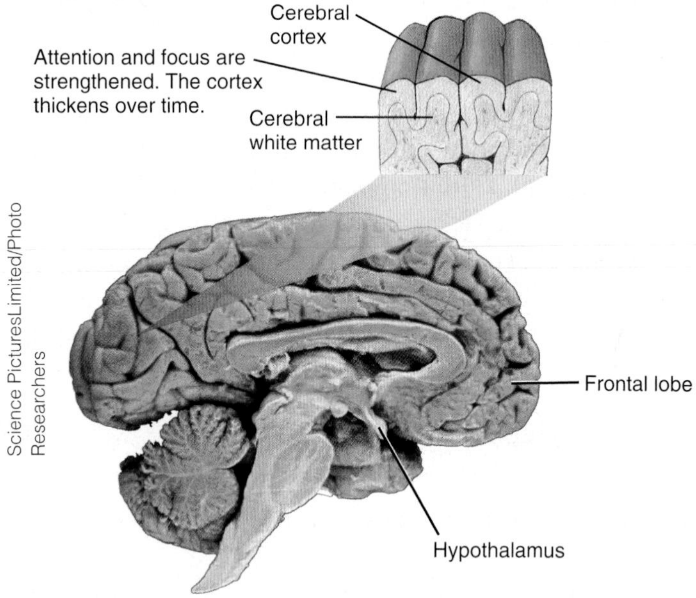


**b.** During meditation, the hypothalamus diminishes the sympathetic response and increases the parasympathetic response. Shutting down the so-called fight-or-flight response in this way allows for deep rest, slower respiration, and increased and more coordinated use of the brain's two hemispheres. At the same time, meditation engages the part of the brain that is responsible for decision making and reasoning, the *frontal lobe*.

of light. I felt the point of consciousness that was myself growing wider, surrounded by waves of light" (Krishna, 1999, pp. 4–5).

**meditation** A group of techniques designed to refocus attention, block out all distractions, and produce an alternate state of consciousness.

This is how spiritual leader Gopi Krishna described his experience with **meditation**. Although most people in the beginning stages of meditation report a simpler, mellow type of relaxation followed by a mild euphoria, some advanced meditators report experiences of profound rapture and joy or strong hallucinations.

The highest functions of consciousness occur in the frontal lobe, particularly in the cerebral cortex. Scientists are seeing increasing evidence that the altered state of consciousness one experiences during meditation occurs when one purposely changes how the prefrontal cortex (the area immediately behind your eyes) functions. Typically, your prefrontal cortex is balancing your working memory, temporal integration, and higher-order thinking, among other tasks. Scientists theorize, based

THE PLANNER

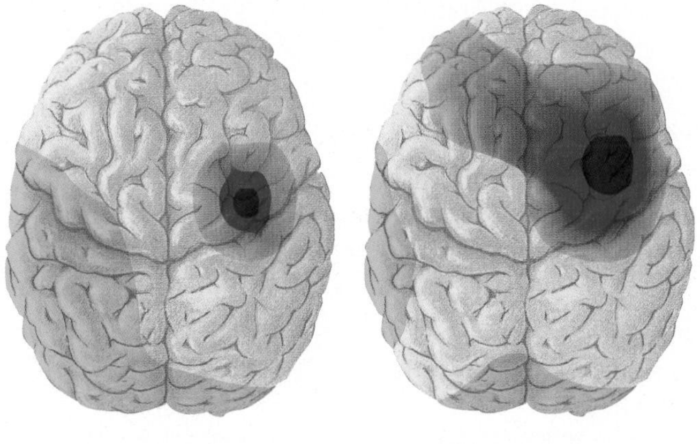


c. Researchers have found that a wider area of the brain responds to sensory stimuli during meditation, suggesting that meditation enhances the coordination between the brain hemispheres (Kilpatrick et al., 2011; Lyubimov, 1992). Note how much the small blue colored area on the right hemisphere before meditation enlarged and spread across to the left hemisphere after meditation.

on brain imaging, that when you focus on a single object, emotion, or word, you diminish the amount of brain cells that must be devoted to these multiple tasks, and they instead become involved in the singular focus of your meditation. This narrow focus allows other areas of the brain to be affected, and since the neurons devoted to time have changed focus, you experience a sense of timelessness and mild euphoria (Cvetkovic & Cosic, 2011; Harrison, 2005). Evidence also shows that meditation can help us cope better with stress (see Chapter 3 and *What a Psychologist Sees*).

## Hypnosis

Relax . . . your eyelids are so very heavy . . . your muscles are becoming more and more relaxed . . . your breathing is becoming deeper and deeper . . . relax . . . your eyes are closing . . . let go . . . relax.

**hypnosis** Trance-like state of heightened suggestibility, deep relaxation, and intense focus.

Hypnotists use suggestions like these to begin **hypnosis**. Once hypnotized, some people can be convinced that they are standing at the edge of the ocean listening to the sound of the waves and feeling the ocean mist on their faces. Invited to eat a delicious apple that is actually an onion, the hypnotized person may relish the flavor. Told they are watching a very funny or sad movie, hypnotized people may begin to laugh or cry at their self-created visions.

From the 1700s to modern times, entertainers and quacks have used (and abused) hypnosis (see *Myth Busters* on the next page), but physicians, dentists, and therapists also have long employed it as a respected clinical tool. Modern scientific research has removed much of the mystery surrounding hypnosis. A number of features characterize the hypnotic state (Jamieson & Hasegawa, 2007; Jensen et al., 2008; Nash & Barnier, 2008):

- Narrowed, highly focused attention (ability to "tune out" competing sensory stimuli)
- Increased use of imagination and hallucinations
- A passive and receptive attitude
- Decreased responsiveness to pain
- Heightened suggestibility, or a greater willingness to respond to proposed changes in perception ("This onion is an apple")

Today, even with available anesthetics, hypnosis is occasionally used in surgery and for the treatment of chronic pain and severe burns (Jensen et al., 2008, 2011; Nash & Barnier, 2008; Nusbaum et al., 2011; Smith, 2011). Hypnosis has found its best use in medical areas, such as dentistry and childbirth, in which patients have a high degree of anxiety, fear, and misinformation. Because tension and anxiety strongly affect pain, any technique that helps the patient relax is medically useful.

In psychotherapy, hypnosis can help patients relax, remember painful memories, and reduce anxiety. Despite the many myths about hypnosis, it has been used with modest success in the treatment of phobias and in helping people to lose weight, stop smoking, and improve study habits (Amundson & Nuttgens, 2008; Golden, 2006; Manning, 2007; Smith 2011).

## MYTH BUSTERS

THE PLANNER

### HYPNOSIS MYTHS AND FACTS

| MYTH | FACT |
|---|---|
| 1. **Faking:** Hypnosis participants are "faking it," and playing along with the hypnotist. | There are conflicting research positions about hypnosis. Although most participants are not consciously faking hypnosis, some researchers believe the effects result from a blend of conformity, relaxation, obedience, suggestion, and role playing (Lynn, Rhue, & Kirsch, 2010; Orne, 2006). Other theorists believe that hypnotic effects result from a special altered state of consciousness (Bob, 2008; Naish, 2006; Bowers & Woody, 1996; Hilgard, 1978, 1992). A group of "unified" theorists suggests that hypnosis is a combination of both relaxation/role playing and a unique alternate state or consciousness. |
| 2. **Forced hypnosis:** People can be hypnotized against their will or hypnotically "brainwashed." | Hypnosis requires a willing, conscious choice to relinquish control of one's consciousness to someone else. The best potential subjects are those who are able to focus attention, are open to new experiences, and are capable of imaginative involvement or fantasy (Carvalho et al., 2008; Green & Lynn, 2011; Lynn, Rhue, & Kirsch, 2010; Terhune, Cardena, & Lindgren, 2011; Wickramasekera, 2008). |
| 3. **Unethical behavior:** Hypnosis can make people behave immorally or take dangerous risks against their will. | Hypnotized people retain awareness and control of their behavior, and they can refuse to comply with the hypnotist's suggestions (Kirsch, Mazzoni, & Montgomery, 2006; Lynn, Rhue, & Kirsch, 2010). |
| 4. **Superhuman strength:** Under hypnosis, people can perform acts of special, superhuman strength. | When nonhypnotized people are simply asked to try their hardest on tests of physical strength, they generally can do anything that a hypnotized person can (Orne, 2006). |
| 5. **Exceptional memory:** Under hypnosis, people can recall things they otherwise could not. | Although the heightened relaxation and focus that hypnosis engenders improves recall for some information, it adds little if anything to regular memory and hypnotized people also are more willing to guess (Erdelyi, 2010; Lynn, Rhue, & Kirsch, 2010; Wagstaff et al., 2007; Wickramasekera, 2008). Because memory is normally filled with fabrication and distortion (Chapter 7), hypnosis generally increases the potential for error. |

Bill Bridges/Getty Images, Inc.

Michael Newman/PhotoEdit

**CONCEPT CHECK** STOP

1. What part of the brain is activated in particular during meditation?

2. Who might benefit from hypnosis?

# Summary

THE PLANNER

## 1 Consciousness, Sleep, and Dreaming 120

- **Consciousness,** an organism's awareness of its own self and surroundings, exists along a continuum, from high awareness (**controlled processing**) to low awareness (sleep, dreaming, anesthesia, and coma). Sleep is a particular **alternate state of consciousness (ASC)**.
- **Controlled processing** requires focused attention and generally interferes with other ongoing activities. **Automatic processing** requires minimal attention and generally does not interfere with other ongoing activities.
- Many physiological functions follow 24-hour **circadian rhythms**. Disruptions in circadian rhythms, as well as long-term sleep deprivation, cause increased fatigue, cognitive and mood disruptions, and other health problems.
- The **electroencephalogram (EEG)** detects and records electrical changes in the nerve cells of the cerebral cortex. People progress through four distinct stages of **non-rapid-eye-movement (NREM) sleep**, with periods of **rapid-eye-movement (REM) sleep** occurring at the end of each sleep cycle, as shown in the diagram. Both REM and NREM sleep are important for our biological functioning.

Scientific study of sleep and dreaming • Figure 5.3

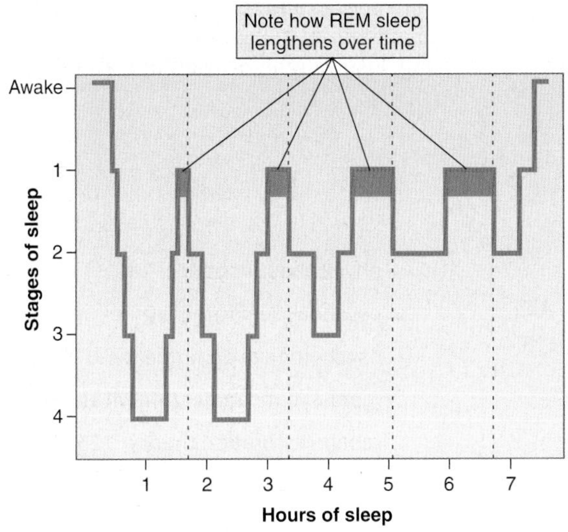


- The **adaptation/protection theory** proposes that sleep evolved to conserve energy and to provide protection from predators. The **repair/restoration theory** suggests that sleep helps us recuperate from the day's events. The **growth/development theory** argues that we use sleep for growth. The **learning/memory theory** says that sleep is used for consolidation of memories. Three major theories for why we dream are Freud's **wish-fulfillment view**, the **activation–synthesis hypothesis**, and the **cognitive view**.
- **Dyssomnias** are problems in the amount, timing, and quality of sleep; they include **insomnia**, **sleep apnea**, and **narcolepsy**. **Parasomnias** are abnormal disturbances occurring during sleep; they include **nightmares**, **night terrors**, **sleepwalking**, and **sleeptalking**. Natural sleep aids, based on behavior therapy, are helpful ways to treat sleep disorders and avoid problems associated with medication.

## 2 Psychoactive Drugs 130

- **Psychoactive drugs** influence the nervous system in a variety of ways. Alcohol affects neural membranes throughout the entire nervous system. Most psychoactive drugs act in a more specific way, by either enhancing a particular neurotransmitter's effect (an **agonist drug**, shown in the diagram) or inhibiting it (an **antagonist drug**). Drugs can interfere with neurotransmission at any of four stages: production or synthesis; storage and release; reception; or removal.

Cocaine: An agonist drug in action • Figure 5.10

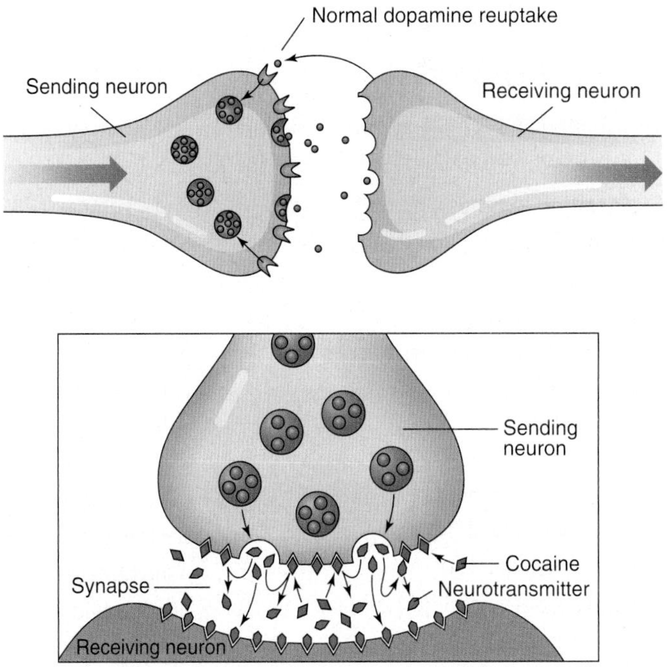

- The term **drug abuse** refers to drug-taking behavior that causes emotional or physical harm to oneself or others. **Addiction** refers to a condition in which a person feels compelled to use a specific drug. **Psychological dependence** refers to the mental desire or craving to achieve a drug's effects. **Physical dependence** refers to biological changes that make a drug necessary for minimum daily functioning, as shown in the diagram. Repeated use of a drug can produce decreased sensitivity, or **tolerance**. Sometimes, using one drug increases tolerance for another (**cross-tolerance**).

**How opiates create physical dependence • Figure 5.11**

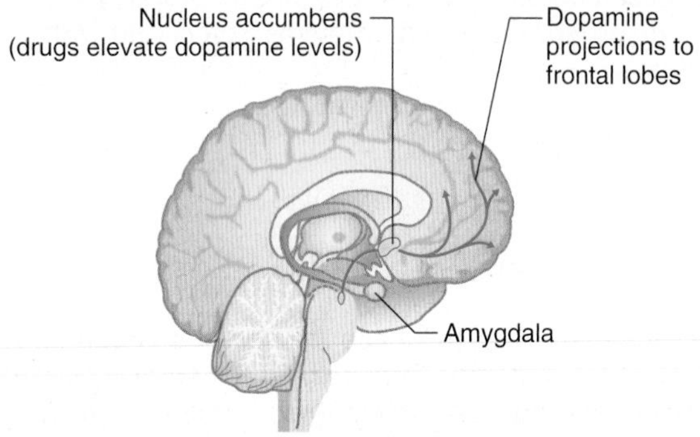


- Psychologists divide psychoactive drugs into four categories: **depressants** (such as alcohol, barbiturates, Rohypnol, and Ketamine), **stimulants** (such as caffeine, nicotine, cocaine, amphetamine, methamphetamine, and Ecstasy), **opiates** (such as morphine, heroin, and codeine), and **hallucinogens** (such as marijuana and LSD). Almost all psychoactive drugs may cause serious health problems and, in some cases, even death.

## 3 Meditation and Hypnosis 137

- The term **meditation** refers to techniques designed to refocus attention, block out distractions, and produce an alternate state of consciousness. Some followers believe that meditation offers a more enlightened form of consciousness, and researchers have verified that it can produce dramatic changes in basic physiological processes, as shown in the diagram.

**What a Psychologist Sees: Benefits of Meditation**

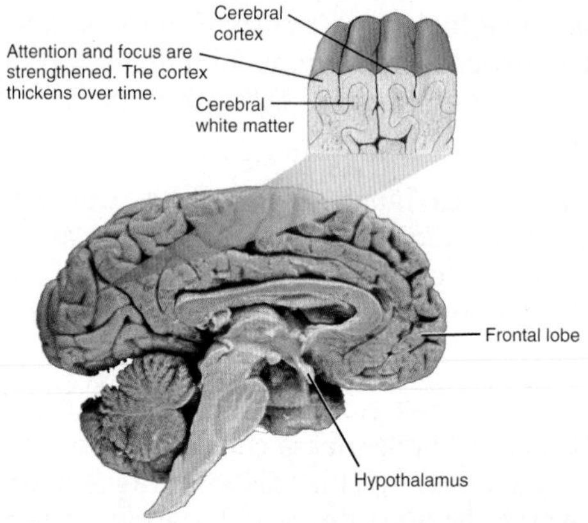


- Modern research has removed the mystery surrounding **hypnosis**, a trancelike state of heightened suggestibility, deep relaxation, and intense focus. It is used in surgery and medicine, and it is especially useful in medical areas in which patients have a high degree of anxiety, fear, and misinformation. In psychotherapy, hypnosis can help patients relax, remember painful memories, and reduce anxiety. It is also used to help with a number of behavioral issues, such as efforts to quit smoking, lose weight, or overcome phobias.

# Key Terms

**RETRIEVAL PRACTICE** Write a definition for each term before turning back to the referenced page to check your answer.

- activation–synthesis hypothesis 126
- adaptation/protection theory 124
- addiction 130
- agonist drug 130
- alternate state of consciousness (ASC) 120
- antagonist drug 130
- automatic processing 120
- circadian rhythms 121
- cognitive view of dreaming 126
- consciousness 120
- controlled processing 120
- cross-tolerance 132
- depressant 133
- drug abuse 130
- dyssomnia 128
- electroencephalograph (EEG) 123
- growth/development theory 125
- hallucinogen 135
- hypnosis 139
- insomnia 128
- learning/memory theory 125
- meditation 138
- narcolepsy 128
- night terrors 129
- nightmare 129
- non-rapid-eye-movement (NREM) sleep 124
- opiate 135
- parasomnia 129
- physical dependence 130
- psychoactive drug 130
- psychological dependence 130
- rapid-eye-movement (REM) sleep 124
- repair/restoration theory 125
- sleep apnea 129
- sleep deprivation 122
- sleeptalking 129
- sleepwalking 129
- stimulant 134
- tolerance 130
- wish-fulfillment view 126
- withdrawal 130

# Critical and Creative Thinking Questions

1. Do you believe that people have an unconscious mind? If so, how does it affect thoughts, feelings, and behavior?
2. Which of the three main theories of dreams do you most agree with, and why?
3. Do you think marijuana use should be legal? Why or why not?
4. Why might hypnosis help treat people who suffer from chronic pain?

# What is happening in these pictures?

These photos show a kitten during REM and NREM sleep.

Dorling Kindersley/Getty Images, Inc.

Neo Vision/Getty Images, Inc.

**Think Critically**

1. In which stage of sleep is the kitten in each photo, and how do you know?
2. Why might REM sleep serve an important adaptive function for cats?

# Self-Test

**RETRIEVAL PRACTICE** Completing this self-test and comparing your answers with those in Appendix C provide immediate feedback and helpful practice for exams. Additional interactive self-tests are available at www.wiley.com/college/carpenter.

1. *Consciousness* is defined in this text as _____.
   a. ordinary and extraordinary wakefulness
   b. an organism's awareness of its own self and surroundings
   c. mental representations of the world in the here and now
   d. any mental state that requires thinking and processing of sensory stimuli

2. The woman in this photo is most likely _____.
   a. sleepwalking and sleeptalking
   b. employing controlled processing
   c. at the highest end of the continuum of awareness
   d. enjoying automatic processing

3. *Circadian rhythms* are _____.
    a. patterns that repeat themselves on a twice-daily schedule
    b. physical and mental changes associated with the cycle of the moon
    c. rhythmical processes in your brain
    d. biological changes that occur on a 24-hour cycle

4. Identify the main areas of the brain involved in the operation of circadian rhythms.
    a. hypothalamus
    b. pineal gland
    c. suprachiasmatic nucleus

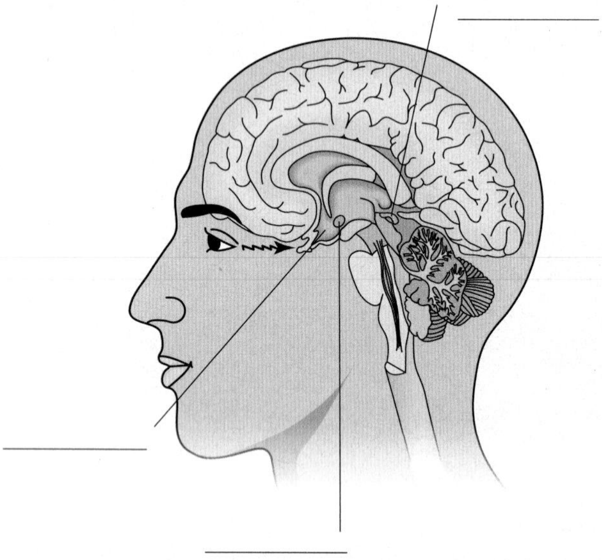

5. With what type of device were the data shown in this diagram recorded?
    a. EKG
    b. PET scan
    c. EEG
    d. EMG

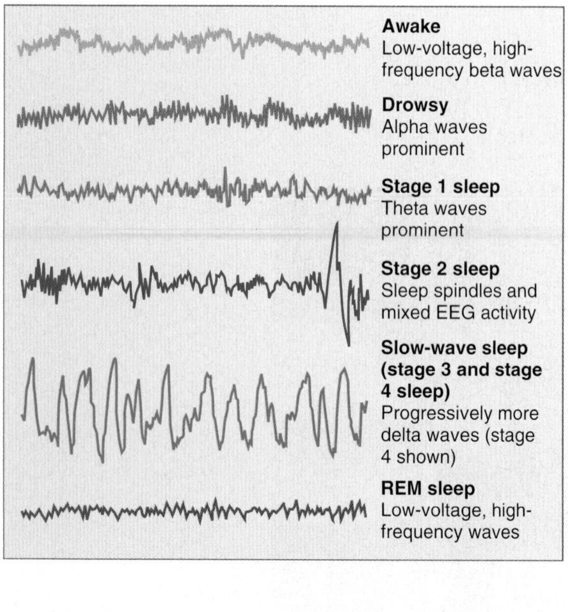


6. The sleep stage marked by irregular breathing, eye movements, high-frequency brain waves, and dreaming is called _____ sleep.
    a. beta
    b. hypnologic
    c. REM
    d. transitional

7. As shown in this graph, both human and nonhuman animals vary in their average number of sleep hours, which is best explained by the _____ theory of sleep.
    a. growth/development
    b. repair/restoration
    c. adaptation/protection
    d. learning/memory

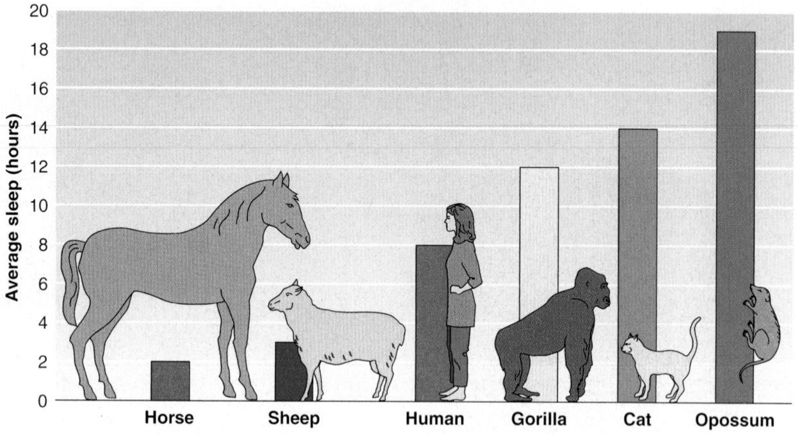


8. Which of the following people is clearly experiencing insomnia?
    a. Joan frequently cannot fall asleep the night before a final exam.
    b. Cliff regularly sleeps less than eight hours per night.
    c. Consuela persistently has difficulty falling or staying asleep.
    d. All of these persons are clearly experiencing insomnia.

**9.** The puppy in this photo was very hungry but is now sleeping next to his bowl because he most likely suffers from _____.

a. sleep apnea

b. a parasomnia

c. narcolepsy

d. insomnia

**10.** *Psychoactive drugs* _____.

a. change conscious awareness, mood, or perception

b. are addictive, mind altering, and dangerous to your health

c. are illegal unless prescribed by a medical doctor

d. all of these options

**11.** Match the four major categories of psychoactive drugs with the correct photo:

a. depressants, photo _____

b. stimulants, photo _____

c. opiates, photo _____

d. hallucinogens, photo _____

**12.** *Depressants* include all of the following EXCEPT _____.

a. downers such as sedatives, barbiturates, antianxiety drugs

b. alcohol

c. tobacco

d. Rohypnol

**13.** Alternate states of consciousness (ASCs) can be achieved in which of the following ways?

a. during sleep and dreaming

b. via chemical channels

c. through hypnosis and meditation

d. all of these options

**14.** _____ is a group of techniques designed to refocus attention and produce an alternate state of consciousness.

a. Parasomnia

b. Scientology

c. Parapsychology

d. Meditation

**15.** _____ is an alternate state of heightened suggestibility characterized by deep relaxation and intense focus.

a. Meditation

b. Amphetamine psychosis

c. Hypnosis

d. Daydreaming

**THE PLANNER** ✓

Review your Chapter Planner on the chapter opener and check off your completed work.

# 6 Learning

What image or *visualization* comes to mind when you hear the word "learning"? Because most non-psychologists generally picture students sitting in a classroom learning how to read or do math calculations, or a small child learning to ride a bike, we chose instead to start this chapter with a photo of excited gamblers. Why? Because we want you to see what a psychologist sees. And we want you to ask the questions a psychologist might ask, such as: Why do people spend long hours throwing dice, pushing buttons on slot machines, and standing in line to buy lottery tickets—especially since they generally lose more often than they win?

We also could have started with a photo of nonhuman animals, like service dogs for the blind, who

open and close doors, help their owners dress and undress, and clearly differentiate between staircases, escalators, and elevator. And then we could have asked you to imagine how these dogs perform such amazing acts.

When attempting to understand either human or nonhuman animal behaviors, the important thing to note is that people aren't born knowing how to throw dice, nor are guide dogs born knowing how to open and close doors. Psychologists would say these are *learned* behaviors. In this chapter, we begin with a focus on two of the most basic forms of learning—*classical* and *operant conditioning*. Then we look at *social-cognitive learning* and the *biological factors* involved in learning. Throughout the chapter we explore how learning theories and concepts impact our everyday lives.

## CHAPTER OUTLINE



## CHAPTER PLANNER ✓

- ❑ Study the picture and read the opening story.
- ❑ Scan the Learning Objectives in each section: p. 148 ❑ p. 155 ❑ p. 163 ❑ p. 167 ❑
- ❑ Read the text and study all figures and visuals. Answer any questions.

**Analyze key features**

- ❑ Process Diagrams p. 149 ❑ p. 153 ❑
- ❑ What a Psychologist Sees p. 150 ❑ p. 158 ❑
- ❑ Study Organizers p. 151 ❑ p. 159 ❑
- ❑ Psychology InSight p. 154 ❑ p. 162 ❑ p. 166 ❑
- ❑ Myth Busters, p. 155
- ❑ Applying Psychology, p. 161
- ❑ Psychological Science, p. 165
- ❑ Stop: Answer the Concept Checks before you go on: p. 154 ❑ p. 162 ❑ p. 166 ❑ p. 169 ❑

**End of chapter**

- ❑ Review the Summary and Key Terms.
- ❑ Answer the Critical and Creative Thinking Questions.
- ❑ Answer What is happening in this picture?
- ❑ Complete the Self-Test and check your answers.

# Classical Conditioning

## LEARNING OBJECTIVES

**RETRIEVAL PRACTICE** While reading the upcoming sections, respond to each Learning Objective in your own words. Then compare your responses with those in Appendix B.

1. **Describe** Pavlov and Watson's contributions to our understanding of learning.
2. **Describe** the three steps in classical conditioning.
3. **Summarize** the six principles of classical conditioning.

As you can see from our opening examples of gambling and service dogs, psychologists define **learning** in a broad way that emphasizes a relatively permanent change, as opposed to the short-term "learning" that often occurs when you're passively listening to lectures or studying a text. This relative permanence applies not only to useful behaviors (guiding the blind, reading this text) but also to bad habits, like gambling or texting while driving a car. The good news is: What is learned can be unlearned.

**learning** The relatively permanent change in behavior or mental processes caused by experience.

We begin this chapter with a study of one of the earliest forms of learning, *classical conditioning*, made famous by Pavlov's salivating dogs.

## Beginnings of Classical Conditioning

Why does your mouth water when you see a large slice of chocolate cake or a juicy steak? The answer to this question was accidentally discovered in the laboratory of Russian physiologist Ivan Pavlov (1849–1936). Pavlov's work focused on the role of saliva in digestion, and one of his experiments involved measuring salivary responses in dogs, using a tube attached to the dogs' salivary glands.

One of Pavlov's students noticed that many dogs began salivating at the sight of the food or the food dish, the smell of the food, or even the sight of the person who delivered the food long before receiving the actual food. This "unscheduled" salivation was intriguing. Pavlov recognized that an involuntary reflex (salivation) that occurred before the appropriate stimulus (food) was presented could not be inborn and biological. It had to have been acquired through experience—through *learning*.

Excited by their accidental discovery, Pavlov and his students conducted several experiments. Their most famous method involved sounding a tone on a tuning fork just before food was placed in the dogs' mouths. After several pairings of tone and food, the dogs would salivate on hearing the tone alone. Pavlov and others went on to show that many things can be conditioned stimuli for salivation if they are paired with food—a bell, a buzzer, a light, and even the sight of a circle or triangle drawn on a card.

**classical conditioning** Learning through involuntarily paired associations; it occurs when a neutral stimulus (NS) becomes paired (associated) with an unconditioned stimulus (US) to elicit a conditioned response (CR). (Also known as Pavlovian conditioning.)

The type of learning that Pavlov described came to be known as **classical conditioning**. To understand classical conditioning, you first need to realize that **conditioning** is basically just another word for learning. Next, you need to know that classical conditioning is a three-step process (*before, during*, and *after conditioning*) (**Figure 6.1**).

***Step 1*** Before conditioning, Pavlov's dogs did not salivate at extraneous stimuli, like the sight of the food, food dish, or the person delivering the food. The initial salivary reflex was *inborn* and biological, which consisted of an **unconditioned stimulus (US)** that normally elicits or produces an unlearned, biological response (in this case *food*). Because this response is also unlearned, it is called an **unconditioned response (UR)** (in this case *salivation*).

**unconditioned stimulus (US)** An unlearned stimulus (e.g., food) that naturally and automatically elicits an unconditioned response (UR) (e.g., salivation).

**unconditioned response (UR)** An unlearned, naturally occurring response (e.g., salivation) to an unconditioned stimulus (US) (e.g., food).

THE PLANNER

## Pavlov's classical conditioning • Figure 6.1

Although Pavlov's initial experiment used a metronome, a ticking instrument designed to mark exact time, his best known method (depicted here) involved presenting a neutral stimulus (NS) (a tone) just before an unconditioned stimulus (US) (food in the dog's mouth). After repeated pairings of the tone and food, the neutral stimulus (NS) became a contioned stimulus (CS), producing a conditioned response (CR) (the dog's salivation). Note also how Pavlov's classic experiment compares with the modern example of the man learning to salivate at the sight of the pizza box.

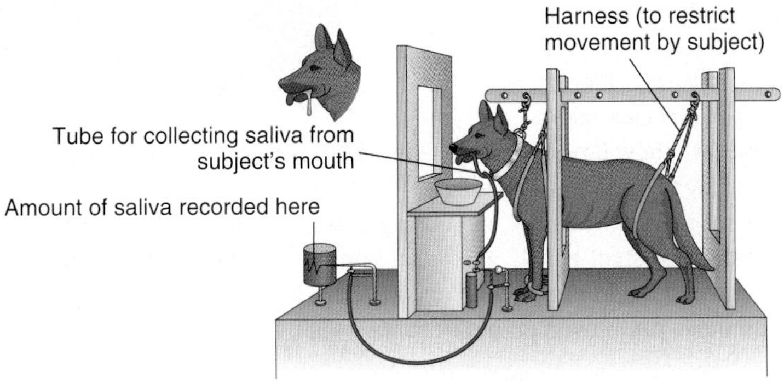


**1 Before conditioning**
The neutral stimulus (NS) produces no relevant response. The unconditioned (unlearned) *stimulus* (US) elicits the unconditioned *response* (UR).

**2 During conditioning**
The neutral stimulus (NS) is repeatedly paired with the unconditioned (unlearned) *stimulus* (US) to produce the unconditioned *response* (UR).

**3 After conditioning**
The neutral stimulus (NS) has become a conditioned (learned) stimulus (CS). This CS now produces a conditioned (learned) *response* (CR), which is usually similar to the previously unconditioned (unlearned) response (UR).

**Summary**
An originally neutral stimulus (NS) becomes a conditioned stimulus (CS), which elicits a conditioned response (CR).

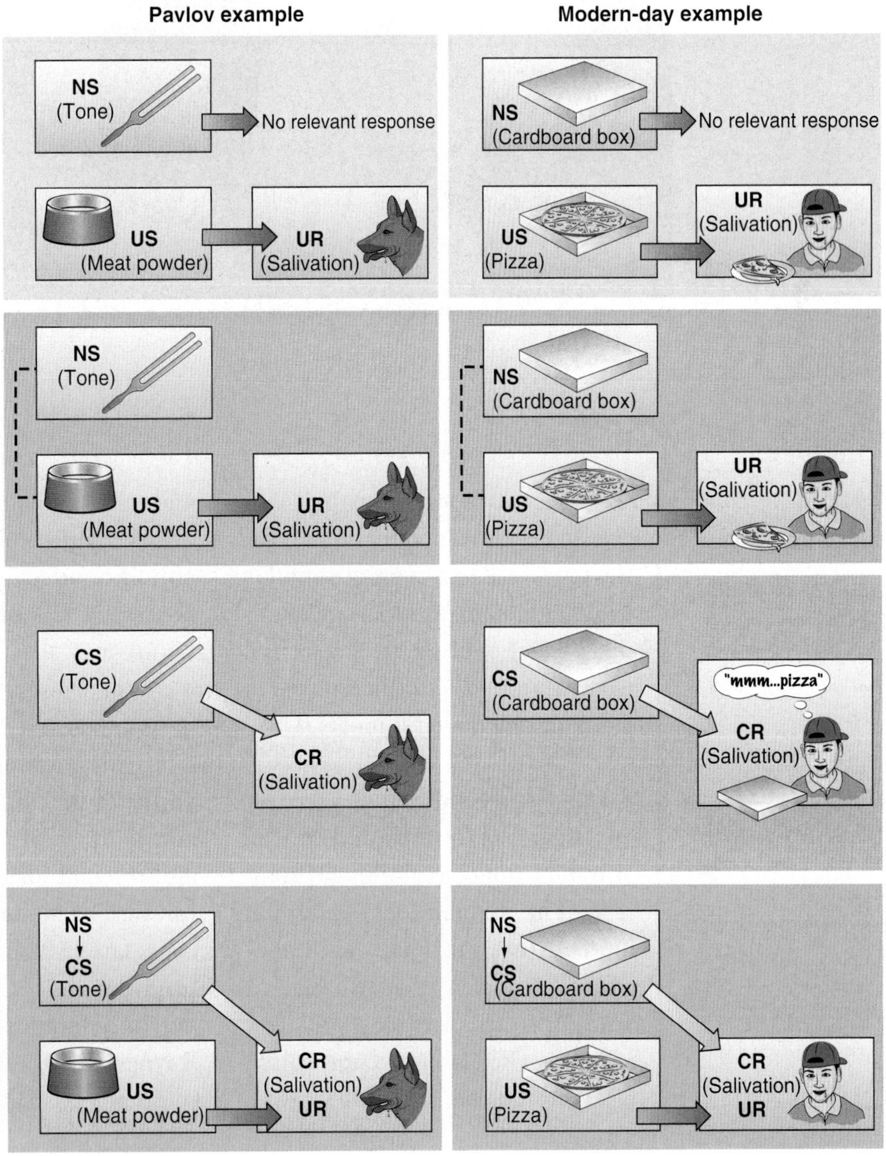

# WHAT A PSYCHOLOGIST SEES

THE PLANNER

## Conditioning Fear

Does simply glancing at this photo of cockroaches (**Figure a**) create feelings of intense, irrational fear? If so, your fear of cockroaches may stem from hearing one or both of your parents scream at the sight of a cockroach, and it would be called a *conditioned emotional response* (*CER*) in the context of classical conditioning.

Watson and Rayner's famous "Little Albert" study demonstrated how some fears can originate through conditioning. In this study, a healthy 11-month-old child, later known as "Little Albert," was first allowed to play with a white laboratory rat. Like most infants, Albert was curious and reached for the rat, showing no fear. Knowing that infants are naturally frightened by loud noises, Watson stood behind Albert and again put the rat near him, but when the infant reached for the rat, Watson banged a steel bar with a hammer. The loud noise frightened Albert and made him cry (**Figure b**). The rat was paired with the loud noise only seven times before Albert expressed fear of the rat even without the noise.

a.

Alamy

Using classical conditioning terms, we would say that the white rat (a neutral stimulus/NS) was paired with the loud noise (an unconditioned stimulus/US) to produce a conditioned emotional response (CER) in Albert, fear of the rat (**Figure c**).

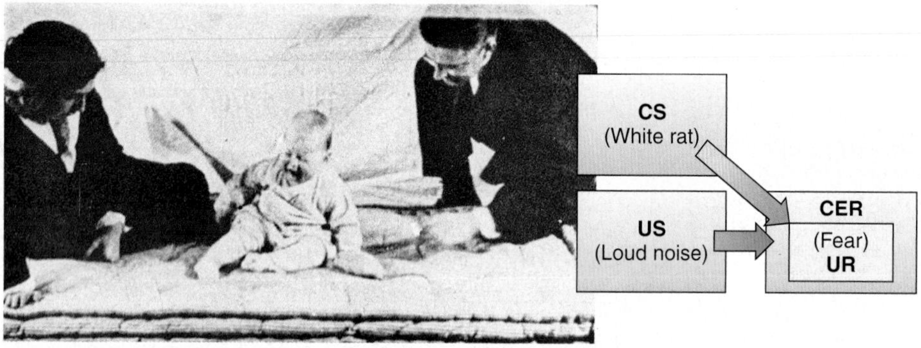


b.

c.

Benjamin Harris

***Step 2*** During conditioning, a tone was repeatedly paired with the presentation of the food. In classical conditioning terms, the tone is called a **neutral stimulus (NS),** a stimulus that does not normally evoke or bring out a response, whereas the food is again referred to as an unconditioned stimulus (US).

**neutral stimulus (NS)** An unlearned stimulus (e.g., tone) that does not elicit a response.

**conditioned stimulus (CS)** A learned (previously neutral) stimulus (e.g., tone) that elicits a conditioned response (CR) (e.g., salivation) as a result of repeated pairings with an unconditioned stimulus (US) (e.g., food).

***Step 3*** After conditioning (repeated pairings of the NS and US), the stimulus (tone) is no longer neutral. The human or nonhuman animal has formed an association between the two paired stimuli, so the previously neutral stimulus (NS) (tone) is now called a **conditioned stimulus (CS),** and the biological response it elicits (salivation) is called a **conditioned response (CR).** This process is visually summarized in Figure 6.1.

**conditioned response (CR)** A learned reaction to a previously neutral (but now conditioned) stimulus (CS) (e.g., tone).

What does a salivating dog have to do with your life? Classical conditioning is the most fundamental way that all animals, including humans, learn most new responses, emotions, and attitudes. The feelings of excitement and/or compulsion to gamble, your love for your parents (or significant other), and your drooling at the sight of chocolate cake or pizza are largely the result of classical conditioning.

In a famous experiment, John Watson and Rosalie Rayner (1920) demonstrated how the emotion of fear could be classically conditioned (*What a Psychologist Sees*).

Although this study remains a classic in psychology, the research procedures used by Watson and Rayner violated several ethical guidelines for scientific research (Chapter 1). They not only deliberately created a serious fear in a child, but they also ended their experiment without *extinguishing* (removing) it. In addition, the researchers have been criticized because they did not measure Albert's fear objectively. Their subjective evaluation raises doubt about the degree of fear conditioned (Paul & Blumenthal, 1989).

Despite such criticisms, John B. Watson made important and lasting contributions to psychology. He emphasized strictly *observable* behaviors and founded the new approach known as *behaviorism*, which explains behavior as a result of observable *stimuli* (in the environment) and observable *responses* (behavioral actions). In addition, Watson's study of Little Albert showed us that many of our likes, dislikes, prejudices, and fears are examples of a **conditioned emotional response (CER)**. In Chapter 14, you'll discover how Watson's research that *produced* Little Albert's fears later led to powerful clinical tools for *eliminating* extreme, irrational fears known as *phobias*, exaggerated and irrational fears of a specific object or situation (Cal et al., 2006; Field, 2006; Schachtman & Reilly, 2011; Schweckendiek et al., 2011).

## Understanding Classical Conditioning

In this section, we'll briefly discuss six important principles of classical conditioning: *acquisition*, *stimulus generalization*, *stimulus discrimination*, *extinction*, *spontaneous recovery*, and *higher-order conditioning* (**Study Organizer 6.1**). These processes complicate classical conditioning, but they help us understand its fundamental nature.

*"I don't care if she is a tape dispenser. I love her."*

Which of the six basic principles of classical conditioning best explain(s) this cartoon?

Answer: Stimulus generalization



**Study Organizer 6.1** Six principles of classical conditioning — THE PLANNER

| Process | Description | Example |
|---|---|---|
| Acquisition | Neutral stimulus (NS) and unconditioned stimulus (US) are paired; neutral stimulus (NS) becomes a conditioned stimulus (CS), eliciting a conditioned response (CR) | You learn to fear (CR) a dentist's office (CS) by associating it with a reflexive response to a painful tooth extraction (US).   |
| Stimulus generalization | Conditioned response (CR) is elicited not only by the conditioned stimulus (CS) but also by stimuli similar to the conditioned stimulus (CS) | You learn to fear most dentists' offices and other places that smell like them. |
| Stimulus discrimination | Certain stimuli similar to the conditioned stimulus (CS) do not elicit the conditioned response (CR) | You learn that your physician's office is not associated with the painful tooth extraction (US) |
| Extinction | Conditioned stimulus (CS) is presented alone, without the unconditioned stimulus (US); eventually the conditioned stimulus (CS) no longer elicits the conditioned response (CR) | You return to your dentist's office for routine checkups, with no extraction, and your fear (CR) gradually disappears. |
| Spontaneous recovery | Sudden reappearance of a previously extinguished conditioned response (CR) | While watching a movie depicting oral surgery, your previous fear (CR) suddenly and temporarily returns. |
| Higher-order conditioning | Neutral stimulus (NS) becomes a conditioned stimulus (CS) through repeated pairing with a previously conditioned stimulus (CS) | The sign outside your dentist's office, an originally neutral stimulus (NS), becomes a conditioned stimulus (CS), associated with the previous conditioned stimulus (CS) of the dentist's office, and you experience fear whenever you see just the sign. |

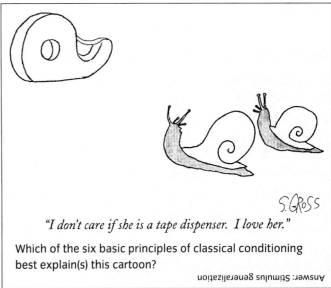

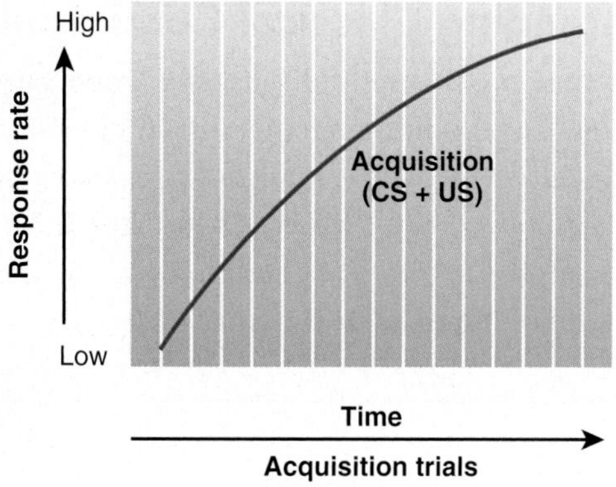


## The process of acquisition • Figure 6.2

During the acquisition phase, the subject shows increasing response to the stimulus with repeated trials or exposure.

After Pavlov's original (accidental) discovery of classical conditioning, he and his associates were interested in expanding their understanding of the basic experiment we've just described. They wanted to go beyond the basic **acquisition** phase, a term describing general classical conditioning when a neutral stimulus (NS) is consistently paired with an unconditioned stimulus (US) so that the NS comes to elicit a conditioned response (CR) (**Figure 6.2**).

In addition to the basic acquisition phase of classical conditioning, Pavlov and his associates also discovered that other stimuli similar to the neutral stimuli often produced a similar conditioned response (CR), a phenomenon known as **stimulus generalization**. The more the stimulus resembles the conditioned stimulus, the stronger the conditioned response (Hovland, 1937). For example, after first conditioning dogs to salivate at the sound of low-pitched tones, Pavlov later demonstrated that the dogs would also salivate in response to higher-pitched tones. Similarly, after conditioning, the infant in Watson and Rayner's experiment ("Little Albert") feared not only rats but also a rabbit, a dog, and a bearded Santa Claus mask.

Eventually, through the process of **stimulus discrimination** (a term that refers to a learned response to a specific stimulus, but not to other similar stimuli), Albert presumably learned to recognize differences between rats and other stimuli. As a result, he probably overcame his fear of Santa Claus, even if he remained afraid of white rats. Similarly, Pavlov's dogs learned to distinguish between the tone that signaled food and those that did not.

Classical conditioning, like all learning, is only *relatively* permanent. Most responses that are learned through classical conditioning can be weakened or suppressed through *extinction*. **Extinction** *is the gradual disappearance of a conditioned response (CR)*. It occurs when the unconditioned stimulus (US) is repeatedly withheld whenever the conditioned stimulus (CS) is presented, which gradually weakens the previous association. When Pavlov repeatedly sounded the tone without presenting food, the dogs' salivation gradually declined. Similarly, if you have a classically conditioned fear of cats and later start to work as a veterinary assistant, your fear will gradually diminish. Can you see the usefulness of this information if you're trying to get over a destructive love relationship? Rather than thinking, "I'll always be in love with this person," remind yourself that given time and repeated contact in a nonloving situation, your feelings will gradually lessen.

Keep in mind, extinction is not *unlearning* (Bouton, 1994). A behavior becomes *extinct* when the response rate decreases and the person or animal no longer responds to the stimulus. It does not mean the person or animal has "erased" the previously learned connection between the stimulus and the response. In fact, if the stimulus is reintroduced, the conditioning is much faster the second time. Furthermore, Pavlov found that if he allowed several hours to pass after the extinction procedure and then presented the tone again, the salivation would spontaneously reappear (**Figure 6.3**).

## Acquisition, extinction, and spontaneous recovery • Figure 6.3

During acquisition, the strength of the conditioned response (CR) rapidly increases and then levels off near its maximum. During extinction, the CR declines erratically until it is extinguished. After a "rest" period in which the organism is not exposed to the conditioned stimulus (CS), spontaneous recovery occurs, and the CS once again elicits a (weakened) CR.

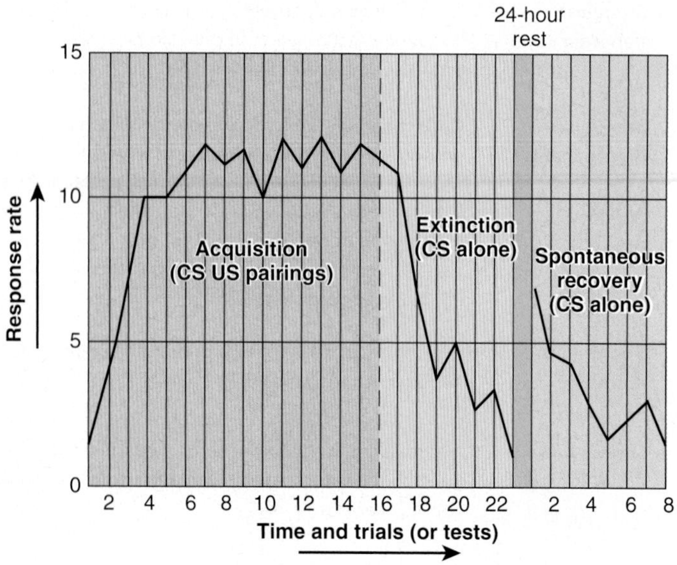

PROCESS DIAGRAM

## Higher-order conditioning • Figure 6.4

THE PLANNER

Children are not born salivating upon seeing the McDonald's golden arches. So why do they beg their parents to take them to "Mickey D's" after simply seeing an ad showing the golden arches? It is because of *higher-order conditioning*, which occurs when a neutral stimulus (NS) becomes a conditioned stimulus (CS) through repeated pairings with a previously conditioned stimulus (CS).

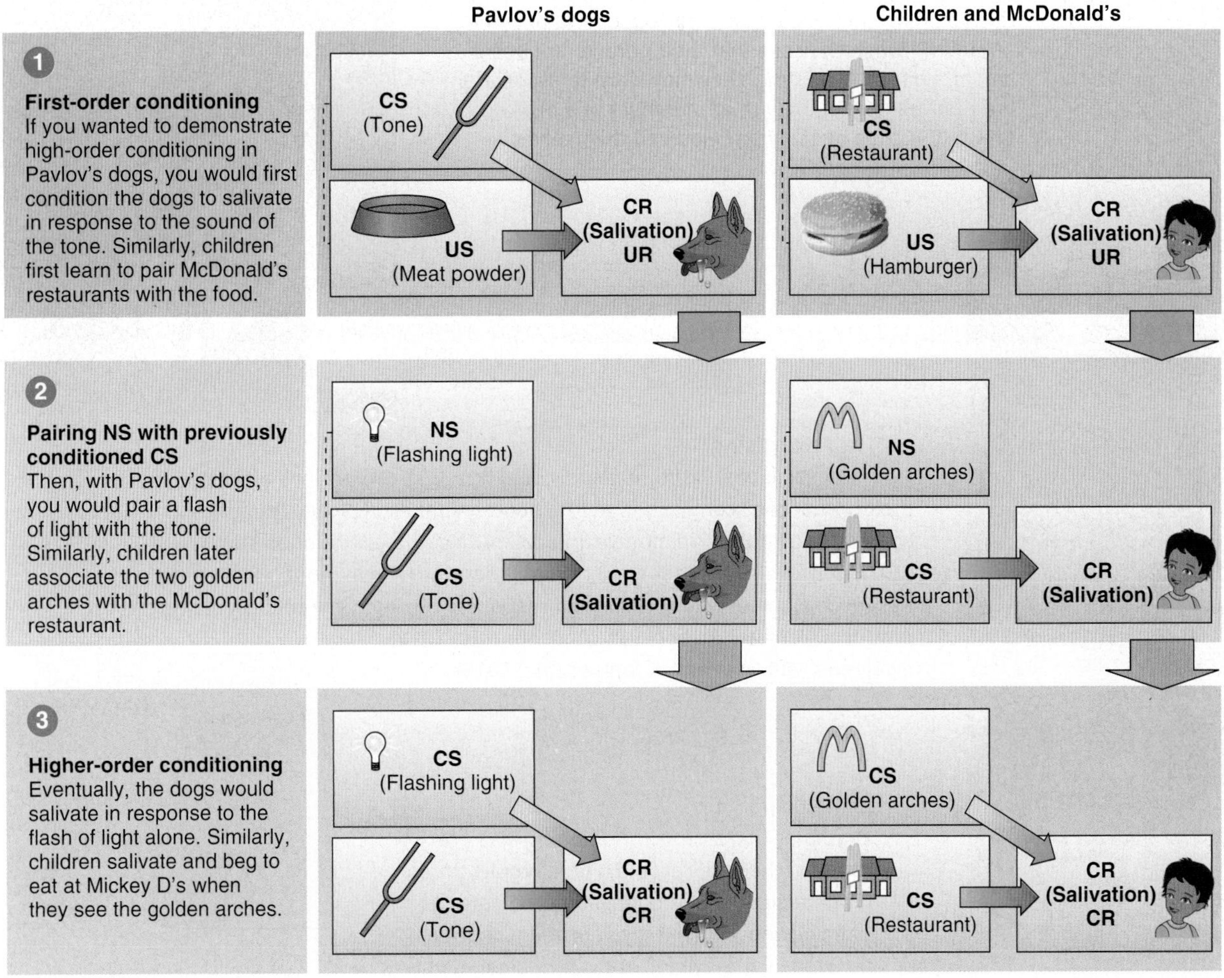

Pavlov's dogs
Children and McDonald's
1
First-order conditioning
If you wanted to demonstrate high-order conditioning in Pavlov's dogs, you would first condition the dogs to salivate in response to the sound of the tone. Similarly, children first learn to pair McDonald's restaurants with the food.
CS (Tone)
US (Meat powder)
CR (Salivation) UR
CS (Restaurant)
US (Hamburger)
CR (Salivation) UR
2
Pairing NS with previously conditioned CS
Then, with Pavlov's dogs, you would pair a flash of light with the tone. Similarly, children later associate the two golden arches with the McDonald's restaurant.
NS (Flashing light)
CS (Tone)
CR (Salivation)
NS (Golden arches)
CS (Restaurant)
CR (Salivation)
3
Higher-order conditioning
Eventually, the dogs would salivate in response to the flash of light alone. Similarly, children salivate and beg to eat at Mickey D's when they see the golden arches.
CS (Flashing light)
CS (Tone)
CR (Salivation) CR
CS (Golden arches)
CS (Restaurant)
CR (Salivation) CR

This **spontaneous recovery** helps explain why you might suddenly feel excited at the sight of a former girlfriend or boyfriend, even though years have passed (and extinction has occurred). It also explains why couples who've recently broken up sometimes misinterpret a sudden flare-up of feelings and return to unhappy relationships. Furthermore, if a conditioned stimulus is reintroduced after extinction, the conditioning occurs much faster the second time around—a phenomenon known as *reconditioning*.

Both spontaneous recovery and reconditioning help underscore why it can be so difficult for us to break bad habits (such as eating too much chocolate cake) or internalize new beliefs (such as egalitarian gender or racial beliefs). The phenomenon of **higher-order conditioning** (**Figure 6.4**) further expands and complicates our learned habits and associations.

For more examples of how classical conditioning impacts everyday life, see **Figure 6.5** on the next page.

# Psychology InSight

## Classical conditioning in everyday life • Figure 6.5

Just as children learn to beg to go to McDonald's through higher-order classical conditioning, they may also develop prejudice. Advertising and medicine also are influenced by classical conditioning.

Randy Olsen/NG Image Collection

**a. Prejudice**

How did the child holding the KKK sign in this photo develop prejudice at such an early age? As you can see in this diagram, children are naturally upset and become afraid (UR) when they see that their parents are upset and afraid (US). Over time, they may learn to associate their parents' reaction with all members of a disliked group (CS), thus becoming prejudiced themselves.

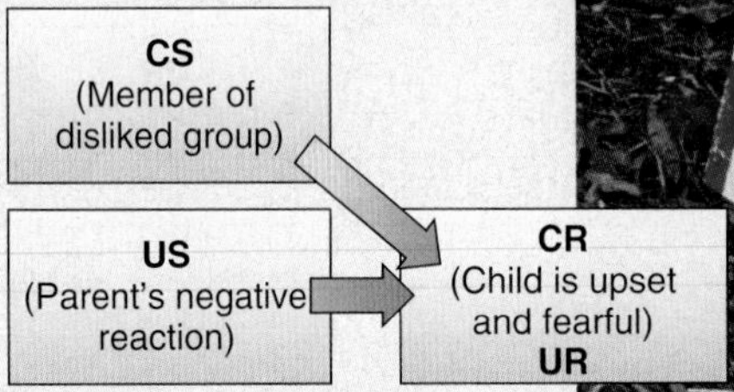

Bill Aron/PhotoEdit

**b. Advertising**

Magazine ads, TV commercials, and business promotions often pair their products or company logo (the neutral stimulus/NS) with pleasant images, such as attractive models and celebrities (the conditioned stimulus/CS). Through higher-order conditioning, these attractive models then trigger desired responses (the conditioned response/CR), such as purchasing their products (Fennis & Stroebe, 2010; Lefrançois, 2012; Sweldens, Van Osselaer, & Janiszewski, 2010).

**c. Medicine**

Examples of classical conditioning also are found in the medical field. For example, a treatment designed for alcohol-addicted patients pairs alcohol with a nausea-producing drug. Afterward, just the smell or taste of alcohol makes the person sick. Some, but not all, patients have found this treatment helpful.

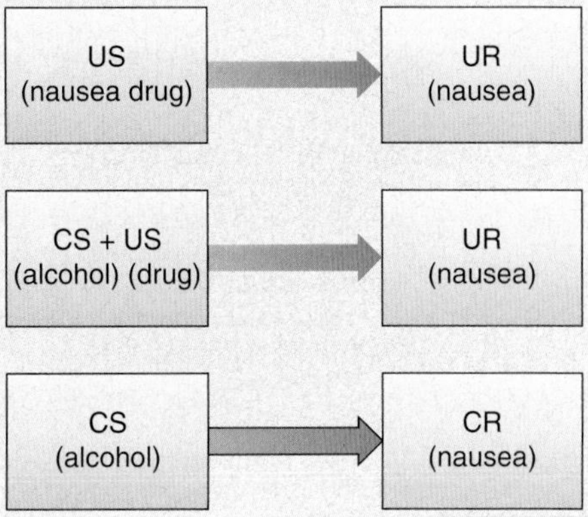


### CONCEPT CHECK

STOP

1. **Why** did Pavlov's dogs salivate before being presented with any food?
2. **What** circumstances might produce a phobia?
3. **How** do advertisers take advantage of higher-order conditioning?

# Operant Conditioning

## LEARNING OBJECTIVES

**RETRIEVAL PRACTICE** While reading the upcoming sections, respond to each Learning Objective in your own words. Then compare your responses with those in Appendix B.

1. **Describe** Thorndike's and Skinner's contributions to operant conditioning research.
2. **Explain** how positive and negative reinforcement influence behavior.
3. **Explain** how punishment influences behavior.

As we've just seen, classical conditioning is based on what happens *before* we *involuntarily* respond—something happens to us and we learn a new response. In contrast, **operant conditioning** is based on what happens *after* we *voluntarily* perform a behavior—we do something and learn from the consequences (see **Figure 6.6**). The learner "operates" on the environment and produces consequences that influence whether the behavior will be repeated. For example, if your friends smile and laugh when you tell a joke, you are likely to joke more. If they frown, groan, or ridicule you, you are likely to joke less. Try the *Myth Busters* to see what you know about learned behavior.

**operant conditioning** Learning through voluntary responses and its consequences (also known as instrumental or Skinnerian conditioning).

**MYTH BUSTERS**

**TRUTH OR CONSEQUENCES**

**TRUE OR FALSE?**

___ 1. The best way to maintain a desired behavior is to reward every response.

___ 2. Punishment is a very effective way to change long-term behavior.

___ 3. Negative reinforcement is another type of punishment.

___ 4. Prejudiced and superstitious people are born that way.

___ 5. New learning changes your actual brain structure.

Answers: All but one of these are false. Detailed answers are provided in this chapter and in Lilienfeld et al., (2010).

## Classical versus operant conditioning • Figure 6.6

Classical conditioning is based on involuntary behavior, whereas operant conditioning is based on voluntary behavior.

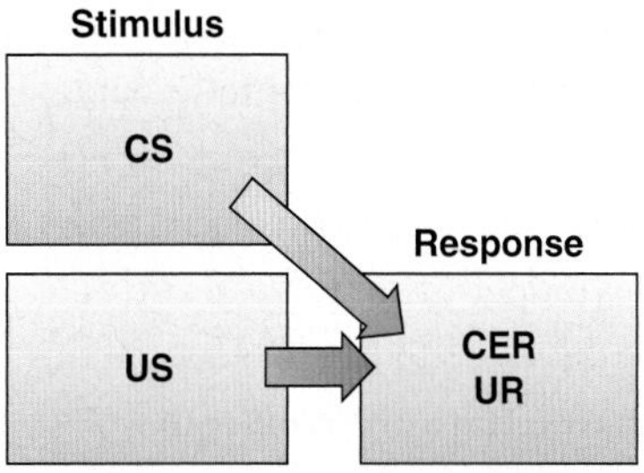


**a. Classical conditioning**
In classical conditioning, the subject is *passive*. Something happens to you, and you learn a new response due to pairing of CS with US.

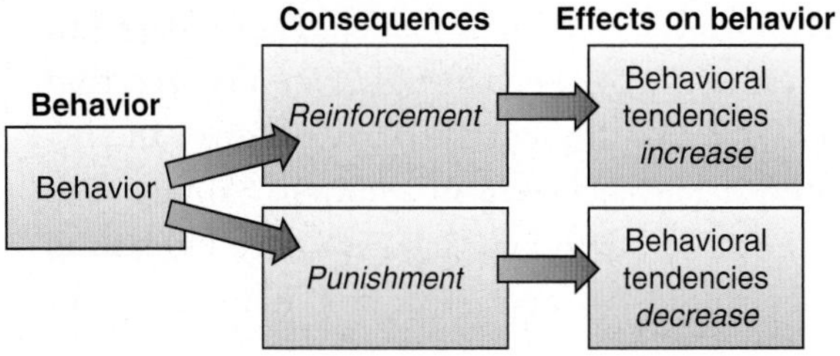


**b. Operant conditioning**
In operant conditioning, the subject is *active*. You do something—you "operate" on the environment; then your behavioral tendencies increase or decrease as a result of consequences—either reinforcement or punishment.

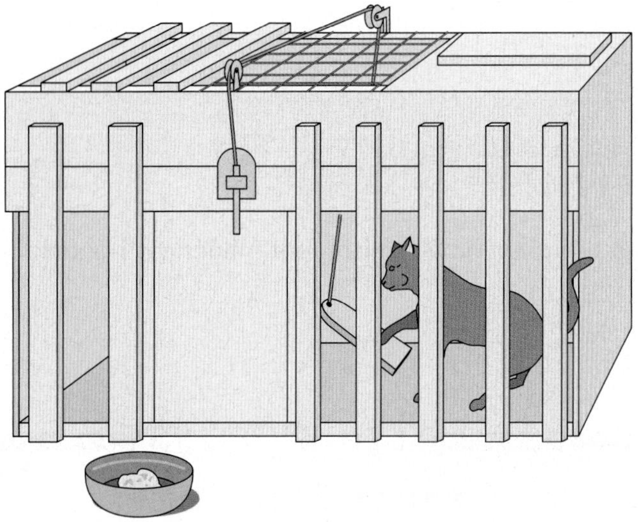

**Thorndike box • Figure 6.7**

In one famous experiment, Thorndike put a cat inside a specially built puzzle box. When the cat stepped on a pedal inside the box (at first, through trial-and-error), the door opened and the cat could get out to eat. With each additional success, the cat's actions became more purposeful, and it soon learned to open the door immediately (from Thorndike, 1898).

Note that *consequences* are the heart of operant conditioning. In classical conditioning, consequences are irrelevant—Pavlov's dogs still got to eat whether they salivated or not. But in operant conditioning, the organism voluntarily performs a behavior (an **operant**) that produces a consequence—either reinforcement or punishment. **Reinforcement** strengthens the response, making it more likely to recur. **Punishment** weakens the response, making it less likely to recur.

> **reinforcement** A consequence that strengthens a response and makes it more likely to recur.
>
> **punishment** A consequence that weakens a response and makes it less likely to recur.

## Beginnings of Operant Conditioning

Edward Thorndike (1874–1949), a pioneer of operant conditioning, determined that the frequency of a behavior is modified by its consequences. He developed the **law of effect** (Thorndike, 1911), a first step in understanding how consequences can modify active, voluntary behaviors (**Figure 6.7**).

> **law of effect** Thorndike's rule that the probability of an action being repeated is strengthened when followed by a pleasant or satisfying consequence.

B. F. Skinner (1904–1990) extended Thorndike's law of effect to more complex behaviors. He emphasized that reinforcement and punishment always occur after the behavior of interest has occurred. In addition, Skinner cautioned that the only way to know how we have influenced someone's behavior is to check whether it increases or decreases. Sometimes, he noted, we think we're reinforcing or punishing, when we're actually doing the opposite. For example, a professor may think she is encouraging shy students to talk by repeatedly praising them each time they speak up in class. But what if shy students are embarrassed by this attention? If so, they may decrease the number of times they talk in class. It's important to note that what is reinforcing or punishing for one person may not be so for another.

## Reinforcement

Reinforcers, which strengthen a response, can be a powerful tool in all parts of our lives. Psychologists group them into two types, primary and secondary. **Primary reinforcers** satisfy an intrinsic, unlearned biological need (food, water, sex). **Secondary reinforcers** are not intrinsic; the value of this reinforcer is learned (money, praise, attention). Each type of reinforcer can produce **positive reinforcement** or **negative reinforcement** (**Table 6.1**), depending on whether certain stimuli are added or taken away.

It's easy to confuse negative reinforcement with punishment, but the two concepts are actually completely opposite. Reinforcement (either positive or negative) strengthens a behavior, whereas punishment weakens a behavior. If the terminology seems confusing, it may help to think of positive and negative reinforcement in the mathematical sense (that is, in terms of something being added [+] or taken away [–]) rather than in terms of good and bad.

When you make yourself do some necessary but tedious task—say, paying bills—before letting yourself go

**How reinforcement strengthens and increases behavior Table 6.1**

| | **Positive reinforcement**<br>**Adds to (+) and strengthens behavior** | **Negative reinforcement**<br>**Takes away (–) and strengthens behavior** |
|---|---|---|
| **Primary reinforcers**<br>Satisfy *biological* needs | You hug your baby and he smiles at you. The "addition" of his smile strengthens the likelihood that you will hug him again.<br>You do a favor for a friend and she buys you lunch in return, which strengthens your tendency to do more favors for your friend in the future. | Your baby is crying, so you hug him and he stops crying. The "removal" of crying strengthens the likelihood that you will hug him again when he cries.<br>You take an aspirin for your headache, which takes away the pain and encourages you to take more aspirin again in the future. |
| **Secondary reinforcers**<br>Satisfy *learned* needs | You increase profits and receive $200 as a bonus, which encourages you to do more things to increase profits.<br>You study hard and receive a good grade on your psychology exam, which strengthens your tendency to study hard in the future. | After high sales, your boss says you won't have to work on weekends, which encourages you to do more things to increase high sales.<br>Professor says you won't have to take the final exam because you did so well on your unit exams, which encourages you to do well on future unit exams. |

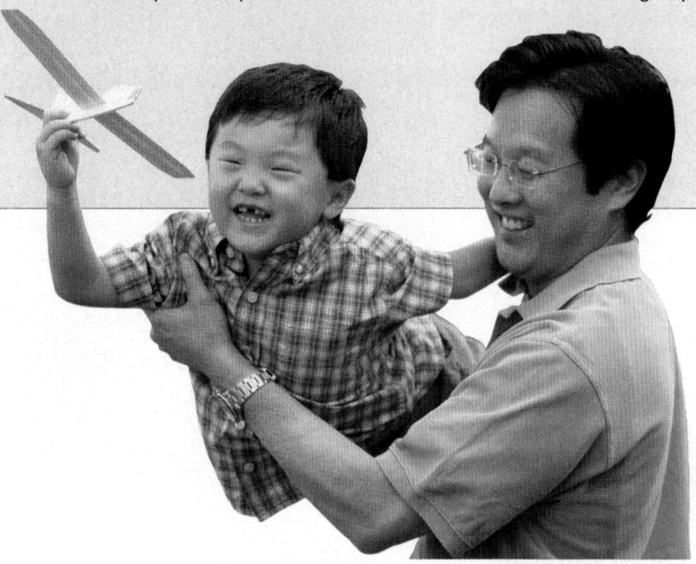

Blend Images/Superstock

to the movies, you are not only using positive reinforcement, you're also using the **Premack principle**. This principle is named after psychologist David Premack, who believes any naturally occurring, high-frequency response can be used to reinforce and increase low-frequency responses (Lefrançois, 2012; Poling, 2010). Recognizing that you love to go to movies, you intuitively tie your less desirable low-frequency activity (paying the bills) to your high-frequency or highly desirable behavior (going to the movies).

What are the best circumstances for using reinforcement? It depends on the desired outcome. To make this decision, you need to understand various **schedules of reinforcement** (Terry, 2003). This term refers to the rate or interval at which responses are reinforced. Although there are numerous schedules of reinforcement, the most important distinction is whether they are continuous or partial. When Skinner was training his animals, he found that learning was most rapid if the response was reinforced every time it occurred—a procedure called **continuous reinforcement**.

As you have probably noticed, real life seldom provides continuous reinforcement. Yet your behavior persists because your efforts are occasionally rewarded. Most everyday behavior is rewarded on a **partial (or intermittent) schedule of reinforcement**, which involves reinforcing only

# WHAT A PSYCHOLOGIST SEES THE PLANNER

## Partial Reinforcement Keeps 'em Coming Back

© poba/iStockphoto

Recall back at the start of the chapter when we asked you think about why so many people spend so much time gambling. Now can you see how the compulsion to keep gambling, in spite of significant losses, is evidence of the strong resistance to extinction, which occurs with partial (or intermittent) schedules of reinforcement?

Machines in Nevada casinos, by law, have payout rates of at least 75%—that is, for every dollar spent, a player "wins" $0.75. Different machines are programmed different ways. Some meet the percentage by giving very large, but infrequent, payouts to a lucky winner. Others give frequent smaller payouts to all players. In either case, people are reinforced just often enough to keep them coming back, always hoping the "partial" reinforcement will lead to more. The same compulsion leads people to go on buying lottery tickets even though the odds of winning are very low.

A more positive example of partial reinforcement occurs when parents use it to maintain their children's desirable behaviors, such as toothbrushing and bed making. After the child has initially learned these behaviors with continuous reinforcement, occasional, partial reinforcement is the most efficient way to maintain the behavior. (Chart adapted from Skinner, 1961.)

age fotostock/Superstock

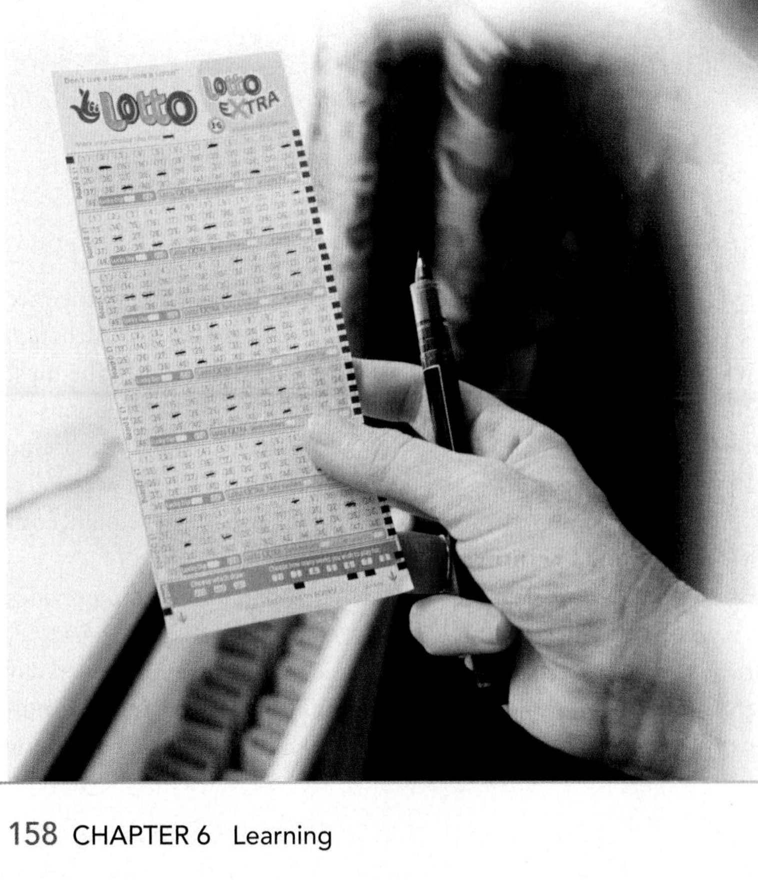


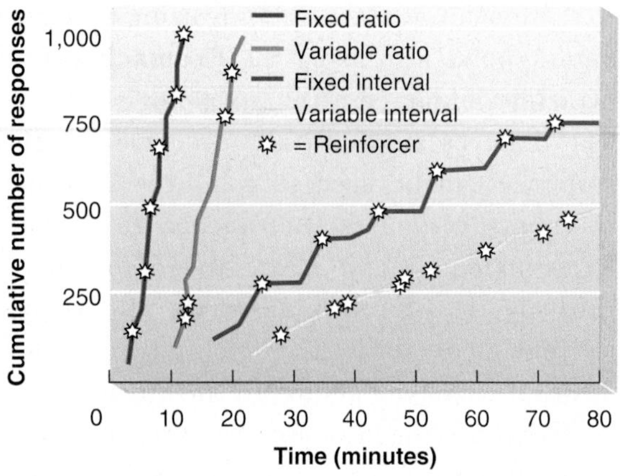

## Study Organizer 6.2 Four schedules of partial reinforcement

THE PLANNER

| | | Definitions | Response rates | Example |
|---|---|---|---|---|
| Ratio schedules (response based) | Fixed ratio (FR) | Reinforcement occurs after a predetermined number of responses | High rate of response, but a brief drop-off just after reinforcement | Car wash employee receives $10 for every 3 cars washed. In a laboratory, a rat receives a food pellet every time it presses the bar 7 times.  |
| | Variable ratio (VR) | Reinforcement occurs after a varying number of responses | High response rate, no pause after reinforcement, and very resistant to extinction | Slot machines are designed to pay out after an average number of responses (maybe every 10 times), but any one machine may pay out on the first respones, then seventh, then the twentieth.  |
| Interval schedules (time based) | Fixed interval (FI) | Reinforcement occurs after a fixed period of time | Responses tend to increase as the time for the next reinforcer is near, but drop off after reinforcement and during interval | You receive a monthly paycheck. Rat's behavior is reinforced with a food pellet when (or if) it presses a bar after 20 seconds have elapsed.  |
| | Variable interval (VI) | Reinforcement occurs after varying periods of time | Relatively low response rates, but they are steady because respondents cannot predict when reward will come | Your professor gives pop quizzes at random times through the course. Rat's behavior is reinforced with a food pellet after a response and a variable unpredictable interval of time.  |

some responses, not all (Miltenberger, 2011). (See *What a Psychologist Sees*.)

Once a task is well learned, it is important to move to a partial schedule of reinforcement. Why? Because under partial schedules, behavior is more resistant to extinction. There are four schedules of partial reinforcement: **fixed ratio (FR)**, **variable ratio (VR)**, **fixed interval (FI)**, and **variable interval (VI)** (see **Study Organizer 6.2**).

Each of the four schedules of partial reinforcement is important for *maintaining* behavior. But how would you teach someone to play the piano or to speak a foreign language? For new and complex behaviors such as these, which aren't likely to occur naturally, **shaping** is an especially valuable tool. Skinner believed that shaping explains a variety of abilities that each of us possesses, from eating with a fork to playing a musical instrument, to driving a stick-shift car (Chance, 2009; Lefrançois, 2012). Parents, athletic coaches, teachers, and animal trainers all use shaping techniques (**Figure 6.8**).

**shaping** Reinforcement by a series of successively improved steps leading to desired response.

### Shaping • Figure 6.8

Momoko, a female monkey, is famous in Japan for her water-skiing, deep-sea diving, and other amazing abilities. Can you describe how her animal trainers used the successive steps of shaping to teach her these skills? First, they reinforced Momoko (with a small food treat) for standing or sitting on the water ski. Then they reinforced her each time she accomplished a successive step in learning to water-ski.

Kurita Kaku/Getty Images

**How punishment weakens and decreases behavior Table 6.2**

| **Positive punishment**<br>**Adds stimulus (+)**<br>**and weakens the behavior** | **Negative punishment**<br>**Takes stimulus away (–) and**<br>**weakens the behavior** |
|---|---|
| You must run four extra laps in your gym class because you were late, which discourages you from being late in the future. | You're excluded from gym class because you were late, which discourages you from being late in the future. |
| A parent adds chores following a child's poor report card, which discourages the child from getting poor grades in the future. | A parent takes away a teen's cell phone following a poor report card. The teen stops getting a poor report card. |
| Your boss complains about your performance, and your poor performance decreases. | Your boss reduces your expense account after a poor performance, and your poor performance decreases. |

## Punishment

Unlike reinforcement, punishment *decreases* the strength of a response. As with reinforcement, there are two kinds of punishment—positive and negative (Miltenberger, 2011; Skinner, 1953).

**Positive punishment** is the addition (+) of a stimulus that decreases (or weakens) the likelihood of the response occurring again. **Negative punishment** is the taking away (–) of a reinforcing stimulus, which decreases (or weakens) the likelihood of the response occurring again (**Table 6.2**). (To check your understanding of the principles of reinforcement and punishment, see **Figure 6.9**.)

Any process that adds or takes away something and causes a behavior to decrease is punishment. Thus, if parents ignore all the A's on their child's report card ("taking away" encouraging comments) and ask repeated questions about the B's and C's, they may unintentionally be punishing the child's excellent grade achievement and weakening the likelihood of future A's.

### The Skinner box application • Figure 6.9

To test his behavioral theories, Skinner used an animal, usually a pigeon or a rat, and an apparatus that has come to be called a *Skinner Box*.

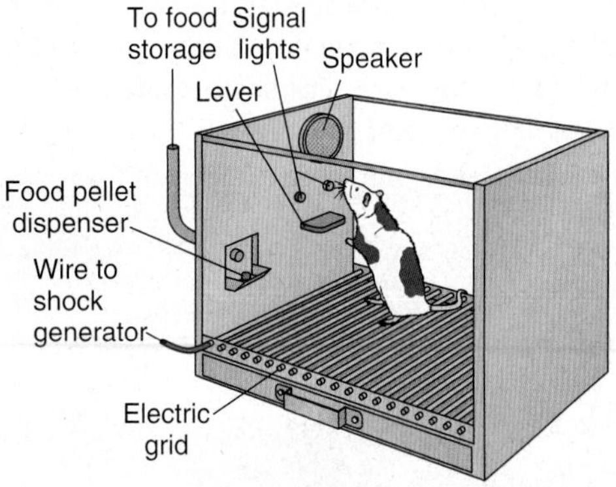


**a.** In Skinner's basic experimental design, an animal such as a rat received a food pellet each time it pushed a lever, and the number of responses was recorded. An electric grid on the cage floor could be used to deliver small electric shocks. The food pellets and shocks could be used to give the animal reinforcement or punishment.

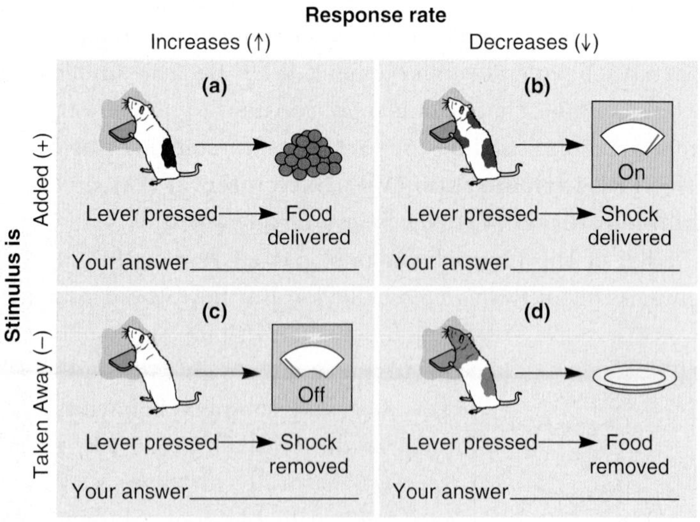


**b.** Are these examples of positive reinforcement, negative reinforcement, positive punishment, or negative punishment? Fill in the name of appropriate learning principle in the spaces provided.

Answers: (a) positive reinforcement, (b) positive punishment, (c) negative reinforcement, (d) negative punishment

To be effective, punishment should be clear, direct, immediate, and consistent. However, in the real world, this is extremely hard to do. Police officers cannot stop every driver every time they speed. To make matters worse, when punishment is not immediate, during the delay the behavior is likely to be reinforced on a partial schedule, which makes it highly resistant to extinction. For example, although gamblers lose far more than they win, the fact that they occasionally win keeps them "hanging in there."

Even if punishment immediately follows the misbehavior, the recipient may learn what not to do but not necessarily what he or she should do. It's much more efficient to teach someone by giving him or her clear examples of correct behavior, rather than simply punishing the incorrect behavior.

In sum, reinforcement and punishment are effective ways of altering behavior in laboratory settings, but evidence from real life suggests these two approaches are not equally desirable. Reinforcement is generally more ethical than punishment, and far easier to administer. Punishment is too often misused; it has at least seven serious drawbacks, shown in *Applying Psychology*.

For more examples of how operant conditioning applies in everyday life, see **Figure 6.10**, on the next page.

## Applying Psychology

THE PLANNER

# Problems with Punishment

1. **Passive aggressiveness** For the recipient, punishment often leads to frustration, anger, and eventually aggression. But most of us have learned from experience that retaliatory aggression toward a punisher (especially one who is bigger and more powerful) is usually followed by more punishment. We therefore tend to control our impulse toward open aggression and instead resort to more subtle techniques, such as showing up late or forgetting to mail a letter for someone. This is known as *passive aggressiveness* (Girardi et al., 2007; Johnson, 2008).
2. **Avoidance behavior** No one likes to be punished, so we naturally try to avoid the punisher. If every time you come home a parent or spouse starts yelling at you, you will delay coming home or find another place to go.
3. **Inappropriate modeling** Have you ever seen a parent spank or hit his or her child for hitting another child? Ironically, the punishing parent may unintentionally serve as a "model" for the same behavior he or she is attempting to stop.
4. **Temporary suppression versus elimination** Punishment generally suppresses the behavior only temporarily, while the punisher is nearby.
5. **Learned helplessness** Why do some people stay in abusive relationships? Research shows that if you repeatedly fail in your attempts to control your environment, you acquire a general sense of powerlessness or *learned helplessness* and you may make no further attempts to escape (Bargai, Ben-Shakhar, & Salev, 2007; Diaz-Berciano et al., 2008; Kim, 2008; Shea, 2008).
6. **Rewarded aggression** Because punishment often produces a decrease in undesired behavior, at least for the moment, the punisher is in effect rewarded for applying punishment.
7. **Perpetuated aggression** A vicious circle may be established in which both the punisher and recipient are reinforced for inappropriate behavior—the punisher for punishing and the recipient for being fearful and submissive. This side effect partially explains the escalation of violence in family abuse and bullying (Anderson, Buckley, & Carnagey, 2008; Fang & Corso, 2007; Huesman, Dubow, & Boxer, 2011). In addition to fear and submissiveness, the recipient also might become depressed and/or respond with his or her own form of aggression.

Photo Researchers, Inc.

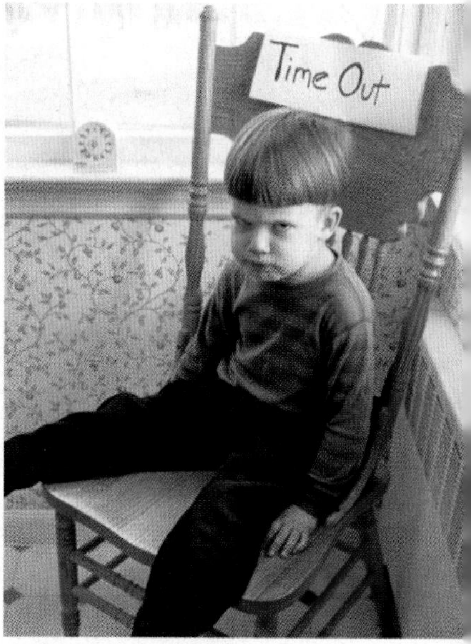


**Think Critically**

1. Why do you think drivers quickly slow down when they see a police car, and then quickly resume their previous speed once the police officer is out of sight?
2. Given all the problems associated with punishment, why is it so often used?

# Psychology InSight

## Operant conditioning in everyday life

• **Figure 6.10**

Reinforcement and punishment shape behavior in many aspects of our lives.

© Image Source/Alamy Limited

**a. Prejudice and discrimination**
Although prejudice and discrimination show up early in life, children are not born believing others are inferior. What reinforcement might these boys receive for teasing this girl? Can you see how operant conditioning, beginning at an early age, can promote prejudice and discrimination?

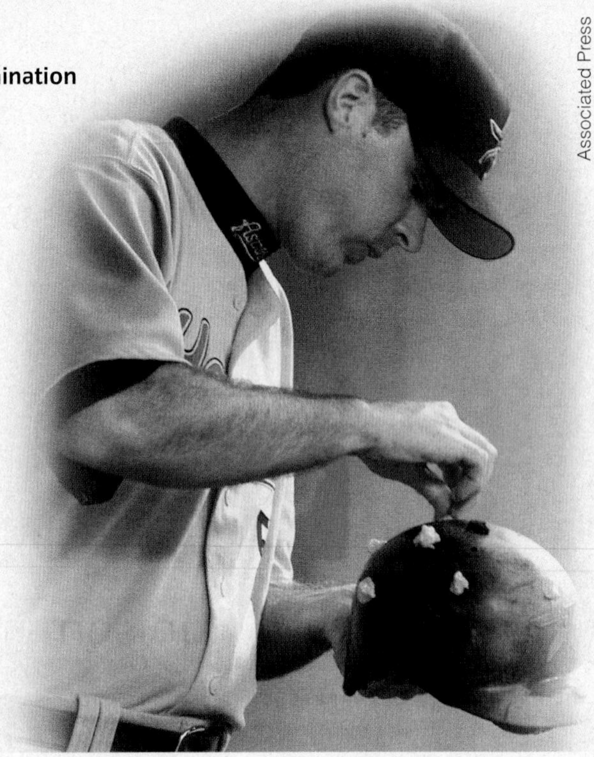

Associated Press

**b. Superstition**
Like prejudice and discrimination, we are not born with superstitious beliefs, they are learned—partly through operant conditioning. Like this baseball player placing gum on his helmet after hitting a home run, many superstitions may have developed from accidental reinforcements—from wearing "something old" during one's wedding to knocking on wood for good fortune.

Robert Goldstein/Science Photo Library/Photo Researchers, Inc.

**c. Biofeedback**
**Biofeedback** relies on operant conditioning to treat ailments such as anxiety and chronic pain. Participants connected to electrodes watch a monitor screen, a series of flashing lights, or listen to different beeps, which all can be used to display changes in their internal bodily functions. For example, in the treatment of chronic pain, participants might work on reducing their muscle tension, and "feedback" from the machines would help them gauge their progress as they try various relaxation strategies.

### CONCEPT CHECK — STOP

1. **What** is the difference between classical conditioning and operant conditioning?
2. **What** is the difference between negative reinforcement and punishment?
3. **What** are some problems with punishment?

# Cognitive-Social Learning

## LEARNING OBJECTIVES

**RETRIEVAL PRACTICE** While reading the upcoming sections, respond to each Learning Objective in your own words. Then compare your responses with those in Appendix B.

1. **Summarize** the cognitive-social theory of learning.
2. **Describe** insight learning and latent learning.
3. **Explain** the principles of observational learning.

So far, we have examined learning processes that involve associations between a stimulus and an observable behavior. Although some behaviorists believe that almost all learning can be explained in such stimulus–response terms, other psychologists feel that there is more to learning than can be explained solely by operant and classical conditioning. **Cognitive-social learning theory** (also called cognitive-social learning or cognitive-behavioral theory) incorporates the general concepts of conditioning, but rather than relying on a simple S–R (stimulus and response) model, this theory emphasizes the interpretation or thinking that occurs within the organism: S–O–R (stimulus–organism–response).

**cognitive-social learning theory** A perspective that emphasizes the roles of thinking and social learning in behavior.

According to this view, animals (including humans) have attitudes, beliefs, expectations, motivations, and emotions that affect learning. Furthermore, both human and nonhuman animals are social creatures that are capable of learning new behaviors through observation and imitation of others. We begin with a look at the *cognitive* part of cognitive-social theory, followed by an examination of the *social* aspects of learning.

## Insight and Latent Learning

Early behaviorists likened the mind to a "black box" whose workings could not be observed directly, but German psychologist Wolfgang Köhler wanted to look inside the box. He believed that there was more to learning—especially learning to solve a complex problem—than responding to stimuli in a trial-and-error fashion. In a series of experiments, he placed a banana outside the reach of a chimpanzee and watched to see how the chimp solved the problem (**Figure 6.11**). One caged chimp had to use a stick to extend its reach. Köhler noticed that the chimp did not solve this problem in a random trial-and-error fashion but, instead, seemed to sit and think about the situation for a while. Then, in a flash of **insight**, the chimp picked up the stick and maneuvered the banana to within its grasp (Köhler, 1925).

**insight** The sudden understanding of a problem that implies the solution.

Another of Köhler's chimps, an intelligent fellow named Sultan, was put in a similar situation. This time there were two sticks available to him, and the banana was placed even farther away, too far to reach with a single stick. Sultan seemingly lost interest in the banana, but he continued to play with the sticks. When he later discovered that the two sticks could be interlocked, he instantly used the now longer stick to pull the banana to within his grasp. Köhler designated this type of learning

**Cognitive-social learning • Figure 6.11**

One of Köhler's chimps demonstrated insight learning when he stacked boxes to solve the problem of how to get the banana. (Note how the chimp in the background is engaged in observational learning, our next topic.)

**insight learning** because some internal mental event that he could only describe as "insight" went on between the presentation of the banana and the use of the stick to retrieve it.

Like Köhler, Edward C. Tolman (1898–1956) believed that previous researchers underestimated animals' cognitive processes and cognitive learning. He noted that, when allowed to roam aimlessly in an experimental maze with no food reward at the end, rats seemed to develop a **cognitive map**, or mental representation of the maze.

To test the idea of cognitive learning, Tolman allowed one group of rats to aimlessly explore a maze with no reinforcement. A second group was reinforced with food whenever they reached the end of the maze. The third group was not rewarded during the first 10 days of the trial, but starting on day 11 they found food at the end of the maze.

As expected from simple operant conditioning, the first and third groups were slow to learn the maze, whereas the second group, which had reinforcement, showed fast, steady improvement. However, when the third group started receiving reinforcement (on the 11th day), their learning quickly caught up to the group that had been reinforced every time (Tolman & Honzik, 1930). This showed that the nonreinforced rats had been thinking and building cognitive maps of the area during their aimless wandering and that their **latent learning** only showed up when there was a reason to display it (the food reward).

**latent learning** Hidden learning that exists without behavioral signs.

Cognitive maps and latent learning are not limited to rats. For example, a chipmunk will pay little attention to a new log in its territory (after initially checking it for food). When a predator comes along, however, the chipmunk heads directly for and hides beneath the log. Recent experiments provide additional clear evidence of latent learning and the existence of internal, cognitive maps in both human and nonhuman animals (Gómez-Laplaza & Gerlai, 2010; Lahav & Mioduser, 2008; Lew, 2011) (**Figure 6.12**).

## Observational Learning

In addition to classical and operant conditioning and cognitive processes (such as insight and latent learning), we also learn many things through **observational learning**. From birth to death, observational learning is very important to our

**observational learning** Learning a new behavior or information by watching others (also known as social learning or modeling).

**Is this learning? • Figure 6.12**

This child often rides through her neighborhood for fun, without a specific destination. Could she tell her parents where the nearest mailbox is located? What type of cognitive maps have you created?

# Psychological Science

THE PLANNER

## Bandura's Bobo Doll

Bandura's "Bobo doll" study is considered a classic in psychology. Wanting to know whether children learn to be aggressive by watching others be aggressive, Bandura and his colleagues set up several experiments. They allowed children to watch a live or televised adult model punch, throw, and hit a large inflated Bobo doll (**Figure a**). Later, the children were allowed to play in the same room with the same toys. As Bandura hypothesized, children who had seen the live or televised aggressive model were much more aggressive with the Bobo doll than children who had not seen the modeled aggression (**Figure b**). In other words, "Monkey see, monkey do."

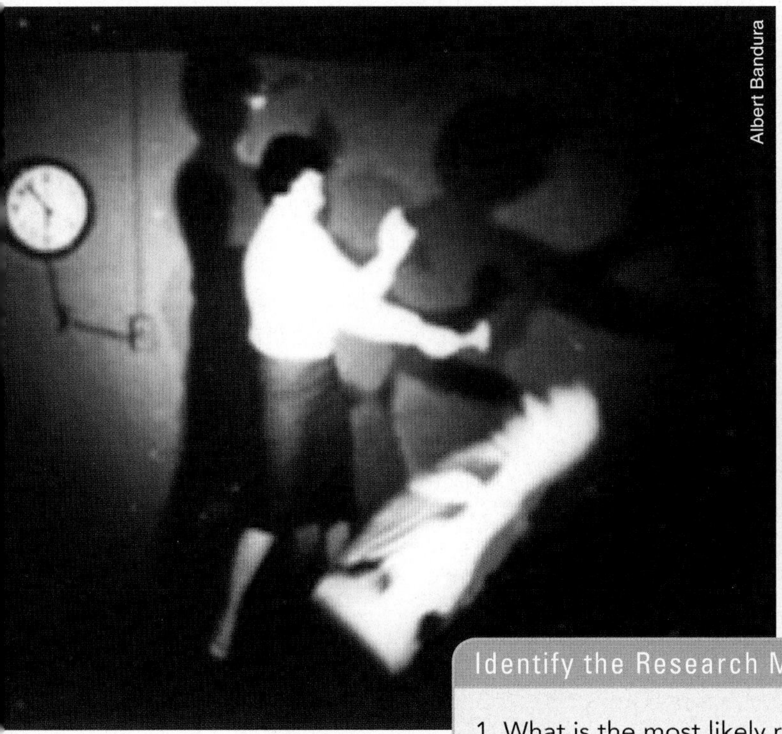


a.

b.

### Identify the Research Method

1. What is the most likely research method used in Bandura's Bobo doll study?
2. If you chose
   - the experimental method, label the IV, DV, experimental group, and control group.
   - the descriptive method, is this a naturalistic observation, survey, or case study?
   - the correlational method, is this a positive, negative, or zero correlation?
   - the biological method, identify the specific research tool (e.g., brain dissection, CT scan).

(Check your answers in Appendix C.)

biological, psychological, and social survival (the *biopsychosocial model*). Watching others helps us avoid dangerous stimuli in our environment, teaches us how to think and feel, and shows us how to act and interact socially (Chance, 2009).

As shown in the *Psychological Science* feature, some of the most compelling examples of observational learning come from the work of Albert Bandura and his colleagues (Bandura, 2007; Bandura, Ross, & Ross, 1961; Bandura & Walters, 1963; Huesmann & Kirwil, 2007).

# Psychology InSight

## Four key factors in observational learning • Figure 6.13

A child wanting to become a premier ballerina would need to incorporate these four factors to maximize learning.
a. Attention
Observational learning requires attention. This is why teachers insist on having students watch their demonstrations.
b. Retention
To learn new behaviors, we need to carefully note and remember the model's directions and demonstrations.
Erik Isakson/Getty Images, Inc.
c. Reproduction
Observational learning requires that we imitate the model.
d. Reinforcement
We are more likely to repeat a modeled behavior if the model is reinforced for the behavior, for example, with applause or other recognition.

According to Bandura, observational learning requires *attention*, *retention*, *reproduction*, and *reinforcement* (**Figure 6.13**). The children in Bandura's Bobo doll study demonstrated each of these four processes by paying *attention* to the adult model who hit the Bobo doll, *retaining* the memory of the model's behavior, *reproducing* the aggressive acts, and, although it wasn't directly mentioned, they expected to be *reinforced* for their aggression, like the adult models were in the actual experiment.

### CONCEPT CHECK

STOP

1. **How** is the cognitive-social theory of learning different from classical and operant conditioning?
2. **What** evidence is there that humans and nonhuman animals form cognitive maps?
3. **What** conclusions can be drawn from Bandura's Bobo doll study?

# Biology of Learning

## LEARNING OBJECTIVES

**RETRIEVAL PRACTICE** While reading the upcoming sections, respond to each Learning Objective in your own words. Then compare your responses with those in Appendix B.

1. **Explain** how an animal's environment might affect learning and behavior.
2. **Describe** the biological foundation for empathy.
3. **Summarize** the role of evolution in learning.

So far in this chapter, we have considered how external forces—from reinforcing events to our observations of others—affect learning. But we also know that for changes in behavior to persist over time, lasting biological changes must occur within the organism. In this section, we will examine neurological and evolutionary influences on learning.

### Neuroscience and Learning

Each time we learn something, either consciously or unconsciously, that experience creates new synaptic connections and alterations in a wide network of brain structures, including the cortex, cerebellum, hippocampus, hypothalamus, thalamus, and amygdala (Fu & Zuo, 2011; Kalat, 2013; Lozano, 2011; Meyers, Xue-Lian, & Constantinidis, 2012; Podgorski, Dunfield, & Haas, 2012).

Evidence that learning changes brain structure first emerged in the 1960s from studies of animals raised in enriched versus deprived environments. Compared with rats raised in a stimulus-poor environment, those raised in a colorful, stimulating "rat Disneyland" had a thicker cortex, increased nerve growth factor (NGF), more fully developed synapses, more dendritic branching, and improved performance on many tests of learning and memory (Diniz et al., 2010; Harati et al., 2011; Lores-Arnaiz et al., 2007; Shoji & Mizoguchi, 2011).

Admittedly, it is a big leap from rats to humans, but research suggests that the human brain also responds to environmental conditions (**Figure 6.14**). For example, older adults who are exposed to stimulating environments generally perform better on intellectual and perceptual tasks than those who are in restricted environments (Daffner, 2010; Merrill & Small, 2011; Schaie, 1994, 2008).

### Mirror Neurons and Imitation

Recent research has identified another neurological influence on learning processes, particularly imitation.

**Daily enrichment • Figure 6.14**

For humans and nonhuman animals alike, environmental conditions play an important role in enabling learning. How might a classroom rich with stimulating toys, games and books foster intellectual development in young children?

PhotoDisc Green/Getty Images

Top-Pet-Pics/Alamy

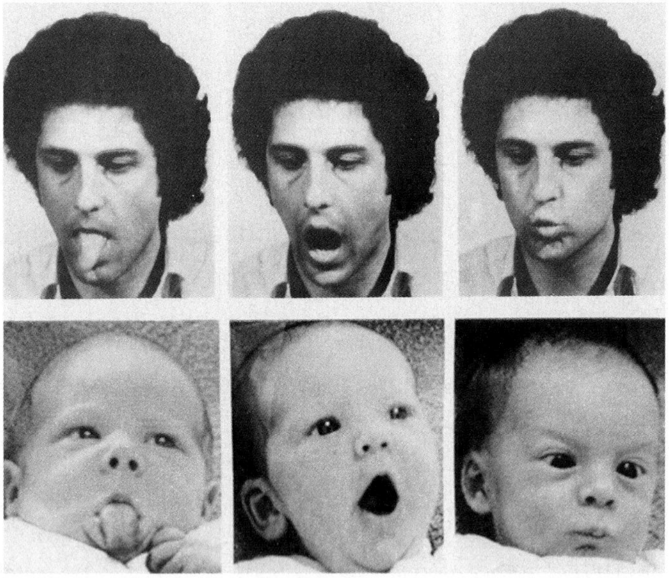
A.N. Meltzoff & M.K. Moore, "Imitation of facial and manual gestures by human neonates." Science, 1977, 198, 75–78

**Imitation • Figure 6.15**

In a series of well-known studies, Andrew Meltzoff and M. Keith Moore (1977, 1985, 1994) found that newborns could imitate such facial movements as tongue protrusion, mouth opening, and lip pursing.

When an adult models a facial expression, even very young infants will respond with a similar expression (**Figure 6.15**). At 9 months, infants will imitate facial actions a full day after seeing them (Heimann & Meltzoff, 1996).

How can children so easily imitate the facial expressions of others? Using fMRIs and other brain-imaging techniques, researchers have identified specific **mirror neurons** believed to be responsible for human empathy and imitation (Ahlsén, 2008; Baird, Scheffer, & Wilson, 2011; Caggiano et al., 2011; Jacob, 2008). These neurons are found in several key areas of the brain, and they help us identify with what others are feeling and to imitate their actions. When we see another person in pain, one reason we empathize and "share their pain" may be that our mirror neurons are firing. Similarly, if we watch others smile, our mirror neurons make it harder for us to frown.

Mirror neurons were first discovered by neuroscientists who implanted wires in the brains of monkeys to monitor areas involved in planning and carrying out movement (Ferrari, Rozzi, & Fogassi, 2005; Rizzolatti et al., 2002, 2006). When these monkeys moved and grasped an object, specific neurons fired, but they also fired when the monkeys simply observed another monkey performing the same or similar tasks.

Mirror neurons in humans also fire when we perform a movement or watch someone else perform. Have you noticed how spectators at an athletic event sometimes slightly move their arms or legs in synchrony with the athletes? Mirror neurons may be the underlying biological mechanism for this imitation. They also might help explain the emotional deficits of children and adults with autism or schizophrenia who often misunderstand the verbal and nonverbal cues of others (Enticott et al., 2012; Kana, Wadsworth, & Travers, 2011; Rizzolatti & Fabbri-Destro, 2010.)

Although scientists are excited about the promising links between mirror neurons and human emotions, imitation, language, learning, and learning disabilities, we do not yet know how they develop. In addition, other researchers have suggested that more research is needed to justify the claims for the role of mirror neurons (Baird, Scheffer, & Wilson, 2011). Stay tuned.

## Evolution and Learning

Humans and other animals are born with various innate reflexes and instincts. Although these biological tendencies help ensure evolutionary survival, they are inherently inflexible. Only through learning are we able to react to important environmental cues—such as spoken words and written symbols—that our innate reflexes and instincts do not address. Thus, from an evolutionary perspective, learning is an adaptation that enables organisms to survive and prosper in a constantly changing world.

Because animals can be operantly conditioned to perform a variety of novel behaviors (like waterskiing), learning theorists initially believed that the fundamental laws of conditioning would apply to almost all species and all behaviors. However, researchers have identified several biological constraints that limit the generality of conditioning principles. These include biological preparedness and instinctive drift.

Years ago, a young woman named Rebecca unsuspectingly bit into a Butterfinger candy bar filled with small, wiggling maggots. Horrified, she ran gagging and screaming to the bathroom. Many years later, Rebecca still feels nauseated when she sees a Butterfinger candy bar (but, fortunately, she doesn't feel similarly nauseated by the sight of her boyfriend, who bought her the candy).

Rebecca's graphic (and true!) story illustrates an important evolutionary process. When a food or drink is associated with nausea or vomiting, that particular food or

© Wildlife/Alamy

**Taste aversion in the wild • Figure 6.16**

Researchers, Gustavson and Garcia (1974), used classical conditioning to teach coyotes not to eat sheep. They began by lacing freshly killed sheep with a chemical that caused extreme nausea and vomiting in the coyotes that ate the tainted meat. Eventually the conditioned coyotes ran away from even the sight or smell of sheep.

drink can become a conditioned stimulus (CS) that triggers a conditioned **taste aversion**. Like other classically conditioned responses, taste aversions develop involuntarily (**Figure 6.16**).

Can you see why this automatic response would be adaptive? If one of our cave-dwelling ancestors became ill after eating a new plant, it would increase his or her chances for survival if he or she immediately developed an aversion to that plant—but not to other family members who might have been present at the time. Similarly, perhaps because of the more "primitive" evolutionary threat posed by snakes, darkness, spiders, and heights, people tend to develop phobias of these stimuli and situations far more easily than of guns, knives, and electrical outlets. We apparently inherit a built-in (innate) readiness to form associations between certain stimuli and responses. This is known as **biological preparedness**.

**biological preparedness** The built-in (innate) readiness to form associations between certain stimuli and responses.

Laboratory experiments have provided general support for both taste aversion and biological preparedness. For example, John Garcia and his colleague Robert Koelling (1966) produced taste aversion in lab rats by pairing flavored water (NS) and a drug (US) that produced gastrointestinal distress (UR). After being conditioned and then recovering from the illness, the rats refused to drink the flavored water (CS) because of the conditioned taste aversion. Remarkably, however, Garcia later discovered that only certain neutral stimuli could produce the nausea. Pairings of a noise (NS) or a shock (NS) with the nausea-producing drug (US) produced no taste aversion. Garcia suggested that when we are sick to our stomachs, we have a natural, evolutionary tendency to attribute it to food or drink. Being biologically prepared to quickly associate nausea with food or drink is adaptive because it helps us avoid that or similar food or drink in the future (Domjan, 2005; Garcia, 2003; Kardong, 2008; Swami, 2011).

Just as Garcia couldn't produce noise–nausea associations, other researchers have found that an animal's natural behavior pattern can interfere with operant conditioning. For example, the Brelands (1961) tried to teach a chicken to play baseball. Through shaping and reinforcement, the chicken first learned to pull a loop that activated a swinging bat and then learned to actually hit the ball. But instead of running to first base, it would chase the ball as if it were food. Regardless of the lack of reinforcement for chasing the ball, the chicken's natural behavior took precedence. This biological constraint is known as **instinctive drift**.

**instinctive drift** The tendency of some conditioned responses (CRs) to shift (or drift) back toward an innate response pattern.

**CONCEPT CHECK** STOP

1. **What** changes occur in rats' brains when they are raised in an enriched (versus deprived) environment?
2. **What** evidence is there for mirror neurons?
3. **Why** is taste aversion evolutionarily adaptive?

# Summary

## 1 Classical Conditioning 148

- **Learning** is a relatively permanent change in behavior or mental processes caused by experience. Pavlov discovered a fundamental form of **conditioning** (learning) called **classical conditioning**, in which a **neutral stimulus (NS)** becomes associated with an **unconditioned stimulus (US)** to elicit a **conditioned response (CR)**. In the "Little Albert" experiment, Watson and Rayner demonstrated how many of our likes, dislikes, prejudices, and fears are examples of a **conditioned emotional response (CER)**, as shown in the diagram.

What a Psychologist Sees: Conditioning Fear

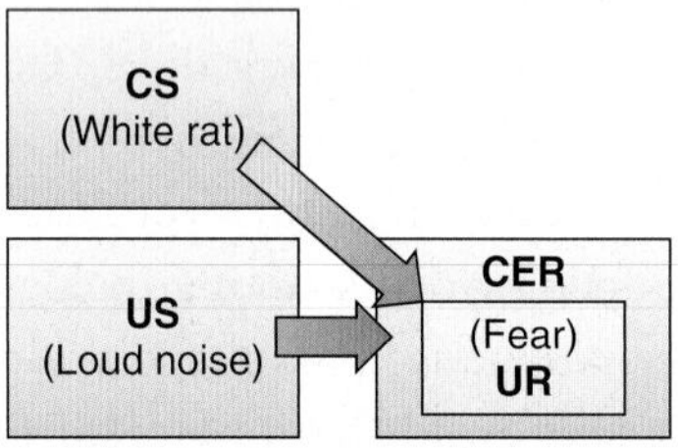


- **Acquisition** is the initial phase of classical conditioning. **Stimulus generalization** occurs when an event similar to the originally conditioned stimulus triggers the same conditioned response. With experience, animals learn to distinguish between an original conditioned stimulus and similar stimuli—**stimulus discrimination**. Most learned behaviors can be weakened through extinction. **Extinction** is a gradual weakening or suppression of a previously conditioned response. However, if a conditioned stimulus is reintroduced after extinction, an extinguished response may spontaneously recover. **Higher-order conditioning** occurs when the NS becomes a CS through repeated pairings with a previous CS.

## 2 Operant Conditioning 155

- In **operant conditioning**, an organism performs a behavior that produces either reinforcement or punishment. **Reinforcement** strengthens the response, while **punishment** weakens the response, as shown in the diagram.

Classical versus operant conditioning • Figure 6.6

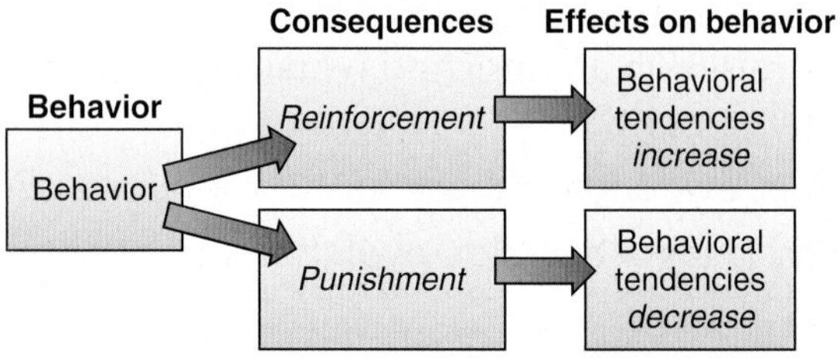


- Thorndike developed the **law of effect**, a first step in understanding how consequences can modify voluntary behaviors. Skinner extended Thorndike's law of effect to more complex behaviors.
- Reinforcers can be either primary or secondary, and reinforcement can be either positive or negative.
- **"Schedules of reinforcement"** refers to the rate or interval at which responses are reinforced. Most behavior is rewarded on one of four **partial schedules of reinforcement**: **fixed ratio (FR)**, **variable ratio (VR)**, **fixed interval (FI)**, or **variable interval (VI)**.
- Organisms learn complex behaviors through **shaping**, in which **reinforcement** is delivered for successive approximations of the desired response.
- **Punishment** weakens the response. To be effective, punishment must be immediate and consistent. Even then, the recipient may learn only what not to do. Punishment can have serious side effects: passive aggressiveness, avoidance behavior, inappropriate modeling, temporary suppression, learned helplessness, rewarded aggression, and perpetuated aggression.

## 3 Cognitive-Social Learning 163

- **Cognitive-social learning theory** emphasizes cognitive and social aspects of learning. Köhler discovered that animals sometimes learn through sudden **insight**, rather than through trial and error, as shown in the diagram. Tolman provided evidence of **latent learning** and internal **cognitive maps**.

Cognitive-social learning • Figure 6.11

- Bandura's research found that children who watched an adult behave aggressively toward an inflated Bobo doll were later more aggressive than those who had not seen the aggression. According to Bandura, **observational learning** requires attention, retention, motor reproduction, and reinforcement.

## 4 Biology of Learning 167

- Learning creates structural changes in the brain. Early evidence for such changes came from research on animals raised in enriched environments, such as the one shown in the photo, versus deprived environments.
- Learning is an evolutionary adaptation that enables organisms to survive and prosper in a constantly changing world. Researchers have identified biological constraints that limit the generality of conditioning principles: **biological preparedness** (such as **taste aversion**) and **instinctive drift**.

**Daily enrichment • Figure 6.14**

# Key Terms

**RETRIEVAL PRACTICE** Write a definition for each term before turning back to the referenced page to check your answer.

- acquisition 152
- biofeedback 162
- biological preparedness 169
- classical conditioning 148
- cognitive map 164
- cognitive-social learning theory 163
- conditioned emotional response (CER) 151
- conditioned response (CR) 150
- conditioned stimulus (CS) 150
- conditioning 148
- continuous reinforcement 157
- extinction 152
- fixed interval (FI) 159
- fixed ratio (FR) 159
- higher-order conditioning 153
- insight 163
- insight learning 164
- instinctive drift 169
- latent learning 164
- law of effect 156
- learning 148
- mirror neurons 168
- negative punishment 160
- negative reinforcement 156
- neutral stimulus (NS) 150
- observational learning 164
- operant 156
- operant conditioning 155
- partial (or intermittent) schedule of reinforcement 157
- positive punishment 160
- positive reinforcement 156
- Premack principle 157
- primary reinforcer 156
- punishment 156
- reinforcement 156
- schedule of reinforcement 157
- secondary reinforcer 156
- shaping 159
- spontaneous recovery 153
- stimulus discrimination 152
- stimulus generalization 152
- taste aversion 169
- unconditioned response (UR) 148
- unconditioned stimulus (US) 148
- variable interval (VI) 159
- variable ratio (VR) 159

# Critical and Creative Thinking Questions

1. How might Watson and Rayner, who conducted the famous "Little Albert" study, have designed a more ethical study of conditioned emotional responses?
2. What are some examples of ways in which observational learning has benefited you in your life? Are there instances in which observational learning has worked to your disadvantage?
3. Do you have any taste aversions? How would you use information in this chapter to remove them?
4. Most classical conditioning is involuntary. Considering this, is it ethical for politicians and advertisers to use classical conditioning to influence our thoughts and behavior? Why or why not?

# What is happening in this picture?

Chris Fitzgerald/CandidatePhotos/NewsCom

For political candidates, kissing babies seems to be as critical as pulling together a winning platform and airing persuasive campaign ads.

**Think Critically**

1. What principle of learning explains why politicians kiss babies?
2. What other stimuli or symbols might elicit similar effects?

# Self-Test

**RETRIEVAL PRACTICE** Completing this self-test and comparing your answers with those in Appendix C provides immediate feedback and helpful practice for exams. Additional interactive self-tests are available at www.wiley.com/college/carpenter.

**1.** _____ *conditioning* occurs when a neutral stimulus becomes associated with an unconditioned stimulus to elicit a conditioned response.

a. Reflex
b. Instinctive
c. Classical
d. Basic

**2.** The process of learning associations between environmental stimuli and behavioral responses is known as _____.

a. maturation
b. contiguity learning
c. conditioning
d. latent learning

**3.** In this photo of John Watson's demonstration of classical conditioning with Little Albert, the UNCONDITIONED STIMULUS was _____.

a. symptoms of fear
b. a rat
c. a bath towel
d. a loud noise

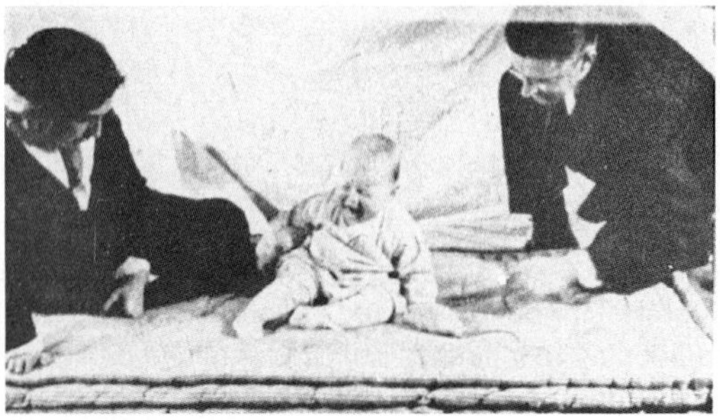

**4.** A baby is bitten by a small dog and then is afraid of all small animals. This is an example of _____.

a. stimulus discrimination
b. extinction
c. reinforcement
d. stimulus generalization

**5.** *Extinction* _____.

a. is a gradual weakening or suppression of a previously conditioned behavior
b. occurs when a CS is repeatedly presented without the US
c. is a weakening of the association between the CS and the US
d. all of these options

**6.** When a neutral stimulus is paired with a previously conditioned stimulus to become a conditioned stimulus as well, this is called _____ conditioning.

a. operant
b. classical
c. higher-order
d. secondary

**7.** If you wanted to use higher-order conditioning to get Little Albert to fear Barbie dolls, you would present a Barbie doll with the _____.

a. loud noise
b. original unconditioned response
c. white rat
d. original conditioned response

8. Which of the following is an example of the use of classical conditioning in everyday life?
   a. Treating alcoholism with a drug that causes nausea when alcohol is consumed
   b. Using attractive models to sell products
   c. The development of prejudice
   d. All of these options

9. Learning in which voluntary responses are controlled by their consequences is called _____.
   a. self-efficacy
   b. classical conditioning
   c. operant conditioning
   d. all of these options

10. The diagram shows the _____ box, used to demonstrate _____.
   a. Thorndike; reinforcement
   b. Thorndike; the law of effect
   c. Skinner; reinforcement and punishment
   d. Skinner; the law of effect

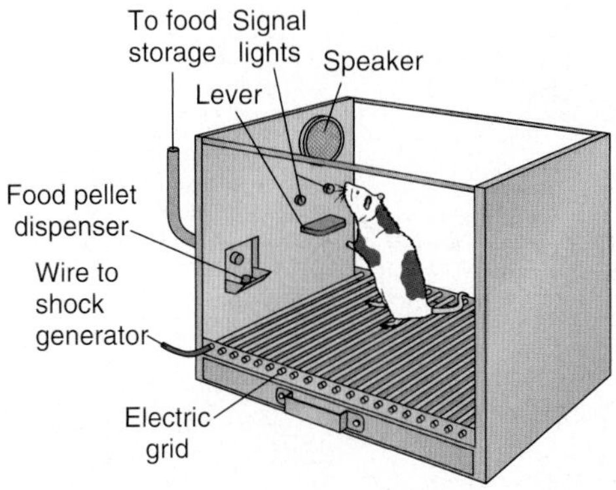


11. _____ strengthens a response and makes it more likely to recur, whereas _____ weakens a response and makes it less likely to recur.
   a. Positive reinforcement; positive punishment
   b. Negative reinforcement; negative punishment
   c. Negative reinforcement; positive punishment
   d. all of these options

12. Learning new behavior or information by watching others is known as _____.
   a. social learning
   b. observational learning
   c. modeling
   d. all of the above

13. In Köhler's *insight* experiment, the chimpanzee in this diagram _____.
   a. used trial and error to reach a banana placed just out of reach
   b. turned its back on the banana out of frustration
   c. sat for a while, then used a stick to bring the banana within reach
   d. didn't like bananas

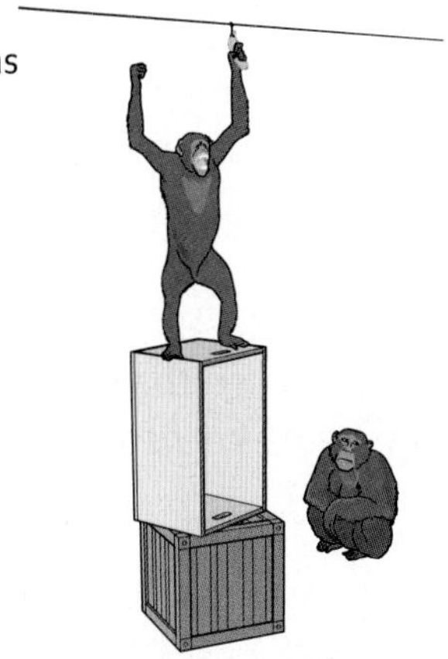

14. The imitation shown by the infant in these photos may be the result of _____.
   a. classical conditioning
   b. mirror neurons
   c. operant conditioning
   d. all of the above

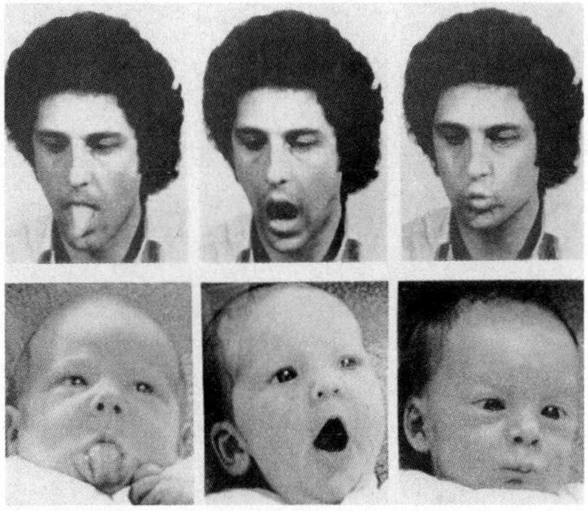

15. Garcia and his colleagues taught coyotes to avoid sheep by pairing a nausea-inducing drug with freshly killed sheep eaten by the coyotes. This is an example of _____.
   a. classical conditioning
   b. operant conditioning
   c. positive punishment
   d. negative punishment

**THE PLANNER** ✓

Review your Chapter Planner on the chapter opener and check off your completed work.

# 7 Memory

When you visualize the topic of memory, what pictures come to mind? Do you think of scrapbooks and people taking photos to capture their memories? Consider this true story of Elizabeth Loftus, one of modern psychology's most famous memory scientists:

When Elizabeth was 14, her mother drowned in their pool. As she grew older, the details surrounding her mother's death became increasingly vague. Decades later, a relative told Elizabeth that she had been the one to find the body. Despite her initial shock, the memories slowly started coming back.

*I could see myself, a thin, dark-haired girl, looking into the flickering blue-and-white pool. My mother, dressed in her nightgown, is floating face down. I start screaming. I remember the police cars, their lights flashing, and the stretcher with the clean, white blanket tucked*

*in around the edges of the body. The memory had been there all along, but I just couldn't reach it.*

(Loftus & Ketcham, 1994, p. 45)

How could a child forget finding her mother's body? How can we remember our second grade teacher's name but forget the name of someone we just met? In this chapter, you'll discover answers to these and other important questions about memory.

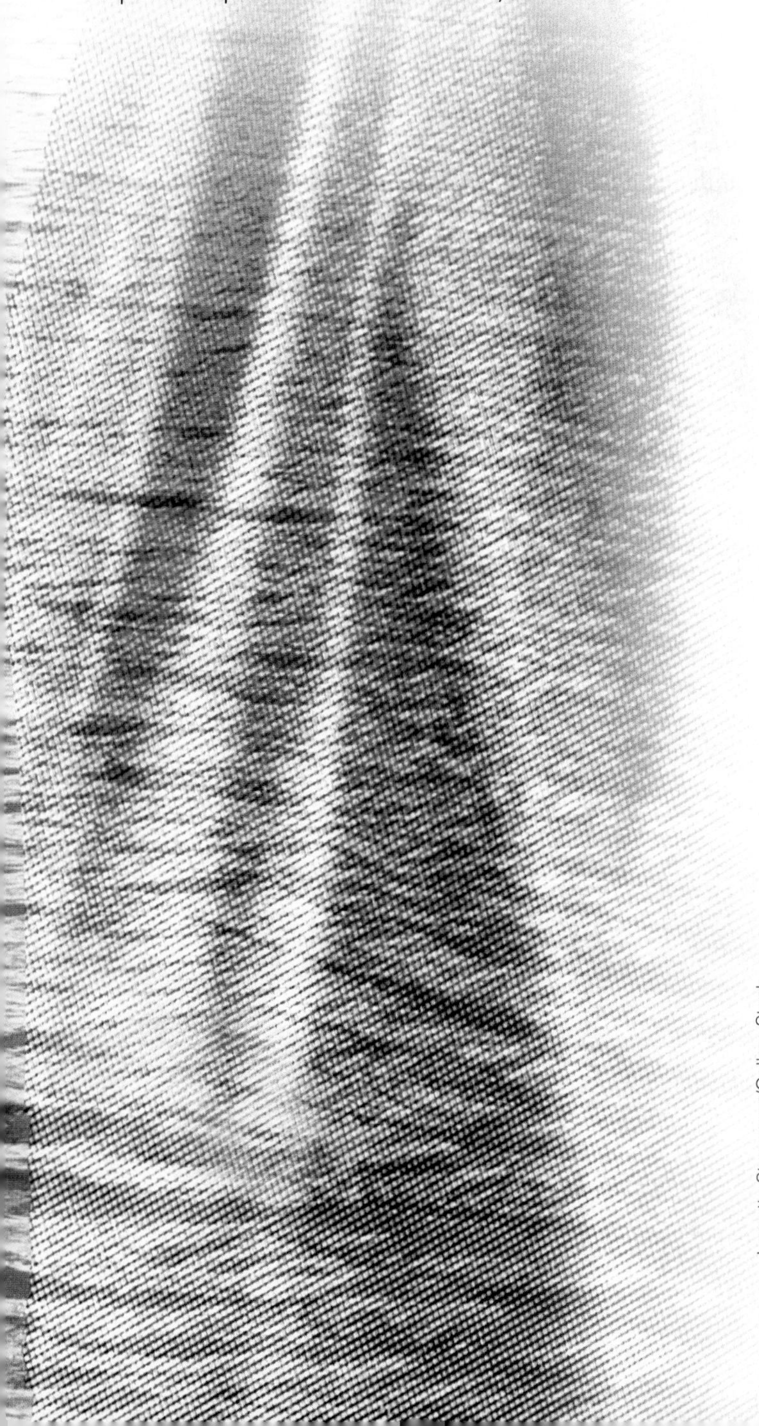

Jeanette Stevenson/Gallery Stock

## CHAPTER OUTLINE



## CHAPTER PLANNER ✓

- ❑ Study the picture and read the opening story.
- ❑ Scan the Learning Objectives in each section: p. 176 ❑ p. 185 ❑ p. 189 ❑ p. 192 ❑
- ❑ Read the text and study all figures and visuals. Answer any questions.

**Analyze key features**

- ❑ Process Diagram, p. 176
- ❑ Applying Psychology p. 178 ❑ p. 182 ❑ p. 193 ❑
- ❑ What a Psychologist Sees, p. 179
- ❑ Study Organizers p. 181 ❑ p. 187 ❑
- ❑ Psychology InSight, p. 184
- ❑ Psychological Science, p. 186
- ❑ Stop: Answer the Concept Checks before you go on. p. 185 ❑ p. 188 ❑ p. 192 ❑ p. 195 ❑

**End of chapter**

- ❑ Review the Summary and Key Terms.
- ❑ Answer the Critical and Creative Thinking Questions.
- ❑ Answer What is happening in this picture?
- ❑ Complete the Self-Test and check your answers.

# The Nature of Memory

## LEARNING OBJECTIVES

**RETRIEVAL PRACTICE** While reading the upcoming sections, respond to each Learning Objective in your own words. Then compare your responses with those in Appendix B.

1. **Review** the principles of the major memory models.
2. **Describe** the purpose of sensory memory.
3. **Explain** how short-term memory (STM) operates as our working memory.
4. **Describe** the various types of long-term memory (LTM).
5. **Explain** how organization, elaborative rehearsal, and retrieval cues improve long-term memory.

Memory allows us to learn from our experiences and to adapt to ever-changing environments—without it, we would have no past or future. Yet our memories are also highly fallible. Although some people think of **memory** as a gigantic library or an automatic tape recorder, our memories are not exact recordings of events. Instead, memory is a *constructive process* through which we actively organize and shape information as it is being processed, stored, and retrieved. As you might expect, this construction often leads to serious errors and biases, which we'll discuss throughout this chapter.

**memory** An internal record or representation of some prior event or experience.

## Memory Models

To understand memory (and its constructive nature), you first need a model of how it operates. Over the years, psychologists have developed numerous models for memory. According to the **encoding, storage, and retrieval (ESR) model**, the barrage of information that we encounter every day goes through three basic operations: **encoding**, **storage**, and **retrieval**. Each of these processes represents a different function closely analogous to the parts and functions of a computer (**Figure 7.1**).

**encoding** Processing information into the memory system.

**storage** Retaining information over time.

**retrieval** Recovering information from memory storage.

PROCESS DIAGRAM

### Encoding, storage, and retrieval • Figure 7.1

THE PLANNER

The encoding, storage, and retrieval model of memory can be compared to a computer's information processing system.

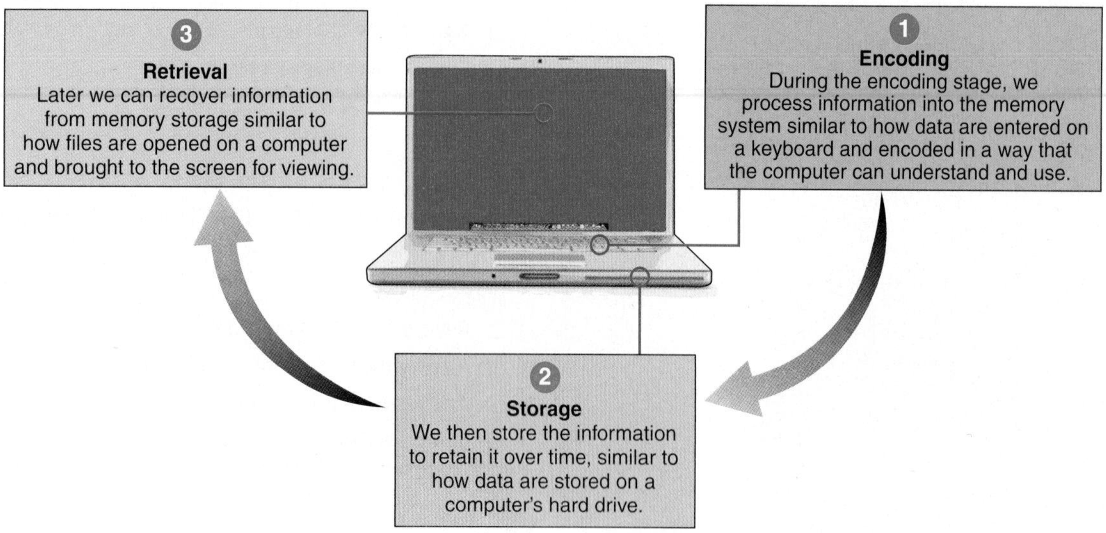

1
Encoding
During the encoding stage, we process information into the memory system similar to how data are entered on a keyboard and encoded in a way that the computer can understand and use.
2
Storage
We then store the information to retain it over time, similar to how data are stored on a computer's hard drive.
3
Retrieval
Later we can recover information from memory storage similar to how files are opened on a computer and brought to the screen for viewing.

## The traditional three-stage memory model • Figure 7.2

Each "box" represents a separate memory system that differs in purpose, duration, and capacity. When information is not transferred from sensory memory or short-term memory, it is assumed to be lost. Information stored in long-term memory can be retrieved and sent back to short-term memory for use.

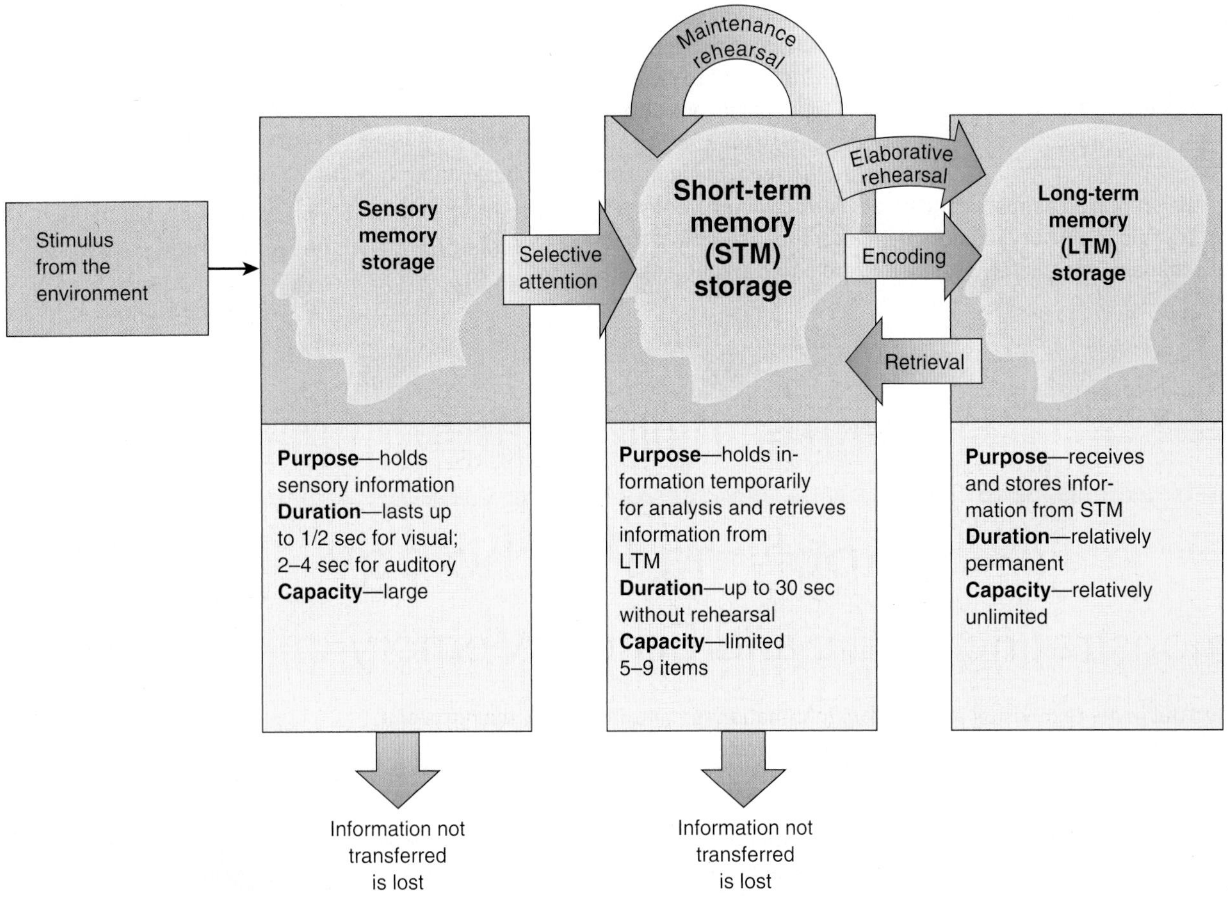


The *storage* stage also involves *information processing* in three interacting memory systems—*sensory, short-term,* and *long-term*—which is generally considered a separate model, known as the **three-stage memory model** (Atkinson & Shiffrin, 1968; Moulin, 2011). According to this model, three different storage "boxes," or memory stages, hold and process information. Each stage has a different purpose, duration, and capacity (**Figure 7.2**). The three-stage memory model remains the leading paradigm in memory research because it offers a convenient way to organize the major findings. Let's discuss the model in more detail.

## Sensory Memory

Everything we see, hear, touch, taste, and smell must first enter our **sensory memory**. Once it's entered, the information remains in sensory memory just long enough to locate relevant bits of data and transfer it on to the next stage of memory. For visual information, known as *iconic memory*, the visual image (icon) stays in sensory memory only about one-half of a second

> **sensory memory** This first memory stage holds sensory information. It has a relatively large capacity, but duration is only a few seconds.

## How do researchers test iconic sensory memory? • Figure 7.3

| | | | |
|---|---|---|---|
| K | Z | R | A |
| Q | B | T | P |
| S | G | N | Y |

In an early study of iconic sensory memory, George Sperling (1960) flashed an arrangement of letters like these for 1/20 of a second. Most people, he found, could recall only 4 or 5 letters. But when instructed to report just the top, middle, or bottom row, depending on whether they heard a high, medium, or low tone, they reported almost all the letters correctly. Apparently, all 12 letters are held in sensory memory right after they are viewed, but only those that are immediately attended to are noted and processed.

(**Figure 7.3**). Auditory information (what we hear) lasts about the same length of time as visual information, one-quarter to one-half of a second, but a weaker "echo," or *echoic memory*, of this auditory information is held up to four seconds (Inui et al., 2010; Neisser 1967; Radvansky, 2011). Both iconic and echoic memory are demonstrated in *Applying Psychology*.

Early researchers believed that sensory memory had an unlimited capacity. However, later research suggests that sensory memory does have limits and that stored images are fuzzier than once thought (Goldstein, 2010; Grondin, Ouellet, & Roussel, 2004; Moulin, 2011).

# Applying Psychology

THE PLANNER

## Demonstrating Iconic and Echoic Memory

As pointed out in the text, visual and auditory information remains in sensory memory for a very short time. Here are some easy ways to demonstrate this phenomenon.

**a. Iconic memory**
To demonstrate the duration of visual, or *iconic memory*, swing a flashlight in a dark room. Because the image, or icon, lingers for a fraction of a second after the flashlight is moved, you see the light as a continuous stream, as in this photo, rather than as a succession of individual points.

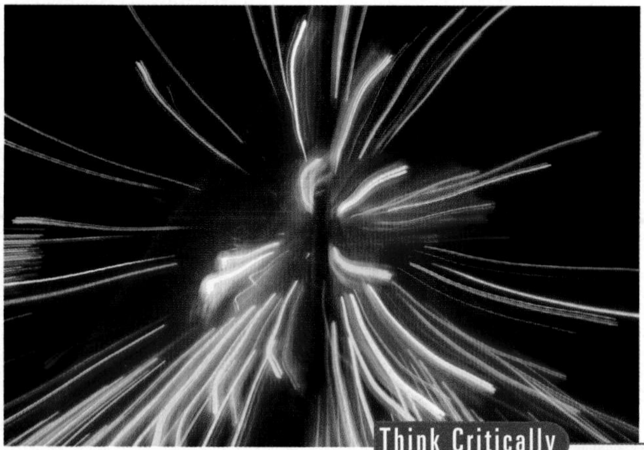

Stephen St. John/NG Image Collection

**b. Echoic memory**
Think back to times when someone asked you a question while you were deeply absorbed in a task. Did you ask "What?" and then immediately find you could answer them without hearing their repeated response? Now you know why. A weaker "echo" of auditory information can last up to four seconds.

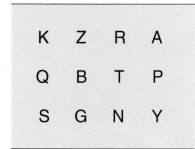

Blue Jean Images/Getty Images

**Think Critically**

1. What do you think would happen if we did not possess iconic or echoic memory?
2. What might happen if visual or auditory sensations lingered not for seconds but for minutes?

## WHAT A PSYCHOLOGIST SEES — THE PLANNER

### Chunking in Chess

What do you see when you observe the arrangement of pieces on a chessboard? To the inexpert eye, a chess game in progress looks like a random assembly of game pieces. Accordingly, novice chess players can remember the positions of only a few pieces when a chess game is underway. But expert players generally remember all the positions. To the experts, the scattered pieces form meaningful patterns—classic arrangements that recur often. Just as you group the letters of this sentence into meaningful words and remember them long enough to understand the meaning of the sentence, expert chess players group the chess pieces into easily recalled patterns (or chunks) (Huffman, Matthews, & Gagne, 2001; Waters & Gobet, 2008).

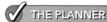

James P. Blair/NG Image Collection

## Short-Term Memory (STM)

The second stage of memory processing, **short-term memory (STM)**, temporarily stores and processes sensory stimuli. If the information is important, STM organizes and sends this information along to long-term memory (LTM). STM also retrieves stored memories from LTM. Otherwise, information decays and is lost.

**short-term memory (STM)** This second memory stage temporarily stores sensory information and decides whether to send it on to long-term memory (LTM). Its capacity is limited to five to nine items, and its duration is about 30 seconds.

The *capacity* and *duration* of STM are limited (Bankó & Vidnyánsky, 2010; Kareev, 2000; Moulin, 2011). To extend the *capacity* of STM, you can use a technique called **chunking** (Boucher & Dienes, 2003; Glicksohn & Cohen, 2011; Miller, 1956; Perlman et al., 2010), as demonstrated in *What a Psychologist Sees*. Have you noticed that credit card, Social Security, and telephone numbers are all grouped into three or four units separated by hyphens? This is because it's easier to remember numbers in "chunks" rather than as a string of single digits.

**chunking** The act of grouping separate pieces of information into a single unit (or chunk).

**maintenance rehearsal** Repeating information to maintain it in short-term memory (STM).

You can extend the *duration* of your STM almost indefinitely by consciously "juggling" the information, a process called **maintenance rehearsal**. You are using maintenance

## How working memory might "work" • Figure 7.4

Working memory involves active processing of incoming sensory information, followed by active encoding and retrieval of information from LTM. Note the "tools" the "worker" uses during this processing.

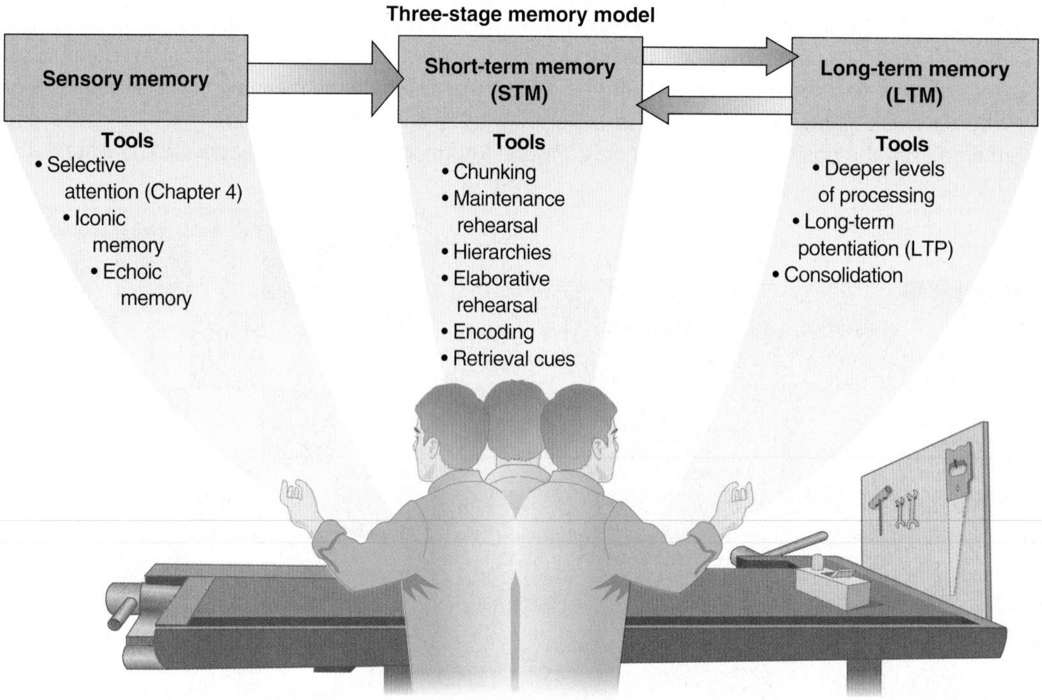


rehearsal when you look up a phone number and repeat it over and over until you punch or press in the number.

People who are good at remembering names also know to take advantage of maintenance rehearsal. They repeat the name of each person they meet, aloud or silently, to keep it active in STM. They also make sure that other thoughts (such as their plans for what to say next) don't intrude.

As you can see in **Figure 7.4**, short-term memory is more than just a passive, temporary "holding area." Most current researchers (Baddeley, 1992, 2007; Berti, 2010; Buchsbaum, Padmanabhan, & Berman, 2011; Camos & Barrouillet, 2011) realize that active processing of information also occurs in STM.

Because short-term memory (STM) is active, or *working*, some researchers prefer the term **working memory**. It helps to visually picture STM as a "workbench," with a "worker" at the bench who selectively attends to certain sensory information, and also sends and retrieves it from long-term memory (LTM). The worker also manipulates the incoming, transferred, and retrieved information. All our conscious thinking occurs in this "working memory," and the manipulation helps explain some of our memory errors and false constructions described in this chapter.

## Long-Term Memory (LTM)

Once information is transferred from STM, it is organized and integrated with other information in **long-term memory (LTM)**. LTM serves as a storehouse for information that must be kept for long periods. When we need the information, it is sent back to STM for our conscious use. Compared with sensory memory and short-term memory, long-term memory has relatively unlimited *capacity* and *duration* (Klatzky, 1984). But, just as with any other possession, the better we label and arrange our memories, the more readily we'll be able to retrieve them.

> **long-term memory (LTM)** This third memory stage stores information for long periods. Its capacity is limitless; its duration is relatively permanent.

How do we store the vast amount of information that we collect over a lifetime? Several types of LTM exist (see **Study Organizer 7.1**).

**Explicit/declarative memory** refers to intentional learning or conscious knowledge. If asked to remember your phone number or your mother's name, you can state (*declare*) the answers directly (*explicitly*).

Explicit/declarative memory can be further subdivided into two parts. *Semantic* memory is memory for general knowledge, rules, events, facts, and specific information. It is our mental encyclopedia. In contrast, *episodic* memory is like a mental diary. It records the major events (*episodes*) in our lives. Some of our episodic memories are short-lived, whereas others can last a lifetime.

Have you ever wondered why most adults can recall almost nothing of the years before age 3? Research suggests that a concept of "self," sufficient language development, and growth of the frontal lobes of the cortex (along with other structures) may be necessary before these early events can be encoded and retrieved many years later (Bauer & Lukowski, 2010; Morris, 2007; Prigatano & Gray, 2008; Rose et al., 2011).

**explicit/declarative memory** The subsystem within long-term memory that consciously stores facts, information, and personal life experiences.

**implicit/nondeclarative memory** The subsystem within long-term memory that consists of unconscious procedural skills, simple classically conditioned responses (Chapter 6), and priming.

**Implicit/nondeclarative memory** refers to unintentional learning or unconscious knowledge. Try telling someone else how you tie your shoelaces without demonstrating the actual behavior. Because your memory of this skill is unconscious and hard to describe (*declare*) in words, this type of memory is sometimes referred to as *nondeclarative*.

Implicit/nondeclarative memory consists of *procedural* motor skills, like tying your shoes or riding a bike, as well as *classically conditioned memory* responses, such as fears or taste aversions.

Implicit/nondeclarative memory also includes *priming*, where prior exposure to a stimulus (*prime*) facilitates or inhibits the processing of new information (McKoon & Ratcliff, 2012; Schmitz & Wentura, 2012; Woollams et al., 2008). Priming often occurs even when we do not consciously remember being exposed to the prime. For example, you might feel nervous being home alone after reading a Stephen King novel, and watching a romantic movie might kindle your own romantic feelings.

**Study Organizer 7.1** Types of long-term memory ✓ THE PLANNER

LTM is divided and subdivided into various types of memory.

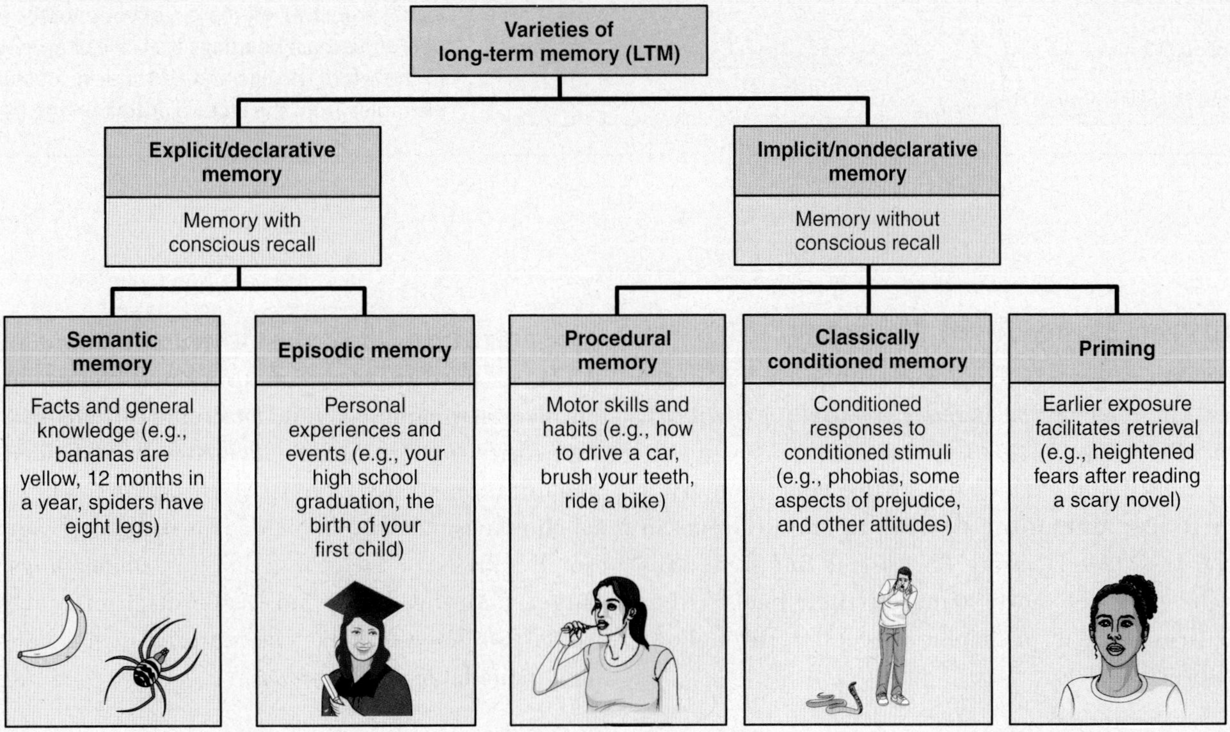

# Applying Psychology

THE PLANNER

## Mnemonic Devices

These three mnemonics improve memory by tagging information to physical locations (*method of loci*), organizing information into main and subsidiary topics (an *outline*), and using familiar information to remember the unfamiliar (*acronyms*).

**a. Method of loci**
Greek and Roman orators developed the *method of loci* to keep track of the many parts of their long speeches. Orators would imagine the parts of their speeches attached to places in a courtyard. For example, if an opening point in a speech was the concept of *justice*, they might visualize a courthouse placed in the first corner of their garden. As they mentally walked around their garden during their speech, they would encounter, in order, each of the points to be made. Using this same example, the second point that the orator might plan to make would be something about the prison system, and the third point would involve the role of government.

| Outline | Details and examples from lecture and text |
|---|---|
| 1. Nature of Memory | |
| a. Memory Models | |
| b. Sensory Memory | |
| c. Short-Term Memory (STM) | |

**b. Outline organization**
When listening to lectures and/or reading the text, draw a vertical line approximately three inches from the left margin of your notebook paper. Write main headings from the chapter outline to the left of the line and add specific details and examples from the lecture or text on the right.

## Improving Long-Term Memory

Several processes can be employed to improve long-term memory. These include *organization, rehearsal* (or *repetition*), and *retrieval*.

One additional "trick" for giving your memory a boost is to use **mnemonic devices** to encode items in a special way (see *Applying Psychology*). But some students find they get more "bang for their buck" using the well-researched principles discussed throughout this chapter.

**Organization** To successfully encode information for LTM, we need to *organize* material into hierarchies. This involves arranging a number of related items into broad categories that are further divided and subdivided. (This organization strategy for LTM is similar to the strategy of chunking material in STM.) For instance, by grouping small subsets of ideas together (as subheadings under larger, main headings, and within diagrams, tables, and so on), we hope to make the material in this book more understandable and *memorable*.

**c. Acronyms**
To use the *acronym method*, create a new code word from the first letters of the items you want to remember. For example, to recall the names of the Great Lakes, think of *HOMES* on a *great lake* (*H*uron, *O*ntario, *M*ichigan, *E*rie, *S*uperior). Visualizing homes on each lake also helps you remember your code word "homes."

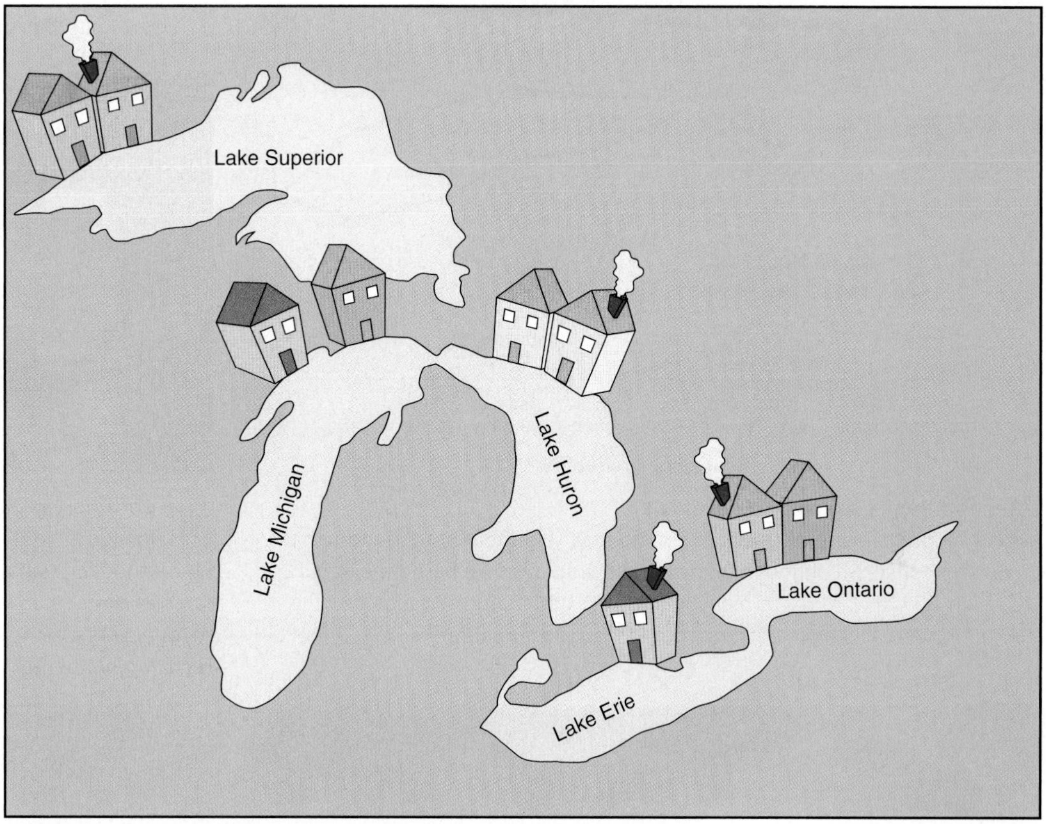


**Think Critically**

1. How could you use the method of loci to remember the types of long-term memory described in Study Organizer 7.1 on page 181?
2. How could you use the acronym method to remember the first six words in the list on page 193?

Admittedly, organization takes time and work. But you'll be happy to know that some memory organization and filing is done automatically while you sleep (Fogel & Smith, 2011; Mograss, Guillem, & Godbout, 2008; Sara, 2010; Verleger et al., 2011). (Unfortunately, despite claims to the contrary, research shows that we can't recruit our sleeping hours to memorize new material, such as a foreign language.)

**Rehearsal** Like organization, *rehearsal* also improves encoding for both STM and LTM. If you need to hold information in STM for longer than 30 seconds, you can simply keep repeating it (maintenance rehearsal). But storage in LTM requires *deeper levels of processing*, called **elaborative rehearsal**.

**elaborative rehearsal** The process of linking new information to previously stored material.

The immediate goal of elaborative rehearsal is to *understand*—not to memorize. Fortunately, this attempt to understand is one of the best ways to encode new information into long-term memory (Moulin, 2011).

# Psychology InSight Retrieval cues • **Figure 7.5**

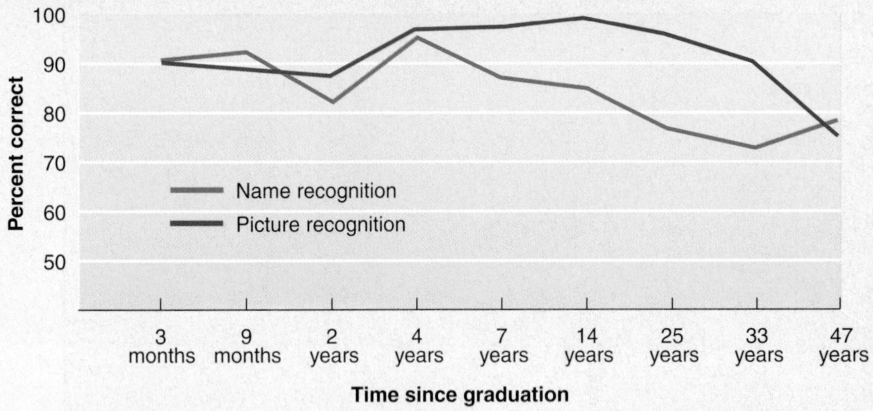


**a. Recognition (a specific retrieval cue)**
A recognition task offers specific cues and only requires you to identify the correct response. So although many people would have a hard time naming their high school classmates many years later, both name and picture recognition remains high, even many years after graduation.

Seth Poppel/Yearbook Library

Cyndy Ord/Getty Images, Inc.

▲ Considering these two photos of Mariah Carey taken in high school and years after graduation, if her former high-school classmates didn't know she had become a famous singer, can you see how they could more easily *recognize* her in the later photo and still have difficulty *recalling* her name?

Antonio M. Rosario/Photographer's Choice/Getty Images

**b. Recall (a general retrieval cue)**
Can you *recall* from memory the names of the eight planets in our solar system? If not, it's probably because *recall*, like an essay question on an exam, requires retrieval with only general, nonspecific cues. Note how much easier it is if you're provided these specific, recognition-type, retrieval cues: Mer-, Ven-, Mar-, Jup-, Sat-, Ura-, Nep-.

**retrieval cue** A clue or prompt that helps stimulate recall and retrieval of a stored piece of information from long-term memory.

**Retrieval** Finally, effective *retrieval* is critical to improving long-term memory. There are two types of **retrieval cues** (**Figure 7.5**). *Specific* cues require you only to *recognize* the correct response. *General* cues require you to *recall* previously learned material by searching through all possible matches in LTM—a much more difficult task.

Whether cues require recall or only recognition is not all that matters. Imagine that while house hunting, you walk into a stranger's kitchen and are greeted with the unmistakable smell of freshly baked bread. Instantly, the aroma transports you to your grandmother's kitchen, where you spent many childhood afternoons doing your homework. You find yourself suddenly thinking of the mental shortcuts your grandmother taught you to help you learn your multiplication tables. You hadn't thought about these little tricks for years, but somehow a whiff of baking bread brought them back to you. Why?

In this imagined episode, you have stumbled upon the **encoding specificity principle** (Tulving & Thompson, 1973). In most cases, we're able to remember better when we attempt to recall information in the *same* context in which it was learned. Have you noticed that you tend to do better on exams when you take them in the same seat and classroom in which you originally studied the material? This happens because the matching location acts as a retrieval cue for the information.

**encoding specificity principle** Retrieval of information is improved when the conditions of recovery are similar to the conditions that existed when the information was encoded.

People also remember information better if their moods during learning and retrieval match (Howe & Malone, 2011; Nouchi & Hyodo, 2007). This phenomenon, called *mood congruence*, occurs because a given mood tends to evoke memories that are consistent with that mood. When you're sad (or happy or angry), you're more likely to remember events and circumstances from other times when you were sad (or happy or angry).

Finally, as generations of coffee-guzzling university students have discovered, if you learn something while under the influence of a drug, such as caffeine, you will remember it more easily when you take that drug again than at other times (Nasehi et al., 2010; Patti et al., 2010; Rezayat et al., 2010). This is called *state-dependent retrieval*.

**CONCEPT CHECK** STOP

1. **What** does it mean to say that memory is a constructive process?
2. **What** happens to information after it leaves sensory memory?
3. **How** can you extend the capacity and duration of short-term memory?
4. **What** is the difference between semantic and episodic memory?
5. **What** is the difference between recognition and recall?

# Forgetting

## LEARNING OBJECTIVES

**RETRIEVAL PRACTICE** While reading the upcoming sections, respond to each Learning Objective in your own words. Then compare your responses with those in Appendix B.

1. **Describe** Ebbinghaus's research on learning and forgetting.
2. **Review** the five key theories of why we forget.
3. **Explain** the six most important factors that contribute to forgetting.

Psychologists have long been interested in how and why we forget. Hermann Ebbinghaus first introduced the experimental study of learning and forgetting in 1885 (*Psychological Science* on the next page).

Psychologists have developed several theories to explain forgetting and have identified a number of factors that can interfere with the process of forming memories. We discuss some of these theories and factors in this section.

## Theories of Forgetting

If you couldn't forget, your mind would be filled with meaningless data, such as what you ate for breakfast every morning of your life. Similarly, think of the incredible pain you would continuously endure if you couldn't distance yourself from tragedy through forgetting. The ability to forget is essential to the proper functioning of memory. But what about those times when forgetting is an inconvenience or even dangerous?

Psychological Science

THE PLANNER

# How Quickly We Forget

Using himself as a subject, Ebbinghaus calculated how long it took to learn a list of three-letter *nonsense syllables* such as *SIB* and *RAL*. He found that one hour after he knew a list perfectly, he remembered only 44% of the syllables. A day later, he recalled 35% and a week later, only 21%.

Depressing as these findings may seem, keep in mind that meaningful material is much more memorable than nonsense syllables. Even so, we all forget some of what we have learned.

On a more cheerful note, after some time passed and he had forgotten the list, Ebbinghaus found that *relearning* a list took less time than the initial learning did. This research suggests that we often retain some memory for things that we have learned, even when we seem to have forgotten them completely.

More recent research based on Ebbinghaus's discoveries has found that there is an ideal time to practice something you have learned. Practicing too soon is a waste of time, and if you practice too late you will already have forgotten what you learned. The ideal time to practice is when you are about to forget.

Polish psychologist Piotr Wozniak used this insight to create a software program called SuperMemo. The program can be used to predict the future state of an individual's memory and help the person schedule reviews of learned information at the optimal time. So far the program has been applied mainly to language learning, helping users retain huge amounts of vocabulary. But Wozniak hopes that someday programs like SuperMemo will tell people when to wake and when to exercise, help them remember what they have read and whom they have met, and remind them of their goals (Wolf, 2008).

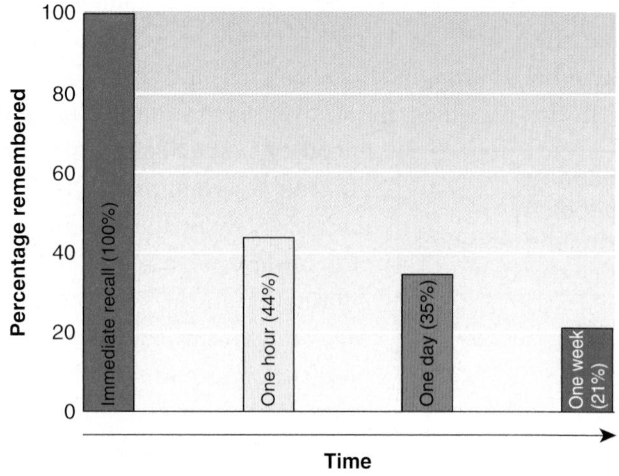


## Identify the Research Method

1. What is the most likely research method used in Ebbinghaus's study?
2. If you chose
   - the experimental method, label the IV, DV, experimental group, and control group.
   - the descriptive method, is this a naturalistic observation, survey, or case study?
   - the correlational method, is this a positive, negative, or zero correlation?
   - the biological method, identify the specific research tool (e.g., brain dissection, CT scan).

(Check your answers in Appendix C.)

There are five major theories that explain why forgetting occurs (**Study Organizer 7.2**): *decay, interference, motivated forgetting, encoding failure*, and *retrieval failure*. Each theory focuses on a different stage of the memory process or a particular type of problem in processing information.

In *decay theory*, memory is processed and stored in a physical form—for example, in a network of neurons. Connections between neurons probably deteriorate over time, leading to forgetting. This theory explains why skills and memory degrade if they go unused ("use it or lose it").

In *interference theory*, forgeting is caused by two competing memories, particularly memories with similar qualities. There are two types of interference: **retroactive interference** and **proactive interference** (**Figure 7.6**).

*Motivated forgetting theory* is based on the idea that we forget some information for a reason. According to Freudian theory, people forget unpleasant or anxiety-producing information either consciously or unconsciously.

In *encoding failure theory*, our sensory memory receives the information and passes it to STM. But during STM, we may decide there is no need to remember the precise details, so we do not fully encode it and pass it on for proper storage in LTM. For example, few people can correctly recognize the details on the U.S. penny (the location of the mint date, the direction of Lincoln's head).

## Study Organizer 7.2 Why we forget: Five key theories

THE PLANNER

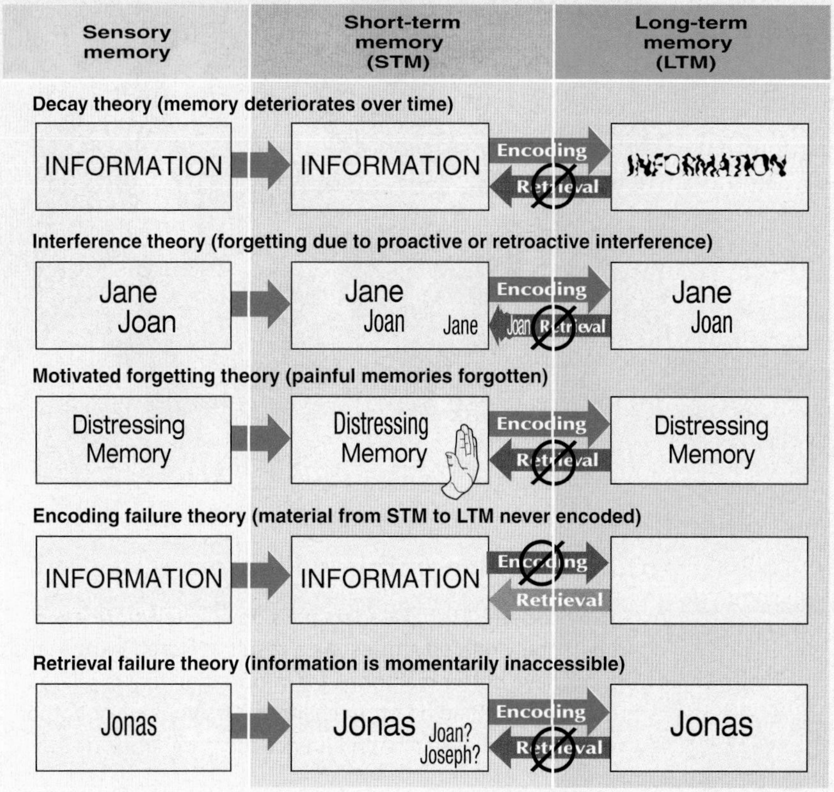


Which of the five theories of forgetting best applies to this cartoon?

*"I forget to drink."*

According to *retrieval failure theory*, memories stored in LTM aren't forgotten. They're just momentarily inaccessible. For example, the **tip-of-the-tongue (TOT) phenomenon**—the feeling that a word or event you are trying to remember will pop out at any second—is known to result from interference, faulty cues, and high emotional arousal.

## Factors Involved in Forgetting

Since Ebbinghaus's original research, scientists have discovered numerous factors that contribute to forgetting. Six of the most important are the *misinformation effect*, the *serial position effect*, *source amnesia*, the *sleeper effect*, *spacing of practice*, and *culture*.

### Two types of interference • Figure 7.6

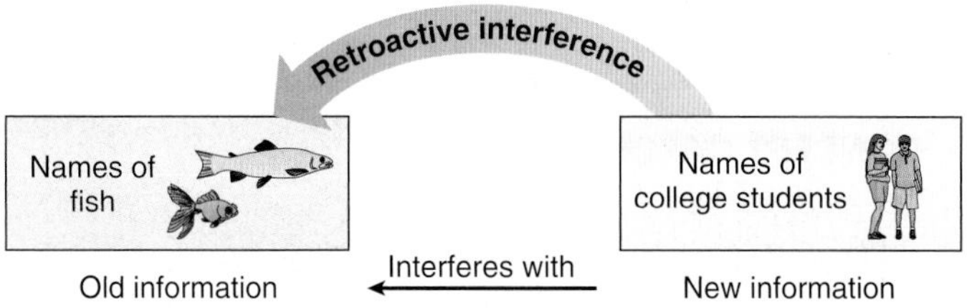


**a. Retroactive** (backward-acting) **interference** occurs when new information interferes with old information. This example comes from a story about an absent-minded icthyology professor (fish specialist) who refused to learn the names of his college students. Asked why, he said, "Every time I learn a student's name, I forget the name of a fish!"

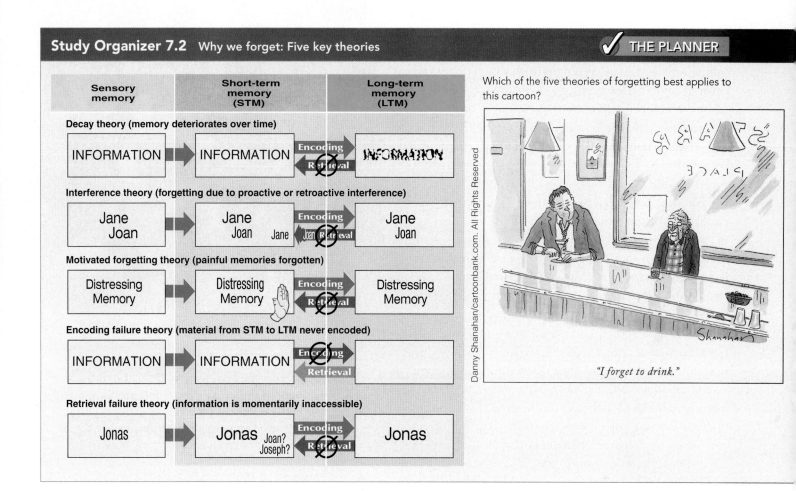

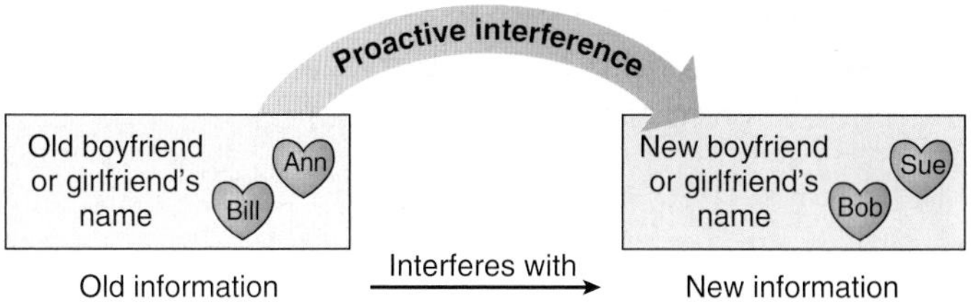


**b. Proactive** (forward-acting) **interference** occurs when old information interferes with new information. Have you ever been in trouble because you used an old romantic partner's name to refer to your new partner? You now have a guilt-free explanation—proactive interference.

Many people (who haven't studied this chapter or taken a psychology course) believe that when they're recalling an event, they're remembering it as if it were an "instant replay." However, as you know, our memories are highly fallible and filled with personal "constructions" that we create during encoding and storage. Research on the **misinformation effect** shows that information that occurs *after an event* may further alter and revise those constructions.

For example, in one study subjects watched a film of a car driving through the countryside, and were then asked to estimate how fast the car was going when it passed the barn. Although there was no actual barn in the film, subjects were six times more likely to report having seen one than those who were not asked about a barn (Loftus, 1982). Other experiments have created false memories by showing subjects doctored photos of themselves taking a completely fictitious hot-air balloon ride, or by asking subjects to simply imagine an event, such as having a nurse remove a skin sample from their finger. In these and similar cases, a large number of subjects later believed that misleading information was correct and that fictitious or imagined events actually occurred (Challies et al., 2011; Hess et al., 2012; Mazzoni & Vannucci, 2007; Pérez-Mata & Diges, 2007).

In addition to problems with the misinformation effect, when study participants are given lists of words to learn, they remember some words better than others depending on where they occurred in the list, known as the **serial position effect**. They remember the words at the beginning (*primacy effect*) and the end of the list (*recency effect*) better than those in the middle, which are quite often forgotten (Bonk & Healy, 2010; Overstreet & Healy, 2011).

The reasons for this effect are complex, but they do have interesting real-life implications. For example, a potential employer's memory for you might be enhanced if you are either the first or last candidate interviewed.

Each day we read, hear, and process an enormous amount of information, and it's easy to confuse "who said what to whom" and in what context. Forgetting the true source of a memory is known as **source amnesia** (Kleider et al., 2008; Leichtman, 2006; Paterson, Kemp, & Ng, 2011).

When we first hear something from an unreliable source, we tend to disregard that information in favor of a more reliable source. However, as the source of the information is forgotten (source amnesia), the unreliable information is no longer discounted. This is called the **sleeper effect** (Appel & Richter, 2007; Ecker, Lewandowsky, & Apai, 2011; Nabi, Moyer-Gusé, & Byrne, 2007).

If we try to memorize too much information at once (as when students "cram" before an exam), we're not likely to learn and remember as much as we would with more distributed study (Karpicke & Bauernschmidt, 2011; Simmons, 2012). **Distributed practice** refers to spacing your learning periods, with rest periods between sessions. Cramming is called **massed practice** because the time spent learning is massed into long, unbroken intervals.

Finally, as illustrated in **Figure 7.7**, cultural factors can play a role in how well people remember what they have learned.

Nat Farbman/Time & Life Pictures/Getty Imag

**Culture and memory • Figure 7.7**

In many societies, tribal leaders pass down vital information through orally related stories. As a result, children living in these cultures have better memories for information that is related through stories than do other children. Can you think of other ways in which culture might influence memory?

### CONCEPT CHECK STOP

1. **How** does previous learning affect relearning?
2. **What** is the difference between proactive and retroactive interference?
3. **What** is the relationship between source amnesia and the sleeper effect?

# Biological Bases of Memory

## LEARNING OBJECTIVES

**RETRIEVAL PRACTICE** While reading the upcoming sections, respond to each Learning Objective in your own words. Then compare your responses with those in Appendix B.

1. **Describe** two kinds of biological changes that occur when we learn something new.
2. **Explain** the effect of hormones on memory.
3. **Identify** some brain areas involved in memory.
4. **Explain** how injury and disease can affect memory.

A number of biological changes occur when we learn something new. Among them are neuronal and synaptic changes and hormonal changes. We discuss these changes in this section, along with the questions of where memories are located and what causes memory loss.

## Neuronal and Synaptic Changes

We know that learning modifies the brain's neural networks (Chapters 2 and 6). As you learn to play tennis, for example, repeated practice builds specific neural "pathways" that make it easier and easier for you to get the ball over the net. This **long-term potentiation (LTP)** happens in at least two ways.

**long-term potentiation (LTP)** A long-lasting increase in neural excitability believed to be a biological mechanism for learning and memory.

First, as early research with rats raised in "enriched" environments showed (Rosenzweig, Bennett, & Diamond, 1972), repeated stimulation of a synapse can strengthen the synapse by causing the dendrites to grow more spines (Chapter 6). This results in more synapses, more receptor sites, and more sensitivity.

Second, when learning and memory occur there is a measurable change in the amount of neurotransmitter released, which thereby increases the neuron's efficiency in message transmission. Research with *Aplysia* (sea slugs) clearly demonstrates this effect (**Figure 7.8**).

Further evidence comes from research with genetically engineered "smart mice," which have extra receptors for a neurotransmitter named NMDA (N-methyl-d-aspartate). These mice performed significantly better on memory tasks than did normal mice (Tang et al., 2001; Tsien, 2000).

Although it is difficult to generalize from sea slugs and mice, research on long-term potentiation (LTP) in humans has been widely supportive (Berger et al., 2008; Kullmann & Lamsa, 2011; Shin et al., 2010).

### How does a sea slug learn and remember? • Figure 7.8

After repeated squirting with water, followed by a mild shock, the sea slug, *Aplysia*, releases more neurotransmitters at certain synapses. These synapses then become more efficient at transmitting signals that allow the slug to withdraw its gills when squirted. As a critical thinker, can you explain why this ability might provide an evolutionary advantage?

Dr. Bill Rudman/Australian Museum

## Hormonal Changes and Emotional Arousal

When stressed or excited, we naturally produce hormones that arouse the body, such as *epinephrine* and *cortisol* (Chapter 3). These hormones in turn affect the amygdala (a brain structure involved in emotion), which then stimulates the hippocampus and cerebral cortex (parts of the brain that are important for memory storage). Research has shown that direct injections of epinephrine or cortisol, or electrical stimulation of the amygdala, will increase the encoding and storage of new information (Hamilton & Gotlib, 2008; Jurado-Berbel et al., 2010; van Stegeren, 2008). However, prolonged or excessive stress and emotional arousal can increase, as well as interfere with, memory (Baucom et al., 2012; Heffelfinger & Newcomer, 2001; Quas et al., 2012).

The powerful effect of hormones on memory also can be seen in what are known as *flashbulb memories*—vivid images of circumstances associated with surprising or strongly emotional events (Brown & Kulik, 1977). In such situations, we secrete fight-or-flight hormones when we initially hear of the event and then replay the event in our minds again and again, which makes for stronger memories. Despite their intensity, flashbulb memories are not as accurate as you might think (Kvavilashvili et al., 2010; Lanciano, Curei, & Semin, 2010; Talarico & Rubin, 2007). For example, when asked how he heard the news of the September 11, 2001 attacks, President George W. Bush's answers contained several errors, which illustrates that flashbulb memories, like all other memories, are not perfect recordings of events (Greenberg, 2004).

## Where Are Memories Located?

Early memory researchers believed that memory was *localized*, or stored in a particular brain area. Later research suggests that, in fact, memory tends to be localized not in a single area but in many separate areas throughout the brain (**Figure 7.9**).

Today, research techniques are so advanced that we can experimentally induce and measure memory-related brain changes as they occur—on-the-spot reporting! For example, James Brewer and his colleagues (1998) used functional magnetic resonance imaging (fMRI) to locate areas of the brain responsible for encoding memories of pictures. They showed 96 pictures of indoor and outdoor scenes to participants while scanning their brains, and then later tested participants' ability to recall the pictures. Brewer and his colleagues identified the *right prefrontal cortex*

### The brain and memory • Figure 7.9

Damage to any one of these areas can affect encoding, storage, and retrieval of memories.

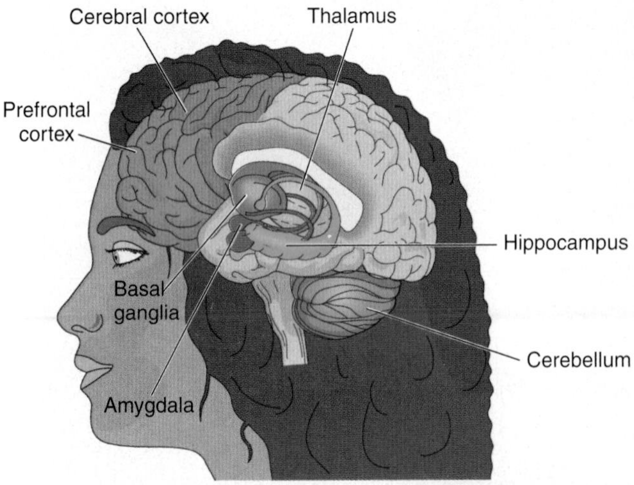


| | |
|---|---|
| Amygdala | Emotional memory (Gerber et al., 2008; Hamilton & Gotlib, 2008; Murty et al., 2011) |
| Basal ganglia and cerebellum | Creation and storage of the basic memory trace and implicit (nondeclarative) memories (such as skills, habits, and simple classical conditions responses) (Foerde & Shohamy, 2011; Stoodley, 2012; Thompson, 2005) |
| Hippocampal formation (hippocampus and surrounding area) | Memory recognition; implicit, explicit, spatial, episodic memory; declarative long-term memory; sequences of events (Gimbel & Brewer, 2011; Murty et al., 2011) |
| Thalamus | Formation of new memories and spatial and working memory (Aggleton et al., 2010; Carlesimo et al., 2011) |
| Cortex | Encoding of explicit (declarative) memories; storage of episodic and semantic memories; skill learning; printing; working memory (Depue, 2012; Herholz, Halpern, & Zatorre, 2012; Thompson, 2005) |

**Think Critically**

1. What effect might damage to the amygdala have on a person's relationships with others?
2. What effect do you think damage to the thalamus might have on a person's day-to-day functioning?

## Two types of amnesia • Figure 7.10

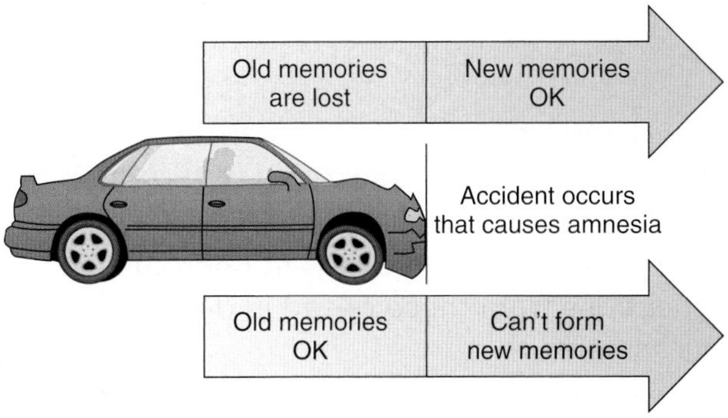


**a. Retrograde amnesia**
The person loses memories of events that occurred *before* the accident, yet has no trouble remembering things that happened afterward (old, "retro" memories are lost).

**b. Anterograde amnesia**
The person cannot form new memories for events that occur *after* the accident. Anterograde amnesia also may result from a surgical injury or from diseases such as chronic alcoholism.

and the *parahippocampal cortex* as being the most active during the encoding of the pictures. These are only two of several brain regions involved in memory storage.

## Biological Causes of Memory Loss

The leading cause of neurological disorders—including memory loss—among Americans between the ages of 15 and 25 is *traumatic brain injury*. These injuries most commonly result from car accidents, falls, blows, and gunshot wounds.

Traumatic brain injury happens when the skull suddenly collides with another object. The compression, twisting, and distortion of the brain inside the skull all cause serious and sometimes permanent damage to the brain. The frontal and temporal lobes often take the heaviest hit because they directly impact against the bony ridges inside the skull.

Loss of memory as a result of brain injury is called *amnesia*. Two major types of amnesia are **retrograde amnesia** and **anterograde amnesia** (**Figure 7.10**). Usually, retrograde amnesia is temporary. Unfortunately, anterograde amnesia is most often permanent, but patients show surprising abilities to learn and remember implicit/nondeclarative tasks (such as procedural motor skills).

Another way to understand these two forms of a amnesia is to compare them to a computer. Retrograde amnesia would be like having your latest stored data (memories) on your hard drive erased. In contrast, anterograde amnesia would be similar to disconnecting your keyboard form the hard drive—no new data (memories) could be entered or stored.

Like traumatic brain injuries, disease can alter the physiology of the brain and nervous system, affecting memory processes. For example, **Alzheimer's disease (AD)** is a progressive mental deterioration that occurs most commonly in old age (**Figure 7.11**). The most noticeable early symptoms are disturbances in memory, which become progressively worse until, in the final stages, the person fails to recognize loved ones, needs total nursing care, and ultimately dies.

**Alzheimer's disease (AD)** A progressive mental deterioration characterized by severe memory loss.

## The effect of Alzheimer's disease on the brain • Figure 7.11

**a. Normal brain**
Note the high amount of red and yellow color (signs of brain activity) in the positron emission tomography scans of the normal brain.

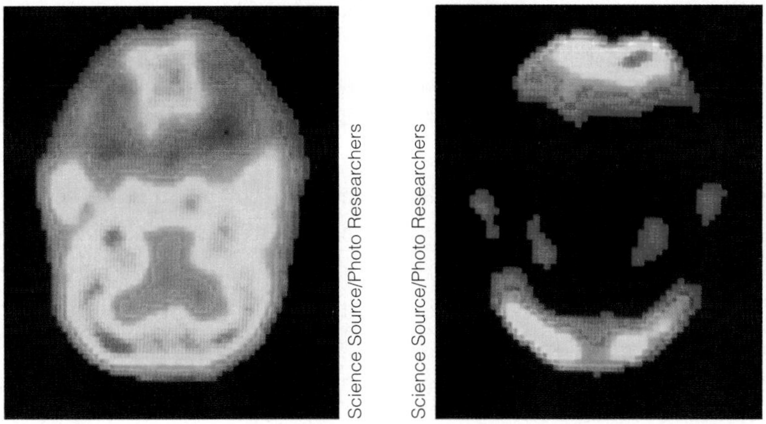

Science Source/Photo Researchers

Science Source/Photo Researchers

**b. Brain of an Alzheimer's disease patient**
The reduced activity in the brain of the Alzheimer's disease patient is evident. The loss is most significant in the temporal and parietal lobes, which indicates that these areas are particularly important for storing memories.

Alzheimer's does not attack all types of memory equally. A hallmark of the disease is an extreme decrease in *explicit/declarative memory* (Irish et al., 2011; Libon et al., 2007; Satler et al., 2007). Alzheimer's patients fail to recall facts, information, and personal life experiences, yet they still retain some *implicit/nondeclarative* memories, such as simple classically conditioned responses and procedural tasks, like brushing their teeth.

Brain autopsies of people with Alzheimer's show unusual *tangles* (structures formed from degenerating cell bodies) and *plaques* (structures formed from degenerating axons and dendrites). Hereditary Alzheimer's generally strikes its victims between the ages of 45 and 55. Some experts believe that the cause of Alzheimer's is primarily genetic, but others think that genetic makeup may make some people more susceptible to environmental triggers (Bekris et al., 2010; Ertekin-Taner, 2007; Persson et al., 2008; Sillén et al., 2011).

**CONCEPT CHECK** STOP

1. **Why** would animals raised in enriched environments develop different neuronal connections from those raised in deprived environments?
2. **How** do hormones contribute to flashbulb memories?
3. **How** do scientists know what parts of the brain are involved in memory?
4. **What** is the difference between retrograde and anterograde amnesia?

# Memory Distortions

## LEARNING OBJECTIVES

**RETRIEVAL PRACTICE** While reading the upcoming sections, respond to each Learning Objective in your own words. Then compare your responses with those in Appendix B.

1. **Explain** why our memories sometimes become distorted.
2. **Describe** problems with eyewitnesses in the criminal justice system.
3. **Distinguish** between false and repressed memories.

*One of my first memories would date, if it were true, from my second year. I can still see, most clearly, the following scene, in which I believed until I was about fifteen. I was sitting in my pram, which my nurse was pushing in the Champs-Élysées, when a man tried to kidnap me. I was held in by the strap fastened round me while my nurse bravely tried to stand between the thief and me. She received various scratches, and I can still see vaguely those on her face. Then a crowd gathered, a policeman with a short cloak and a white baton came up, and the man took to his heels. I can still see the whole scene, and can even place it near the tube station. When I was about fifteen, my parents received a letter from my former nurse saying that she had been converted to the Salvation Army. She wanted to confess her past faults, and in particular to return the watch she had been given as a reward on this occasion. She had made up the whole story, faking the scratches. I, therefore, must have heard, as a child, the account of this story, which my parents believed, and projected it into the past in the form of a visual memory, which was a memory of a memory, but false* (Piaget, 1962, pp. 187–188).

This is a self-reported childhood memory of Jean Piaget, a brilliant and world-famous cognitive and developmental psychologist. Why did Piaget create such a strange and elaborate memory for something that never happened?

There are several reasons why we shape, rearrange, and distort our memories. One of the most common is our need for *logic* and *consistency*. When we're initially forming new memories or sorting through the old ones, we fill in missing pieces, make "corrections," and rearrange information to make it logical and consistent with our previous experiences. If Piaget's beloved nurse said someone attempted to kidnap him, it was only logical for the boy to "remember" the event. Try the memory test in *Applying Psychology* to see how accurate your memory is.

## Applying Psychology

THE PLANNER

# A Memory Test

Carefully read through all the words in the following list.

| | |
|---|---|
| Bed | Drowse |
| Awake | Nurse |
| Tired | Sick |
| Dream | Lawyer |
| Wake | Medicine |
| Snooze | Health |
| Snore | Hospital |
| Rest | Dentist |
| Blanket | Physician |
| Doze | Patient |
| Slumber | Stethoscope |
| Nap | Curse |
| Peace | Clinic |
| Yawn | Surgeon |

Now cover the list and write down all the words you remember. Number of correctly recalled words:

21 to 28 words = excellent memory

16 to 20 words = better than most

12 to 15 words = average

8 to 11 words = below average

7 or fewer words = you might need a nap

How did you do? Do you have a good or excellent memory? Did you recall seeing the words *sleep* and *doctor*? Look back over the list. These words are not there. However, over 65% of students commonly report seeing these words. Why? As mentioned in the introduction to this chapter, memory is not a faithful duplicate of an event; it is a *constructive process*. We actively shape and build on information as it is encoded and retrieved.

**Think Critically**

1. Did you remember the words *bed* and *surgeon*? If so, why?
2. How might constructive memories create misunderstandings at work and in our everyday relationships?

We also shape and construct our memories for the sake of *efficiency*. We summarize, augment, and tie new information in with related memories in LTM. Similarly, when we need to retrieve the stored information, we leave out seemingly unimportant elements or misremember the details.

Despite all their problems and biases, our memories are normally quite accurate and serve us well in most situations. Our memories have evolved to encode, store, and retrieve vital information. Even while sleeping we process and store important memories. However, when faced with tasks like remembering precise details in a scholarly text, the faces and names of potential clients, or where we left our house keys, our brains are simply not as well equipped.

## Memory and the Criminal Justice System

When our memory errors involve the criminal justice system, they may lead to wrongful judgments of guilt or innocence and even life-or-death decisions.

In the past, one of the best forms of trial evidence a lawyer could have was an *eyewitness*—"I was there; I saw it with my own eyes." Unfortunately for lawyers, research has identified several problems with eyewitness testimony (Loftus, 2000, 2001, 2007, 2011; Paterson, Kemp, & Ng, 2011; Rubinstein, 2008). In fact, researchers have demonstrated that it is relatively easy to create false memories (Eslick et al., 2011; Frenda, Nichols, & Loftus, 2011; Loftus & Cahill, 2007; Strange et al., 2011; Zaragoza et al., 2011).

Do you recall our earlier discussion of the misinformation effect and how the experimenters created a false memory of a barn (page 188)? Participants were first asked to watch a film of a car driving through the countryside. Later, those who were asked to estimate how fast the car was going when it passed the barn (actually nonexistent) were six times as likely to later report that they had seen a barn in the film than participants who hadn't been asked about a barn (Loftus, 1982).

Problems with eyewitness recollections are so well established and important that judges now allow expert testimony on the unreliability of eyewitness testimony and routinely instruct jurors on its limits (Pezdek, 2012; Rubinstein, 2008). If you serve as a member of a jury or listen to accounts of crimes in the news, remind yourself of these problems. Also, keep in mind that research participants in eyewitness studies generally report their inaccurate memories with great self-assurance and strong conviction (Douglass & Pavletic, 2012; Jaeger, Cox, & Dobbins, 2012). Eyewitnesses to an actual crime may similarly identify—with equally high confidence—an innocent person as the perpetrator. In one experiment, participants watched people committing a staged crime. Only an hour later, 20% of the eyewitnesses identified innocent people from mug shots, and a week later, 8% identified innocent people in a lineup (Brown, Deffenbacher, & Sturgill, 1977). What memory processes might have contributed to the eyewitnesses' errors? (See **Figure 7.12**.)

## False Versus Repressed Memories

Like eyewitness testimony, false memories can have serious legal and social implications. Do you recall the opening story of Elizabeth, who suddenly remembered finding her mother's drowned body decades after it had happened? Elizabeth's recovery of these gruesome childhood memories, although painful, initially brought great relief. It also seemed to explain why she had always been fascinated by the topic of memory.

Then, her brother called to say there had been a mistake! The relative who told Elizabeth that she had been the one to discover her mother's body later remembered—and other relatives confirmed—that it had actually been Aunt Pearl, not Elizabeth Loftus. Loftus, an expert on memory distortions, had unknowingly created her own *false memory*.

Creating false memories may be somewhat common, but can we recover true memories that are buried in childhood? **Repression** is the supposed unconscious coping mechanism by which we prevent anxiety-provoking thoughts from reaching consciousness. According to some research, repressed memories are *actively* and *consciously* "forgotten" in an effort to avoid the pain of their retrieval (Anderson et al., 2004; Boag, 2012). Others suggest that some memories are so painful that they exist only in an *unconscious* corner of the brain, making them inaccessible to the individual (Haaken, 2010; Mancia & Baggott, 2008). In these cases, therapy supposedly would be necessary to unlock the hidden memories.

This is a complex and controversial topic in psychology. No one doubts that some memories are forgotten and later recovered. What is questioned is the concept of *repressed memories* of painful experiences (especially childhood sexual abuse) and their storage in the unconscious mind (Klein, 2012; Lambert, Good, & Kirk, 2010; Loftus & Cahill, 2007).

### Eyewitnesses and police lineups • Figure 7.12

As humorously depicted in this cartoon, officials now recommend that suspects should never "stand out" from the others in a lineup. Witnesses also are cautioned to not assume that the real criminal is in the lineup, and they should never "guess" when asked to make an identification.

*"Thank you, gentlemen—you may all leave except for No. 3."*

Critics suggest that most people who have witnessed or experienced a violent crime, or are adult survivors of childhood sexual abuse, have intense, persistent memories. They have trouble *forgetting*, not remembering. Some critics also wonder whether therapists sometimes inadvertently create false memories in their clients during therapy. Some worry that if a clinician even suggests the possibility of abuse, the client's own *constructive processes* may lead him or her to create a false memory. The client might start to incorporate portrayals of abuse from movies and books into his or her own memory, forgetting their original sources and eventually coming to see them as reliable.

This is not to say that all psychotherapy clients who recover memories of sexual abuse (or other painful incidents) have invented those memories. Indeed, the repressed memory debate has grown increasingly bitter, and research on both sides is hotly contested. The stakes are high because lawsuits and criminal prosecutions of sexual abuse are sometimes based on recovered memories of childhood sexual abuse. As researchers continue exploring the mechanisms underlying delayed remembering, we must be careful not to ridicule or condemn people who recover true memories of abuse. In the same spirit, we must protect the innocent from wrongful accusations that come from false memories. We look forward to a time when we can justly balance the interests of the victim with those of the accused.

**CONCEPT CHECK** STOP

1. **How** can our desire for logic, consistency, and efficiency thwart accurate memory?
2. **What** are some strategies used in the criminal justice system to improve eyewitness testimony?
3. **What** are the two sides of the debate on repressed memories of childhood sexual abuse?

# Summary

THE PLANNER

## 1 The Nature of Memory 176

- **Memory** is an internal representation of some prior event or experience. Major perspectives on memory include the **encoding, storage, and retrieval (ESR) model** and the **three-stage memory model**.
- In the ESR model shown in the diagram, information enters memory in three stages: **encoding**, **storage**, and **retrieval**. In contrast, the three-stage memory model proposes that information is stored and processed in **sensory memory**, **short-term memory (STM)**, and **long-term memory (LTM)**; these differ in purpose, duration, and capacity.
- **Chunking** and **maintenance rehearsal** improve STM's duration and capacity. Researchers think of STM as our **working memory**.
- **LTM** is an almost unlimited storehouse for information that must be kept for long periods. The two major types of **LTM** are **explicit/declarative memory** and **implicit/nondeclarative memory**. Organization and **elaborative rehearsal** improve encoding. **Retrieval cues** help stimulate retrieval of information from LTM. According to the **encoding specificity principle**, retrieval is improved when conditions of recovery are similar to encoding conditions.

**Encoding, storage, and retrieval • Figure 7.1**

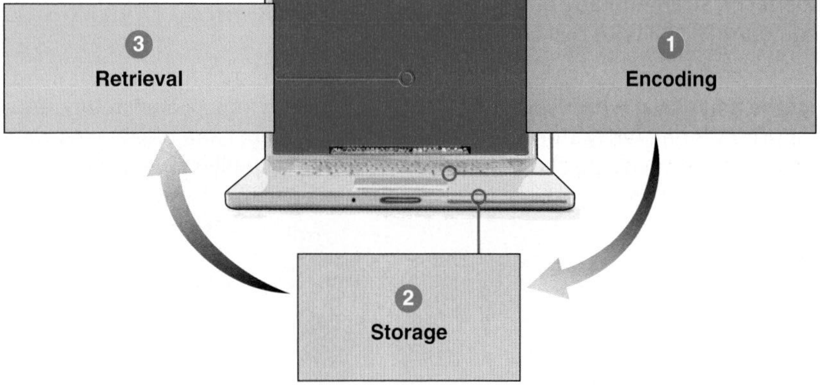

## 2 Forgetting 185

- Researchers have proposed that we forget information through decay, **retroactive** and **proactive interference**, motivated forgetting, encoding failure, and retrieval failure.
- Early research by Ebbinghaus showed that we tend to forget newly learned information quickly, but that we relearn the information more readily the second time, as shown in the graph.

**Psychological Science: How Quickly We Forget**

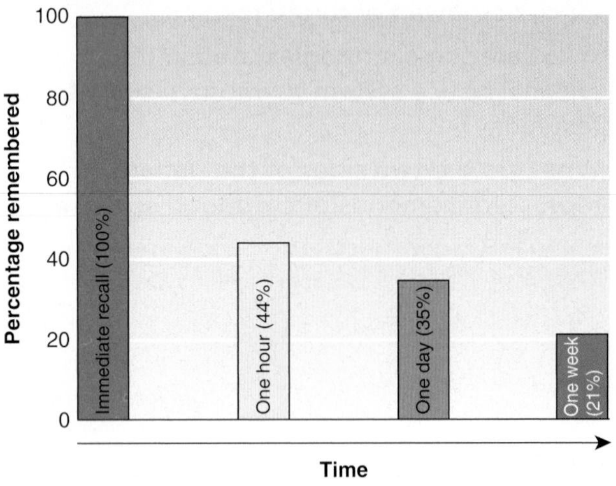


- Six factors that contribute to forgetting are the **misinformation effect**, the **serial position effect** (primacy and recency effects), **source amnesia**, the **sleeper effect**, spacing of practice (**distributed** versus **massed practice**), and culture.

## 3 Biological Bases of Memory 189

- Learning modifies the brain's neural networks through **long-term potentiation (LTP)**, strengthening particular synapses and affecting neurons' ability to release their neurotransmitters.
- Stress hormones affect the amygdala, which stimulates brain areas that are important for memory storage. Heightened arousal increases the encoding and storage of new information. Secretion of fight-or-flight hormones can contribute to "flashbulb" memories.
- Research using advanced techniques has indicated that several brain regions are involved in memory storage, as shown in the diagram.

**The brain and memory • Figure 7.9**

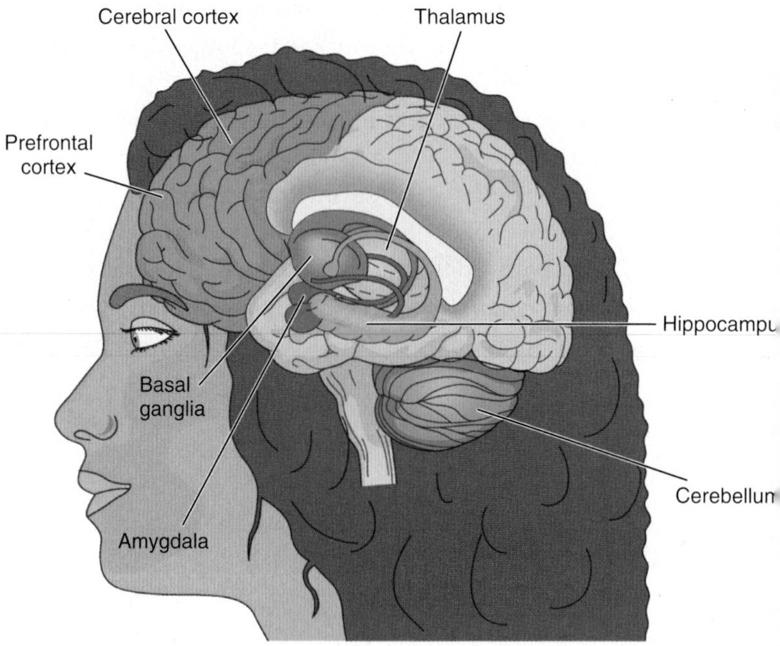


- Two major types of amnesia are **retrograde** and **anterograde amnesia**. Traumatic brain injuries and disease, such as **Alzheimer's disease (AD)**, can cause memory loss.

## 4 Memory Distortions 192

- People shape, rearrange, and distort memories in order to create logic, consistency, and efficiency. Despite all their problems and biases, our memories are normally accurate and usually serve us well.

- When memory errors involve the criminal justice system, they can have serious legal and social consequences. Problems with eyewitness recollections, as illustrated humorously in this cartoon, are well established. Judges often allow expert testimony on the unreliability of eyewitnesses.
- Memory **repression** (especially of childhood sexual abuse) is a complex and controversial topic. Critics note that most people who have witnessed or experienced a violent or traumatic event have intense, persistent memories. Critics worry that if a clinician suggests the possibility of abuse, the client's *constructive processes* may lead him or her to create a false memory of being abused. Researchers continue to explore delayed remembering.

**Eyewitnesses and police lineups • Figure 7.12**

*"Thank you, gentlemen—you may all leave except for No. 3."*

# Key Terms

**RETRIEVAL PRACTICE** Write a definition for each term before turning back to the referenced page to check your answer.

- Alzheimer's disease (AD) 191
- anterograde amnesia 191
- chunking 179
- distributed practice 188
- elaborative rehearsal 183
- encoding 176
- encoding specificity principle 185
- encoding, storage, and retrieval (ESR) model 176
- explicit/declarative memory 181
- implicit/nondeclarative memory 181
- long-term memory (LTM) 180
- long-term potentiation (LTP) 189
- maintenance rehearsal 179
- massed practice 188
- memory 176
- misinformation effect 188
- mnemonic devices 182
- proactive interference 186
- repression 194
- retrieval 176
- retrieval cue 184
- retroactive interference 186
- retrograde amnesia 191
- sensory memory 177
- serial position effect 188
- short-term memory (STM) 179
- sleeper effect 188
- source amnesia 188
- storage 176
- three-stage memory model 177
- tip-of-the-tongue (TOT) phenomenon 187
- working memory 180

# Creative and Critical Thinking Questions

1. If you were forced to lose one type of memory—sensory, short-term, or long-term—which would you select? Why?
2. Why might students do better on a test if they take it in the same seat and classroom where they originally studied the material?
3. What might be the evolutionary benefit of heightened (but not excessive) arousal enhancing memory?
4. Why might advertisers of shoddy services or products benefit from "channel surfing" if the television viewer is skipping from news programs to cable talk shows to infomercials?
5. As an eyewitness to a crime, how could you use information in this chapter to improve your memory for specific details? If you were a juror, what would you say to the other jurors about the reliability of eyewitness testimony?

# What is happening in this picture?

Associated Press

Many people feel they can remember events such as the September 11, 2001, terrorist attacks, shown here, or the explosion of the space shuttle *Challenger*, or the capture and death of Osama bin Laden with perfect clarity.

**Think Critically**

1. What are such memories called?
2. What biological process accounts for this kind of intense memory?
3. Are memories like this impervious to distortion?

# Self-Test

**RETRIEVAL PRACTICE** Completing this self-test and comparing your answers with those in Appendix C provides immediate feedback and helpful practice for exams. Additional interactive, self-tests are available at www.wiley.com/college/carpenter.

1. In a computer model of memory, (a) ______ would happen at the keyboard, (b) ______ on the monitor, and (c) ______ on the hard drive.

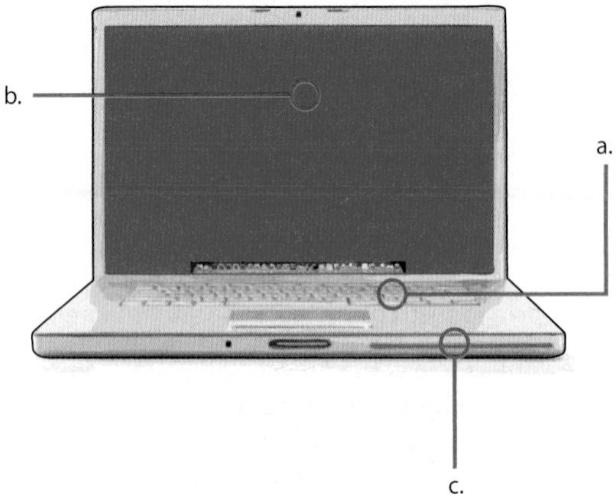


2. Label the three-stage memory model in the correct sequence.

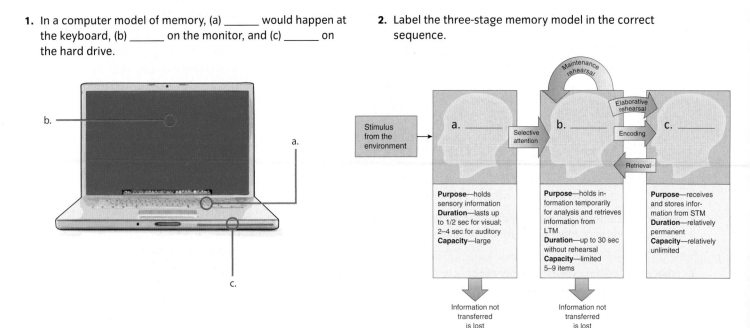

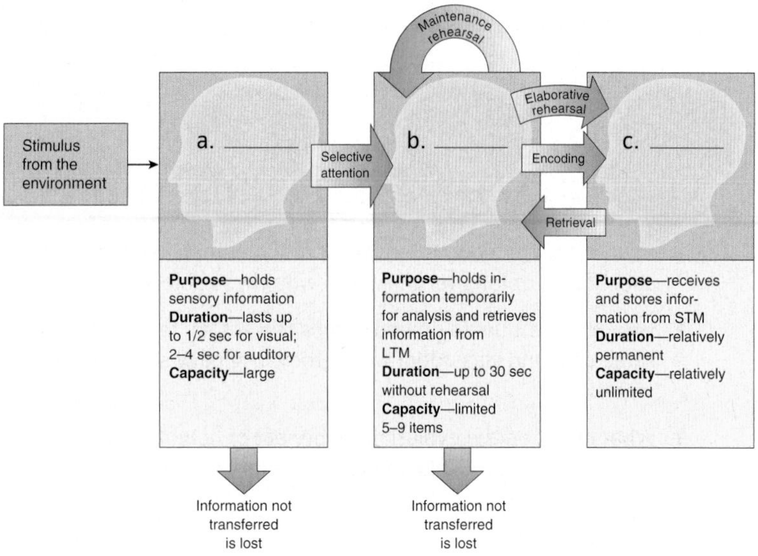

3. The following descriptions are characteristic of _____: information lasts only a few seconds or less, and a relatively large (but not unlimited) storage capacity.

a. perceptual processes     c. working memory

b. short-term storage     d. sensory memory

4. _____ is the process of grouping separate pieces of information into a single unit.

a. Chunking     c. Collecting

b. Cheating     d. Dual-coding

5. Label the two major systems of long-term memory (LTM).

```
        ┌─────────────────────────┐
        │   Varieties of          │
        │ long-term memory (LTM)  │
        └─────────────────────────┘
          │                     │
┌──────────────────┐   ┌──────────────────┐
│ a. _____   │   │ b. _____   │
│ Memory with      │   │ Memory without   │
│ conscious recall │   │ conscious recall │
└──────────────────┘   └──────────────────┘
```

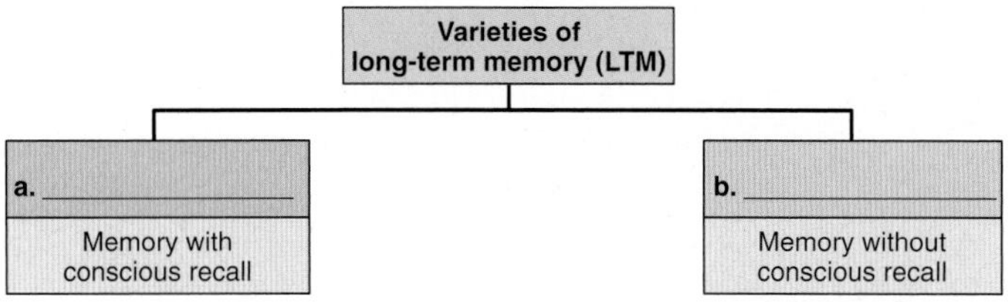

6. _____ devices improve memory by encoding items in a special way.

a. Eidetic imagery     c. Reverberating circuit

b. Mnemonic     d. ECS

7. In answering this question, the correct multiple-choice option may serve as a _____ for recalling accurate information from your long-term memory.

a. specificity code     c. retrieval cue

b. priming pump     d. flashbulb stimulus

8. The encoding specificity principle says that information retrieval is improved when _____.

a. both maintenance and elaborative rehearsal are used

b. reverberating circuits consolidate information

c. conditions of recovery are similar to encoding conditions

d. long-term potentiation is accessed

9. List the 5 major theories of forgetting:

a. _____

b. _____

c. _____

d. _____

e. _____

10. Distributed practice is a learning technique in which _____.

a. subjects are distributed across equal study sessions

b. learning periods alternate with nonlearning rest periods

c. learning decays faster than it can be distributed

d. several students study together, distributing various subjects according to their individual strengths

11. The long-lasting increase in neural excitability believed to be a biological mechanism for learning and memory is called _____.

a. maintenance rehearsal

b. adrenaline activation

c. long-term potentiation

d. the reverberating response

12. Label the two types of amnesia.

a. _____

Old memories are lost | New memories OK

Accident occurs that causes amnesia

Old memories OK | Can't form new memories

b. _____

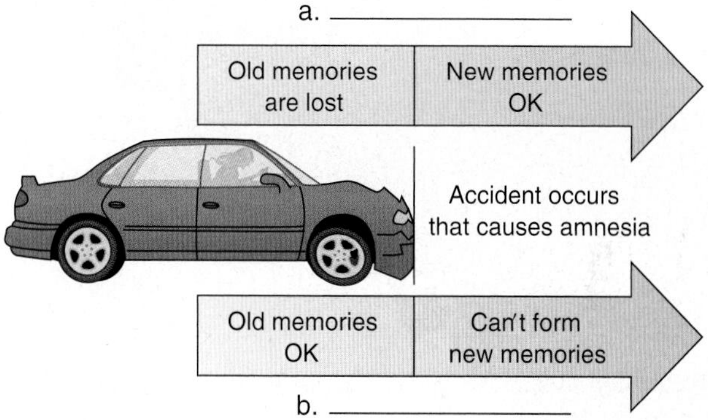

13. A progressive mental deterioration characterized by severe memory loss that occurs most commonly in the elderly is called _____.

a. retrieval loss

b. prefrontal cortex deterioration

c. Alzheimer's disease

d. age-related amnesia

14. Researchers have demonstrated that it is _____ to create false memories.

a. relatively easy     c. rarely possible

b. moderately difficult     d. never possible

15. _____ memories are related to anxiety-provoking thoughts or events that are supposedly prevented from reaching consciousness.

a. Suppressed     c. Motivated

b. Flashback     d. Repressed

**THE PLANNER** ✓

Review your Chapter Planner on the chapter opener and check off your completed work.

# 8 Thinking, Language, and Intelligence

What do you imagine or visualize when you think of intelligence? Many of us think of Nobel Prize winners, great inventors, or chess champions. But what about professional skateboarder Danny Way who rocketed down a 120-foot ramp at almost 50 miles an hour and leapt a 61-foot gap across the Great Wall of China? Success as a skateboarder obviously requires intelligence—perhaps of a different kind than people generally associate with being "smart."

Intelligence is a complex topic. We begin with an exploration of the mental processes involved in thinking, problem solving, and creativity. Then we look at the world of language—its components, development, and interrelationship with thought. We close with a review of how we define and measure intelligence. Along the way, you'll discover that how we think and use language are key aspects of what is generally referred to as intelligence, and why the three topics are combined into this one chapter.

Mike Ehrmann/Wire Image/Getty Images, Inc.

## CHAPTER OUTLINE



## CHAPTER PLANNER ✓

- ❑ Study the picture and read the opening story.
- ❑ Scan the Learning Objectives in each section: p. 202 ❑ p. 208 ❑ p. 213 ❑ p. 218 ❑
- ❑ Read the text and study all figures and visuals. Answer any questions.

**Analyze key features**

- ❑ Psychology InSight, p.203
- ❑ Process Diagram, p. 204
- ❑ Applying Psychology, p. 207
- ❑ What a Psychologist Sees p. 209 ❑ p. 220 ❑
- ❑ Study Organizer, p. 211
- ❑ Psychological Science, p. 222
- ❑ Stop: Answer the Concept Checks before you go on: p. 207 ❑ p. 213 ❑ p. 217 ❑ p. 223 ❑

**End of chapter**

- ❑ Review the Summary and Key Terms.
- ❑ Answer the Critical and Creative Thinking Questions.
- ❑ Answer What is happening in this picture?
- ❑ Complete the Self-Test and check your answers.

# Thinking

## LEARNING OBJECTIVES

**RETRIEVAL PRACTICE** While reading the upcoming sections, respond to each Learning Objective in your own words. Then compare your responses with those in Appendix B.

1. **Describe** the roles of mental images and concepts in thinking.
2. **Describe** the three stages of problem solving.
3. **Identify** the barriers to problem solving.
4. **Explain** the characteristics associated with creativity.

Thinking, language, and intelligence are closely related facets of **cognition**. Every time we take in information and mentally act on it, we're thinking. And these thought processes are both localized and distributed throughout our brains in networks of neurons. For example, during problem solving or decision making, our brains are most active in the **prefrontal cortex**. This region associates complex ideas; makes plans; forms, initiates, and allocates attention; and supports multitasking. In addition to the localization of thinking processes, the prefrontal cortex also links to other areas of the brain, such as the limbic system (Chapter 2), to synthesize information from several senses (Anderson et al., 2011; Banich & Compton, 2011; Carlson, 2011; Sacchetti, Sacco, & Strata, 2007). Now that we know where thinking occurs, we need to discuss its basic components.

**cognition** Mental activities involved in acquiring, storing, retrieving, and using knowledge.

### Mental imagery • Figure 8.1

Some of our most creative and inspired moments come when we're forming and manipulating mental images. This climber is probably visualizing her next move, and her ability to do so is critical to her success.

Blend Images/Superstock

## Cognitive Building Blocks

Imagine yourself lying relaxed in the warm, gritty sand on an ocean beach. Do you see palms swaying in the wind? Can you smell the salty sea and taste the dried salt on your lips? Can you hear children playing in the surf? What you've just created is a **mental image** (**Figure 8.1**), a mental representation of a previously stored sensory experience, which includes visual, auditory, olfactory, tactile, motor, and gustatory imagery (McKellar, 1972). We all have a mental space where we visualize and manipulate our sensory images (Borst et al., 2011; Moulton & Kosslyn, 2011; Schifferstein & Hilscher, 2010).

In addition to mental images, our thinking also involves forming **concepts**, or mental representations of a group or category (Smith, 1995). Concepts can be concrete (*car, concert*) or abstract (*intelligence, pornography*). They are essential to thinking and communication because they simplify and organize information. Normally, when you see a new object or encounter a new situation, you relate it to your existing conceptual structure and categorize it according to where it fits. For example, if you see a metal box with four wheels driving on the highway, you know it is a car, even if you've never seen that particular model before. How do we learn concepts? They develop through the environmental interactions of three major building blocks: *prototypes*, *artificial concepts*, and *hierarchies* (**Figure 8.2**).

## Psychology InSight Concepts • **Figure 8.2**

When forming concepts, we most often use prototypes, artificial concepts, and hierarchies to simplify and categorize information. For example, when we encounter a bird, we fit it into our existing concept of a bird.

Stephen St. John/NG Image Collection

**a. Prototypes**
When initially learning about the world, a young child develops a general, natural concept based on a typical representative, or **prototype,** of *bird* after a parent points out a number of examples. Once the child develops the prototype of a *bird*, he or she then is able to quickly classify all flying animals, such as this robin, correctly.

**b. Artificial concepts**
We create **artificial concepts** from logical rules or definitions. When an example doesn't quite fit the prototype, like this penguin, we must review our artificial concept of a bird: warm-blooded animals that fly, have wings and beaks, and lay eggs. Although this penguin doesn't fly, it has wings, a beak, and lays eggs. So it must be a bird.

Gordon Wiltsie/NG Image Collection

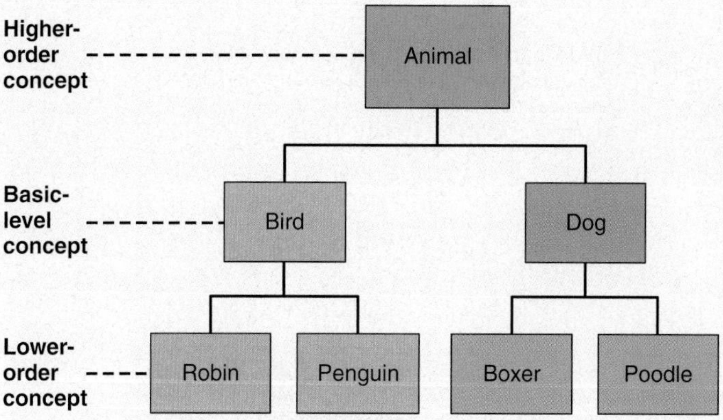


**c. Hierarchies**
Creating **hierarchies,** or subcategories within broader concepts, makes mastering new material faster and easier. Note, however, that we tend to begin with basic-level concepts (the middle row on the diagram) when we first learn something (Rosch, 1978). For example, a child develops the basic-level concept for bird before learning the higher-order concept *animal* or the lower-order concept *robin.*

## Solving Problems

Several years ago in Los Angeles, a 12-foot-high tractor-trailer got stuck under a bridge that was 6 inches too low. After hours of towing, tugging, and pushing, the police and transportation workers were stumped. Then a young boy happened by and asked, "Why don't you let some air out of the tires?" It was a simple, creative suggestion—and it worked.

Our lives are filled with problems, some simple, some difficult. In all cases, **problem solving** requires moving from a given state (the problem) to a goal state (the solution), a process that usually involves three steps:

PROCESS DIAGRAM

## Three steps to the goal • Figure 8.3

THE PLANNER

There are three stages of problem solving that help you attain your goal, for example, moving to a new home.

### 1 Preparation

Begin by clarifying the problem using these three steps in preparation.

• Define the ultimate goal.

Move to a new home close to work.

• Outline your limits and/or desires.

✓ Must allow pets.
✓ Must be close enough to walk.
✓ I prefer a house to an apartment building.
✓ Fireplaces are nice.

• Separate the negotiable from the nonnegotiable.

✓ Must allow pets.
✓ Must be close enough to walk.
* I prefer a house to an apartment building.
* Fireplaces are nice.

### 2 Production

Next, test your possible paths and solutions with one or both of these methods.

• Use an **algorithm**, a logical step-by-step procedure that, if followed correctly, will eventually solve the problem. But algorithms may take a long time—especially for complex problems.

Look at every ad in the paper and call all of those that allow pets.

• Use a **heuristic**, a simple rule for problem solving that does not guarantee a solution, but offers a likely shortcut to it.

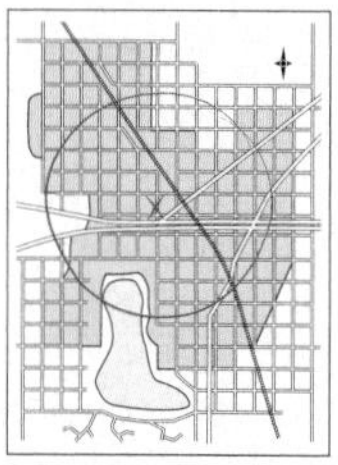

Work backwards from the solution—start by drawing a 1-mile radius around work to narrow the search.

### 3 Evaluation

Did your possible solutions solve the problem?

• If no, then you must return to the production and/or preparation stages.

• If yes, then take action to achieve your goal.

*preparation*, *production*, and *evaluation* (Bourne, Dominowski, & Loftus, 1979) (**Figure 8.3**).

In addition to these three steps, we also sometimes solve problems with a sudden flash of **insight**, like Köhler's chimps who stacked boxes to reach the bananas (Chapter 6). On other occasions, we may mentally "set aside" our problem for a while, a so-called **incubation period**, and then find that the solution comes to mind without further conscious thought.

## Barriers to Problem Solving

Why are some problems so difficult to solve? It may be because we often stick to problem-solving strategies (**mental sets**) that worked in the past, rather than trying new, possibly more effective ones (**Figure 8.4**). Or we fail to let our inventive instincts run free, thinking of objects as functioning only in their prescribed, customary way—a phenomenon called **functional fixedness**.

When a child uses soft cushions to build a fort, or you use a table knife instead of a screwdriver to tighten a screw, you both have successfully avoided functional fixedness. Similarly, the individual who discovered a way to retrofit diesel engines to allow them to use discarded restaurant oil as fuel has overcome functional fixedness—and may become a very rich man! See **Figure 8.5** for practice with functional fixedness.

Other barriers to effective problem solving stem from our tendency to ignore important information. Have you ever caught yourself agreeing with friends who support your political opinions, and discounting conflicting opinions? This inclination to seek confirmation for our preexisting beliefs and to overlook contradictory evidence is known as the **confirmation bias** (Christandl, Fetchenhauer, & Hoelzl, 2011; Nickerson, 1998; Reich, 2004).

British researcher Peter Wason (1968) first demonstrated the confirmation bias. He asked participants to generate a list of numbers that conformed to the same rule that applied to this set of numbers:

2 4 6

Hypothesizing that the rule was "numbers increasing by two," most participants generated sets such as (4, 6, 8) or (1, 3, 5). Each time, Wason assured them that their sets of numbers conformed to the rule but that the rule "numbers increasing by two" was incorrect. The problem was that the participants were searching only for information that confirmed their hypothesis. Proposing a series such as (1, 3, 4) would have led them to reject their initial hypothesis and discover the correct rule: "numbers in increasing order of magnitude."

In addition to *functional fixedness* and the *confirmation bias*, another barrier to problem solving sometimes results from *heuristics* (see again Figure 8.3). For example, when we use the **availability heuristic**, we take a mental shortcut based on the ease with which we can recall other similar instances or events. In other words, we conclude "if we can easily think of it, it must be important" (Fortune & Goodie, 2012; Kahneman & Tversky, 1973; Miller et al., 2012; Tversky & Kahneman, 1974, 1993). Someone who easily recalls images of his or her grandfather smoking cigarettes his entire life

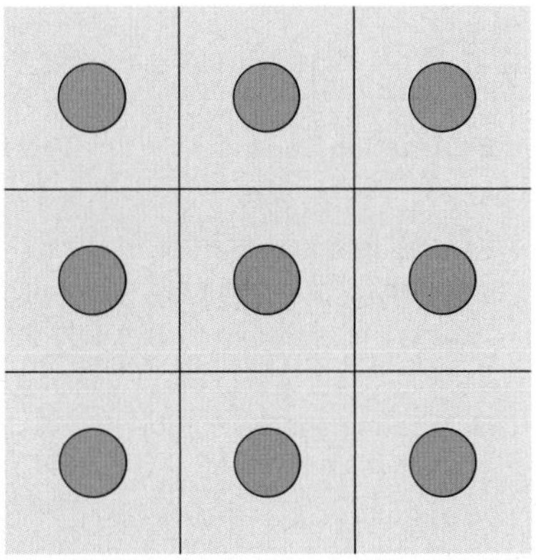

### The nine-dot problem • Figure 8.4

Without lifting your pencil, can you draw no more than four lines to connect all nine dots? If not, it may be because of *mental sets*—you're trying to use problem-solving strategies that have worked well for you in the past. Try "thinking outside the box," then compare your answer to the solution presented at the end of the chapter.

### Overcoming functional fixedness • Figure 8.5

Can you use these supplies to mount the candle on a wall so that it can be lit in a normal way—without toppling over? The solution is at the end of the chapter.

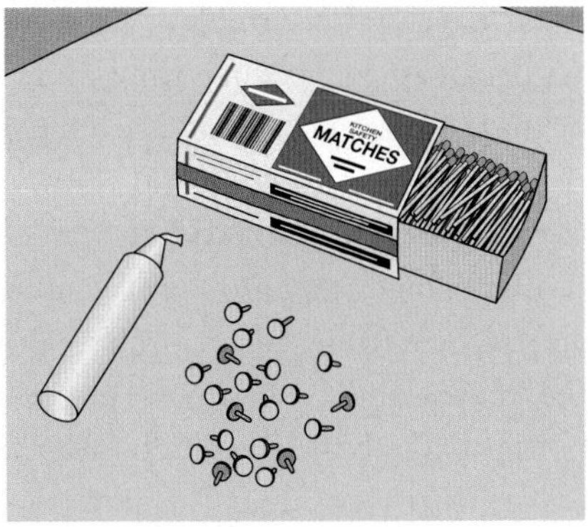

**Three elements of creative thinking Table 8.1**

| | Explanations | Thomas Edison examples |
|---|---|---|
| **Originality** | Seeing unique or different solutions to a problem | After noting that electricity passing through a conductor produces a glowing red or white heat, Edison imagined using this light for practical applications. |
| **Fluency** | Generating a large number of possible solutions | Edison tried literally hundreds of different materials to find one that would heat to the point of glowing white without burning up. |
| **Flexibility** | Shifting with ease from one type of problem-solving strategy to another | When he couldn't find a long-lasting material, Edison tried heating it in a vacuum—thereby creating the first light bulb. |

Courtesy U. S. Department of the Interior, National Parks Service, Edison National Historic Site

and living to be 100 may fail to recognize the uniqueness of this one personal example, and thereby ignore or distort the statistically based true dangers of smoking. Similarly, sensationalistic media coverage of occasional instances of bullying on a school bus or kidnapping may create an erroneous availability heuristic, which encourages many parents to needlessly drive their children to and from school.

Like the availability heuristic, the **representativeness heuristic** also sometimes hinders problem solving. Using this heuristic, we estimate the probability of something based on how well the circumstances match (*represent*) our prototype (Fisk, Burg, & Holden, 2006; Read & Grushka-Cockayne, 2011). For example, if John is six feet, five inches tall, we may guess that he is an NBA basketball player rather than, say, a bank president. But in this case, the representative heuristic ignores *base rate information*—the probability of a characteristic occurring in the general population. In the United States, bank presidents outnumber NBA players by about 50 to 1. So despite his height, John is much more likely to be a bank president.

Before we go on, it's important to note that although the availability and representativeness heuristics are presented here as "barriers" to problem solving, they more often help us because they reduce time and effort. They allow us to "make inferences that are fast, frugal, and accurate" (Todd & Gigerenzer, 2000, p. 736). A prime example is the **recognition heuristic**, in which judgments are made by relying on one single cue (recognition), while ignoring other information. For example, when students in the United States and Germany were presented with two sample cities (e.g., San Diego and San Antonio) and asked to choose which city had the largest population, both Germans and Americans scored well because even when they had little information they chose the more recognizable city as being the larger of the two, which was correct at the time (Gigerenzer, 1996; Gigerenzer & Goldstein, 2011).

## Creativity

What makes a person creative? Conceptions of creativity are subject to cultural relevance and to whether a solution or performance is considered useful at the time. In general, three characteristics are associated with **creativity**: *originality*, *fluency*, and *flexibility*. Thomas Edison's invention of the light bulb offers a prime example of each of these characteristics (**Table 8.1**).

**creativity** The ability to produce valued outcomes in a novel way.

Most tests of creativity focus on **divergent thinking**, a type of thinking where many possibilities are developed from a single starting point (Baer, 1994). For example, in the Unusual Uses Test, people are asked to think of as many uses as possible for an object, such as a brick. In the Anagrams Test, people are asked to reorder the letters in a word to make as many new words as possible. To test your own creativity, try the activities in *Applying Psychology*.

A classic example of divergent thinking is the decision of Xiang Yu, a Chinese general in the third century B.C., to crush his troops' cooking pots and burn their ships. One might think that no general in his right mind would make such a decision, but Xiang Yu explained that his purpose was to focus the troops on moving forward, as they had no hope of retreating. His divergent thinking was rewarded with victory on the battlefield.

# Applying Psychology

THE PLANNER

## Are You Creative?

Everyone exhibits a certain amount of creativity in some aspects of life. Even when doing ordinary tasks, like planning an afternoon of errands, you are being somewhat creative. Similarly, if you've ever tightened a screw with a penny or used a thick book on a chair as a booster seat for a child, you've found creative solutions to problems.

Would you like to test your own creativity?

- Find 10 coins and arrange them in the configuration shown here. By moving only two coins, form two rows that each contain six coins. The solution is at the end of the chapter.
- In five minutes, see how many words you can make using the letters in the word *hippopotamus*.
- In five minutes, list all the things you can do with a paper clip.

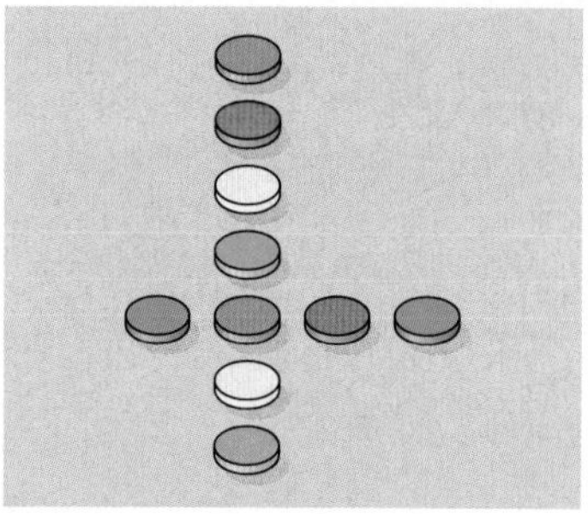

**Think Critically**

1. How did you do? If you had trouble with any of these tasks, can you identify which of the three characteristics of creativity (originality, fluency, or flexibility) best explains your lack of a solution? How might you use this information to improve your overall creativity?
2. Creativity is usually associated with art, poetry, and the like. Can you think of other areas in which creativity is highly valued?

One prominent theory of creativity is Robert J. Sternberg and Todd Lubart's **investment theory** (1992, 1996). According to this theory, creative people tend to "buy low" in the realm of ideas, championing ideas that others dismiss (much like a bold entrepreneur might invest in low-priced, unpopular stocks, believing that their value will rise). Once their creative ideas are highly valued, they "sell high" and move on to another unpopular but promising idea. Investment theory also suggests that creativity requires the coming together of six interrelated resources (Mieg, 2011; Sternberg, 2010; Sternberg & Lubart, 1996). These resources are summarized in **Table 8.2**. One way to improve your personal creativity is to study this list and then strengthen yourself in those areas you think need improvement.

**CONCEPT CHECK** STOP

1. **What** is the difference between prototypes and artificial concepts?
2. **When** are heuristics more appropriate for problem solving than algorithms?
3. **What** is an example of the availability heuristic?
4. **How** does investment theory explain creativity?

**Resources of creative people Table 8.2**

| | |
|---|---|
| **Intellectual ability** | Enough intelligence to see problems in a new light |
| **Knowledge** | Sufficient basic knowledge of the problem to effectively evaluate possible solutions |
| **Thinking style** | Novel ideas and ability to distinguish between the worthy and worthless |
| **Personality** | Willingness to grow and change, take risks, and work to overcome obstacles |
| **Motivation** | Sufficient motivation to accomplish the task and more internal than external motivation |
| **Environment** | An environment that supports creativity |

Michael Travis/Corbis

Which resources best explain Lady Gaga's phenomenal success?

# Language

## LEARNING OBJECTIVES

**RETRIEVAL PRACTICE** While reading the upcoming sections, respond to each Learning Objective in your own words. Then compare your responses with those in Appendix B.

1. **Identify** the building blocks of language.
2. **Describe** the prominent theories of how language and thought interact.
3. **Describe** the major stages of language development.
4. **Review** the evidence that nonhuman animals are able to learn and use language.

Using **language** enables us to mentally manipulate symbols, thereby expanding our thinking. Whether it's spoken, written, or signed, language also allows us to communicate our thoughts, ideas, and feelings.

To produce language, we first build words using **phonemes** [FO-neems] and **morphemes** [MOR-feems]. Then we string words into sentences using rules of **grammar** (*syntax* and *semantics*) (**Figure 8.6**).

What happens in our brains when we produce and comprehend language? Recall from Chapter 2 that for most of us our language centers are located in the left frontal lobe, with *Broca's area* linked to speech production and *Wernicke's area* being important for language comprehension (*What a Psychologist Sees*). Keep in mind, however, that in our everyday conversations both areas are active at the same time, along with other parts of the brain.

**language** A form of communication using sounds and symbols combined according to specified rules.

**phoneme** The smallest basic unit of speech or sound. The English language has about 40 phonemes.

**morpheme** The smallest meaningful unit of language, formed from a combination of phonemes.

**grammar** Rules that specify how phonemes, morphemes, words, and phrases should be combined to express meaningful thoughts. These rules include syntax and semantics.

**Building blocks of language • Figure 8.6**

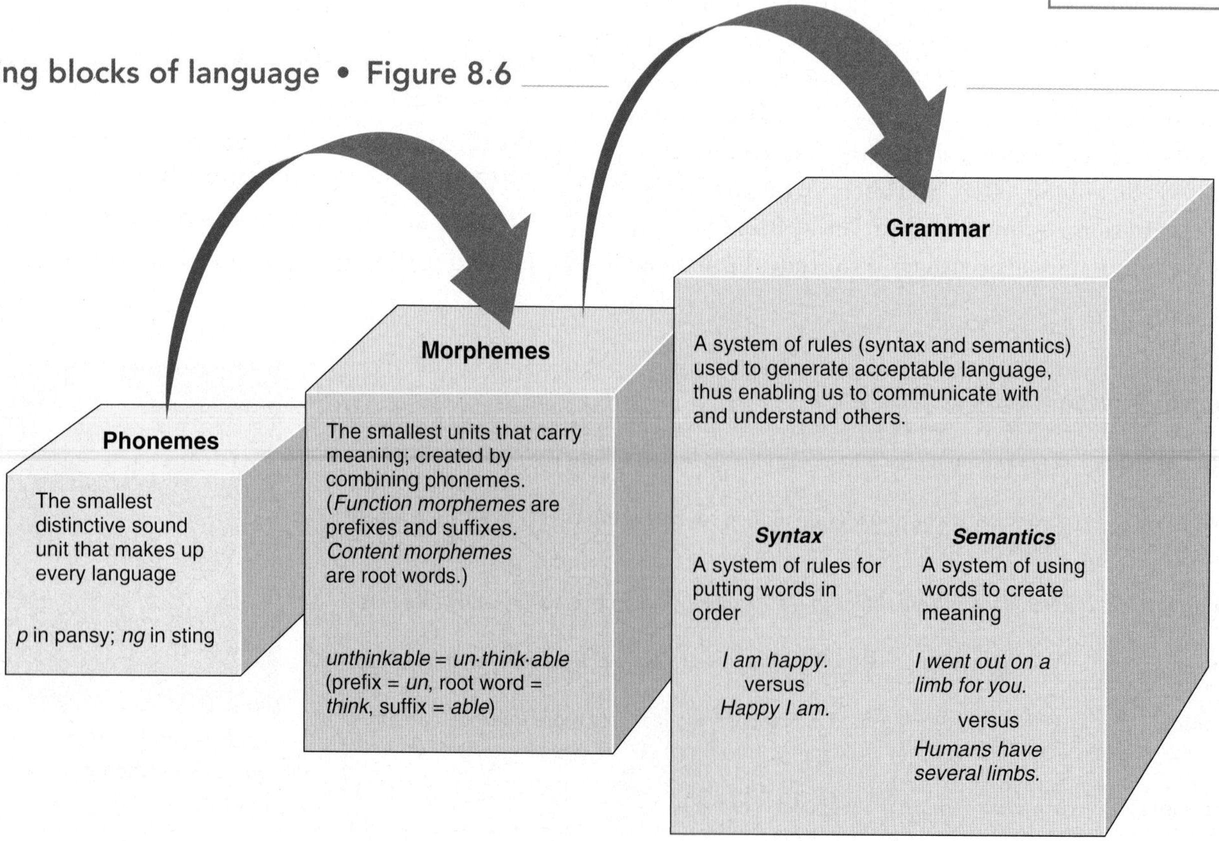

Phonemes
The smallest distinctive sound unit that makes up every language
*p* in pansy; *ng* in sting
Morphemes
The smallest units that carry meaning; created by combining phonemes. (*Function morphemes* are prefixes and suffixes. *Content morphemes* are root words.)
*unthinkable* = *un·think·able* (prefix = *un*, root word = *think*, suffix = *able*)
Grammar
A system of rules (syntax and semantics) used to generate acceptable language, thus enabling us to communicate with and understand others.
Syntax
A system of rules for putting words in order
*I am happy.* versus *Happy I am.*
Semantics
A system of using words to create meaning
*I went out on a limb for you.* versus *Humans have several limbs.*

## WHAT A PSYCHOLOGIST SEES

THE PLANNER

### Language and the Brain

Language, just like our thought processes, is both localized and distributed throughout our brain (**Figure a**). For example, the amygdala is active when we engage in a special type of language—cursing or swearing. Why? Recall from Chapter 2 that the amygdala is linked to emotions, especially fear and rage. So it's logical that the brain regions activated by swearing or hearing swear words would be the same as those for fear and aggression.

Scientists can track brain activity through a colored *positron emission tomography (PET) scan*. Injection of the radioactive isotope oxygen-15 into the bloodstream of the subject makes areas of the brain with high metabolic activity "light up" in red and orange on the scan (**Figures b–d**).

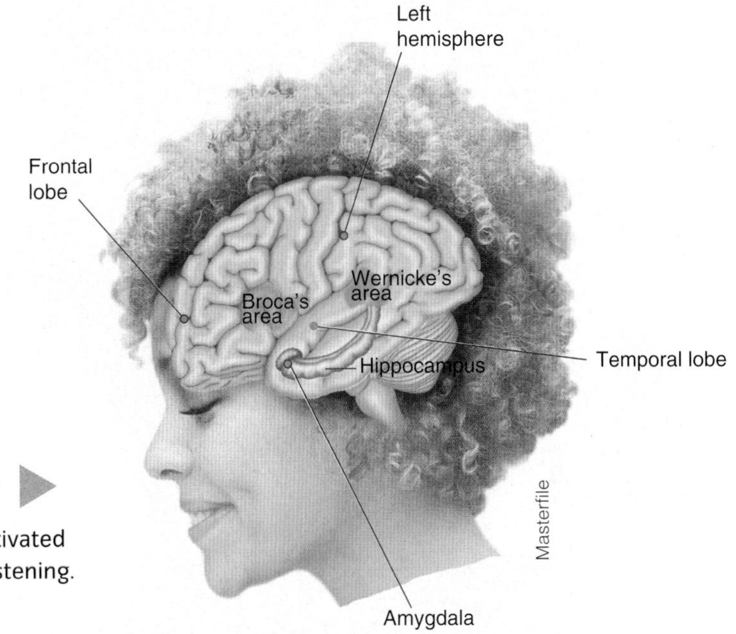


a. Broca's area is responsible for speech generation, and Wernicke's area is responsible for language comprehension. Other parts of the brain also are activated during different types of language generation and listening.

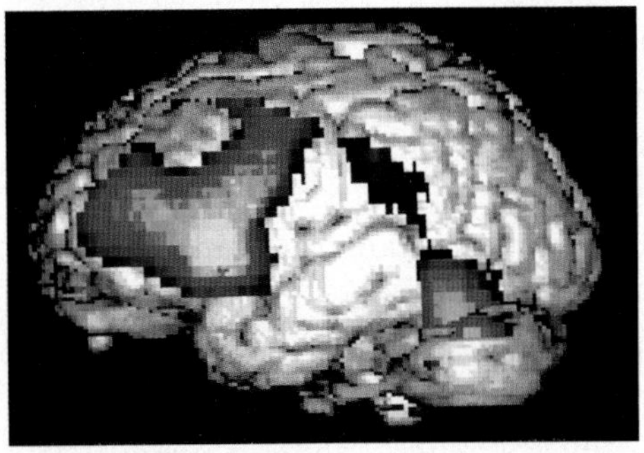

WDCN/Univ. College London/ Photo Researchers

b. Language generated in the frontal lobe (center left) has its cognition checked in the temporal lobe (lower right).

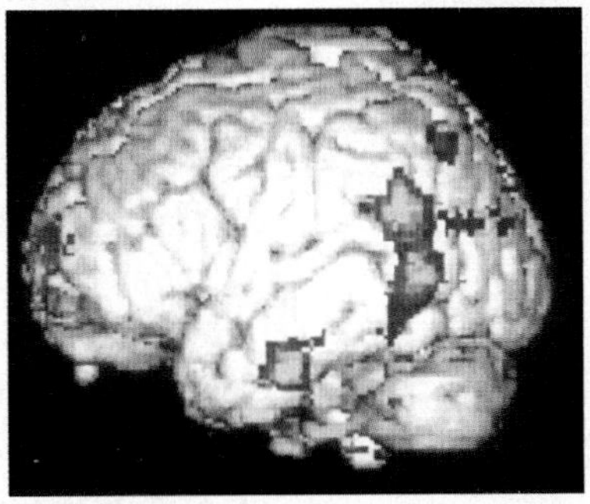

WDCN/Univ. College London/ Photo Researchers

c. Working out the meaning of heard words makes areas of the temporal lobe light up.

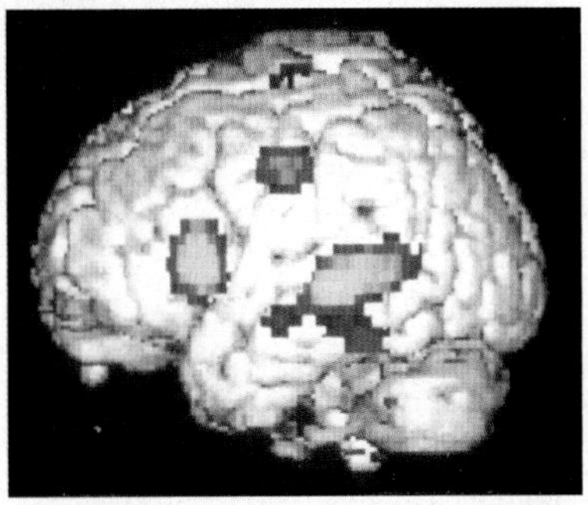

WDCN/Univ. College London/ Photo Researchers

d. Repeating words increases activity in Broca's area and Wernicke's area, as well as a motor region responsible for pronouncing words (reddish area at the top).

## Language and Thought

Does the fact that you speak English instead of Spanish—or Chinese instead of Swahili—determine how you reason, think, and perceive the world? Linguist Benjamin Whorf (1956) believed so. As evidence for his **linguistic relativity hypothesis**, Whorf offered a now classic example: Because Inuits (previously known as Eskimos) supposedly have many words for snow (*apikak* for "first snow falling," *pukak* for "snow for drinking water," and so on), they can reportedly perceive and think about snow differently from English speakers, who have only one word—*snow*.

Though intriguing, Whorf's hypothesis has not fared well. He apparently exaggerated the number of Inuit words for snow (Pullum, 1991) and ignored the fact that English speakers have a number of terms to describe various forms of snow, such as *slush*, *sleet*, *hard pack*, and *powder*.

Myrleen Ferguson Cafe/PhotoEdit

**Can you identify this emotion? • Figure 8.7**

Infants as young as 2.5 months can nonverbally express emotions, such as joy, surprise, or anger.

Other research has directly contradicted Whorf's theory. For example, Eleanor Rosch (1973) found that although people of the Dani tribe in New Guinea possess only two color names—one indicating cool, dark colors, and the other describing warm, bright colors—they discriminate among multiple hues as well as English speakers do.

Whorf apparently was mistaken in his belief that language *determines* thought. But there is no doubt that language *influences* thought (Deutscher, 2010; Jarvis, 2011; Walther et al., 2011). People who speak both Chinese and English report that the language they're currently using affects their sense of self (Berry et al., 2011; Matsumoto, 2010). When using Chinese, they tend to conform to Chinese cultural norms; when speaking English, they tend to adopt Western norms.

Our words also influence the thinking of those who hear them. That's why companies avoid *firing* employees; instead, employees are *outplaced* or *nonrenewed*. Similarly, the military uses terms like *preemptive strike* to cover the fact that they attacked first and *tactical redeployment* to refer to a retreat. And research has shown that consumers who receive a *rebate* are less likely to spend the money than those who receive a *bonus* (Epley, 2008).

## Language Development

From birth, a child communicates through facial expressions, eye contact, and body gestures (**Figure 8.7**). Babies only hours old begin to "teach" their caregivers when and how they want to be held, fed, and played with.

Eventually, children also communicate verbally, progressing through several distinct stages of language acquisition. These stages are summarized in **Study Organizer 8.1**. By age 5, most children have mastered basic grammar and typically use about 2,000 words (a level of mastery considered adequate for getting by in any given culture). Past this point, vocabulary and grammar gradually improve throughout life (Owens, 2011; Rowe & Levine, 2011).

Some theorists believe that language capability is innate, primarily a matter of maturation. Noam Chomsky (1968, 1980) suggests that children are "prewired" with a neurological ability known as a **language acquisition device (LAD)** that enables a child to analyze language and to extract the basic rules of grammar. This mechanism needs only minimal exposure to adult speech to unlock its potential. As evidence for this nativist position, Chomsky observes that children everywhere progress through the same stages of language development at about the same ages. He also notes that babbling is the same in all languages and that deaf babies babble just like hearing babies.

Nurturists argue that the nativist position doesn't fully explain individual differences in language development. They hold that children learn language through a complex system of rewards, punishments, and imitation. For example, parents smile and encourage any vocalizations from a very young infant. Later, they respond even more enthusiastically when the infant babbles "mama" or "dada." In this way, parents unknowingly use *shaping* (Chapter 6) to help babies learn language (**Figure 8.8**).

## Study Organizer 8.1 Language acquisition

THE PLANNER

### Prelinguistic stage

**Birth to 12 months**

iStockphoto

| FEATURES | EXAMPLES |
|---|---|
| Crying (reflexive in newborns) becomes more purposeful | hunger cry, anger cry, and pain cry |
| **Cooing** (vowel-like sounds) at 2–3 months | "ooooh," "aaaah" |
| **Babbling** (consonants added) at 4–6 months | "bahbahbah," "dahdahdah" |

### Linguistic stage

**12 months to 2 years**

iStockphoto

| FEATURES | EXAMPLES |
|---|---|
| Babbling resembles language of the environment, and child understands sounds relate to meaning | |
| Speech consists of one-word utterances | "Mama," "juice," "up" |
| Expressive ability more than doubles once words are joined into short phrases | "Daddy milk," "no night-night!" |
| **Overextension** (using words to include objects that do not fit the word's meaning) | all men = "Daddy"<br>all furry animals = "doggy" |

**2 years to 5 years**

iStockphoto

| FEATURES | EXAMPLES |
|---|---|
| **Telegraphic speech** (like telegrams, omits nonessential connecting words) | "Me want cookie"<br>"Grandma go bye-bye?" |
| Vocabulary increases at a phenomenal rate | |
| Child acquires a wide variety of grammar rules | adding *–ed* for past tense, adding s to form plurals |
| **Overgeneralization** (applying basic rules of grammar even to cases that are exceptions to the rule) | "I goed to the zoo" "Two mans" |

© image100/Age Fotostock America, Inc.

### Nature or nurture? • Figure 8.8

Both sides of the nature-nurture controversy have staunch supporters. However, most psychologists believe that language acquisition is a combination of both biology (nature) and environment (nurture) (Hoff, 2009; Plomin, De Fries, & Fulker, 2007). Can you see how both might contribute to this child's pretend phone conversation?

Ron Cohn/The Gorilla Foundation/koko.org

**Signing • Figure 8.9**

According to her teacher, Penny Patterson, Koko has used ASL to converse with others, talk to herself, joke, express preferences, and even lie (Linden, 1993; Patterson, 2002).

## Can Humans Talk with Nonhuman Animals?

Without question, nonhuman animals communicate, sending warnings, signaling sexual interest, sharing locations of food sources, and so on. But can nonhuman animals master the complexity of human language? Since the 1930s, many language studies have attempted to answer this question by probing the language abilities of chimpanzees, gorillas, and other animals (e.g., Berwick et al., 2011; Call, 2011; Lyn et al., 2011; May-Collado, 2010).

One of the most successful early studies was conducted by Beatrice and Allen Gardner (1969), who recognized chimpanzees' manual dexterity and ability to imitate gestures. The Gardners used American Sign Language (ASL) with a chimp named Washoe. By the time Washoe was 4 years old, she had learned 132 signs and was able to combine them into simple sentences such as "Hurry, gimme toothbrush" and "Please tickle more." The famous gorilla Koko also uses ASL to communicate; she reportedly uses more than 1,000 words (**Figure 8.9**).

In another well-known study, a chimp named Lana learned to use symbols on a computer to get things she wanted, such as food, a drink, a tickle from her trainers, and having her curtains opened (Rumbaugh et al., 1974) (**Figure 8.10**).

Dolphins are also the subject of interesting language research (e.g., May-Collado, 2010). Communication with dolphins is done by means of hand signals or audible commands transmitted through an underwater speaker system. In one typical study, trainers gave dolphins commands made up of two-to five-word sentences, such as "Big ball—square—return," which meant that they should go get the big ball, put it in the floating square, and return to the trainer (Herman, Richards, & Woltz, 1984). By varying the syntax (for example, the order of the words) and specific content of the commands, the researchers showed that dolphins are sensitive to these aspects of language.

Psychologists disagree about how to interpret these findings on apes and dolphins. Most psychologists believe that nonhuman animals communicate but that their ideas are severely limited. For example, some critics claim that apes and dolphins are unable to convey subtle meanings, use language creatively, or communicate at an abstract level (Jackendoff, 2003; Siegala & Varley, 2008).

Others propose that these animals do not truly understand language but are simply operantly conditioned (Chapter 6) to imitate symbols to receive rewards (Savage-Rumbaugh, 1990; Terrace, 1979). Finally, others suggest that data regarding animal language has not always been well documented (Font & Carazo, 2010; Willingham, 2001; Wynne, 2007).

Proponents of animal language respond that apes can use language creatively and have even coined some words

**Computer-aided communication • Figure 8.10**

Apes lack the necessary anatomical structures to vocalize the way humans do. For this reason, language research with chimps and gorillas has focused on teaching the animals to use sign language or to "speak" by pointing to symbols on a keyboard. Do you think this amounts to using language the same way humans do?

Michael Nichols /NG Image Collection

of their own. For example, Koko signed "finger bracelet" to describe a ring and "eye hat" to describe a mask (Patterson & Linden, 1981). Proponents also argue that, as demonstrated by the dolphin studies, animals can be taught to understand basic rules of sentence structure.

Still, the gap between human and nonhuman animals' language is considerable. Current evidence suggests that, at best, nonhuman animal language is less complex, less creative, and has fewer rules than any language used by humans.

**CONCEPT CHECK** STOP

1. **What** is the difference between phonemes and morphemes?
2. **What** is an example of language influencing thought?
3. **Why** do some psychologists believe that language is an innate ability, whereas others believe that it is learned through imitation and reinforcement?

# Intelligence

## LEARNING OBJECTIVES

**RETRIEVAL PRACTICE** While reading the upcoming sections, respond to each Learning Objective in your own words. Then compare your responses with those in Appendix B.

1. **Review** the history of theorizing about single versus multiple intelligences.
2. **Describe** the components of emotional intelligence.
3. **Explain** how intelligence is measured.

Many people equate intelligence with "book smarts." For others, what is intelligent depends on the characteristics and skills that are valued in a particular social group or culture (Berry et al., 2011; Hunt, 2011; Sternberg, Jarvin, & Grigorenko, 2011). For example, the Mandarin word that corresponds most closely to the word "intelligence" is a character meaning "good brain and talented" (Matsumoto, 2000). The word is also associated with traits like imitation, effort, and social responsibility (Keats, 1982).

Even among Western psychologists there is debate over the definition of intelligence. In this discussion, we rely on a formal definition of **intelligence** developed by psychologist David Wechsler (pronounced "WEX-ler") (1944, 1977).

**intelligence** The global capacity to think rationally, act purposefully, and deal effectively with the environment.

### Do We Have One or Many Intelligences?

One of the central debates in research on intelligence concerns whether intelligence is a single ability or a collection of many specific abilities.

In the 1920s, British psychologist Charles Spearman first observed that high scores on separate tests of mental abilities tend to correlate with each other. Spearman (1923) thus proposed that intelligence is a single factor, which he termed **general intelligence (*g*)**. He believed that *g* underlies all intellectual behavior, including reasoning, solving problems, and performing well in all areas of cognition. Spearman's work laid the foundations for today's standardized intelligence tests (Matlin, 2013; Wright, 2011).

About a decade later, L. L. Thurstone (1938) proposed seven primary mental abilities: verbal comprehension, word fluency, numerical fluency, spatial visualization, associative memory, perceptual speed, and reasoning. J. P. Guilford (1967) later expanded this number, proposing that as many as 120 factors were involved in the structure of intelligence.

Around the same time, Raymond Cattell (1963, 1971) reanalyzed Thurstone's data and argued against the idea of multiple intelligences. He believed that two subtypes of *g* exist:

- **Fluid intelligence (*gf*)** refers to innate, inherited reasoning abilities, memory, and speed of information processing. Fluid intelligence is relatively independent of education and experience, and like other biological capacities it declines with age (Jost et al., 2011; Murray et al., 2011; Rozencwajg et al., 2005).
- **Crystallized intelligence (*gc*)** refers to the store of knowledge and skills gained through experience and education (Hunt, 2011; Sternberg & Kaufman, 2012). Crystallized intelligence tends to increase over the lifespan.

**Gardner's multiple intelligences Table 8.3**

| Type of intelligence | Possible careers |
|---|---|
| **Linguistic**<br>Language, such as speaking, reading a book, writing a story | Novelist, journalist, teacher |
| **Spatial**<br>Mental maps, such as figuring out how to pack multiple presents in a box or how to draw a floor plan | Engineer, architect, pilot |
| **Bodily/kinesthetic**<br>Body movement, such as dancing, soccer, and football | Athlete, dancer, ski instructor |
| **Intrapersonal**<br>Understanding oneself, such as setting achievable goals or recognizing self-defeating emotions | Increased success in almost all careers |
| **Logical/mathematical**<br>Problem solving or scientific analysis, such as following a logical proof or solving a mathematical problem | Mathematician, scientist, engineer |
| **Musical**<br>Musical skills, such as singing or playing a musical instrument | Singer, musician, composer |
| **Interpersonal**<br>Social skills, such as managing diverse groups of people | Salesperson, manager, therapist, teacher |
| **Naturalistic**<br>Being attuned to nature, such as noticing seasonal patterns or using environmentally safe products | Biologist, naturalist |
| **Spiritual/existential**<br>Attunement to meaning of life and death and other conditions of life | Philosopher, theologian |

©AP/Wide World Photos

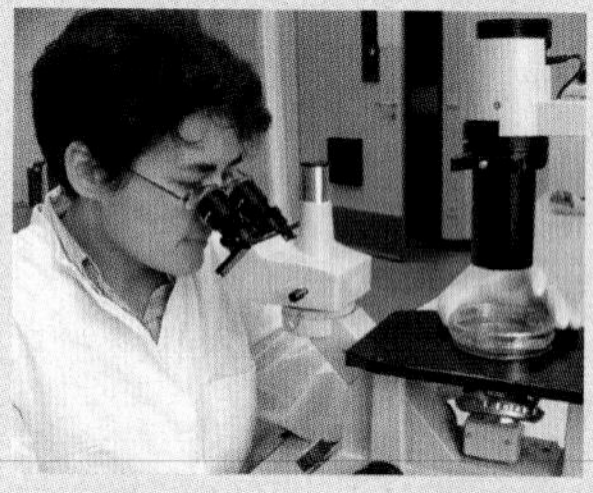

iStockphoto

Kevin Winter/AMA2010/Getty Images, Inc.

Source: Adapted from Gardner 1983, 1999, 2008, 2011.

Today there is considerable support for the concept of *g* as a measure of academic smarts. However, many contemporary cognitive theorists believe that intelligence is not a single general factor but a collection of many separate specific abilities.

One of these cognitive theorists, Howard Gardner, believes that people have many kinds of intelligences. The fact that brain-damaged patients often lose some intellectual abilities while retaining others suggests to Gardner that different intelligences are located in discrete areas throughout the brain. According to **Gardner's theory of multiple intelligences** (1983, 1999, 2008, 2011), people have different profiles of intelligence because they are stronger in some areas than others (**Table 8.3**). They also use their intelligences differently to learn new material, perform tasks, and solve problems.

Robert **Sternberg's triarchic theory of successful intelligence** also involves multiple abilities. As discussed in **Table 8.4**, Sternberg theorized that three separate, learned aspects of intelligence exist: (1) *analytic*, (2) *creative*, and (3) *practical* (Sternberg, 1985, 2007, 2012).

Sternberg (1985, 1999) emphasizes the process underlying thinking rather than just the product. He also stresses the importance of applying mental abilities to real-world situations rather than testing mental abilities in isolation (e.g., Sternberg, 2005; Sternberg & Hedlund, 2002). In short, Sternberg avoids the traditional idea of intelligence as an innate form of "book smarts," and instead emphasizes successful intelligence as the learned ability to adapt to, shape, and select environments in order to accomplish personal and societal goals.

**Sternberg's triarchic theory of successful intelligence Table 8.4**

| | Analytical intelligence | Creative intelligence | Practical intelligence |
|---|---|---|---|
| **Sample skills** | Good at analysis, evaluation, judgment, and comparison skills | Good at invention, coping with novelty, and imagination skills | Good at application, implementation, execution, and utilization skills |
| **Methods of assessment** | Intelligence tests assess the meaning of words based on context, and how to solve number-series problems | Open-ended tasks, writing a short story, drawing a piece of art, solving a scientific problem requiring insight | Tasks requiring solutions to practical, personal problems |

## Emotional Intelligence

In addition to the multiple intelligences proposed by Gardner and Sternberg, Daniel Goleman's research (1995, 2000, 2008) and best-selling books have popularized the concept of **emotional intelligence (EI)**, based on original work by Peter Salovey and John Mayer (1990).

According to the theory, emotional intelligence involves knowing and managing one's emotions, empathizing with others, and maintaining satisfying relationships. In other words, an emotionally intelligent person successfully combines the three components of emotions (cognitive, physiological, and behavioral). Proponents of EI have suggested ways in which the close collaboration between emotion and reason may promote personal well-being and growth (Salovey et al., 2000) (**Figure 8.11**).

Popular accounts such as Goleman's have suggested that traditional measures of human intelligence ignore a crucial range of abilities that characterize people who excel in real life: self-awareness, impulse control, persistence, zeal and self-motivation, empathy, and social deftness. Goleman also proposes that many societal problems, such as domestic abuse and youth violence, can be attributed to a low EI. Therefore, he argues, EI should be fostered and taught to everyone.

Critics fear that a handy term like EI invites misuse, but their strongest reaction is to Goleman's proposals for teaching EI. For example, Paul McHugh, director of psychiatry at Johns Hopkins University, suggests that Goleman is "presuming that someone has the key to the right emotions to be taught to children. We don't even know the right emotions to be taught to adults" (cited in Gibbs, 1995, p. 68).

### How do we develop emotional intelligence? • Figure 8.11

The mother in this photo appears to be empathizing with her young daughter and helping her to recognize and manage her own emotions. According to Goleman, this type of modeling and instruction is vital to the development of emotional intelligence.

Digital Vision/Getty Images

## Measuring Intelligence

Different IQ tests approach the measurement of intelligence from different perspectives. However, most are designed to predict grades in school. Let's look at the most commonly used IQ tests.

The **Stanford-Binet Intelligence Scale** is loosely based on the first IQ tests developed in France around the turn of the last century by Alfred Binet. In the United States, Lewis Terman (1916) developed the Stanford-Binet (at Stanford University) to test the intellectual ability of U.S.-born children ages 3 to 16. The test is revised periodically—most recently in 2003. The test is administered individually and consists of such tasks as copying geometric designs, identifying similarities, and repeating number sequences.

In the original version of the Stanford-Binet, results were expressed in terms of a mental age. For example, if a 7-year-old's score equaled that of an average 8-year-old, the child was considered to have a mental age of eight. To determine the child's **intelligence quotient (IQ)**, mental age was divided by the child's chronological age (actual age in years) and multiplied by 100.

> **intelligence quotient (IQ)** Once a formula (mental age ÷ chronological age × 100), the term IQ is now used to describe intelligence test scores.

The most widely used intelligence test, the **Wechsler Adult Intelligence Scale (WAIS)**, was developed by David Wechsler in the early 1900s. He later created a similar test for school-aged children. Like the Stanford-Binet, Wechsler's tests yield an overall intelligence score, but they have separate scores for verbal intelligence (such as vocabulary comprehension, and knowledge of general information) and performance (such as arranging pictures to tell a story or arranging blocks to match a given pattern) (**Table 8.5**). The advantages of Wechsler's test are that different abilities can be evaluated either separately or together and the test can be used with non-English speakers (the verbal portion is omitted).

**Items similar to those on the Wechsler Adult Intelligence Scale (WAIS) Table 8.5**

| Test | Description | Example |
|---|---|---|
| **Verbal scale** | | |
| Information | Taps general range of information | On which continent is France? |
| Comprehension | Tests understanding of social conventions and ability to evaluate past experience | Why do people need birth certificates? |
| Arithmetic | Tests arithmetic reasoning through verbal problems | How many hours will it take to drive 150 miles at 50 miles per hour? |
| Similarities | Asks in what way certain objects or concepts are similar; measures abstract thinking | How are a calculator and a typewriter alike? |
| Digit span | Tests attention and rote memory by orally presenting series of digits to be repeated forward or backward | Repeat the following numbers backward: 2 4 3 5 1 8 6 |
| Vocabulary | Tests ability to define increasingly difficult words | What does repudiate mean? |
| **Performance scale** | | |
| Digit symbol | Tests speed of learning through timed coding tasks in which numbers must be associated with marks of various shapes | Shown: 1 2 3 4 Fill in: 4 2 1 3  |
| Picture completion | Tests visual alertness and visual memory through presentation of an incompletely drawn figure; the missing part must be discovered and named | Tell me what is missing:  |
| Block design | Tests ability to perceive and analyze patterns presenting designs that must be copied with blocks | Assemble blocks to match this design:  |
| Picture arrangement | Tests understanding of social situations through a series of comic-strip-type pictures that must be arranged in the right sequence to tell a story | Put these pictures in the right order:   |
| Object assembly | Tests ability to deal with part/whole relationships by presenting puzzle pieces that must be assembled to form a complete object | Assemble the pieces into a complete object: 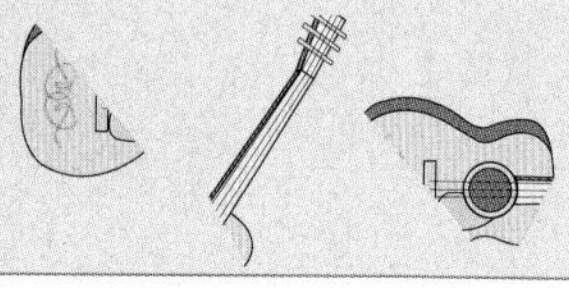 |

## The distribution of scores on the Stanford-Binet Intelligence Test • Figure 8.12

Sixty-eight percent of children score within one standard deviation (16 points) above or below the national average, which is 100 points. About 16% score above 116, and about 16% score below 84.

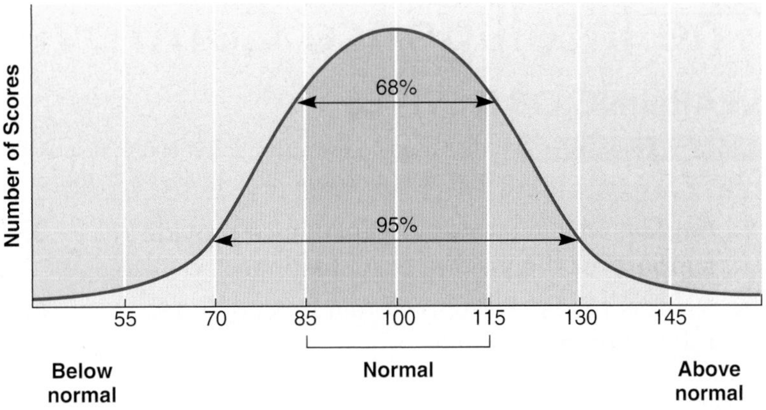


Today, most intelligence test scores are expressed as a comparison of a single person's score to a national sample of similar-aged people (**Figure 8.12**). Even though the actual IQ is no longer calculated using the original formula comparing mental and chronological ages, the term *IQ* remains as a shorthand expression for intelligence test scores.

What makes a good test? How are the tests developed by Binet and Wechsler any better than those published in popular magazines and presented on television programs? To be scientifically acceptable, all psychological tests must fulfill three basic requirements:

- **Standardization** Intelligence tests (as well as personality, aptitude, and most other tests) must be standardized in two senses (Gregory, 2011; Plucker & Esping, 2013, Wright, 2011). First, every test must have *norms*, or average scores, developed by giving the test to a representative sample of people (a diverse group of people who resemble those for whom the test is intended). Second, testing procedures must be standardized. All test takers must be given the same instructions, questions, and time limits, and all test administrators must follow the same objective score standards.
- **Reliability** To be trustworthy, a test must be consistent, or reliable. Reliability is usually determined by retesting subjects to see whether their test scores change significantly (Gregory, 2011). Retesting can be done via the **test-retest method**, in which participants' scores on two separate administrations of the same test are compared, or via the **split-half method**, which involves splitting a test into two equivalent parts (e.g., odd and even questions) and determining the degree of similarity between the two halves.
- **Validity** Validity is the ability of a test to measure what it is designed to measure. The most important type of validity is **criterion-related validity**, or the accuracy with which test scores can be used to predict another variable of interest (known as the criterion). Criterion-related validity is expressed as the *correlation* (Chapter 1) between the test score and the criterion. If two variables are highly correlated, then one variable can be used to predict the other. Thus, if a test is valid, its scores will be useful in predicting an individual's behavior in some other specified situation. One example of this is using intelligence test scores to predict grades in college.

**standardization** Establishment of the norms and uniform procedures for giving and scoring a test.

**reliability** A measure of the consistency and stability of test scores when a test is readministered.

**validity** The ability of a test to measure what it was designed to measure.

Can you see why a test that is standardized and reliable but not valid is worthless? For example, a test for skin sensitivity may be easy to standardize (the instructions specify exactly how to apply the test agent), and it may be reliable (similar results are obtained on each retest). But it certainly would not be valid for predicting college grades.

### CONCEPT CHECK

STOP

1. **What** is the difference between Gardner's and Sternberg's theories of intelligence?
2. **How** would someone with low emotional intelligence behave?
3. **Why** is criterion-related validity important?

# The Intelligence Controversy

## LEARNING OBJECTIVES

**RETRIEVAL PRACTICE** While reading the upcoming sections, respond to each Learning Objective in your own words. Then compare your responses with those in Appendix B.

1. **Explain** why extremes in intelligence provide support for the validity of IQ testing.
2. **Review** research on how brain functioning is related to intelligence.
3. **Describe** how genetics and environment interact to shape intelligence.
4. **Summarize** the controversy over whether or not IQ tests are culturally biased.

Psychologists have long debated several important questions related to intelligence: What causes some people to be more intelligent than others? What factors—environmental or hereditary—influence an individual's intelligence? Are IQ tests culturally biased? These questions, and the controversies surrounding them, are discussed in this section.

## Extremes in Intelligence

One of the best methods for judging the validity of a test is to compare people who score at the extremes. Despite the uncertainties discussed in the previous section, intelligence tests provide one of the major criteria for assessing mental ability at the extremes—specifically, for diagnosing **intellectual disability** and **mental giftedness**.

The clinical label *intellectually disabled* (previously referred to as *mentally retarded*) is applied when someone is significantly below average in intellectual functioning and has significant deficits in adaptive functioning (such as communicating, living independently, social or occupational functioning, or maintaining safety and health) (American Psychiatric Association, 2000).

Fewer than 3% of people are classified as having an intellectual disability (**Table 8.6**). Of this group, 85% have only mild intellectual disability and many become self-supporting, integrated members of society. Furthermore, people can score low on some measures of intelligence and still be average or even gifted in others. The most dramatic examples are people with **savant syndrome** (**Figure 8.13**).

**savant syndrome** A condition in which a person with generally limited mental abilities exihibits exceptional skill or brilliance in some limited field.

Some forms of intellectual disability stem from genetic abnormalities, such as Down syndrome, fragile-X syndrome, and phenylketonuria (PKU). Other causes are environmental, including prenatal exposure to alcohol and other drugs, extreme deprivation or neglect in early life, and brain damage from accidents. However, in many cases, there is no known cause of the intellectual disability.

At the other end of the intelligence spectrum are people with especially high IQs (typically defined as being in the top 1 or 2%).

**Degrees of intellectual disability Table 8.6**

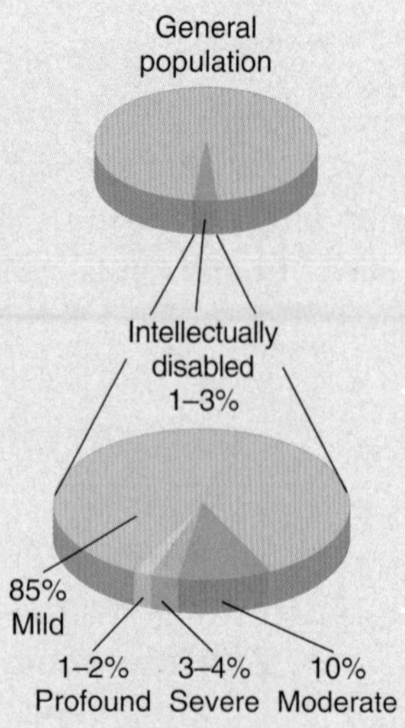


| Level of disability | IQ scores | Characteristics |
|---|---|---|
| **Mild (85%)** | 50–70 | Usually able to become self-sufficient; may marry, have families, and secure full-time jobs in unskilled occupations |
| **Moderate (10%)** | 35–49 | Able to perform simple unskilled tasks; may contribute to a certain extent to their livelihood |
| **Severe (3–4%)** | 20–34 | Able to follow daily routines, but with continual supervision; with training, may learn basic communication skills |
| **Profound (1–2%)** | below 20 | Able to perform only the most rudimentary behaviors, such as walking, feeding themselves, and saying a few phrases |

© Justin Sutcliffe/Redux Pictures

## An unusual form of intelligence • Figure 8.13

People with *savant syndrome* generally score very low on IQ tests (usually between 40 and 70), yet they demonstrate exceptional skills or brilliance in specific areas, such as rapid calculation, art, memory, or musical ability (Meyer, 2011; Olson, 2010; Pring et al., 2008). Derek Paravicini, a musical savant, pictured here, was born premature, blind, and with a severe learning disability. In spite of all of these challenges, he plays the concert piano entirely by ear and has a repertoire of thousands of memorized pieces.

In 1921, Lewis Terman identified 1,500 gifted children—affectionately nicknamed the "Termites"—with IQs of 140 or higher and tracked their progress through adulthood. The number who became highly successful professionals was many times the number a random group would have provided (Leslie, 2000; Terman, 1954). Those who were most successful tended to have extraordinary motivation, and they also had someone at home or school who was especially encouraging (Goleman, 1980). There were some notable failures, and the "Termites" became alcoholics, got divorced, and committed suicide at close to the national rate (Campbell & Feng, 2011; Leslie, 2000; Terman, 1954). In sum, a high IQ is no guarantee of success in every endeavor; it only offers more intellectual opportunities.

## The Brain's Influence on Intelligence

A basic tenet of neuroscience is that all mental activity (including intelligence) results from neural activity in the brain. Most recent research on the biology of intelligence has focused on brain functioning. For example, neuroscientists have found that people who score highest on intelligence tests also respond more quickly on tasks involving perceptual judgments (Bowling & Mackenzie, 1996; Posthuma et al., 2001; Sternberg, Jarvin, & Grigorenko, 2011).

In addition to a faster response time, research using positron emission tomography (PET) scans to measure brain activity (Chapter 1) suggests that intelligent brains work smarter, or more efficiently, than less-intelligent brains (Jung & Haier, 2007; Neubauer et al., 2004; Posthuma et al., 2001) (**Figure 8.14**).

## Do intelligent brains work more efficiently? • Figure 8.14

When given problem-solving tasks, low-IQ brains (PET scans on the left) show more activity (red and yellow indicate more brain activity) in relevant brain areas than high-IQ brains (PET scans on right). This research suggests that lower-IQ brains actually work harder, although less efficiently, than higher-IQ brains.

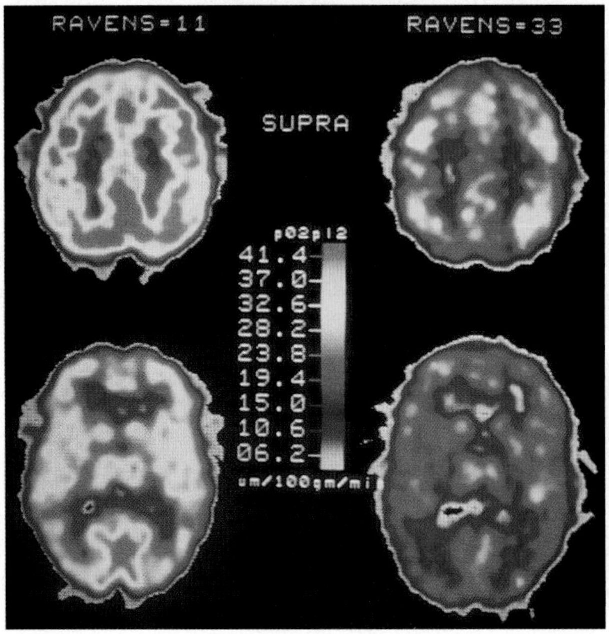


Courtesy Richard J. Haier, University of California-Irvine

Does size matter? It makes logical sense that bigger brains would be smarter—after all, humans have larger brains than less intelligent species, such as dogs. (Some animals, such as whales and dolphins, have larger brains than humans, but our brains are larger relative to our body size.) In fact, brain-imaging studies have found a significant correlation between brain size (adjusted for body size) and intelligence (Christensen et al., 2008; Deary et al., 2007; Ivanovic et al., 2004; Stelmack, Knott, & Beauchamp, 2003). On the other hand, Albert Einstein's brain was no larger than normal (Witelson, Kigar, & Harvey, 1999). In fact, some of Einstein's brain areas were actually smaller than average, but the area responsible for processing mathematical and spatial information was 15% larger than average.

## Genetic and Environmental Influences on Intelligence

Similarities in intelligence between family members are due to a combination of hereditary (shared genetic material) and environmental factors (similar living arrangements and experiences). Researchers who are interested in the role of heredity in intelligence often focus on identical (monozygotic) twins because they share 100% of their genetic material, as described in *What a Psychologist Sees*.

The long-running Minnesota Study of Twins, an investigation of identical twins raised in different homes and reunited only as adults (Bouchard, 1994, 1999; Bouchard et al., 1998; Johnson et al., 2007), found that genetic

# WHAT A PSYCHOLOGIST SEES

THE PLANNER

## Family Studies of Intelligence

PhotoDisc, Inc./Getty Images

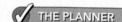

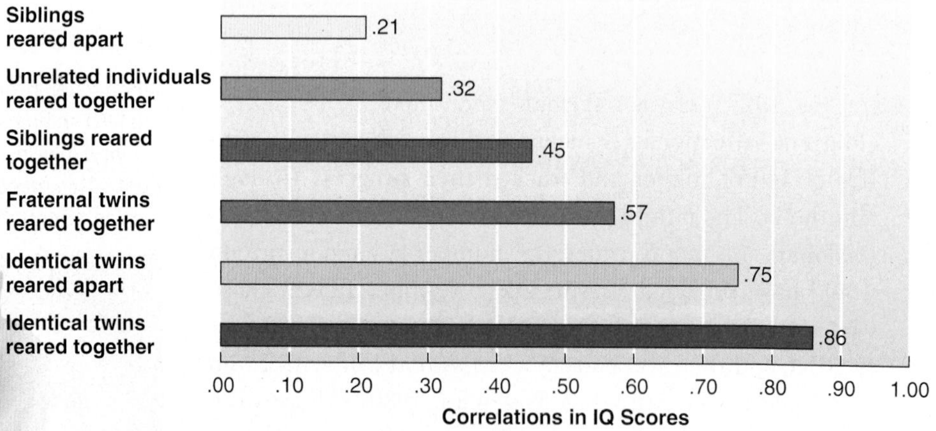


Why are some people intellectually gifted while others struggle? As identical twin studies demonstrate, genetics play an important role. In the figure here, note the higher correlations between identical twins' IQ scores compared to the correlations between all other pairs (Bouchard & McGue, 1981; Bouchard et al., 1998; McGue et al., 1993). Although heredity equips each of us with innate intellectual capabilities, the environment significantly influences whether a person will reach his or her full intellectual potential (Dickens & Flynn, 2001; Sangwan, 2001). For example, early malnutrition can retard a child's brain development, which in turn affects the child's curiosity, responsiveness to the environment, and motivation for learning—all of which can lower the child's IQ. The reverse is also true: An enriching early environment can set the stage for intellectual growth.

factors appear to play a surprisingly large role in the IQ scores of identical twins.

However, such results are not conclusive. Adoption agencies tend to look for similar criteria in their choice of adoptive parents. Therefore, the homes of these "reared apart" twins were actually quite similar. In addition, these twins also shared the same nine-month prenatal environment, which also might have influenced their brain development and intelligence (White, Andreasen, & Nopoulos, 2002).

**Are IQ tests culturally biased?** One of the most controversial issues in psychology involves group differences in intelligence test scores and what they really mean. In 1969, Arthur Jensen sparked a heated debate when he argued that genetic factors are "strongly implicated" as the cause of ethnic differences in intelligence. A book by Richard J. Herrnstein and Charles Murray titled *The Bell Curve: Intelligence and Class Structure in American Life* reignited this debate in 1994 when the authors claimed that African Americans score below average in IQ because of their "genetic heritage."

Are intelligence tests biased? Although there is no clear answer in this debate, psychologists have made several points:

- Selectively highlighting IQ scores from one group (African American) is deceptive because race and ethnicity, like intelligence itself, are almost impossible to define. Depending on the definition that you use, there are between 3 and 300 races, and no race is pure in a biological sense (Fujimura et al., 2010; Navarro, 2008; Sternberg & Grigorenko, 2008). Furthermore, like President Barack Obama, Tiger Woods, and Mariah Carey, many people today self-identify as multiracial.
- Environmental factors may override genetic potential and later affect IQ test scores. Like plants that come from similar seeds but are placed in poor versus enriched soil, minority children more often grow up in lower socioeconomic conditions, which may hamper their true intellectual potential (**Figure 8.15**).
- Cultural factors may influence test scores. In some ethnic groups, a child who excels in school is ridiculed for trying to be different from his or her classmates. Moreover, if children's own language and dialect do not match their education system or the IQ tests they take, they are obviously at a disadvantage (Cathers-Schiffman & Thompson, 2007; García & Náñez, 2011; Johnson, Brett, & Deary, 2010; Sidhu, Malhi, & Jerath, 2010; Sternberg & Grigorenko, 2008).
- Group distributions of scores overlap considerably. Therefore, any group differences in IQ scores have little relevance for judging individuals. For example, many individual African Americans receive higher IQ scores then many individual European Americans (Garcia & Stafford, 2000; Myerson et al., 1998; Reifman, 2000).

## Genetics versus environment • Figure 8.15

Even when you begin with the exact same package of seeds (genetic inheritance), the average height of corn plants in the fertile soil will be greater than those in the poor soil (environmental influences). The same may be true for intelligence. Therefore, no conclusions can be drawn about possible genetic contributions to the differences between groups.

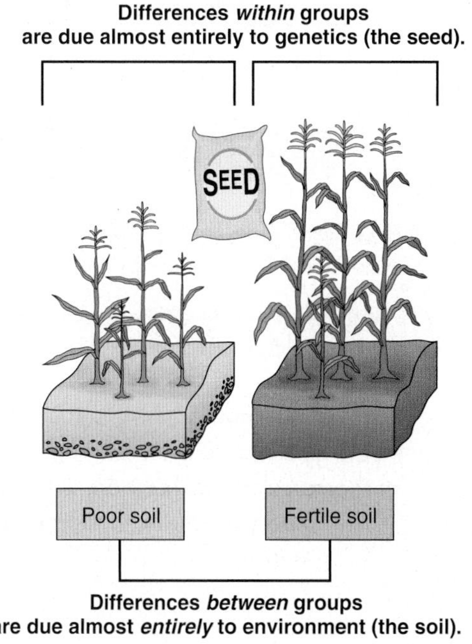
Differences *within* groups
are due almost entirely to genetics (the seed).
SEED
Poor soil
Fertile soil
Differences *between* groups
are due almost *entirely* to environment (the soil).

THE PLANNER

## Stereotype Threat

In the first study of stereotype threat, Claude Steele and Joshua Aronson (1995) recruited African American and European American Stanford University students (with similar ability levels) to complete a "performance exam" that supposedly measured intellectual abilities. The exam's questions were similar to those on the Graduate Record Exam (GRE). Results showed that African American students performed far below European American students. In contrast, when the researchers told students it was a "laboratory task," there was no difference between African American and European American scores.

Subsequent work showed that stereotype threat occurs because members of stereotyped groups are anxious that they will fulfill their group's negative stereotype. This anxiety in turn hinders their performance on tests. Some people cope with stereotype threat by **disidentifying**, telling themselves they don't care about the test scores (Major et al., 1998). Unfortunately, this attitude lessens motivation, decreasing performance.

Stereotype threat affects many social groups, including African Americans, women, Native Americans, Latinos, low-income people, elderly people, and white male athletes (e.g., Bates, 2007; Keller & Bless, 2008; Owens & Massey, 2011). This research helps explain some group differences in intelligence and achievement tests. As such, it underscores why relying solely on such tests to make critical decisions affecting individual lives—for example, in hiring, college admissions, or clinical application—is unwarranted and possibly even unethical.

Reuters/Larry Downing/Landov

Early research suggested that having Barack Obama as a positive role model improved academic performance by African Americans—thus offsetting the stereotype threat (e.g., Marx, Ko, & Friedman, 2009). However, later studies found either no relationship between test performance and positive thoughts about Obama or mixed results (Aronson et al., 2009; Smith, 2012).

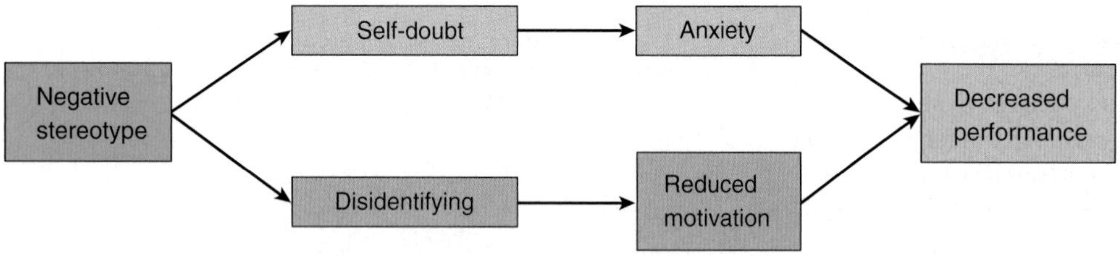


### Identify the Research Method

1. What is the most likely research method used in this study of stereotype threat by Steele and Aronson (1995)?
2. If you chose
   - the experimental method, label the IV, DV, experimental group, and control group.
   - the descriptive method, is this a naturalistic observation, survey, or case study?
   - the correlational method, is this a positive, negative, or zero correlation?
   - the biological method, identify the specific research tool (e.g., brain dissection, CT scan).

(Check your answers in Appendix C.)

- Traditional IQ tests do not measure many of our multiple intelligences (Gardner, 2002, 2011; Manly et al., 2004; Naglieri & Ronning, 2000; Rutter, 2007; Sternberg, 2007, 2012; Sternberg & Grigorenko, 2008).
- **Stereotype threat** can cause some group members to doubt their abilities, which may, in turn, significantly reduce their test scores (Keller & Bless, 2008; Owens & Massey, 2011; Shapiro, 2011) (see *Psychological Science*).

**stereotype threat** Negative stereotypes about minority groups cause some members to doubt their abilities.

- Intelligence (as measured by IQ tests) is not a fixed trait. Around the world, IQ scores have increased over the last half century. This well-established phenomenon, known as the *Flynn effect*, may be due to improved nutrition, better public education, more proficient test-taking skills, and rising levels of education for a greater percentage of the world's population (Flynn, 1987, 2007, 2010; Huang & Hauser, 1998; Mingroni, 2004; Pietschnig, Voracek, & Formann, 2011).

The ongoing debate over the nature of intelligence and its measurement highlights the complexities of studying cognition. In this chapter, we've explored three cognitive processes: *thinking*, *language*, and *intelligence*. As you've seen, all three processes are greatly affected by numerous interacting factors.

**CONCEPT CHECK** STOP

1. **What** abilities and limitations would you expect in someone with severe intellectual disability?
2. **Why** might more-intelligent people show less activity in cognitive-processing areas than less-intelligent people?
3. **How** have twin studies improved researchers' understanding of intelligence?
4. **What** is stereotype threat, and why does it occur?

# Summary

THE PLANNER

## 1 Thinking 202

- Thinking is a central aspect of **cognition**. Thought processes are distributed throughout the brain in neural networks. **Mental images** and **concepts** aid our thought processes. There are three major building blocks for concepts—**prototypes, artificial concepts,** and **hierarchies**, shown in the diagram.
- **Problem solving** usually involves three steps: *preparation, production,* and *evaluation*. **Algorithms** and **heuristics** help us produce solutions.
- Barriers to problem solving include **mental sets, functional fixedness, confirmation bias, availability heuristic, representativeness heuristic,** and **recognition heuristic.**
- **Creativity** is the ability to produce valued outcomes in a novel way. Tests of creativity usually focus on **divergent thinking.** One prominent theory of creativity is **investment theory.**

**Concepts: Hierarchies • Figure 8.2**

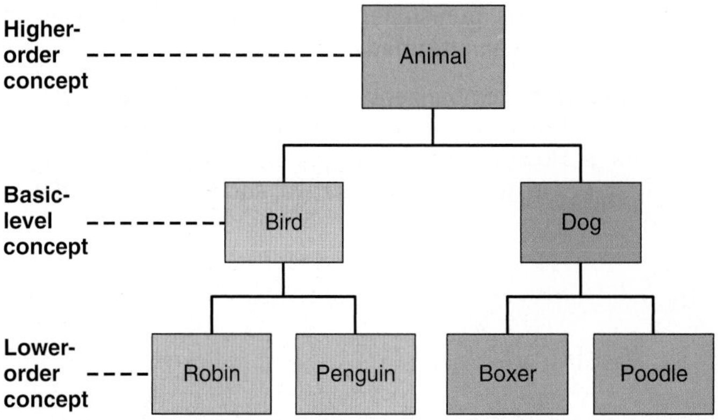

## 2 Language 208

- **Language** supports thinking and enables us to communicate. To produce language, we use **phonemes**, **morphemes**, and **grammar** (syntax and semantics). Several different parts of our brain are involved in producing and listening to language, as shown in the diagram.

### What a Psychologist Sees: Language and the Brain

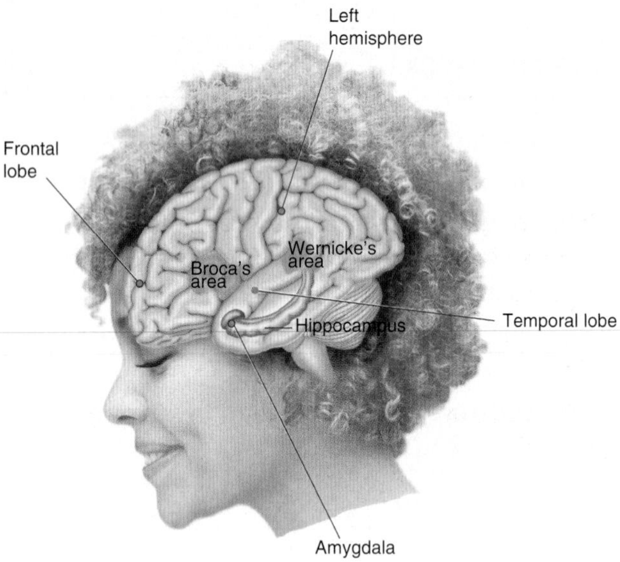


- According to Whorf's **linguistic relativity hypothesis,** language determines thought. Generally, Whorf's hypothesis is not supported. However, language does strongly influence thought.
- Children communicate nonverbally from birth. Their verbal communication proceeds in stages: *prelinguistic* (crying, **cooing, babbling**) and *linguistic* (single utterances, **telegraphic speech,** and acquisition of rules of grammar).
- According to Chomsky, humans are "prewired" with a **language acquisition device (LAD)** that enables language development. Nurturists hold that children learn language through rewards, punishments, and imitation. Most psychologists hold an intermediate view.
- Research with chimpanzees, gorillas, and dolphins suggests that these animals can learn and use basic rules of language. However, critics suggest nonhuman animal language is less complex, less creative, and not as rule-laden as human language.

## 3 Intelligence 213

- There is considerable debate over the meaning of **intelligence**. Here it is defined as the global capacity to think rationally, act purposefully, and deal effectively with the environment.
- Spearman proposed that intelligence is a single factor, which he termed **general intelligence (*g*).** Thurstone and Guilford argued that intelligence included numerous, distinct abilities. Cattell proposed two subtypes of *g*: **fluid intelligence (*gf*)** and **crystallized intelligence (*gc*).** Many contemporary cognitive theorists, including Gardner and Sternberg, believe that intelligence is a collection of many separate specific abilities. Goleman believes that **emotional intelligence (EI)** (the ability to empathize and manage one's emotions and relationships) is just as important as any other kind of intelligence.
- Early intelligence tests involved computing a person's mental age to arrive at an **intelligence quotient (IQ)**. Today, two of the most widely used intelligence tests are the **Stanford-Binet Intelligence Scale** and the **Wechsler Adult Intelligence Scale (WAIS).** Intelligence tests commonly compare the performance of an individual with other individuals of the same age, as shown on the graph.

**The distribution of scores on the Stanford-Binet Intelligence Test • Figure 8.12**

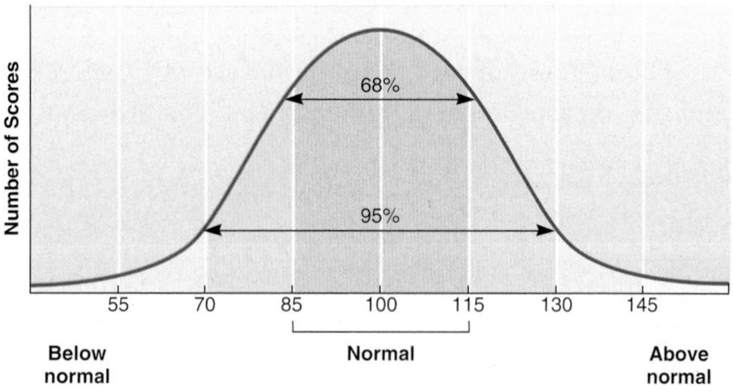


- To be scientifically acceptable, all psychological tests must fulfill three basic requirements: **standardization, reliability,** and **validity**.

## 4 The Intelligence Controversy 218

- Intelligence tests provide one of the major criteria for assessing **intellectual disability** and **mental giftedness**, both of which exist on a continuum. In **savant syndrome**, a person with limited mental abilities is exceptional in some limited field. Studies of people who are intellectually gifted found that they had more intellectual opportunities and tended to excel professionally. However, a high IQ is no guarantee of success in every endeavor.
- Most recent research on the biology of intelligence has focused on brain functioning, not size. Research indicates that intelligent people's brains respond especially quickly and efficiently.

- Both hereditary and environmental factors contribute to intelligence. Researchers interested in the role of heredity in intelligence often focus on identical twins. These studies have found that genetics play an important role in intelligence (as shown in the diagram). However, the environment significantly influences whether a person will reach his or her full intellectual potential.
- In response to claims that IQ tests are culturally biased, psychologists have emphasized that race and ethnicity are difficult to define, environmental and cultural factors may influence test scores, group distribution scores overlap, traditional IQ tests do not measure many multiple intelligences, **stereotype threat** can reduce test scores, and intelligence is not a fixed trait.

**What a Psychologist Sees: Family Studies of Intelligence**

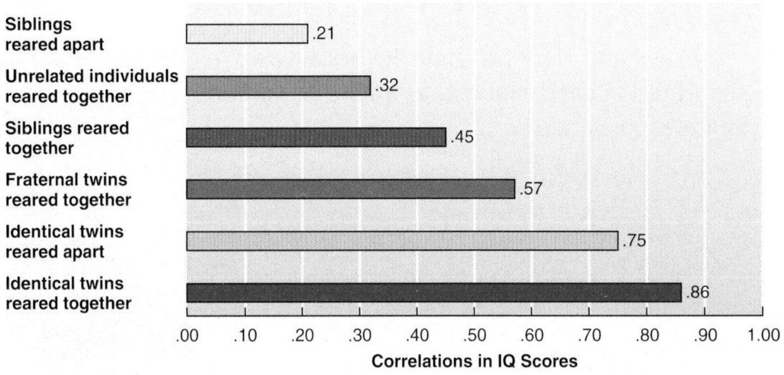


# Key Terms

**RETRIEVAL PRACTICE** Write a definition for each term before turning back to the referenced page to check your answer.

- algorithm 204
- artificial concept 203
- availability heuristic 205
- babbling 211
- cognition 202
- concept 202
- confirmation bias 205
- cooing 211
- creativity 206
- criterion-related validity 217
- crystallized intelligence (*gc*) 213
- disidentifying 222
- divergent thinking 206
- emotional intelligence (EI) 215
- fluid intelligence (*gf*) 213
- functional fixedness 205
- Gardner's theory of multiple intelligences 214
- general intelligence (*g*) 213
- grammar 208
- heuristic 204
- hierarchy 203
- incubation period 205
- insight 205
- intellectual disability 218
- intelligence 213
- intelligence quotient (IQ) 216
- investment theory 207
- language 208
- language acquisition device (LAD) 210
- linguistic relativity hypothesis 209
- mental giftedness 218
- mental image 202
- mental set 205
- morpheme 208
- overextension 211
- overgeneralization 211
- phoneme 208
- prefrontal cortex 202
- problem solving 203
- prototype 203
- recognition heuristic 206
- reliability 217
- representativeness heuristic 206
- savant syndrome 218
- split-half method 217
- standardization 217
- Stanford-Binet Intelligence Scale 215
- stereotype threat 223
- Sternberg's triarchic theory of successful intelligence 214
- telegraphic speech 211
- test-retest method 217
- validity 217
- Wechsler Adult Intelligence Scale (WAIS) 216

# Critical and Creative Thinking Questions

1. During problem solving, do you use primarily algorithms or heuristics? What are the advantages of each?
2. Would you like to be more creative? Can you do anything to acquire more of the resources of creative people?
3. Do you believe that we are born with an innate "language acquisition device" or that language development is a function of our environments?
4. Do you think apes and dolphins have true language? Why or why not?
5. Physicians, teachers, musicians, politicians, and people in many other occupations may become more successful with age and can continue working well into old age. Which kind of general intelligence might explain this phenomenon?
6. If Gardner's and Sternberg's theories of multiple intelligences are correct, what are the implications for intelligence testing and for education?

# What is happening in this picture?

Jerry Levy and Mark Newman, identical twins separated at birth, first met as adults at a firefighter's convention.

T.K. Wanstal/The Image Work

**Think Critically**

1. What factors might explain why they both became firefighters?
2. Does the brothers' choosing the same uncommon profession seem like a case of special "twin-telepathy"?
3. How might the *confirmation bias* contribute to this perception?

# Self-Test

**RETRIEVAL PRACTICE** Completing this self-test and comparing your answers with those in Appendix C provides immediate feedback and helpful practice for exams. Additional interactive, self-tests are available at www.wiley.com/college/carpenter.

**1.** The mental activities involved in acquiring, storing, retrieving, and using knowledge are collectively known as ______.

a. perception  
b. cognition  
c. consciousness  
d. awareness

**2.** Mental representations of previously stored sensory experiences are called ______.

a. illusions  
b. psychoses  
c. mental images  
d. mental propositions

**3.** The ______ stage of problem solving is demonstrated in this diagram.

**4.** This is a logical step-by-step procedure that, if followed, will always produce the solution.

a. algorithm  
b. heuristic  
c. problem-solving set  
d. brainstorming

**5.** ______ is the ability to produce valued outcomes in a novel way.

a. Problem-solving  
b. Incubation  
c. Functional flexibility  
d. Creativity

**6.** ______ is the set of rules that specify how phonemes, morphemes, words, and phrases should be combined to express meaningful thoughts.

a. Syntax  
b. Pragmatics  
c. Semantics  
d. Grammar

**7.** Label the three building blocks of language:

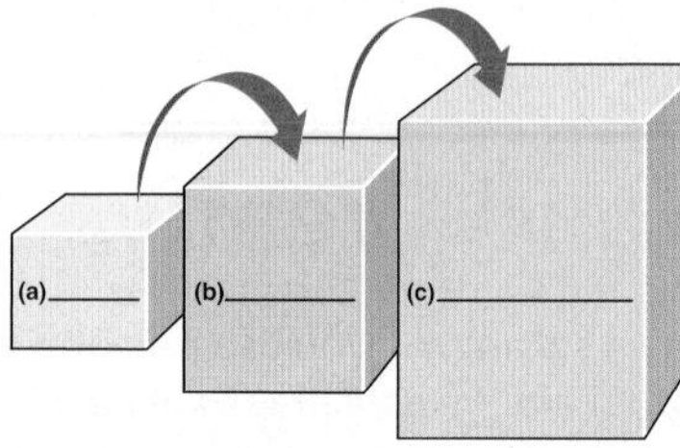


**8.** According to Chomsky, the innate mechanism that enables a child to analyze language is known as ______.

a. telegraphic understanding device (TUD)  
b. language acquisition device (LAD)  
c. language and grammar translator (LGT)  
d. overgeneralized neural net (ONN)

9. The definition of *intelligence* stated in your textbook stresses the global capacity to ______.

a. perform in school and on the job

b. read, write, and make computations

c. perform verbally and physically

d. think rationally, act purposefully, and deal effectively with the environment

10. The IQ test sample shown here is from the ______, the most widely used intelligence test.

a. Wechsler Intelligence Scale for Children

b. Wechsler Adult Intelligence Scale

c. Stanford-Binet Intelligence Scale

d. Binet-Terman Intelligence Scale

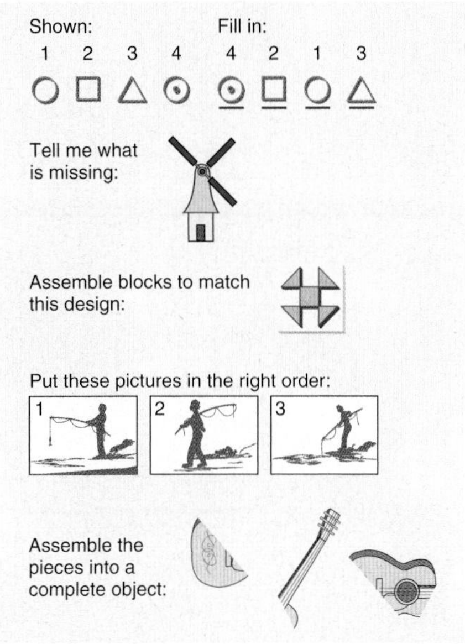


11. The development of uniform procedures for administering and scoring a test is called ______.

a. norming

b. standardization

c. procedural protocol

d. normalization

12. People with intellectual disability who demonstrate exceptional ability in specific areas are called ______.

a. savants

b. idiot geniuses

c. mildly retarded

d. connoisseurs

13. Speed of response and ______ are correlated with IQ scores.

a. gender

b. height

c. brain efficiency

d. all of the above

14. Which of the following groups would be most likely to have similar IQ test scores?

a. identical twins raised apart

b. fraternal twins raised together

c. identical twins raised together

d. step-brothers raised together

15. Stereotype threat affects the IQ scores of which of the following groups?

a. women

b. white male athletes

c. the elderly

d. all of these options

**THE PLANNER** ✓

Review your Chapter Planner on the chapter opener and check off your completed work.

# SOLUTIONS

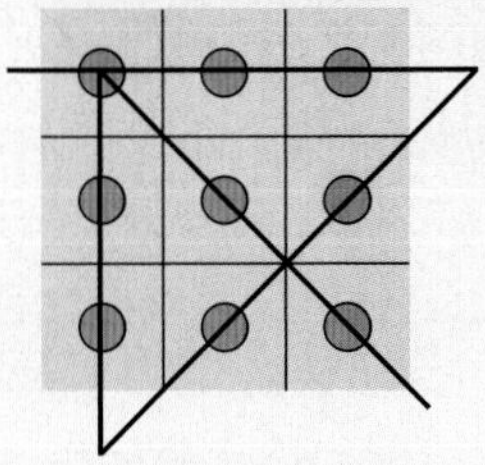

## Nine-dot problem solution

People find this puzzle difficult because they see the arrangement of dots as a square—a mental set that limits possible solutions.

## Candle problem solution

Use the tacks to mount the matchbox tray to the wall. Light the candle and use some melted wax to mount the candle to the matchbox.

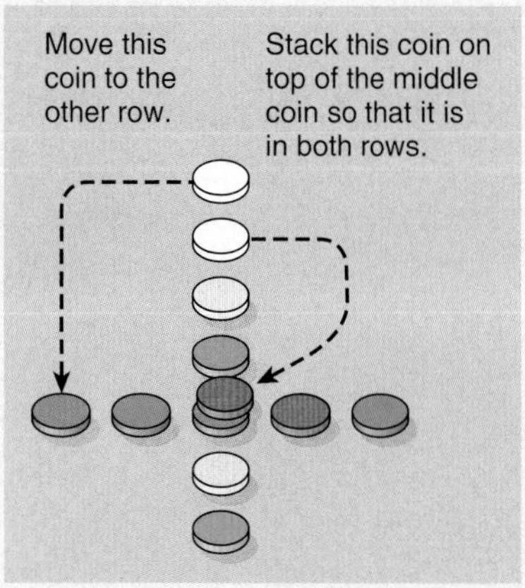

Move this coin to the other row.
Stack this coin on top of the middle coin so that it is in both rows.

## Coin problem solution

# 9 Lifespan Development I

Imagine that you could go back to the moment right before your conception—when your father's sperm met your mother's egg—and could change certain things. Would you still be "you" if you had a different father or mother? What if you chose another hometown or added or subtracted siblings? How might these changes affect who you are today?

As you can see from this brief fantasy, what you are now is a reflection of thousands of contributions from the past, and what you will be tomorrow is still unwritten. Life doesn't stand still. If we live long enough, you and I (and every other human on this planet) will be many people in our lifetime—infant, child, teenager, adult, and senior citizen.

Would you like to know more about yourself at each of these ages? The next two chapters explore the major areas of research in developmental psychology. In this first chapter, we'll study changes in our physical and cognitive development from conception to death. In Chapter 10, we'll examine how our social, moral, and personality development change over our lifespan.

© Agnieszka Kirinicjanow/iStockphoto

## CHAPTER OUTLINE



## CHAPTER PLANNER ✓

- ❑ Study the picture and read the opening story.
- ❑ Scan the Learning Objectives in each section: p. 230 ❑ p. 234 ❑ p. 243 ❑
- ❑ Read the text and study all figures and visuals. Answer any questions.

**Analyze key features**

- ❑ Psychological Science p. 231 ❑ p. 239 ❑
- ❑ Process Diagram, p. 235
- ❑ Psychology InSight, p. 237
- ❑ Study Organizer, p. 245
- ❑ Applying Psychology, p. 248
- ❑ What a Psychologist Sees, p. 249
- ❑ Stop: Answer the Concept Checks before you go on: p. 233 ❑ p. 243 ❑ p. 251 ❑

**End of chapter**

- ❑ Review the Summary and Key Terms.
- ❑ Answer the Critical and Creative Thinking Questions.
- ❑ Answer What is happening in this picture?
- ❑ Complete the Self-Test and check your answers.

# Studying Development

## LEARNING OBJECTIVES

**RETRIEVAL PRACTICE** While reading the upcoming sections, respond to each Learning Objective in your own words. Then compare your responses with those in Appendix B.

1. **Summarize** the three most important debates or questions in developmental psychology.
2. **Contrast** the cross-sectional research design with the longitudinal research design.

We begin our study of human development by focusing on some key theoretical issues and debates, along with several research approaches unique to this field. Keep in mind that development does not stop after childhood. **Developmental psychology** studies how we grow and change throughout the eight major stages of life (**Table 9.1**).

**developmental psychology** The study of age-related changes in behavior and mental processes from conception to death.

**Lifespan development Table 9.1**

| Stage | Approximate age |
|---|---|
| Prenatal | Conception to birth |
| Infancy | Birth to 18 months |
| Early childhood | 18 months to 6 years |
| Middle childhood | 6–12 years |
| Adolescence | 12–20 years |
| Young adulthood | 20–45 years |
| Middle adulthood | 45–60 years |
| Later adulthood | 60 years to death |

### Theoretical Issues

Developmental psychology's three most important debates or questions are about *nature versus nurture*, *stages versus continuity*, and *stability versus change*.

**Nature or nurture** The issue of "nature versus nurture" has been with us since the beginning of psychology (Chapter 1). According to the nature position, human behavior and development are governed by automatic, genetically predetermined signals in a process known as **maturation**. Just as a flower unfolds in accord with its genetic blueprint, we humans crawl before we walk and walk before we run.

**maturation** Development governed by automatic, genetically predetermined signals.

Furthermore, there is an optimal period shortly after birth, one of several **critical periods** during our lifetime, when an organism is especially sensitive to certain experiences that shape the capacity for future development. For example, Konrad Lorenz's (1937) early studies of **imprinting** found that baby geese will attach to (or imprint on) the first suitable moving stimulus they see within a certain critical period—normally the first 36 hours of their lives (**Figure 9.1**). Human children also have critical periods for attachment and development (see *Psychological Science*).

**critical period** A period of special sensitivity to specific types of learning that shapes the capacity for future development.

On the other side of the debate, those who hold a nurturist position argue that development occurs by learning through personal experience and observation of others.

Nina Leen/Time Life Pictures/Getty Images, Inc.

**Imprinting • Figure 9.1**

While baby geese normally imprint on their mother, Lorenz discovered that if he was the first moving creature the geese observed, they would imprint and follow him everywhere.

Psychological Science

THE PLANNER

## Deprivation and Development

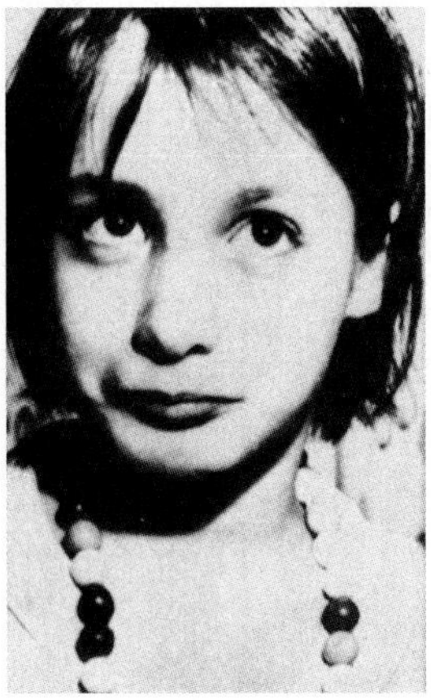

AP/Wide World Photos

What happens if a child is deprived of appropriate stimulation during critical periods of development? Consider the story of Genie, the so-called "wild child." From the time she was 20 months old until authorities rescued her at age 13, Genie was locked alone in a tiny, windowless room. By day, she sat naked and tied to a chair with nothing to do and no one to talk to. At night, she was put in a kind of straitjacket and "caged" in a covered crib. Genie's abusive father forbade anyone to speak to her for those 13 years. If Genie made noise, her father beat her while he barked and growled like a dog.

Genie's tale is a heartbreaking account of the lasting scars from a disastrous childhood. In the years after her rescue, Genie spent thousands of hours receiving special training, and by age 19 she could use public transportation and was adapting well to special classes at school. Genie was far from normal, however. Her intelligence scores were still close to the cutoff for intellectual disability. And although linguists and psychologists worked with her for many years, she never progressed beyond sentences like "Genie go," which suggests that because of her extreme childhood isolation and abuse she missed the necessary *critical period* for language development (Curtiss, 1977; LaPointe, 2005; Rymer, 1993; Saxton, 2010). To make matters worse, she was also subjected to a series of foster home placements, one of which was abusive. Later Genie lived in a home for intellectually disabled adults (Rymer, 1993).

### Identify the Research Method

1. What is the most likely research method used in this study of Genie (Rymer, 1993)?
2. If you chose
   - the experimental method, label the IV, DV, experimental group, and control group.
   - the descriptive method, is this a naturalistic observation, survey, or case study?
   - the correlational method, is this a positive, negative, or zero correlation?
   - the biological method, identify the specific research tool (e.g., brain dissection, CT scan).

(Check your answers in Appendix C.)

**Stages or continuity** A second key debate concerns whether development occurs in stages that are discrete and qualitatively different from one to another, or in a continuous pattern with gradual, but steady and quantitative (measurable) changes (**Figure 9.2**). Stage theorists believe development occurs at different rates, alternating between periods of little change and periods of abrupt, rapid change. Later in this chapter, we discuss stages in physical development and Piaget's stage theory of cognitive development.

Continuity theorists, on the other hand, maintain that development is *continuous*, with new abilities, skills, and knowledge gradually added at a relatively uniform pace. Therefore, adult thinking and intelligence differ quantitatively from a child's.

### Stages versus continuity in development • Figure 9.2

There is an ongoing debate about whether development is better characterized by stages or gradual, continuous development.

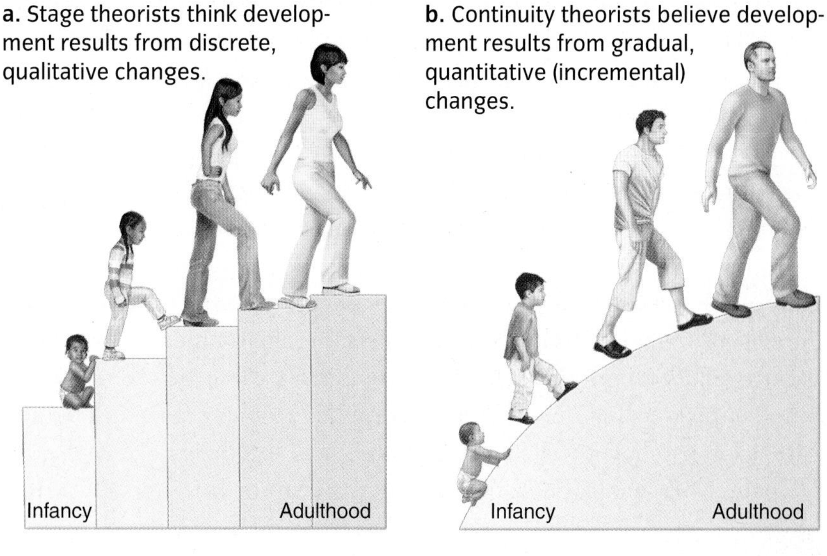


**a.** Stage theorists think development results from discrete, qualitative changes.

**b.** Continuity theorists believe development results from gradual, quantitative (incremental) changes.

## Cross-sectional versus longitudinal research • Figure 9.3

To study development, psychologists may use either a cross-sectional or longitudinal design for their research.

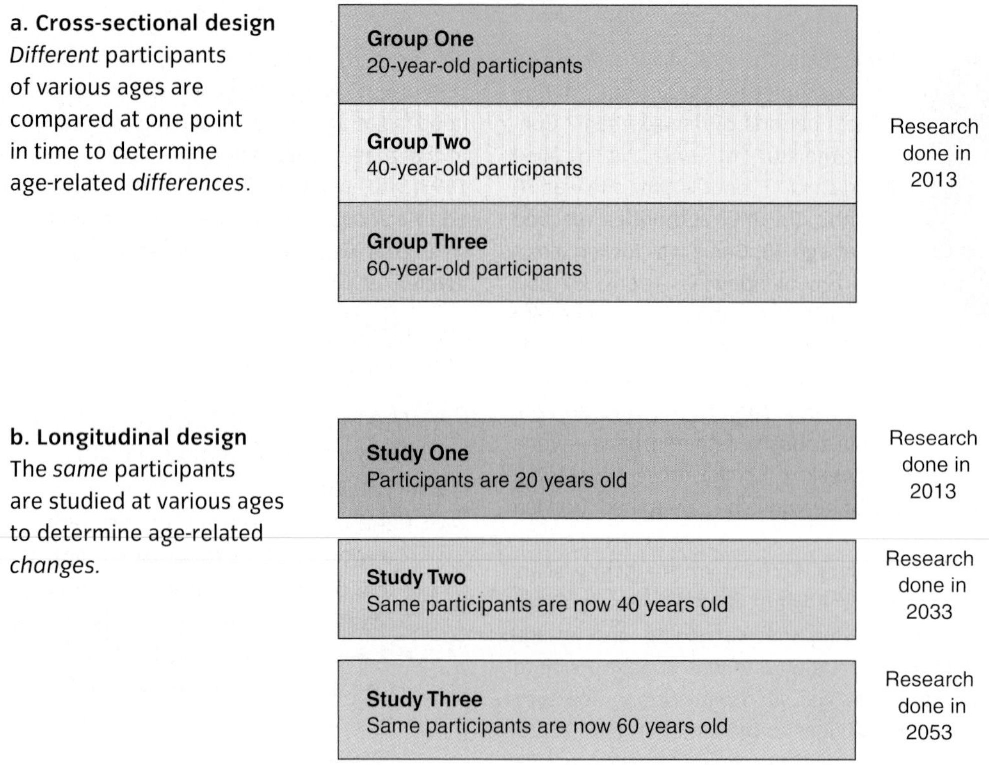
a. Cross-sectional design
*Different* participants of various ages are compared at one point in time to determine age-related *differences*.
Group One
20-year-old participants
Group Two
40-year-old participants
Group Three
60-year-old participants
Research done in 2013
b. Longitudinal design
The *same* participants are studied at various ages to determine age-related *changes*.
Study One
Participants are 20 years old
Research done in 2013
Study Two
Same participants are now 40 years old
Research done in 2033
Study Three
Same participants are now 60 years old
Research done in 2053

**Stability or change** Have you generally maintained your personal characteristics as you matured from infant to adult (stability)? Or does your current personality bear little resemblance to the personality you displayed during infancy (change)? Psychologists who emphasize stability in development hold that measurements of personality taken during childhood are important predictors of adult personality. Of course, psychologists who emphasize change disagree.

Which of these positions is most correct? Most psychologists do not take a hard line either way. Rather, they prefer an *interactionist perspective* and/or the *biopsychosocial model*. For example, in the *nature-nurture* debate, psychologists generally agree that development emerges from unique genetic predispositions *and* from experiences in the environment (Hartwell, 2008; Hudziak, 2008; Rutter, 2007). Note that the "nature versus nurture" debate applies to numerous areas of psychology, such as language and intelligence (Chapter 8), and psychological disorders (Chapter 13).

## Research Approaches

Developmental psychologists typically use all of the research methods discussed in Chapter 1. However, to study changes that happen throughout the entire human lifespan, they need to add two specific techniques, cross-sectional design and longitudinal design. The **cross-sectional design** examines people of various ages (e.g., 20, 40, 60, and 80) all at the same time to see how each age differs at that one specific time of measurement. The **longitudinal design** involves repeated measures of one person or a group of same-aged people over a long period of time to see how the individual or the group change over time (**Figure 9.3**).

**cross-sectional design** Research approach that measures individuals of various ages at one point in time and gives information about age differences.

**longitudinal design** Research approach that measures a single individual or a group of same-aged individuals over an extended period and gives information about age changes.

## Which results are "true"? • Figure 9.4

Cross-sectional studies have shown that reasoning and intelligence reach their peak in early adulthood and then gradually decline. In contrast, longitudinal studies have found that a marked decline does not begin until about age 60.

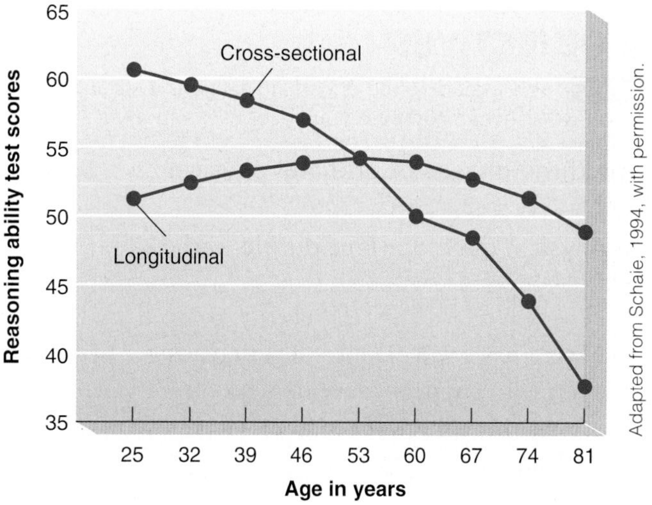


Imagine that you are a developmental psychologist interested in studying intelligence in adults. Which design would you choose—cross-sectional or longitudinal? Before you decide, note the different research results shown in **Figure 9.4**.

Why do the two methods show such different results? Cross-sectional studies sometimes confuse genuine age differences with **cohort effects**—differences that result from specific histories of the age group studied. As shown in the top line in Figure 9.4, the 81-year-olds measured by the cross-sectional design have dramatically lower scores than the 25-year-olds. But is this due to aging or possibly to broad environmental differences, such as less formal education or poorer nutrition? Because the different age groups, called *cohorts*, grew up in different historical periods, the results may not apply to people growing up at other times. With the cross-sectional design, age effects and cohort effects are sometimes hopelessly tangled.

Longitudinal studies also have their limits. They are expensive in terms of time and money, and their results are restricted in generalizability. Because participants often drop out or move away during the extended test period, the experimenter may end up with a self-selected sample that differs from the general population in important ways. Each method of research has its own strengths and weaknesses (**Table 9.2**). Keep these differences in mind when you read the findings of developmental research.

### CONCEPT CHECK

STOP

1. **How** does imprinting support the nature position of the nature-nurture debate?
2. **Which** kind of study provides the most in-depth information, cross-sectional or longitudinal?

**Advantages and disadvantages of cross-sectional and longitudinal research designs Table 9.2**

| Research design | Advantages | Disadvantages |
|---|---|---|
| **Cross-sectional** | • Provides information about age differences<br>• Quick and less expensive<br>• Typically larger sample | • Cohort effects difficult to separate<br>• Restricted generalizability (measures behaviors and mental processes at only one point in time) |
| **Longitudinal** | • Provides information about age changes<br>• Increased confidence in results<br>• More in-depth information per participant | • More expensive and time consuming<br>• Restricted generalizability (typically smaller sample due to participant dropouts over time) |

# Physical Development

## LEARNING OBJECTIVES

**RETRIEVAL PRACTICE** While reading the upcoming sections, respond to each Learning Objective in your own words. Then compare your responses with those in Appendix B.

1. **Describe** the three phases of prenatal physical development.
2. **Summarize** physical development during early childhood.
3. **Describe** the physical changes that occur during adolescence and adulthood.

In this section, we will explore the fascinating processes of physical development from conception through childhood, adolescence, and adulthood.

## Prenatal and Early Childhood

The early years of development are characterized by rapid and unparalleled change. In fact, if you continued to develop at the same rapid rate that marked your first two years of life, you would weigh several tons and be over 12 feet tall as an adult! Thankfully, physical development slows, yet it is important to note that change continues until the very moment of death. Let's look at some of the major physical changes that occur throughout the lifespan.

Your prenatal development began at conception, when your mother's egg, or ovum, united with your father's sperm cell (**Figure 9.5**). At that time, you were a single cell barely 1/175 of an inch in diameter—smaller than the period at the end of this sentence. This new cell, called a **zygote**, then began a process of rapid cell division that resulted in a multimillion-celled infant (you) some nine months later.

The vast changes that occur during the nine months of a full-term pregnancy are usually divided into three stages: the **germinal period**, the **embryonic period**, and the **fetal period** (**Figure 9.6**).

Prenatal growth, as well as growth during the first few years after birth, is *proximodistal* (near to far), with the innermost parts of the body developing before the outermost. Thus, a fetus's arms develop before its hands and fingers. Development also proceeds *cephalocaudally* (head to tail). Thus, a fetus's head is disproportionately large compared to the lower part of its body.

**germinal period** The first stage of prenatal development, which begins with conception and ends with implantation in the uterus (the first two weeks).

**embryonic period** The second stage of prenatal development, which begins after uterine implantation and lasts through the eighth week.

**fetal period** The third stage of prenatal development (eight weeks to birth), characterized by rapid weight gain and fine detailing of organs and systems.

### Conception • Figure 9.5

Millions of sperm are released when a man ejaculates into a woman's vagina.

**a.** Only a few hundred sperm survive the arduous trip up to the egg.

**b.** Although a joint effort is required to break through the outer coating, only one sperm will actually fertilize the egg.

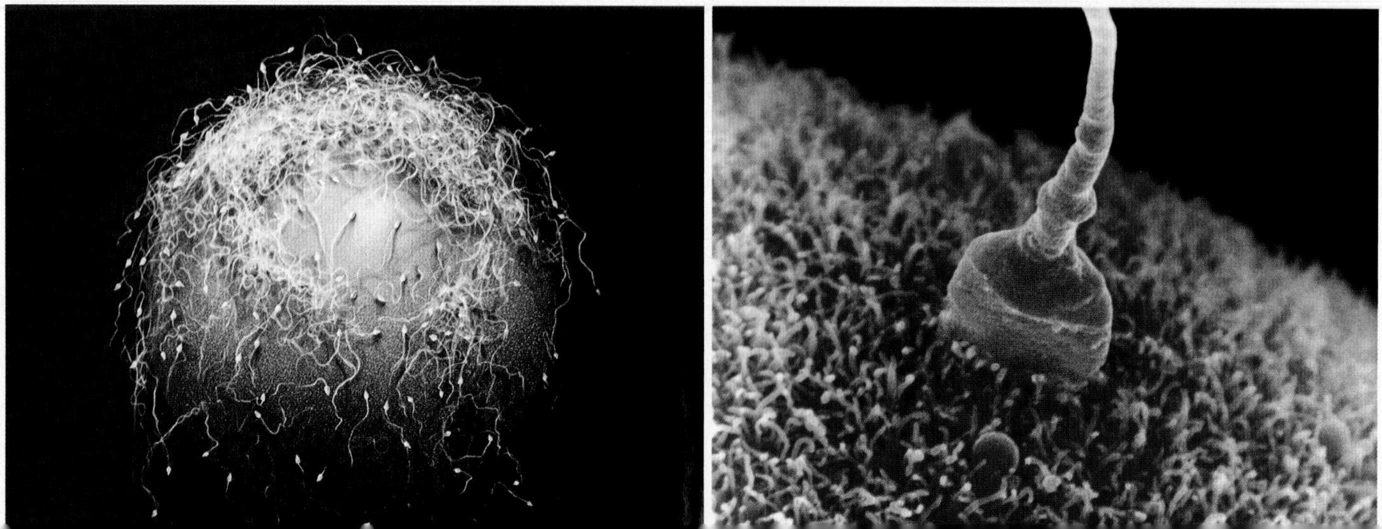

Francis Leroy, Biocosmos/Photo Researchers

David M. Phillips/Photo Researchers

PROCESS DIAGRAM

THE PLANNER

## Prenatal development • Figure 9.6

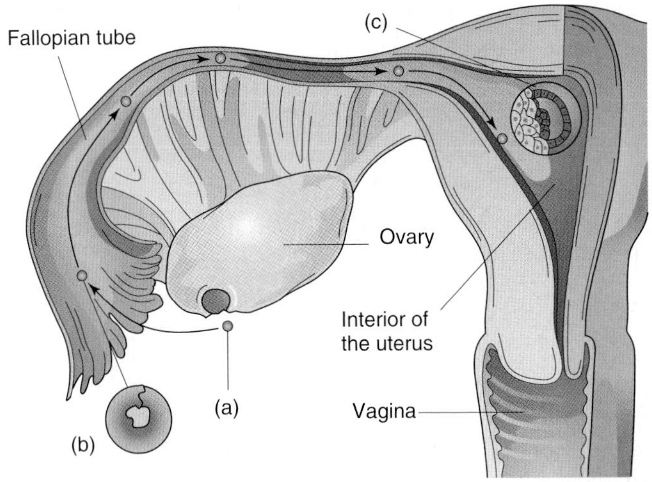


**1 Germinal period: From conception to implantation**
After discharge from either the left or right ovary (a), the ovum travels to the opening of the fallopian tube.

If fertilization occurs (b), it normally takes place in the first third of the fallopian tube. The fertilized ovum is referred to as a zygote.

When the zygote reaches the uterus, it implants itself in the wall of the uterus (c) and begins to grow tendril-like structures that intertwine with the rich supply of blood vessels located there. After implantation, the organism is known as an embryo.

Biophoto Associates/Photo Researchers

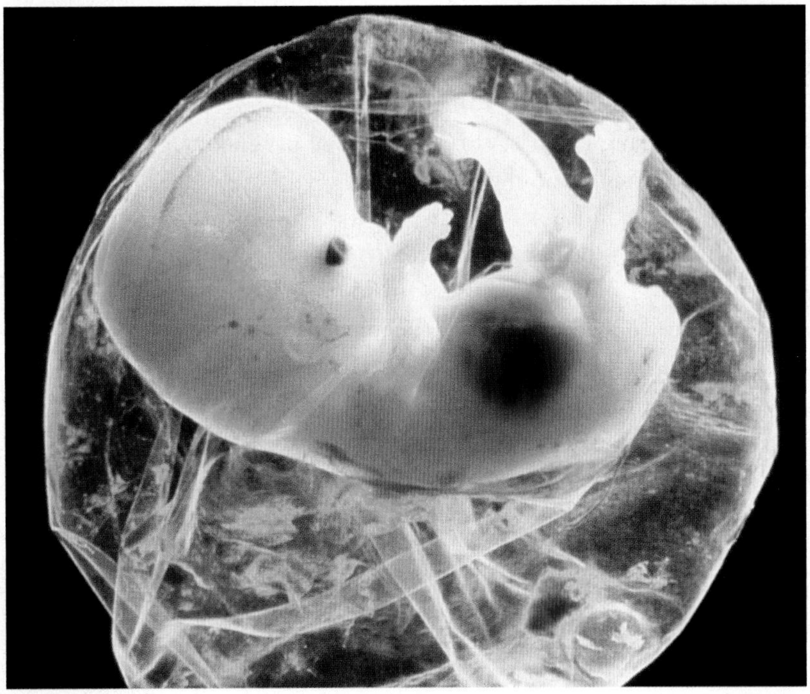

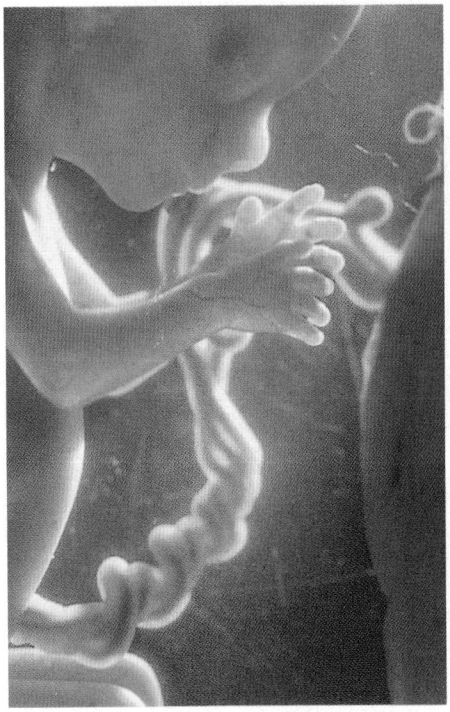

Petit Format/Nestle/Photo Researchers

**2 Embryonic period: From implantation to eight weeks**
At eight weeks, the major organ systems have become well differentiated. Note that at this stage, the head grows at a faster rate than other parts of the body.

**3 Fetal period: From the end of the second month to birth**
At four months, all the actual body parts and organs are established. The fetal stage is primarily a time for increased growth and "fine detailing."

| Sample environmental conditions that endanger the child Table 9.3 | |
|---|---|
| **Maternal factors** | **Possible effects on embryo, fetus, newborn, or young child** |
| **Malnutrition** | Low birth weight, malformations, less developed brain, greater vulnerability to disease |
| **Stress exposure** | Low birth weight, hyperactivity, irritability, feeding difficulties |
| **Exposure to x-rays** | Malformations, cancer |
| **Legal and illegal drugs:** Alcohol, nicotine | Inhibition of bone growth, hearing loss, low birth weight, fetal alcohol syndrome (FAS), intellectual disability, attention deficits in childhood, death |
| **Diseases:** German measles (rubella), herpes, HIV/AIDS, toxoplasmosis | Blindness, deafness, intellectual disability, heart and other malformations, brain infection, spontaneous abortion, premature birth, low birth weight, death |

*Sources:* Abadinsky, 2011; Howell, Coles, & Kable, 2008; Levinthal, 2011; Maisto, Galizio, & Connors, 2011; Whitbourne, 2011.

During pregnancy, the **placenta** (the vascular organ that unites the fetus to the mother's uterus) serves as the link for food and excretion of wastes. It also screens out some, but not all, harmful substances. Environmental hazards such as x-rays or toxic waste, drugs, and diseases such as rubella (German measles) can cross the placental barrier. These influences generally have their most devastating effects during the first three months of pregnancy, making this a *critical period* in development.

The pregnant mother plays a primary role in prenatal development because her health directly influences the child she is carrying (**Table 9.3**). Almost everything she ingests can cross the placental barrier (a better term might be *placental sieve*). However, the father also plays a role (other than just fertilization). Environmentally, the father's smoking may pollute the air the mother breathes, and genetically, he may transmit heritable diseases. In addition, research suggests that alcohol, opiates, cocaine, various gases, lead, pesticides, and industrial chemicals can damage sperm (Baker & Nieuwenhuijsen, 2008; Ferretti et al., 2006; Levy et al., 2011).

Perhaps the most important—and generally avoidable—danger to the fetus comes from drugs, both legal and illegal. Nicotine and alcohol are major **teratogens**, environmental agents that cause damage during prenatal development. Mothers who smoke tobacco or drink alcohol during pregnancy have significantly higher rates of premature births, low-birth-weight infants, and fetal deaths. Their children also show increased behavior and cognitive problems (Abadinsky, 2011; Espy et al., 2011; Larkby et al., 2011).

Alcohol also readily crosses the placenta, affects fetal development, and can result in a neurotoxic syndrome called **fetal alcohol syndrome (FAS)** (**Figure 9.7**). About one in a hundred babies in the United States is born with FAS or other birth defects resulting from the mother's alcohol use during pregnancy (National Organization on Fetal Alcohol Syndrome, 2008; Popova et al., 2011).

Like the prenatal period, early childhood is also a time of rapid physical development. Although

## Fetal alcohol syndrome • Figure 9.7

Prenatal exposure to alcohol can cause facial abnormalities and stunted growth. But the most disabling features of FAS are neurobehavioral problems, ranging from hyperactivity and learning disabilities to intellectual disability, depression, and psychoses (Pellegrino & Pellegrino, 2008; Sowell et al., 2008; Wass, 2008).

David H. Wells/Corbis

# Psychology InSight

Brain development • **Figure 9.8**

The brain undergoes dramatic changes from conception to the first few years of life. Keep in mind, however, that our brains continue to change and develop throughout our lifespan.

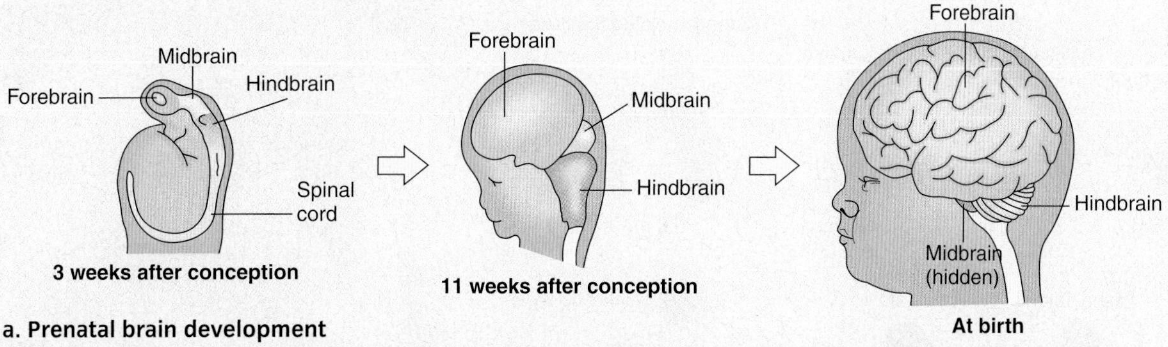


**a. Prenatal brain development**

Recall from Chapter 2 that the human brain is divided into three major sections—the hindbrain, midbrain, and forebrain. Note how at three weeks after conception these three brain sections are one long neural tube, which later becomes the brain and spinal cord.

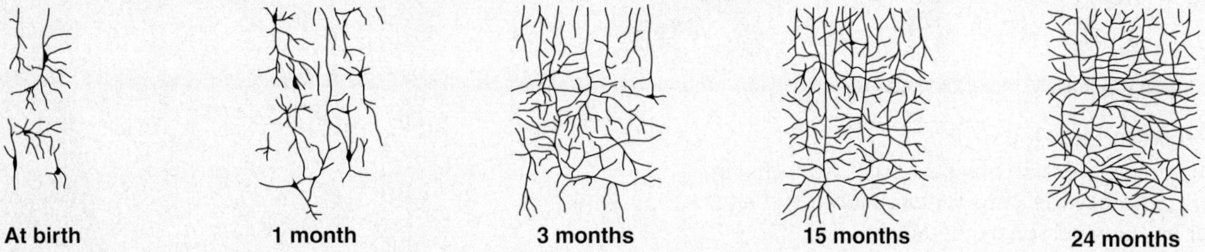


**b. Brain growth in the first two years**

As infants learn and develop, synaptic connections between active neurons strengthen, and dendritic connections become more elaborate. Synaptic pruning (reduction of unused synapses) helps support this process. Myelination, the accumulation of fatty tissue coating the axons of nerve cells, continues until early adulthood.

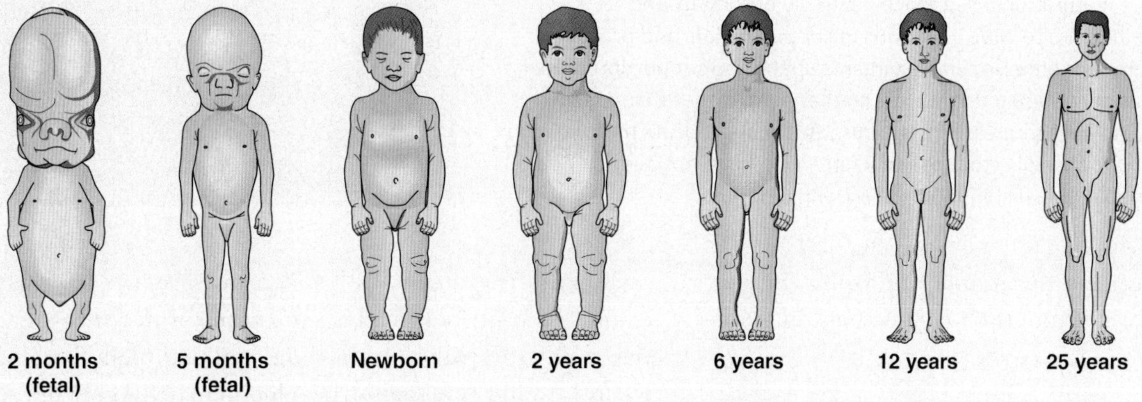


**c. Brain and body changes over your lifespan**

There are dramatic changes in your brain and body proportions as you grow older. At birth, your head was one-forth your total body's size, whereas in adulthood, your head is one-eighth.



Shakespeare described infants as capable of only "mewling and puking in the nurse's arms," they are actually capable of much more. Let's explore three key areas of change in early childhood: *brain*, *motor*, and *sensory/perceptual development*.

**Brain development** The brain and other parts of the nervous system grow faster than any other part of the body during both prenatal development and the first two years of life, illustrated in **Figure 9.8**. Further brain development and learning occur primarily because neurons

## Motor development • Figure 9.9

Motor development, like virtually all aspects of development, results from an inseparable, interaction of genetics, environment, and the individual's unique traits and learning experiences (Adolph & Berger, 2012; Atun-Einy, Berger, & Scher, 2012; Benjamin Neelon et al., 2012).

**a. Milestones in motor development**
The acquisition and progression of motor skills, from chin up to walking up steps, is generally the same for all children, but each child will follow his or her own personal timetable.

Victor Englebert/PhotoResearchers,

**b. Culture and motor development**
A similar maturational sequence generally occurs around the world, despite wide variations in cultural beliefs and practices. For example, some Hopi Indian infants spend a great portion of their first year of life being carried in a cradleboard, rather than crawling and walking freely on the ground. Yet by age 1, their motor skills are very similar to those of infants who have not been restrained in this fashion (Dennis & Dennis, 1940).

grow in size and because the number of axons and dendrites, as well as the extent of their connections, increases (DiPietro, 2000).

**Motor development** Compared to the hidden, internal changes in brain development, the orderly emergence of active movement skills, known as *motor development*, is easily observed and measured. The newborn's first motor abilities are limited to *reflexes*, or involuntary responses to stimulation. For example, the rooting reflex occurs when something touches a baby's cheek: the infant will automatically turn its head, open its mouth, and root for a nipple.

In addition to simple reflexes, the infant soon begins to show voluntary control over the movement of various body parts (**Figure 9.9**). Thus, a helpless newborn who cannot even lift her head is soon transformed into an active toddler capable of crawling, walking, and climbing. Keep in mind that motor development is largely due to natural maturation, but it can also be affected by environmental influences like disease and neglect.

**Sensory and perceptual development** At birth, a newborn can smell most odors and distinguish between sweet, salty, and bitter tastes. Breast-fed newborns also recognize and show preference for the odor and taste of their mother's milk over another mother's (Allam et al., 2010; DiPietro, 2000; Rattaz et al., 2005). In addition, the newborn's sense of touch and pain is highly developed, as evidenced by reactions to heel pricks for blood testing and to circumcision (Williamson, 1997).

The newborn's sense of vision, however, is poorly developed. At birth, an infant is estimated to have vision between 20/200 and 20/600 (Haith & Benson, 1998). Imagine what the infant's visual life is like: the level of detail you see at 200 or 600 feet (if you have 20/20 vision) is what they see at 20 feet. Within the first few months, vision quickly improves, and by six months it is 20/100 or better. At two years, visual acuity reaches a near-adult level of 20/20 (Courage & Adams, 1990).

One of the most interesting findings in infant sensory and perceptual research concerns hearing. Not only can the newborn hear quite well at birth (Matlin & Foley, 1997) but also, during the last few months in the womb, the fetus can apparently hear sounds outside the mother's body (Vaughan, 1996). This raises the interesting possibility of fetal learning, and some have advocated special stimulation for the fetus as a way of increasing intelligence, creativity, and general alertness (e.g., Van de Carr & Lehrer, 1997). For more on infant perception, read *Psychological Science*.

Ψ Psychological Science

THE PLANNER

## How an Infant Perceives the World

Because infants cannot talk or follow directions, researchers have had to create ingenious experiments to measure their perceptual abilities and preferences. One of the earliest experimenters, Robert Fantz (1956, 1963), designed a "looking chamber" to measure how long infants stared at stimuli. Research using this apparatus indicates that infants prefer complex rather than simple patterns and pictures of faces rather than nonfaces.

Researchers also use newborns' heart rates and innate abilities, such as the sucking reflex, to study learning and perceptual development (Bendersky & Sullivan, 2007). To study the sense of smell, researchers measure changes in the newborns' heart rates when odors are presented. Presumably, if they can smell one odor but not another, their heart rates will change in the presence of the first but not the second. Brain scans, such as fMRI, MRI, and CTs, also help scientists study changes in the infant's brain. From research such as this, we now know that the senses develop very early in life.

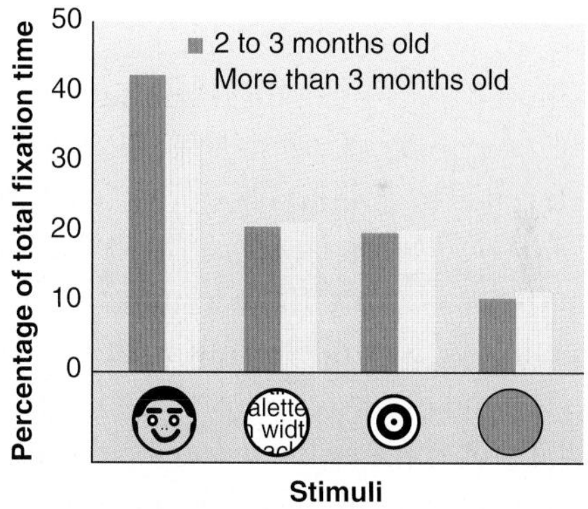


Randy Olson/NG Image Collection

### Identify the Research Method

1. What is the most likely research method used in Fantz's "looking chamber" study?
2. If you chose
   - the experimental method, label the IV, DV, experimental group, and control group.
   - the descriptive method, is this a naturalistic observation, survey, or case study?
   - the correlation method, is this a positive, negative, or zero correlation?
   - the biological method, identify the specific research tool (e.g., brain dissection, CT scan).

(Check your answers in Appendix C.)

**Ready for responsibility? • Figure 9.10**

Adolescence is not a universal concept. Unlike in the United States and other Western nations, some nonindustrialized countries have no need for a slow transition from childhood to adulthood; children simply assume adult responsibilities as soon as possible.

Studies on possible fetal learning have found that newborn infants easily recognize their own mother's voice over that of a stranger (Kisilevsky et al., 2003). They also show a preference for children's stories (such as *The Cat in the Hat* or *The King, the Mice, and the Cheese*) that were read to them while they were still in the womb (DeCasper & Fifer, 1980; Karmiloff & Karmiloff-Smith, 2002; Music, 2011). On the other hand, some experts caution that too much or the wrong kind of stimulation before birth can be stressful for both the mother and fetus. They suggest that the fetus gets what it needs without any special stimulation.

## Adolescence and Adulthood

Changes in height and weight, breast development and menstruation for girls, and a deepening voice and beard growth for boys are important milestones for adolescents. **Puberty**, the period of adolescence when a person becomes capable of reproduction, is a major physical milestone for everyone. It is a clear biological signal of the end of childhood.

**Adolescence** is the loosely defined psychological period of development between childhood and adulthood, commonly associated with puberty. In the United States, it roughly corresponds to the teenage years. The concept of adolescence and its meaning varies greatly across cultures (**Figure 9.10**).

The clearest and most dramatic physical sign of puberty is the **growth spurt**, which is characterized by rapid increases in height, weight, and skeletal growth (**Figure 9.11**), and by significant changes in reproductive structures and sexual characteristics. Maturation

**Adolescent growth spurt • Figure 9.11**

Note the gender differences in height gain during puberty. Most girls are about two years ahead of boys in their growth spurt and are therefore taller than most boys between the ages of 10 and 14.

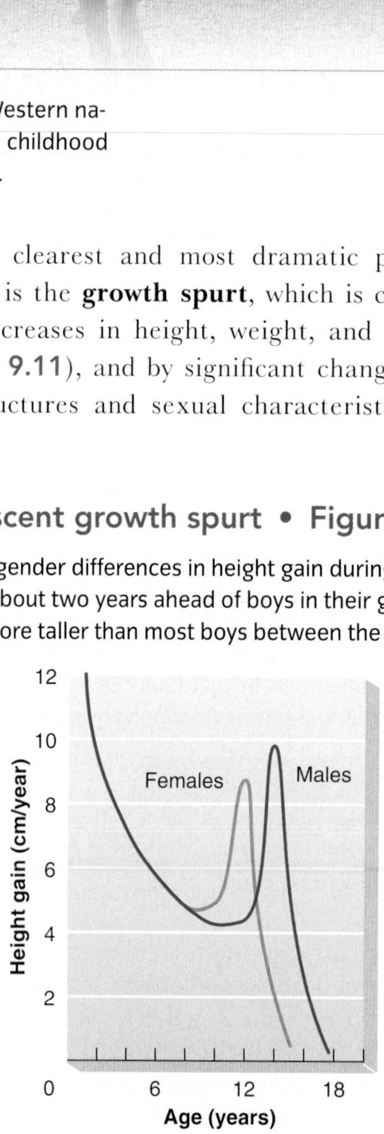


From Tanner, J. N., Whitehouse R. N., and Takaishi, M. Standards from birth to maturity for height, weight, height velocity, and weight velocity: British children, 1965. I. *Archives of Diseases in Childhood*, 41, 454-471, 1966. Reproduced with permission from BMJ Publishing Group Ltd.

## Secondary sex characteristics • Figure 9.12

Complex physical changes in puberty primarily result from hormones secreted from the ovaries and testes, the pituitary gland in the brain, and the adrenal glands near the kidneys.

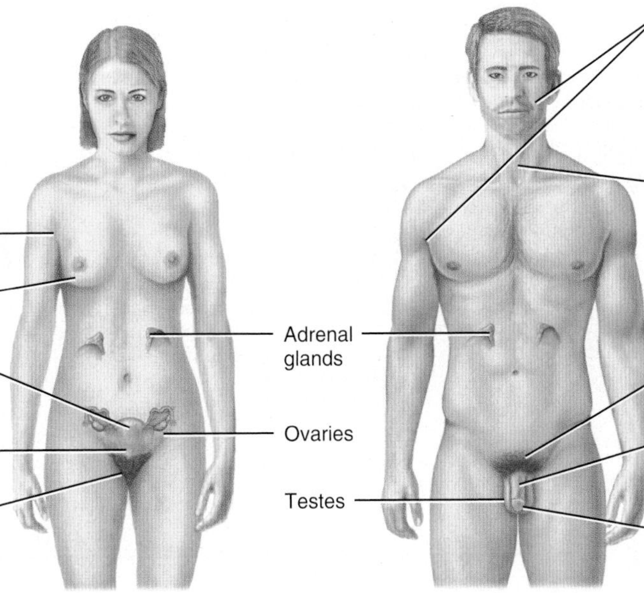


and hormone secretion cause rapid development of the ovaries, uterus, and vagina and the onset of menstruation (**menarche**) in the adolescent female. In the adolescent male, the testes, scrotum, and penis develop, and he experiences his first ejaculation (**spermarche**). The testes and ovaries in turn produce hormones that lead to the development of secondary sex characteristics, such as the growth of pubic hair, deepening of the voice and growth of facial hair in men, and the growth of breasts in women (**Figure 9.12**).

Age-related physical changes that occur after the obvious pubertal changes are less dramatic. Other than some modest increases in height and muscular development during the late teens and early twenties, most individuals experience only minor physical changes until middle age.

For women, **menopause**, the cessation of the menstrual cycle, which occurs somewhere between the ages of 45 and 55, is the second most important life milestone in physical development. The decreased production of estrogen (the dominant female hormone) produces certain physical changes. However, the popular belief that menopause (or "the change of life") causes serious psychological mood swings, loss of sexual interest, and depression is not supported by current research (Matlin, 2012; Moilanen et al., 2010). In fact, most studies find that postmenopausal women report relief, increased libido, and other positive reactions to the end of their menstrual cycles (Chrisler, 2008; Strauss, 2010).

When psychological problems do arise, they may reflect the social devaluation of aging women, not the physiological process of menopause. Given that in Western society women are highly valued for their youth and beauty, a biological process such as aging can be difficult for some women. Women living in youth-oriented cultures with negative stereotypes about aging tend to experience more anxiety and depression during menopause (Mingo, Herman, & Jasperse, 2000; Sampselle et al., 2002; Winterich, 2003).

In contrast to women, men experience a more gradual decline in hormone levels, and most men can father children until their seventies or eighties. Physical changes such as unexpected weight gain, decline in sexual responsiveness, loss of muscle strength, and graying or loss of hair may lead some men (and women as well) to feel depressed and to question their life progress. They often see these alterations as a biological signal of aging and mortality. Such physical and psychological changes in some men are generally referred to as the **male climacteric** (or *andropause*). As you will learn in Chapter 10, however, the popular belief that almost all men (and some women) go through a deeply disruptive midlife crisis, involving serious dissatisfaction with their work and personal relationships, is largely a myth.

## Achievement in later years • Figure 9.13

Contrary to the unfortunate (and untrue) stereotype that, "You can't teach an old dog new tricks," our cognitive abilities generally grow and improve throughout our lifespan, as demonstrated by the achievements of these individuals.

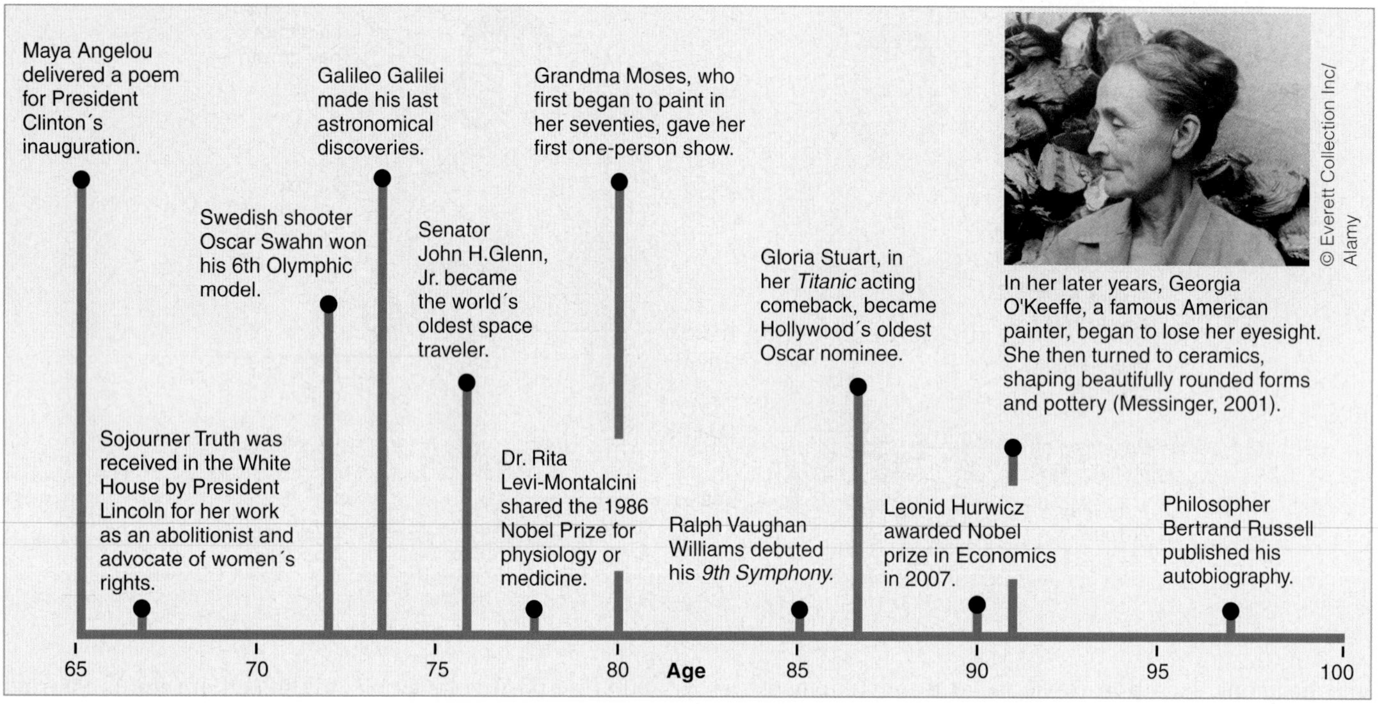

Maya Angelou delivered a poem for President Clinton's inauguration.
Sojourner Truth was received in the White House by President Lincoln for her work as an abolitionist and advocate of women's rights.
Swedish shooter Oscar Swahn won his 6th Olympic model.
Galileo Galilei made his last astronomical discoveries.
Senator John H.Glenn, Jr. became the world's oldest space traveler.
Dr. Rita Levi-Montalcini shared the 1986 Nobel Prize for physiology or medicine.
Grandma Moses, who first began to paint in her seventies, gave her first one-person show.
Ralph Vaughan Williams debuted his *9th Symphony.*
Gloria Stuart, in her *Titanic* acting comeback, became Hollywood's oldest Oscar nominee.
Leonid Hurwicz awarded Nobel prize in Economics in 2007.
In her later years, Georgia O'Keeffe, a famous American painter, began to lose her eyesight. She then turned to ceramics, shaping beautifully rounded forms and pottery (Messinger, 2001).
Philosopher Bertrand Russell published his autobiography.
65
70
75
80
Age
85
90
95
100
© Everett Collection Inc/ Alamy

Adapted from chart of elderly achievers. In The brain: A user's manual. Copyright © 1982 by Diagram Visual Information.

After middle age, most physical changes in development are gradual and occur in the heart and arteries and in the sensory receptors. For example, cardiac output (the volume of blood pumped by the heart each minute) decreases, whereas blood pressure increases due to the thickening and stiffening of arterial walls. Visual acuity and depth perception decline, hearing acuity lessens (especially for high-frequency sounds), and smell sensitivity decreases (Baldwin & Ash, 2011; Heinrich & Schneider, 2011; Morgan & Murphy, 2010; Snyder & Alain, 2007; Whitbourne, 2011).

In addition to these physical losses, television, magazines, movies, and advertisements generally portray aging as a time of balding and graying hair, sagging body parts, poor vision, hearing loss, and, of course, no sex life. Such negative portrayals contribute to our society's widespread **ageism**—prejudice or discrimination based on physical age. However, as advertising companies pursue the revenue of the huge aging baby boomer population, there has been a recent shift toward a more accurate portrayal of aging as a time of vigor, interest, and productivity (**Figure 9.13**).

What about memory problems and inherited genetic tendencies toward Alzheimer's disease and other serious diseases of old age? The public and most researchers have long thought that aging is accompanied by widespread death of neurons in the brain. Although this decline does happen with degenerative disorders like Alzheimer's disease, this is no longer believed to be a part of normal aging (Chapter 2). It is also important to remember that age-related memory problems are not on a continuum with Alzheimer's disease (Wilson et al., 2000). That is, normal forgetfulness does not mean that serious dementia is around the corner.

Aging does seem to take its toll on the speed of information processing (Chapter 7). Decreased speed of processing may reflect problems with encoding (putting information into long-term storage) and retrieval (getting information out of storage). If memory is like a filing system, older people may have more filing cabinets, and it may take them longer to initially file and later retrieve information. Although mental speed declines with age, general information processing and memory ability is largely unaffected by the aging process (Binstock & George, 2011;

Lachman, 2004; Whitbourne, 2011). Despite their concerns about "keeping up with 18-year-olds," older returning students often do as well or better than their younger counterparts in college classes. This superior performance by older adult students is due in part to their generally greater academic motivation, but it also reflects the importance of prior knowledge. Cognitive psychologists have demonstrated that the more people know, the easier it is for them to lay down new memories (Goldstein, 2011; Matlin, 2012). Older students, for instance, generally find this chapter on development easier to master than younger students. Their interactions with children and greater knowledge about life changes create a framework on which to hang new information.

What causes us to age and die? If we set aside contributions from **secondary aging** (changes resulting from disease, disuse, or neglect), we are left to consider **primary aging** (gradual, inevitable age-related changes in physical and mental processes). There are two main theories explaining primary aging and death: *programmed theory* and *damage theory*.

According to the **programmed theory**, aging is genetically controlled. Once the ovum is fertilized, the program for aging and death is set and begins to run. Researcher Leonard Hayflick (1977, 1996) found that human cells seem to have a built-in lifespan. He observed that after doubling about 50 times, laboratory-cultured cells ceased to divide—they reached the Hayflick limit. The other explanation of primary aging is **damage theory**, which proposes that an accumulation of damage to cells and organs over the years ultimately causes death.

Whether aging is genetically controlled or caused by accumulated damage over the years, scientists generally agree that humans appear to have a maximum lifespan of about 110 to 120 years. Although we can try to control secondary aging in an attempt to reach that maximum, so far we have no means to postpone primary aging.

**CONCEPT CHECK** STOP

1. **What** role does the placenta play in prenatal development?
2. **What** are some important milestones in motor development in early childhood?
3. **What** physical changes occur during middle age and late adulthood?

# Cognitive Development

## LEARNING OBJECTIVES

**RETRIEVAL PRACTICE** While reading the upcoming sections, respond to each Learning Objective in your own words. Then compare your responses with those in Appendix B.

1. **Explain** the role of schemas, assimilation, and accommodation in cognitive development.
2. **Describe** the major characteristics of Piaget's four stages of cognitive development.
3. **Compare** Piaget's theory of cognitive development to Vygotsky's.

Just as a child's body and physical abilities change, his or her way of knowing and perceiving the world also grows and changes. This seems intuitively obvious, but early psychologists—with one exception—focused on physical, emotional, language, and personality development. The one major exception was Jean Piaget (pronounced pee–ah–ZHAY).

Piaget demonstrated that a child's intellect is fundamentally different from an adult's. He showed that an infant begins at a cognitively "primitive" level and that intellectual growth progresses in distinct stages, motivated by an innate need to know. Piaget's theory, developed in the 1920s and 1930s, has proven so comprehensive and insightful that it remains the major force in the cognitive area of developmental psychology today.

To appreciate Piaget's contributions, we need to consider three major concepts: schemas, assimilation, and accommodation. **Schemas** are the most basic units of intellect.

**schemas** Cognitive structures or patterns consisting of a number of organized ideas that grow and differentiate with experience.

## Accommodation • Figure 9.14

An infant uses accommodation to make the transition to solid foods.

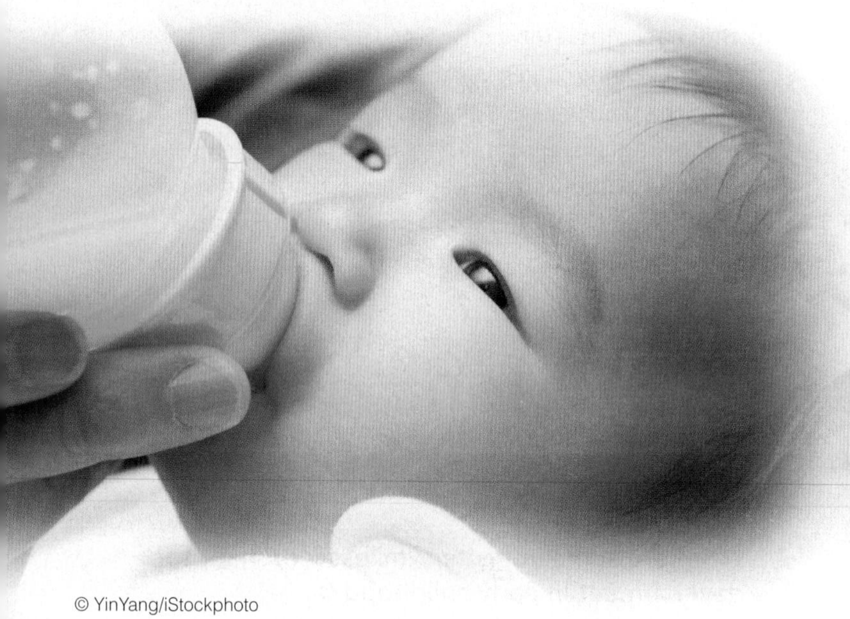

© YinYang/iStockphoto

**a.** For the first few months of life an infant ingests all her food by sucking from a nipple.

© amana images inc./Alamy

**b.** When first offered solid food from a spoon, the infant must learn to accommodate by adjusting her lips and tongue in a way that moves the food off the spoon and into her mouth.

They act as patterns that organize our interactions with the environment, like an architect's drawings or a builder's blueprints.

In the first few weeks of life, for example, the infant apparently has several schemas based on the innate reflexes of sucking, grasping, and so on. These schemas are primarily motor and may be little more than stimulus-and-response mechanisms—the nipple is presented, and the baby sucks. Soon, other schemas emerge. The infant develops a more detailed schema for eating solid food, a different schema for the concepts of "mother" and "father," and so on. Schemas, our tools for learning about the world, expand and change throughout our lives. For example, most adults who once used paper and pencil schedules have developed new schemas for their smart phone's electronic calendar, as well as for reading textbooks on a digital device, and storing computer data in the cloud. With rapid advances in technology, we'll also need new schemas for driverless cars and other devices that we haven't even imagined yet. Stay tuned!

**Assimilation** and **accommodation** are the two major processes by which schemas grow and change over time. Assimilation is the process of absorbing new information into existing schemas. For instance, infants use their sucking schema not only in sucking nipples but also in sucking blankets and fingers.

In accommodation, existing ideas are modified to fit new information. It generally occurs when new information or stimuli cannot be assimilated. New schemas are developed or old schemas are changed to better fit with the new information. An infant's first attempt to eat solid food with a spoon is a good example of accommodation (**Figure 9.14**).

**assimilation** In Piaget's theory, the process of absorbing new information into existing schemas.

**accommodation** In Piaget's theory, the process of adjusting old schemas or developing new ones to better fit with new information.

## Stages of Cognitive Development

According to Piaget, all children go through approximately the same four stages of cognitive development, regardless of the culture in which they live (**Study Organizer 9.1**). Piaget also believed these stages cannot be skipped because skills acquired at earlier stages are essential to mastery at later stages. Let's take a closer look at these four stages: *sensorimotor*, *preoperational*, *concrete operational*, and *formal operational*.

## Study Organizer 9.1 Piaget's four stages of cognitive development

THE PLANNER

| | | |
|---|---|---|
| **Sensorimotor stage (birth to age 2)** | **Limits**<br>• Beginning of stage lacks object permanence (understanding that things continue to exist even when not seen, heard, or felt)<br><br>**Abilities**<br>• Uses senses and motor skills to explore and develop cognitively<br><br>**Example**<br>• Children at this stage like to play with their food. | <br>The Copyright Group/SuperStock |
| **Preoperational stage (ages 2 to 7)** | **Limits**<br>• Cannot perform "operations" (lacks reversibility)<br>• Intuitive thinking versus logical reasoning<br>• Egocentric thinking (inability to consider another's point of view)<br>• Animistic thinking (believing all things are living)<br><br>**Abilities**<br>• Has significant language and thinks symbolically<br><br>**Example**<br>• Children at this stage often believe the moon follows them. | 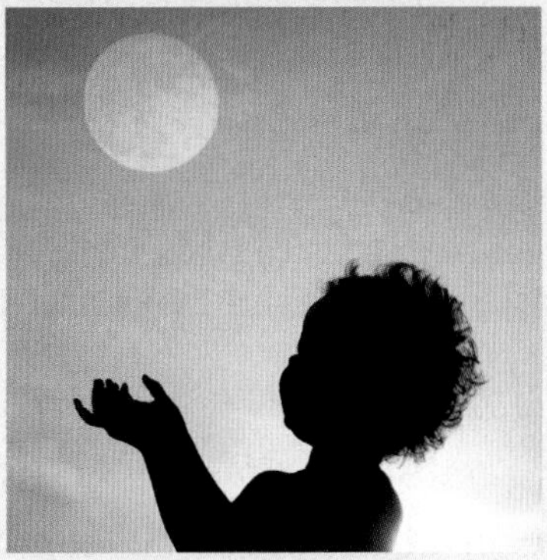<br>© Igor Demchenkov/iStockphoto |
| **Concrete operational stage (ages 7 to 11)** | **Limits**<br>• Cannot think abstractly and hypothetically<br>• Thinking tied to concrete, tangible objects and events<br><br>**Abilities**<br>• Can perform "operations" on concrete objects<br>• Understands conservation (realizing that changes in shape or appearance can be reversed)<br>• Less egocentric<br>• Can think logically about concrete objects and events<br><br>**Example**<br>• Children at this stage begin to question the existence of Santa. | <br>Hill Street Studios/Blend Images/Getty Images, Inc. |
| **Formal operational stage (age 11 and over)** | **Limits**<br>• Adolescent egocentrism at the beginning of this stage, with related problems of the personal fable and imaginary audience<br><br>**Abilities**<br>• Can think abstractly and hypothetically<br><br>**Example**<br>• Children at this stage show great concern for physical appearance. | <br>Roy Melnychuk/Taxi/Getty Images |

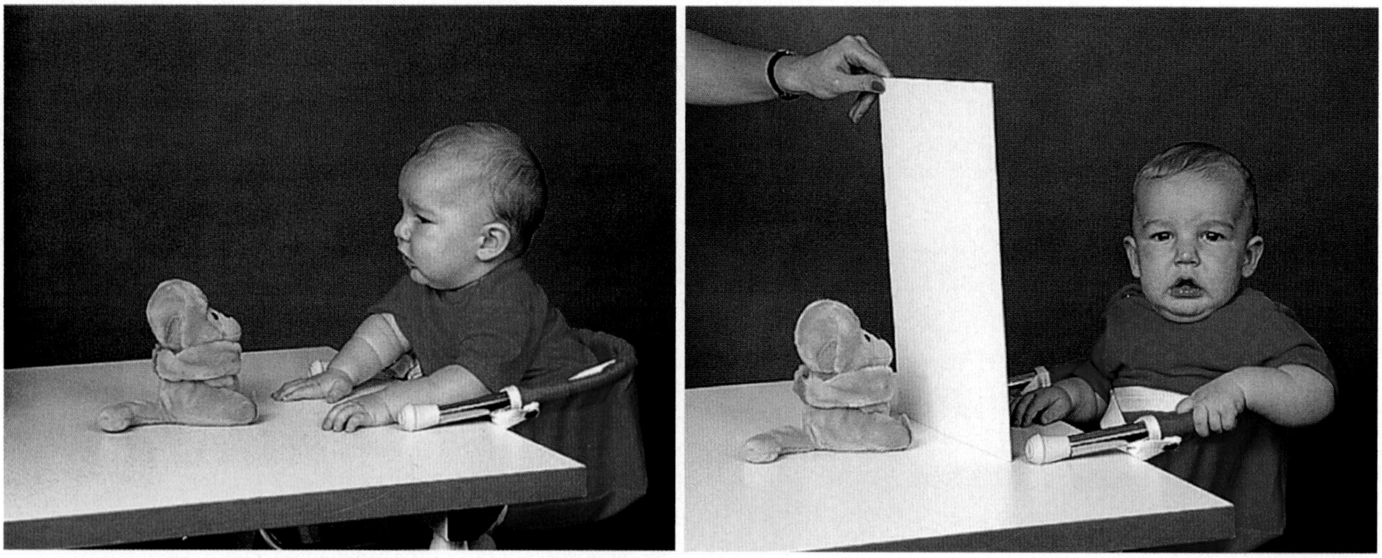

Doug Goodman/Photo Researchers, Inc.

Doug Goodman/Photo Researchers, Inc.

**Object permanence • Figure 9.15**

The child in these two photos seems to believe the toy no longer exists once it is blocked from sight. According to Piaget, young infants lack object permanence—an understanding that objects continue to exist even when they cannot be seen, heard, or touched.

**sensorimotor stage** Piaget's first stage (birth to approximately age 2), in which schemas are developed through sensory and motor activities.

**preoperational stage** Piaget's second stage (roughly ages 2 to 7 years), which is characterized by the ability to employ significant language and to think symbolically, though the child lacks operations (reversible mental processes), and thinking is egocentric and animistic.

**Sensorimotor stage** The **sensorimotor stage** lasts from birth until "significant" language acquisition (about age 2). During this time, children explore the world and develop their schemas primarily through their senses and motor activities—hence the term *sensorimotor*. One important concept that infants acquire during the sensorimotor stage is **object permanence** (**Figure 9.15**).

**Preoperational stage** During the **preoperational stage** (roughly ages 2 to 7), language advances significantly, and the child begins to think symbolically—using symbols, such as words, to represent concepts. Three other qualities characterize this stage: *lack of operations*, *egocentrism*, and *animism*.

Piaget labeled this period "preoperational" because the child lacks **operations**, or reversible mental processes. For instance, if a preoperational boy who has a brother is asked, "Do you have a brother?" he will easily respond, "Yes." However, when asked, "Does your brother have a brother?" he will answer, "No!" To understand that his brother has a brother, he must be able to reverse the concept of "having a brother."

Children at this stage have difficulty understanding that there are points of view other than their own, which is known as **egocentrism**. The preschooler who moves in front of you to get a better view of the TV, or repeatedly asks questions while you are talking on the telephone, is demonstrating egocentrism. Children of this age assume that others see, hear, feel, and think exactly as they do.

Consider the following telephone conversation between a 3-year-old, who is at home, and her mother, who is at work:

Mother: Emma, is that you?
Emma: (Nods silently.)
Mother: Emma, is Daddy there? May I speak to him?
Emma: (Twice nods silently.)

Egocentric preoperational children fail to understand that the phone caller cannot see their nodding head. Preoperational children's egocentrism also sometimes leads them to believe that their "bad thoughts" caused their sibling or parent to get sick or that their misbehavior caused their parents' marital problems. Because they think the world centers on them, they often cannot separate reality from what goes on inside their own heads.

Children in the preoperational stage believe that objects such as the sun, trees, clouds, and bars of soap have motives, feelings, and intentions (for example, "dark clouds are angry" and "soap sinks because it is tired"). **Animism** refers to the belief that all things are living (or animated).

Can preoperational children be taught how to use operations and to avoid egocentric and animistic thinking? Although some researchers have reported success in accelerating the preoperational stage, Piaget did not believe in pushing children ahead of their own developmental schedules. He believed that children should grow at their own pace, with minimal adult interference (Elkind, 1981, 2000). In fact, Piaget thought that Americans were particularly guilty of pushing their children, calling American childhood the "Great American Kid Race."

**Concrete operational stage** Following the preoperational stage, at approximately age 7, children enter the **concrete operational stage**. During this stage, many important thinking skills emerge. However, as the name implies, their thinking tends to be limited to *concrete*, tangible objects and events. Unlike the preoperational child, children in this stage are less egocentric in their thinking and become capable of true logical thought. As most parents know, children at this stage stop believing in Santa Claus because they logically conclude that one man can't deliver presents to everyone in one night.

**concrete operational stage** Piaget's third stage (roughly ages 7 to 11), in which the child can perform mental operations on concrete objects and understand reversibility and conservation, though abstract thinking is not yet present.

Because they understand the concept of *reversibility*, concrete operational children also can now successfully perform "operations." They recognize that certain physical attributes (such as volume) remain unchanged, although the outward appearance is altered, a process known as **conservation**. Tests for conservation can identify whether a child is in the preoperational or the concrete operational stage (see *Applying Psychology* on the next page).

**formal operational stage** Piaget's fourth stage (around age 11 and beyond), which is characterized by abstract and hypothetical thinking.

**Formal operational stage** The final stage in Piaget's theory is the **formal operational stage**, which typically begins around age 11. In this stage, children begin to apply their operations to abstract concepts in addition to concrete objects. They also become capable of hypothetical thinking ("What if?"), which allows systematic formulation and testing of concepts.

For example, before filling out applications for part-time jobs, adolescents may think about possible conflicts with school and friends, the number of hours they want to work, and the kind of work for which they are qualified. Formal operational thinking also allows the adolescent to construct a well-reasoned argument based on hypothetical concepts and logical processes. Consider the following argument:

1. If you hit a glass with a feather, the glass will break.
2. You hit the glass with a feather.

What is the logical conclusion? The correct answer, "The glass will break," is contrary to fact and direct experience. Therefore, the child in the concrete operational stage would have difficulty with this task, whereas the formal operational thinker understands that this problem is about abstractions that need not correspond to the real world.

Along with the benefits of this cognitive style come several problems. Adolescents in the early stages of the formal operational period demonstrate a type of egocentrism different from that of the preoperational child. Although adolescents recognize that others have unique thoughts and perspectives, they often fail to differentiate between what they are thinking and what others are thinking. This adolescent egocentrism has two characteristics that may affect social interactions as well as problem solving—the personal fable and the imaginary audience.

Because of their unique form of egocentrism, adolescents may conclude that they alone are having insights or difficulties and that no one else understands or sympathizes with them. David Elkind (1976, 2001, 2007) described this as the formation of a **personal fable**, an intense investment in an adolescent's own thoughts and feelings and a belief that these thoughts are unique. For example, one young woman remembered being very upset in middle school when her mother tried to comfort her over the loss of an important relationship. "I felt like she couldn't possibly know how it felt—no one could. I couldn't believe that anyone had ever suffered like this or that things would ever get better."

# Applying Psychology

THE PLANNER

## Putting Piaget to the Test

If you have access to children in the preoperational or concrete operational stages, try some of the following experiments, which researchers use to test Piaget's various forms of conservation. The equipment is easily obtained, and you will find their responses fascinating. Keep in mind that this should be done as a game. The child should not feel that he or she is failing a test or making a mistake.

| Type of conservation task (average age at which concept is fully grasped) | Your task as experimenter . . . | Child is asked . . . |
|---|---|---|
| **Length** (ages 6–7) | **Step 1** Center two sticks of equal length. Child agrees that they are of equal length. **Step 2** Move one stick.  | **Step 3** "Which stick is longer?" Preoperational child will say that one of the sticks is longer. Child in concrete stage will say that they are both the same length. |
| **Substance amount** (ages 6–7) | **Step 1** Center two identical clay balls. Child acknowledges that the two have equal amounts of clay. **Step 2** Flatten one of the balls.  | **Step 3** "Do the two pieces have the same amount of clay?" Preoperational child will say that the flat piece has more clay. Child in concrete stage will say that the two pieces have the same amount of clay. |
| **Liquid volume** (ages 7–8) | **Step 1** Present two identical glasses with liquid at the same level. Child agrees that liquid is at the same height in both glasses. **Step 2** Pour the liquid from one of the short, wide glasses into the tall, thin one.  | **Step 3** "Do the two glasses have the same amount of liquid?" Preoperational child will say that the tall, thin glass has more liquid. Child in concrete stage will say that the two glasses have the same amount of liquid. |
| **Area** (ages 8–10) | **Step 1** Center two identical pieces of cardboard with wooden blocks placed on them in identical positions. Child acknowledges that the same amount of space is left open on each piece of cardboard. **Step 2** Scatter the blocks on one piece of the cardboard. 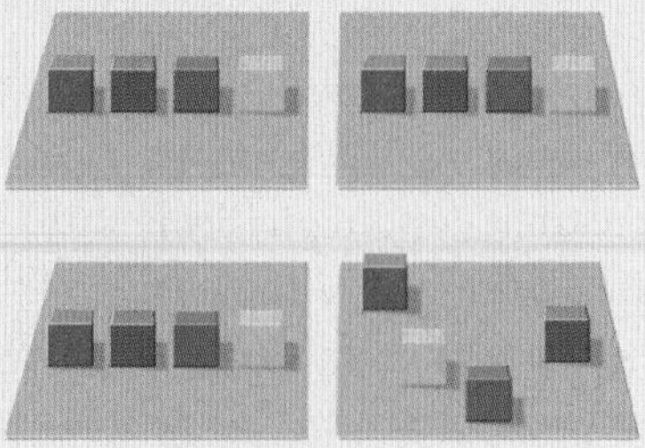 | **Step 3** "Do the two pieces of cardboard have the same amount of open space?" Preoperational child will say that the cardboard with scattered blocks has less open space. Child in concrete stage will say that both pieces have the same amount of open space. |

**Think Critically**

1. Based on their responses, are the children you tested in the preoperational or concrete stage?
2. If you repeat the same tests with each child, do their answers change? Why or why not?

Several forms of risk taking, such as engaging in sexual intercourse without contraception, driving dangerously, and experimenting with drugs, seem to arise from the personal fable (Alberts, Elkind, & Ginsburg, 2007; Flavell, Miller, & Miller, 2002; Hill & Lapsley, 2011; Moshman, 2011). Adolescents have a sense of uniqueness, invulnerability, and immortality. They recognize the dangers of these activities, but "thanks to" their adolescent egocentrism and personal fable they believe the rules and statistical dangers don't apply to them (see *What a Psychologist Sees*).

# WHAT A PSYCHOLOGIST SEES ✓ THE PLANNER

## Brain and Behavior in Adolescence

Adolescents are often criticized for their mood swings, poor decisions, and risky behaviors (**Figure a**), but psychologists now believe this may be due to their less developed frontal lobes (Bava & Tapert, 2010; Christakou, Brammer, & Rubia, 2011; Moshman, 2011). Recall from Chapter 2 that your frontal lobes are largely responsible for higher level cognitive functioning, such as emotional regulation, planning ahead, and reckless behavior.

In contrast to the rapid synaptic growth experienced in the earlier years (**Figures b** and **c**), the adolescent's brain actively destroys (prunes) unneeded connections (**Figure d**). Although it may seem counterintuitive, this pruning actually improves brain functioning by making the remaining neurons more efficient. Full maturity of the frontal lobes occurs around the mid-twenties.

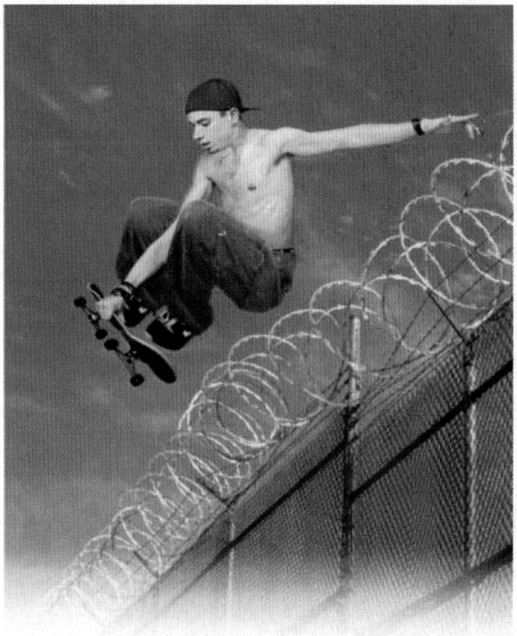

© PCN Photography/Alamy

**a.** Can you see how this type of risk-taking behavior may reflect the personal fable—adolescents' tendency to believe they are unique and special and that dangers don't apply to them?

Growth
Pruning

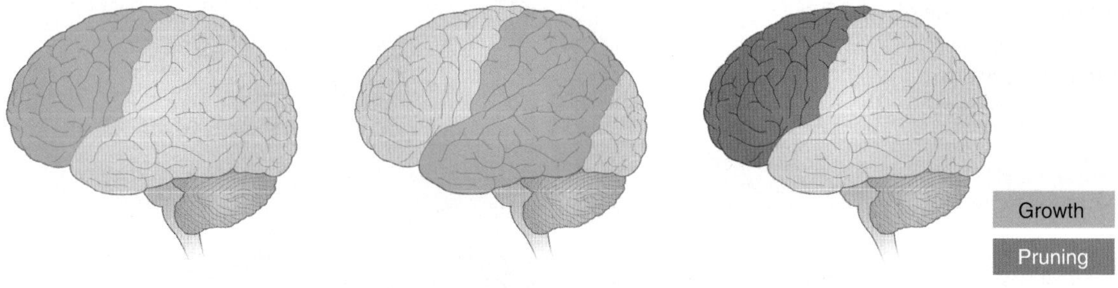

**b.** During early childhood (ages 3–6), the frontal lobes experience a significant increase in the connections between neurons, which helps explain a child's rapid cognitive growth.

**c.** This rapid synaptic growth shifts to the temporal and parietal lobes during the ages of 7 to 15, which corresponds to their significant increases in language and motor skills.

**d.** During the teen years (ages 16–20), synaptic pruning is widespread in the frontal lobes.

In addition to the personal fable, in early adolescence, people tend to believe that they are the center of others' thoughts and attentions, instead of considering that everyone is equally wrapped up in his or her own concerns and plans. In other words, adolescents feel that all eyes are focused on their behaviors. Elkind referred to this as the **imaginary audience**.

If the imaginary audience results from an inability to differentiate the self from others, the personal fable is a product of differentiating too much. Thankfully, these two forms of adolescent egocentrism tend to decrease during later stages of the formal operational period.

## Vygotsky Versus Piaget

As influential as Piaget's account of cognitive development has been, there are other important theories and criticisms of Piaget to consider. For example, Russian psychologist Lev Vygotsky emphasized the sociocultural influences on a child's cognitive development, rather than Piaget's internal schemas. According to Vygotsky, children construct knowledge through their culture, language, and collaborative social interactions with more experienced thinkers (Ferrari, Robinson, & Yasnitsky, 2010; Trawick-Smith & Dziurgot, 2011). Unlike Piaget, Vygotsky believed that adults play an important instructor role in development and that this instruction is particularly helpful when it falls within a child's **zone of proximal development (ZPD)** (**Figure 9.16**).

Having briefly discussed Vygotsky's alternative theory, let's consider two major criticisms of Piaget: underestimated abilities and underestimated genetic and cultural influences. Research shows that Piaget may have underestimated young children's cognitive development (**Figure 9.17**). For example, researchers report that very young infants have a basic concept of how objects move, are aware that objects continue to exist even when screened from view, and can recognize speech sounds (Baillargeon, 2000, 2008; Madole, Oaks, & Rakison, 2011).

Nonegocentric responses also appear in the earliest days of life. For example, newborn babies tend to cry in response to the cry of another baby (Diego & Jones, 2007; Geangu et al., 2010). Also, preschoolers will adapt their speech by using shorter, simpler expressions when talking to 2-year-olds rather than to adults.

Piaget's model, like other stage theories, has also been criticized for not sufficiently taking into account genetic and cultural differences (Cole & Gajdamaschko, 2007; Shweder, 2011; Zelazo, Chandler, & Crone, 2010). During Piaget's time, the genetic influences on cognitive abilities were

### Vygotsky's zone of proximal development (ZPD) • Figure 9.16

Much like the modern concept of instructional scaffolding, Vygotsky suggests that the most effective teaching focuses on tasks in between those that a learner can do without help (the lower limit) and those that he or she cannot do even with help (the upper limit). In this zone of proximal development (ZPD), tasks and skills can be readily developed with the guidance and encouragement of a more knowledgeable person.

© Mary Kate Denny/PhotoEdit

**Are preoperational children always egocentric? • Figure 9.17**

Some toddlers and preschoolers clearly demonstrate empathy for other people. How does this ability to take another's perspective contradict Piaget's beliefs about egocentrism in very young children?

poorly understood, but there has been a rapid explosion of information in this field in recent years. In addition, formal education and specific cultural experiences can also significantly affect cognitive development. Consider the following example from a researcher attempting to test the formal operational skills of a farmer in Liberia (Scribner, 1977):

> Researcher: All Kpelle men are rice farmers. Mr. Smith is not a rice farmer. Is he a Kpelle man?
>
> Kpelle farmer: I don't know the man. I have not laid eyes on the man myself.

Instead of reasoning in the "logical" way of Piaget's formal operational stage, the Kpelle farmer reasoned according to his specific cultural and educational training, which apparently emphasized personal knowledge. Not knowing Mr. Smith, the Kpelle farmer did not feel qualified to comment on him. Thus, Piaget's theory may have underestimated the effect of culture on a person's cognitive functioning.

Despite criticisms, Piaget's contributions to psychology are enormous. As one scholar put it, "assessing the impact of Piaget on developmental psychology is like assessing the impact of Shakespeare on English literature or Aristotle on philosophy—impossible" (Beilin, 1992, p. 191).

**CONCEPT CHECK** STOP

1. **When** does accommodation occur?
2. **How** do egocentrism and animism limit children's thinking during the preoperational stage?
3. **What** criticisms has Piaget's theory received?

# Summary

THE PLANNER

## 1 Studying Development 230

- **Developmental psychology** is the study of age-related changes in behavior and mental processes from conception to death. Development is an ongoing, lifelong process.
- The three most important debates or questions in human development are about *nature versus nurture* (including studies of **maturation** and **critical periods**), *stages versus continuity* (illustrated in the diagram), and *stability versus change*. For each question, most psychologists prefer an interactionist perspective.
- Developmental psychologists use two special techniques in their research: **cross-sectional design** and **longitudinal design**. Although both have valuable attributes, each also has disadvantages. Cross-sectional studies can confuse genuine age differences with **cohort effects**. On the other hand, longitudinal studies are expensive and time-consuming, and their results are restricted in generalizability.

**Stages versus continuity in development • Figure 9.2**

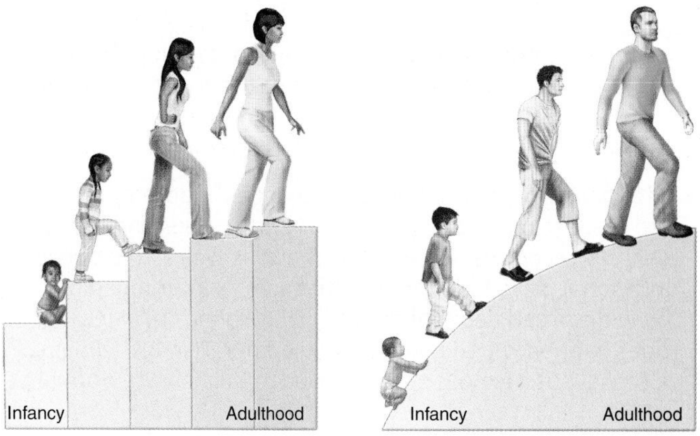

## 2 Physical Development 234

- The early years of development are characterized by rapid change. Prenatal development begins at conception and is divided into three stages: the **germinal period** (shown in the diagram), the **embryonic period**, and the **fetal period**. During pregnancy, the **placenta** serves as the link for food and the excretion of wastes, and it screens out some harmful substances. However, some environmental hazards (**teratogens**), such as alcohol and nicotine, can cross the placental barrier and endanger prenatal development.

**Prenatal development • Figure 9.6**

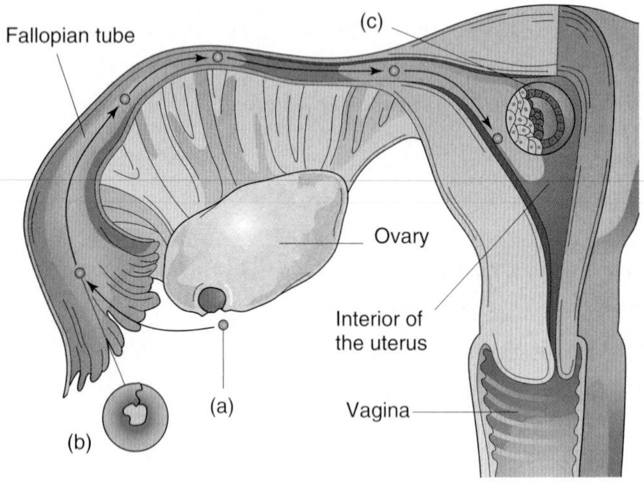


- Early childhood is also a time of rapid physical development, including brain, motor, and sensory/perceptual development.
- During **adolescence**, both boys and girls undergo dramatic changes in appearance and physical capacity. **Puberty** is the period of adolescence when a person becomes capable of reproduction. The clearest and most dramatic physical sign of puberty is the **growth spurt**, characterized by rapid increases in height, weight, and skeletal growth and by significant changes in reproductive structures and sexual characteristics.
- During adulthood, most individuals experience only minor physical changes until middle age. Around age 45–55, women experience **menopause**, the cessation of the menstrual cycle. At the same time, men experience a gradual decline in the production of sperm and testosterone, as well as other physical changes, known as the **male climacteric**.
- After middle age, most physical changes in development are gradual and occur in the heart and arteries and in the sensory receptors. Most researchers, as well as the public in general, have long thought that aging is accompanied by widespread death of neurons in the brain. This decline does happen with degenerative disorders like Alzheimer's disease, but it is no longer believed to be a part of normal aging. Although mental speed declines with age, general information processing and much of memory ability are largely unaffected by the aging process. There are two main theories explaining **primary aging** and death—**programmed theory** and **damage theory**.

## 3 Cognitive Development 243

- In the 1920s and 1930s, Piaget conducted groundbreaking work on children's cognitive development. Piaget's theory remains the major force in the cognitive area of developmental psychology today.
- Three major concepts are central to Piaget's theory: **schemas, assimilation**, and **accommodation**. According to Piaget, all children progress through four stages of cognitive development: the **sensorimotor stage**, the **preoperational stage**, the **concrete operational stage**, and the **formal operational stage**. As they move through these stages, children acquire progressively more sophisticated ways of thinking.
- Piaget's account of cognitive development has been enormously influential, but it has also received significant criticisms. For example, Vygotsky emphasized the sociocultural influences on a child's cognitive development, rather than Piaget's internal schemas. He also believed that adults play an important instructor role in development and that this instruction is particularly helpful when it falls within a child's **zone of proximal development (ZPD),** as shown in the figure. Piaget has also been criticized for underestimating children's abilities, as well as the genetic and cultural influences on cognitive development.

**Vygotsky's zone of proximal development (ZPD) • Figure 9.16**

# Key Terms

**RETRIEVAL PRACTICE** Write a definition for each term before turning back to the referenced page to check your answer.

- accommodation 244
- adolescence 240
- ageism 242
- animism 247
- assimilation 244
- cohort effect 233
- concrete operational stage 247
- conservation 247
- critical period 230
- cross-sectional design 232
- damage theory 243
- developmental psychology 230
- egocentrism 246
- embryonic period 234
- fetal alcohol syndrome (FAS) 236
- fetal period 234
- formal operational stage 247
- germinal period 234
- growth spurt 240
- imaginary audience 250
- imprinting 230
- longitudinal design 232
- male climacteric 241
- maturation 230
- menarche 241
- menopause 241
- object permanence 246
- operations 246
- personal fable 247
- placenta 236
- preoperational stage 246
- primary aging 243
- programmed theory 243
- puberty 240
- schema 243
- secondary aging 243
- sensorimotor stage 246
- spermarche 241
- teratogen 236
- zone of proximal development (ZPD) 250
- zygote 234

# Critical and Creative Thinking Questions

1. Based on what you have learned about the advantages and disadvantages of cross-sectional and longitudinal designs, can you think of a circumstance when each might be preferable over the other?
2. If a mother knowingly ingests a quantity of alcohol that causes her child to develop fetal alcohol syndrome (FAS), is she guilty of child abuse? Why or why not?
3. Based on what you have learned about schemas, what are some new schemas that you developed as part of your transition from high school to college?
4. Based on what you have learned about development during late adulthood, do you think that this period is necessarily a time of physical and mental decline?
5. Piaget's theory states that all children progress through all of the discrete stages of cognitive development in order and without skipping any. Do you agree with this theory? Do you know any children who seem to contradict this theory?

George F. Mobley/NG Image Collection

# What is happening in this picture?

Parents of young children are generally surprised, amused, and even frustrated with their child's repeated and frequent episodes of "playing with their food." They often wonder what motivates this behavior.

**Think Critically**

1. What stage of cognitive development does this child's behavior typify?
2. What schemas might the child build by "exploring" her food in this way?

# Self-Test

**RETRIEVAL PRACTICE** Completing this self-test and comparing your answers with those in Appendix C provides immediate feedback and helpful practice for exams. Additional interactive, self-tests are available at www.wiley.com/college/carpenter.

**1.** The study of age-related changes in behavior and mental processes from conception to death is called ______.

a. thanatology

b. neo-gerontology

c. developmental psychology

d. longitudinal psychology

**2.** Development governed by automatic, genetically predetermined signals is called ______.

a. growth

b. natural progression

c. maturation

d. tabula rasa

**3.** Label the two basic types of research designs:

a. ______________

**Group One**
20-year-old participants

**Group Two**
40-year-old participants

**Group Three**
60-year-old participants

Research done in 2013

b. ______________

**Study One**
Participants are 20 years old
Research done in 2013

**Study Two**
Same participants are now 40 years old
Research done in 2033

**Study Three**
Same participants are now 60 years old
Research done in 2053

**4.** This is the first stage of prenatal development, which begins with conception and ends with implantation in the uterus (the first two weeks).

a. embryonic period

b. zygote stage

c. critical period

d. germinal period

**5.** As shown in the diagram, at birth, an infant's head is ______ its body's size, whereas in adulthood, the head is ______ its body's size.

a. 1/3; 1/4

b. 1/3; 1/10

c. 1/4; 1/10

d. 1/4; 1/8

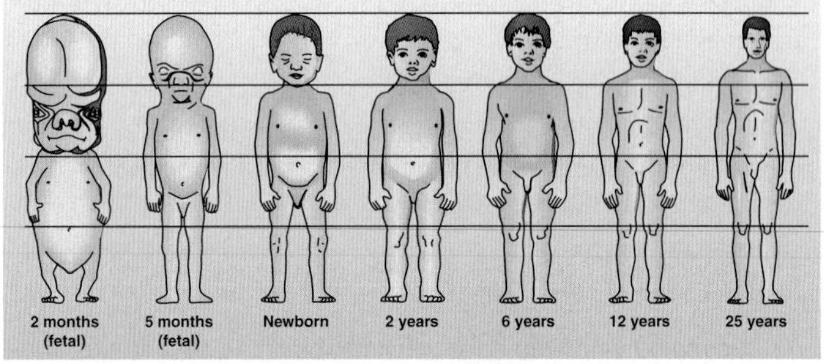


**6.** Which of the following is NOT true regarding infant sensory and perceptual development?

a. Vision is almost 20/20 at birth.

b. A newborn's sense of pain is highly developed at birth.

c. An infant can recognize, and prefers, its own mother's breast milk by smell.

d. An infant can recognize, and prefers, its own mother's breast milk by taste.

**7.** The clearest and most physical sign of puberty is the ______, characterized by rapid increases in height, weight, and skeletal growth.

a. menses

b. spermarche

c. growth spurt

d. age of fertility

**8.** Some employers are reluctant to hire 50- to 60-year-old workers because of a generalized belief that they are sickly and will take too much time off. This is an example of ______.

a. discrimination

b. prejudice

c. ageism

d. all of these options

9. ______ was one of the first scientists to demonstrate that a child's intellect is fundamentally different from an adult's.

a. Baumrind

b. Beck

c. Piaget

d. Elkind

10. ______ occurs when existing schemas are used to interpret new information, whereas ______ involves changes and adaptations of the schemas.

a. Adaptation; accommodation

b. Adaptation; reversibility

c. Egocentrism; postschematization

d. Assimilation; accommodation

11. Label the four stages of Piaget's cognitive development model.

a. ______________________

b. ______________________

c. ______________________

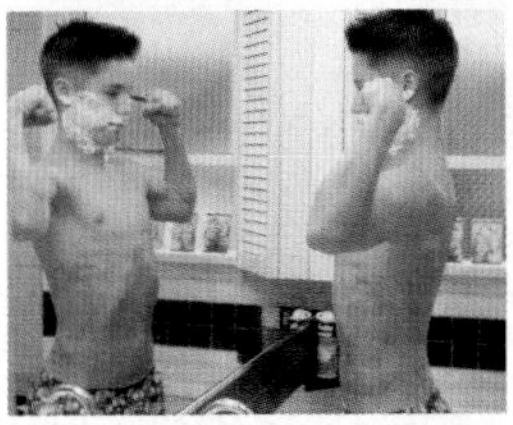

d. ______________________

12. These photos are an example of the lack of ______.

a. sensory permanence

b. perceptual constancy

c. perceptual permanence

d. object permanence

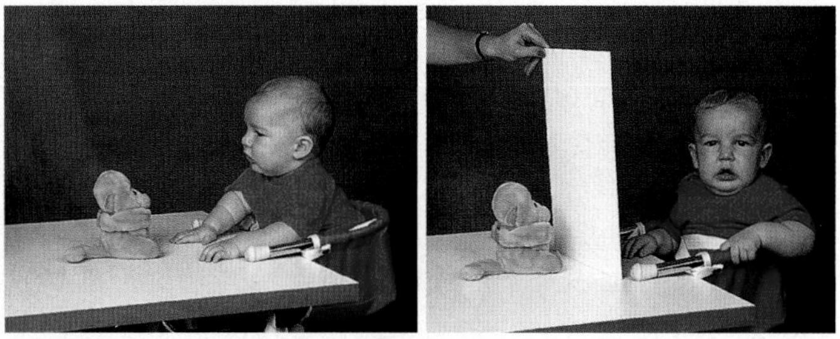

13. The ability to think abstractly or hypothetically occurs in Piaget's ______ stage.

a. egocentric

b. postoperational

c. formal operational

d. concrete operational

14. Believing that their own thoughts and feelings are unique is common in adolescents, who tend to exhibit the early formal operational characteristic called the ______.

a. personal fable

b. special person misconception

c. imaginary audience

d. egocentric fable

15. Critics have suggested that Piaget underestimated ______.

a. the cognitive abilities of infants and young children

b. genetic influences on cognition

c. the effect of cultural experiences on cognition

d. all of these options

**THE PLANNER**

Review your Chapter Planner on the chapter opener and check off your completed work.

# 10 Lifespan Development II

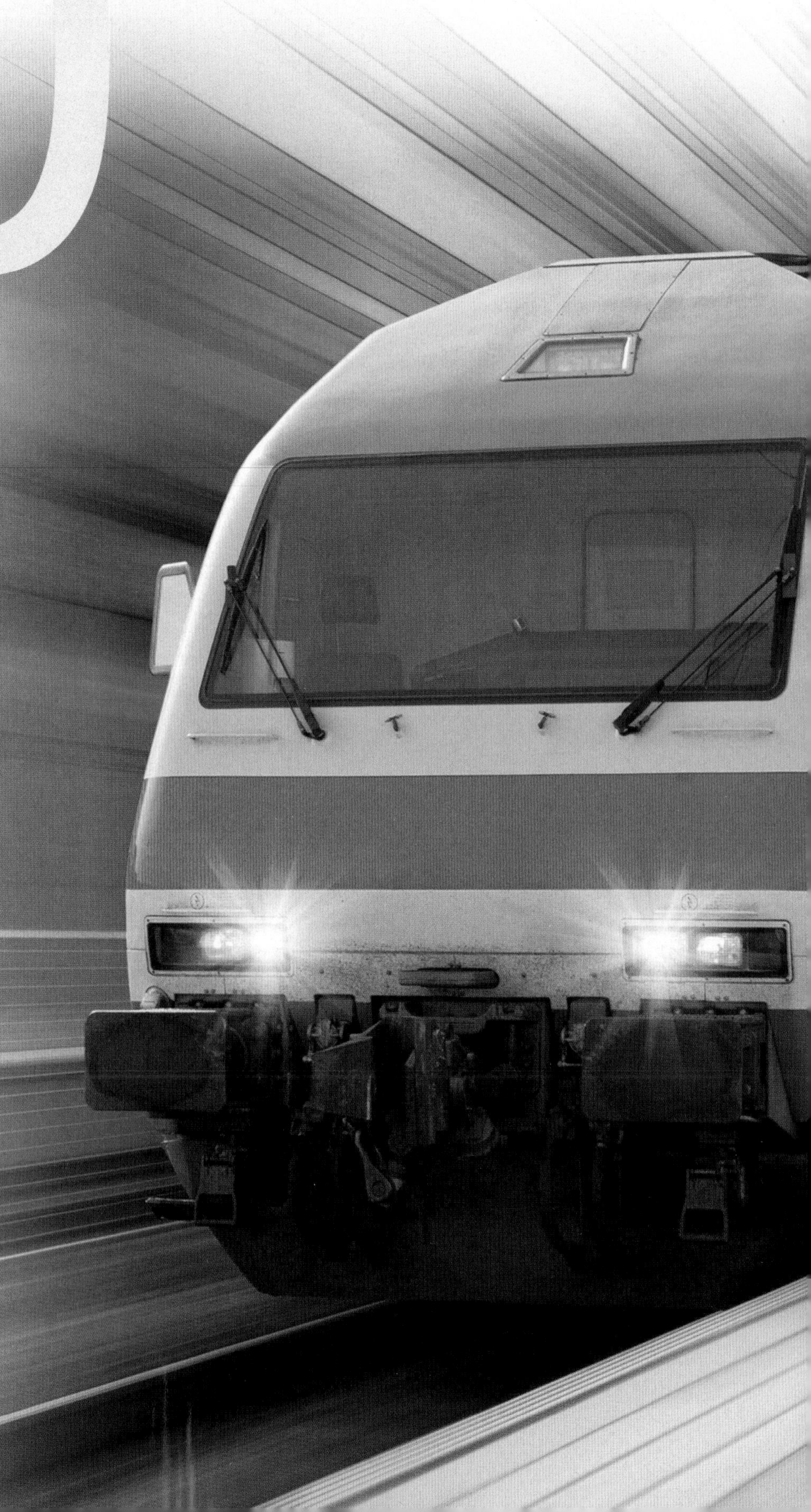

Imagine yourself standing on a bridge over a railroad track and seeing below you a runaway train that's about to kill five people. Because you also are standing next to a switching mechanism, you can easily divert the train onto a spur, thus allowing those five people to survive. But here's the catch: diverting the train will condemn *one* person, who is standing on the spur, to death (Appiah, 2008, 2010). What would you do? Would you allow one person to die in order to save five others? Would your answer be different if that one person were your mother, father, or some other much-loved person? What might lead different people to make different decisions?

It's unlikely that you will ever have to make such a gruesome choice. Yet everyone encounters moral dilemmas from time to time. How we approach moral dilemmas—as well as many other events and circumstances throughout our lives—reflects several key facets of our lifespan development. Chapter 9 explored lifespan changes in physical and cognitive development. In this chapter, we'll look at social, moral, and personality development, along with how sex, gender, and culture influence development, and the special challenges of adulthood.

© Oleksiy Mark/iStockphoto

## CHAPTER OUTLINE



## CHAPTER PLANNER ✓

- ❑ Study the picture and read the opening story.
- ❑ Scan the Learning Objectives in each section:
  p. 258 ❑ p. 266 ❑ p. 272 ❑
- ❑ Read the text and study all figures and visuals.
  Answer any questions.

**Analyze key features**

- ❑ What a Psychologist Sees, p. 258
- ❑ Psychology InSight, p. 259
- ❑ Applying Psychology p. 260 ❑ p. 273 ❑
- ❑ Study Organizer, p. 261
- ❑ Process Diagrams p. 262 ❑ p. 265 ❑
- ❑ Psychological Science, p. 274
- ❑ Myth Busters, p. 276
- ❑ Stop: Answer the Concept Checks before you go on.
  p. 266 ❑ p. 271 ❑ p. 278 ❑

**End of chapter**

- ❑ Review the Summary and Key Terms.
- ❑ Answer the Critical and Creative Thinking Questions.
- ❑ Answer What is happening in this picture?
- ❑ Complete the Self-Test and check your answers.

# Social, Moral, and Personality Development

## LEARNING OBJECTIVES

**RETRIEVAL PRACTICE** While reading the upcoming sections, respond to each Learning Objective in your own words. Then compare your responses with those in Appendix B.

1. **Describe** how attachment influences our social development.
2. **Describe** Baumrind's four different parenting styles.
3. **Summarize** the central characteristics of Kohlberg's theory of moral development.
4. **Identify** Erikson's eight stages of psychosocial development.

In addition to physical and cognitive development (Chapter 9), developmental psychologists study social, moral, and personality development by looking at how social forces and individual differences affect development over the lifespan. In this section, we focus on three major facets of development that shed light on how we affect each other: attachment, Kohlberg's stages of moral development, and Erikson's psychosocial stages.

## Social Development

An infant arrives in the world with a multitude of behaviors that encourage a strong bond of **attachment** with primary caregivers.

**attachment** A strong affectional bond with special others that endures over time.

In studying attachment behavior, researchers are often divided along the lines of the nature-nurture controversy. Those who advocate the nativist, or innate, position cite John Bowlby's work (1969, 1989, 2000). He proposed that newborn infants are biologically equipped with verbal and nonverbal behaviors (such as crying, clinging, and smiling) and with "following" behaviors (such as crawling and walking after the caregiver) that elicit instinctive nurturing responses from the caregiver.

Konrad Lorenz's (1937) studies of imprinting and attachment (Chapter 9), along with Harry Harlow and his colleagues' experiment with **contact comfort** in rhesus monkeys (*What a Psychologist Sees*), further support the biological argument for attachment.

# WHAT A PSYCHOLOGIST SEES

THE PLANNER

## The Power of Touch

Parents around the world tend to kiss, nuzzle, comfort, and respond to their children with lots of physical contact **(Figure a)**. Psychologists know that contact comfort, the pleasurable tactile sensations provided by a soft and cuddly "parent," is a powerful contributor to attachment.

Harry Harlow and his colleagues (1950, 1959, 1966, 1971) investigated the variables that might affect attachment. They created two types of wire-framed surrogate (substitute) "mother" monkeys: one covered by soft terry cloth and one left uncovered (**Figure b**). The infant monkeys were fed by either the cloth or the wire mother, but they otherwise had access to both mothers. The researchers found that monkeys "reared" by a cloth mother clung frequently to the soft material of their surrogate mother and developed greater emotional security and curiosity than did monkeys assigned to the wire mother.

Jodi Cobb/NG Image Collection

a.

Nina Leen/Time Life Pictures/Getty Images

b.

# Psychology InSight

Research on infant attachment • **Figure 10.1**

For most children, parents are the earliest and most important factor in a child's social development, and the attachment between parent and child is of particular interest to developmental psychologists.

**a. Strange situation procedure**

Mary Ainsworth and her colleagues (1967, 1978, 2010) found significant differences in the typical levels of attachment between infants and their mothers using a technique called the **strange situation procedure**, in which they observed how infants responded to the presence or absence of their mother and a stranger.

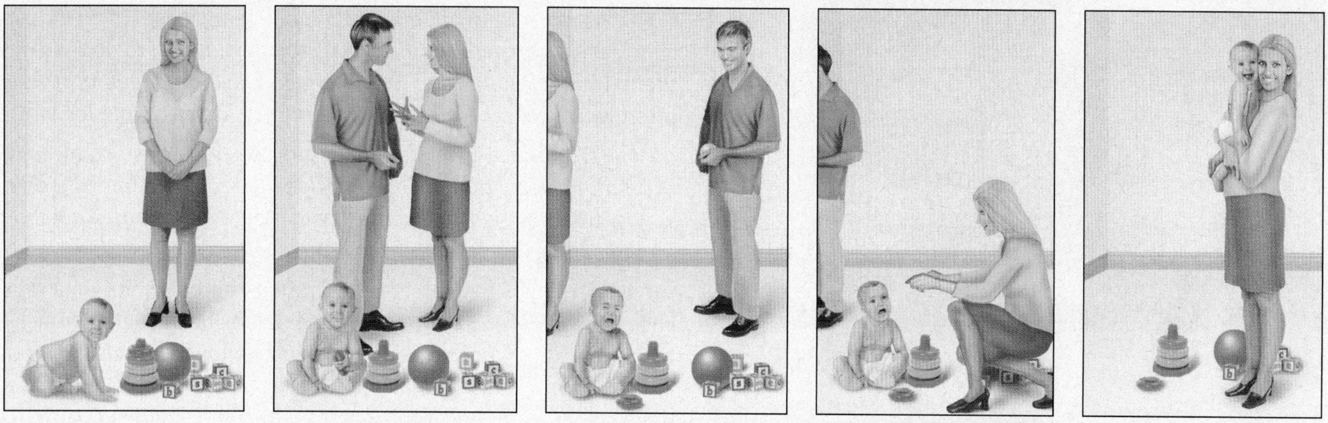

1. The baby plays while the mother is nearby.
2. A stranger enters the room, speaks to the mother, and approaches the child.
3. The mother leaves and the stranger stays in the room with an unhappy baby.
4. The mother returns and the stranger leaves.
5. The baby is reunited with the mother.

**b. Degrees of attachment**

Using the strange situation procedure, Ainsworth found that children could be divided into three groups: *Secure, anxious/avoidant*, and *anxious/ambivalent*. A later psychologist, Mary Main, added a fourth category, *disorganized/disoriented* (Main & Solomon, 1986, 1990).

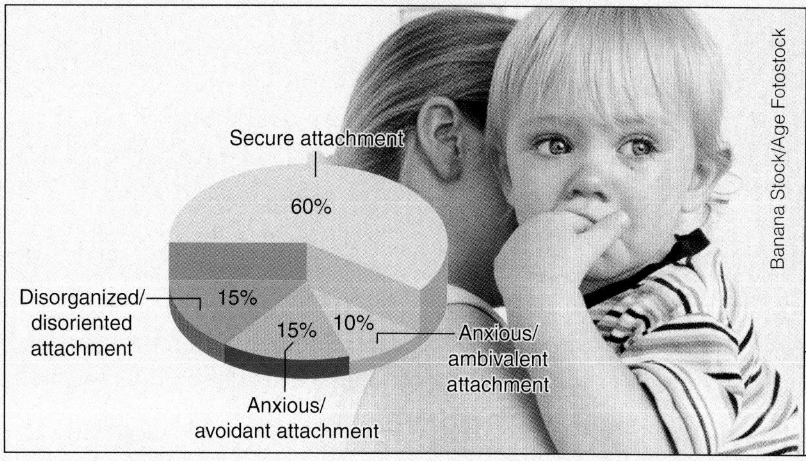


Banana Stock/Age Fotostock

| | |
|---|---|
| **Secure**<br>Seeks closeness with mother when stranger enters. Uses her as a safe base from which to explore, shows moderate distress on separation from her, and is happy when she returns. | |
| **Anxious/ambivalent**<br>Infant becomes very upset when mother leaves the room and shows mixed emotions when she returns. | |
| **Anxious/avoidant**<br>Infant does not seek closeness or contact with the mother and shows little emotion when the mother departs or returns. | |
| **Disorganized/disoriented**<br>Infant exhibits avoidant or ambivalent attachment, often seeming either confused or apprehensive in the presence of the mother. | |

Developmental psychologist Mary Ainsworth and her colleagues (1967, 1978) have found significant differences in the typical levels of attachment between infants and their mothers (**Figure 10.1**).

What if a child does not form an attachment? Research shows that infants raised in impersonal surroundings (such as in institutions that do not provide the stimulation and love of a regular caregiver) or under abusive conditions

# Applying Psychology

THE PLANNER

## What's Your Romantic Attachment Style?

Andrew O'Toole/Getty Images

Thinking of your current and past romantic relationships, place a check next to those statements that best describe your feelings.

_____ **1.** *I find it relatively easy to get close to others and am comfortable depending on them and having them depend on me. I don't often worry about being abandoned or about someone getting too close.*

_____ **2.** *I am somewhat uncomfortable being close. I find it difficult to trust partners completely or to allow myself to depend on them. I am nervous when anyone gets close, and love partners often want me to be more intimate than is comfortable for me.*

_____ **3.** *I find that others are reluctant to get as close as I would like. I often worry that my partner doesn't really love me or won't stay with me. I want to merge completely with another person, and this desire sometimes scares people away.*

According to research, 55% of adults agree with item 1 (secure attachment), 25% choose number 2 (anxious/avoidant attachment), and 20% choose item 3 (anxious/ambivalent attachment) (adapted from Fraley & Shaver, 1997; Hazan & Shaver, 1987). Note that the percentages for these adult attachment styles do not include the disorganized/disoriented attachment pattern of infants.

**Think Critically**

1. Do your responses match your attachment experiences as a child?
2. Does this affect your present relationship?

suffer from a number of problems. They seldom cry, coo, or babble; they become rigid when picked up; and they have few language skills. They also tend to form shallow or anxious relationships. Some appear forlorn, withdrawn, and uninterested in their caretakers, whereas others seem insatiable in their need for affection (Zeanah, 2000). Finally, they also tend to show intellectual, physical, and perceptual disabilities; increased susceptibility to infection; and neurotic "rocking" and isolation behaviors. In some cases, they die from lack of attachment (Bowlby, 1973, 1982, 2000; Duniec & Raz, 2011; Gunnar, 2010; Spitz & Wolf, 1946; Zeanah, 2000).

In addition to varying levels of infant attachment to parents, researchers also have found that these early patterns of attachment tend to be related to our later adult styles of romantic love (Graham, 2011; Hatfield & Rapson, 2010; Madey & Rogers, 2009; Nosko et al., 2011; Sprecher & Fehr, 2011; Tatkin, 2011). If we developed a secure, anxious/ambivalent, anxious/avoidant, or disorganized/disoriented style as infants, we tend to follow these same patterns in our adult approach to intimacy and affection. You can check your own romantic attachment style in *Applying Psychology*. However, keep in mind that it's always risky to infer causation from correlation. Moreover, even if early attachment experiences are correlated with our later relationships, they do not determine them. Throughout life, we can learn new social skills and different approaches to all our relationships.

## Parenting Styles

How much of our personality comes from the way our parents treat us as we're growing up? Researchers since the 1920s have studied the effects of different methods of childrearing on children's behavior, development, and mental health. For example, when researchers did follow up studies of children with the four levels of attachment we just discussed, they found that infants with a secure attachment style had caregivers who were sensitive and responsive to their signals of distress, happiness, and fatigue (Ainsworth, 1967; Ainsworth et al., 1978; Gini et al., 2007; Higley, 2008; Völker, 2007). Anxious/avoidant infants had caregivers who were aloof and distant, and anxious/ambivalent infants had inconsistent caregivers who alternated between strong affection and indifference. Follow-up studies found that, over time, securely attached children were the most sociable, emotionally aware, enthusiastic, cooperative, persistent, curious, and competent (Bar-Haim et al., 2007; Brown & Whiteside, 2008; Flaherty & Sadler, 2011; Johnson, Dweck, & Chen, 2007).

Other studies done by Diana Baumrind (1980, 1991, 1995) found that parenting styles could be reliably divided into three broad patterns, *permissive*, *authoritarian*, and *authoritative*, which could be identified by their degree of *control/demandingness* (*C*) and *warmth/responsiveness* (*W*) (**Study Organizer 10.1**).

## Moral Development

Developing a sense of right and wrong, or morality, is a part of psychological development. Consider the following situation in terms of what you would do:

> *In Europe, a woman was near death from a special kind of cancer. There was one drug that doctors thought might save her. It was a form of radium that a druggist in the same town had recently discovered. The drug was expensive to make, but the druggist was charging 10 times what the drug cost him to make. He paid $200 for the radium and charged $2,000 for a small dose of the drug. The sick woman's husband, Heinz, went to everyone he knew to borrow the money, but he could gather together only about $1,000, half of what it cost. He told the druggist that his wife was dying and asked him to sell it cheaper or let him pay later. But the druggist said, "No, I discovered the drug, and I'm going to make money from it." So Heinz got desperate and broke into the man's store to steal the drug for his wife. (Kohlberg, 1964, pp. 18–19)*

Was Heinz right to steal the drug? What do you consider moral behavior? Is morality "in the eye of the beholder," or are there universal truths and principles? Whatever your answer, your ability to think, reason, and respond to Heinz's dilemma reportedly demonstrates your current level of moral development.

One of the most influential researchers in moral development was Lawrence Kohlberg (1927–1987). He presented what he called "moral stories" like the Heinz dilemma to people of all ages, and on the basis of his findings, he developed a model of moral development (1964, 1984).

**Study Organizer 10.1** Parenting styles — THE PLANNER

| Parenting style | Description | Example | Effect on children |
|---|---|---|---|
| **Permissive-neglectful** (permissive-indifferent) (low C, low W) | Parents make few demands, with little structure or monitoring, and show little interest or emotional support; may be actively rejecting. | "I don't care about you—or what you do." | Children tend to have poor social skills and little self-control (being overly demanding and disobedient). |
| **Permissive-indulgent** (low C, high W) | Parents set few limits or demands, but are highly involved and emotionally connected. | "I care about you—and you're free to do what you like!" | Children often fail to learn respect for others and tend to be impulsive, immature, and out of control. |
| **Authoritarian** (high C, low W) | Parents are rigid and punitive, while also being low on warmth and responsiveness. | "I don't care what you want. Just do it my way, or else!" | Children tend to be easily upset, moody, aggressive, and often fail to learn good communication skills. |
| **Authoritative** (high C, high W) | Parents generally set and enforce firm limits, while also being highly involved, tender, and emotionally supportive. | "I really care about you, but there are rules and you need to be responsible." | Children become self-reliant, self-controlled, high achieving, and emotionally well-adjusted; also seem more content, goal oriented. friendly, and socially competent. |

*Sources:* Areepattamannil, 2010; Celada, 2011; Driscoll, Russell, & Crockett, 2008; Martin & Fabes, 2009; McKinney, Donnelly, & Renk, 2008; Topham et al., 2011.

## Kohlberg's stages of moral development • Figure 10.2

THE PLANNER

Kohlberg believed that individuals progress through three levels and six stages of moral development.

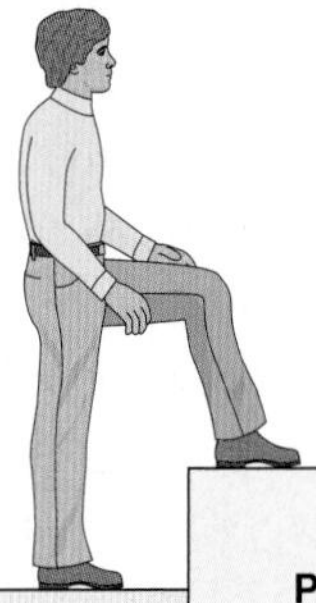

### PRECONVENTIONAL LEVEL

(Stages 1 and 2—birth to adolescence)
Moral judgment is *self-centered.* What is right is what one can get away with, or what is personally satisfying. Moral understanding is based on rewards, punishments, and the exchange of favors.

**1 Punishment-obedience orientation**

Focus is on self-interest—obedience to authority and avoidance of punishment. Because children at this stage have difficulty considering another's point of view, they also ignore people's intentions.

**2 Instrumental-exchange orientation**

Children become aware of others' perspectives, but their morality is based on reciprocity—an equal exchange of favors.

### CONVENTIONAL LEVEL

(Stages 3 and 4—adolescence and young adulthood)
Moral reasoning is *other-centered.* Conventional societal rules are accepted because they help ensure the social order.

**3 Good-child orientation**

Primary moral concern is being nice and gaining approval, and judges others by their intentions—"His heart was in the right place."

**4 Law-and-order orientation**

Morality based on a larger perspective—societal laws. Understanding that if everyone violated laws, even with good intentions, there would be chaos.

### POSTCONVENTIONAL LEVEL

(Stages 5 and 6—adulthood)
Moral judgments based on *personal standards for right and wrong.* Morality also defined in terms of abstract principles and values that apply to all situations and societies.

**5 Social contact orientation**

Appreciation for the underlying purposes served by laws. Societal laws are obeyed because of the "social contract," but they can be morally disobeyed if they fail to express the will of the majority or fail to maximize social welfare.

**6 Universal-ethics orientation**

"Right" is determined by universal ethical principles (e.g., nonviolence, human dignity, freedom) that *all* religions or moral authorities might view as compelling or fair. These principles apply whether or not they conform to existing laws.

*Sources:* Adapted from Kohlberg, L. "Stage and Sequence: The Cognitive Developmental Approach to Socialization," in D. A. Goslin, *The Handbook of Socialization Theory and Research.* Chicago: Rand McNally, 1969, p. 376 (Table 6.2).

What is the right answer to Heinz's dilemma? Kohlberg was interested not in whether participants judged Heinz to be right or wrong but in the reasons they gave for their decisions. On the basis of participants' responses, Kohlberg proposed three broad levels in the evolution of moral reasoning, each composed of two distinct stages (**Figure 10.2**). Individuals at each stage and level may or may not support Heinz's stealing of the drug, but their reasoning changes from level to level.

Kohlberg believed that, like Piaget's stages of cognitive development (Chapter 9), his stages of moral development are universal and invariant. That is, they supposedly exist in all cultures, and everyone goes through each of the stages in a predictable fashion. The age trends that are noticed tend to be rather broad.

**Preconventional level** (Stages 1 and 2—birth to adolescence). At the **preconventional level**, moral judgment is self-centered. What is right is what one can get away with or what is personally satisfying. Moral understanding is based on rewards, punishments, and the exchange of favors. This level is called "preconventional" because children have not yet accepted society's (conventional) rule-making processes.

> **preconventional level** Kohlberg's first level of moral development, in which morality is based on rewards, punishment, and the exchange of favors.
>
> **conventional level** Kohlberg's second level of moral development, in which moral judgments are based on compliance with the rules and values of society.

**Conventional level** (Stages 3 and 4—adolescence and young adulthood). At the **conventional level**, moral reasoning advances from being self-centered to other-centered. The individual personally accepts conventional societal rules because they help ensure social order, and he or she judges morality in terms of compliance with these rules and values.

**Postconventional level** (Stages 5 and 6—adulthood). At the **postconventional level**, individuals develop personal standards for right and wrong. They also define morality in terms of abstract principles and values that apply to all situations and societies (**Figure 10.3**). A 20-year-old who judges the "discovery" and settlement of North America by Europeans as immoral because it involved the theft of land from native peoples is thinking in postconventional terms.

> **postconventional level** Kohlberg's highest level of moral development, in which individuals develop personal standards for right and wrong, and they define morality in terms of abstract principles and values that apply to all situations and societies.

**Assessing Kohlberg's theory** Kohlberg has been credited with enormous insights and contributions, but his theories have also been the focus of three major areas of criticism:

***Moral reasoning versus behavior*** Are people who achieve higher stages on Kohlberg's scale really more moral than others, or do they just "talk a good game"? Some studies show a positive correlation between higher stages of reasoning and higher levels of moral behavior

**Postconventional moral reasoning • Figure 10.3**

In the postconventional level of moral reasoning, when laws are consistent with interests of the majority, they are obeyed because of the "social contract." However, laws can be morally disobeyed if they fail to express the will of the majority or fail to maximize social welfare.

Jacom Stephens/Vetta/Getty Images

(Gasser & Malti, 2011; Gini, Pozzoli, & Hauser, 2011; Langdon, Clare, & Murphy, 2011), but others have found that situational factors are better predictors of moral behavior (**Figure 10.4**) (Bandura, 1986, 1991, 2008; Kaplan, 2006; Satcher, 2007; Slovácková & Slovácek, 2007). For example, research participants are more likely to steal when they are told the money comes from a large company rather than from individuals (Greenberg, 2002). And both men and women will tell more sexual lies during casual relationships than during close relationships (Williams, 2001).

***Possible gender bias*** Researcher Carol Gilligan criticized Kohlberg's model because on his scale women tend to be classified at a lower level of moral reasoning than men. Gilligan suggested that this difference occurred because Kohlberg's theory emphasizes values more often held by men, such as rationality and independence, while de-emphasizing common female values, such as concern for others and belonging (Gilligan, 1977, 1990, 1993; Kracher & Marble, 2008). Most follow-up studies of Gilligan's theory, however, have found few, if any, gender differences in level or type of moral reasoning (Bateman & Valentine, 2010; Fumagalli et al., 2010; Hyde, 2007; Mercadillo et al., 2011; Smith, 2007).

***Cultural differences*** Cross-cultural studies confirm that children from a variety of cultures generally follow Kohlberg's model and progress sequentially from his first level, the preconventional, to his second, the conventional (Rest et al., 1999; Snarey, 1985, 1995). At the same time, studies find differences among cultures (Jensen, 2011; LePage et al., 2011; Rai & Fiske, 2011). For example, cross-cultural comparisons of responses to Heinz's moral dilemma show that Europeans and Americans tend to consider whether they like or identify with the victim in questions of morality. In contrast, Hindu Indians consider social responsibility and personal concerns to be separate issues (Miller & Bersoff, 1998). Researchers suggest that the difference reflects the Indians' broader sense of social responsibility.

In India, Papua New Guinea, and China, as well as in Israeli kibbutzim, people don't choose between the rights of the individual and the rights of society (as the top levels of Kohlberg's model require). Instead, most people seek a compromise solution that accommodates both interests (Killen & Hart, 1999; Miller & Bersoff, 1998). Thus, Kohlberg's standard for judging the highest level of morality (the postconventional) may be more applicable to cultures that value individualism over community and interpersonal relationships.

*"I swear I wasn't looking at smut—I was just stealing music."*

**Morality gap? • Figure 10.4**

What makes people who normally behave ethically willing to steal intellectual property such as music or software?

## Personality Development

Like Piaget and Kohlberg, Erik Erikson (1902–1994) developed a stage theory of development. He identified eight **psychosocial stages** of social development (**Figure 10.5**), each marked by a psychosocial crisis or conflict related to a specific developmental task. The name given to each stage reflects a specific crisis encountered at that stage and the two possible outcomes. For example, the crisis or task of most young adults is *intimacy versus isolation*. This age group's developmental task is developing deep, meaningful relationships with others. Those who don't meet this developmental challenge risk social isolation. Erikson believed that the more successfully we overcome each psychosocial crisis, the better chance we have to develop in a healthy manner (Erikson, 1950).

**psychosocial stages** The eight developmental stages, each involving a crisis that must be successfully resolved, that individuals pass through in Erikson's theory of psychosocial development.

Many psychologists agree with Erikson's general idea that psychosocial crises contribute to personality development (Conzen, 2010; Fukase & Okamoto, 2010; Markstrom & Marshall, 2007; Zhang & He, 2011). However, Erikson also has his critics (Beyers & Seiffge-Krenke, 2010; Spano

## Erikson's eight stages of psychosocial development • Figure 10.5

THE PLANNER

Erikson identified eight stages of development, each of which is associated with a psychosocial crisis.

Steve Raymer/NG Image Collection

**1 Trust versus mistrust (birth–age 1)**

Infants learn to *trust* or *mistrust* their caregivers and the world based on whether or not their needs—such as food, affection, safety—are met.

Prof. Karen Huffman

**2 Autonomy versus shame and doubt (ages 1–3)**

Toddlers start to assert their sense of independence (*autonomy*). If caregivers encourage this self-sufficiency, the toddler will learn to be independent versus feelings of *shame* and *doubt*.

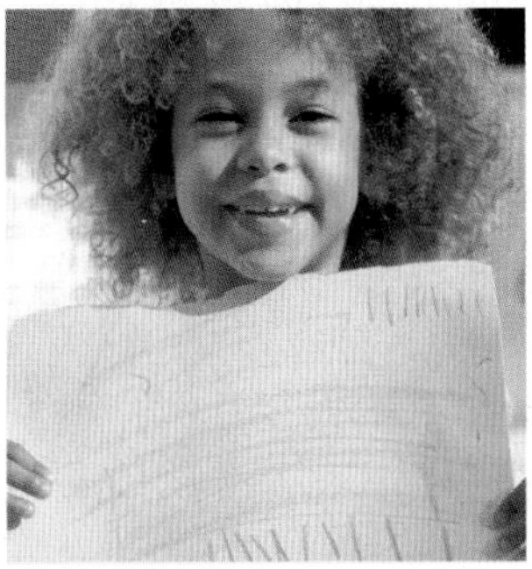

Dynamic Graphics, Inc. Creatas

**3 Initiative versus guilt (ages 3–6)**

Preschoolers learn to *initiate* activities and develop self-confidence and a sense of social responsibility. If not, they feel irresponsible, anxious, and *guilty*.

PhotoDisc/Getty Images, Inc.

**4 Industry versus inferiority (ages 6–12)**

Elementary school-aged children who succeed in learning new, productive life skills, develop a sense of pride and competence (industry). Those who fail to develop these skills feel inadequate and unproductive (*inferior*).

Randy Olson/NG Image Collection

**5 Identity versus role confusion (ages 12–20)**

Adolescents develop a coherent and stable self-definition (identity) by exploring many roles and deciding who or what they want to be in terms of career, attitudes, and so on. Failure to resolve this **identity crisis** may lead to apathy, withdrawal, and/or *role confusion*.

Pablo Corral Vega/NG Image Collection

**6 Intimacy versus isolation (early adulthood)**

Young adults form lasting, meaningful relationships, which help them develop a sense of connectedness and *intimacy* with others. If not, they become psychologically *isolated*.

IT Stock

**7 Generativity versus stagnation (middle adulthood)**

The challenge for middle-aged adults is to be nurturant of the younger generation. Failing to meet this challenge leads to self-indulgence and a sense of *stagnation*.

Richard Olsenius/NG Image Collection

**8 Ego integrity versus despair (late adulthood)**

During this stage, older adults reflect on their past. If this reflection reveals a life well-spent, the person experiences self-acceptance and satisfaction (*ego integrity*). If not, he or she experiences regret and deep dissatisfaction (*despair*).

et al., 2010). First, Erikson's psychosocial stages are difficult to test scientifically. Second, the labels he used to describe the eight stages may not be entirely appropriate cross-culturally. For example, in individualistic cultures, *autonomy* is highly preferable to *shame and doubt*. But in collectivist cultures, the preferred resolution might be *dependence* or *merging relations* (Berry et al., 2011).

Despite their limits, Erikson's stages have greatly contributed to the study of North American and European psychosocial development. By suggesting that development continues past adolescence, Erikson's theory has encouraged further research.

**CONCEPT CHECK** STOP

1. **How** do the securely attached, avoidant, and anxious/ambivalent attachment styles differ?
2. **How** is the behavior of children of permissive/indulgent parents different from the behavior of children with authoritative parents?
3. **What** criticisms have been leveled at Kohlberg's theory of moral development?
4. **What** are some cross-cultural limitations to Erikson's psychosocial theory?

# How Sex, Gender, and Culture Affect Development

## LEARNING OBJECTIVES

**RETRIEVAL PRACTICE** While reading the upcoming sections, respond to each Learning Objective in your own words. Then compare your responses with those in Appendix B.

1. **Describe** how sex and gender differences are related to physical, cognitive, personality, and social development.
2. **Explain** how individualistic versus collectivistic cultures shape personality development.

Imagine for a moment what your life would be like if you were a member of the other sex. Would you think differently? Would you be more or less sociable and outgoing? Would your career plans or friendship patterns change? Most people believe that whether we are male or female has a strong impact on many facets of development. But why is that? Why is it that the first question most people ask after a baby is born is, "Is it a girl or a boy?"

**sex** Biological maleness and femaleness, including chromosomal sex. Also, activities related to sexual behaviors, such as masturbation and intercourse.

**gender** The psychological and sociocultural meanings added to biological maleness or femaleness.

**gender roles** The societal expectations for normal and appropriate male and female behavior.

## Sex and Gender Influences on Development

In this section, we will explore how our development is affected by **sex** (a biological characteristic determined at the moment of conception) and by **gender** and **gender roles** (cultural meanings that accompany biological sex).

**Sex differences** Physical anatomy is the most obvious biological sex difference between men and women (**Figure 10.6**). Men and women also differ in secondary sex characteristics, such as facial hair and breast growth; signs of reproductive capability, such as menstruation and ejaculation of sperm; and physical responses to middle age or the end of reproduction, such as *menopause* and the *male climacteric*. There are also several functional and structural differences in the brains of men and women (Allen et al., 2007; Hyde & DeLamater, 2011; Matlin, 2012; Valla & Ceci, 2011). These result partly from the influence of prenatal sex hormones on the developing fetal brain.

**Gender differences** In addition to biological sex differences, scientists have found numerous gender differences, which affect our cognitive, social, and personality development.

For example, females tend to score higher on tests of verbal skills, whereas males score higher on math and visuospatial tests (Castelli, Corazzini, & Geminiani, 2008; Hunt, 2011; Olszewski-Kubilius & Lee, 2011; Reynolds et al., 2008; van der Sluis et al., 2008). Keep in mind

## Major physical differences between the sexes • Figure 10.6

The average man is 35 pounds heavier, has less body fat, and is 5 inches taller than the average woman. Men tend to have broader shoulders, slimmer hips and slightly longer legs in proportion to their height. There are also physical differences in the brain and in the muscular and skeletal systems of men and women.

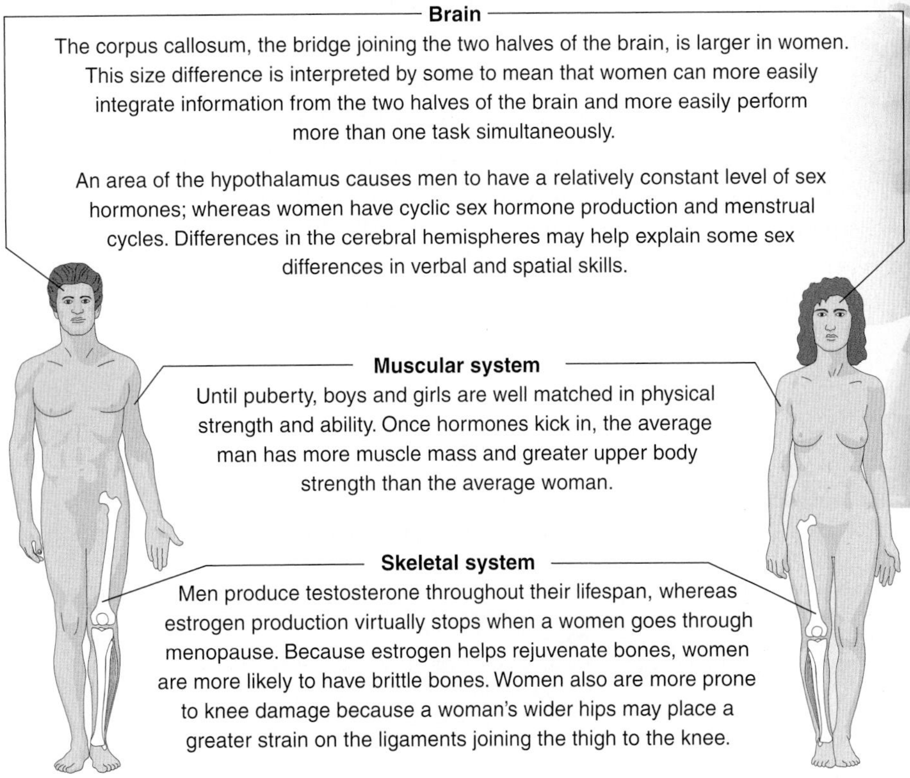

Brain
The corpus callosum, the bridge joining the two halves of the brain, is larger in women. This size difference is interpreted by some to mean that women can more easily integrate information from the two halves of the brain and more easily perform more than one task simultaneously.
An area of the hypothalamus causes men to have a relatively constant level of sex hormones; whereas women have cyclic sex hormone production and menstrual cycles. Differences in the cerebral hemispheres may help explain some sex differences in verbal and spatial skills.
Muscular system
Until puberty, boys and girls are well matched in physical strength and ability. Once hormones kick in, the average man has more muscle mass and greater upper body strength than the average woman.
Skeletal system
Men produce testosterone throughout their lifespan, whereas estrogen production virtually stops when a women goes through menopause. Because estrogen helps rejuvenate bones, women are more likely to have brittle bones. Women also are more prone to knee damage because a woman's wider hips may place a greater strain on the ligaments joining the thigh to the knee.

Jon Feingersh/Iconica/Getty Images, Inc.

*Source:* Miracle, Tina S.; Miracle, Andrew W.; Baumeister, Roy F., Human Sexuality: Meeting Your Basic Needs, 1st Edition, © 2003, p. 302. Adapted by permission of Pearson Education, Inc., Upper Saddle River, NJ.

that these differences are statistically small and represent few meaningful differences.

Some researchers suggest that these differences in cognitive ability may reflect biological factors, including structural differences in the cerebral hemispheres, hormones, or the degree of hemispheric specialization.

One argument against this biological model, however, is that male–female differences in verbal ability and math scores have declined in recent years (Brown & Josephs, 1999; Halpern, 1997, 2000; Lizarraga & Ganuza, 2003). However, the gap has not been narrowed with regard to male and female math scores on the SAT. Men with the highest overall IQ scores still tend to outscore women on this one measure.

Like cognitive ability, aggressive behavior also differs slightly between the genders. For example, young boys are more likely to engage in mock fighting and rough-and-tumble play, and as adolescents and adults, they are far more likely to commit aggressive crimes (Campbell & Muncer, 2008; Giancola & Parrott, 2008; Hay et al., 2011; King, 2012). But gender differences are clearer for physical aggression (like hitting) than for other forms of aggression. Early research suggested that females were more likely to engage in more indirect and relational forms of aggression, such as spreading rumors and ignoring or excluding someone (Fisher, 2011; Radcliff & Joseph, 2011; Willer & Cupach, 2011). But more recent studies have not found such clear differences (Marsee, Weems, & Taylor, 2008; Shahim, 2008).

Some researchers believe that biological factors cause gender differences in aggression—a nativist position. Several studies have linked the male hormone testosterone to aggressive behavior (Carré, McCormick, & Hariri, 2011;

Hermans, Ramsey, & van Honk, 2008; Victoroff et al., 2011). Other studies have found that aggressive men have disturbances in their levels of serotonin, a neurotransmitter that is inversely related to aggression (Berman, Tracy, & Coccaro, 1997; Holtzworth-Munroe, 2000; Nelson & Chiavegatto, 2001). In addition, studies on identical twins have found that genetic factors account for about 50% of aggressive behavior (Burt, 2011; Cadoret, Leve, & Devor, 1997; Segal & Bouchard, 2000).

Other researchers take a more nurturist position. They suggest that gender differences in aggressiveness result from environmental experiences with social dominance and pressures that encourage "sex-appropriate" behaviors and skills (Bosson & Vandello, 2011; Richardson & Hammock, 2007; Wallis, 2011). **Table 10.1** summarizes the gender differences between men and women.

**Gender-role development** By age 2, children are well aware of gender roles. They recognize that boys "should" be strong, independent, aggressive, dominant, and achieving, whereas girls "should" be soft, dependent, passive, emotional, and "naturally" interested in children (Collins, 2011; King, 2012; Matlin, 2012; Renzetti, Curran, & Kennedy-Bergen, 2006). The existence of similar gender roles in many cultures suggests that evolution and biology may play a role. However, most research emphasizes two major theories of gender-role development: *social learning* and *cognitive developmental* (**Figure 10.7**).

**Research-supported sex and gender differences Table 10.1**

| Behavior | More often shown by men | More often shown by women |
|---|---|---|
| **Sexual** | • Begin masturbating sooner in life cycle and higher overall occurrence rates | • Begin masturbating later in life cycle and lower overall occurrence rates |
| | • Start sexual life earlier and have first orgasm through masturbation | • Start sexual life later and have first orgasm from partner stimulation |
| | • More likely to recognize their own sexual arousal | • Less likely to recognize their own sexual arousal |
| | • More orgasm consistency with sexual partner | • Less orgasm consistency with sexual partner |
| **Touching** | • Touched, kissed, and cuddled less by parents | • Touched, kissed, and cuddled more by parents |
| | • Less physical contact with other men and respond more negatively to being touched | • More physical contact with other women and respond more positively to being touched |
| | • More likely to initiate both casual and intimate touch with sexual partner | • Less likely to initiate either casual or intimate touch with sexual partner |
| **Friendship** | • Larger number of friends and express friendship by shared activities | • Smaller number of friends and express friendship by shared communication about self |
| **Personality** | • More aggressive from a very early age | • Less aggressive from a very early age |
| | • More self-confident of future success | • Less self-confident of future success |
| | • Attribute success to internal factors and failures to external factors | • Attribute success to external factors and failures to internal factors |
| | • Achievement more task oriented; motives are mastery and competition | • Achievement more socially directed with emphasis on self-improvement; higher work motives |
| | • More self-validating | • More dependent on others for validation |
| | • Higher self-esteem | • Lower self-esteem |
| **Cognitive abilities** | • Slightly superior in math and visuospatial skills | • Slightly superior in verbal skills |

*Sources:* Crooks & Baur, 2011; Hunt, 2011; Hyde & DeLamater, 2011; King, 2012; Masters & Johnson, 1961, 1966, 1970; Matlin, 2012.

## Gender-role development • Figure 10.7

Social-learning theory focuses on a child's passive process of learning about gender through observation, rewards, and punishments, whereas cognitive-developmental theory emphasizes a child's active role in building a gender schema.

Elizabeth Crews/The Image Works

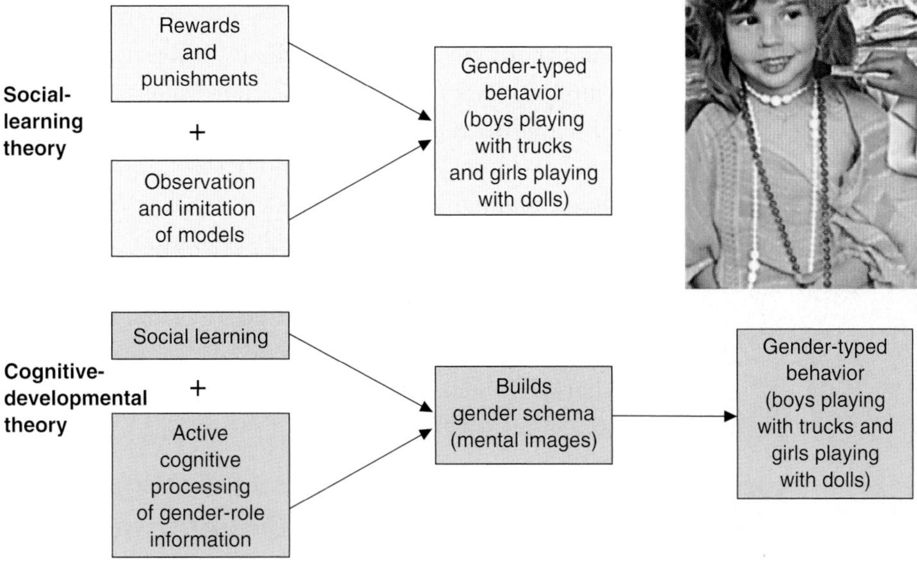


*Social-learning theorists* emphasize the power of the immediate situation and observable behaviors on gender-role development. They suggest that girls learn how to be "feminine" and boys learn how to be "masculine" in two major ways: (1) They receive rewards or punishments for specific gender-role behaviors and (2) they watch and imitate the behavior of others, particularly the same-sex parent (Bandura, 1989, 2000, 2006, 2008; Collins, 2011; Fulcher, 2011). A boy who puts on his father's tie or baseball cap wins big, indulgent smiles from his parents. But what would happen if he put on his mother's nightgown or lipstick? Parents, teachers, and friends generally reward or punish behaviors according to traditional gender-role expectations. Thus, a child "socially learns" what it means to be male or female.

According to the *cognitive-developmental theory*, social learning is part of gender-role development, but it's much more than a passive process of receiving rewards or punishments and modeling others. Instead, cognitive developmentalists argue that children actively observe, interpret, and judge the world around them (Bem, 1981, 1993; Hollander, Renfrow, & Howard, 2011; Leaper, 2011). As children process information about the world, they also create internal rules governing correct behaviors for boys versus girls. On the basis of these rules, they form **gender schemas** (mental images) of how they should act (**Figure 10.8**).

## Developing gender schemas • Figure 10.8

How would social-learning theory and cognitive-developmental theory explain why children tend to choose stereotypically "appropriate" toys?

© Romell/Masterfile

Ranald Mackechnie/Stone/Getty Images

Androgyny • Figure 10.9

For both children and adults, it is generally more difficult for males to express so-called female traits like nurturance and sensitivity than it is for women to adopt such traditionally male traits as assertiveness and independence (Kimmel, 2000; Leaper, 2000; Wood et al., 1997). Can you see how being more androgynous may help many couples better meet the demands of modern life?

**Androgyny** One way to overcome rigid or destructive gender-role stereotypes is to express both the "masculine" and "feminine" traits found in each individual—for example, being assertive and aggressive when necessary but also gentle and nurturing. Combining characteristics in this way is known as **androgyny** [an-DRAW-juh-nee].

Researchers have found that masculine and androgynous individuals generally have higher self-esteem and creativity, are more socially competent and motivated to achieve, and exhibit better overall mental health than those with traditional feminine traits (Choi, 2004; Stoltzfus et al., 2011; Wall, 2007; Woo & Oie, 2006).

What makes androgyny and masculinity adaptive? Research shows that traditional masculine characteristics (analytical thinking, independence) are more highly valued than traditional feminine traits (affectivity, cheerfulness) (**Figure 10.9**). For example, in business a good manager is generally perceived as having predominantly masculine traits (Johnson et al., 2008; Stoltzfus et al., 2011). Also, when college students in 14 countries were asked to describe their "current self" and their "ideal self," the ideal self-descriptions for both men and women contained more masculine than feminine qualities (Williams & Best, 1990).

Recent studies show that gender roles are becoming less rigidly defined (Cotter, Hermsen, & Vanneman, 2011; Hollander, Renfrow, & Howard, 2011). Asian American and Mexican American groups show some of the largest changes toward androgyny, and African Americans remain among the most androgynous of all ethnic groups (Denmark, Rabinovitz, & Sechzer, 2005; Duval, 2006; Renzetti et al., 2006).

## Cultural Influences on Development

Our development is rooted in the concept of *self*—how we define and understand ourselves. Yet the very concept of self reflects our culture. In **individualistic cultures**, the needs and goals of the individual are emphasized over the needs and goals of the group. When asked to complete the statement "I am...," people from individualistic cultures tend to respond with personality traits ("I am shy"; "I am outgoing") or their occupation ("I am a teacher"; "I am a student").

**individualistic culture** Culture in which the needs and goals of the individual are emphasized over the needs and goals of the group.

In **collectivistic cultures**, however, the opposite is true. The person is defined and understood primarily by looking at his or her place in the social unit (Berry et al., 2011; Dierdorff, Bell, & Belohlav, 2011; McCrae, 2004). Relatedness, connectedness, and interdependence are valued, as opposed to separateness, independence, and individualism. When asked to complete the statement "I am...," people from collectivistic cultures tend to mention their families or nationality ("I am a daughter"; "I am Chinese").

**collectivistic culture** Culture in which the needs and goals of the group are emphasized over the needs and goals of the individual.

If you are North American or western European, you are more likely to be individualistic than collectivistic

**A worldwide ranking of cultures Table 10.2**

| Individualistic cultures | Intermediate cultures | Collectivistic cultures |
|---|---|---|
| United States | Israel | Hong Kong |
| Australia | Spain | Chile |
| Great Britain | India | Singapore |
| Canada | Argentina | Thailand |
| Netherlands | Japan | West Africa region |
| New Zealand | Iran | El Salvador |
| Italy | Jamaica | Taiwan |
| Belgium | Arab region | South Korea |
| Denmark | Brazil | Peru |
| France | Turkey | Costa Rica |
| Sweden | Uruguay | Indonesia |
| Ireland | Greece | Pakistan |
| Norway | Philippines | Colombia |
| Switzerland | Mexico | Venezuela |

How might these two groups differ in their physical, socioemotional, cognitive, and personality development?



(**Table 10.2**). And you may find the concept of a self defined in terms of others almost contradictory. A core selfhood probably seems intuitively obvious to you. Recognizing that over 70% of the world's population lives in collectivistic cultures, however, may improve your cultural sensitivity and prevent misunderstandings (Singelis et al., 1995). For example, North Americans generally define *sincerity* as behaving in accordance with one's inner feelings, whereas Japanese see it as behavior that conforms to a person's role expectations (carrying out one's duties) (Yamada, 1997). Can you see how Japanese behavior might appear insincere to a North American and vice versa?

**CONCEPT CHECK** STOP

1. **What** is the difference between sex and gender?
2. **How** is the concept of "self" different in individualistic versus collectivistic cultures?

# Developmental Challenges Through Adulthood

## LEARNING OBJECTIVES

**RETRIEVAL PRACTICE** While reading the upcoming sections, respond to each Learning Objective in your own words. Then compare your responses to those in Appendix B.

1. **Describe** the factors that ensure realistic expectations for marriage and long-term committed relationships.
2. **Explain** the factors that affect life satisfaction during the adult working years and retirement.
3. **Describe** the three basic concepts about death and dying that people learn to understand through the course of development.

In this section, we will explore three of the most important developmental tasks that people face during adulthood: developing a loving, committed relationship with another person; finding rewarding work and a satisfying retirement; and coping with death and dying.

## Committed Relationships

One of the most important tasks faced during adulthood is that of establishing some kind of continuing, loving sexual relationship with another person. Yet navigating such partnerships is often very challenging. For example, approximately half of all marriages in the United States end in divorce, with serious implications for both adult and child development (Barczak et al., 2010; Hakvoort et al., 2011; Li, 2008; Osler et al., 2008; Steiner et al., 2011). For the adults, both spouses generally experience emotional as well as practical difficulties and are at high risk for depression and physical health problems. In many cases, these problems are present even before the marital disruption.

Realistic expectations are a key ingredient in successful relationships (Beachkofsky, 2010; Cheever, 2010; Gottman & Levenson, 2002; Hall & Adams, 2011). Yet many people harbor unrealistic expectations about marriage and the roles of husband and wife, opening the door to marital problems. Answer the questions in *Applying Psychology* to find out if your expectations are realistic.

Finding a loving, committed partner can be difficult, but it's equally important to study and avoid potential violence in your relationships. Dating and partner violence, child abuse, and elder abuse are tragically common, and research in this area is notoriously limited because this type of violence usually occurs in private and victims are reluctant to report it out of shame, powerlessness, or fear of reprisal (Buzawa, Buzawa, & Stark, 2012; Fife & Schrager, 2011; Lewin & Herron, 2007; McGuinness & Schneider, 2007).

What we do know is that violence is more likely in families and relationships experiencing financial problems, substance abuse, mental disorders, and/or social isolation (Abadinsky, 2011; Koss, White, & Kazdin, 2011; Raphel, 2008; Siever, 2008). These external factors, combined with the fact that many abusers lack good communication and interpersonal skills and their victims generally have no one to turn to for help or feedback, may lead to increasing anxiety and frustration that explodes into violence. In fact, one of the clearest identifiers of abuse potential is *impulsivity* (Gansler et al., 2011; Venables et al., 2011).

This impulsivity is related not only to external factors, such as economic stress and social isolation, but also to parts of the brain, such as the amygdala, prefrontal cortex, and hypothalamus, which are closely related to the expression and control of aggression. Head injuries, strokes, dementia, schizophrenia, alcoholism, and abuse of stimulant drugs have all been linked to these three areas and to aggressive outbursts. And low levels of the neurotransmitters serotonin and GABA (gamma-aminobutyric acid) are associated with irritability, hypersensitivity to provocation, and impulsive rage (Lee, Chong, & Coccaro, 2011; Levinthal, 2011; Livingston, 2011; Takahashi et al., 2011).

How can we reduce this type of violence and aggression? Treatment with antianxiety and serotonin-enhancing drugs like fluoxetine (Prozac) may lower the amount of impulsive violence. However, it's important to note that the primary goal of most abusers is the ultimate power and control over their victims and the relationship, and treating only biological factors within the abuser may detract from the serious and potentially fatal consequences for

# Applying Psychology

THE PLANNER

## Are Your Relationship Expectations Realistic?

To evaluate your own expectations, answer the following questions about traits and factors common to happy marriages and committed long-term relationships (Beachkofsky, 2010; Gonzaga, Carter, Buckwalter, 2010; Gottman & Levenson, 2002; Gottman & Notarius, 2000; Marks et al., 2008; Rauer, 2007):

**1. Established "love maps"**

Yes__ No__ *Do you believe that emotional closeness "naturally" develops when two people have the right chemistry?*

In successful relationships, both partners are willing to share their feelings and life goals. This sharing leads to detailed "love maps" of each other's inner emotional life and the creation of shared meaning in the relationship.

**2. Shared power and mutual support**

Yes__ No__ *Have you unconsciously accepted the imbalance of power promoted by many TV sitcoms, or are you willing to fully share power and to respect your partner's point of view, even if you disagree?*

The common portrayal of husbands as "head of household" and wives as the "little women" who secretly wield the true power may help create unrealistic expectations for marriage.

**3. Conflict management**

Yes__ No__ *Do you expect to "change" your partner or to be able to resolve all your problems?*

Successful couples work hard (through negotiation and accommodation) to solve their solvable conflicts, to accept their unsolvable ones, and to know the difference.

**4. Similarity**

Yes__ No__ *Do you believe that "opposites attract?"*

Although we all know couples who are very different but are still happy, similarity (in values, beliefs, religion, and so on) is one of the best predictors of long-lasting relationships (Chapter 15).

**5. Supportive social environment**

Yes__ No__ *Do you believe that "love conquers all"?*

Unfortunately, several environmental factors can overpower or slowly erode even the strongest love. These include age (younger couples have higher divorce rates), money and employment (divorce is higher among the poor and unemployed), parents' marriages (divorce is higher for children of divorced parents), length of courtship (longer is better), and premarital pregnancy (no pregnancy is better, and waiting a while after marriage is even better).

**6. Positive emphasis**

Yes__ No__ *Do you believe that an intimate relationship is a place where you can indulge your bad moods and openly criticize one another?*

Think again. Positive emotions, positive mood, and positive behavior toward one's partner are vitally important to a lasting, happy relationship.

Andersen Ross/Blend Images/ Getty Images

## Psychological Science

THE PLANNER

# The Power of Resilience

How people fare in the face of violent, impoverished, or neglectful situations has been studied throughout the world (e.g., Burt & Masten, 2010; Dimitry, 2012; Easterbrooks et al., 2011; Masten & Wright, 2010). For example, research on children and adolescents in the Middle East, who have grown up with numerous long-standing armed conflicts and wars, has found that the number of conflict-related traumatic experiences is closely associated with a higher number of mental, behavioral, and emotional problems (Dimitry, 2012).

Despite the odds, some children are able to survive and even prosper in harsh circumstances (**Figure a**). Ann Masten at the University of Minnesota and Douglas Coatsworth at the University of Miami (1998) identified specific traits and environmental conditions of **resiliency**, the ability to adapt effectively in the face of threats, which seem to account for the resilient child's success (**Figure b**). Good intellectual functioning, for example, may help resilient children solve problems or protect themselves from adverse conditions (Anderson & Bang, 2012; Irvin, 2012; Masten & Narayan, 2012).

a.

Akram Saleh/Getty Images

b.

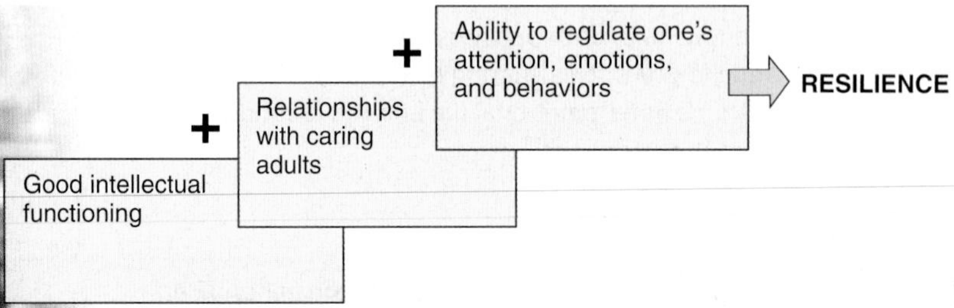


### Identify the Research Method

1. What is the most likely research method used for the Masten and Coatsworth (1998) study of resilience?
2. If you chose
   - the experimental method, label the IV, DV, experimental group, and control group.
   - the descriptive method, is this a naturalistic observation, survey, or case study?
   - the correlational method, is this a positive, negative, or zero correlation?
   - the biological method, identify the specific research tool (e.g., brain dissection, CT scan).

(Check your answers in Appendix C.)

the victims (Christensen, 2011; Lepistö, Luukkaala, & Paavilainen, 2011; van Marle, 2010).

Why don't victims immediately report the abuse to authorities or simply leave? First, abuse generally involves numerous events that follow a cyclical and escalating pattern. In the beginning, perpetrators can be devoted and caring partners or parents, but when disagreements happen the abuser responds with increasing levels of intimidation, bullying, and violence, while the victim learns that the only way to calm the situation is to respond with increasing levels of compliance and subservience. Second, abuse occurs in many forms (physical, verbal, and emotional), which makes it harder to identify and report. Finally, domestic violence is much more difficult to report and prosecute than attacks by strangers.

Despite these obstacles, if you or someone you know is involved in violence, trained counselors are available 24 hours a day (1-800-799-SAFE). For insight into how some children survive these and other adverse situations, see *Psychological Science*.

## Work and Retirement

Work affects our health, our friendships, where we live, and even our leisure activities. Too often, however, career choices are made based on dreams of high income. In a 1995 survey

Blend Images/SuperStock, Inc.

## Satisfaction after retirement • Figure 10.10

Active involvement is a key ingredient to a fulfilling old age.

conducted by the Higher Education Research Institute, nearly 74% of college freshmen said that being "very well-off financially" was "very important" or "essential." Seventy-one percent felt the same way about raising a family.

Work and career are a big part of adult life and self-identity, but the large majority of men and women in the United States choose to retire sometime in their sixties. What helps people successfully navigate this important life change? According to the **activity theory of aging**, successful aging appears to be most strongly related to good health, control over one's life, social support, and participation in community services and social activities (Burr, Santo, & Pushkar, 2011; Fernandez-Ballesteros et al., 2011; Kubicek et al., 2011; Reynolds et al., 2008) (**Figure 10.10**).

**activity theory of aging** Theory that successful aging is fostered by a full and active commitment to life.

The activity theory of aging has largely displaced the older notion that successful aging entails a natural and graceful withdrawal from life, known as **disengagement theory** (Achenbaum & Bengtson, 1994; Cummings & Henry, 1961; Feldman & Beehr, 2011; Marshall & Bengtson, 2011; Neugarten, Havighurst, & Tobin, 1968; Riebe et al., 2005; Sanchez, 2006).

**disengagement theory** Theory that successful aging is characterized by mutual withdrawal between the elderly and society.

**socioemotional selectivity theory** A natural decline in social contact as older adults become more selective with their time.

An influential modern theory, **socioemotional selectivity theory**, helps explain the predictable decline in social contact that almost everyone experiences as they move into their older years (Carstensen et al., 2011; Charles & Carstensen, 2007; Yeung, Wong, & Lok, 2011). According to this theory, we don't naturally withdraw from society in our later years—we deliberately decrease our total number of social contacts in favor of familiar people who provide emotionally meaningful interactions (**Figure 10.11**).

## Socioemotional selectivity • Figure 10.11

During infancy, emotional connection is essential to our survival. During childhood, adolescence, and early adulthood, information gathering is critical and the need for emotional connection declines. During late adulthood, emotional satisfaction is again more important—we tend to invest our time in those who can be counted on in times of need.

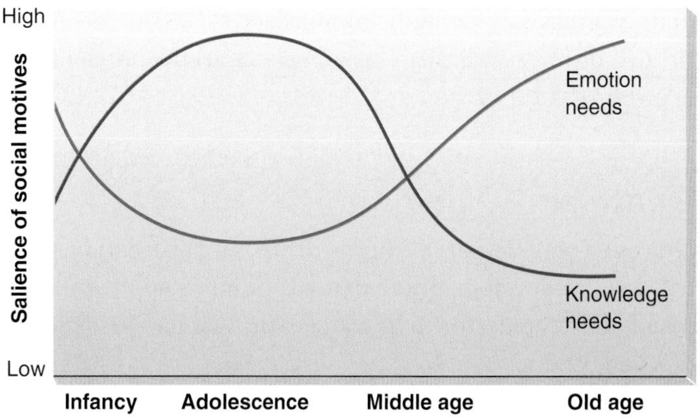


*Source: Annual Review of Gerontology and Geriatrics*, Volume 17, 1997: Focus on Emotion and Adult Development, Carstensen, Gross, & Fung, 1997, Springer Publishing Company, LLC.

# MYTH BUSTERS

THE PLANNER

## MYTHS OF DEVELOPMENT

A number of popular beliefs about age-related crises are not supported by research. The popular idea of a *midlife crisis* began largely as a result of Gail Sheehy's national best-seller *Passages* (1976). Sheehy drew on the theories of Daniel Levinson (1977, 1996) and psychiatrist Roger Gould (1975), as well as her own interviews. She popularized the idea that almost everyone experiences a "predictable" crisis" at about age 35 for women and 40 for men. Middle age often *is* a time of reexamining one's values and lifetime goals. However, Sheehy's book led many people to automatically expect a midlife crisis with drastic changes in personality and behavior. Research suggests that a severe reaction or crisis may actually be quite rare and not typical of what most people experience during middle age (Freund & Ritter, 2009; Lilienfeld et al., 2010; Whitbourne & Mathews, 2009).

Many people also believe that when the last child leaves home, most parents experience an *empty nest syndrome*—a painful separation and time of depression for the mother, the father, or both parents. Again, research suggests that the empty nest syndrome may be an exaggeration of the pain experienced by a few individuals and an effort to downplay positive reactions (Clay, 2003; Mitchell & Lovegreen, 2009). For example, one major benefit of the empty nest may be an increase in marital satisfaction, as shown in the graph. Moreover, parent-child relationships do continue once the child leaves home. As one mother said, "The empty nest is surrounded by telephone wires" (Troll, Miller, & Atchley, 1979).

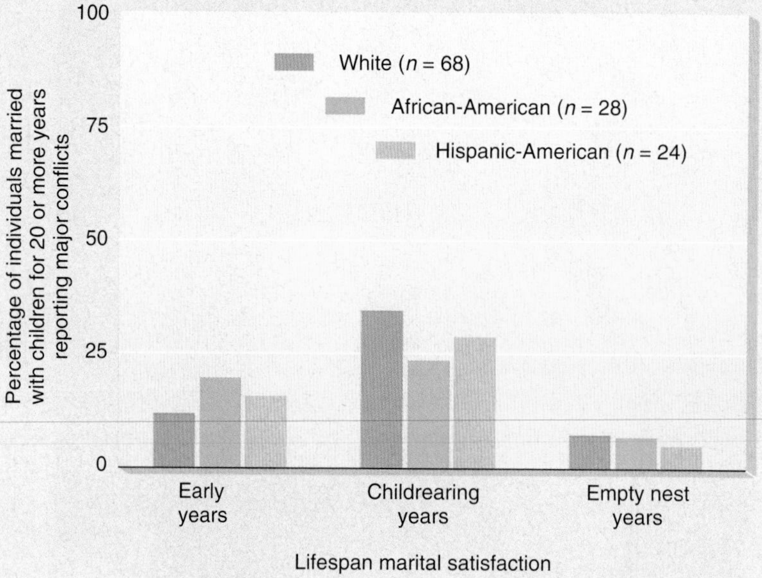


*Source:* From Mackey, R.A., and O'Brien, B.A. (1998). Marital conflict management: Gender and ethnic differences. *Social work*, 43 (2), 128-141. By permission of Oxford University Press.

As we've seen, there are losses and stress associated with the aging process—although much less than most people think (see *Myth Busters*). Perhaps the greatest challenge for the elderly, at least in the United States, is the ageism they encounter. In societies that value older people as wise elders or keepers of valued traditions, the stress of aging is much less than in societies that view them as mentally slow and socially useless. In cultures like that in the United States in which youth, speed, and progress are strongly emphasized, a loss or decline in any of these qualities is deeply feared and denied (Powell, 1998).

## Death and Dying

One unavoidable part of life is death. In this section, we will look at developmental changes in how people understand and prepare for their own deaths and for the deaths of loved others.

As adults, we understand death in terms of three basic concepts: (1) *permanence*—once a living thing dies, it cannot be brought back to life; (2) *universality*—all living things eventually die; and (3) *nonfunctionality*—all living functions, including thought, movement, and vital signs, end at death.

Research shows that permanence, the notion that death cannot be reversed, is the first and most easily understood concept (**Figure 10.12**). Understanding of universality comes slightly later, and by the age of 7, most children have mastered nonfunctionality and have an adultlike understanding of death.

Although parents may fear that discussing death with children and adolescents will make them unduly anxious, those who are offered open, honest discussions of death have an easier time accepting it (Buckle & Fleming, 2011; Corr, Nabe, & Corr, 2009; Kastenbaum, 2007; Talwar, Harris, & Schleifer, 2011).

The same is true for adults—in fact, avoiding thoughts and discussion of death and associating aging with death

Susan Van Etten/PhotoEdit

**How do children understand death? • Figure 10.12**

Preschoolers seem to accept the fact that the dead person cannot get up again, perhaps because of their experiences with dead butterflies and beetles found while playing outside (Furman, 1990). Later, they begin to understand all that death entails and that they, too, will someday die.

contribute to ageism (Atchley, 1997). Moreover, the better we understand death and the more wisely we approach it, the more fully we can live. Since the late 1990s, right-to-die and death-with-dignity advocates have brought death out in the open, and mental health professionals have suggested that understanding the psychological processes of death and dying may play a significant role in good adjustment. Cultures around the world interpret and respond to death in widely different ways: "Funerals are the occasion for avoiding people or holding parties, for fighting or having sexual orgies, for weeping or laughter, in a thousand combinations" (Metcalf & Huntington, 1991) (**Figure 10.13**).

Confronting our own death is the last major crisis we face in life. What is it like? Is there a "best" way to prepare to die? Is there such a thing as a "good" death? After spending hundreds of hours at the bedsides of the terminally ill, Elisabeth Kübler-Ross developed a highly controversial stage theory of the psychological processes surrounding death (1983, 1997, 1999). She proposed that most people go through five sequential stages when facing death:

- *Denial* of the terminal condition ("This can't be true; it's a mistake!")
- *Anger* ("Why me? It isn't fair!")
- *Bargaining* ("God, if you let me live, I'll dedicate my life to you!")
- *Depression* ("I'm losing everyone and everything I hold dear.")
- *Acceptance* ("I know that death is inevitable and my time is near.")

Critics of the stage theory of dying stress that the five-stage sequence has not been scientifically validated and that each person's death is a unique experience (Kastenbaum, 2007; Konigsberg, 2011; Lilienfeld et al., 2010; O'Rourke, 2010). Others worry that popularizing such a stage theory will cause more avoidance and stereotyping of the dying ("He's just in the anger stage right now."). Kübler-Ross (1983, 1997, 1999) agrees that not all people go through the same stages in the same way and regrets that anyone would use her theory as a model for a "good" death.

**Culture influences our response to death • Figure 10.13**

In October 2006, a dairy truck driver took over a one-room Amish schoolhouse in Pennsylvania, killed and gravely injured several young girls, then shot himself. Instead of responding in rage, his Amish neighbors attended his funeral. Amish leaders urged forgiveness for the killer and called for a fund to aid his wife and three children. Rather than creating an on-site memorial, the schoolhouse was razed, to be replaced by pasture. What do you think of this response? The fact that many Americans were offended, shocked, or simply surprised by the Amish reaction illustrates how strongly culture affects our emotions, beliefs, and values.

Mel Evans/©AP/Wide World Photos

In spite of the potential abuses, Kübler-Ross's theory has provided valuable insights and spurred research into a long-neglected topic. **Thanatology**, the study of death and dying, has become a major topic in human development. Thanks in part to thanatology research, the dying are being helped to die with dignity by the **hospice** movement, which has created special facilities and trained staff and volunteers to provide loving support for the terminally ill and their families (Casarett, 2011; Claxton-Oldfield et al., 2011; DeSpelder & Strickland, 2007; Kumar, Markert, & Patel, 2011).

**CONCEPT CHECK** STOP

1. **Why** is it difficult for victims of violent relationships to leave?
2. **What** is the socioemotional selectivity theory?
3. **What** are the stages of Kübler-Ross's theory of the psychological processes surrounding death?

# Summary

THE PLANNER

## 1 Social, Moral, and Personality Development 258

- Nativists believe that **attachment** is innate, whereas nurturists believe that it is learned. Studies of imprinting and the experiment with rhesus monkeys support the biological argument for attachment. Harlow and his colleagues' research with monkeys raised by cloth or wire "mothers" found that **contact comfort** might be the most important factor in attachment.
- Ainsworth and her colleagues examined how infants coped when faced with the **strange situation procedure**. Ainsworth's study of attachment in human infants found four key patterns: *securely attached, anxious/ambivalent, anxious/avoidant,* or *disorganized/disoriented attachment*, shown in the diagram. In addition to attachment patterns, Baumrind's three parenting styles, *permissive, authoritarian,* and *authoritative*, also affect a child's social development.

**Research on infant attachment: Degrees of attachment • Figure 10.1**

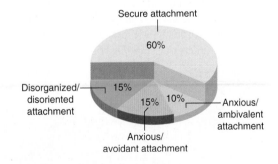

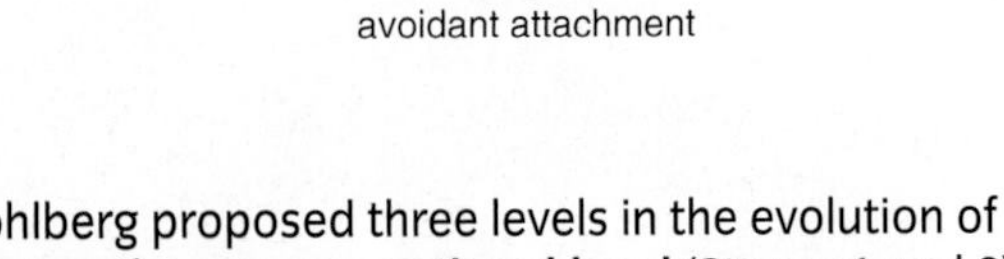


- Kohlberg proposed three levels in the evolution of moral reasoning: the **preconventional level** (Stages 1 and 2), the **conventional level** (Stages 3 and 4), and the **postconventional level** (Stages 5 and 6). Kohlberg's theory has been criticized for possibly measuring only moral reasoning and not moral behavior, as well as for possible gender and **cultural bias**.
- Erikson identified eight **psychosocial stages** of development, each marked by a crisis (such as the adolescent **identity crisis**) related to a specific developmental task. Some critics argue that Erikson's stages are difficult to test scientifically, and that his theory does not hold true across all cultures. However, many psychologists agree with Erikson's general idea that psychosocial crises do contribute to personality development.

## 2 How Sex, Gender, and Culture Affect Development 266

- In addition to biological **sex** differences, scientists have found numerous **gender** differences relevant to cognitive, social, and personality development (for example, in cognitive ability and aggressive behavior) and have proposed both biological and environmental explanations for these differences. Most research emphasizes two major theories of gender-role development: social learning (shown in the diagram) and cognitive developmental. One way to overcome gender-role stereotypes is to combine "masculine" and "feminine" traits **(androgyny)**.

**Gender-role development • Figure 10.7**

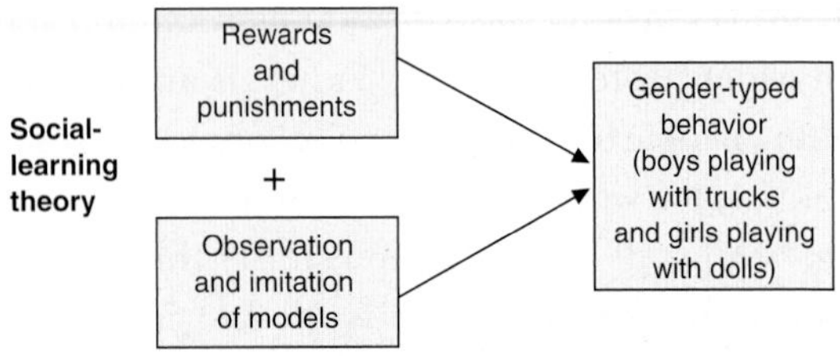


- Our culture shapes how we define and understand ourselves. In **individualistic cultures**, the needs and goals of the individual are emphasized; in **collectivistic cultures**, the person is defined and understood primarily by looking at his or her place in the social unit.

## 3 Developmental Challenges Through Adulthood 272

- Developing realistic expectations and avoiding violence and abuse are vitally important to good relationships and social development. However, **resiliency** helps some children survive and even prosper in harsh, abusive environments.
- The **activity theory of aging** proposes that successful aging depends on having control over one's life, having social support, and participating in community services and social activities. The older **disengagement theory** has been largely abandoned. According to the **socioemotional selectivity theory**, people don't naturally withdraw from society in late adulthood but deliberately decrease the total number of their social contacts in favor of familiar people who provide emotionally meaningful interactions, as shown in the graph.

**Socioemotional selectivity • Figure 10.11**

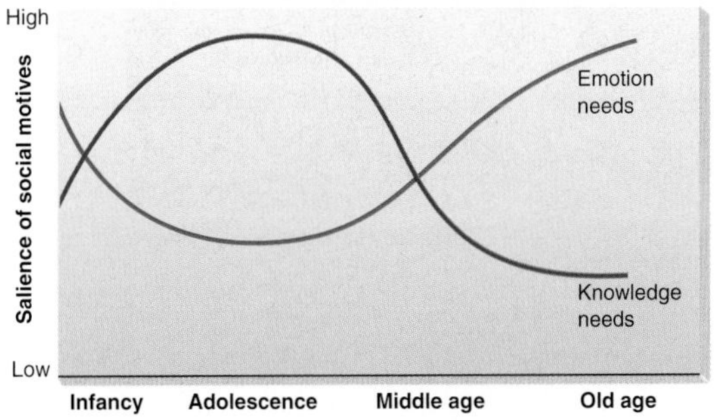


- Adults understand death in terms of three basic concepts: *permanence, universality*, and *nonfunctionality*. Children most easily understand the concept of permanence; understanding of universality and nonfunctionality comes later. For children, adolescents, and adults, understanding the psychological processes of death and dying is important to good adjustment. Kübler-Ross proposed that most people go through five sequential stages when facing death: *denial, anger, bargaining, depression*, and *acceptance*. But she has emphasized that her theory should not be interpreted as a model for a "good" death.

# Key Terms

**RETRIEVAL PRACTICE** Write a definition for each term before turning back to the referenced page to check your answer.

- activity theory of aging 275
- androgyny 270
- attachment 258
- collectivistic culture 270
- contact comfort 258
- conventional level 263
- disengagement theory 275
- gender 266
- gender role 266
- gender schema 269
- hospice 278
- identity crisis 265
- individualistic culture 270
- postconventional level 263
- preconventional level 263
- psychosocial stages 264
- resiliency 274
- sex 266
- socioemotional selectivity theory 275
- strange situation procedure 259
- thanatology 278

# Critical and Creative Thinking Questions

1. Were Harlow and his colleagues acting ethically when they separated young rhesus monkeys from their mothers and raised them with either a wire or cloth "mother"? Why or why not?
2. According to Erikson's psychosocial theory, what developmental crisis must you now resolve in order to successfully move on to the next stage? Are there any earlier crises that you feel you may not have successfully resolved?
3. Can you think of instances where you have adopted traditional gender roles? What about times where you have expressed more androgyny?
4. Would you like to know ahead of time that you were dying? Or would you prefer to die suddenly with no warning? Briefly explain your choice.

# What is happening in this picture?

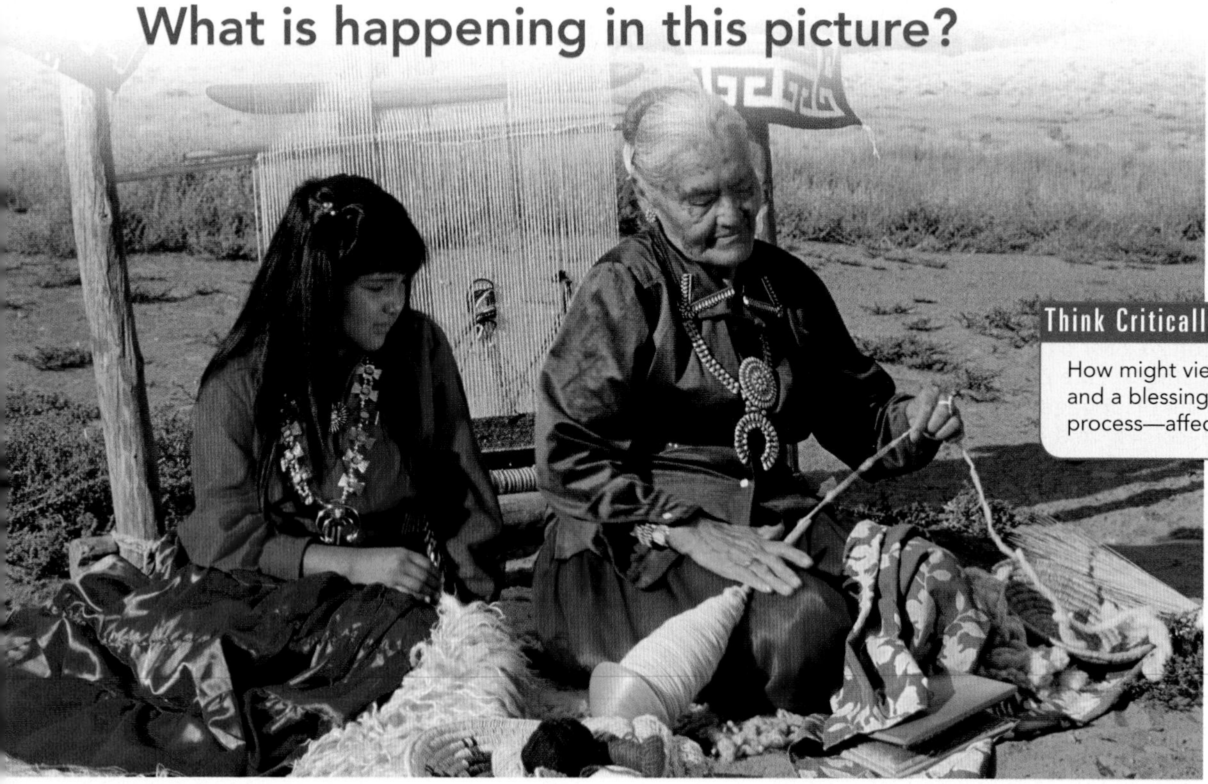

David W. Hamilton/Getty Images

Native Americans generally revere and respect the elder members of their tribe.

**Think Critically**

How might viewing old age as an honor and a blessing—as opposed to a dreaded process—affect the experience of aging?

# Self-Test

**RETRIEVAL PRACTICE** Completing this self-test and comparing your answers with those in Appendix C provides immediate feedback and helpful practice for exams. Additional interactive, self-tests are available at www.wiley.com/college/carpenter.

**1.** ______ is a strong affectional bond with special others that endures over time.

a. Bonding  c. Love
b. Attachment  d. Intimacy

**2.** In Ainsworth's studies on infant attachment, the ______ infants sought their mothers' comfort while also squirming to get away when the mother returned to the room.

a. anxious/ambivalent  c. insecurely attached
b. caregiver-based  d. dependent

**3.** Kohlberg believed his stages of moral development to be ______.

a. universal and invariant
b. culturally bound, but invariable within a culture
c. universal, but variable within each culture
d. culturally bound and variable

**4.** Once an individual has accepted and complied with the rules and values of society, that person has advanced to the ______ level of moral development.

a. instrumental-exchange  c. conventional
b. social-contract  d. postconventional

**5.** According to Erikson, humans progress through eight stages of psychosocial development. Label the CORRECT sequence on the photos for the "successful" completion of the first four stages.

Stage 1 ______

Stage 2 ______

Stage 3 ______

Stage 4 ______

6. Label the CORRECT sequence on the photos for the "successful" completion of the second four stages of Erikson's eight stages of psychosocial development.

Stage 5 _____

Stage 6 _____

Stage 7 _____

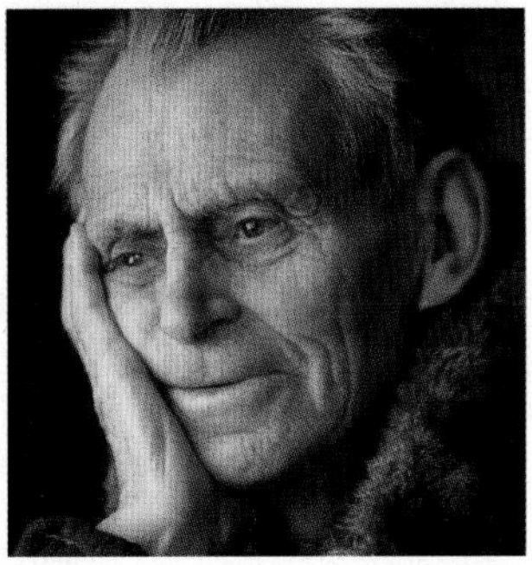

Stage 8 _____

7. Your _____ is related to societal expectations for normal and appropriate male or female behavior.
   a. gender identity
   b. gender role
   c. sexual identity
   d. sexual orientation

8. On the basis of internal rules governing correct behaviors for boys versus girls, children form _____ of how they should act.
   a. sexual identities
   b. gender identities
   c. gender schemas
   d. any of these options

9. Cultures in which the needs and goals of the group are emphasized over the needs and goals of the individual are _____.
   a. collectivistic cultures
   b. individualistic cultures
   c. worldwide cultures
   d. all of the above

10. Which of the following is a factor in resiliency?
    a. good intellectual functioning
    b. relationships with caring adults
    c. ability to regulate attention, emotions, and behavior
    d. all of these options

11. The _____ theory of aging suggests that successful adjustment is fostered by a full and active commitment to life.
    a. activity
    b. commitment
    c. engagement
    d. life-affirming

12. The _____ theory of aging suggests that adjustment to retirement is fostered by mutual withdrawal between the elderly and society.
    a. disenchantment
    b. withdrawal
    c. disengagement
    d. death-preparation

13. Which of the following concepts is considered permanent, universal, and nonfunctional?
    a. androgyny
    b. resiliency
    c. gender schema
    d. death

14. Which of the following is TRUE about a child's understanding of death?
    a. Preschoolers understand that death is permanent.
    b. Preschoolers may not understand that death is universal.
    c. Children understand that death is nonfunctional by the age of 7.
    d. All of these options are true.

15. Which of the following is the correct sequence for Kübler-Ross's stage theory of dying?
    a. denial; anger; bargaining; depression; acceptance
    b. anger; denial; depression; bargaining; acceptance
    c. denial; bargaining; anger; depression; acceptance
    d. denial; depression; anger; acceptance

**THE PLANNER** ✓

Review your Chapter Planner on the chapter opener and check off your completed work.

# 11 Motivation and Emotion

Imagine yourself in a cap and gown celebrating your own college graduation. Will you experience the same emotions as the people shown in this photo? What propels students to work hard and sacrifice so much for long-term accomplishments, like a college degree? Do most people share similar motivations and emotions? Consider these quotes from two of our most famous Americans:

*I have a dream that my four little children will one day live in a nation where they will not be judged by the color of their skin, but by the content of their character. . . .*

*Dr. Martin Luther King, Jr.* (1929–1968), world famous civil rights leader, author, clergyman, and winner of many awards and honors, including the Nobel Peace prize, Time Person of the Year, and second on Gallup's list of the most widely admired people of the twentieth century.

*Knowledge is love and light and vision. . . . Life is an exciting business, and most exciting when it is lived for others. No pessimist ever discovered the secret of the stars, or sailed to an uncharted land, or opened a new doorway for the human spirit.*

*Helen Keller* (1880–1968), world-famous author, activist, suffragist, speaker, and winner of many awards, including being listed as one of Gallup's most widely admired people of the twentieth century.

What do you think motivated Dr. King and Helen Keller to fight so hard to pursue their dreams, in spite of serious illnesses and setbacks? This chapter reveals what psychologists have discovered about why we do what we do, as well as what our emotions are made of and why we often feel as we do. To share these discoveries, read on!

## CHAPTER OUTLINE

Ariel Skelley/Blend Images/Getty Images

## CHAPTER PLANNER

- ❑ Study the picture and read the opening story.
- ❑ Scan the Learning Objectives in each section: p. 284 ❑ p. 289 ❑ p. 299 ❑
- ❑ Read the text and study all figures and visuals. Answer any questions.

**Analyze key features**

- ❑ Study Organizers p. 284 ❑ p. 301 ❑
- ❑ Myth Busters p. 285 ❑ p. 297 ❑
- ❑ Process Diagrams p. 286 ❑ p. 295 ❑
- ❑ Applying Psychology, p. 287
- ❑ Psychology InSight, p. 298
- ❑ Psychological Science, p. 302
- ❑ What a Psychologist Sees, p. 304
- ❑ Stop: Answer the Concept Checks before you go on. p. 289 ❑ p. 298 ❑ p. 307 ❑

**End of chapter**

- ❑ Review the Summary and Key Terms.
- ❑ Answer the Critical and Creative Thinking Questions.
- ❑ Answer What is happening in this picture?
- ❑ Complete the Self-Test and check your answers.

# Theories of Motivation

## LEARNING OBJECTIVES

**RETRIEVAL PRACTICE** While reading the upcoming sections, respond to each Learning Objective in your own words. Then compare your responses with those in Appendix B.

1. **Summarize** the three biologically based theories of motivation.
2. **Explain** how incentives, attributions, and expectations affect motivation.
3. **Describe** Maslow's hierarchy of needs.

We begin our study of motivation by examining several theories of motivation that fall into three general categories—biological, psychological, and biopsychosocial (**Study Organizer 11.1**). While studying these theories, try to identify which theory best explains your personal behaviors, such as going to college or choosing a major. Research in *motivation* attempts to answer such "what," "how," and "why" questions. Take the *Myth Buster* quiz to see what you already know about motivation.

## Biological Theories

Many theories of **motivation** focus on inborn biological processes that control behavior. Among these biologically oriented theories are *instinct*, *drive-reduction*, and *arousal* theories.

**motivation** A set of factors that activate, direct, and maintain behavior, usually toward a goal.

In the earliest days of psychology, researchers like William McDougall (1908) proposed that humans had numerous "instincts," such as repulsion, curiosity, and

**Study Organizer 11.1** Six major theories of motivation — THE PLANNER

James A. Sugar/NG Image Collection

Which of the six theories of motivation best explains this behavior?

| Theory | Description |
|---|---|
| **BIOLOGICAL** | |
| 1. Instinct | Motivation results from innate, biological instincts, which are unlearned responses found in almost all members of a species. |
| 2. Drive reduction | Motivation begins with biological need (a lack or deficiency) that elicits a *drive* toward behavior that will satisfy the original need and restore homeostasis. |
| 3. Optimal arousal | Organisms are motivated to achieve and maintain an optimal level of arousal. |
| **PSYCHOLOGICAL** | |
| 4. Incentive | Motivation results from external stimuli that "pull" the organism in certain directions. |
| 5. Cognitive | Motivation is affected by expectations and attributions, or how we interpret or think about our own or others' actions. |
| **BIOPSYCHOSOCIAL** | |
| 6. Maslow's hierarchy of needs | Lower needs like hunger and safety must be satisfied before advancing to higher needs (such as belonging and self-actualization). |

## MYTH BUSTERS

THE PLANNER

### MOTIVATION AND EMOTION MYTHS

**TRUE OR FALSE?**

- Being either too excited or too relaxed can interfere with test performance.
- People who are high achievers prefer moderately difficult tasks (not too hard and not too easy).
- Getting paid for your hobbies may reduce your overall creativity and enjoyment.
- Smiling can make you feel happy and frowning can create negative feelings.
- Polygraph tests are reliable lie detectors.
- Lie detector tests may be fooled by biting your tongue.

Answers: One of these is false. Check your text for details.

self-assertiveness. Other researchers later added their favorite instincts, and by the 1920s, the list of recognized instincts had become impossibly long. One researcher found listings for over 10,000 human instincts (Bernard, 1924).

In addition, the label *instinct* led to unscientific, circular explanations—"men are aggressive because they are instinctively aggressive" or "women are maternal because they have a natural maternal instinct." However, in recent years, a branch of biology called sociobiology has revived the case for **instincts** when strictly defined (**Figure 11.1**).

**instincts** Behavioral patterns that are unlearned, always expressed in the same way, and universal in a species.

In the 1930s, the concepts of drive and drive reduction began to replace the theory of instincts. According to **drive-reduction theory** (Hull, 1952), when biological needs (such as food, water, and oxygen) are unmet, a state of tension (known as a *drive*) is created,

### Instincts • Figure 11.1

a. Instinctual behaviors are obvious in many animals. Birds build nests, bears hibernate, and salmon swim upstream to spawn.

Paul Sutherland/NG Image Collection

b. Sociobiologists such as Edward O. Wilson (1975, 1978) believe that humans also have instincts, like competition or aggression, which are genetically transmitted from one generation to the next.

Stockbyte/Getty Images, Inc.

PROCESS DIAGRAM

## Drive-reduction theory • Figure 11.2

THE PLANNER

When we are hungry or thirsty, the disruption of our normal state of equilibrium creates a drive that motivates us to search for food or water. Once action is taken and the need is satisfied, homeostasis is restored and our motivation is also decreased.

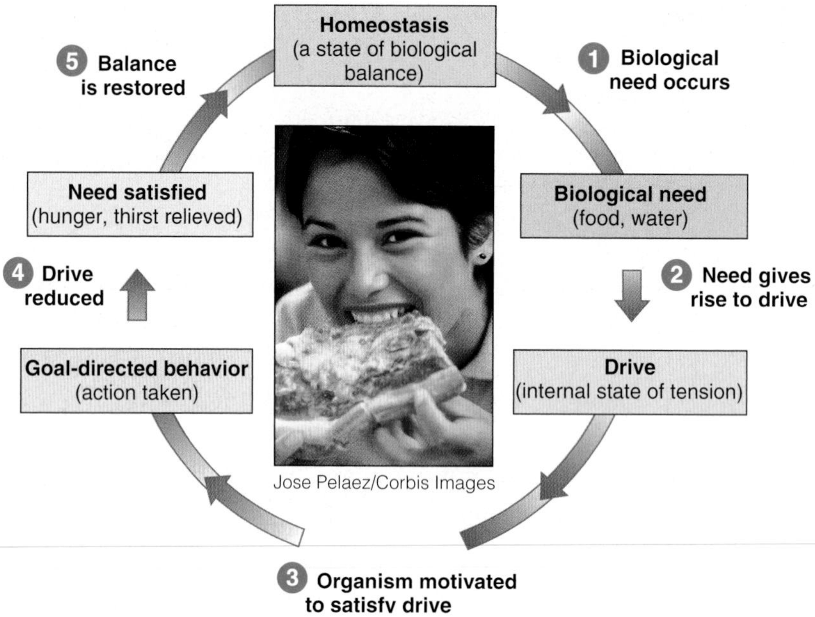


Jose Pelaez/Corbis Images

and the organism is motivated to reduce it. Drive-reduction theory is based largely on the biological concept of **homeostasis**, a term that literally means "standing still" (**Figure 11.2**).

In addition to our obvious biological needs, humans and other animals are innately curious and require a certain amount of novelty and complexity from the environment.

**homeostasis** A body's tendency to maintain a relatively stable state, such as a constant internal temperature.

According to **optimal-arousal theory**, organisms are motivated to achieve and maintain an optimal level of arousal that maximizes their performance. Either too much or too little arousal diminishes performance (**Figure 11.3**). The optimal amount of arousal may vary from person to person (see *Applying Psychology*).

## Optimal level of arousal • Figure 11.3

Our need for stimulation (the arousal motive) suggests that behavior efficiency increases as we move from deep sleep to increased alertness. However, once we pass the maximum level of arousal, our performance declines. Can you see how the optimal alertness of the young man in this photo would clearly increase his learning and memory during class lectures?

© Joshua Hodge Photography/iStockphoto

## Applying Psychology

THE PLANNER

# Sensation Seeking

Marco Simoni/Getty Images

What motivates people to bungee jump over deep canyons or white-water raft down dangerous rivers? According to research, these "high-sensation seekers" may be biologically "prewired" to need a higher than usual level of stimulation (Zuckerman, 1979, 1994, 2004).

To sample the kinds of questions that are asked on tests for sensation seeking, circle the choice (A or B) that BEST describes you:

1. **A** I would like a job that requires a lot of traveling.
   **B** I would prefer a job in one location.
2. **A** I get bored seeing the same old faces.
   **B** I like the comfortable familiarity of everyday friends.
3. **A** The most important goal of life is to live it to the fullest and experience as much as possible.
   **B** The most important goal of life is to find peace and happiness.
4. **A** I would like to try parachute jumping.
   **B** I would never want to try jumping out of a plane, with or without a parachute.
5. **A** I prefer people who are emotionally expressive even if they are a bit unstable.
   **B** I prefer people who are calm and even-tempered.

*Source*: Zuckerman, M. (1978, February). The search for high sensation, *Psychology Today*, pp. 38-46.

Research suggests that four distinct factors characterize sensation seeking (Legrand et al., 2007; Wallerstein, 2008; Zuckerman, 2004, 2008):

1. Thrill and adventure seeking (skydiving, driving fast, or trying to beat a train)
2. Experience seeking (travel, unusual friends, drug experimentation)
3. Disinhibition ("letting loose")
4. Susceptibility to boredom (lower tolerance for repetition and sameness)

Being very high or very low in sensation seeking might cause problems in relationships with individuals who score toward the other extreme. This is true not just between partners or spouses but also between parent and child and therapist and patient. There might also be job difficulties for high-sensation seekers in routine clerical or assembly-line jobs or for low-sensation seekers in highly challenging and variable occupations.

**Think Critically**

1. If your answers to the brief quiz above indicate that you are a high-sensation seeker, what do you do to satisfy that urge, and what can you do to make sure it doesn't get out of control?
2. If you are low in sensation seeking, has this trait interfered with some aspect of your life? If so, what could you do to improve your functioning in this area?

## Psychological Theories

Instinct and drive-reduction theories explain some motivations, but why do we continue to eat after our biological need is completely satisfied? Or why does someone work overtime when his or her salary is sufficient to meet all basic biological needs? These questions are better answered by psychosocial theories that emphasize incentives and cognition.

Unlike drive-reduction theory, which states that internal factors *push* people in certain directions, **incentive theory** maintains that external stimuli *pull* people toward desirable goals or away from undesirable ones. Most of us initially eat

Don Smetzer/PhotoEdit

**Expectancies as psychological motivation • Figure 11.4**

What expectations might these students have that would motivate them to master a new language?

because our hunger "pushes" us (drive-reduction theory). But the sight of apple pie or ice cream too often "pulls" us toward continued eating (incentive theory).

According to **cognitive theories**, motivation is directly affected by **attributions**, or how we interpret or think about our own and others' actions. If you receive a high grade in your psychology course, for example, you can interpret that grade in several ways. You earned it because you really studied. You "lucked out." Or the textbook was exceptionally interesting and helpful (our preference!). People who attribute their successes to personal ability and effort tend to work harder toward their goals than people who attribute their successes to luck (Beacham et al., 2011; Houtz et al., 2007; Martinko, Harvey, & Dasborough, 2011; Weiner, 1972, 1982).

**Expectancies**, or what we believe or assume will happen, are also important to motivation (Reinhard & Dickhäuser, 2011; Schunk, 2008) (**Figure 11.4**). If you anticipate that you will receive a promotion at work, you're more likely to work overtime for no pay than if you expect no promotion.

## Biopsychosocial Theories

Research in psychology generally emphasizes either biological or psychosocial factors (nature or nurture). But in the final analysis, biopsychosocial factors almost always provide the best explanation. Theories of motivation are no exception. One researcher who recognized this was Abraham Maslow (1954, 1970, 1999). Maslow believed that we all have numerous needs that compete for fulfillment but that some needs are more important than others. For example, your need for food and shelter is generally more important than college grades.

Maslow's **hierarchy of needs** prioritizes needs, with survival needs at the bottom (needs that must be met before others) and social, spiritual needs at the top (**Figure 11.5**).

**hierarchy of needs** Maslow's theory of motivation that lower motives (such as physiological and safety needs) must be met before going on to higher needs (such as belonging and self-actualization).

Maslow's hierarchy of needs seems intuitively correct—a starving person would first look for food, then love and friendship, and then self-esteem. This prioritizing and the concept of **self-actualization** are important contributions to the study of motivation (Frick, 2000; Harper, Harper, & Stills, 2003). But critics argue that parts of Maslow's theory are poorly researched and biased toward Western individualism. Furthermore, his theory presupposes a homogeneous life experience and people sometimes seek to satisfy higher-level needs even when their lower-level needs have not been met (Cullen & Gotell, 2002; Kress et al., 2011; Neher, 1991).

## Maslow's hierarchy of needs • Figure 11.5

Maslow's theory of motivation suggests we all share a compelling need to "move up"—to grow, improve ourselves, and ultimately become "self-actualized."

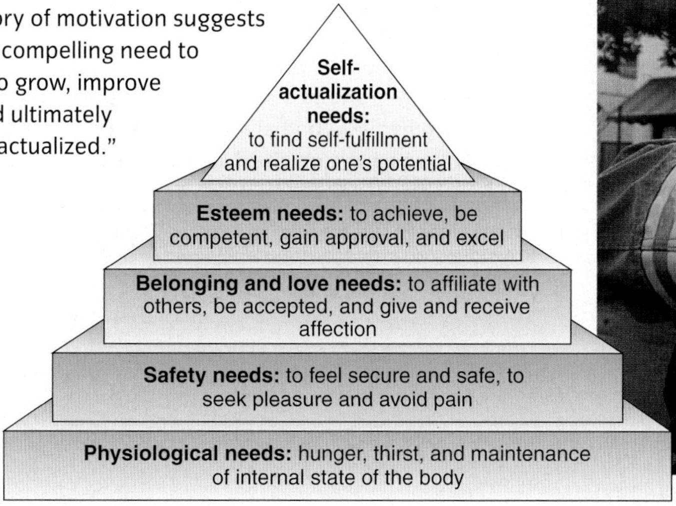

Nam Y Huh/©AP/Wide World Photos

Which of Maslow's five levels of need are on display in this photo? Can you see how the needs of the rescuer clearly differ from those of the person who is being rescued?

**CONCEPT CHECK** STOP

1. **Why** are modern sociobiological theories of instincts more scientifically useful than older "instinct" theories?
2. **What** four factors are associated with sensation seeking?
3. **What** criticisms have been made of Maslow's biopsychosocial (hierarchy of needs) theory?

# Motivation and Behavior

## LEARNING OBJECTIVES

**RETRIEVAL PRACTICE** While reading the upcoming sections, respond to each Learning Objective in your own words. Then compare your responses with those in Appendix B.

1. **Describe** how internal (biological) and external (psychosocial) factors direct hunger and eating as well as how they are involved in serious eating disorders.
2. **Explain** why some people are more highly achievement motivated than others.
3. **Summarize** what happens to the human body during sexual activity.
4. **Compare** intrinsic and extrinsic motivation.

Why do people put themselves in dangerous situations? Why do salmon swim upstream to spawn? Behavior results from many motives. For example, we discuss the need for sleep in Chapter 5, and we look at aggression, altruism, and interpersonal attraction in Chapter 15. Here, we will focus on three basic motives: hunger, achievement, and sexuality. Then we'll turn to a discussion of how different kinds of motivation affect our intrinsic interests and performance.

### Hunger and Eating

What motivates hunger? Is it your growling stomach? Or is it the sight of a juicy hamburger or the smell of a freshly baked cinnamon roll?

**The stomach** Early hunger researchers believed that the stomach controls hunger, contracting to send hunger signals when it is empty. Today, we know that it's more complicated. As dieters who drink lots of water to keep their stomachs feeling full have been disappointed to discover, sensory input from an empty stomach is not essential for feeling hungry. In fact, humans and nonhuman animals without stomachs continue to experience hunger.

However, there is a connection between the stomach and feeling hungry. Receptors in the stomach and intestines detect levels of nutrients, and specialized pressure receptors in the stomach walls signal feelings of emptiness or *satiety* (fullness or satiation). The stomach and other parts of the gastrointestinal tract also release chemical signals that play a role in hunger (Moran & Daily, 2011; Nicolaidis, 2011; Nogueiras & Tschöp, 2005; Steinert & Beglinger, 2011).

**Biochemistry** Like the stomach, the brain and other parts of the body produce numerous neurotransmitters, hormones, enzymes, and other chemicals that affect hunger and satiety (e.g., Arumugam et al., 2008; Cooper et al., 2011; Cummings, 2006; Wardlaw & Hampl, 2007). Research in this area is complex because of the large number of known (and unknown) bodily chemicals and the interactions among them. It's unlikely that any one chemical controls our hunger and eating. Other internal factors, such as *thermogenesis* (the heat generated in response to food ingestion), also play a role (Acheson et al., 2011; Ping-Delfos &Soares, 2011; Subramanian & Vollmer, 2002).

**The brain** In addition to its chemical signals, particular brain structures also influence hunger and eating. Let's look at the hypothalamus, which helps regulate eating, drinking, and body temperature.

Early research suggested that one area of the hypothalamus, the lateral hypothalamus (LH), stimulates eating, while another area, the ventromedial hypothalamus (VMH), creates feelings of satiation, signaling the animal to stop eating. When the VMH area was destroyed in rats, researchers found that the rats overate to the point of extreme obesity (**Figure 11.6**). In contrast, when the LH area was destroyed, the animals starved to death if they were not force-fed.

## How the brain affects eating • Figure 11.6

Several areas of the brain are involved in the regulation of hunger.

**a.** This diagram shows a section of the human brain, including the ventromedial hypothalamus (VMH) and the lateral hypothalamus (LH), which are involved in the regulation of hunger.

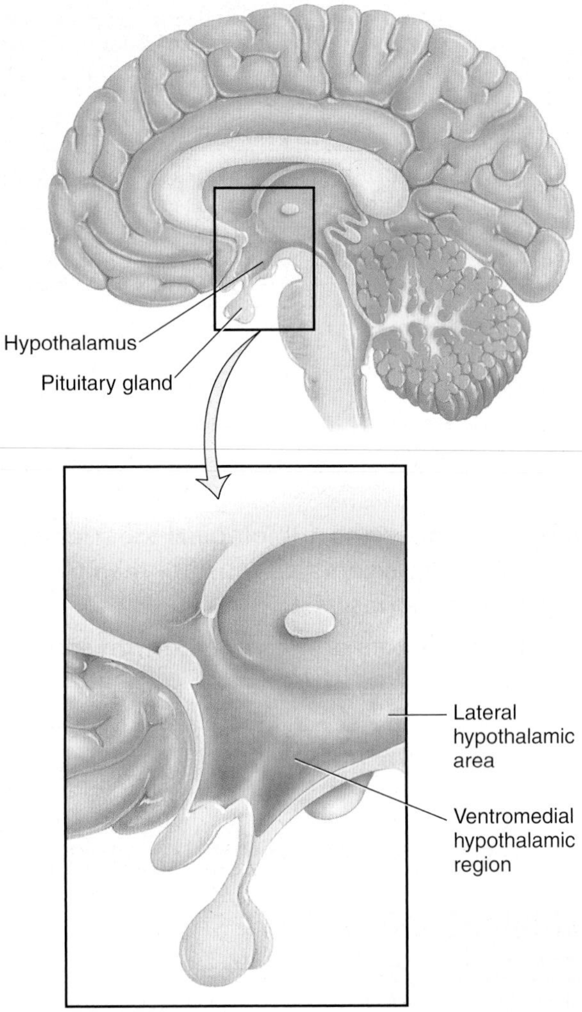


**b.** After the ventromedial area of the hypothalamus of the rat on the left was destroyed, its body weight tripled. A rat of normal weight is shown on the right for comparison.

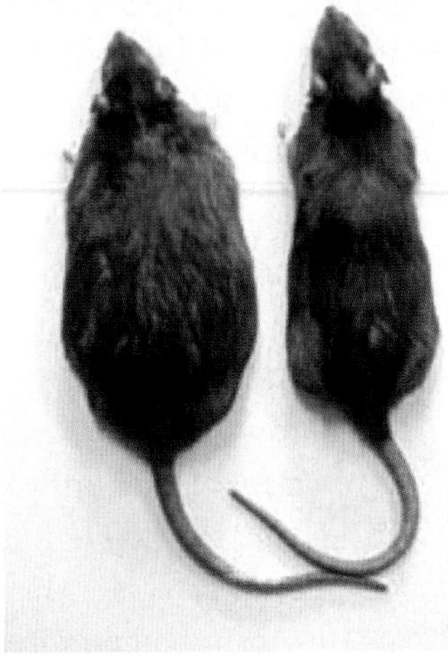

Olivier Voisin/Photo Researchers

Later research, however, showed that the LH and VMH areas are not simple on–off switches for eating. For example, lesions (damage) to the VMH make animals picky eaters—they reject food that doesn't taste good. The lesions also increase insulin secretion, which may cause overeating (Challem et al., 2000). Today, researchers know that the hypothalamus plays an important role in hunger and eating, but it is not the brain's "eating center." In fact, hunger and eating, like virtually all behavior, are influenced by numerous neural circuits that run throughout the brain (Berthoud & Morrison, 2008; Malik, McGlone & Dagher, 2011; van der Laan et al., 2011).

**Psychosocial factors** The internal motivations for hunger we've discussed (the stomach, biochemistry, the brain) are powerful. But psychosocial factors—for example, spying a dessert cart or a McDonald's billboard, or even simply noticing that it's almost lunchtime—can be equally important stimulus cues for hunger and eating.

Another important psychosocial influence on when, what, where, and why we eat is cultural conditioning. North Americans, for example, tend to eat dinner at around 6 P.M., whereas people in Spain and South America tend to eat around 10 P.M. When it comes to *what* we eat, have you ever eaten rat, dog, or horse meat? If you are a typical North American, this might sound repulsive to you, yet most Hindus would feel a similar revulsion at the thought of eating meat from cows.

In sum, there are numerous biological and psychosocial factors involved in the regulation of hunger and eating (**Figure 11.7**), and researchers are still struggling to discover and explain how all these processes work together.

## Key mechanisms in hunger regulation • Figure 11.7

Different parts of your body communicate with your brain to trigger feelings of hunger.

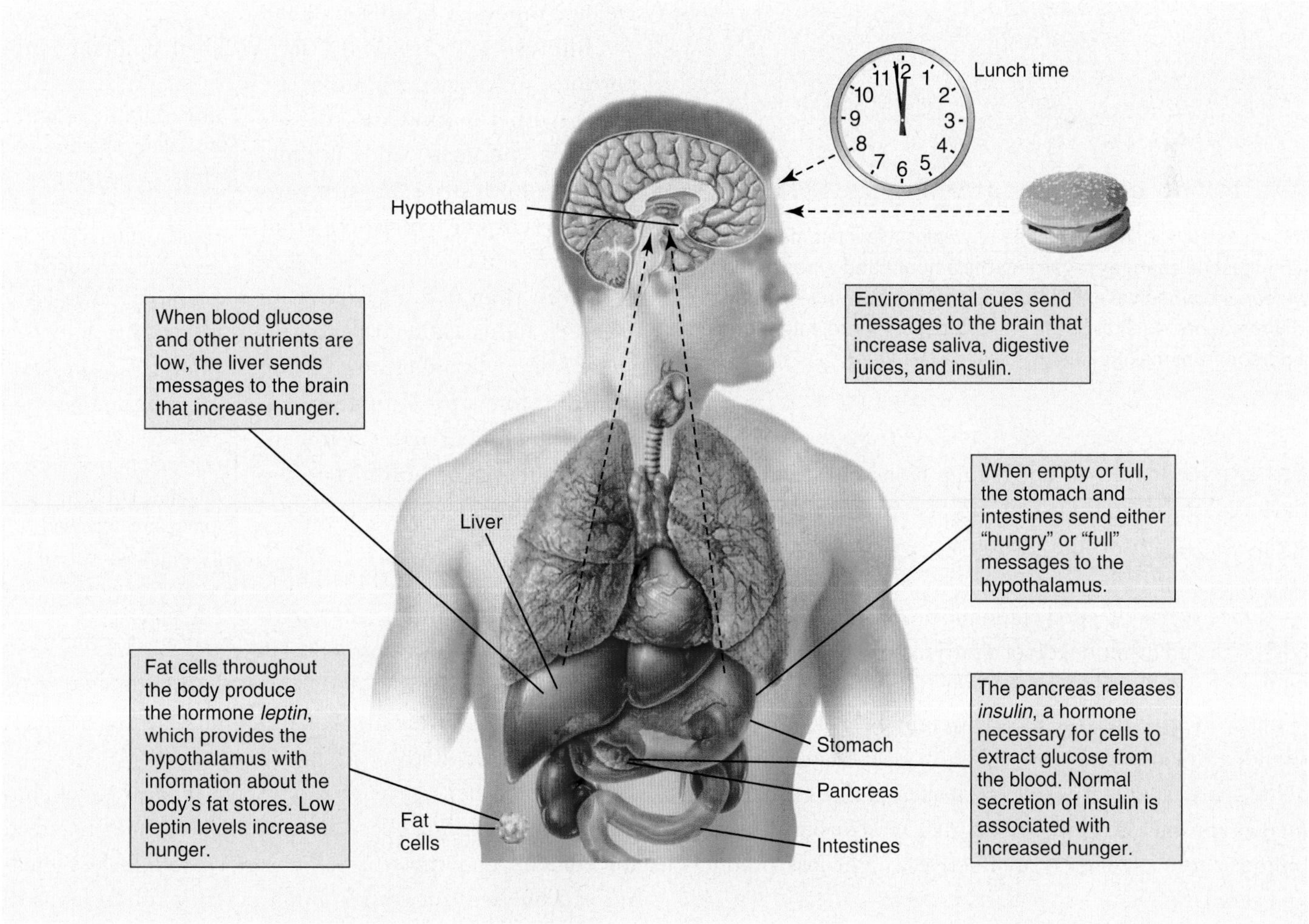

Lunch time
Hypothalamus
Environmental cues send messages to the brain that increase saliva, digestive juices, and insulin.
When blood glucose and other nutrients are low, the liver sends messages to the brain that increase hunger.
When empty or full, the stomach and intestines send either "hungry" or "full" messages to the hypothalamus.
Liver
Fat cells throughout the body produce the hormone *leptin*, which provides the hypothalamus with information about the body's fat stores. Low leptin levels increase hunger.
The pancreas releases *insulin*, a hormone necessary for cells to extract glucose from the blood. Normal secretion of insulin is associated with increased hunger.
Stomach
Pancreas
Fat cells
Intestines

Christian Thomas/Getty Images

## A fattening environment • Figure 11.8

To successfully lose (and maintain) weight, we must make permanent lifestyle changes regarding the amount and types of foods we eat and when we eat them. Can you see how our everday environments, such as in the workplace seen here, might prevent a person from making healthy lifestyle changes?

**Eating disorders** The same biopsychosocial forces that explain hunger and eating also play a role in three serious eating disorders: *obesity*, *anorexia nervosa*, and *bulimia nervosa*.

Obesity has reached epidemic proportions in the United States and other developed nations. Well over half of all adults in the United States meet the current criteria for clinical **obesity** (having a body weight 15% or more above the ideal for one's height and age). Each year, billions of dollars are spent treating serious and life-threatening medical problems related to obesity, and consumers spend billions more on largely ineffective weight-loss products and services.

For Americans, controlling weight is a particularly difficult task. We are among the most sedentary people of all nations, and we've become accustomed to "supersized" cheeseburgers, "Big Gulp" drinks, and huge servings of dessert (Carels et al., 2008; Fisher & Kral, 2008; Herman & Polivy, 2008). We've also learned that we should eat three meals a day (whether we're hungry or not); that "tasty" food requires lots of salt, sugar, and fat; and that food is an essential part of all social gatherings (**Figure 11.8**).

However, we all know some people who can eat anything they want and still not add pounds. This may be a result of their ability to burn calories more effectively (thermogenesis), a higher metabolic rate, or other factors. Adoption and twin studies indicate that genes also play a role. Heritability for obesity ranges between 30 and 70% (Andersson & Walley, 2011; Johnson, 2011; Lee et al., 2008). Unfortunately, identifying the genes for obesity is difficult. Researchers have isolated over 2,000 genes that contribute to normal and abnormal weight (Camarena et al., 2004; Costa, Brennen, & Hochgeschwender, 2002; Devlin, Yanovski, & Wilson, 2000).

Interestingly, as obesity has reached epidemic proportions, we've seen a similar rise in two other eating disorders—**anorexia nervosa** and **bulimia nervosa**. Both disorders are serious and chronic conditions that require treatment.

**anorexia nervosa** An eating disorder characterized by a severe loss of weight resulting from self-imposed starvation and an obsessive fear of obesity.

**bulimia nervosa** An eating disorder involving the consumption of large quantities of food (bingeing), followed by vomiting, extreme exercise, or laxative use (purging).

More than 50% of women in Western industrialized countries show some signs of an eating disorder, and approximately 2% meet the clinical criteria for anorexia nervosa or bulimia nervosa (Porzelius et al., 2001; Wade, Keski-Rahkonen, & Hudson, 2011; Welch et al., 2011). These disorders are found at all socioeconomic levels. A few men also develop eating disorders, although the incidence is rarer in men (Jacobi et al., 2004; Mond & Arrighi, 2011; Raevuori et al., 2008).

Anorexia nervosa is characterized by an overwhelming fear of becoming obese, a distorted body image, a need for control, and the use of dangerous weight-loss measures. The resulting extreme malnutrition often leads to emaciation, osteoporosis, bone fractures, interruption of menstruation, and loss of brain tissue. A significant percentage of individuals with anorexia nervosa ultimately die of the disorder (Huas et al., 2011; Kaye, 2008; Rosling et al., 2011).

Occasionally, the person suffering from anorexia nervosa succumbs to the desire to eat and gorges on food, then vomits or takes laxatives. However, this type of bingeing and purging is more characteristic of bulimia nervosa. Individuals with this disorder also show impulsivity in other areas, sometimes engaging in excessive shopping, alcohol abuse, or petty shoplifting (Claes et al., 2011; Kaye, 2008; Vaz-Leal et al., 2011). The vomiting associated with bulimia nervosa causes dental damage, severe damage to the throat and stomach, cardiac arrhythmias, metabolic deficiencies, and serious digestive disorders.

There are many suspected causes of anorexia nervosa and bulimia nervosa. Some theories focus on physical causes, such as hypothalamic disorders, low levels of various neurotransmitters, and genetic or hormonal disorders. Other theories emphasize psychosocial factors, such as a need for perfection, a perceived loss of control, being teased about body weight, destructive thought patterns, depression, dysfunctional families, distorted body image, and sexual abuse (e.g., Caqueo-Urizar et al., 2011; Fairburn et al., 2008; Kaye, 2008; Sachdev et al., 2008).

Cultural perceptions and stereotypes about weight and eating also play important roles in eating disorders (Eddy et al., 2007; George & Franko, 2010; Herman & Polivy, 2008). For instance, Asian and African Americans report fewer eating and dieting disorders and greater body satisfaction than do European Americans (Taylor et al., 2007; Wang et al., 2011; Whaley, Smith, & Hancock, 2011). Similarly, Mexican students report less concern about their own weight and more acceptance of obese people than do other North American students (Crandall & Martinez, 1996).

Although social pressures for thinness certainly contribute to eating disorders, anorexia nervosa has also been found in nonindustrialized areas like the Caribbean island of Curaçao (Ferguson, Winegard, & Winegard, 2011; Hoek et al., 2005). On that island, being overweight is socially acceptable, and the average woman is considerably heavier than the average woman in North America. However, some women there still have anorexia nervosa. This research suggests that both culture and biology help explain eating disorders. Regardless of the causes, it is important to recognize the symptoms of anorexia and bulimia (**Table 11.1**) and to seek therapy if the symptoms apply to you.

**DSM-IV-TR[a] symptoms of anorexia nervosa and bulimia nervosa Table 11.1**

Tony Freeman/PhotoEdit

| Symptoms of anorexia nervosa | Symptoms of bulimia nervosa |
| --- | --- |
| • Body weight below 85% of normal for one's height and age | • Normal or above-normal weight |
| • Intense fear of becoming fat or gaining weight, even though underweight | • Recurring binge eating |
| • Disturbance in one's body image or perceived weight | • Eating an amount of food that is much larger than most people would consume |
| • Self-evaluation unduly influenced by body weight | • Feeling a lack of control over eating |
| • Denial of seriousness of abnormally low body weight | • Purging behavior (vomiting or misuse of laxatives or diuretics) |
| • Absence of menstrual period in women | • Excessive exercise to prevent weight gain |
| • Purging behavior (vomiting or misuse of laxatives or diuretics) | • Fasting to prevent weight gain |
| | • Self-evaluation unduly influenced by body weight |

In anorexia nervosa, body image is so distorted that even a skeletal, emaciated body is perceived as fat.

[a]DSM-IV-TR = *Diagnostic and Statistical Manual of Mental Disorders,* fourth edition, revised.

## Achievement

Do you wonder what motivates Olympic athletes to work so hard for a gold medal? Or what about someone like Thomas Edison, who patented more than 1,000 inventions? What drives some people to high achievement?

**achievement motivation** A desire to excel, especially in competition with others.

The key to understanding what motivates high-achieving individuals lies in what psychologist Henry Murray (1938) identified as a high need for achievement (nAch), or **achievement motivation**.

Several traits distinguish people who have high achievement motivation (McClelland, 1958, 1987, 1993; Senko, Durik, & Harackiewicz, 2008; Quintanilla, 2007):

- *Preference for moderately difficult tasks* People high in nAch (need for achievement) avoid tasks that are too easy because they offer little challenge or satisfaction. They also avoid extremely difficult tasks because the probability of success is too low.
- *Competitiveness* High-achievement-oriented people are more attracted to careers and tasks that involve competition and an opportunity to excel.
- *Preference for clear goals with competent feedback* High-achievement-oriented people tend to prefer tasks with a clear outcome and situations in which they can receive feedback on their performance. They also prefer criticism from a harsh but competent evaluator to criticism from one who is friendlier but less competent.
- *Responsibility* People with high nAch prefer being personally responsible for a project so that they can feel satisfied when the task is well done.
- *Persistence* High-achievement-oriented people are more likely to persist at a task when it becomes difficult. In one study, 47% of high nAch individuals persisted on an "unsolvable task" until time was called, compared with only 2% of people with low nAch.
- *More accomplished* People who have high nAch scores do better than others on exams, earn better grades in school, and excel in their chosen professions.

Achievement orientation appears to be largely learned in early childhood, primarily through interactions with parents (**Figure 11.9**). Highly motivated children tend to have parents who encourage independence and frequently reward successes (Katz, Kaplan, & Buzukashvily, 2011;

### Future achiever • Figure 11.9

A study by Richard de Charms and Gerald Moeller (1962) found a significant correlation between the achievement themes in children's literature and the industrial accomplishments of various countries.

Michael Newman/PhotoEdit

## Masters and Johnson's sexual response cycle • Figure 11.10

THE PLANNER

Masters and Johnson identified a typical 4-stage pattern of sexual response. Note that this simplified description does not account for individual variation and should not be used to judge what's "normal."

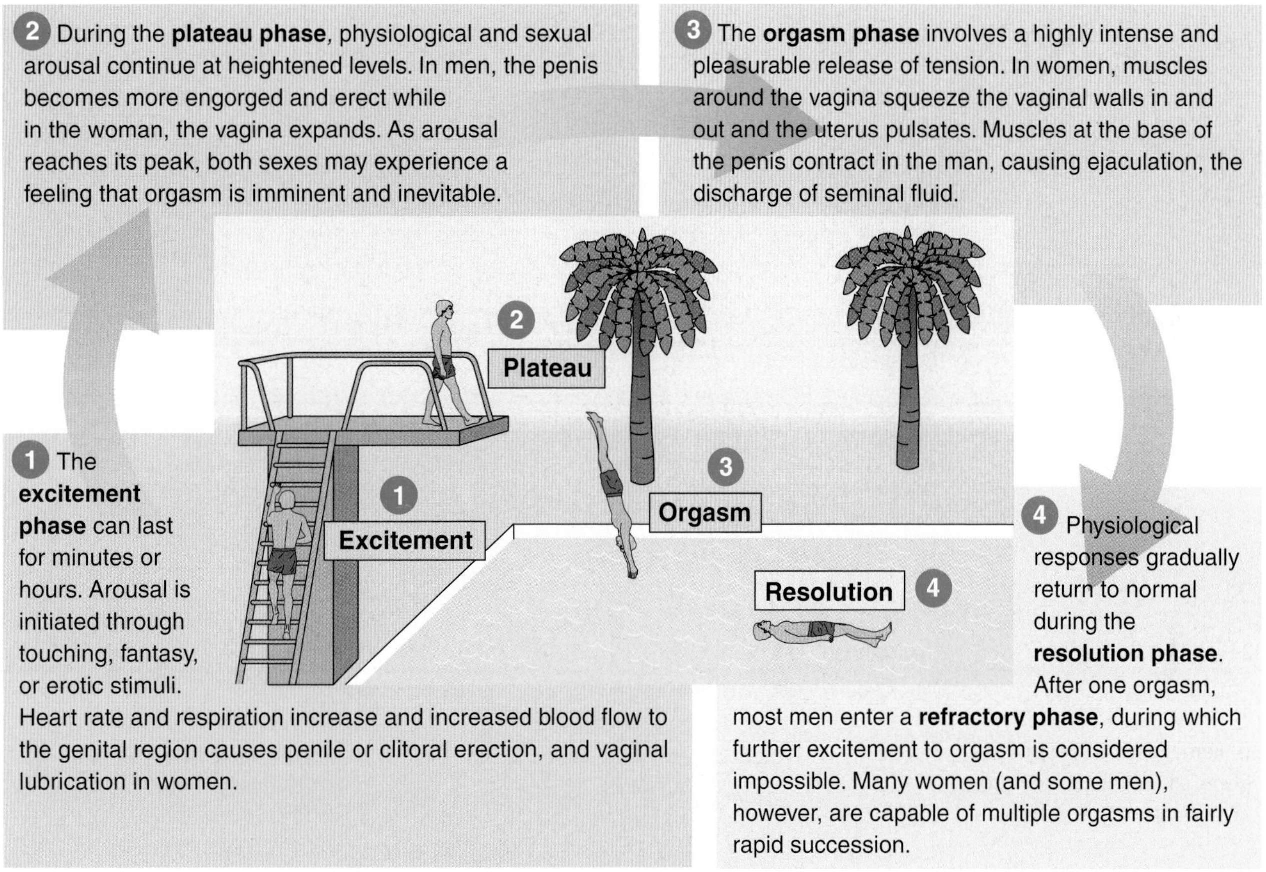

2 During the **plateau phase**, physiological and sexual arousal continue at heightened levels. In men, the penis becomes more engorged and erect while in the woman, the vagina expands. As arousal reaches its peak, both sexes may experience a feeling that orgasm is imminent and inevitable.
3 The **orgasm phase** involves a highly intense and pleasurable release of tension. In women, muscles around the vagina squeeze the vaginal walls in and out and the uterus pulsates. Muscles at the base of the penis contract in the man, causing ejaculation, the discharge of seminal fluid.
2 Plateau
1 Excitement
3 Orgasm
Resolution 4
1 The **excitement phase** can last for minutes or hours. Arousal is initiated through touching, fantasy, or erotic stimuli. Heart rate and respiration increase and increased blood flow to the genital region causes penile or clitoral erection, and vaginal lubrication in women.
4 Physiological responses gradually return to normal during the **resolution phase**. After one orgasm, most men enter a **refractory phase**, during which further excitement to orgasm is considered impossible. Many women (and some men), however, are capable of multiple orgasms in fairly rapid succession.

Maehr & Urdan, 2000). Our cultural values also affect achievement needs (Hofer et al., 2010; Vitoroulis et al., 2011; Xu & Barnes, 2011).

## Sexuality

Obviously, there is strong motivation to engage in sexual behavior: it's essential for the survival of our species, and it's also pleasurable. But **sexuality** includes much more than reproduction. For most humans (and some other animals), a sexual relationship fulfills many needs, including the need for connection, intimacy, pleasure, and the release of sexual tension.

William Masters and Virginia Johnson (1966) were the first to conduct laboratory studies on what happens to the human body during sexual activity. They attached recording devices to male and female volunteers and monitored or filmed their physical responses as they moved from nonarousal to orgasm and back to nonarousal. They labeled the bodily changes during this series of events a **sexual response cycle** (**Figure 11.10**). Researchers have further characterized

**sexual response cycle** Masters and Johnson's description of the four-stage bodily response to sexual arousal, which consists of excitement, plateau, orgasm, and resolution.

## Differences in male and female sexual response patterns • Figure 11.11

Although the overall pattern of sexual response is similar in both sexes, there is more variation in specific patterns among women.

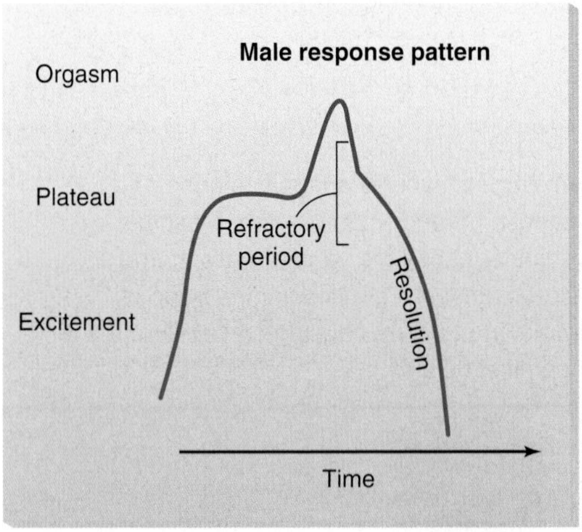


**a.** Immediately after orgasm, men generally enter a refractory period, which lasts from several minutes up to a day.

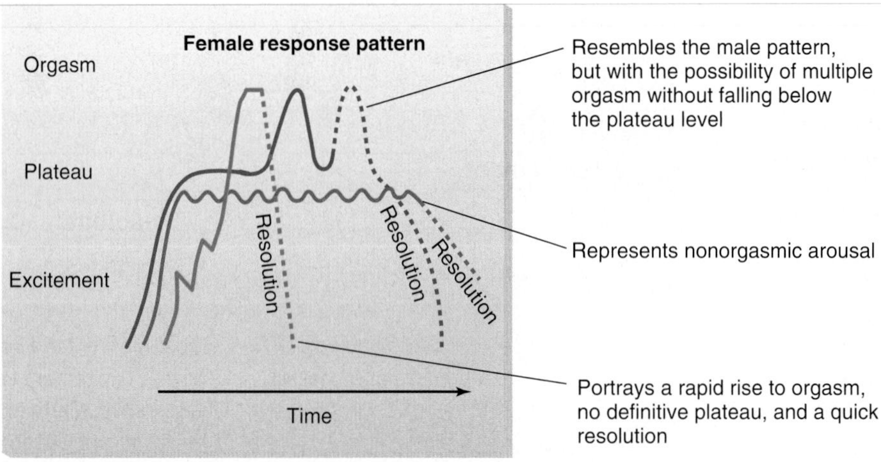


**b.** Female sexual responses generally follow one or more of three basic patterns.

Graphs used by permission: Master and Johnson's Sexual Response Cycle from Masters, W. H., & Johnson, V. E. (1966). *Human sexual response*. Boston: Little, Brown.

differences between sexual response patterns in men and women (**Figure 11.11**).

**Sexual orientation** Of course, an important part of people's sexuality is the question of to whom they are sexually attracted. What leads some people to be homosexual and others to be heterosexual? Unfortunately, the roots of **sexual orientation** are poorly understood. However, most studies suggest that genetics and biology play the dominant role (Bailey, Dunne, & Martin, 2000; Bao & Swaab, 2011; Ellis et al., 2008; Hines, 2011; Jannini et al., 2010; LeVay, 2011).

> **sexual orientation** Primary erotic attraction toward members of the same sex (homosexual, gay or lesbian), both sexes (bisexual), or the other sex (heterosexual).

Studies on identical and fraternal twins and adopted siblings found that if one identical twin was gay, 48 to 65% of the time so was the second twin (Hyde, 2005; Långstrom et al., 2010; Moutinho, Pereira, & Jorge, 2011). The rate was 26 to 30% for fraternal twins and 6 to 11% for brothers and sisters who were adopted into one's family. Estimates of homosexuality in the general population run between 2 and 10%. (Note that the actual percentages may be much larger, given the prejudice and misconceptions surrounding sexual orientation—a topic in our next section.)

Research with rats and sheep hints that prenatal hormone levels may also affect fetal brain development and sexual orientation (Bagermihl, 1999; Roselli, Reddy, & Kaufman, 2011). However, the effects of hormones on human sexual orientation are unknown. Furthermore, no well-controlled study has ever found a difference in adult hormone levels between heterosexuals and gays and lesbians (Hall & Schaeff, 2008; Hines, 2011; LeVay, 2003, 2011).

Scientific research has disproved several widespread myths and misconceptions about homosexuality (Bergstrom-Lynch, 2008; Boysen & Vogel, 2007; Drucker, 2010; LeVay, 2003, 2011) (see *Myth Busters*). Although mental health authorities long ago discontinued labeling homosexuality as a mental illness, it continues to be a divisive societal issue. Gays, lesbians, bisexuals, and transgendered people often confront **sexual prejudice**, and many endure verbal and physical attacks; disrupted family and peer relationships; and high rates of anxiety, depression, and suicide (Espelage et al., 2008; Hyde & DeLamater, 2011; Newcomb & Mustanski, 2010; Rivers, 2011). Sexual prejudice is a socially reinforced phenomenon, not an individual pathology (as the older term *homophobia* implies).

> **sexual prejudice** Negative attitudes toward an individual because of her or his sexual orientation.

## Extrinsic Versus Intrinsic Motivation

Should parents reward children for getting good grades? Many psychologists are concerned about the widespread practice of giving external, or extrinsic, rewards to motivate behavior (e.g., Anderman & Dawson, 2011; Deci & Moller, 2005; Gunderman & Kamer, 2011). They're concerned that providing such **extrinsic motivation** will seriously affect the individual's personal, **intrinsic motivation.** Participation in sports and hobbies, like swimming or playing guitar, is usually intrinsically motivated. Unfortunately, for many students and workers, going to school or going to work is primarily extrinsically motivated.

> **extrinsic motivation** Motivation based on obvious external rewards or threats of punishment.
>
> **intrinsic motivation** Motivation resulting from personal enjoyment of a task or activity.

## MYTH BUSTERS

THE PLANNER

### SEXUAL ORIENTATION

**DISPROVED "THEORIES"**

***Myth #1: Seduction theory*** Gays and lesbians were seduced as children by adults of their own sex.

***Myth #2: "By default" theory*** Gays and lesbians were unable to attract partners of the other sex or have experienced unhappy heterosexual experiences.

***Myth #3: Poor parenting theory*** Sons become gay because of domineering mothers and weak fathers. Daughters become lesbians because of weak or absent mothers and having only fathers as their primary role model.

***Myth #4: Modeling theory*** Children raised by gay and lesbian parents usually end up adopting their parents' sexual orientation.

Lisa Poole/© AP/Wide World Photos

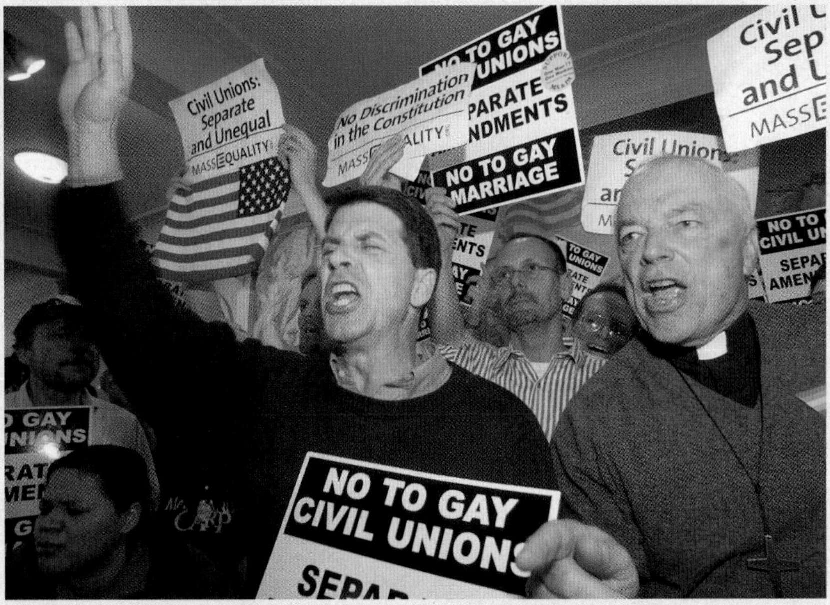


Debate, legislation, and judicial action surrounding gay marriage have brought attitudes about homosexuality into full public view.

# Psychology InSight

## Extrinsic versus intrinsic motivation

**• Figure 11.12**

When we perform a task for no ulterior purpose, we are intrinsically motivated ("because I like it"; "because it's fun"). In contrast, we are extrinsically motivated by rewards like praise or payment. Some types of extrinsic rewards actually decrease enjoyment and hamper performance.

**a.** Researchers found that a group of preschoolers spent more time drawing when they were intrinsically motivated (no reward) than when they were either promised an extrinsic reward or were given an unexpected reward when they were done (Lepper, Greene, & Nisbett, 1973).

© Christopher Futcher/iStockphoto

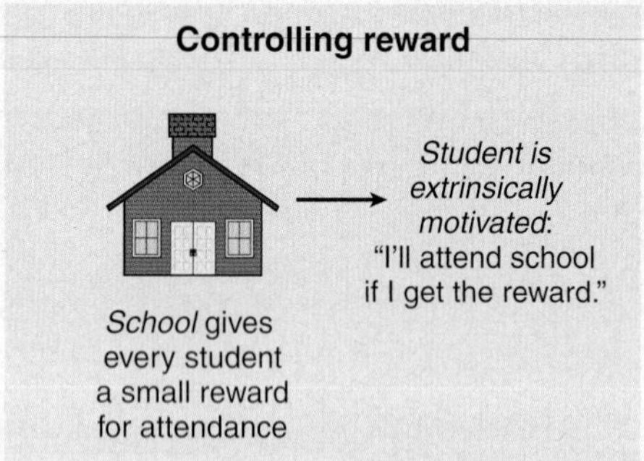

Controlling reward
Student is extrinsically motivated: "I'll attend school if I get the reward."
School gives every student a small reward for attendance

Approval reward
Student is extrinsically motivated: "I'll get good grades to get their approval."
Parents: "We'll be very happy if you get A's like our neighbor's boy."

**b.** If extrinsic rewards are used to control or gain approval—for example, when schools pay all students for simple attendence or when parents give children approval or privileges for achieving good grades—they often decrease motivation (Eisenberger & Armeli, 1997; Eisenberger & Rhoades, 2002; Houlfort, 2006).

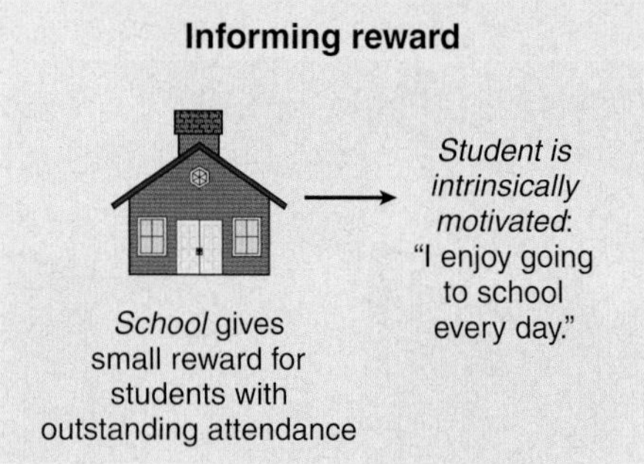

**c.** Extrinsic rewards can be motivating if they are used to produce intrinsic motivation. For example, when superior performance is recognized with an award or a special "no strings attached" treat (Deci, 1995), it may increase enjoyment.

Research has shown that people who are given extrinsic rewards (money, praise, or other incentives) for an intrinsically satisfying activity, such as watching TV, playing cards, or even engaging in sex, often lose enjoyment and interest and may decrease the time spent on the activity (Hennessey & Amabile, 1998; Kohn, 2000; Moneta & Siu, 2002) (**Figure 11.12**).

### CONCEPT CHECK

STOP

1. **What** are the symptoms of anorexia nervosa?
2. **What** traits characterize people with high-achievement motivation?
3. **What** is sexual prejudice?
4. **Why** does providing extrinsic rewards sometimes decrease enjoyment and hamper performance?

# Components and Theories of Emotion

## LEARNING OBJECTIVES

**RETRIEVAL PRACTICE** While reading the upcoming sections, respond to each Learning Objective in your own words. Then compare your responses with those in Appendix B.

1. **Describe** the biological, cognitive, and behavioral components of emotion.
2. **Compare** the three major theories of emotion.
3. **Explain** cultural similarities and differences in emotion.
4. **Review** the problems with relying on polygraph testing as a "lie detector."

Emotions play an important role in our lives. They color our dreams, memories, and perceptions (Kalat & Shiota, 2012). When they are disordered, they contribute significantly to psychological problems. But what do we mean by the term **emotion**? In everyday usage, we describe emotions in terms of feeling states—we feel "thrilled" when our political candidate wins an election, "defeated" when our candidate loses, and "miserable" when our loved ones reject us. Obviously, what you and I mean by these terms, or what we individually experience with various emotions, can vary greatly among individuals.

**emotion** A subjective feeling that includes arousal (heart pounding), cognitions (thoughts, values, and expectations), and expressions (frowns, smiles, and running).

## Three Components of Emotion

Psychologists define and study emotion according to three basic components—*biological*, *cognitive*, and *behavioral*.

**Biological (arousal) component** Internal physical changes occur in our bodies whenever we experience an emotion. Imagine walking alone on a dark street when someone jumps from behind a stack of boxes and starts running toward you. How would you respond? Like most people, you would undoubtedly interpret the situation as threatening and would run. Your predominant emotion, fear, would involve several physiological reactions, such as increased heart rate and blood pressure, perspiration, and goose bumps (piloerection). Such biological reactions are controlled by certain brain structures and by the autonomic branch of the nervous system (ANS).

Our emotional experiences appear to result from important interactions between several areas of the brain, most particularly the *cerebral cortex* and *limbic system* (Langenecker et al., 2005; LeDoux, 2002; Panksepp, 2005). As we discuss in Chapter 2, the cerebral cortex, the outermost layer of the brain, serves as our body's ultimate control and information-processing center, including our ability to recognize and regulate our emotions.

Studies of the limbic system, located in the innermost part of the brain, have shown that one area of the limbic system, the *amygdala*, plays a key role in emotion—especially fear. It sends signals to the other areas of the brain, causing increased heart rate and all the other physiological reactions related to fear.

Interestingly, emotional arousal sometimes occurs without conscious awareness. According to psychologist Joseph LeDoux (1996, 2002, 2007), when sensory inputs arrive in the *thalamus* (our brain's sensory switchboard), it sends separate messages up to the cortex (which "thinks" about the stimulus) and the amygdala (which immediately activates the body's alarm system). Although this dual pathway occasionally leads to "false alarms," such as mistaking a stick for a snake, LeDoux believes it is a highly adaptive warning system essential to our survival. He states that "the time saved by the amygdala in acting on the thalamic interpretation, rather than waiting for the cortical input, may be the difference between life and death."

As important as the brain is to emotion, it is the *autonomic nervous system* (ANS, Chapter 2) that produces

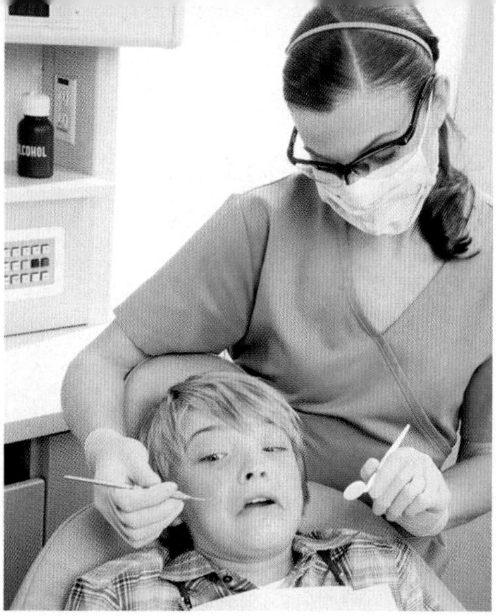

Upper Cut Images/SuperStock, Inc.

| Sympathetic | | Parasympathetic |
|---|---|---|
| Pupils dilated | **Eyes** | Pupils constricted |
| Decreased saliva | **Mouth** | Increased saliva |
| Vessels constricted (skin cold and clammy) | **Skin** | Vessels dilated (normal blood flow) |
| Respiration increased | **Lungs** | Respiration normal |
| Increased heart rate | **Heart** | Decreased heart rate |
| Increased epinephrine and norepinephrine | **Adrenal glands** | Decreased epinephrine and norepinephrine |
| Decreased motility | **Digestion** | Increased motility |

**Emotion and the autonomic nervous system • Figure 11.13**

During emotional arousal, the sympathetic branch of the autonomic nervous system prepares the body for fight or flight. (The hormones epinephrine and norepinephrine keep the system under sympathetic control until the emergency is over.) The parasympathetic branch returns the body to a more relaxed state (homeostasis).

the obvious signs of arousal. These largely automatic responses result from interconnections between the ANS and various glands and muscles (**Figure 11.13**).

**Cognitive (thinking) component** Emotional reactions are very individual: what you experience as intensely pleasurable may be boring or aversive to another. To study the cognitive (thought) component of emotions, psychologists typically use self-report techniques, such as paper-and-pencil tests, surveys, and interviews. However, people are sometimes unable or unwilling to accurately describe (or remember) their emotional states. For these reasons, our cognitions about our own and others' emotions are difficult to measure scientifically. This is why many researchers supplement participants' reports on their emotional experiences with methods that assess emotional experience indirectly (for example, by measuring physiological response or behavior).

**Behavioral (expressive) component** Emotional expression is a powerful form of communication, and facial expressions may be our most important form of emotional communication. Researchers have developed sensitive techniques to measure subtleties of feeling and to differentiate honest expressions from fake ones. Perhaps most interesting is the difference between the **social smile** and the **Duchenne smile** (named after French anatomist Duchenne de Boulogne, who first described it in 1862) (**Figure 11.14**). In a false, social smile, our voluntary cheek muscles are pulled back, but our eyes are unsmiling. Smiles of real pleasure, on the other hand, use the muscles not only around the cheeks but also around the eyes.

**Duchenne smile • Figure 11.14**

People who show a Duchenne, or real, smile (**a**) and laughter elicit more positive responses from strangers and enjoy better interpersonal relationships and personal adjustment than those who use a social smile (**b**) (Keltner, Kring, & Bonanno, 1999; Prkachin & Silverman, 2002).

a.

Courtesy Karen Huffman

b.

Courtesy Karen Huffman

The Duchenne smile illustrates the importance of nonverbal means of communicating emotion. We all know that people communicate in ways other than speaking or writing. However, few people recognize the full importance of nonverbal signals. Imagine yourself as an interviewer. Your first job applicant greets you with a big smile, full eye contact, a firm handshake, and an erect, open posture. The second applicant doesn't smile, looks down, offers a weak handshake, and slouches. Whom do you think you will hire?

Psychologist Albert Mehrabian would say that you're much less likely to hire the second applicant due to his or her "mixed messages." His research suggests that when we're communicating feelings or attitudes, and the verbal and nonverbal dimensions don't match, the receiver trusts the predominant form of communication, which is about 93% nonverbal (the way the words are said and facial expression) versus the literal meaning of the words (Mehrabian, 1968, 1971, 2007).

Unfortunately, Mehrabian's research is often overgeneralized, and many people misquote him as saying that "over 90% of communication is nonverbal." Clearly, if a police officer says, "Put your hands up," his or her verbal words might carry 100% of the meaning. However, when we're confronted with a mismatch between verbal and nonverbal, it is safe to say that we pay far more attention to the nonverbal because we believe that it more often tells us what someone is really thinking or feeling. The importance of nonverbal communication, particularly facial expressions, is further illustrated by the popularity of "smileys," and other emotional symbols, in our everyday e-mail and text messages.

## Three Major Theories of Emotion

Researchers generally agree on the three components of emotion (biological, cognitive, and behavioral), but there is less agreement on *how* we become emotional (Kalat & Shiota, 2012). The major competing theories are the *James-Lange, Cannon-Bard*, and *Schachter and Singer's two-factors* (**Study Organizer 11.2**).

Imagine you're walking in the woods and suddenly see a coiled snake on the path next to you. What emotion would you experience? Most people would say they would be very afraid, but why? Common sense tells us that our hearts pound and we tremble when we're afraid, or that

**Study Organizer 11.2** Comparing three major theories of emotion    ✓ THE PLANNER

**Stimulus**

**James-Lange theory**
Perception of an environmental stimulus (snake) triggers physiological arousal, which we experience as a felt emotion (fear).

⟹ [Physiological arousal] ⟹ [Emotion: **FEAR**]

**Cannon-Bard theory**
Perception of an environmental stimulus (snake) triggers the thalamus to send a simultaneous message that activates a physiological arousal at the same time as the felt emotion (fear).

⟹ [Thalamus relays information] ⟹ [Physiological arousal]
⟹ [Emotion: **FEAR**]

**Schachter and Singer's two-factor theory**
Perception of an environmental stimulus (snake) triggers physiological arousal, which we cognitively appraise and label ("I'm afraid"). Then, the appraisal (label) is felt as the emotion (fear).

⟹ [Physiological arousal] ⟹ [Label ("I'm scared")] ⟹ [Emotion: **FEAR**]

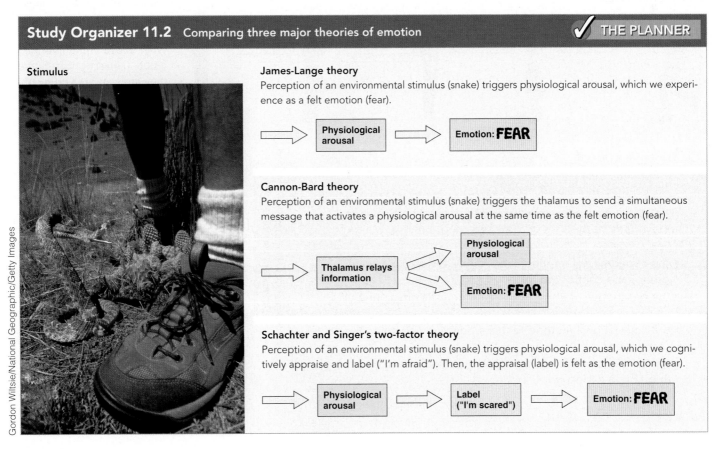

Gordon Wiltsie/National Geographic/Getty Images

Psychological Science

THE PLANNER

# Schachter and Singer's Classic Study

In their classic study (1962), Schachter and Singer gave research participants injections of epinephrine, a hormone/neurotransmitter that produces feelings of arousal, or saline shots (a placebo). How subjects responded suggested that arousal could be labeled as happiness or anger, depending on the context. Schachter and Singer's research demonstrated that emotion is determined by two factors: physiological arousal and cognitive appraisal (labeling).

Participants were first told that this was a study of how certain vitamins affect visual skills, and then they were asked to give permission to be injected with a small shot of the vitamin "Suproxin." This injection actually was a shot of the hormone/neurotransmitter epinephrine (or adrenalin), which triggers feelings of arousal such as racing heart, flushed skin, and trembling hands. Those participants who gave permission were then divided into four groups and given injections (**Figure a**).

Each participant was then placed in a room with either a "happy" or an "angry" trained confederate (who was actually an accomplice of the experimenter). Both the participant and confederate were told that before they could take the supposed vision test, they needed to complete a questionnaire and allow time for the drug to take effect. As predicted, when participants were placed in a room with the happy, euphoric confederate, they rated themselves as happy. When placed with the angry, irritated confederate, they reported feeling angry. Although the subjects were feeling similar effects of the epinephrine, they labeled those effects differently (happy or angry) depending on the external cues in their environment (**Figure b**). This effect on emotions was most apparent in the groups that were not told about the true effects of the injections, who engaged in more acts of euphoria (**Figure c**). Why? Those in the informed group correctly attributed their arousal to the drug rather than to the environment.

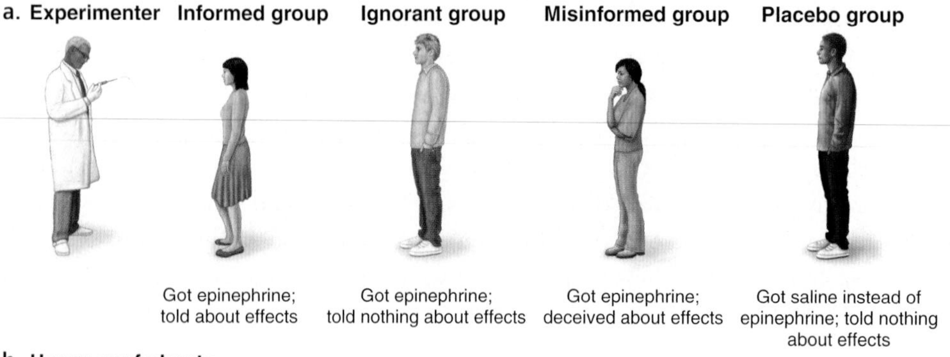

Ignorant group

Placebo group

Got epinephrine; told nothing about effects

Got saline instead of epinephrine; told nothing about effects

b. Happy confederate
I feel strange, it must be the injection
I feel strange, sort of "happy" like that fella
I feel strange. I guess I feel very happy like that guy over there
I feel strange, sort of happy like that fella


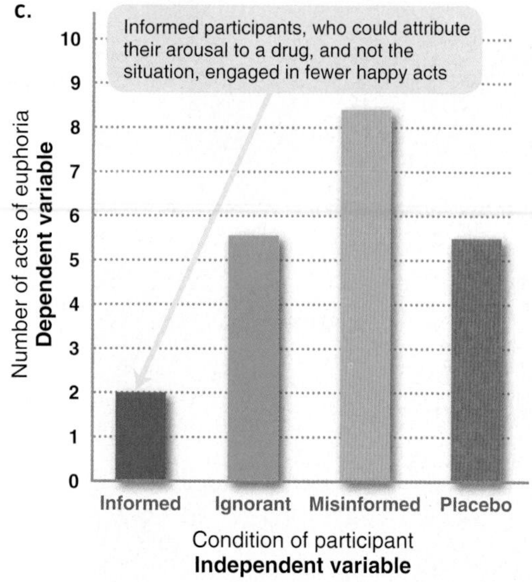


Informed participants, who could attribute their arousal to a drug, and not the situation, engaged in fewer happy acts

## Identify the Research Method

1. What is the most likely research method used for Schacter and Singer's study?
2. If you chose
   - the experimental method, label the IV, DV, experimental group, and control group.
   - the descriptive method, is this a naturalistic observation, survey, or case study?
   - the correlational method, is this a positive, negative, or zero correlation?
   - the biological method, identify the specific research tool (e.g., brain dissection, CT scan).

(Check your answers in Appendix C.)

we cry when we're sad. But according to the **James-Lange theory** felt emotions begin with physiological arousal of the autonomic nervous system (ANS) (Chapter 2). This arousal (e.g., pounding heart, breathlessness, trembling all over) then causes us to experience the emotion we call "fear." Contrary to popular opinion, James wrote: "We feel sorry because we cry, angry because we strike, afraid because we tremble" (James, 1890).

In contrast, the **Cannon-Bard theory** proposed that arousal and emotion occur separately, but simultaneously. Following perception of an emotion-provoking stimulus, the thalamus (a subcortical brain structure) sends two simultaneous messages: one to the ANS, which causes physiological arousal, and one to the brain's cortex, which causes awareness of the felt emotion.

Finally, **Schachter and Singer's two-factor theory** suggests that our emotions start with physiological arousal followed by a conscious, cognitive appraisal. We then look to external cues from the environment and to others around us to find a label and explanation for the arousal. Therefore, if we cry at a wedding, we label our emotion as joy or happiness. If we cry at a funeral, we label the emotion as sadness (see *Psychological Science*).

Schachter and Singer's two-factor theory may have important practical implications. Depending on the cues present in our environment, we apparently can interpret the exact same feelings of arousal in very different ways. For example, if you're shy or afraid of public speaking, try interpreting your feelings of nervousness as the result of too much coffee or the heating in the room. Some research on this type of "misattribution of arousal" has had positive outcomes (Chafin, Christenfeld, & Gerin, 2008; Inzlicht & Al-Khindi, 2012; Thornton & Tizard, 2010).

Later research on emotion has supported or discounted parts of all three of these major theories and added some new ideas. For example, as discussed earlier, in the behavioral component of emotion, facial expressions (like the Duchenne smile) help us express our emotions. And, according to the **facial-feedback hypothesis**, movements of our facial muscles produce and/or intensify our subjective experience of emotion (**Figure 11.15**). More specifically, sensory input (seeing a snake) is first routed to subcortical areas of the brain that activate facial movements. These facial changes then initiate and intensify emotions (Adelmann & Zajonc, 1989; Ceschi & Scherer, 2001; Dimberg & Söderkvist, 2011; Neal & Chartrand, 2011).

Other research emphasizes specific pathways between the *thalamus, amygdala,* and *cerebral cortex* to help explain differing emotional responses, such as the almost instantaneous emotional response to some stimuli versus our slower-developing responses (Debiec & LeDoux, 2006; Lazarus, 1991, 1998; LeDoux, 1996, 2003, 2007;

## Testing the facial-feedback hypothesis • Figure 11.15

Hold a pen or pencil between your teeth with your mouth open. Spend about 30 seconds in this position. How do you feel? According to research, pleasant feelings are more likely when teeth are showing than when they are not. *Source*: Adapted from Strack, Martin, & Stepper, 1988.

Mark Owens/John Wiley & Sons, Inc.

Mark Owens/John Wiley & Sons, Inc.

# WHAT A PSYCHOLOGIST SEES

THE PLANNER

## Fast and Slow Pathways for Fear

A hiker who stepped near a snake on the path would have an almost instantaneous response to jump away to safety. Only a few seconds later would she really be consciously aware of what kind of danger the snake posed. The hiker would just feel lucky to have escaped a snake bite, but psychologists know that this two-phase fear response is a result of two different pathways for sensory data in the brain.

When sensory input arrives at the thalamus, the thalamus sends it along a fast route directly to the amygdala (the red line), as well as along a slower, more indirect route to the visual cortex (the blue line). Can you see how this speedy, direct route allows us to quickly respond to a feared stimulus (like the snake) even before we're consciously aware of our emotions or behaviors? In contrast, the indirect route, involving the visual cortex, provides more detailed information that allows us to consciously evaluate the danger of this particular snake and our most appropriate response.

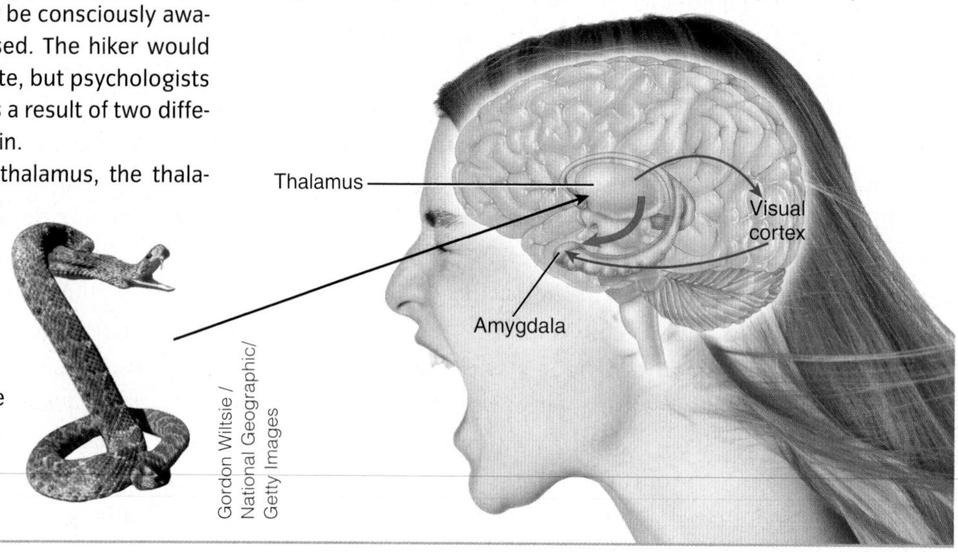


Gordon Wiltsie / National Geographic/ Getty Images

Masterfile

Zajonc, 1984) (see *What a Psychologist Sees*). Both our fast-acting and slower-developing emotional responses seem to depend primarily on pathways between the *thalamus*, which serves as a relay station for incoming information, the *amygdala*, which directly influences emotions, and the *cerebral cortex*, which is responsible for higher order mental processes (Chapter 2).

## Culture, Evolution, and Emotion

Are emotions the same across all cultures? Given the seemingly vast array of emotions within our own culture, it may surprise you to learn that some researchers believe that all our feelings can be condensed into 7 to 10 culturally universal emotions (**Table 11.2**). These researchers hold that other emotions, such as love, are

**The basic human emotions Table 11.2**

| Carroll Izard | Paul Ekman and Wallace Friesen | Robert Plutchik | Silvan Tomkins |
|---|---|---|---|
| Fear | Fear | Fear | Fear |
| Anger | Anger | Anger | Anger |
| Disgust | Disgust | Disgust | Disgust |
| Surprise | Surprise | Surprise | Surprise |
| Joy | Happiness | Joy | Enjoyment |
| Shame | – | – | Shame |
| Contempt | Contempt | – | Contempt |
| Sadness | Sadness | Sadness | – |
| Interest | – | Anticipation | Interest |
| Guilt | – | – | – |
| – | – | Acceptance | – |
| – | – | – | Distress |

## Plutchik's wheel of emotions • Figure 11.16

Robert Plutchik (1984, 1994, 2000) suggested that primary emotions (inner circle) combine to form secondary emotions (located outside the circle). Plutchik also found that emotions that lie next to each other are more alike than those that are farther apart. Can you provide examples of this phenomenon from your own life?

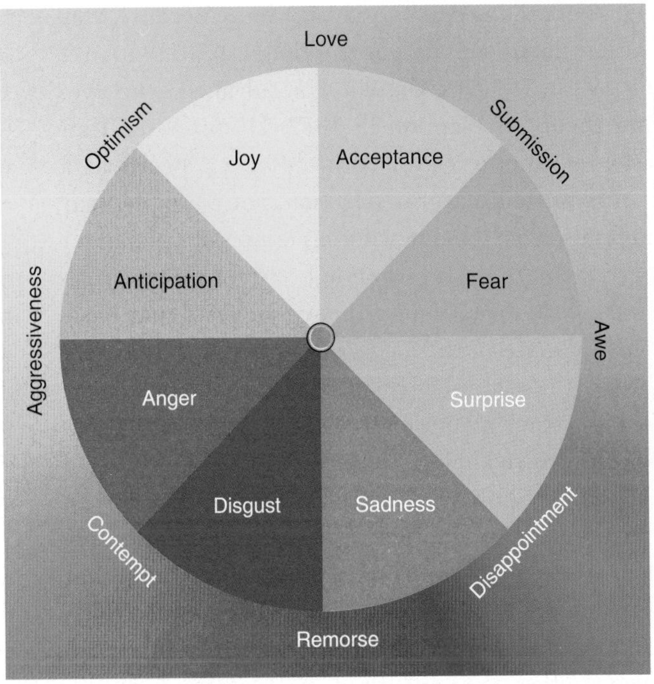


Plutchik, Emotion: A Psychoevolutionary Synthesis, 1st Edition, © 1979, Adapted by permission of Pearson Education, Inc., Upper Saddle River, NJ.

simply combinations of primary emotions with variations in intensity (**Figure 11.16**).

Some research indicates that people in all cultures express and recognize the basic emotions in essentially the same way (Berry et al., 2011; Ekman, 1993, 2004; Matsumoto, 1992, 2000, 2010; Said, Haxby, & Todorov, 2011). In other words, across cultures, a frown is recognized as a sign of displeasure and a smile as a sign of pleasure (**Figure 11.17**).

## Can you identify these emotions? • Figure 11.17

Most people can reliably identify at least six basic emotions—happiness (shown in the photo on the left), surprise, anger, sadness, fear, and disgust. On occasion, our facial expressions don't seem to match our presumed emotions, as is the case in the photo on the right. This woman has just won an important award, yet she looks sad rather than happy.

Rubberball/Getty Images, Inc.

Kevork Djansezian/©AP/Wide World Photos

From an evolutionary perspective, the idea of universal facial expressions makes adaptive sense because they signal others about our current emotional state (Ekman & Keltner, 1997). Charles Darwin first advanced the evolutionary theory of emotion in 1872. He proposed that expression of emotions evolved in different species as a part of survival and natural selection. For example, fear helps animals avoid danger, whereas expressions of anger and aggression are useful when fighting for mates or resources. Modern evolutionary theory suggests that basic emotions originate in the *limbic system*. Given that higher brain areas (the cortex) developed later than the subcortical limbic system, evolutionary theory proposes that basic emotions evolved before thought.

Studies with infants provide further support for an evolutionary basis for emotions. For example, infants only a few hours old show distinct expressions of emotion that closely match adult facial expressions (Field et al., 1982; Meltzoff & Moore, 1977, 1985, 1994). And all infants, even those who are born deaf and blind, show similar facial expressions in similar situations (Field et al., 1982; Gelder et al., 2006). In addition, a recent study showed that families may have characteristic facial expressions, shared even by family members who have been blind from birth (Peleg et al., 2006). This collective evidence points to a strong biological, evolutionary basis for emotional expression and decoding.

So how do we explain cultural differences in emotions? Although we all seem to share similar facial expressions for some emotions, each culture has its own **display rules** governing how, when, and where to express emotions (Ekman, 1993, 2004; Fok et al., 2008; Koopmann-Holm & Matsumoto, 2011; Matsumoto & Hwang, 2011). For instance, parents pass along their culture's specific display rules when they respond angrily to some emotions in their children, when they are sympathetic to others, and when they simply ignore an expression of emotion. Public physical contact is also governed by display rules. North Americans and Asians are less likely than people in other cultures to touch each other. For example, only the closest family and friends might hug in greeting or farewell. In contrast, Latin Americans and Middle Easterners often embrace and hold hands as a sign of casual friendship (Axtell, 2007).

## The Polygraph as a Lie Detector

We've discussed the three major theories of emotion, and how emotions are affected by culture and evolution. Now, we turn our attention to one of the hottest, and most controversial, topics in emotion research—the **polygraph**.

> **polygraph** An instrument that measures sympathetic arousal (heart rate, respiration rate, blood pressure, and skin conductivity) to detect emotional arousal, which in turn supposedly reflects lying versus truthfulness.

Traditional polygraph tests are based on the theory that when people lie, they feel guilty and anxious. Special sensors supposedly detect these testable emotions by measuring sympathetic and parasympathetic nervous system responses (Grubin, 2010) (**Figure 11.18**).

The problem is that lying is only loosely related to anxiety and guilt. Some people become nervous even when telling the truth, whereas others remain calm when deliberately lying. A polygraph cannot tell which emotion is being felt (nervousness, excitement, sexual arousal, etc.) or whether a response is due to emotional arousal or something else. One study found that people could affect the outcome of a polygraph by about 50% simply by pressing their toes against the floor or biting their tongues (Honts & Kircher, 1994). In fact, although proponents claim that polygraph tests are 90% accurate or better, actual tests show error rates ranging between 25 and 75% (DeClue, 2003; Handler et al., 2009; Iacono, 2008).

For these reasons, most judges and scientists have serious reservations about using polygraphs as lie detectors (DeClue, 2003). Scientific controversy and public concern led the U.S. Congress to pass a bill that severely restricts the use of polygraphs in the courts, in government, and in private industry.

Tens of millions to hundreds of millions of dollars have been spent on new and improved lie-detection techniques (Kluger & Masters, 2006). Perhaps the most promising new technique is the use of brain scans like the functional magnetic resonance imaging (fMRI).

Unfortunately, each of these new lie-detection techniques has serious shortcomings and unique problems. Researchers have questioned their reliability and validity, while civil libertarians and judicial scholars question their ethics and legalities.

## Polygraph testing • Figure 11.18

Polygraph testing is based on the assumption that when we lie we feel guilty, fearful, or anxious.

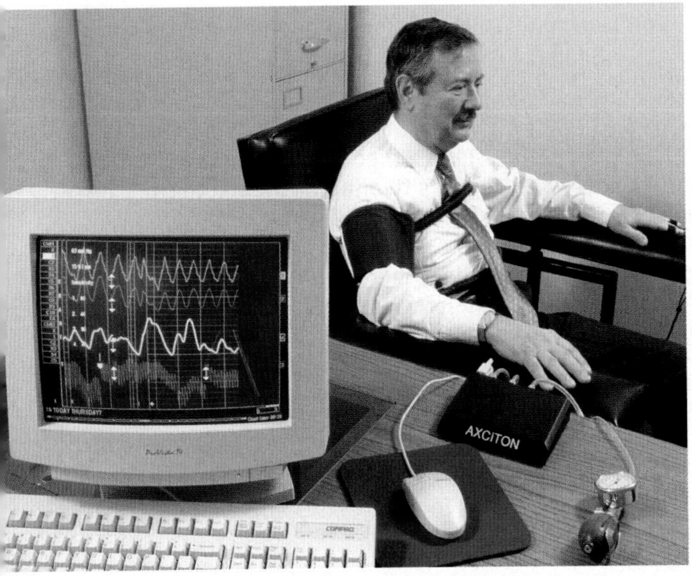
AXCITON

a. During a standard polygraph test, a band around the person's chest measures breathing rate, a cuff monitors blood pressure, and finger electrodes measure sweating, or galvanic skin response (GSR).

b. Note how the GSR rises sharply in response to the question, "Have you ever taken money from this bank?"

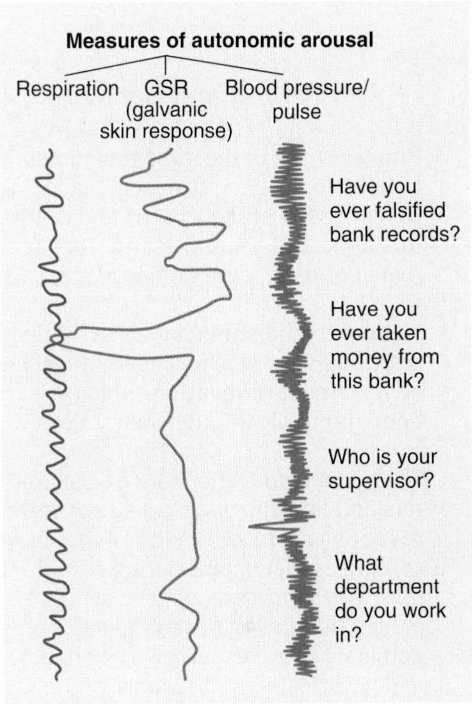
Measures of autonomic arousal
Respiration
GSR (galvanic skin response)
Blood pressure/ pulse
Have you ever falsified bank records?
Have you ever taken money from this bank?
Who is your supervisor?
What department do you work in?

### CONCEPT CHECK

1. **How** does the autonomic nervous system (ANS) respond to fearful stimuli?
2. **What** are the differences between the James-Lange theory and the Cannon-Bard theory of emotion?
3. **Why** do some researchers believe that the basic emotions are consistent from one culture to the next?
4. **What** is the connection between lying and emotional arousal?

# Summary

THE PLANNER

Maslow's hierarchy of needs • Figure 11.5

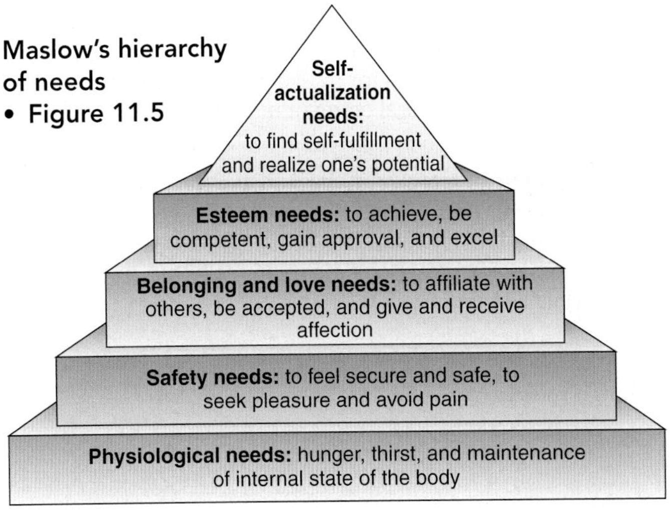
Self-actualization needs: to find self-fulfillment and realize one's potential
Esteem needs: to achieve, be competent, gain approval, and excel
Belonging and love needs: to affiliate with others, be accepted, and give and receive affection
Safety needs: to feel secure and safe, to seek pleasure and avoid pain
Physiological needs: hunger, thirst, and maintenance of internal state of the body

## 1 Theories of Motivation 284

- The three major biological theories of **motivation** involve **instincts**, drives (produced by the body's need for **homeostasis**), and arousal, or need for novelty and complexity. For people who are high in sensation seeking, the need for stimulation is especially high.
- Psychological theories emphasize the role of incentives, attributions, and expectancies in cognition.
- Maslow's **hierarchy of needs** theory (shown in the diagram) takes a biopsychosocial approach. It prioritizes needs, with survival needs at the bottom and social and spiritual needs at the top. Although the theory has made important contributions, some critics argue that it is poorly researched and biased toward Western individualism.

## 2 Motivation and Behavior 289

- Hunger is one of the strongest motivational drives, and both biological (the stomach, biochemistry, the brain) and psychosocial (stimulus cues and cultural conditioning) factors affect hunger and eating. These same factors play a role in **obesity**, **anorexia nervosa**, and **bulimia nervosa**.

- The key to understanding what motivates high-achieving individuals lies in a high need for achievement (nAch), or **achievement motivation**, which is learned in early childhood, primarily through interactions with parents.

- The human motivation for sex is extremely strong. Masters and Johnson first studied and described the **sexual response cycle**, the series of physiological and sexual responses that occurs during sexual activity. Other sex research has focused on the roots of **sexual orientation**, and most studies suggest that genetics and biology play the dominant role. Sexual orientation remains a divisive issue, and gays, lesbians, bisexuals, and transgendered people often confront **sexual prejudice**.

- Providing **extrinsic motivation** (money, praise, or other incentives) for an intrinsically satisfying activity can undermine people's enjoyment and interest (**intrinsic motivation**) for the activity, as shown in the graph. This is especially true when extrinsic motivation is used to control—for example, when parents give children money or privileges for achieving good grades.

**Extrinsic versus intrinsic motivation • Figure 11.12**

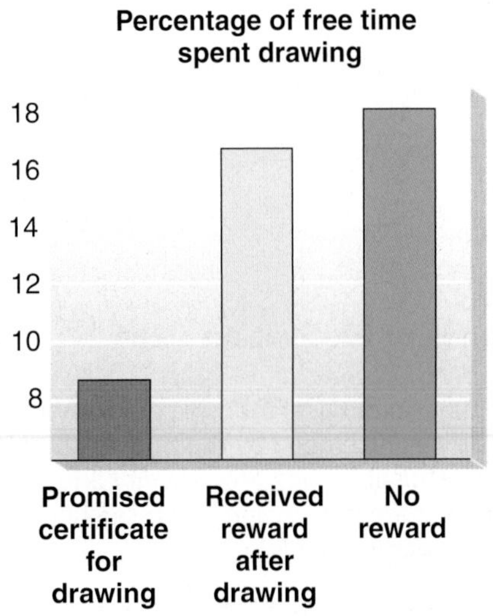


## 3 Components and Theories of Emotion 299

- **Emotion** has biological (brain and autonomic nervous system), cognitive (thoughts), and behavioral (expressive) components.

- The major theories of how the components of emotion interact are the **James-Lange theory, Cannon-Bard theory** and **Schachter and Singer's two-factor theory.** Each emphasizes different sequences or aspects of the three elements, and each has its limits and contributions to our understanding of emotion. Other research emphasizes how different pathways in the brain trigger faster and slower emotional responses, as shown in the diagram.

**What a Psychologist Sees: Fast and Slow Pathways for Fear**

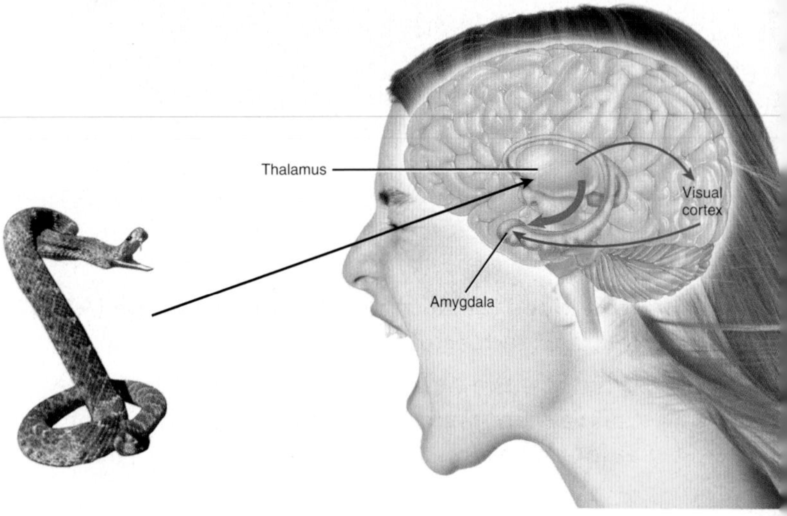


- Some researchers believe that there are 7 to 10 basic, universal emotions that are shared by people of all cultures, and that people in all cultures express and recognize the basic emotions in essentially the same way, supporting the evolutionary theory of emotion. Studies with infants provide further support. Although we all seem to share similar facial expressions for some emotions, **display rules** for emotions (as well as for public physical contact, another aspect of the behavioral element of emotion) vary across cultures.

- **Polygraph** tests attempt to detect lying by measuring physiological signs of guilt and anxiety. However, because lying is only loosely related to these emotions, most judges and scientists have serious reservations about using polygraphs as lie detectors.

# Key Terms

**RETRIEVAL PRACTICE** Write a definition for each term before turning back to the referenced page to check your answer.

- achievement motivation 294
- anorexia nervosa 292
- attribution 288
- bulimia nervosa 292
- Cannon-Bard theory 303
- cognitive theory 288
- display rules 306
- drive-reduction theory 285
- Duchenne smile 300
- emotion 299
- excitement phase 295
- expectancy 288
- extrinsic motivation 297
- facial-feedback hypothesis 303
- hierarchy of needs 288
- homeostasis 286
- incentive theory 287
- instinct 285
- intrinsic motivation 297
- James-Lange theory 303
- motivation 284
- obesity 292
- optimal-arousal theory 286
- orgasm phase 295
- plateau phase 295
- polygraph 306
- refractory period 295
- resolution phase 295
- Schachter and Singer's two-factor theory 303
- self-actualization 288
- sexual orientation 296
- sexual prejudice 297
- sexual response cycle 295
- sexuality 295
- social smile 300

# Critical and Creative Thinking Questions

1. Like most Americans, you may have difficulty controlling your weight. Using information from this chapter, can you identify the factors or motives that best explain your experience?
2. How can you restructure elements of your personal, work, or school life to increase intrinsic versus extrinsic motivation?
3. If you were going out on a date with someone or applying for an important job, how might you use the four theories of emotion to increase the chances that things will go well?
4. Have you ever felt depressed after listening to a friend complain about his problems? How might the facial-feedback hypothesis explain this?
5. Why do you think people around the world experience and express the same basic emotions? What evolutionary advantages might help explain these similarities?

# What is happening in this picture?

Curiosity is an important aspect of the human experience.

Tetra Images/SuperStock, Inc.

**Think Critically**

1. Which of the six theories of motivation best explains this behavior?
2. How do you think animal motivation, like this kitten's, might differ from that of humans?

# Self-Test

**RETRIEVAL PRACTICE** Completing this self-test and comparing your answers with those in Appendix C provides immediate feedback and helpful practice for exams. Additional interactive self-tests are available at www.wiley.com/college/carpenter.

**1.** *Motivation* is BEST defined as ______.

a. a set of factors that activate, direct, and maintain behavior, usually toward a goal

b. the physiological and psychological arousal that occurs when a person really wants to achieve a goal

c. what makes you do what you do

d. the conscious and unconscious thoughts that focus a person's behaviors and emotions in the same direction toward a goal

**2.** This diagram illustrates the ______ theory, in which motivation decreases once homeostasis occurs.

a. drive-induction

b. heterogeneity

c. drive-reduction

d. biostability

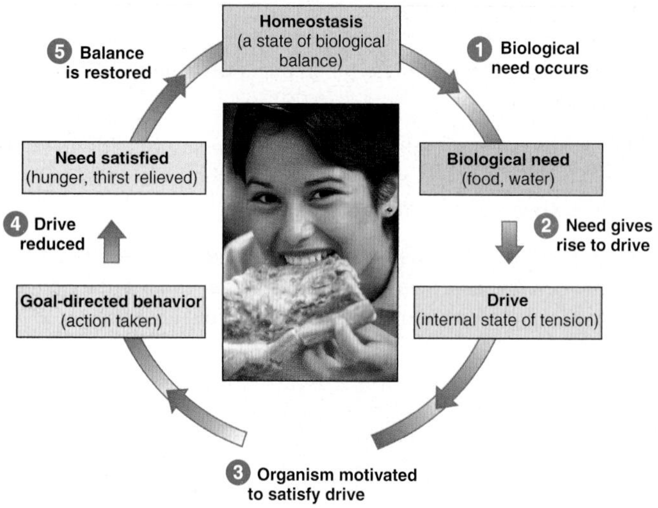


**3.** ______ is the body's tendency to maintain a relatively stable state for internal processes.

a. Homeostasis

b. Heterogeneity

c. Drive-induction

d. Biostability

**4.** The ______ theory says people are "pulled" by external stimuli to act a certain way.

a. cognitive

b. incentive

c. Maslow's hierarchy of needs

d. drive reduction

**5.** According to ______, illustrated in the diagram, some motives have to be satisfied before a person can advance to fulfilling higher motives.

a. Freud's psychosexual stages of development

b. Kohlberg's moral stages of development

c. Erikson's psychosocial stages of development

d. Maslow's hierarchy of needs

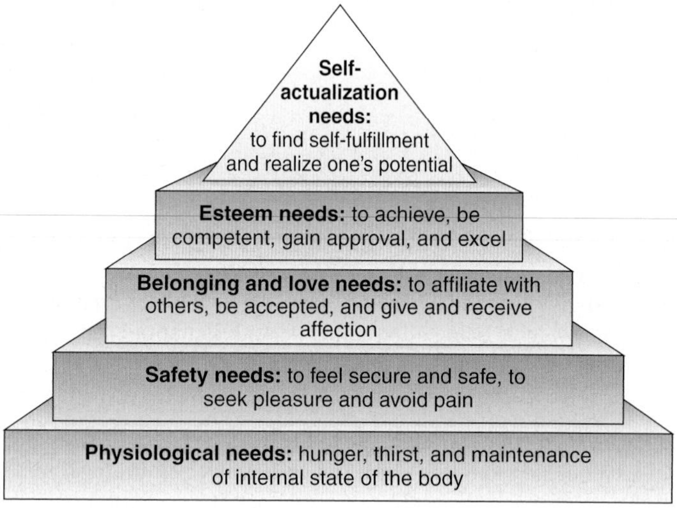


**6.** The feeling of fullness you get once you have ingested enough food is called ______.

a. bloating

b. homeostasis

c. stimulus overload

d. satiety

**7.** ______ involves the consumption of large quantities of food followed by self-induced vomiting, the use of laxatives, or extreme exercise.

a. Anorexia nervosa

b. The binge-purge syndrome

c. Bulimia nervosa

d. Pritikin dieting

**8.** According to your textbook, the desire to excel, especially in competition with others, is known as ______.

a. drive-reduction theory

b. intrinsic motivation

c. achievement motivation

d. all of the above

**9.** Label the correct sequence of events in Masters and Johnson's sexual response cycle on the diagram.

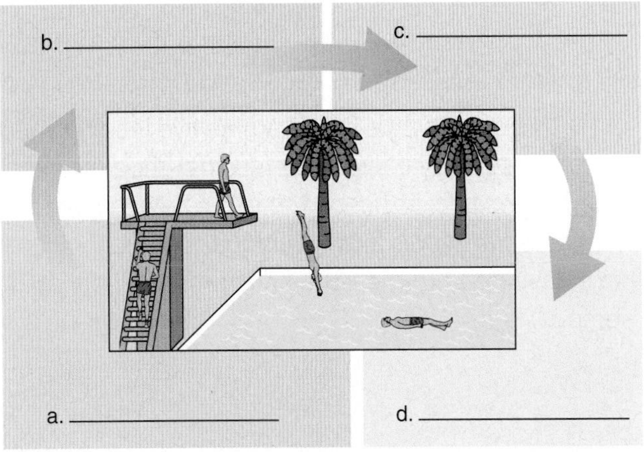
b.
c.
a.
d.

**10.** ______ is the term for negative attitudes toward someone based on his or her sexual orientation.

a. Heterophobia
b. Sexual prejudice
c. Heterosexism
d. Sexual phobia

**11.** Extrinsic motivation is based on ______.

a. the desire for rewards or threats of punishment
b. the arousal motive
c. the achievement motive
d. all of these options

**12.** The three components of emotion are ______.

a. cognitive, biological, and behavioral
b. perceiving, thinking, and acting
c. positive, negative, and neutral
d. active/passive, positive/negative, and direct/indirect

**13.** In a ______ smile, as shown in the photo, the cheek muscles are pulled back and the muscles around the eyes also contract.

a. Duchenne
b. Madonna
c. Mona Lisa
d. da Vinci

**14.** According to ______, we look to external rather than internal cues to understand emotions, as shown in the diagram.

a. the Cannon-Bard theory
b. the James-Lange theory
c. the facial feedback hypothesis
d. Schachter and Singer's two-factor theory

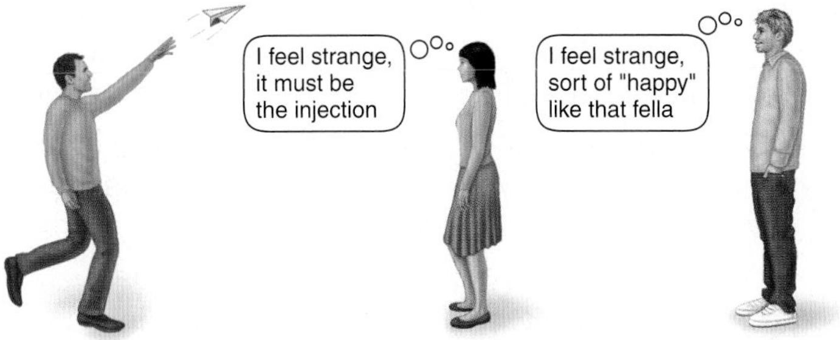
I feel strange, it must be the injection
I feel strange, sort of "happy" like that fella

**15.** Researchers believe all our feelings can be condensed to ______ culturally universal emotions.

a. 2 to 3
b. 5 to 6
c. 7 to 10
d. 11 to 15

**THE PLANNER**

Review your Chapter Planner on the chapter opener and check off your completed work.

# 12 Personality

When you hear the word "personality," do you picture famous TV or movie stars with "great personalities," like Steven Colbert or Whoopi Goldberg? When you meet someone for the first time, do you sometimes instantly like or dislike them based on their personality? What about your "virtual" friends on Facebook or Twitter? Did you decide you liked their personality solely based on their photos or writings?

Most of us spend a lot of time thinking and talking about personality. We want to understand and categorize others so we can better predict how they might react under certain situations. We also want to know more about our own personalities and how our friends and acquaintances perceive us.

To gain a greater understanding and appreciation of personality, psychologists conduct research in a variety of areas, including our genetic and biological characteristics, our lifespan developmental experiences, and our typical patterns of information processing and social interactions. As you'll happily discover in this chapter, modern personality research provides invaluable insights into the people around you, as well as increased self-knowledge.

In this chapter, we first examine five leading theories of personality (*psychoanalytic/psychodynamic, trait, humanistic, social-cognitive, biological*), and then we discuss the tools and techniques psychologists have developed to measure and compare individual personalities.

## CHAPTER OUTLINE



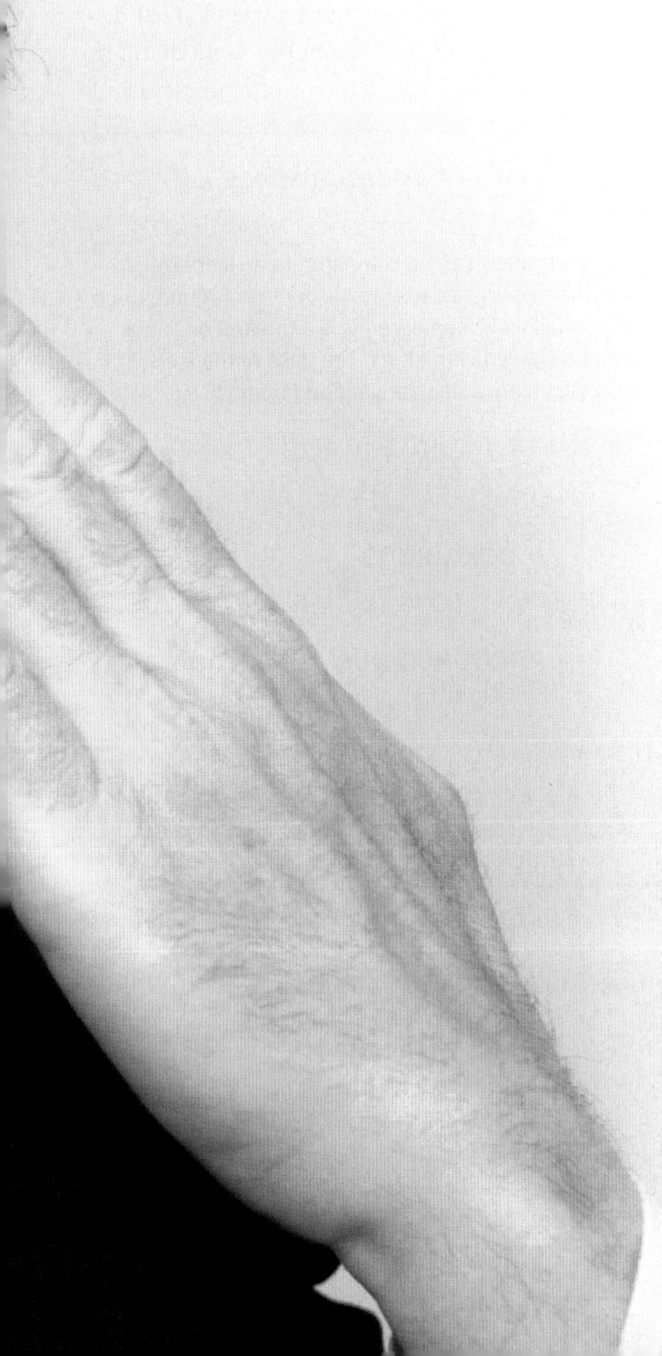

Charles Sykes/©AP/Wide World Photos

## CHAPTER PLANNER ✓

❑ Study the picture and read the opening story.

❑ Scan the Learning Objectives in each section:
p. 314 ❑ p. 319 ❑ p. 323 ❑ p. 326 ❑ p. 328 ❑ p. 330 ❑

❑ Read the text and study all figures and visuals. Answer any questions.

**Analyze key features**

❑ Study Organizer, p. 316

❑ Process Diagram, p. 317

❑ Psychological Science, p. 321

❑ Applying Psychology, p. 322

❑ Psychology InSight, p. 323

❑ Myth Busters, p. 331

❑ What a Psychologist Sees, p. 333

❑ Stop: Answer the Concept Checks before you go on.
p. 319 ❑ p. 322 ❑ p. 325 ❑ p. 327 ❑ p. 329 ❑ p. 333 ❑

**End of chapter**

❑ Review the Summary and Key Terms.

❑ Answer the Critical and Creative Thinking Questions.

❑ Answer What is happening in this picture?

❑ Complete the Self-Test and check your answers.

# Psychoanalytic/Psychodynamic Theories

## LEARNING OBJECTIVES

**RETRIEVAL PRACTICE** While reading the upcoming sections, respond to each Learning Objective in your own words. Then compare your responses with those in Appendix B.

1. **Identify** Freud's most basic and controversial contributions to the study of personality.
2. **Explain** how Adler's, Jung's, and Horney's theories differ from Freud's thinking.
3. **Explore** the major criticisms of Freud's psychoanalytic theories.

The ways you are different from other people and what patterns of behavior are typical of you make up your **personality**. You might qualify as an "extrovert," for example, if you are talkative and outgoing most of the time. Or you may be described as "conscientious" if you are responsible and self-disciplined most of the time. (Keep in mind that *personality* is not the same as *character*, which refers to your ethics, morals, values, and integrity.)

**personality** Relatively stable and enduring patterns of thoughts, feelings, and actions.

One of the earliest theories of personality was Sigmund Freud's psychoanalytic perspective, which emphasized unconscious processes and unresolved past conflicts. We will examine Freud's theories in some detail and then briefly discuss three of his most influential early followers.

## Freud's Psychoanalytic Theory

Working from about 1890 until he died in 1939, Freud developed a theory of personality that has been one of the most influential— and most controversial—theories in all of science (Burger, 2011; Cautin, 2011; Cordón, 2012; Dufresne, 2007). Let's examine some of Freud's most basic and debatable concepts.

**conscious** Freud's term for thoughts or motives that a person is currently aware of or is remembering.

**preconscious** Freud's term for thoughts or motives that are just beneath the surface of awareness and can be easily brought to mind.

**unconscious** Freud's term for thoughts or motives that lie beyond a person's normal awareness, which still exert great influence.

Freud called the mind the "psyche" and asserted that it contains three **levels of consciousness,** or awareness: the **conscious**, the **preconscious**, and the **unconscious** (**Figure 12.1**). Using the Freudian idea of "levels of consciousness," we can say that at this moment your conscious mind is focusing on this text. However, your preconscious may include feelings of hunger and thoughts of friends you need to contact. Any repressed sexual desires, aggressive impulses, or irrational thoughts and feelings are reportedly stored in your unconscious.

Freud believed that most psychological disorders originate from repressed memories and instincts (sexual and aggressive) that are hidden in the unconscious

### Freud's three levels of consciousness • Figure 12.1

Freud's levels of awareness can be compared to an iceberg, where the tip of the iceberg represents the conscious mind, open to view. The area shallowly submerged is like the preconscious mind, which can be viewed with effort. The base of the iceberg is like the unconscious mind, completely hidden from view.

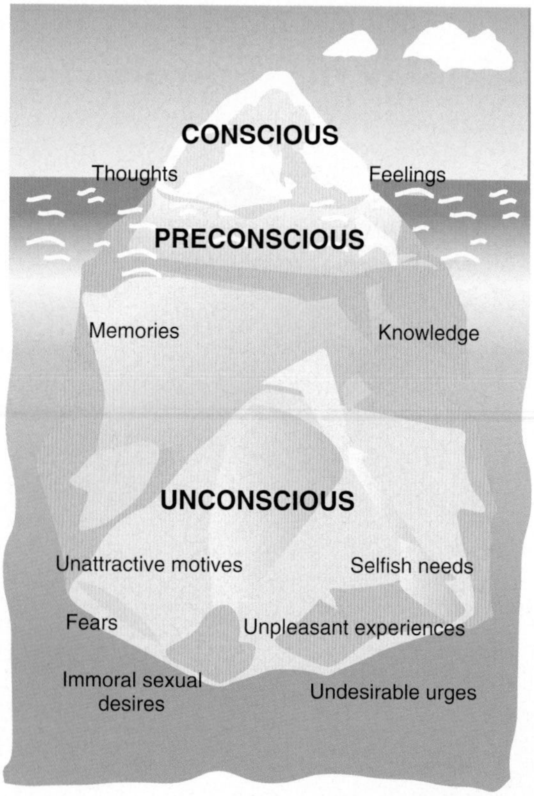

*"Good morning, beheaded—uh, I mean beloved."*

**Freudian slips • Figure 12.2**

Freud believed that a small slip of the tongue (known as a "Freudian slip") can reflect unconscious feelings that we normally keep hidden.

(**Figure 12.2**). To treat these disorders, Freud developed *psychoanalysis* (Chapter 14).

In addition to proposing that the mind functions at three levels of consciousness, Freud also thought that personality was composed of three mental structures: *id, ego,* and *superego* (**Figure 12.3**).

According to Freud, the **id** is made up of innate, biological instincts and urges. It is immature, impulsive, and irrational. The id is also totally unconscious and serves as the reservoir of mental energy. When its primitive drives build up, the id seeks immediate gratification to relieve the tension—a concept known as the **pleasure principle**. In other words, the id is like a newborn baby. It wants what it wants when it wants it.

As a child grows older, the second part of the psyche—the ego—develops. The **ego** is responsible for planning, problem solving, reasoning, and controlling the potentially destructive energy of the id. In Freud's system, the ego corresponds to the *self*—our conscious identity of ourselves as persons.

One of the ego's tasks is to channel and release the id's energy in ways that are compatible with the external world. Thus, the ego is responsible for delaying gratification when necessary. Contrary to the id's pleasure principle, the ego operates on the **reality principle** because it can understand and deal with objects and events in the "real world."

The final part of the psyche to develop is the **superego**, a set of ethical rules for behavior. The superego develops from internalized parental and societal standards. It constantly strives for perfection and is therefore as unrealistic as the id. Some Freudian followers have suggested that the superego operates on the **morality principle** because violating its rules results in feelings of guilt. When the ego fails to satisfy both the id and the superego, anxiety slips into conscious awareness. Because anxiety is uncomfortable, people avoid it through **defense mechanisms**.

**defense mechanisms** In Freudian theory, the ego's protective method of reducing anxiety by distorting reality and self-deception.

**Freud's personality structure • Figure 12.3**

According to Freud, personality is composed of three structures—the id, ego, and superego. Note how the ego is primarily conscious and preconscious, whereas the id is entirely unconscious.

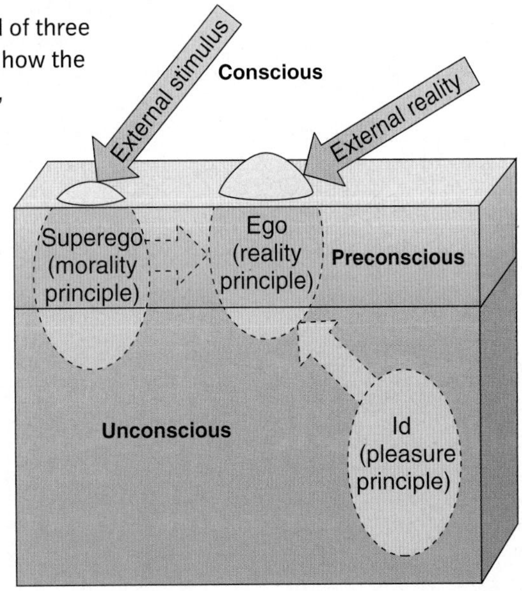

## Why do we use defense mechanisms? • Figure 12.4

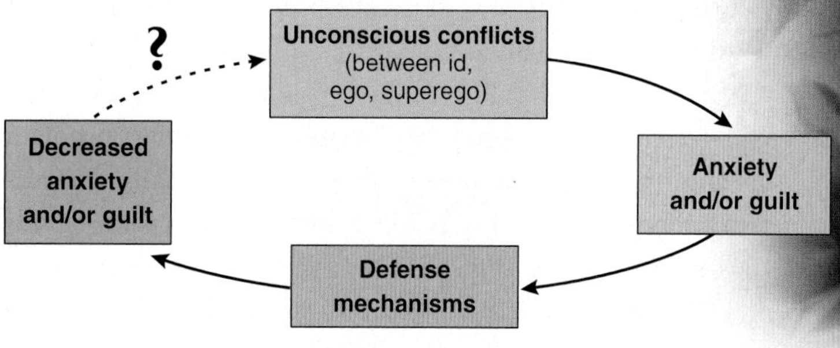


**a.** Freud believed that unconscious conflicts create anxiety and/or guilt, which are relieved by defense mechanisms. The problem is that defense mechanisms distort reality, and the decreased anxiety and/or guilt may lead to their overuse—thereby creating a negative cycle of self-deception.

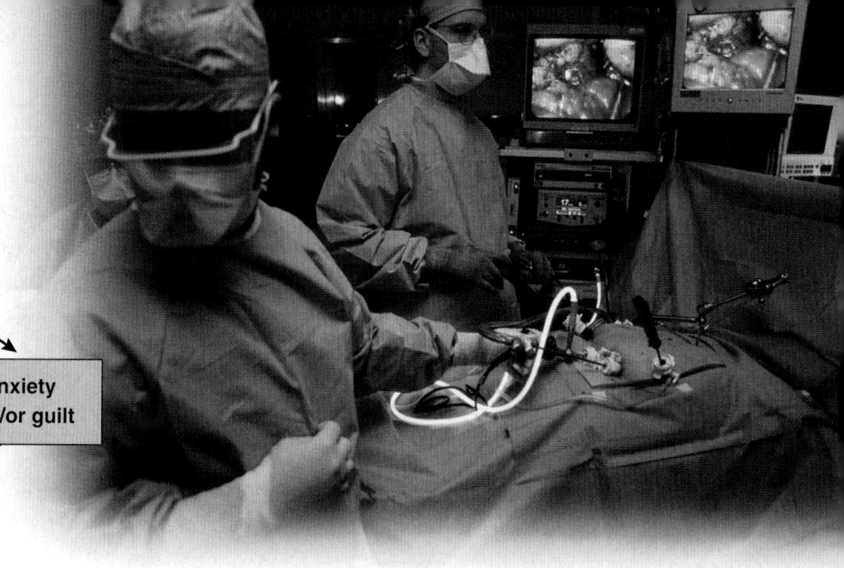

Karen Kasmauski NGS Image S

**b.** During a gruesome surgery, for example, physicians and nurses may intellectualize the procedure as an unconscious way of dealing with their personal anxieties.

**Figure 12.4** illustrates the resolution of unconscious conflicts through defense mechanisms and provides an example of **intellectualization**. Note that defense mechanisms can be healthy and helpful if we use them in moderation or on a temporary basis—like the physicians in Figure 12.4b. Unfortunately, they also help us develop and perpetuate some of our most dangerous habits. For example, an alcoholic who uses his paycheck to buy drinks (a message from the id) may feel very guilty (a response from the superego). He may reduce this conflict by telling himself that he deserves a drink because he works so hard. This is an example of the defense mechanism **rationalization**.

Although Freud described many kinds of defense mechanisms (**Study Organizer 12.1**), he believed that repression was the most important. **Repression** is the mechanism by which the ego prevents the most unacceptable, anxiety-provoking thoughts from entering consciousness.

**Study Organizer 12.1** Sample psychological defense mechanisms — THE PLANNER

| Defense mechanism | Description | Example |
|---|---|---|
| **Repression** | Preventing painful or unacceptable thoughts from entering consciousness | Forgetting the details of a tragic accident |
| **Sublimation** | Redirecting socially unacceptable impulses into acceptable activities | Redirecting aggressive impulses by becoming a professional fighter |
| **Denial** | Refusing to accept an unpleasant reality | Alcoholics refusing to admit their addiction |
| **Rationalization** | Creating a socially acceptable excuse to justify unacceptable behavior | Justifying cheating on an exam by saying "everyone else does it" |
| **Intellectualization** | Ignoring the emotional aspects of a painful experience by focusing on abstract thoughts, words, or ideas | Emotionless discussion of your divorce while ignoring underlying pain |
| **Projection** | Transferring unacceptable thoughts, motives, or impulses to others | Becoming unreasonably jealous of your mate while denying your own attraction to others |
| **Reaction formation** | Not acknowledging unacceptable impulses and overemphasizing their opposite | Promoting a petition against adult bookstores even though you are secretly fascinated by pornography |
| **Regression** | Reverting to immature ways of responding | Throwing a temper tantrum when a friend doesn't want to do what you'd like |
| **Displacement** | Redirecting impulses from original source toward a less threatening person or object | Yelling at a coworker after being criticized by your boss |

PROCESS DIAGRAM

THE PLANNER

## Freud's five psychosexual stages of development • Figure 12.5

Psychosexual development

| | Name of stage (Approximate age) | Erogenous zone (Key conflict or developmental task) | | Supposed symptoms of fixation and/or regression |
|---|---|---|---|---|
| 1 | **Oral** (0–18 months) | **Mouth** (Weaning from breast or bottle) | 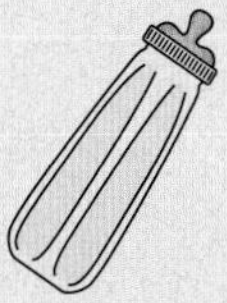 | Overindulgence reportedly contributes to gullibility ("swallowing" anything), dependence, and passivity. Underindulgence leads to aggressiveness, sadism, and a tendency to exploit others. Freud also believed orally fixated adults may orient their lives around their mouths—overeating, becoming alcoholic, smoking, or talking a great deal. |
| 2 | **Anal** (18 months–3 years) | **Anus** (Toilet training) |  | Fixation or regression supposedly leads to highly controlled and compulsively neat (anal-retentive) personality, or messy, disorderly, rebellious, and destructive (anal-expulsive) personality. |
| 3 | **Phallic** (3–6 years) | **Genitals** (Overcoming the Oedipus complex by identifying with same-sex parent) | 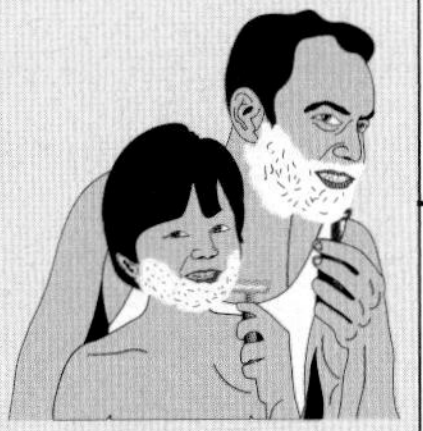 | According to Freud, unresolved, sexual longing for the opposite-sex parent can lead to long-term resentment and hostility toward the same-sex parent. Freud also believed that boys develop an **Oedipus complex,** or attraction to their mothers. He thought that young girls develop an attachment to their fathers and harbor hostile feelings toward their mothers, whom they blame for their lack of a penis. According to Freud most girls never overcome **penis envy** or give up their rivalry with their mothers, which leads to enduring moral inferiority. |
| 4 | **Latency** (6 years–puberty) | **None** (Interacting with same-sex peers) |  | The latency stage is a reported period of sexual "dormancy." Children do not have particular psychosexual conflicts that must be resolved during this period. |
| 5 | **Genital** (puberty–adult) | **Genitals** (Establishing intimate relationships with the opposite sex) |  | Unsuccessful outcomes at this stage supposedly may lead to sexual relationships based only on lustful desires, not on respect and commitment. |

**psychosexual stages** In Freudian theory, the five developmental periods (oral, anal, phallic, latency, and genital) during which particular kinds of pleasures must be gratified if personality development is to proceed normally.

Although defense mechanisms are now an accepted part of modern psychology, other Freudian ideas are more controversial. For example, according to Freud, strong biological urges residing within the id push all children through five universal **psychosexual stages** (**Figure 12.5**). The term *psychosexual* reflects Freud's belief that children experience sexual feelings from birth (in different forms from those of adolescents or adults).

Freud held that if a child's needs are not met or are overindulged at one particular psychosexual stage, the child may become *fixated*, and a part of his or her personality will remain stuck at that stage. Furthermore, under stress, individuals supposedly may return (or *regress*) to a stage at which earlier needs were frustrated or overly gratified.

## Psychodynamic/Neo-Freudian Theories

Some initial followers of Freud later rebelled and proposed theories of their own; they became known as **neo-Freudians**.

Alfred Adler (1870–1937) was the first to leave Freud's inner circle. Instead of seeing behavior as motivated by unconscious forces, he believed that it is purposeful and goal-directed. According to Adler's **individual psychology**, we are motivated by our goals in life—especially our goals of obtaining security and overcoming feelings of inferiority.

Adler believed that almost everyone suffers from an **inferiority complex**, or deep feelings of inadequacy and incompetence that arise from our feelings of helplessness as infants. According to Adler, these early feelings result in a "will-to-power" that can take one of two paths. It can either cause children to strive to develop superiority over others through dominance, aggression, or expressions of envy, or—more positively—it can cause children to develop their full potential and creativity and to gain mastery and control in their lives (Adler, 1964, 1998) (**Figure 12.6**).

Another early Freud follower turned dissenter, Carl Jung (pronounced "yoong"), developed **analytical psychology.** Like Freud, Jung (1875–1961) emphasized unconscious processes, but he believed that the unconscious contains positive and spiritual motives as well as sexual and aggressive forces.

Jung also thought that we have two forms of unconscious mind: the personal unconscious and the collective unconscious. The *personal unconscious* is created from our individual experiences, whereas the *collective unconscious* is identical in each person and is inherited (Jung, 1946, 1959, 1969). The collective unconscious consists of primitive images and patterns of thought, feeling, and behavior that Jung called **archetypes** (**Figure 12.7**).

Because of archetypal patterns in the collective unconscious, we perceive and react in certain predictable ways. One set of archetypes refers to gender roles (Chapter 10). Jung claimed that both males and females have patterns for feminine aspects of personality (*anima*) and masculine aspects of personality (*animus*), which allow us to express both masculine and feminine personality traits and to understand the opposite sex.

Like Adler and Jung, psychoanalyst Karen Horney (pronounced "HORN-eye") was an influential follower of Freud's who later came to reject major aspects of Freudian theory. She is credited with having developed a creative blend of Freudian, Adlerian, and Jungian theory, along with the first feminist critique of Freud's theory (Horney, 1939, 1945). Horney emphasized women's positive traits and suggested that most of Freud's ideas about female personality reflected male bias and misunderstanding. For example, she proposed that women's everyday experience with social inferiority led to power envy versus Freud's idea of biological penis envy.

### An upside to feelings of inferiority? • Figure 12.6

Adler suggested that the will-to-power could be positively expressed through social interest—identifying with others and cooperating with them for the social good. Can you explain how these volunteers might be fulfilling their will-to-power interest?

Keith Morris/Alamy

Roger Wood/Corbis

### Archetypes in the collective unconscious • Figure 12.7

According to Jung, the collective unconscious is the ancestral memory of the human race, which supposedly explains the similarities in religion, art, symbolism, and dream imagery across cultures, such as the repeated symbol of the snake in ancient Egyptian tomb painting. Can you think of other explanations?

Horney also emphasized that personality development depends largely on social relationships—particularly on the relationship between parent and child. She believed that when a child's needs are not met by nurturing parents, he or she may develop lasting feelings of helplessness and insecurity. How people respond to this so-called **basic anxiety**, in Horney's view, sets the stage for later adult psychological health. She believed that everyone copes with this basic anxiety in one of three ways—we move toward, away from, or against other people—and that psychological health requires a balance among these three styles.

In sum, Horney proposed that our adult personalities are shaped by our childhood relationship with our parents—not by fixation or regression at some stage of psychosexual development, as Freud argued.

## Evaluating Psychoanalytic Theories

Before going on, it's important to consider the major criticisms of Freud's psychoanalytic theories:

- *Inadequate empirical support* Many psychoanalytic concepts—such as the psychosexual stages—cannot be empirically tested.
- *Overemphasis on sexuality, biology, and unconscious forces* Modern psychologists believe Freud underestimated the role of learning and culture in shaping personality.
- *Sexism* Many psychologists (beginning with Karen Horney) reject Freud's theories as derogatory toward women.

In response to the claim of inadequate empirical support, numerous modern studies have supported certain Freudian concepts like the defense mechanisms and the fact that a lot of our information processing occurs outside our conscious awareness (e.g., automatic processing, Chapter 5, implicit memories, Chapter 7). Moreover, many contemporary clinicians still value Freud's insights on childhood experiences and unconscious influences on personality development (Burger, 2011; Celes, 2010; De Sousa, 2011; Kring et al., 2012; Schülein, 2007; Siegel, 2010).

However, today there are few Freudian purists. Instead, modern psychodynamic theorists and psychoanalysts tend to place less emphasis on sexual instincts and more on sociocultural influences on personality development (Cautin, 2011; Cordón, 2012; Diem-Wille, 2011; Knekt et al., 2008; Tryon, 2008; Westen, 1998; Young-Bruehl & Schwartz, 2011).

**CONCEPT CHECK** STOP

1. **How** do the conscious, preconscious, and unconscious shape personality in Freud's view?
2. **What** is the difference between the personal unconscious and the collective unconscious?
3. **What** is an example of sexism in Freud's psychoanalytic theory?

# Trait Theories

## LEARNING OBJECTIVES

**RETRIEVAL PRACTICE** While reading the upcoming sections, respond to each Learning Objective in your own words. Then compare your responses with those in Appendix B.

1. **Explain** how early trait theorists approached the study of personality.
2. **Identify** the Big Five personality traits.
3. **Summarize** the major critiques of trait theory.

Stop for a moment and think about the key personality characteristics of your best friend. For example, you might say: *He's a great guy who's a lot of fun to be with. But I sometimes get tired of his constant jokes and pranks. On the other hand, he does listen well and will be serious when I need him to be.*

When describing another's personality, people use terms that refer to that person's most frequent and typical characteristics ("fun," "constant jokes and pranks," "listens well"). These unique and defining characteristics are the foundation for the *trait approach*, which seeks to discover what characteristics form the core of human personality.

## Early Trait Theorists

An early study of dictionary terms found almost 4,500 words that described personality **traits** (Allport & Odbert, 1936). Faced with this enormous list, Gordon Allport (1937) believed that the best way to understand personality was to arrange a person's unique personality traits into a hierarchy, with the most pervasive or important traits at the top.

**traits** Relatively stable and consistent characteristics that can be used to describe someone.

Later psychologists reduced the list of possible personality traits using a statistical technique called **factor analysis**, in which large arrays of data are grouped into more basic units (factors). Raymond Cattell (1950, 1965, 1990) condensed the list of traits to 30 to 35 basic characteristics. Hans Eysenck (1967, 1982, 1990) reduced the list even further. He described personality as a relationship among three basic types of traits: *extroversion–introversion*, *neuroticism* (tendency toward insecurity, anxiety, guilt, and moodiness), and *psychoticism* (exhibiting signs of being out of touch with reality).

## Modern Trait Theory

Factor analysis was also used to develop the most promising modern trait theory, the **five-factor model (FFM)** (Chamorro-Premuzic, 2011; Costa & McCrae, 2011; Soto et al., 2011).

**five-factor model (FFM)** The trait theory of personality composed of openness, conscientiousness, extroversion, agreeableness, and neuroticism.

Combining previous research findings and the long list of possible personality traits, researchers discovered that five traits came up repeatedly, even when different tests were used.

These five major dimensions of personality are often dubbed the **Big Five** (**Figure 12.8**). A handy way to remember the five factors is to note that the first letters of each of the five-factor model spell the word *ocean*. The Big Five are:

**O** *Openness* People who rate high in this factor are original, imaginative, curious, open to new ideas, artistic, and interested in cultural pursuits. Low scorers tend to be conventional, down-to-earth, narrower in their interests, and not artistic. Interestingly, critical thinkers tend to score higher than others on this factor (Clifford, Boufal, & Kurtz, 2004).

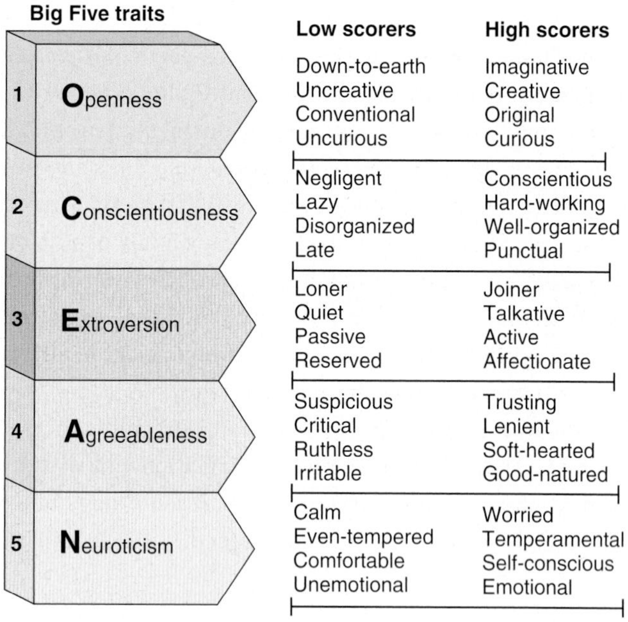


**The five-factor model (FFM) • Figure 12.8**

Different adjectives may describe your personality depending on whether you score high or low in each of the Big Five traits.

**C** *Conscientiousness* This factor ranges from responsible, self-disciplined, organized, and achieving at the high end to irresponsible, careless, impulsive, lazy, and undependable at the other.

**E** *Extroversion* This factor contrasts people who are sociable, outgoing, talkative, fun loving, and affectionate at the high end with introverted individuals who tend to be withdrawn, quiet, passive, and reserved at the low end.

**A** *Agreeableness* Individuals who score high in this factor are good-natured, warm, gentle, cooperative, trusting, and helpful, whereas low scorers are irritable, argumentative, ruthless, suspicious, uncooperative, and vindictive.

**N** *Neuroticism (or emotional stability)* People who score high in neuroticism are emotionally unstable and prone to insecurity, anxiety, guilt, worry, and moodiness. People at the other end are emotionally stable, calm, even-tempered, easygoing, and relaxed.

## Evaluating Trait Theories

The five-factor model (FFM) is the first to achieve the major goal of trait theory—to describe and organize personality characteristics using the smallest number of traits. There also is strong cross-cultural research support for the FFM (see *Psychological Science*).

Trait theories have successfully identified specific personality traits, which allow us to readily compare one person to another. This identification in turn has led to the development of several useful personality measures that will be discussed at the end of this chapter.

Although trait theories, particularly the FFM, have proven to be successful at *describing* personality, critics argue that they fail to offer *causal* explanations for why people develop specific traits (Chamorro-Premuzic, 2011; Cheung, van de Vivjer, & Leong, 2011; Funder, 2000; Furguson et al., 2011). And although trait theories have documented fairly high levels of personality stability over the lifespan, they have failed to identify which characteristics last a lifetime and which are most likely to change

## Ψ Psychological Science

THE PLANNER

# Mate Preferences and the Five-Factor Model (FFM)

Psychologist David Buss and his colleagues (1989, 1990, 2003) surveyed more than 10,000 men and women from 37 countries in order to determine what traits men and women valued in a mate. They found a surprising level of agreement in the characteristics that men and women value in a mate, shown in the figure. Note how most of the FFM personality traits are found at the top of each list, which shows that both men and women prefer dependability (conscientiousness), emotional stability (low neuroticism), pleasing disposition (agreeableness), and sociability (extroversion) to the alternatives.

These shared preferences for certain personality traits may reflect an evolutionary advantage to people who are more conscientious, extroverted, and agreeable—and less neurotic. The evolutionary approach also is confirmed by other cross-cultural studies (Figueredo et al., 2011; McCrae, 2011; Ortiz et al., 2007; Rushton & Irwing, 2011) and comparative studies with dogs, chimpanzees, and other species (Adams, 2011; Gosling, 2008; Mehta & Gosling, 2006; Wahlgren & Lester, 2003).

Seymore/Reportage/Getty Images

| ♂ What Men Want in a Mate | ♀ What Women Want in a Mate |
|---|---|
| 1. Mutual attraction—love | 1. Mutual attraction—love |
| 2. Dependable character | 2. Dependable character |
| 3. Emotional stability and maturity | 3. Emotional stability and maturity |
| 4. Pleasing disposition | 4. Pleasing disposition |
| 5. Good health | 5. Education and intelligence |
| 6. Education and intelligence | 6. Sociability |
| 7. Sociability | 7. Good health |
| 8. Desire for home and children | 8. Desire for home and children |
| 9. Refinement, neatness | 9. Ambition and industriousness |
| 10. Good looks | 10. Refinement, neatness |

From Buss et al., "*International Preferences in Selecting Mates*," Journal of Cross-Cultural Psychology, 21, pp. 5-47 © 1990. Reprinted by Permission of SAGE Publications.

### Identify the Research Method

1. What is the most likely research method conducted by David Buss and his colleagues (1989, 1990, 2003a, 2003b)?
2. If you chose
   - the experimental method, label the IV, DV, experimental group, and control group.
   - the descriptive method, is this a naturalistic observation, survey, or case study?
   - the correlational method, is this a positive, negative, or zero correlation?
   - the biological method, identify the specific research tool (e.g., brain dissection, CT scan).

(Check your answers in Appendix C.)

Jonathan Ernst/Getty Images, Inc.

**Can personality change? • Figure 12.9**

Prior to serving jail time for dog fighting, NFL quarterback Michael Vick relied almost exclusively on his natural athletic ability. Out of jail, he became a great teammate—the first to the field and the last to leave, and in 2010, he was named NFL's "Comeback Player of the Year" (Wilner, 2011).

(Kandler et al., 2012; McCrae, 2011; Mottus, Johnson, & Deary, 2012) (**Figure 12.9**).

Like all theories, trait theories have their limits. However, their overarching goal of identifying and describing the most basic traits that define our personality has largely been met. Furthermore, these traits have proven useful in predicting real-life behaviors (see *Applying Psychology*).

**CONCEPT CHECK** STOP

1. **What** is the purpose of factor analysis?
2. **What** dimensions of personality are central to the five-factor model?
3. **What** limitation does trait theory have when looking at human development over the lifespan?

# Applying Psychology

THE PLANNER

## Personality and Your Career

Are some people better suited for certain jobs than others? According to psychologist John Holland's *personality-job-fit theory*, a match (or "good fit") between our individual personality and our career choice is a major factor in determining job satisfaction (Holland, 1985, 1994). Research shows that a good fit between personality and occupation helps increase subjective well-being, job success, and job satisfaction. In other words, people tend to be happier and like their work when they're well matched to their jobs (Borchers, 2007; Donohue, 2006; Gottfredson & Duffy, 2008; Kieffer, Schinka, & Curtiss, 2004). Check this table to see what job would be a good match for your personality.

| Personality characteristics | Holland personality type | Matching/congruent occupations |
|---|---|---|
| Shy, genuine, persistent, stable, conforming, practical | 1. *Realistic:* Prefers physical activities that require skill, strength, and coordination | Mechanic, drill press operator, assembly-line worker, farmer |
| Analytical, original, curious, independent | 2. *Investigative:* Prefers activities that involve thinking, organizing, and understanding | Biologist, economist, mathematician, news reporter |
| Sociable, friendly, cooperative, understanding | 3. *Social:* Prefers activities that involve helping and developing others | Social worker, counselor, teacher, clinical psychologist |
| Conforming, efficient, practical, unimaginative, inflexible | 4. *Conventional:* Prefers rule-regulated, orderly, and unambiguous activities | Accountant, bank teller, file clerk, manager |
| Imaginative, disorderly, idealistic, emotional, impractical | 5. *Artistic:* Prefers ambiguous and unsystematic activities that allow creative expression | Painter, musician, writer, interior decorator |
| Self-confident, ambitious, energetic, domineering | 6. *Enterprising:* Prefers verbal activities with opportunities to influence others and attain power | Lawyer, real estate agent, public relations specialist, small business manager |

Adapted and reproduced with special permission of the publisher, Psychological Assessment Resources, Inc., 16204 North Florida Avenue, Lutz, Florida 33549, from the _Dictionary of Holland Occupational Codes, Third Edition_, by Gary D. Gottfredson, Ph.D., and John L. Holland, Ph.D., Copyright 1982, 1989, 1996. Further reproduction is prohibited without permission from PAR, Inc.

# Humanistic Theories

## LEARNING OBJECTIVES

**RETRIEVAL PRACTICE** While reading the upcoming sections, respond to each Learning Objective in your own words. Then compare your responses with those in Appendix B.

1. **Explain** the importance of the self in Rogers's theory of personality.
2. **Describe** how Maslow's hierarchy of needs affects personality.
3. **Identify** the pros and cons of humanistic theories.

Humanistic theories of personality emphasize each person's internal feelings, thoughts, and sense of basic worth. In contrast to Freud's generally negative view of human nature, humanists believe that people are naturally good (or, at worst, neutral) and that they possess a natural tendency toward **self-actualization**, the inborn drive to develop all one's talents and capabilities.

According to this view, our personality and behavior depend on how we perceive and interpret the world, not on traits, unconscious impulses, or rewards and punishments. Humanistic psychology was developed largely by Carl Rogers and Abraham Maslow.

### Rogers's Theory

To psychologist Carl Rogers (1902–1987), the most important component of personality is the *self*—what a person comes to identify as "I" or "me." Today, Rogerians (followers of Rogers) use the term **self-concept** to refer to all the information and beliefs you have regarding your own nature, unique qualities, and typical behaviors.

**self-concept** Rogers's term for all the information and beliefs that individuals have about their own nature, qualities, and behavior.

Rogers believed that poor mental health and maladjustment developed from a mismatch, or incongruence, between the self-concept and actual life experiences (**Figure 12.10**).

## Psychology InSight

Personality congruence and mental health • **Figure 12.10**

According to Carl Rogers, mental health and adjustment are related to the degree of congruence between a person's self-concept and life experiences. He argued that self-esteem—how we feel about ourselves—is particularly dependent on this congruence.

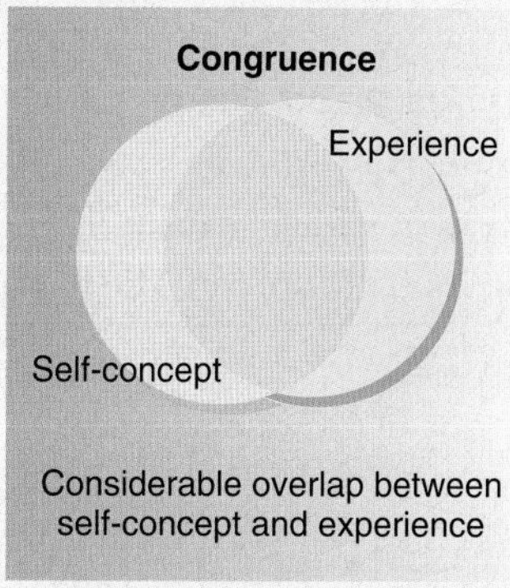


**a. Well-adjusted individual**
Note the considerable overlap between self-concept and experience. Rogers believed fully functioning people willingly integrate life experiences into their self-concept, which allows them to grow and profit from what life can teach them.

**b. Poorly adjusted individual**
According to Rogers, maladjusted people tend to deny or distort life experiences that are incongruent (or don't match) their self-concept, thus blocking their chances for personal growth and fulfillment.

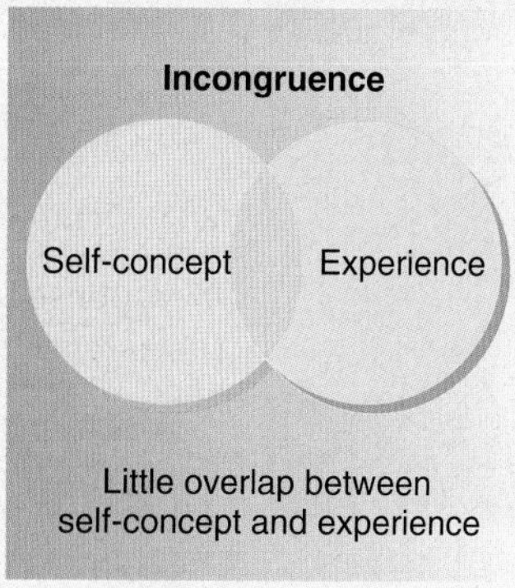


**c. Family effects on congruence**
Can you see how an artistic child would likely have higher self-esteem if her family valued art highly than if they did not?

© Masterfile

Rogers believed that mental health, congruence, and self-esteem are part of our innate, biological capacities. Like plants that grow in cracks in the cement, we are born into the world with an innate need to survive, grow, and enhance ourselves. Therefore, Rogers believed that we should trust our feelings to guide us toward mental health and happiness.

Then why do some people have low self-concepts and poor mental health? Rogers believed that these outcomes generally result from early childhood experiences with parents and other adults who make their love and acceptance *conditional* and contingent on behaving in certain ways and expressing only certain feelings.

If a child learns over time that his negative feelings and behaviors (which we all have) are totally unacceptable and unlovable, his self-concept and self-esteem may become distorted. He may always doubt the love and approval of others because they don't know "the real person hiding inside."

To help children develop to their fullest personality and life potential, adults need to create an atmosphere of **unconditional positive regard**—a warm and caring environment in which a child is never disapproved of as a person.

**unconditional positive regard** Rogers's term for a caring, nonjudgmental attitude toward a person with no contingencies attached.

Some people mistakenly believe that unconditional positive regard means that we should allow people to do whatever they please. But humanists separate the value of the person from his or her behaviors. They accept the person's positive nature while discouraging destructive or hostile behaviors. Humanistic psychologists believe in guiding children to control their behavior so that they can develop a healthy self-concept and healthy relationships with others (**Figure 12.11**).

## Maslow's Theory

Like Rogers, Abraham Maslow believed that there is a basic goodness to human nature and a natural tendency toward *self-actualization*. He saw personality development as a natural progression from lower to higher levels—a basic hierarchy of needs. As newborns, we focus on physiological needs (hunger, thirst), and then as we grow and develop, we move on up through four higher levels (**Figure 12.12**).

According to Maslow, self-actualization is the inborn drive to develop all one's talents and capacities. It involves understanding one's own potential, accepting oneself and others as unique individuals, and taking a problem-centered approach to life situations (Maslow, 1970). Self-actualization is an ongoing process of growth rather than an end product or accomplishment.

Maslow believed that only a few, rare individuals, such as Albert Einstein, Mohandas Gandhi, and Eleanor Roosevelt, become fully self-actualized. However, he saw self-actualization as part of every person's basic hierarchy of needs. (See Chapter 11 for more information on Maslow's theory.)

David Laurens/PhotoAlto/Corbis

### Unconditional positive regard • Figure 12.11

In response to a child who is angry and hits his younger sister, the parent acknowledges that it is the behavior that is unacceptable, and not the child: "I know you're angry with your sister, but we don't hit. And you won't be able to play with her for a while unless you control your anger."

## Maslow's hierarchy of needs • Figure 12.12

Although our natural movement is upward from physical needs toward the highest level, *self-actualization*, Maslow also believed we sometimes "regress" toward a lower level—especially under stressful conditions. For example, during national disasters, people first rush to stockpile food and water (physiological needs), and then often clamor for a strong leader to take over and make things right (safety needs).

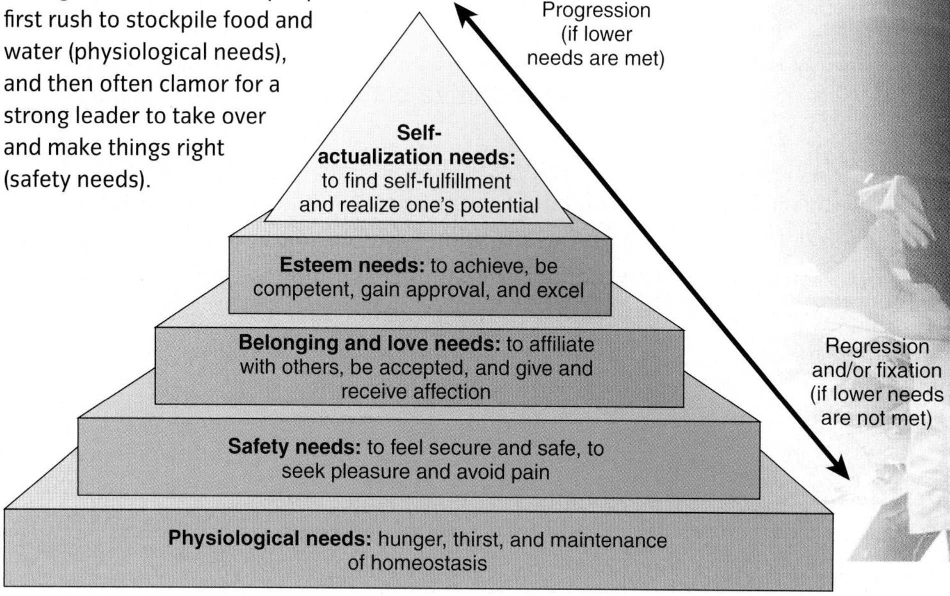

Douglas R. Clifford/©AP/Wide World Photos

## Evaluating Humanistic Theories

Humanistic psychology was extremely popular during the 1960s and 1970s. It was seen as a refreshing new perspective on personality after the negative determinism of the psychoanalytic approach and the mechanical nature of learning theories (Chapter 6). Although this early popularity has declined, humanistic theories have provided valuable insights useful for personal growth and self-understanding. Their theories also play a major role in contemporary counseling and psychotherapy, as well as in modern childrearing, education, and managerial practices.

At the same time, humanistic theories have also been criticized (e.g., Cervone & Pervin, 2010; Chamorro-Premuzic, 2011; Funder, 2000). Three of the most important criticisms are these:

1. *Naïve assumptions* Some critics suggest that humanistic theories are unduly optimistic and overlook the negative aspects of human nature. For example, how would humanists explain immoral politicians, the spread of terrorism, and humankind's ongoing history of war and murder?
2. *Poor testability and inadequate evidence* Like many psychoanalytic terms and concepts, humanistic concepts (such as unconditional positive regard and self-actualization) are difficult to define operationally and to test scientifically.
3. *Narrowness* Like trait theories, humanistic theories have been criticized for merely describing personality rather than explaining it. For example, where does the motivation for self-actualization come from? To say that it is an "inborn drive" doesn't satisfy those who favor using experimental research and hard data to learn about personality.

### CONCEPT CHECK

STOP

1. **How** are self-concept and self-esteem linked in Rogers's theory?
2. **What** is self-actualization?
3. **What** criticism of humanistic theories is also a weakness of trait theories?

# Social-Cognitive Theories

## LEARNING OBJECTIVES

**RETRIEVAL PRACTICE** While reading the upcoming sections, respond to each Learning Objective in your own words. Then compare your responses with those in Appendix B.

1. **Explain** Bandura's concept of self-efficacy and Rotter's concept of locus of control.
2. **Summarize** the attractions and criticisms of the social-cognitive perspective on personality.

As you've just seen, psychoanalytic/psychodynamic, trait, and humanistic theories all tend to focus on internal, personal factors in personality development. In contrast, today's *social-cognitive* theories emphasize the influence of *social* interactions with the environment and *cognitions*—our thoughts, feelings, expectations, and values.

## Bandura's and Rotter's Approaches

Albert Bandura (also discussed in Chapter 6) has played a major role in reintroducing thought processes into personality theory. Cognition, or thought, is central to his concept of **self-efficacy** (Bandura, 1997, 2000, 2006, 2008, 2011).

**self-efficacy** Bandura's term for the learned belief that one is capable of producing desired results, such as mastering new skills and achieving personal goals.

According to Bandura, if you have a strong sense of self-efficacy, you believe you can generally succeed, regardless of past failures and current obstacles. Your self-efficacy will in turn affect which challenges you choose to accept and the effort you expend in reaching goals. However, Bandura emphasized that self-efficacy is always specific to the situation—it does not necessarily carry across situations. For example, self-defense training significantly improves women's belief that they could escape from or disable a potential assailant or rapist, but it does not lead them to feel more capable in all areas of their lives (Weitlauf et al., 2001).

**reciprocal determinism** Bandura's belief that cognitions, behaviors, and the environment interact to produce personality.

Finally, according to Bandura's concept of **reciprocal determinism,** self-efficacy beliefs will affect how others respond to you, influencing your chances for success (**Figure 12.13**). Thus, a cognition ("I can succeed") will affect behaviors ("I will work hard and ask for a promotion"), which in turn will affect the environment ("My employer recognized my efforts and promoted me").

Julian Rotter's theory is similar to Bandura's in that it suggests that learning experiences create **cognitive expectancies** that guide behavior and influence the environment (Rotter, 1954, 1990). According to Rotter, your

**Bandura's theory of reciprocal determinism • Figure 12.13**

According to Albert Bandura, personality is determined by a three-way, reciprocal (interaction) of the internal characteristics of the person, the external environment and the person's behavior.

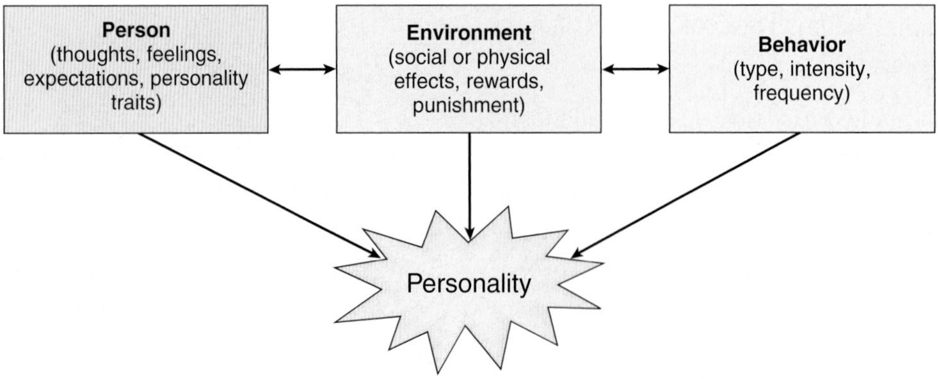

## Locus of control and achievement • Figure 12.14

Despite the sarcasm of this cartoon, research links a perception of control with higher achievement, greater life satisfaction, and better overall mental health (Baumeister & Tierney, 2011; Burns, 2008; Infurna et al., 2011; You, Hong, & Ho, 2011; Zampieri & Pedrosa de Souza, 2011).

*"We're encouraging people to become involved in their own rescue."*



behavior or personality is determined by (1) what you expect to happen following a specific action and (2) the reinforcement value attached to specific outcomes.

To understand your personality and behavior, Rotter would use personality tests that measure your internal versus external **locus of control** (Chapter 3). Rotter's tests ask people to respond to statements such as, "People get ahead in this world primarily by luck and connections rather than by hard work and perseverance," and, "When someone doesn't like you, there is little you can do about it." As you may suspect, people with an external locus of control think that environment and external forces have primary control over their lives, whereas people with an internal locus of control think that they can control events in their lives through their own efforts (**Figure 12.14**).

## Evaluating Social-Cognitive Theories

The social-cognitive perspective holds several attractions. First, it offers testable, objective hypotheses and operationally defined terms, and it relies on empirical data. For example, how do we decide in the real world which risks are worth taking and which are not? Modern research using functional magnetic resonance imaging (fMRI) documents specific areas of the brain that differ between people with risk-averse and risk-seeking personalities (Cox et al., 2010).

Second, social-cognitive theories emphasize how the environment affects, and is affected by, individuals. Walter Mischel, a social-cognitive theorist, ignited the classic person–situation debate when he suggested that our personalities change according to a given situation (Mischel, 1968, 1984, 2004; Mischel & Shoda, 2008). Given that most of us behave differently in different situations—a shy person who never speaks up in large crowds may be very talkative at home or with friends—Mischel initially argued that personality traits may be illusory because they are relatively poor predictors of actual behavior. Although he later proposed an interactionist solution, saying that behavior emerges from an interplay between personality characteristics and situational forces, the debate continues. Critics suggest that social-cognitive theories focus too much on situational influences and fail to adequately acknowledge the stability of personality, as well as unconscious, emotional, and biological influences (Chamorro-Premuzic, 2011; Cervone & Pervin, 2010; Westen, 1998, 2007).

### CONCEPT CHECK

STOP

1. **Why** might self-efficacy beliefs affect actual achievement, according to Bandura?
2. **What** are the two viewpoints in the person–situation debate?

# Biological Theories

## LEARNING OBJECTIVES

**RETRIEVAL PRACTICE** While reading the upcoming sections, respond to each Learning Objective in your own words. Then compare your responses with those in Appendix B.

1. **Summarize** the roles that brain structures and neurochemistry play in personality.
2. **Explain** the limitations of biological theories.
3. **Describe** how the biopsychosocial model integrates different theories of personality.

In this section, we explore how inherited biological factors influence our personalities. We conclude with a discussion of how all theories of personality ultimately interact within the *biopsychosocial model*.

## Three Major Contributors to Personality

Hans Eysenck, the trait theorist discussed earlier in the chapter, was one of the earliest to propose that personality traits are biologically based—at least in part. And modern research agrees that certain brain areas may contribute to some personality characteristics. For instance, modern research in social-cognitive theory has used fMRI brain scans and identified specific areas of the brain believed to be associated with the trait of risk taking (Cox et al., 2010). Earlier research also found increased electroencephalographic (EEG) activity in the left frontal lobes of the brain is associated with sociability (or extroversion), whereas greater EEG activity in the right frontal lobes is associated with shyness (introversion) (Tellegen, 1985).

A major limitation of research on brain structures and personality is the difficulty in identifying which structures are uniquely connected with particular personality traits. Neurochemistry seems to offer more precise data on how biology influences personality (Kagan, 1998).

For example, sensation seeking (Chapter 11) has consistently been linked with levels of monoamine oxidase (MAO), an enzyme that regulates levels of neurotransmitters such as dopamine (Lee, 2011; Zuckerman, 1994, 2004). Dopamine also seems to be correlated with novelty seeking and extroversion (Dalley et al., 2007; Linnet et al., 2011; Marusich et al., 2011; Nemoda, Szekely, & Sasvari-Szekely, 2011).

How can neurochemistry have such effects? Studies suggest that high-sensation seekers and extroverts tend to experience less physical arousal than introverts from the same stimulus (Figueredo et al., 2011; Munoz & Anastassiou-Hadjicharalambous, 2011). Extroverts' low arousal apparently motivates them to seek out situations that will elevate their arousal. Moreover, it is believed that a higher arousal threshold is genetically transmitted. In other words, personality traits like sensation seeking and extroversion may be inherited.

This recognition that genetic factors also have an important influence on personality has contributed to the relatively recent field, called **behavioral genetics** (Congdon & Canli, 2008; Kandler, Riemann, & Kampfe, 2009). Researchers in this area attempt to determine the extent to which behavioral differences among people are due to genetics as opposed to environment (Chapter 2).

One way to measure genetic influences is to compare similarities in personality between identical twins and fraternal twins. For example, studies of the heritability of the five-factor model personality traits suggest that genetic factors contribute about 40 to 50% of personality (Bouchard, 1997, 2004; Eysenck, 1967, 1990; Hopwood et al., 2011; McCrae et al., 2004; Plomin, 1990; Schermer et al., 2011; Veselka, Schermer, & Vernon, 2011).

In addition to twin studies, researchers compare the personalities of parents with those of their biological children and their adopted children. Studies of extroversion and neuroticism have found that parents' traits correlate moderately with those of their biological children and hardly at all with those of their adopted children (Bouchard, 1997; McCrae et al., 2000).

## Evaluating Biological Theories

Modern research in biological theories has provided exciting insights, and it has established clear links between some personality traits and various brain areas, neurotransmitters, and/or genetics. However, it's important to keep in mind that personality traits are never the result of a single

## Multiple influences on personality • Figure 12.15

Research indicates that three major factors influence personality: genetic (inherited) factors; nonshared environmental factors (how each individual's genetic factors react and adjust to his or her environment); and shared environmental factors (parental and family experiences).

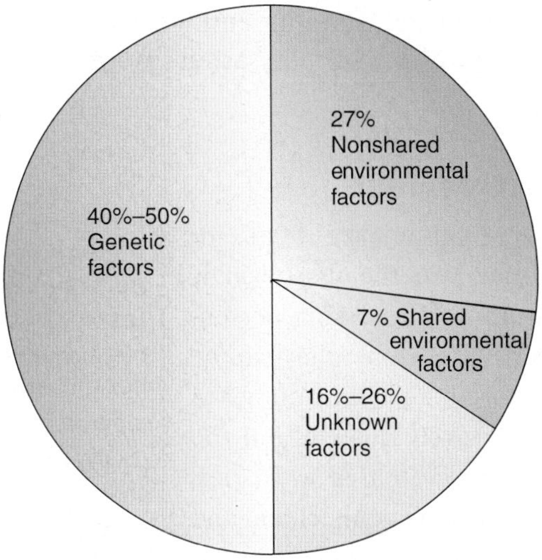


*Source:* Bouchard, 1997, 2004; Eysenck, 1967, 1990; Jang et al., 2006; McCrae et al., 2004; Plomin, 1990; Weiss, Bates, & Luciano, 2008

Rosanne Olson/Stone/Getty Images

biological process. For example, studies do show a strong inherited basis for personality, but researchers are careful not to overemphasize genetics (Beauchamp et al., 2011; Burger, 2011; Chadwick, 2011). Some believe the importance of the unshared environment (aspects of the environment that differ from one individual to another, even within a family) has been overlooked. Others fear that research on "genetic determinism" could be misused to "prove" that an ethnic or a racial group is inferior, that male dominance is natural, or that social progress is impossible.

In sum, there is no doubt that biological studies have produced valuable results. However, like all the previous theories discussed in this chapter, it's also clear that much more research is needed before we can fully understand the nature and causes of personality.

## Biopsychosocial Model

No one personality theory explains everything we need to know about personality. Each theory offers different insights into how a person develops the distinctive set of characteristics we call "personality." That's why instead of adhering to any one theory, many psychologists believe in the biopsychosocial approach, or the idea that several factors—biological, psychological, and social—overlap in their contributions to personality (**Figure 12.15**) (Mischel, Shoda, & Ayduk, 2008).

### CONCEPT CHECK

STOP

1. **How** can traits like sensation seeking and extroversion be related to neurochemistry?
2. **What** would be an argument against overemphasizing genetics as the basis for personality?
3. **Why** is the biopsychosocial model important to research on personality?

# Personality Assessment

## LEARNING OBJECTIVES

**RETRIEVAL PRACTICE** While reading the upcoming sections, respond to each Learning Objective in your own words. Then compare your responses with those in Appendix B.

1. **Summarize** the four major methods psychologists use to measure personality.
2. **Describe** the benefits and limitations of each method of personality assessment.

Numerous methods have been used over the decades to assess personality. Modern personality assessments are used by clinical and counseling psychologists, psychiatrists, and others for diagnosing psychotherapy patients and for assessing their progress in therapy. Personality assessment is also used for educational and vocational counseling and to aid businesses in making hiring decisions. Personality assessments can be grouped into a few broad categories: interviews, observations, objective tests, and projective tests.

## Interviews and Observation

We all use informal "interviews" to get to know other people. When first meeting someone, we usually ask about his or her job, academic interests, family, or hobbies. Psychologists also use interviews. Unstructured interviews are often used for job selection and for diagnosing psychological problems. In an unstructured format, interviewers get impressions and pursue hunches or let the interviewee expand on information that promises to disclose personality characteristics. In structured interviews, the interviewer asks specific questions so that the interviewee's responses can be evaluated more objectively (and compared with others' responses).

In addition to interviews, psychologists also assess personality by directly and methodically observing behavior. The psychologist looks for examples of specific behaviors and follows a careful set of evaluation guidelines. For instance, a psychologist might arrange to observe a troubled client's interactions with his or her family. Does the client become agitated by the presence of certain family members and not others? Does he or she become passive and withdrawn when asked a direct question? Through careful observation, the psychologist gains valuable insights into the client's personality, as well as family dynamics (**Figure 12.16**).

## Objective Tests

**Objective personality tests**, or inventories, are the most widely used method of assessing personality, for two reasons: they can be administered to a large number of people relatively quickly, and the tests can be evaluated in a standardized fashion. (An older, and no longer used, method of assessing personality is described in *Myth Busters*.)

**objective personality tests** Standardized questionnaires that require written, self-report responses, usually to multiple-choice or true-false questions.

Some objective tests measure one specific personality trait, such as sensation seeking (Chapter 11) or locus of control. However, psychologists in clinical, counseling, and industrial settings often wish to assess a range of personality traits. To do so, they generally use multi-trait (or *multiphasic*) inventories.

### Behavioral observation • Figure 12.16

**How might careful observation help a psychologist better understand a troubled client's personality and family dynamics?**

David De Lossy/Getty Images

## MYTH BUSTERS

### PERSONALITY AND BUMPS ON THE HEAD?

In the 1800s, if you wanted to have your personality assessed, you would go to a phrenologist, who would determine your personality by measuring the bumps on your skull and comparing the measurements with a chart that associated different areas of the skull with particular traits, such as *sublimity* (ability to squelch natural impulses, especially sexual) and *ideality* (ability to live by high ideals). What traits might be measured if we still believed in phrenology today?

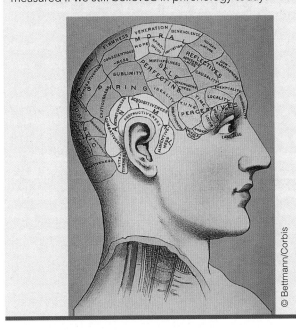

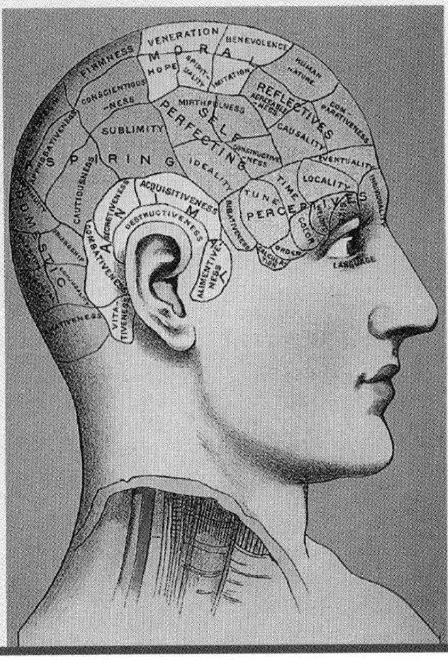


© Bettmann/Corbis

The most widely studied and clinically used multi-trait test is the **Minnesota Multiphasic Personality Inventory (MMPI)**—or its revision, the MMPI-2 (Butcher, 2000, 2011; Butcher & Perry, 2008). This test consists of over 500 statements that participants respond to with *True*, *False*, or *Cannot Say*. The following are examples of the kinds of statements found on the MMPI:

> My stomach frequently bothers me.
> I have enemies who really wish to harm me.
> I sometimes hear things that other people can't hear.
> I would like to be a mechanic.
> I have never indulged in any unusual sex practices.

Did you notice that some of these questions are about very unusual, abnormal behavior? Although there are many "normal" questions on the full MMPI, the test is designed primarily to help clinical and counseling psychologists diagnose psychological disorders. MMPI test items are grouped into 10 clinical scales, each measuring a different disorder (**Table 12.1**). There are also a number of validity scales designed to reflect the extent to which respondents (1) distort their answers (for example, to fake psychological disturbances or to appear more psychologically healthy than they really are), (2) do not understand the items, and (3) are being uncooperative.

**Subscales of the MMPI-2 Table 12.1**

| Clinical scales | Typical interpretations of high scores | Validity scales | Typical interpretations of high scores |
|---|---|---|---|
| 1. Hypochondriasis | Numerous physical complaints | 1. L (lie) | Denies common problems, projects a "saintly" or false picture |
| 2. Depression | Seriously depressed and pessimistic | | |
| 3. Hysteria | Suggestible, immature, self-centered, demanding | 2. F (confusion) | Answers are contradictory |
| 4. Psychopathic deviate | Rebellious, nonconformist | 3. K (defensiveness) | Minimizes social and emotional complaints |
| 5. Masculinity–femininity | Interests like those of other sex | | |
| 6. Paranoia | Suspicious and resentful of others | 4. ? (cannot say) | Many items left unanswered |
| 7. Psychasthenia | Fearful, agitated, brooding | | |
| 8. Schizophrenia | Withdrawn, reclusive, bizarre thinking | | |
| 9. Hypomania | Distractible, impulsive, dramatic | | |
| 10. Social introversion | Shy, introverted, self-effacing | | |

There are other objective personality measures that are less focused on psychological disorders. A good example is the NEO Personality Inventory-Revised, which assesses the dimensions comprising the five-factor model, and the Myers-Briggs Type Indicator (MBTI), which assesses four dimensions derived from Carl Jung's theory of personality: extroversion-introversion (EI), sensing-intuition (SI), thinking-feeling (TF), and judging-perceiving (JP).

Personality tests like the MMPI are often confused with *career inventories* or vocational interest tests. Career counselors use these latter tests (along with aptitude and achievement tests) to help people identify occupations and careers that match their unique traits, values, and interests.

## Projective Tests

Unlike objective tests, **projective tests** use unstructured stimuli that can be perceived in many ways. As the name implies, projective tests supposedly allow each person to project his or her own unconscious conflicts, psychological defenses, motives, and personality traits onto the test materials. Because respondents may be unable (or unwilling) to express their true feelings if asked directly, the ambiguous stimuli reportedly provide an indirect "psychological x-ray" of important unconscious processes (Hogan, 2006). Two of the most widely used projective tests are the **Rorschach Inkblot Test** and the **Thematic Apperception Test (TAT)** (see *What a Psychologist Sees*).

**projective tests** Psychological tests that use ambiguous stimuli, such as inkblots or drawings, which allow the test taker to project his or her unconscious thoughts onto the test material.

## Are Personality Measurements Accurate?

Let's evaluate the strengths and the challenges of each of the four methods of personality assessment: interviews, observation, objective tests, and projective tests.

**Interviews and observations** Both interviews and observations can provide valuable insights into personality, but they are time-consuming and expensive. Furthermore, raters of personality tests frequently disagree in their evaluations of the same individuals. Interviews and observations also involve unnatural settings, and, in fact, the very presence of an observer can alter a subject's behavior.

**Objective tests** Tests like the MMPI-2 provide specific, objective information about a broad range of personality traits in a relatively short period. However, they are also the subject of at least three major criticisms:

1. *Deliberate deception and social desirability bias* Some items on personality inventories are easy to "see through," so respondents may intentionally, or unintentionally, fake particular personality traits. In addition, some respondents want to look good and will answer questions in ways that they perceive are socially desirable. (The validity scales of the MMPI-2 are designed to help prevent these problems.)
2. *Diagnostic difficulties* When inventories are used for diagnosis, overlapping items sometimes make it difficult to pinpoint a diagnosis (Graham, 1991). In addition, clients with severe disorders sometimes score within the normal range, and normal clients sometimes score within the elevated range (Borghans et al., 2011; Weiner, 2008).
3. *Cultural bias and inappropriate use* Some critics think that the standards for "normalcy" on objective tests fail to recognize the impact of culture. For example, Latinos generally score higher than respondents from North American and western European cultures on the masculinity–femininity scale of the MMPI-2 (Dana, 2005; Ketterer, 2011; Mundia, 2011). However, this tendency reflects traditional gender roles and cultural training more than any individual personality traits.

**Projective tests** Although projective tests are extremely time-consuming to administer and interpret, their proponents suggest that because they are unstructured, respondents may be more willing to talk honestly about sensitive topics. Critics point out, however, that the *reliability* and *validity* (Chapter 8) of projective tests is among the lowest of all tests of personality (Garb et al., 2005; Gacono et al., 2008; Wood et al., 2011).

As you can see, each of these methods has its limits. Psychologists typically combine the results from various methods to create a full picture of an individual's personality.

# WHAT A PSYCHOLOGIST SEES

THE PLANNER

## Projective Tests

When you listen to a piece of music or look at a picture, you might say the music is sad or that the people in the picture look happy, but not everyone would have the same interpretation. Some psychologists believe that these differing interpretations reveal important things about each individual's personality. Therefore, your responses to specially selected images or stimuli can be used in a certain type of personality test, called a *projective test*. Responses to projective tests reportedly reflect unconscious parts of the personality that "project" onto the stimuli.

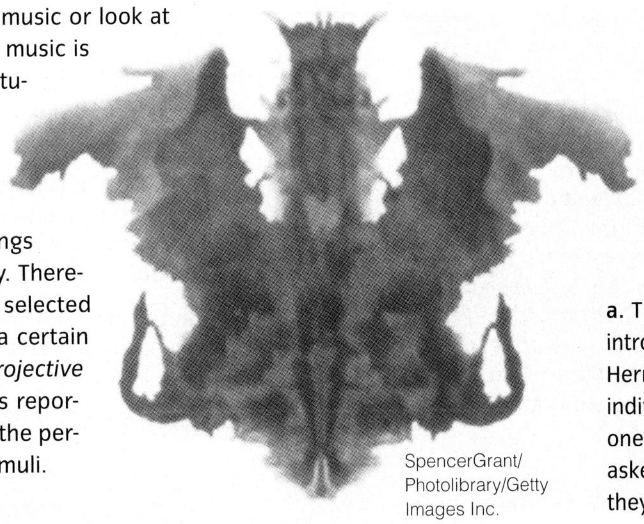

SpencerGrant/Photolibrary/Getty Images Inc.

**a.** The Rorschach Inkblot Test was introduced in 1921 by Swiss psychiatrist Hermann Rorschach. With this technique, individuals are shown 10 inkblots like the one shown here, one at a time, and are asked to report what figures or objects they see in each of them.

**b.** Created by personality researcher Henry Murray in 1938, the Thematic Apperception Test (TAT) consists of a series of ambiguous black-and-white pictures that are shown to the test taker, who is asked to create a story related to each. Can you think of two different stories that a person might create for the picture of the man and woman here? How might a psychologist interpret each story?

**CONCEPT CHECK** STOP

1. **How** do structured and unstructured interviews differ?
2. **Why** are objective personality tests like the MMPI used so widely?

# Summary

THE PLANNER

## 1 Psychoanalytic/Psychodynamic Theories 314

- Freud, the founder of psychodynamic theory, believed that the mind contained three **levels of consciousness** (as shown in the diagram): the **conscious**, the **preconscious**, and the **unconscious**. He believed that most psychological disorders originate from unconscious memories and instincts. Freud also asserted that personality was composed of the **id**, the **ego**, and the **superego**. When the ego fails to satisfy both the id and the superego, anxiety slips into conscious awareness, which triggers **defense mechanisms**. In addition, Freud believed all children go through five **psychosexual stages**, which affect personality development.

Freud's three levels of consciousness • Figure 12.1

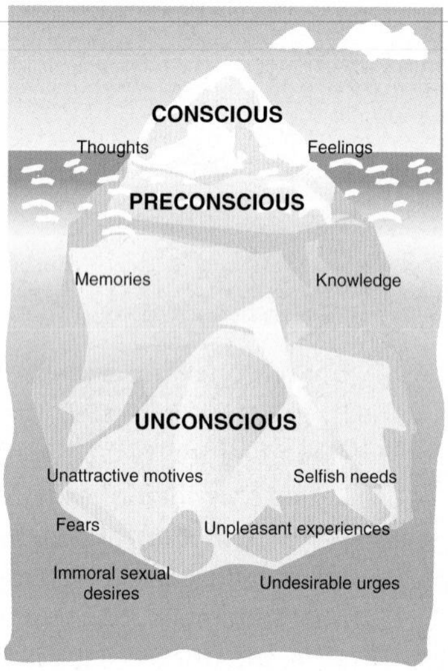


- **Neo-Freudians** such as Adler, Jung, and Horney were influential followers of Freud who later came to reject major aspects of Freudian theory.

## 2 Trait Theories 319

- Psychologists define **personality** as an individual's relatively stable and enduring patterns of thoughts, feelings, and actions.
- Allport believed that the best way to understand personality was to arrange a person's unique personality **traits** into a hierarchy. Cattell and Eysenck later reduced the list of possible personality traits using **factor analysis**.
- The **five-factor model (FFM)** identifies major dimensions of personality, or the **Big Five**, shown in the diagram.

The five-factor model (FFM) • Figure 12.8

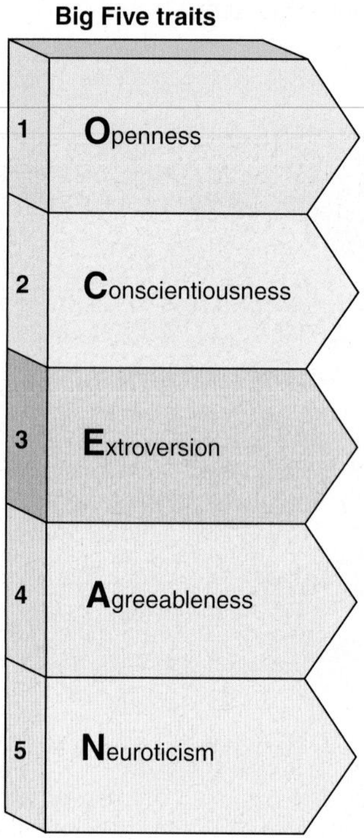

## 3 Humanistic Theories 323

- According to Rogers, mental health and self-esteem are related to the degree of congruence between our **self-concept** and life experiences (shown in the diagrams). He argued that poor mental health results when young children do not receive **unconditional positive regard** from caregivers.

**Personality congruence and mental health**
**• Figure 12.10**

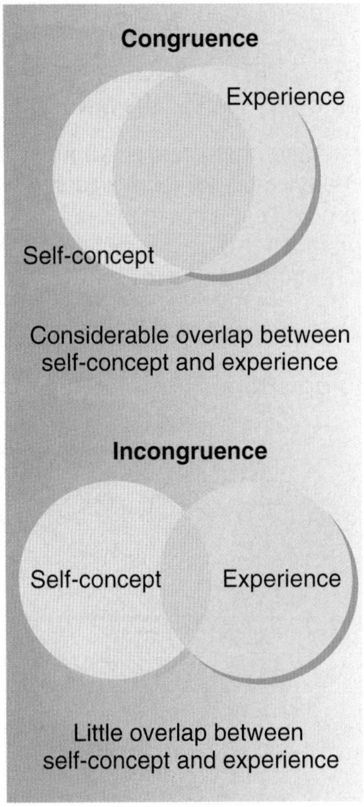


- Maslow saw personality as the quest to fulfill basic physiological needs and to move toward the highest level of **self-actualization.**

## 4 Social-Cognitive Theories 326

- Cognition is central to Bandura's concept of self-efficacy. According to Bandura, **self-efficacy** affects which challenges we choose to accept and the effort we expend in reaching goals. His concept of **reciprocal determinism** (see diagram) states that self-efficacy beliefs will also affect others' responses to us.

**Bandura's theory of reciprocal determinism**
**• Figure 12.13**

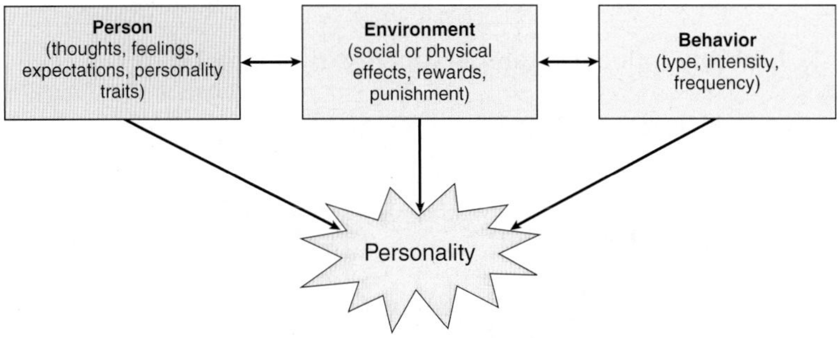


- Rotter's theory suggests that learning experiences create **cognitive expectancies** that guide behavior and influence the environment. Rotter believed that having an internal versus external **locus of control** affects personality and achievement.

## 5 Biological Theories 328

- There is evidence that certain brain areas may contribute to personality. However, neurochemistry seems to offer more precise data on how biology influences personality. Research in **behavioral genetics** indicates that genetic factors also strongly influence personality.
- Instead of adhering to any one theory of personality, many psychologists believe in the biopsychosocial approach—the idea that several factors overlap in their contributions to personality, as shown in the chart.

**Multiple influences on personality • Figure 12.15**

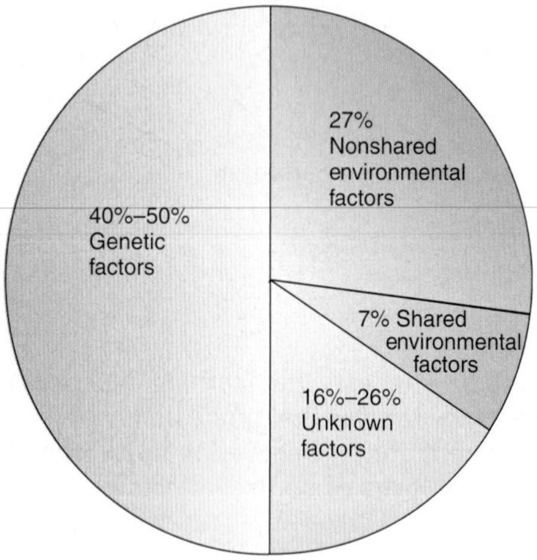


## 6 Personality Assessment 330

- In an unstructured interview format, interviewers get impressions and pursue hunches or let the interviewee expand on information that promises to disclose personality characteristics. In structured interviews, the interviewer asks specific questions so that the interviewee's responses can be evaluated more objectively. Psychologists also assess personality by directly observing behavior.
- **Objective personality tests** are widely used because they can be administered broadly and relatively quickly and because they can be evaluated in a standardized fashion. To assess a range of personality traits, psychologists use multitrait inventories, such as the **MMPI**.
- **Projective tests** use unstructured stimuli that can be perceived in many ways. Projective tests, such as the **Rorschach Inkblot Test**, shown in the photo, supposedly allow each person to project his or her own unconscious conflicts, psychological defenses, motives, and personality traits onto the test materials.

**What a Psychologist Sees: Projective Tests**

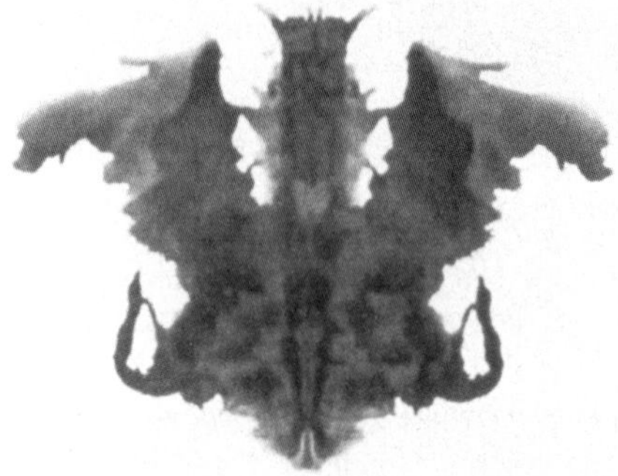

# Key Terms

**RETRIEVAL PRACTICE** Write a definition for each term before turning back to the referenced page to check your answer.

- analytical psychology 318
- archetype 318
- basic anxiety 319
- behavioral genetics 328
- Big Five 320
- cognitive expectancy 326
- conscious 314
- defense mechanism 315
- ego 315
- factor analysis 320
- five-factor model (FFM) 320
- id 315
- individual psychology 318
- inferiority complex 318
- intellectualization 316
- levels of consciousness 314
- locus of control 327
- Minnesota Multiphasic Personality Inventory (MMPI) 331
- morality principle 315
- neo-Freudian 318
- objective personality test 330
- Oedipus complex 317
- penis envy 317
- personality 314
- pleasure principle 315
- preconscious 314
- projective test 332
- psychosexual stages 317
- rationalization 316
- reality principle 315
- reciprocal determinism 326
- repression 316
- Rorschach Inkblot Test 332
- self-actualization 323
- self-concept 323
- self-efficacy 326
- superego 315
- Thematic Apperception Test (TAT) 332
- trait 320
- unconditional positive regard 324
- unconscious 314

# Critical and Creative Thinking Questions

1. If scientists have so many problems with Freud, why do you think his theories are still popular with the public? Should psychologists continue to discuss his theories (and include them in textbooks)? Why or why not?
2. How do you think you would score on each of the "Big Five" personality dimensions?
3. In what ways is Adler's individual psychology more optimistic than Freudian theory?
4. How do Bandura's and Rotter's social-cognitive theories differ from biological theories of personality?
5. Which method of personality assessment (interviews, behavioral observation, objective testing, or projective testing) do you think is likely to be most informative? Can you think of circumstances where one kind of assessment might be more effective than the others?

# What is happening in this picture?

William Thomas Gain/Getty Images

Undertaking what seems like a purely physical hobby may hold keys to more complex biopsychosocial interactions within us.

**Think Critically**

1. How would Freud interpret this woman's aggressive behavior toward her "assailant"?
2. How would Bandura's social-cognitive concept of self-efficacy explain the benefits of taking a self-defense course?
3. What biological factors might explain why some people choose activities that provide high levels of physiological arousal while others avoid such situations?

# Self-Test

**RETRIEVAL PRACTICE** Completing this self-test and comparing your answers with those in Appendix C provides immediate feedback and helpful practice for exams. Additional interactive, self-tests are available at www.wiley.com/college/carpenter.

**1.** ______ is defined as relatively stable and enduring patterns of thoughts, feelings, and actions.

a. Character

b. Trait

c. Temperament

d. Personality

**2.** Label Freud's three levels of consciousness.

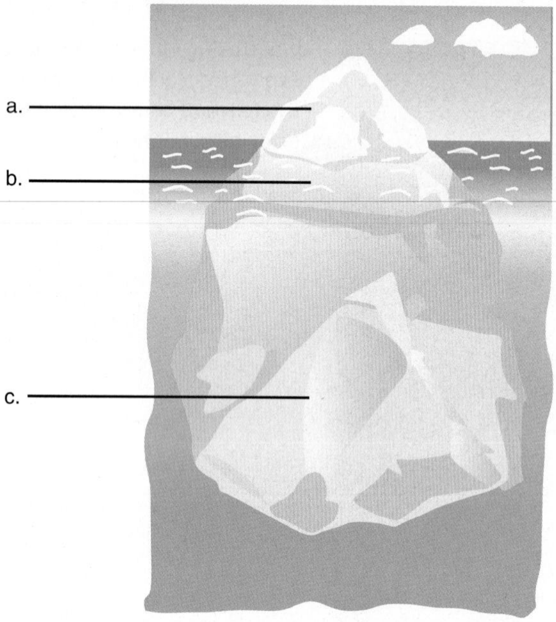


**3.** According to Freud, the three mental structures that form personality are the ______.

a. unconscious, preconscious, and conscious

b. oral, anal, and phallic

c. Oedipus, Electra, and sexual/aggressive

d. id, ego, and superego

**4.** Used excessively, defense mechanisms can be dangerous because they ______.

a. block intellectualization

b. become ineffective

c. distort reality

d. become fixated

**5.** Label Freud's psychosexual stages of development in correct sequence.

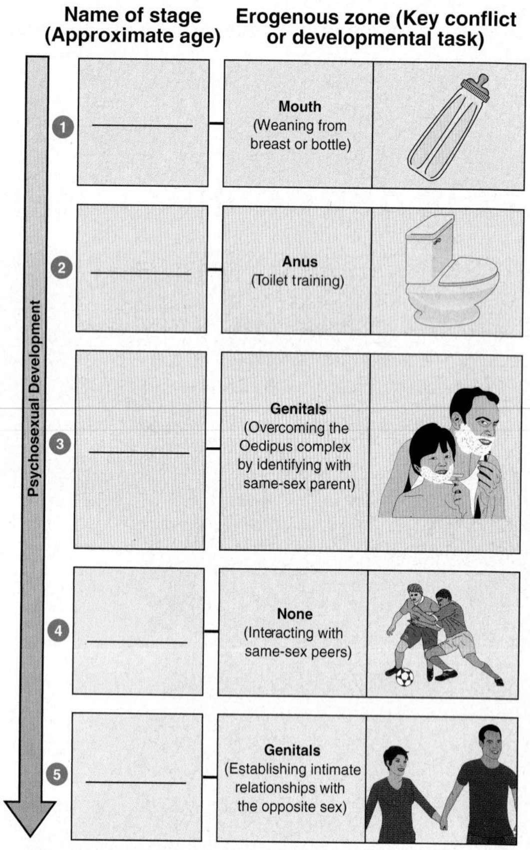


**6.** Three of the most influential neo-Freudians were ______.

a. Plato, Aristotle, and Descartes

b. Dr. Laura, Dr. Phil, and Dr. Ruth

c. Pluto, Mickey, and Minnie

d. Adler, Jung, and Horney

**7.** ______ is a statistical technique that groups large arrays of data into more basic units.

a. MMPI

b. FFM

c. Factor analysis

d. Regression analysis

8. Label the list of personality traits in the five-factor model.

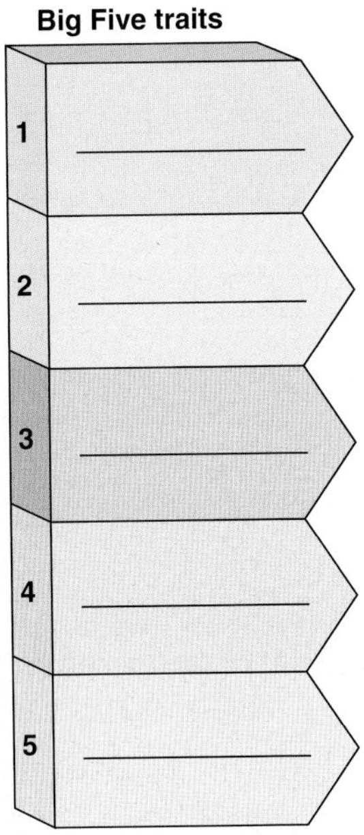


| | Low scorers | High scorers |
|---|---|---|
| 1 | Down-to-earth<br>Uncreative<br>Conventional<br>Uncurious | Imaginative<br>Creative<br>Original<br>Curious |
| 2 | Negligent<br>Lazy<br>Disorganized<br>Late | Conscientious<br>Hard-working<br>Well-organized<br>Punctual |
| 3 | Loner<br>Quiet<br>Passive<br>Reserved | Joiner<br>Talkative<br>Active<br>Affectionate |
| 4 | Suspicious<br>Critical<br>Ruthless<br>Irritable | Trusting<br>Lenient<br>Soft-hearted<br>Good-natured |
| 5 | Calm<br>Even-tempered<br>Comfortable<br>Unemotional | Worried<br>Temperamental<br>Self-conscious<br>Emotional |

9. *Unconditional positive regard* is a Rogerian term for ______.
   a. accepting any and all behavior as a positive manifestation of self-actualization
   b. caring, non-judgmental attitude toward a person without attaching any contingencies
   c. nonjudgmental listening
   d. phenomenological congruence

10. The personality theorist who believed in the basic goodness of individuals and their natural tendency toward self-actualization was ______.
   a. Karen Horney
   b. Alfred Adler
   c. Abraham Maslow
   d. Carl Jung

11. According to Bandura, ______ involves a person's belief about whether he or she can successfully engage in behaviors related to personal goals.
   a. self-actualization
   b. self-esteem
   c. self-efficacy
   d. self-congruence

12. ______ believed personality is affected by cognitive expectancies that guide behavior and influence the environment.
   a. Maslow
   b. Bandura
   c. Adler
   d. Rotter

13. ______ appear(s) to have the largest influence (40 to 50%) on personality.
   a. Nonshared environment
   b. Shared environments
   c. Genetics
   d. Unknown factors

14. The most widely researched and clinically used self-report personality test is the ______.
   a. MMPI-2
   b. Rorschach Inkblot Test
   c. TAT
   d. SVII

15. The photo shows an example of a(n) ______ test.
   a. projective
   b. ambiguous stimuli
   c. Rorschach Inkblot
   d. All of these options

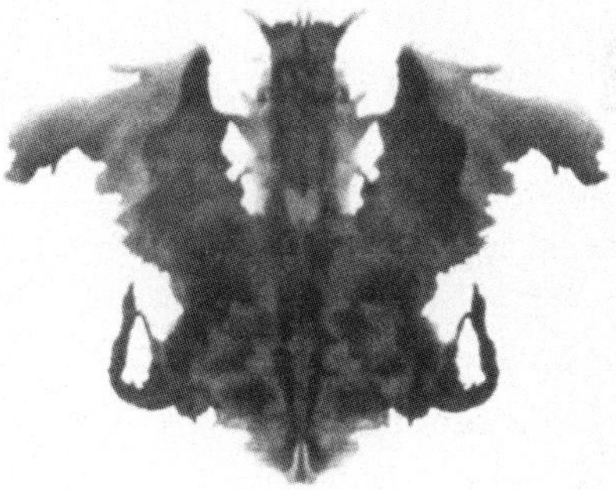

**THE PLANNER** ✓

Review your Chapter Planner on the chapter opener and check off your completed work.

# 13 Psychological Disorders

If someone asked you to visualize and describe a psychologically disordered person, what would you say? Would your imagination match this true story?

> *Mary's troubles first began in adolescence. She was frequently truant, and her grades declined sharply. During family counseling sessions, it was discovered that Mary had prostituted herself for drug money. . . . She idealized new friends, but when they disappointed her, she angrily cast them aside. . . . Mary's problems, coupled with a preoccupation with inflicting pain on herself (by cutting and burning) and persistent thoughts of suicide, led to her admittance to a psychiatric hospital at age 26 (Kring et al., 2012, pp. 465–466).*

Mary has severe psychological problems. Was there something in her early background to explain her later behavior? Is there something medically wrong with her? What about less severe forms of abnormal behavior?

In this chapter, we discuss how psychological disorders are identified, explained, and classified, and explore five major categories of psychological disorders. We also look at how gender and culture affect mental disorders.

© Hexvivo/iStockphoto

## CHAPTER OUTLINE



## CHAPTER PLANNER

- ❑ Study the picture and read the opening story.
- ❑ Scan the Learning Objectives in each section: p. 342 ❑ p. 348 ❑ p. 352 ❑ p. 355 ❑ p. 361 ❑ p. 364 ❑
- ❑ Read the text and study all figures and visuals. Answer any questions.

**Analyze key features**

- ❑ Psychology InSight, p. 342
- ❑ Myth Busters p. 343 ❑ p. 353 ❑ p. 356 ❑
- ❑ What a Psychologist Sees, p. 344
- ❑ Study Organizers p. 346 ❑ p. 349 ❑
- ❑ Process Diagram, p. 351
- ❑ Applying Psychology, p. 354
- ❑ Psychological Science, p. 365
- ❑ Stop: Answer the Concept Checks before you go on: p. 347 ❑ p. 351 ❑ p. 355 ❑ p. 360 ❑ p. 363 ❑ p. 367 ❑

**End of chapter**

- ❑ Review the Summary and Key Terms.
- ❑ Answer the Critical and Creative Thinking Questions.
- ❑ Answer What is happening in this picture?
- ❑ Complete the Self-Test and check your answers.

# Studying Psychological Disorders

## LEARNING OBJECTIVES

**RETRIEVAL PRACTICE** While reading the upcoming sections, respond to each Learning Objective in your own words. Then compare your responses with those in Appendix B.

1. **List** the four criteria for identifying abnormal behavior.
2. **Review** how perspectives of abnormal behavior have changed throughout history.
3. **Explain** the purposes and criticisms of the *Diagnostic and Statistical Manual* (*DSM*).

On the continuum ranging from normal to **abnormal behavior**, people can be unusually healthy or extremely disturbed.

Mental health professionals generally agree on four criteria for abnormal behavior: (often called the four "D's") *deviance, dysfunction, distress,* and *danger* (**Figure 13.1**).

**abnormal behavior** Patterns of emotion, thought, and action that are considered pathological (diseased or disordered) for one or more of these four reasons: deviance, dysfunction, distress, or danger.

## Psychology InSight

Four criteria for abnormal behavior • **Figure 13.1**

Behavior is considered abnormal if it meets one or more of these four criteria (Hansell & Damour, 2008).

Corbis Flirt/Alamy

**a. Deviance**
Behaviors, thoughts, or emotions may be considered abnormal when they deviate from a society or culture's norms or values. For example, it's normal to be a bit concerned if friends are whispering, but abnormal if you're equally concerned when strangers are whispering.

Ian West/Alamy

**b. Dysfunction**
When someone's dysfunction interferes with his or her daily functioning, such as drinking to the point that you cannot hold a job, stay in school, or maintain relationships, it would be considered abnormal behavior.

Peter Dazeley/Photographer's Choice/Getty Image, Inc.

**c. Distress**
Behaviors, thoughts, or emotions that cause significant personal distress may qualify as abnormal. Self-abuse relationship problems, and suicidal thoughts all indicate serious personal\distress and unhappiness.

Digital Vision/Getty Images

**d. Danger**
If someone's thoughts, emotions, or behaviors present a danger to self or others, such as engaging in road rage to the point of physical confrontation, it would be considered abnormal.

Rather than being fixed categories both "abnormal" and "normal" behaviors exist along a continuum (Hansell & Damour, 2008), with normal behavior being more frequent and abnormal behavior more rare.

## MYTH BUSTERS

THE PLANNER

### FIVE COMMON MYTHS ABOUT MENTAL ILLNESS

- ***Myth #1: People with psychological disorders act in bizarre ways and are very different from normal people.***

  **Fact:** This is true for only a small minority of individuals and during a relatively small portion of their lives. In fact, sometimes even mental health professionals find it difficult to distinguish normal from abnormal individuals without formal screening.

- ***Myth #2: Mental disorders are a sign of personal weakness.***

  **Fact:** Psychological disorders are a function of many factors, such as exposure to stress, genetic disposition, and family background. Mentally disturbed individuals can't be blamed for their illness any more than we blame people who develop cancer or other illnesses.

- ***Myth #3: Mentally ill people are often dangerous and unpredictable.***

  **Fact:** Only a few disorders, such as some psychotic and antisocial personality disorders, are associated with violence. The stereotype that connects mental illness and violence persists because of prejudice and selective media attention.

- ***Myth #4: A person who has been mentally ill never fully recovers.***

  **Fact:** With therapy, the vast majority of people who are diagnosed as mentally ill eventually improve and lead normal, productive lives. Moreover, the extreme symptoms of some mental disorders are generally only temporary.

Hoberman Collection/Alamy

**Is this behavior abnormal?** Eccentric? Yes. Mentally disordered? Probably not.

- ***Myth #5: Most mentally ill individuals can work at only low-level jobs.***

  **Fact:** Like all illnesses, mental disorders are complex, and their symptoms, severity, and prognoses differ for each individual. Although some people are seriously disabled by their disorder, others are very productive and high functioning. For example, U.S. President Abraham Lincoln, British Prime Minister Winston Churchill, scientist Isaac Newton, painter Vincent Van Gogh, and author Charles Dickens are all famous achievers who suffered from serious mental disorders at various times throughout their careers.

*Sources:* Barlow & Durand, 2012; *Famous People with Schizophrenia*, 2011; *Famous People with Mental illness*, 2011; Kring et al., 2012 Lilienfeld et al., 2010.

However, as we consider these criteria, remember that no single criterion is adequate for identifying all forms of abnormal behavior. Furthermore, when considering the two criteria of deviance and danger, it's important to note that judgments of deviancy vary historically and cross-culturally, and that the public generally overestimates the danger posed by the mentally ill (see *Myth Busters*).

## Explaining Abnormality

What causes abnormal behavior? Historically, evil spirits and witchcraft have been blamed (Goodwin, 2012; Millon, 2004; Petry, 2011; Riva et al., 2011). Stone Age people, for example, believed that abnormal behavior stemmed from demonic possession; the "therapy" was to bore a hole in the skull so that the evil spirit could escape. During the European Middle Ages, a troubled person was sometimes treated with exorcism, an effort to drive the Devil out through prayer, fasting, noise-making, beating, and drinking terrible-tasting brews. During the fifteenth century, many believed that some individuals chose to consort with the Devil. Many of these supposed witches were tortured, imprisoned for life, or executed.

As the Middle Ages ended, special mental hospitals called *asylums* began to appear in Europe. Initially designed to provide quiet retreats from the world and to protect society (Barlow & Durand, 2012; Coleborne & Mackinnon, 2011; Freckelton, 2011; Millon, 2004), the asylums unfortunately became overcrowded, inhumane prisons.

Improvement came in 1792 when Philippe Pinel, a French physician, was placed in charge of a Parisian asylum. Believing that inmates' behavior was caused by underlying physical illness, he insisted they be unshackled and removed from their unlighted, unheated cells.

# WHAT A PSYCHOLOGIST SEES ✓ THE PLANNER

## Seven Psychological Perspectives on Abnormal Behavior

Everyone sometimes makes an impulse purchase, and most people are reluctant to discard some possessions that are of questionable value. But when the acquisition of and inability to discard worthless items becomes extreme, it can interfere with basic aspects of living, such as cleaning, cooking, sleeping on a bed, and moving around one's home (**Figure a**). This abnormal behavior is associated with several psychological disorders, but it is most commonly found in people who have obsessive-compulsive disorder, or OCD (an anxiety disorder discussed later in this chapter).

Each of the seven major perspectives in psychology emphasizes different factors believed to contribute to abnormal behavior such as hoarding, but in practice they overlap (**Figure b**).

WR Publishing/Alamy

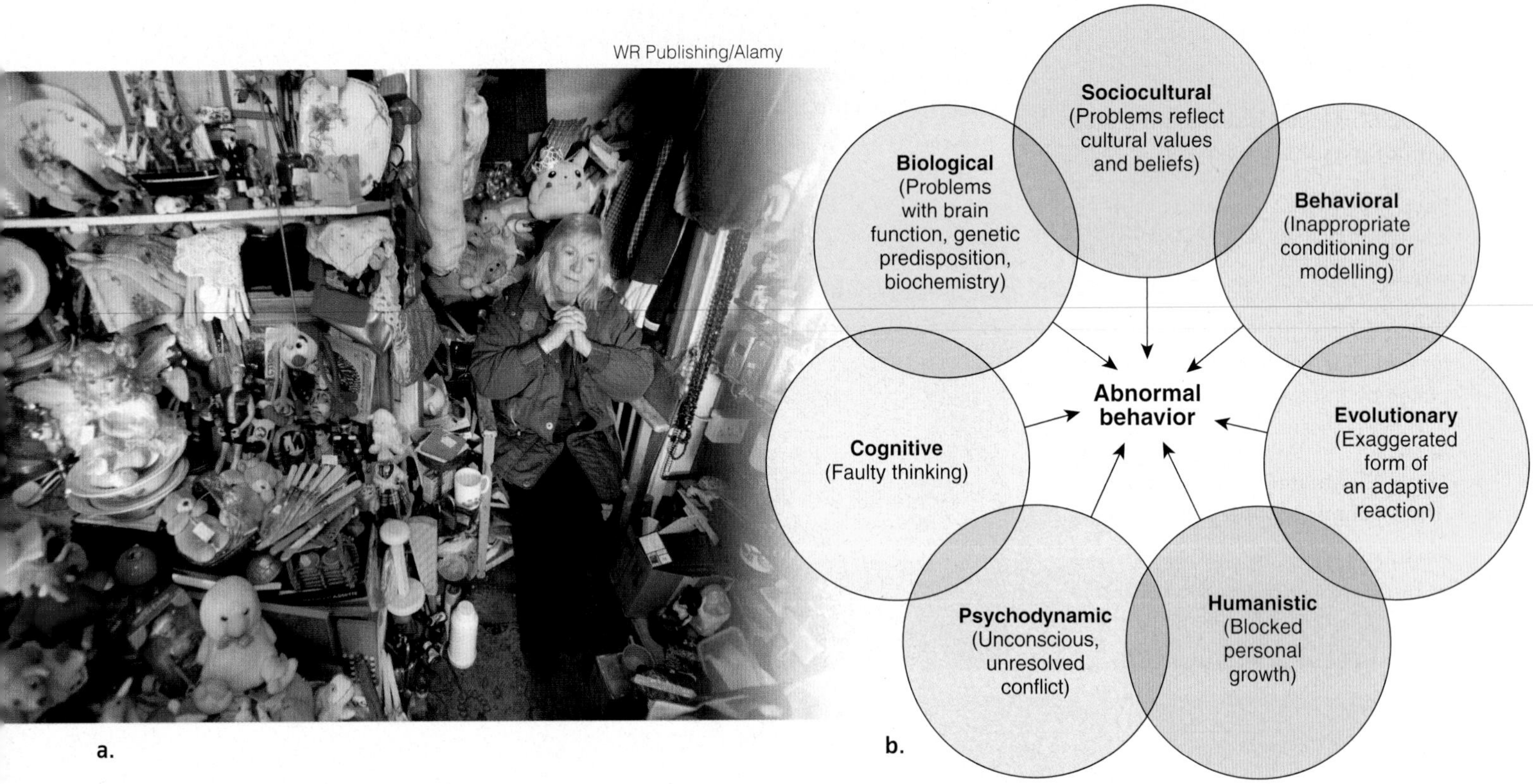


a. b.

Many inmates improved so dramatically that they could be released. Pinel's **medical model** eventually gave rise to the modern specialty of **psychiatry**.

Unfortunately, when we label people "mentally ill," we may create new problems. One of the most outspoken critics of the medical model is psychiatrist Thomas Szasz (1960, 2000, 2004). Szasz believed that the medical model encourages people to believe that they have no responsibility for their actions. He contended that mental illness is a myth used to label individuals who are peculiar or offensive to others (Cresswell, 2008). Furthermore, labels can become self-perpetuating—that is, the person can begin to behave according to the diagnosed disorder.

Despite these potential dangers, the medical model—and the concept of mental illness—remains a founding principle of psychiatry. In contrast, psychology offers a multifaceted approach to explaining abnormal behavior, as described in *What a Psychologist Sees*.

> **medical model** The perspective that diseases (including mental illness) have physical causes that can be diagnosed, treated, and possibly cured.
>
> **psychiatry** The branch of medicine that deals with the diagnosis, treatment, and prevention of mental disorders.

## Classifying Abnormal Behavior

Without a clear, reliable system for classifying the wide range of psychological disorders, scientific research on them would be almost impossible, and communication among mental health professionals would be seriously impaired. Fortunately, mental health specialists share a uniform classification system, the ***Diagnostic and Statistical Manual (DSM)*** (American Psychiatric Association, 2000). This manual has been updated and revised several times, and the latest, fifth edition is scheduled for publication in 2013 (Regier et al., 2011).

***Diagnostic and Statistical Manual (DSM)*** A reference book developed by the American Psychiatric Association that classifies mental disorders on the basis of symptoms, and provides a common language and standard criteria useful for communication and research.

Each revision of the *DSM* has expanded the list of disorders and changed the descriptions and categories to reflect both the latest in scientific research and changes in the way abnormal behaviors are viewed within our social context (Blashfield, Flanagan, & Raley, 2010; Sass, 2011). For example, take the terms **neurosis** and **psychosis.** In previous editions, the term *neurosis* reflected Freud's belief that all neurotic conditions arise from unconscious conflicts (Chapter 12). Now, conditions that were previously grouped under the heading *neurosis* have been formally redistributed as anxiety disorders, somatoform disorders, and dissociative disorders.

Unlike neurosis, the term *psychosis* is still listed in the current edition of the *DSM* because it remains useful for distinguishing the most severe mental disorders, such as schizophrenia and some mood disorders.

What about the term *insanity*? Where does it fit in? **Insanity** is a legal term indicating that a person cannot be held responsible for his or her actions, or is judged incompetent to manage his or her own affairs, because of mental illness. In the law, the definition of mental illness rests primarily on a person's inability to tell right from wrong (**Figure 13.2**). For psychologists, insanity is not the same as abnormal behavior. It's important to keep in mind that modern research does suggest that brain disease and damage can dramatically affect thoughts, emotions, and behaviors.

Steve Ueckert/AP/Wide World Photos

### The insanity plea—guilty of a crime or mentally ill? • Figure 13.2

Texas mother Andrea Yates was found not guilty by reason of insanity for drowning her five children in the bathtub in 2001. Yates was committed to a state mental hospital, where she will remain until she is no longer deemed a threat. The insanity plea is used in less than 1% of all cases that reach trial and is successful in only a fraction of those (Claydon, 2012; Green et al., 2011; Slobogin, 2006; West & Lichtenstein, 2006).

## Understanding and Evaluating the *DSM*

To understand a disorder, we must first name and describe it. The *DSM* identifies and describes the symptoms of approximately 400 disorders, which are grouped into 17 categories (**Study Organizer 13.1**). Note that we focus on only the first 5 of the 17 categories in this chapter. Also, keep in mind that people may be diagnosed with more than one disorder at a time, which is referred to as **comorbidity**.

> **comorbidity** Co-occurrence of two or more disorders in the same person at the same time, as when a person suffers from both depression and alcoholism.

### Study Organizer 13.1 Subcategories of mental disorders in the *DSM*

THE PLANNER

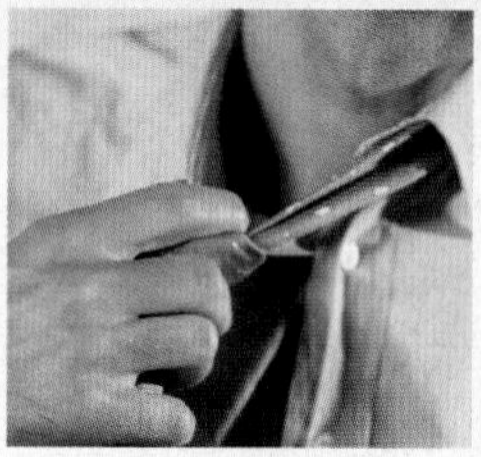

Corbis Images

Anxiety disorders

© Aldo Murillo/iStockphoto

Eating disorders

© gabyjalbert/iStockphoto

Impulse control disorders

1. **Anxiety disorders** Problems associated with severe anxiety, such as *phobias, obsessive-compulsive disorder,* and *posttraumatic stress disorder.*
2. **Mood disorders** Problems associated with severe disturbances of mood, such as *major depressive disorder, mania,* or alternating episodes of the two (*bipolar disorder*).
3. **Schizophrenia and other psychotic disorders** A group of disorders characterized by major disturbances in perception, language and thought, emotion, and behavior.
4. **Dissociative disorders** Disorders in which the normal integration of consciousness, memory, or identity is suddenly and temporarily altered, such as *dissociative amnesia* and *dissociative identity disorder.*
5. **Personality disorders** Problems related to lifelong maladaptive personality traits, including *antisocial personality disorder* (violation of others' rights with no sense of guilt) or *borderline personality disorder* (impulsivity and instability in mood and relationships).
6. **Sleep disorders** Serious disturbances of sleep, such as *insomnia* (too little sleep), *sleep terrors,* or *hypersomnia* (too much sleep). (See Chapter 5.)
7. **Substance-related disorders** Problems caused by alcohol, tobacco, and other drugs. (See Chapter 5.)
8. **Intellectual disability** Problems caused by limitations in intellectual functioning and adaptive behavior. (See Chapter 8.)
9. **Eating disorders** Problems related to food, such as *anorexia nervosa* and *bulimia nervosa.* (See Chapter 11.)
10. **Somatoform disorders** Problems related to unusual preoccupation with physical health or physical symptoms with no physical cause.
11. **Factitious disorders** Conditions in which physical or psychological symptoms are intentionally produced in order to assume a patient's role.
12. **Sexual and gender identity disorders** Sex or gender identity, such as unsatisfactory sexual activity, finding unusual objects or situations arousing, or gender identity confusion.
13. **Impulse control disorders (not elsewhere classified)** Problems related to *kleptomania* (impulsive stealing), *pyromania* (setting of fires), and *pathological gambling.*
14. **Adjustment disorders** Problems involving excessive emotional reaction to specific stressors such as divorce, family discord, or economic concerns.
15. **Disorders first diagnosed in infancy, childhood, or early adolescence** Problems that appear before adulthood, such as *autism, learning disorder,* and *attention deficit disorder.*
16. **Delirium, dementia, amnestic, and other cognitive disorders** Problems caused by known damage to the brain, including *Alzheimer's disease, Huntington's disease,* and physical trauma to the brain.
17. **Other conditions that may be a focus of clinical attention** Problems related to physical or sexual abuse, relational problems, occupational problems, and so forth.

* Subcategories 1–5 are discussed in this chapter. Subcategories 6–9 are discussed elsewhere in this text. Subcategories 10–17 are not directly covered in this text.

*Source:* DSM- IV-TR (2000), American Psychiatric Association, NICHY (2012).

In addition to naming and categorizing specific disorders, the *DSM* also takes into account both the person and his or her life situation (**Figure 13.3**).

Classification and diagnosis of mental disorders are essential to scientific study. Without such a system, we could not effectively identify and diagnose the wide variety of disorders, predict their future courses, or suggest appropriate treatment. The *DSM* also facilitates communication among professionals and patients, and it serves as a valuable educational tool.

Unfortunately, the *DSM* does have its limitations and potential problems. For example, critics suggest that it may be casting too wide a net and *overdiagnosing*. Given that physicians and psychologists can only be compensated for treatment by insurance companies if each client is assigned a specific *DSM* code number, can you see how compilers of the *DSM* may be encouraged to add more diagnoses?

© Mediaphotos/iStockphoto

**Mental health and the economy • Figure 13.3**

During the recent economic crisis in the United States, many job seekers routinely lined up to apply for the few available openings. Unemployment and poverty appear to be strongly linked to psychological dysfunction.

In addition, the *DSM* has been criticized for a potential *cultural bias*. It does provide a culture-specific section and a glossary of culture-bound syndromes, such as *amok* (Indonesia), *genital retraction syndrome* (Asia), and *windigo psychosis* (First Nations' cultures), which we discuss later in this chapter (p. 364). However, the overall classification still reflects a western European and American perspective (Eap et al., 2010; Flaskerud, 2010; Gellerman & Lu, 2011; Nabar, 2011).

Perhaps the most troubling criticism of the *DSM* is the over reliance on the medical model and how it may unfairly label people. Consider the classic study conducted by David Rosenhan (1973) when he and his colleagues presented themselves to several hospital admissions offices with complaints of hearing voices (a classic symptom of schizophrenia). Aside from making this single false complaint and providing alternate names and occupations, they answered all questions truthfully. Not surprisingly (given their reports of hearing strange voices), Rosenhan and his colleagues were all diagnosed with mental disorders and admitted to the hospital. Once admitted, however, the "patients" acted completely normal for the duration of their stays, yet none of these individuals was ever recognized by hospital staff as phony. Can you see how once a person has been diagnosed, it becomes all too easy to overlook an individual's actual behavior in favor of the prior mental illness label?

**CONCEPT CHECK** STOP

1. **Why** isn't any single criterion adequate for classifying abnormal behavior?
2. **How** does psychology diverge from the medical model?
3. **How** might the *DSM* both underdiagnose and overdiagnose mental illness?

# Anxiety Disorders

## LEARNING OBJECTIVES

**RETRIEVAL PRACTICE** While reading the upcoming sections, respond to each Learning Objective in your own words. Then compare your responses with those in Appendix B.

1. **Describe** the symptoms of generalized anxiety disorder (GAD), panic disorder, phobias, and obsessive-compulsive disorder (OCD).
2. **Summarize** how psychological, biological, and sociocultural factors contribute to anxiety disorders.

The most frequently occurring mental disorders in the general population are **anxiety disorders**, which are diagnosed twice as often in women as in men (National Institute of Mental Health, 2011; Zerr, Holly, & Pina, 2011). Fortunately, they are also among the easiest disorders to treat and have one of the best chances for recovery (Chapter 14).

**anxiety disorder** A type of abnormal behavior characterized by unrealistic, irrational fear.

## Four Major Anxiety Disorders

Symptoms of anxiety, such as rapid breathing, dry mouth, and increased heart rate, plague all of us during stressful moments. But some people experience anxiety that is so intense and chronic that it seriously disrupts their lives. They feel threatened, unable to cope, unhappy, and insecure in a world that seems dangerous and hostile. In this section, we consider four anxiety disorders: *generalized anxiety disorder (GAD), panic disorder, phobias,* and *obsessive-compulsive disorder (OCD).* The fifth anxiety disorder, *posttraumatic stress disorder (PTSD),* is discussed in Chapter 3 (**Study Organizer 13.2**). Although we cover these disorders separately, people often have more than one anxiety disorder (Corr, 2011; Ibiloglu & Caykoylu, 2011).

**Generalized anxiety disorder** Sufferers of **generalized anxiety disorder (GAD)** experience fear and worries that are not focused on any one specific threat, which is why it's referred to as *generalized.* Their anxiety also is chronic, uncontrollable, and lasts at least six months. Because of persistent muscle tension and autonomic fear reactions, people with this disorder may develop headaches, heart palpitations, dizziness, and insomnia, making it even harder to cope with normal daily activities. This disorder affects twice as many women as it does men (Horwath & Gould, 2011).

**Panic disorder** Sudden, but brief, attacks of intense apprehension that cause trembling, dizziness, and difficulty breathing are symptoms of **panic disorder**. Panic attacks generally happen after frightening experiences or prolonged stress (and sometimes even after exercise). Panic disorder is diagnosed when several apparently spontaneous panic attacks lead to a persistent concern about future attacks. A common complication of panic disorder is agoraphobia, discussed in the next section (Cully & Stanley, 2008; Horwath & Gould, 2011).

**Phobias** **Phobias** involve a strong, irrational fear and avoidance of objects or situations that are usually considered harmless (fear of elevators or fear of going to the dentist, for example). Although the person recognizes that the fear is irrational, the experience is still one of overwhelming anxiety, and a full-blown panic attack may follow. The fourth edition of the *DSM* divides phobic disorders into three broad categories: agoraphobia, specific phobias, and social phobias.

People with *agoraphobia* restrict their normal activities because they fear having a panic attack in crowded, enclosed, or wide-open places where they would be unable to receive help in an emergency. In severe cases, people with agoraphobia may refuse to leave the safety of their homes.

A *specific phobia* is a fear of a specific object or situation, such as needles, rats, spiders, or heights. Claustrophobia (fear of closed spaces) and acrophobia (fear of heights) are the specific phobias most often treated by therapists. People with specific phobias generally recognize that their

**Study Organizer 13.2** Major anxiety disorders ✓ THE PLANNER

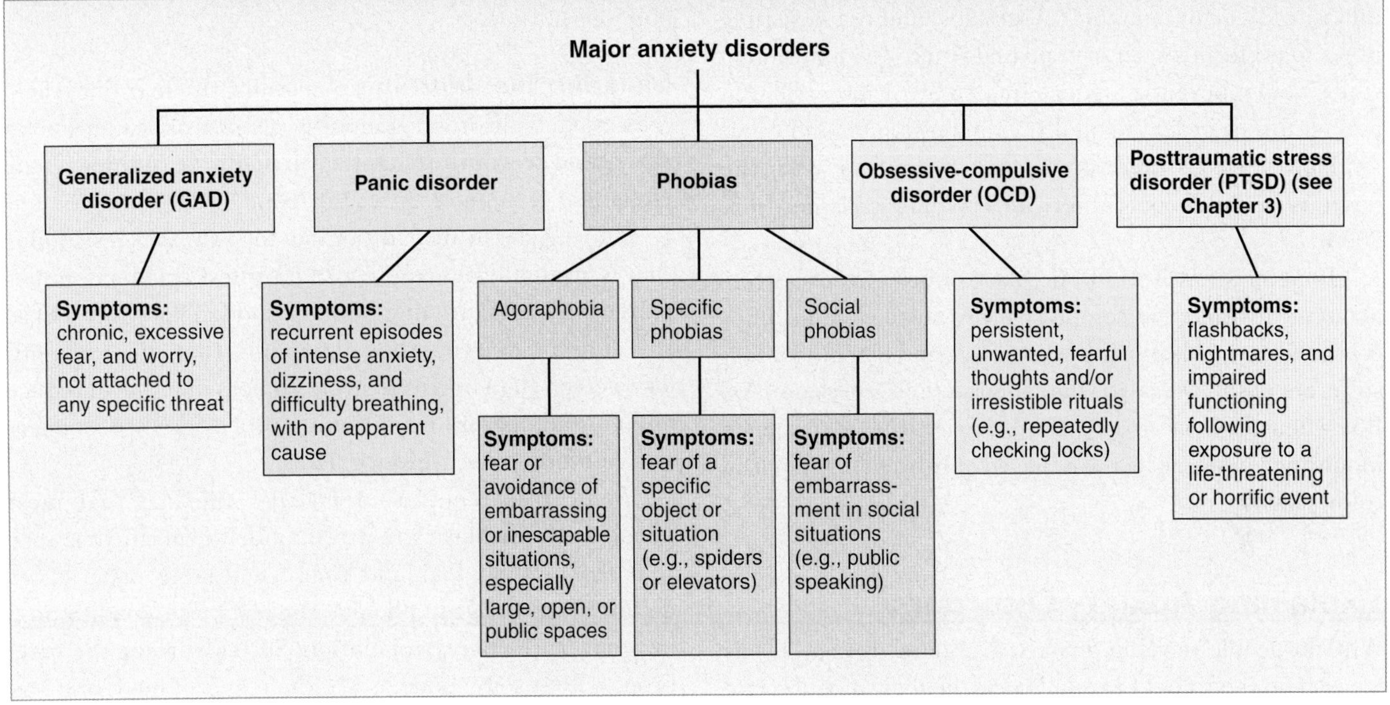
Major anxiety disorders
Generalized anxiety disorder (GAD)
Symptoms: chronic, excessive fear, and worry, not attached to any specific threat
Panic disorder
Symptoms: recurrent episodes of intense anxiety, dizziness, and difficulty breathing, with no apparent cause
Phobias
Agoraphobia
Symptoms: fear or avoidance of embarrassing or inescapable situations, especially large, open, or public spaces
Specific phobias
Symptoms: fear of a specific object or situation (e.g., spiders or elevators)
Social phobias
Symptoms: fear of embarrass-ment in social situations (e.g., public speaking)
Obsessive-compulsive disorder (OCD)
Symptoms: persistent, unwanted, fearful thoughts and/or irresistible rituals (e.g., repeatedly checking locks)
Posttraumatic stress disorder (PTSD) (see Chapter 3)
Symptoms: flashbacks, nightmares, and impaired functioning following exposure to a life-threatening or horrific event

fears are excessive and unreasonable, but they are unable to control their anxiety and will go to great lengths to avoid the feared stimulus (**Figure 13.4**).

People with *social phobias* are irrationally fearful of embarrassing themselves in social situations. Fear of public speaking and of eating in public are the most common social phobias. The fear of public scrutiny and potential humiliation may become so pervasive that normal life is impossible (Acarturk et al., 2008; Alden & Regambal, 2011; Moitra et al., 2011).

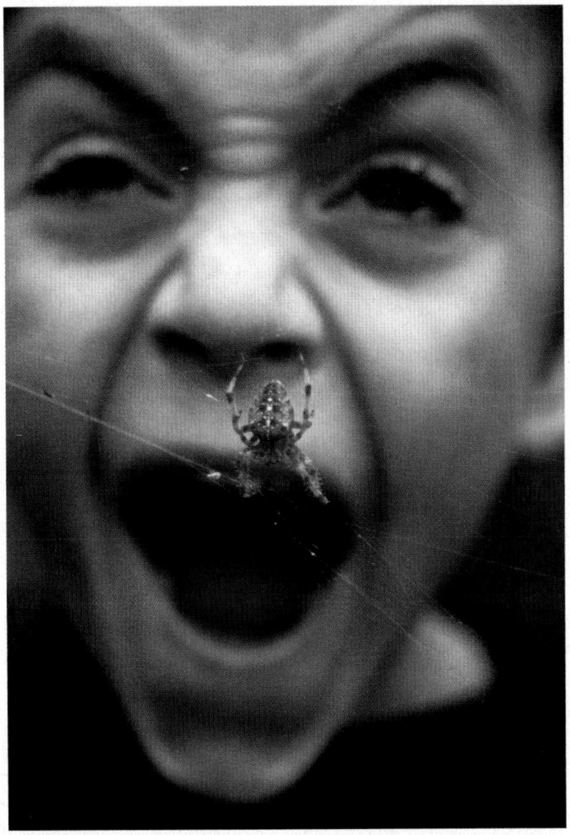

Mark Clarke/Photo Researchers

## Spider phobia • Figure 13.4

Can you imagine being so frightened by a spider that you would try to jump out of a speeding car to get away from it? This is how a person suffering from a phobia might feel.

**Obsessive-compulsive disorder (OCD)** **Obsessive compulsive disorder (OCD)** involves persistent, unwanted, fearful thoughts (obsessions) and/or irresistible urges to perform an act or repeated rituals (compulsions), which help relieve the anxiety created by the obsession. In adults, this disorder is equally common in men and women. However, it is more prevalent among boys when the onset is in childhood (American Psychiatric Association, 2000).

Imagine what it would be like to worry so obsessively about germs that you compulsively wash your hands hundreds of times a day until they are raw and bleeding. Most sufferers of OCD realize that their actions are senseless. But when they try to stop the behavior, they experience mounting anxiety, which is relieved only by giving in to the compulsions.

## Explaining Anxiety Disorders

Why do people develop anxiety disorders? Research has focused on the roles of psychological, biological, and sociocultural processes (the *biopsychosocial model*) (**Figure 13.5**).

Psychological contributions to anxiety disorders are primarily in the form of faulty cognitive processes and maladaptive learning.

**Faulty cognitions** People with anxiety disorders have habits of thinking, or cognitive habits, that make them prone to fear. They tend to be hypervigilant—they constantly scan their environment for signs of danger and ignore signs of safety. They also tend to magnify ordinary threats and failures and to be hypersensitive to others' opinions of them.

**Maladaptive learning** According to learning theorists, anxiety disorders generally result from conditioning and social learning (Chapter 6) (Cully & Stanley, 2008; Hollander & Simeon, 2011; Lovibond, 2011).

During classical conditioning, for example, a stimulus that is originally neutral (e.g., a harmless spider) becomes paired with a frightening event (a sudden panic attack) so that it becomes a conditioned stimulus that elicits anxiety. The person then begins to avoid spiders in order to reduce anxiety (an operant conditioning process known as negative reinforcement) (**Figure 13.6**).

Some researchers contend that the fact that most people with phobias cannot remember a specific instance that led to their fear and that frightening experiences do not always trigger phobias suggests that conditioning may not be the only explanation. Social learning theorists propose that some phobias are the result of modeling and imitation.

Phobias may also be learned vicariously (indirectly). In one study, rhesus monkeys viewed videos showing another monkey apparently experiencing extreme fear of a toy snake, a toy rabbit, a toy crocodile, and flowers (Cook & Mineka, 1989). The "viewing" monkeys were later afraid of the snake and crocodile but not of the rabbit or flowers, suggesting that phobias have both learned and biological components.

**Anxiety disorders and the biopsychosocial model • Figure 13.5**

The biopsychosocial model takes into account the wide variety of factors than can contribute to anxiety disorders.

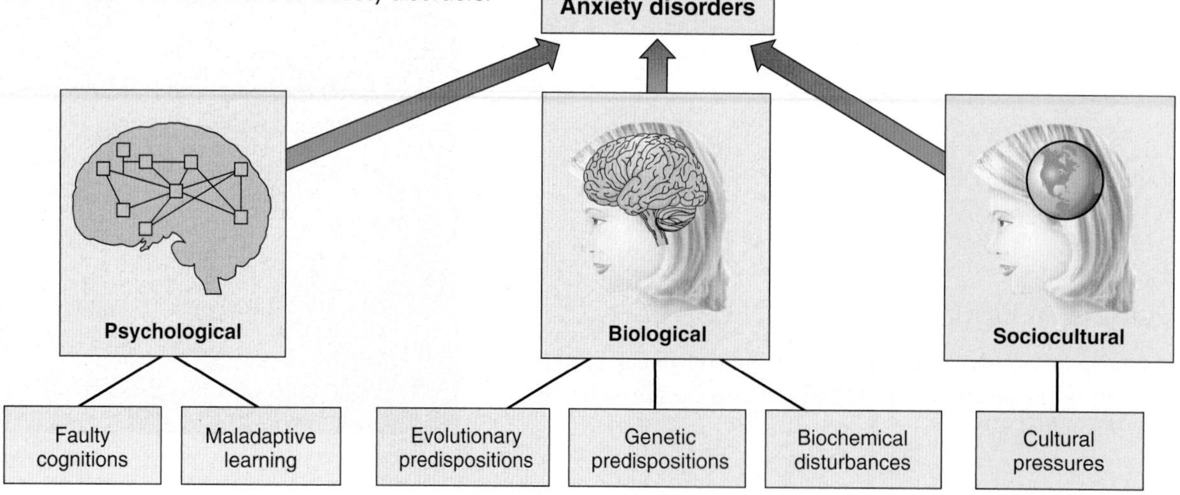

PROCESS DIAGRAM

THE PLANNER

## Conditioning and phobias • Figure 13.6

Classical conditioning combined with operant conditioning can lead to phobias. Take the example of Little Albert's classically conditioned fear of rats, discussed in Chapter 6.

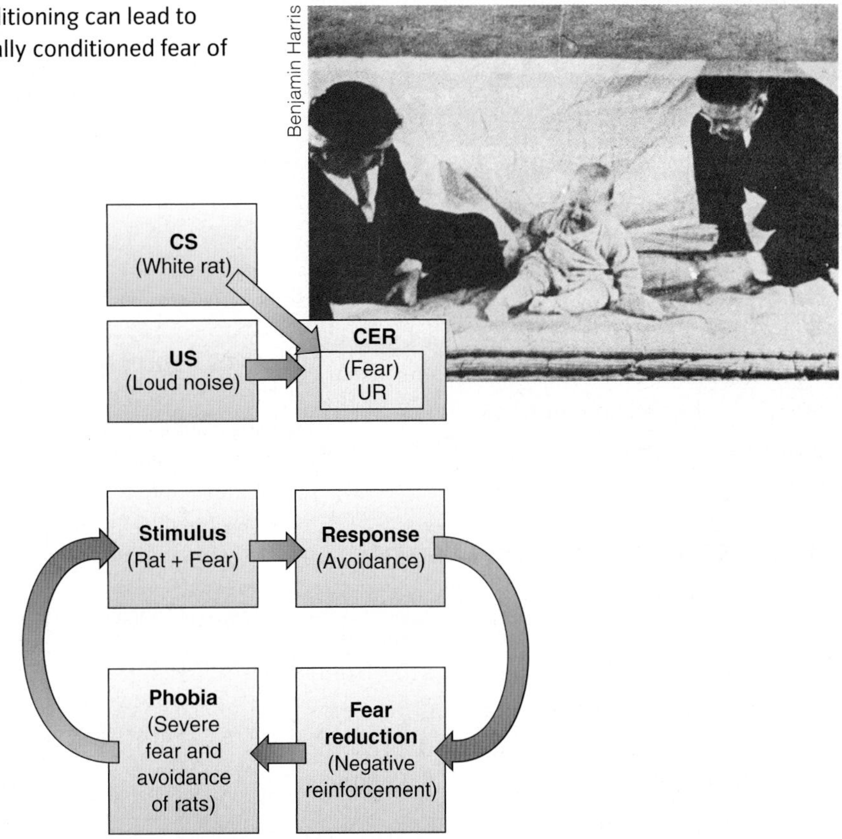


1. **Classical conditioning**
   By pairing the white rat with a loud noise, experimenters used classical conditioning to teach Little Albert to fear rats.

2. **Operant conditioning**
   Later Albert might have learned to avoid rats through operant conditioning. Avoiding rats would reduce Albert's fear, giving unintended reinforcement to his fearful behavior.

3. **Phobia**
   Over time, this cycle of fear and avoidance could produce an even more severe fear, or phobia.

**Biological factors** Some researchers believe that phobias reflect an evolutionary predisposition to fear that which was dangerous to our ancestors (Boyer & Bergstrom, 2011; Glover, 2011; Mineka & Oehlberg, 2008; Walker et al., 2008). Some people with panic disorder also seem to be genetically predisposed toward an overreaction of the autonomic nervous system, further supporting the argument for a biological component. In addition, stress and arousal seem to play a role in panic attacks, and drugs such as caffeine or nicotine and even hyperventilation can trigger an attack, all suggesting a biochemical disturbance.

**Sociocultural factors** In addition to psychological and biological components, sociocultural factors can contribute to anxiety. There has been a sharp rise in anxiety disorders in the past 50 years, particularly in Western industrialized countries. Can you see how our increasingly fast-paced lives—along with our increased mobility, decreased job security, and decreased family support—might contribute to anxiety? Unlike the dangers that humans faced in our evolutionary history, today's threats are less identifiable and immediate, which may lead some people to become hypervigilant and predisposed to anxiety disorders.

Further support for sociocultural influences on anxiety disorders is that anxiety disorders can have dramatically different forms in other cultures. For example, in a collectivist twist on anxiety, the Japanese have a type of social phobia called *taijin kyofusho* (TKS), which involves morbid dread of doing something to embarrass others. This disorder is quite different from the Western version of social phobia, which centers on a fear of criticism.

**CONCEPT CHECK** STOP

1. **How** do generalized anxiety disorders and phobias differ?
2. **What** are examples of the faulty cognitions that characterize anxiety disorders?

# Mood Disorders

## LEARNING OBJECTIVES

**RETRIEVAL PRACTICE** While reading the upcoming sections, respond to each Learning Objective in your own words. Then compare your responses with those in Appendix B.

1. **Explain** how major depressive disorder and bipolar disorder differ.
2. **Summarize** research on the biological and psychological factors that contribute to mood disorders.

As the name implies, **mood disorders** (also known as affective disorders) are characterized by extreme disturbances in emotional states. There are two main types of mood disorders—*major depressive disorder* and *bipolar disorder.*

## Understanding Mood Disorders

We all feel "blue" sometimes, especially following the loss of a job, end of a relationship, or death of a loved one. People suffering from **major depressive disorder,** however, may experience a lasting and continuously depressed mood without a clear trigger (Alloy et al., 2011; Feliciano, Segal, & Vair, 2011).

**major depressive disorder** Long-lasting depressed mood that interferes with the ability to function, feel pleasure, or maintain interest in life (Swartz & Margolis, 2004).

**bipolar disorder** Repeated episodes of mania (unreasonable elation and hyperactivity) and depression.

Clinically depressed people are so deeply sad and discouraged that they often have trouble sleeping, are likely to lose (or gain) weight, and may feel so fatigued that they cannot go to work or school or even comb their hair and brush their teeth. They may sleep constantly, have problems concentrating, and feel so profoundly sad and guilty that they consider suicide. Seriously depressed individuals have a hard time thinking clearly or recognizing their own problems.

When depression is *unipolar,* the depressive episode eventually ends, and the person returns to a "normal" emotional level. People with **bipolar disorder,** however, rebound to the opposite state, known as *mania* (**Figure 13.7**).

### Mood disorders • Figure 13.7

If major depressive disorders and bipolar disorders were depicted on a graph, they might look something like this.

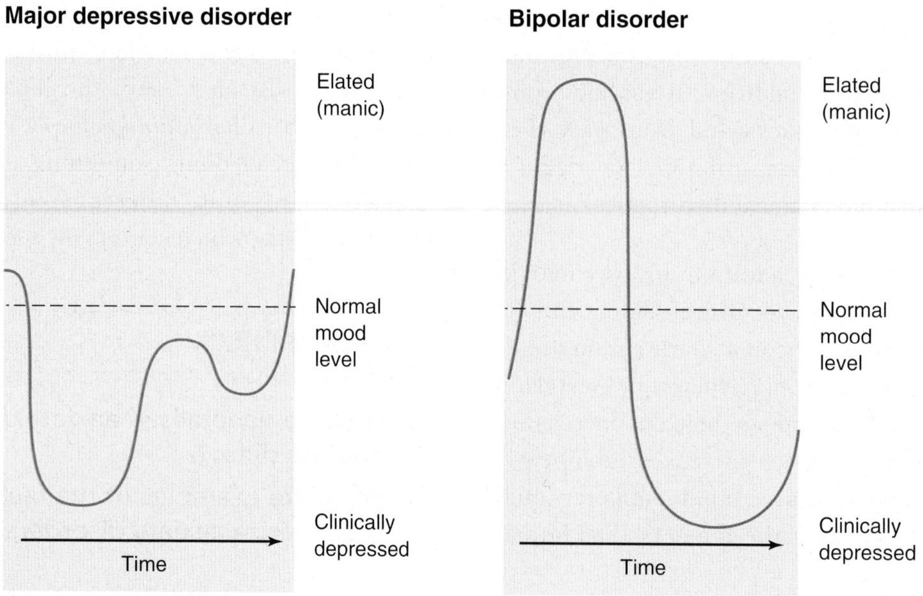

## MYTH BUSTERS

THE PLANNER

### DANGER SIGNS FOR SUICIDE

Suicide is a serious danger associated with severe depression, bipolar disorder, and several other disorders. Because of the shame and secrecy surrounding the suicidal person, there are many misconceptions and stereotypes. Can you correctly identify which of the following is true or false?

1. People who talk about suicide are not likely to commit suicide.
2. Suicide usually takes place with little or no warning.
3. Suicidal people are fully intent on dying.
4. Children of parents who attempt suicide are at greater risk of committing suicide.
5. Suicidal people remain so forever.
6. Men are more likely than women to actually kill themselves by suicide.
7. When a suicidal person has been severely depressed and seems to be "snapping out of it," the danger of suicide decreases substantially.
8. Only depressed people commit suicide.
9. Thinking about suicide is rare.
10. Asking a depressed person about suicide will push him or her over the edge and cause a suicidal act that would not otherwise have occurred.

Now, compare your responses to the experts' answers and explanations (Bailey et al., 2011; Baldessarini & Tondo, 2008; Borges et al., 2010; Brent & Melhem, 2008; Callanan & Davis, 2012; Dutra et al., 2008; Kring et al., 2012; Lilienfeld et al., 2010; Murray, 2009; Schneidman, 1981).

1. and **2. False.** Up to three-quarters of those who take their own lives talk about it and give warnings about their intentions beforehand. They may say, "If something happens to me, I want you to ..." or "Life just isn't worth living." They also provide behavioral clues, such as giving away valued possessions, withdrawing from family and friends, and losing interest in favorite activities.

3. **False.** Only about 3 to 5% of suicidal people truly intend to die. Most are just unsure about how to go on living. They cannot see their problems objectively enough to realize that they have alternative courses of action. They often gamble with death, arranging it so that fate or others will save them. Moreover, once the suicidal crisis passes, they are generally grateful to be alive.
4. **True.** Children of parents who attempt or commit suicide are at much greater risk of following in their footsteps. As Schneidman (1969) puts it, "The person who commits suicide puts his psychological skeleton in the survivor's emotional closet" (p. 225).
5. **False.** People who want to kill themselves are usually suicidal only for a limited period.
6. **True.** Although women are much more likely to attempt suicide, men are far more likely to actually commit suicide. Men are also more likely to use stronger methods, such as guns versus pills.
7. **False.** When people are first coming out of a depression, they are actually at greater risk! This is because they now have the energy to actually commit suicide.
8. **False.** Suicide rates are highest for people with major depressive disorders. However, suicide is also the leading cause of premature death in people who suffer from schizophrenia. In addition, suicide is a major cause of death in people with anxiety disorders and alcohol and other substance-related disorders. Furthermore, suicide is not limited to people with depression. Poor physical health, serious illness, substance abuse (particularly alcohol), loneliness, unemployment, and even natural disasters may push many over the edge.
9. **False.** Estimates from various studies are that 40 to 80% of the general public have thought about committing suicide at least once in their lives.
10. **False.** Because society often considers suicide a terrible, shameful act, asking directly about it can give the person permission to talk. In fact, not asking might lead to further isolation and depression.

During a manic episode, the person is overly excited, extremely active, and easily distracted. In addition, he or she exhibits unrealistically high self-esteem, an inflated sense of importance, and poor judgment—giving away valuable possessions or going on wild spending sprees. The person also is often hyperactive and may not sleep for days at a time, yet does not become fatigued. Thinking is speeded up and can change abruptly to new topics, showing "rapid flight of ideas." Speech is also rapid ("pressured speech"), and it is difficult for others to get a word in edgewise.

A manic episode may last a few days or a few months, and it generally ends abruptly. The ensuing depressive episode generally lasts three times as long as the manic episode. The lifetime risk for bipolar disorder is low—somewhere between 0.5 and 1.6%—but it can be one of the most debilitating and lethal disorders, with a suicide rate between 10 and 20% among sufferers (Bender & Alloy, 2011; Gutiérrez-Rojas, Jurado, & Gurpegui, 2011; Klimes-Dougan et al., 2008). See the *Myth Busters* to learn the truth about some common misconceptions on suicide.

## Explaining Mood Disorders

Biological factors appear to play a significant role in both major depressive disorder and bipolar disorder. Recent research suggests that structural brain changes may contribute to these mood disorders. Other research points to imbalances of the neurotransmitters serotonin, norepinephrine, and dopamine (Barton et al., 2008; Lopez-Munoz & Alamo, 2011; Pu et al., 2011; Wiste et al., 2008). Drugs that alter the activity of these neurotransmitters also decrease the symptoms of depression (and are therefore called antidepressants) (Chuang, 1998).

Some research indicates that mood disorders may be inherited. For example, when one identical twin has a mood disorder, there is about a 50% chance that the other twin will also develop the illness (Boardman, Alexander, & Stallings, 2011; Elder & Mosack, 2011; Faraone, 2008). However, it is important to remember that relatives generally have similar environments as well as similar genes.

Finally, the evolutionary perspective suggests that moderate depression may be a normal and healthy adaptive response to a very real loss (such as the death of a loved one). The depression helps us to step back and reassess our goals (Nettle, 2011; Panksepp & Wyatt, 2011; Surbey, 2011; Varga, 2011). Clinical, severe depression may just be an extreme version of this generally adaptive response.

Psychosocial theories of depression focus on environmental stressors and disturbances in the person's interpersonal relationships, thought processes, self-concept, and a history of abuse and assault (Betancourt et al., 2011;

## Applying Psychology

THE PLANNER

# How Can You Tell If Someone Is Suicidal?

If you believe someone is contemplating suicide, act on your beliefs. Stay with the person if there is any immediate danger. Encourage him or her to talk to you rather than withdraw. Show the person that you care, but do not give false reassurances that "everything will be okay." This type of response may make the suicidal person feel *more* alienated. Instead, openly ask if the person is feeling hopeless and suicidal. Do not be afraid to discuss suicide with people who feel depressed or hopeless, fearing that you will just put ideas into their heads. The reality is that people who are left alone or who are told they can't be serious about suicide often attempt it.

If you suspect someone is imminently suicidal, it is vitally important that you help the person obtain counseling. Get immediate help! Consider calling the police for emergency intervention, the person's therapist, and/or the toll-free 24/7 hotline 1-800-SUICIDE. Most cities also have walk-in centers that provide emergency counseling.

In addition, share your suspicions with parents, friends, or others who can help in a suicidal crisis. To save a life, you may have to betray a secret when someone confides in you. If you feel you can't do this alone, get another person to help you.

Even people who enjoy fame and success may be at risk for suicide. British author J. K. Rowling, creator of the *Harry Potter* book series, revealed that she had "suicidal thoughts" while suffering from clinical depression in the mid-1990s.

Chip Somodevilla/Getty Images

Cheung, Gilbert, & Irons, 2004; Harkness et al., 2010; Mendelson, Turner, & Tandon, 2011; Mota et al., 2012; Neri et al., 2012). The psychoanalytic explanation sees depression as anger (stemming from feelings of rejection) turned inward when an important relationship is lost. The humanistic school says that depression results when a person demands perfection of him- or herself or when positive growth is blocked. (Keen, 2011; O'Connell, 2008; Weinstock, 2008).

According to the **learned helplessness** theory (Seligman, 1975, 1994, 2007), depression occurs when people (and other animals) become resigned to the idea that they are helpless to escape from something painful. Learned helplessness may be particularly likely to trigger depression if the person attributes failure to causes that are internal ("my own weakness"), stable ("this weakness is long-standing and unchanging"), and global ("this weakness is a problem in lots of settings") (Alloy et al., 2011; Ball, McGuffin, & Farmer, 2008; Newby & Moulds, 2011; Wise & Rosqvist, 2006).

Keep in mind that suicide is a major danger associated with mood disorders (both major depressive disorder and bipolar disorder). Given that some people who suffer with these disorders are so disturbed that they lose contact with reality, they may fail to recognize the danger signs or to seek help. *Applying Psychology* offers important information for both the sufferers and their family and friends.

**CONCEPT CHECK** STOP

1. **What** are the key characteristics of a manic episode?
2. **How** might major depression be an exaggeration of an evolutionary adaptation?

# Schizophrenia

## LEARNING OBJECTIVES

**RETRIEVAL PRACTICE** While reading the upcoming sections, respond to each Learning Objective in your own words. Then compare your responses with those in Appendix B.

1. **Describe** some common symptoms of schizophrenia.
2. **Compare** the traditional five-subtype system for classifying different types of schizophrenia with the two-group system that has recently emerged.
3. **Summarize** the biological and psychosocial factors that contribute to schizophrenia.

Imagine that your daughter has just left for college and that you hear voices inside your head shouting, "You'll never see her again! She'll die." Or what if you saw live animals in your refrigerator? These experiences have plagued "Mrs. T" for decades (Gershon & Rieder, 1993).

Mrs. T suffers from **schizophrenia** (skit-so-FREE-nee-uh). Schizophrenia is often so severe that it is considered a psychosis, meaning that the person is out of touch with reality. People with schizophrenia have serious problems caring for themselves, relating to others, and holding a job. In extreme cases, people with schizophrenia require institutional or custodial care.

**schizophrenia** A group of psychotic disorders involving major disturbances in perception, language, thought, emotion, and behavior. The individual withdraws from people and reality, often into a fantasy life of delusions and hallucinations.

Schizophrenia is one of the most widespread and devastating mental disorders. Approximately 1 out of every 100 people will develop schizophrenia in his or her lifetime, and approximately half of all people who are admitted to mental hospitals are diagnosed with this disorder (Gottesman, 1991; Kendler et al., 1996; Kessler et al., 1994; Regier et al., 1993). Schizophrenia usually emerges between the late teens and the mid-30s and only rarely prior to adolescence or after age 45. It seems to be equally prevalent in men and women, but it's generally more severe and strikes earlier in men than in women (Chang et al., 2011; Combs et al., 2008; Faraone, 2008; Gottesman, 1991; Lee et al., 2011).

## MYTH BUSTERS

THE PLANNER

### DO PEOPLE WITH SCHIZOPHRENIA HAVE MULTIPLE PERSONALITIES?

As shown in the cartoon and in popular movies, such as *Me, Myself, and Irene*, *schizophrenia* is commonly confused with *multiple personality disorder* (now known as *dissociative identity disorder*, p. 361). This widespread myth persists in part from confusing terminology. Literally translated, schizophrenia means "split mind," referring to a split from reality that shows itself in disturbed perceptions, language, thought, emotions, and/or behavior.

In contrast, dissociative identity disorder (DID) refers to the condition in which two or more distinct personalities exist within the same person at different times. People with schizophrenia have only one personality.

Why does this matter? Confusing schizophrenia with multiple personalities is not only technically incorrect, it also trivializes the devastating effects of both disorders, which may include severe anxiety, social isolation, unemployment, homelessness, substance abuse, clinical depression, and even suicide (Lilienfeld et al., 2010; Preventing Suicide, 2009; Smith et al., 2011).

Dave Parker/CartoonStock

"THANKS FOR CURING MY SCHIZOPHRENIA—WE'RE BOTH FINE NOW!"

Many people confuse schizophrenia with dissociative identity disorder, which is sometimes referred to as *split* or *multiple personality disorder* (see *Myth Busters*). Schizophrenia means "split mind," but when Eugen Bleuler coined the term in 1911, he was referring to the fragmenting of thought processes and emotions, not personalities (Neale, Oltmanns, & Winters, 1983). As we discuss later in this chapter, dissociative identity disorder is the rare condition of having more than one distinct personality.

## Symptoms of Schizophrenia

Schizophrenia is a group of disorders characterized by a disturbance in one or more of the following areas: *perception*, *language*, *thought*, *affect* (emotions), or *behavior*.

**Perception** The senses of people with schizophrenia may be either enhanced or blunted. The filtering and selection processes that allow most people to concentrate on whatever they choose are impaired, and sensory stimulation is jumbled and distorted.

People with schizophrenia may also experience **hallucinations**—most commonly auditory hallucinations (hearing voices and sounds).

**hallucination** False, imaginary sensory perception that occurs without an external stimulus.

On rare occasions, people with schizophrenia will hurt others in response to their distorted perceptions. But a person with schizophrenia is more likely to be self-destructive and suicidal than violent toward others.

**Language and thought** For people with schizophrenia, words lose their usual meanings and associations, logic is impaired, and thoughts are disorganized and bizarre. When language and thought disturbances are mild, the individual jumps from topic to topic. With more severe disturbances, the person jumbles phrases and words together (into a "word salad") or creates artificial words.

The most common—and frightening—thought disturbance experienced by people with schizophrenia is the lack of contact with reality (psychosis).

**Delusions** are also common in people with schizophrenia. We all experience exaggerated thoughts from time to time, such as thinking a friend is trying to avoid us, but the delusions of schizophrenia are much more extreme. For example, if someone falsely believed that the postman who routinely delivered mail to his house every afternoon was a co-conspirator in a plot to kill him, it would, likely qualify as a *delusion of persecution*. In *delusions of grandeur*, people believe that they are someone very important, perhaps Jesus Christ or the queen of England. In *delusions of reference*, unrelated events are given special significance, as when a person believes that a radio program is giving him or her a special message.

**delusion** Mistaken belief based on misrepresentations of reality.

A recent trend among individuals suffering from various mental disorders is to join Internet groups where they can share their experiences with others. For example, people with schizophrenia and other psychotic disorders can now share with fellow sufferers their belief that they are being stalked or that their minds are being controlled by technology. Some experts consider Internet groups that offer peer support to be helpful to the mentally ill, but others believe that they may reinforce troubled thinking and impede treatment. Those who use the sites report feeling relieved by the sense that they are not alone in their suffering, but it should be noted that these sites are not moderated by professionals and pose the danger of actually amplifying the symptoms of mentally ill individuals (Kershaw, 2008).

**Emotion** Changes in emotion usually occur in people with schizophrenia. In some cases, emotions are exaggerated and fluctuate rapidly. At other times, emotions become blunted. Some people with schizophrenia have *flattened affect*—almost no emotional response of any kind.

**Behavior** Disturbances in behavior may take the form of unusual actions that have special meaning. For example, one patient massaged his head repeatedly to "clear it" of unwanted thoughts. People with schizophrenia also may become *cataleptic* and assume a nearly immobile stance for an extended period.

## Types of Schizophrenia

For many years, researchers divided schizophrenia into five subtypes: *paranoid*, *catatonic*, *disorganized*, *undifferentiated*, and *residual* (**Table 13.1**). Although these terms are included in the fourth edition of the *DSM*, critics contend that this system does not differentiate in terms of prognosis, cause, or response to treatment and that the undifferentiated type is merely a catchall for cases that are difficult to diagnose (American Psychiatric Association, 2000).

For these reasons, researchers have proposed an alternative classification system:

1. **Positive schizophrenia symptoms** involve additions to or exaggerations of normal thought processes and behaviors, including bizarre delusions and hallucinations.
2. **Negative schizophrenia symptoms** involve the loss or absence of normal thought processes and behaviors, including impaired attention, limited or toneless speech, flat or blunted affect, and social withdrawal.

**Subtypes of schizophrenia Table 13.1**

| | |
|---|---|
| **Paranoid** | Delusions (persecution and grandeur) and hallucinations (hearing voices) |
| **Catatonic** | Motor disturbances (immobility or agitated, purposeless activity) and echo speech (repeating the speech of others) |
| **Disorganized** | Incoherent speech, flat or exaggerated emotions, and social withdrawal |
| **Undifferentiated** | Varied symptoms that meet the criteria for schizophrenia but do not fall into any of the above subtypes |
| **Residual** | No longer meets the full criteria for schizophrenia but still shows some symptoms |

Positive symptoms are more common when schizophrenia develops rapidly, whereas negative symptoms are more often found in slow-developing schizophrenia. Positive symptoms are associated with better adjustment before the onset and a better prognosis for recovery.

In addition to these two groups, the latest *DSM* suggests adding another dimension to reflect *disorganization of behavior*, including rambling speech, erratic behavior, and inappropriate affect.

## Explaining Schizophrenia

Because schizophrenia comes in many different forms, it probably has multiple biological and psychosocial bases. Let's look at biological contributions first.

Prenatal stress and viral infections, birth complications, immune responses, maternal malnutrition, and advanced paternal age all may contribute to the development of schizophrenia (Markham & Koenig, 2011; Tandon, Keshavan, & Nasrallah, 2008; Weinberger & Harrison, 2011). However, most biological theories of schizophrenia focus on genetics, neurotransmitters, and brain abnormalities.

- *Genetics* Although researchers are beginning to identify specific genes related to schizophrenia, most genetic studies have focused on twins and adoptions (Drew et al., 2011; Faraone, 2008; Owens et al., 2011; Riley & Kendler, 2011). This research indicates that the risk for schizophrenia increases with genetic similarity; that is, people who share more genes with a person who has schizophrenia are more likely to develop the disorder (**Figure 13.8**).

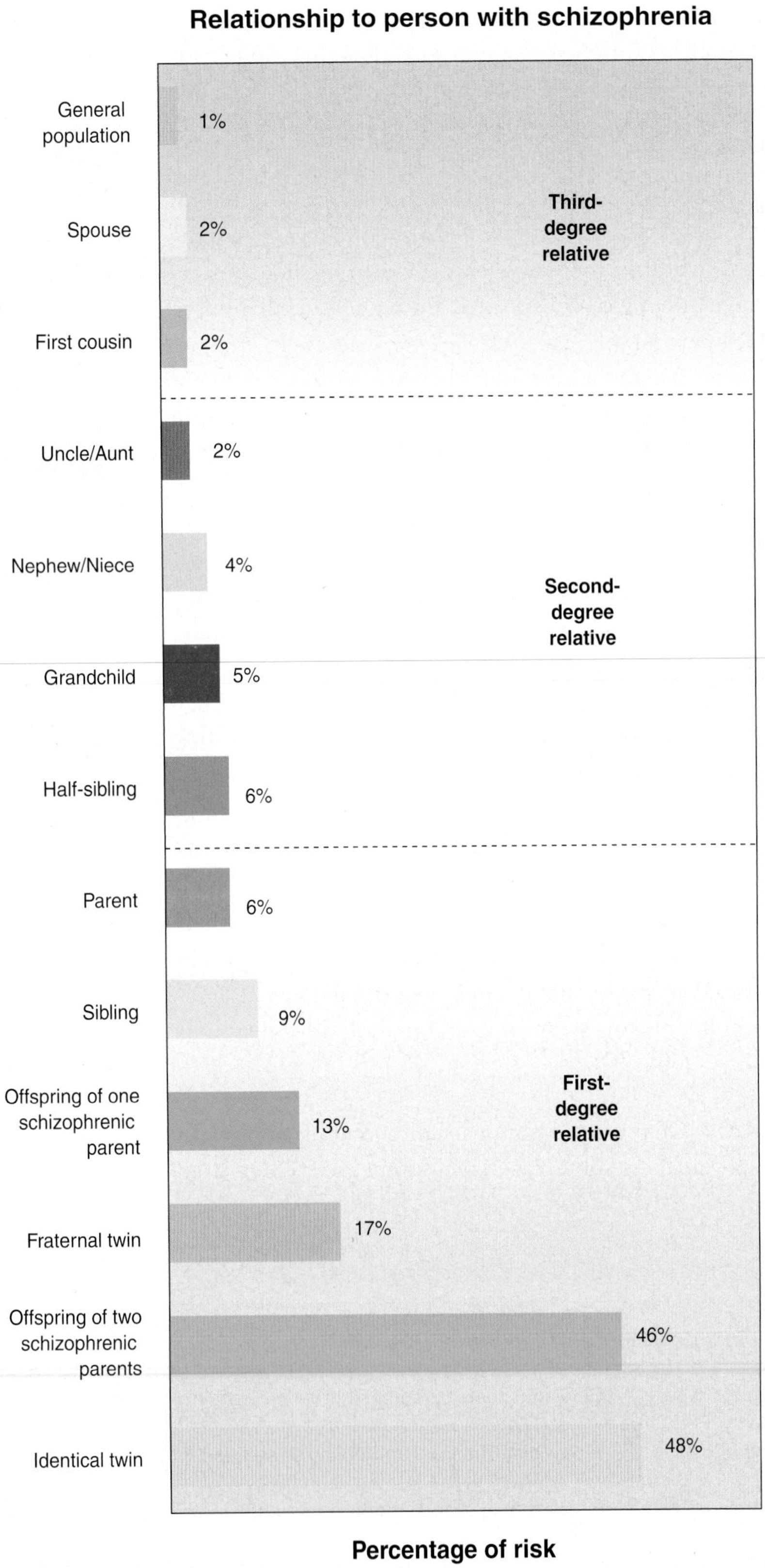


From *Schizophrenia Genesis: The Origins of Madness* by Irving I. Gottesman. © 1991 by Irving I. Gottesman. Used with the permission of Worth Publishers.

**Genetics and schizophrenia • Figure 13.8**

Biological relatives of people with schizophrenia have a higher risk than the general population of developing the disorder during their lifetimes. Closer relatives (who have a more similar genetic makeup) have a greater risk than more distant relatives (Gottesman, 1991, p. 96).

## Brain activity in schizophrenia • Figure 13.9

These positron emission tomography (PET) scans show variations in the brain activity of normal individuals, people with major depressive disorder, and individuals with schizophrenia. Warmer colors (red, yellow) indicate increased activity.

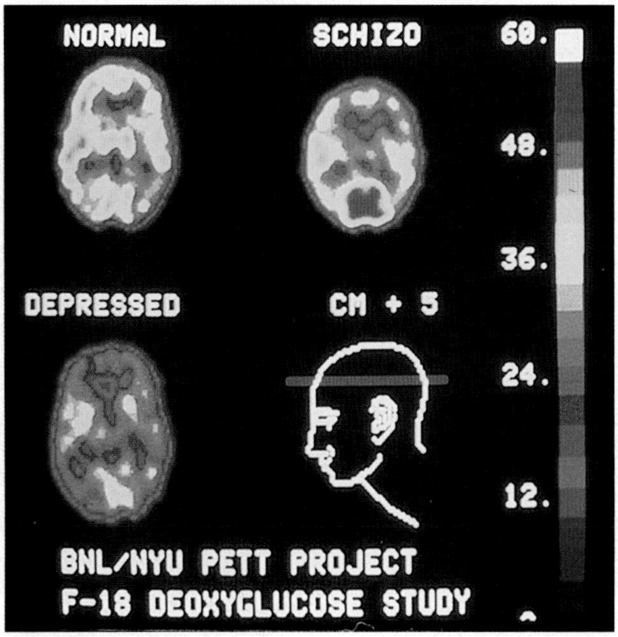


Science Source/Photo Researchers, Inc.

- *Neurotransmitters* Precisely how genetic inheritance produces schizophrenia is unclear. According to the **dopamine hypothesis**, overactivity of certain dopamine neurons in the brain causes schizophrenia (Lodge & Grace, 2011; Miyake et al., 2011; Nord & Farde, 2011; Seeman, 2011). This hypothesis is based on two observations. First, administering amphetamines increases the amount of dopamine and can produce (or worsen) some symptoms of schizophrenia, especially in people with a genetic predisposition to the disorder. Second, drugs that reduce dopamine activity in the brain reduce or eliminate some symptoms of schizophrenia.
- *Brain abnormalities* The third major biological theory for schizophrenia involves abnormalities in brain function and structure. Researchers have found larger cerebral ventricles (fluid-filled spaces in the brain) in some people with schizophrenia (Agartz et al., 2011; Fusar-Poli et al., 2011; Galderisi et al., 2008).

  Also, some people with chronic schizophrenia have a lower level of activity in their frontal and temporal lobes—areas that are involved in language, attention, and memory (**Figure 13.9**). Damage in these regions might explain the thought and language disturbances that characterize schizophrenia. This lower level of brain activity, and schizophrenia itself, may also result from an overall loss of gray matter (neurons in the cerebral cortex) (Crespo-Facorro et al., 2007; Salgado-Pineda et al., 2011; White & Hilgetag, 2011).

Clearly, biological factors play a key role in schizophrenia. But the fact that even in identical twins—who share identical genes—the heritability of schizophrenia is only 48% tells us that nongenetic factors must contribute the remaining percentage. Most psychologists believe that there are at least two possible psychosocial contributors.

According to the **diathesis-stress model** of schizophrenia, stress plays an essential role in triggering schizophrenic episodes in people with an inherited predisposition (or diathesis) toward the disease (Arnsten, 2011; Beaton & Simon, 2011; McGrath, 2011).

Some investigators suggest that communication disorders in family members may also be a predisposing factor for schizophrenia. Such disorders include unintelligible speech, fragmented communication, and parents' frequently sending severely contradictory messages to children. Several studies have also shown greater rates of relapse and worsening of symptoms among hospitalized patients who went home to families that were critical and hostile toward them or overly involved in their lives emotionally (McFarlane, 2011; Roisko et al., 2011).

## The biopsychosocial model and schizophrenia • Figure 13.10

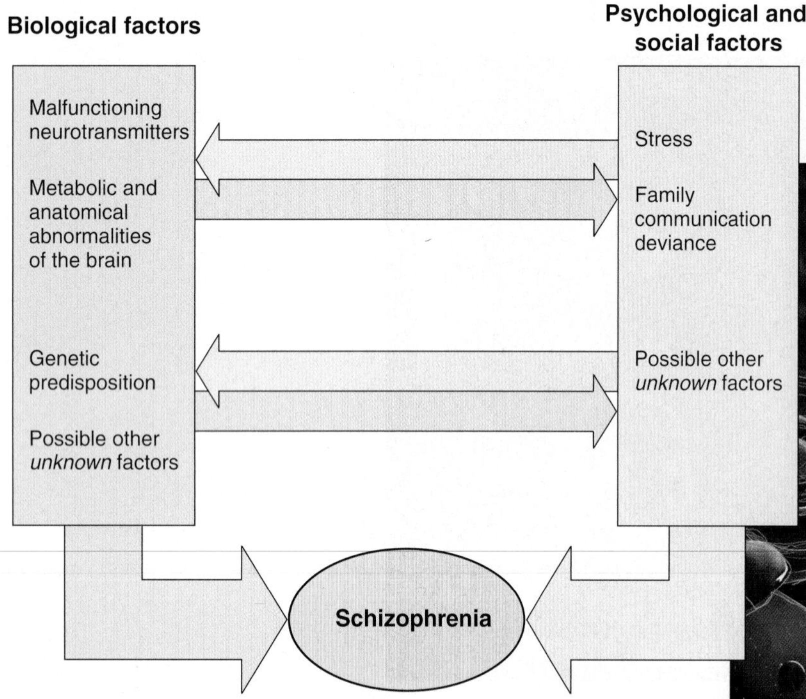


**a.** There is strong evidence linking schizophrenia to biological, psychological, and social factors—the *biopsychosocial model*.

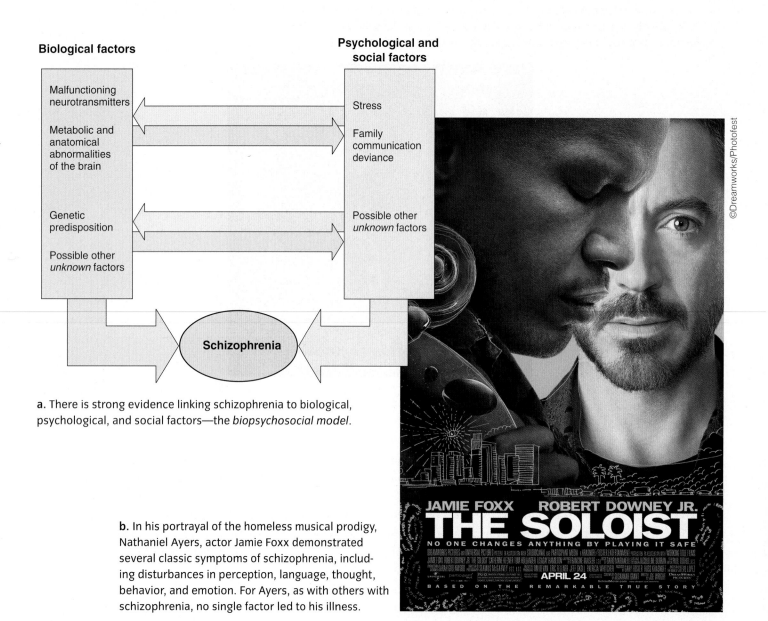

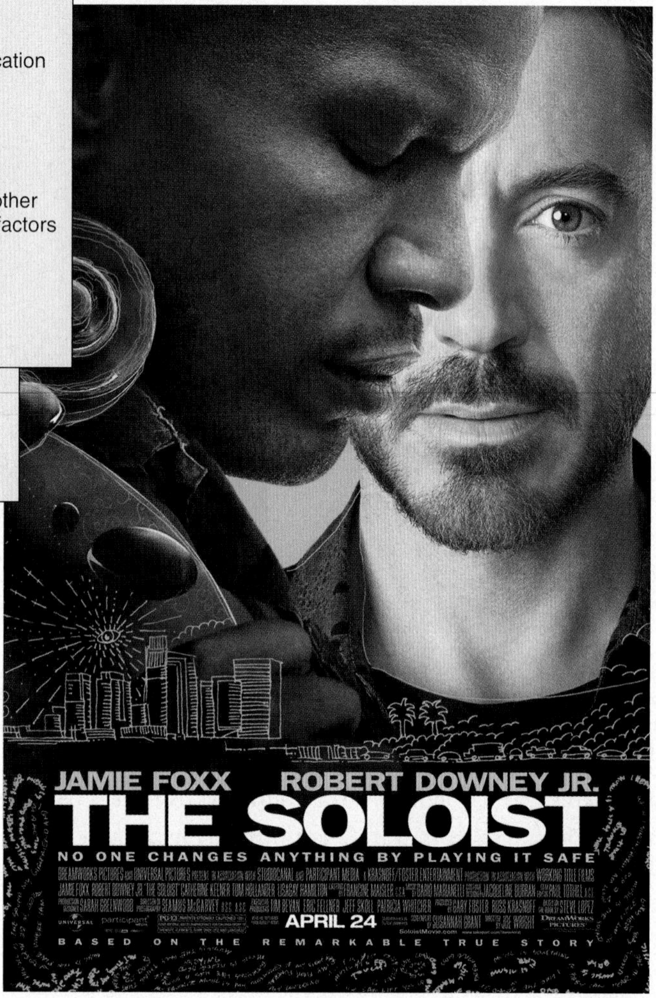


©Dreamworks/Photofest

**b.** In his portrayal of the homeless musical prodigy, Nathaniel Ayers, actor Jamie Foxx demonstrated several classic symptoms of schizophrenia, including disturbances in perception, language, thought, behavior, and emotion. For Ayers, as with others with schizophrenia, no single factor led to his illness.

How should we evaluate the different theories about the causes of schizophrenia? Critics of the dopamine hypothesis and the brain damage theory argue that those theories fit only some cases of schizophrenia. Moreover, with both theories, it is difficult to determine cause and effect. The disturbed-communication theories are also hotly debated, and research is inconclusive. Schizophrenia is probably the result of a combination of known and unknown interacting factors (**Figure 13.10**).

### CONCEPT CHECK

STOP

1. **How** do hallucinations and delusions differ?
2. **How** do positive and negative symptoms of schizophrenia differ?
3. **What** is the diathesis-stress model of schizophrenia?

# Other Disorders

## LEARNING OBJECTIVES

**RETRIEVAL PRACTICE** While reading the upcoming sections, respond to each Learning Objective in your own words. Then compare your responses with those in Appendix B.

1. **Describe** the types of dissociative disorders.
2. **Identify** the major characteristics of personality disorders.

Having discussed anxiety disorders, mood disorders, and schizophrenia, we now explore two additional disorders—*dissociative* and *personality*.

## Dissociative Disorders

The most dramatic psychological disorders are **dissociative disorders**. There are several types of dissociative disorders, but all involve a splitting apart (a *dissociation*) of significant aspects of experience from memory or consciousness. Individuals dissociate from the core of their personality by failing to remember past experiences (*dissociative amnesia*) (**Figure 13.11**), by leaving home and wandering off (*dissociative fugue*), by losing their sense of reality and feeling estranged from the self (*depersonalization disorder*), or by developing completely separate personalities (*dissociative identity disorder*).

**dissociative disorder** Amnesia, fugue, or multiple personalities resulting from avoidance of painful memories or situations.

Unlike most other psychological disorders, the primary cause of dissociative disorders appears to be environmental variables, with little or no genetic influence (Hulette, Freyd, & Fisher, 2011; Sinason, 2011; Waller & Ross, 1997).

The most severe dissociative disorder is **dissociative identity disorder (DID)**—previously known as multiple personality disorder—in which at least two separate and distinct personalities exist within a person at the same time. Each personality has unique memories, behaviors, and social relationships. Transition from one personality to another occurs suddenly and is often triggered by psychological stress. Usually, the original personality has no knowledge or awareness of the alternate personalities, but all of the personalities may be aware of lost periods of time. The disorder is diagnosed more among women than among men. Women with DID also tend to have more identities, averaging 15 or more, compared with men, who average 8 (American Psychiatric Association, 2000).

**Dissociation as an escape • Figure 13.11**

The major force behind all dissociative disorders is the need to escape from anxiety. Imagine witnessing a loved one's death in a horrible car accident. Can you see how your mind might cope by blocking out all memory of the event?

DID is a controversial diagnosis. Some experts suggest that many cases are faked or result from false memories and an unconscious need to please the therapist (Kihlstrom, 2005; Lawrence, 2008; Pope et al., 2007; Stafford & Lynn, 2002). On the other side of the debate are psychologists who accept the validity of DID and provide treatment guidelines (Brown, 2011; Chu, 2011; Dalenberg et al., 2007; Lipsanen et al., 2004; Spiegel & Maldonado, 1999).

## Personality Disorders

What would happen if the characteristics of a personality were so inflexible and maladaptive that they significantly impaired someone's ability to function? This is what happens with **personality disorders**. Several types of personality disorders are included in this category in the fourth edition of the *DSM*, but here we will focus on antisocial personality disorder and borderline personality disorder.

**personality disorder** Inflexible, maladaptive personality traits that cause significant impairment of social and occupational functioning.

**Antisocial personality disorder** The term **antisocial personality disorder** is used interchangeably with the terms *sociopath* and *psychopath*. These labels describe behavior so far outside the ethical and legal standards of society that many consider it the most serious of all mental disorders. Unlike people with anxiety, mood disorders, and schizophrenia, those with this diagnosis feel little personal distress (and may not be motivated to change). Yet their maladaptive traits generally bring considerable harm and suffering to others (Hervé et al., 2004; Kirkman, 2002; Nathan et al., 2003). Although serial killers are often seen as classic examples of antisocial personality disorder (**Figure 13.12**), many sociopaths harm people in less dramatic ways—for example, as ruthless businesspeople and crooked politicians.

The four hallmarks of antisocial personality disorder are *egocentrism* (preoccupation with oneself and insensitivity to the needs of others), *lack of conscience*, *impulsive behavior*, and *superficial charm* (American Psychiatric Association, 2000).

Unlike most adults, individuals with antisocial personality disorder act impulsively, without giving thought to

### Public trials and mental illness • Figure 13.12

In the summer of 2012, many people were surprised that James Holmes (the clean-cut young man in the photo on the left) was accused of being the Colorado movie theater shooter. These horrific murders and injuries, along with dying his hair to match Batman's nemesis—The Joker (photo on the right), led many lay people and mental health professionals to question his sanity. Does intensive media coverage of events like this increase our misconceptions about the mentally ill?

North County Times/ZUMA Press/Newscom

RJ SANGOSTI/AFP/Getty Images/Newscom

the consequences. They are usually poised when confronted with their destructive behavior and feel contempt for anyone they are able to manipulate. They also change jobs and relationships suddenly, and they often have a history of truancy from school and of being expelled for destructive behavior. People with antisocial personalities can be charming and persuasive, and they have remarkably good insight into the needs and weaknesses of other people.

Twin and adoption studies suggest a possible genetic predisposition to antisocial personality disorder (Beaver et al., 2011; Kendler & Prescott, 2006). Biological contributions are also suggested by studies that have found abnormally low autonomic activity during stress, right hemisphere abnormalities, and reduced gray matter in the frontal lobes and biochemical disturbances (Anderson et al., 2011; De Oliveira-Souza et al., 2008; Huesmann & Kirwil, 2007; Schiffer et al., 2011; Sundram et al., 2011).

Evidence also exists for environmental or psychological causes. Antisocial personality disorder is highly correlated with abusive parenting styles and inappropriate modeling (Calkins & Keane, 2009; Henderson, 2010; Torry & Billick, 2011; Wachlarowicz et al., 2012). People with antisocial personality disorder often come from homes characterized by emotional deprivation, harsh and inconsistent disciplinary practices, and antisocial parental behavior. Still other studies show a strong interaction between both heredity and environment (Douglas et al., 2011; Gabbard, 2006; Hudziak, 2008; Philibert et al., 2011).

**Borderline personality disorder** **Borderline personality disorder (BPD)** is among the most commonly diagnosed personality disorders (Ansell & Grilo, 2007; Gunderson, 2011). The core features of this disorder are impulsivity and instability in mood, relationships, and self-image. Originally, the term implied that the person was on the borderline between neurosis and schizophrenia (Kring et al., 2012). The modern conceptualization no longer has this connotation, but BPD remains one of the most complex and debilitating of all the personality disorders.

Mary's story of chronic, lifelong dysfunction, described in the chapter opener, illustrates the serious problems associated with this disorder. People with BPD experience extreme difficulties in relationships. Subject to chronic feelings of depression, emptiness, and intense fear of abandonment, they also engage in destructive, impulsive behaviors, such as sexual promiscuity, drinking, gambling, and eating sprees. In addition, they may attempt suicide and sometimes engage in self-mutilating behavior (Chapman, Leung, & Lynch, 2008; Crowell et al., 2008; Lynam et al., 2011; Muehlenkamp et al., 2011).

People with BPD tend to see themselves and everyone else in absolute terms—perfect or worthless (Mondimore & Kelly, 2011; Scott et al., 2011). They constantly seek reassurance from others and may quickly erupt in anger at the slightest sign of disapproval. The disorder is also typically marked by a long history of broken friendships, divorces, and lost jobs.

Those with BPD frequently have a childhood history of neglect; emotional deprivation; and physical, sexual, or emotional abuse (Christopher et al., 2007; Distel et al., 2011; Gratz et al., 2011). Borderline personality disorder also tends to run in families, and some data suggest it is a result of impaired functioning of the brain's frontal lobes and limbic system, areas that control impulsive behaviors (Leichsenring et al., 2011; Schulze et al., 2011).

Some therapists have had success treating BPD (Fleischhaker et al., 2011; Leichsenring et al., 2011; Mondimore & Kelly, 2011), but people with BPD appear to have a deep well of intense loneliness and a chronic fear of abandonment. Sadly, given their troublesome personality traits, friends, lovers, and even family and therapists often do "abandon" them—thus creating a tragic self-fulfilling prophecy.

**CONCEPT CHECK** STOP

1. **Which** is more important in the development of dissociative disorders, environment or genetics?
2. **How** do personality disorders differ from the other psychological disorders discussed in this chapter?

# How Gender and Culture Affect Abnormal Behavior

## LEARNING OBJECTIVES

**RETRIEVAL PRACTICE** While reading the upcoming sections, respond to each Learning Objective in your own words. Then compare your responses with those in Appendix B.

1. **Identify** the risk factors that might explain gender differences in depression.
2. **Explain** why it is difficult to directly compare mental disorders across cultures.
3. **Outline** strategies for avoiding ethnocentrism in diagnosis.

Among the Chippewa, Cree, and Montagnais-Naskapi Indians in Canada, there is a disorder called *windigo*—or *wiitiko*—*psychosis*, which is characterized by delusions and cannibalistic impulses. Believing that they have been possessed by the spirit of a windigo, a cannibal giant with a heart and entrails of ice, victims become severely depressed (Faddiman, 1997). As the malady begins, the individual typically experiences loss of appetite, diarrhea, vomiting, and insomnia, and he or she may see people turning into beavers and other edible animals. In later stages, the victim becomes obsessed with cannibalistic thoughts and may even attack and kill loved ones in order to devour their flesh (Berreman, 1971).

If you were a therapist, how would you treat this disorder? Does it fit neatly into any of the categories of psychological disorders that you have learned about? We began this chapter discussing the complexities and problems with defining, identifying, and classifying abnormal behavior. Before we close, we need to add two additional confounding factors: gender and culture. In this section, we explore a few of the many ways in which men and women differ in how they experience abnormal behavior. We also look at cultural variations in abnormal behavior.

### Gender Differences

When you picture someone suffering from depression, anxiety, alcoholism, or antisocial personality disorder, what is the gender of each person? Most people tend to visualize a woman for the first two and a man for the last two. There is some truth to these stereotypes.

Research has found considerable gender differences in the prevalence rates of various mental disorders. Let's start with the well-established fact that around the world, the rate of severe depression for women is about double that of men (Dawson et al., 2010; Dillard et al., 2012; Luppa et al., 2012; World Health Organization, 2011). Why is there such a striking gender difference?

Certain risk factors for depression (e.g., genetic predisposition, marital problems, pain and medical illnesses) are common to both men and women. However, poverty is a well-known contributor to many psychological disorders, and women are far more likely to fall into the lowest socioeconomic groups. Women also experience more sexual trauma, partner abuse, and chronic stress in their daily lives, which are all well-known contributing factors in depression (Betancourt et al., 2011; Harkness et al., 2010; Mendelson, Turner, & Tandon, 2010; Mota et al., 2012; Neri et al., 2012; Nolen-Hoeksema, Larson, & Grayson, 2000; Whiffen & Demidenko, 2006).

On the other hand, the most common symptoms of depression, such as crying, low energy, dejected facial expressions, and withdrawal from social activities, are more socially acceptable for women. And the fact that gender differences in depression are more pronounced in cultures with traditional gender roles (e.g., Seedat et al., 2009) suggests that male depression may be underdiagnosed because men are better at hiding and redirecting their emotions (Fields & Cochran, 2011). In our society, men are typically socialized to suppress their emotions and to show their distress by acting out (showing aggression), acting impulsively (driving recklessly and committing petty crimes), and engaging in substance abuse.

Recent research suggests that some gender differences in depression may relate to how women and men tend to internalize or externalize their emotions (see *Psychological Science*).

 THE PLANNER

# Gender Differences in Internalizing Versus Externalizing

Drawing from a national survey of over 43,000 noninstitutionalized people in the United States, researchers identified an interesting *internalizing versus externalizing* pattern that ties together previous research and may better explain gender differences in mental disorders (Eaton et al., 2012).

Using structured interview techniques, researchers found that women ruminate more frequently than men, which means that they are more likely to focus repetitively on their *internal* negative emotions and problems rather than engaging in more external problem-solving strategies. In contrast, men tend to be more disinhibited and more likely to *externalize* their emotions and problems (Dawson et al., 2010; Nolen-Hoeksema, Wisco, & Lyubomirksy, 2008; Tarter et al., 2003; Tompkins et al., 2011; Xu et al., 2012).

The model developed by Eaton and his colleagues (2012) shows that for both men and women the best predictor for certain disorders is the individual's underlying tendency for internalizing or externalizing (**Figure a**). Those who internalize their emotions and problems are more likely to develop inward-focused major depressive disorder and anxiety disorders. In contrast, those who externalize their emotions and problems are more at risk of suffering from outward-focused substance-related disorders and antisocial personality disorder.

Men, who show a greater tendency to externalize their emotions, significantly outnumber women in alcohol and drug dependence, as well as in antisocial personality disorders. The reverse is true for depression and anxiety, in which women greatly outnumber men (**Figure b**).

Understanding the importance of internalizing and externalizing may help mental health professionals treat disorders. For example, in order to prevent depression or anxiety, a therapist could teach coping and cognitive skills that reduce the likelihood of rumination. Rewarding deliberate, planned behaviors versus unintentional, spur-of-the moment ones might be helpful in preventing substance-related disorders (Eaton et al., 2012).

a.

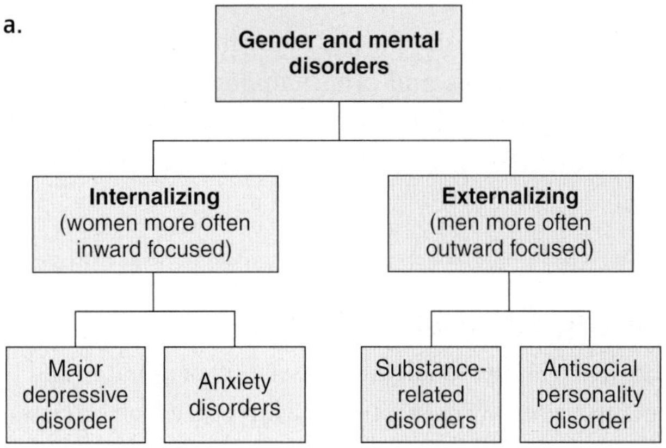


b.

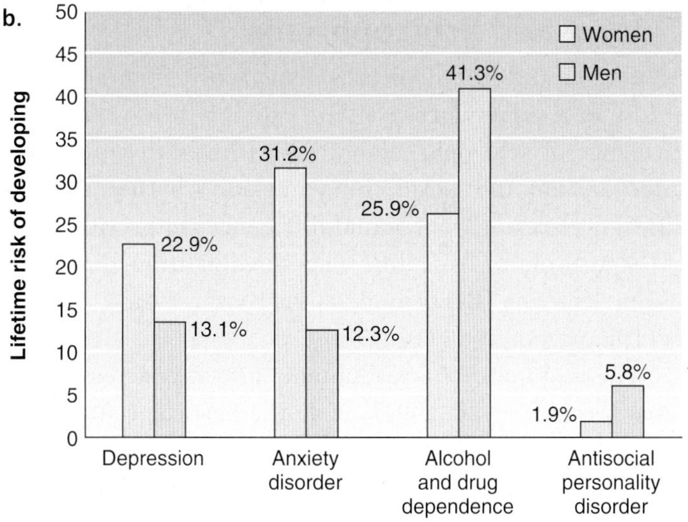


*Source:* Data from Eaton et al., 2012.

## Identify the Research Method

1. What is the most likely research method used for the Eaton et al., 2012 study cited above?
2. If you chose
   - the experimental method, label the IV, DV, experimental group, and control group.
   - the descriptive method, is this a naturalistic observation, survey, or case study?
   - the correlational method, is this a positive, negative, or zero correlation?
   - the biological method, identify the specific research tool (e.g., brain dissection, CT scan).

(Check your answers in Appendix C.)

## What is stressful? • Figure 13.13

a. Some stressors are culturally specific, such as feeling possessed by evil forces or being the victim of witchcraft.

David Alan Harvey/Magnum Photos, Inc.

b. Other stressors are shared by many cultures, such as the unexpected death of a loved one or loss of a job (Al-Issa, 2000; Cechnicki et al., 2011; Neria et al., 2002; Ramsay, Stewart, & Compton, 2012).

Benelux/Zefa/Corbis

## Culture and Schizophrenia

Peoples of different cultures experience mental disorders in a variety of ways. For example, the reported incidence of schizophrenia varies within different cultures around the world. It is unclear whether these differences result from actual differences in prevalence of the disorder or from differences in definition, diagnosis, or reporting (Berry et al., 2011; Butler, Sati, & Abas, 2011; Papageorgiou et al., 2011).

The symptoms of schizophrenia also vary across cultures (Stompe et al., 2003), as do the particular stresses that may trigger its onset (**Figure 13.13**).

Finally, despite the advanced treatment facilities and methods in industrialized nations, the prognosis for people with schizophrenia is actually better in nonindustrialized societies. This may be because the core symptoms of schizophrenia (poor rapport with others, incoherent speech, etc.) make it more difficult to survive in highly industrialized countries. In addition, in most industrialized nations, families and other support groups are less likely to feel responsible for relatives and friends with schizophrenia (Brislin, 2000; Eaton, Chen, & Bromet, 2011).

## Avoiding Ethnocentrism

Most research on psychological disorders originates and is conducted primarily in Western cultures. Such a restricted sampling can limit our understanding of disorders in general and lead to an ethnocentric view of mental disorders.

Fortunately, cross-cultural researchers have devised ways to overcome these difficulties (Matsumoto & Juang, 2008; Triandis, 2007). For example, Robert Nishimoto (1988) has found several **culture-general symptoms** that are useful in diagnosing disorders across cultures (**Table 13.2**).

**Twelve culture-general symptoms of mental health difficulties Table 13.2**

| | | |
|---|---|---|
| Nervous | Trouble sleeping | Low spirits |
| Weak all over | Personal worries | Restless |
| Feel apart, alone | Can't get along | Hot all over |
| Worry all the time | Can't do anything worthwhile | Nothing turns out right |

*Source:* Brislin, 2000.

## Culture-bound disorders • Figure 13.14

Some disorders are fading as remote areas become more Westernized, whereas other disorders (such as anorexia nervosa) are spreading as other countries adopt Western values.

| Puerto Rican and other Latin cultures | Southeast Asian, Malaysian, Indonesian, Thai | West African | Ethiopian | South Chinese and Vietnamese | Westerners |
|---|---|---|---|---|---|
| *Ataque de nervios* ("attack of nerves") | Running amok | Brain fag | Possession by the *Zar* | *Koro* | *Anorexia nervosa* (as other countries become Westernized they're showing some rare cases of anorexia) |
| **Symptoms:** Trembling, heart palpitations, and seizurelike episodes often associated with the death of a loved one, accidents, or family conflict | **Symptoms:** Wild, out-of-control, aggressive behaviors and attempts to injure or kill others | **Symptoms:** "Brain tiredness," a mental and physical response to the challenges of schooling | **Symptoms:** Involuntary movements, mutism, and incomprehensible language | **Symptoms:** Belief that the penis is retracting into the abdomen and that when it is fully retracted, death will result; attempts to prevent the supposed retraction may lead to severe physical damage | **Symptoms:** Occurs primarily among young women; preoccupied with thinness, they exercise excessively and refuse to eat; death can result |

*Source:* Barlow & Durand, 2012; Dhikav et al., 2008; Gaw, 2001; Kring et al., 2012; Laungani, 2007; Tolin et al., 2007.

In addition, Nishimoto found several **culture-bound symptoms**. For example, the Vietnamese Chinese reported "fullness in head," the Mexican respondents had "problems with [their] memory," and the Anglo-Americans reported "shortness of breath" and "headaches." Apparently, people learn to express their problems in ways that are acceptable to others in the same culture (Brislin, 1997, 2000; Butler, Sati, & Abas, 2011; Dhikav et al., 2008; Iwata et al., 2011; Kanayama & Pope, 2011; Marques et al., 2011).

This division between culture-general and culture-bound symptoms also helps us understand depression. Certain symptoms of depression (such as intense sadness, poor concentration, and low energy) seem to exist across all cultures (World Health Organization, 2011). But there is also evidence of some culture-bound symptoms. For example, feelings of guilt are found more often in North America and Europe. And in China, *somatization* (the conversion of depression into bodily complaints) occurs more frequently than it does in other parts of the world (Helms & Cook, 1999; Lawrence et al., 2011; Lim et al., 2011).

Just as there are culture-bound and culture-general symptoms, researchers have found that mental disorders are themselves sometimes culturally bound (**Figure 13.14**). The earlier example of windigo psychosis, a disorder limited to a small group of Canadian Indians, illustrates just such a case.

As you can see, culture has a strong effect on mental disorders. Studying the similarities and differences across cultures can lead to better diagnosis and understanding. It also helps mental health professionals who work with culturally diverse populations understand both culturally general and culturally bound symptoms.

### CONCEPT CHECK

STOP

1. **Why** might depression be frequently overlooked in men?
2. **How** does schizophrenia differ from one culture to another?
3. **Which** symptoms of mental disorders are culture general?

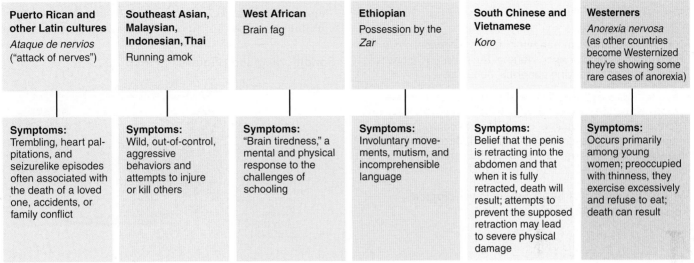

# Summary

THE PLANNER

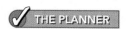

## 1 Studying Psychological Disorders 342

- Criteria for **abnormal behavior** include *deviance*, *dysfunction*, *distress*, and *danger*. None of these criteria alone is adequate for classifying abnormal behavior.
- Superstitious explanations for abnormal behavior were replaced by the **medical model**, which eventually gave rise to the modern specialty of **psychiatry**. In contrast to the medical model, psychology offers a multifaceted approach to explaining abnormal behavior (see diagram).

**What a Psychologist Sees: Seven Psychological Perspectives on Abnormal Behavior**

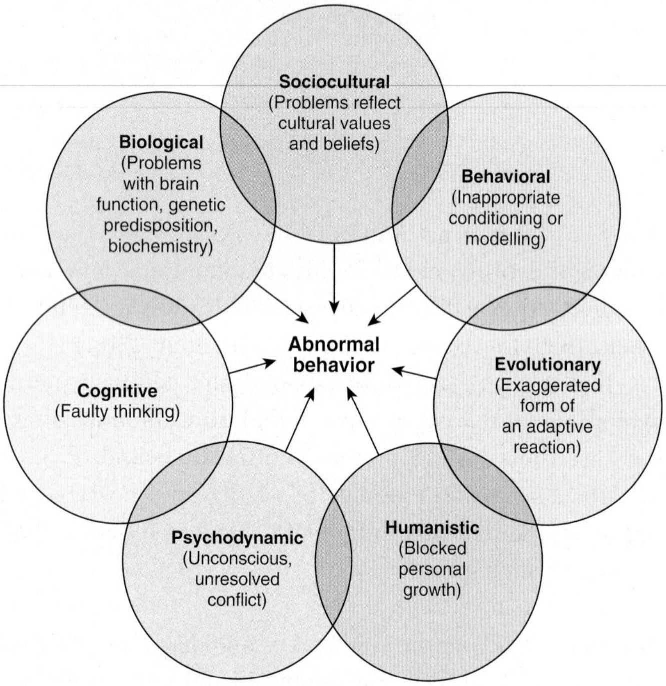


## 2 Anxiety Disorders 348

- Major **anxiety disorders** include **generalized anxiety disorder (GAD), panic disorder, phobias** (including agoraphobia, specific phobias, and social phobias), and **obsessive-compulsive disorder (OCD).**
- Psychological (faulty cognitions and maladaptive learning), biological (evolutionary and genetic predispositions, biochemical disturbances), and sociocultural factors (cultural pressures toward hypervigilance) likely all contribute to anxiety. Classical and operant conditioning also can contribute to phobias, as shown in the diagram.

**Conditioning and phobias • Figure 13.6**

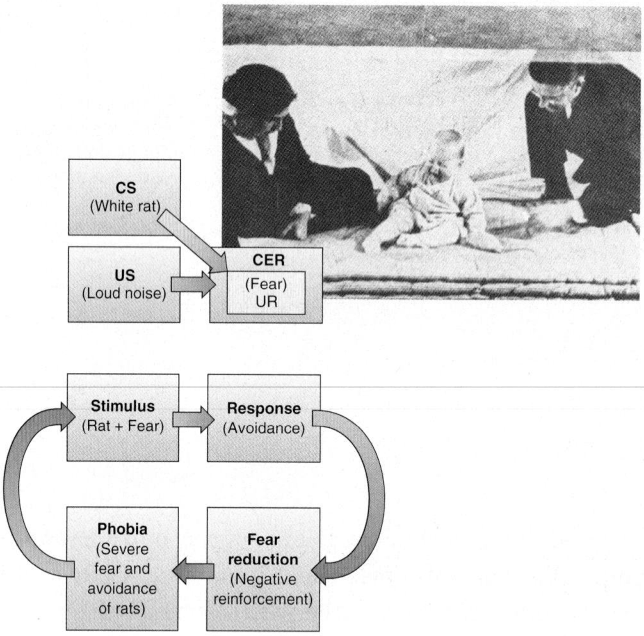


## 3 Mood Disorders 352

- **Mood disorders** are characterized by extreme disturbances in emotional states. People suffering from **major depressive disorder** may experience a lasting depressed mood without a clear trigger. In contrast, people with **bipolar disorder** alternate between periods of depression and mania (hyperactivity and poor judgment), as shown in the diagram.

**Mood disorders • Figure 13.7**

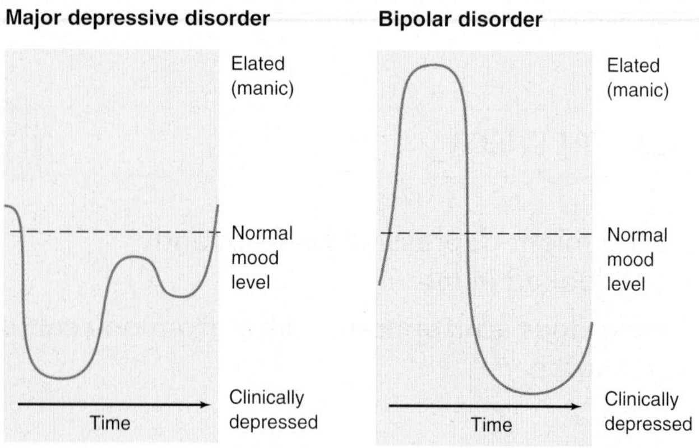

- Biological factors play a significant role in mood disorders. Psychosocial theories of depression focus on environmental stressors and disturbances in interpersonal relationships, thought processes, self-concept, and learning history (including **learned helplessness**).

## 4 Schizophrenia 355

- **Schizophrenia** is a group of disorders, each characterized by a disturbance in perception (including **hallucinations**), language, thought (including **delusions**), emotions, and/or behavior.
- In the past, researchers divided schizophrenia into multiple subtypes. More recently, researchers have proposed focusing instead on **positive schizophrenia symptoms** versus **negative schizophrenia symptoms**. The fourth edition *DSM* also suggests adding another dimension to reflect disorganization of behavior.
- Most biological theories of schizophrenia focus on genetics, neurotransmitters, and brain abnormalities. Psychologists believe that there are also at least two possible psychosocial contributors: stress and communication disorders in families, as shown in the diagram.

**The biopsychosocial model and schizophrenia • Figure 13.10**

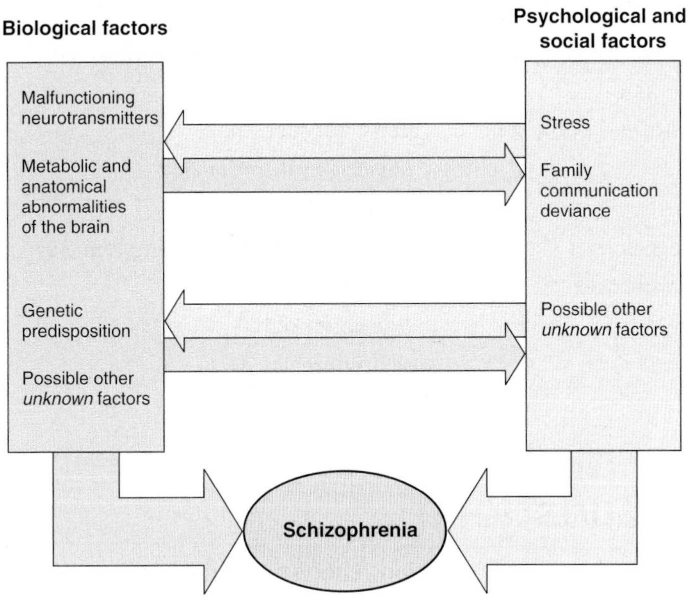


## 5 Other Disorders 361

- **Dissociative disorders** include dissociative amnesia, dissociative fugue, depersonalization disorder, and **dissociative identity disorder (DID).** Environmental variables appear to be the primary cause of dissociative disorders. Dissociation can be a form of escape from a past trauma, such as that shown in the photo.

**Dissociation as an escape • Figure 13.11**

- **Personality disorders** occur when inflexible, maladaptive personality traits cause significant impairment of social and occupational functioning. **Antisocial personality disorder**, is characterized by egocentrism, lack of conscience, impulsive behavior, and superficial charm. The most common personality disorder is **borderline personality disorder (BPD).** Its core features are impulsivity and instability in mood, relationships, and self-image. Although some therapists have success with drug therapy and behavior therapy, prognosis is not favorable.

## 6 How Gender and Culture Affect Abnormal Behavior 364

- Men and women differ in their respective rates and experiences of abnormal behavior. For instance, in the case of depression, modern research suggests that the gender-ratio differences may reflect an underlying predisposition toward internalizing or externalizing their emotions and problems, as shown in the diagram.
- Peoples of different cultures experience mental disorders in a variety of ways. For example, the reported incidence of schizophrenia varies within different cultures around the world, as do the disorder's symptoms, triggers, and prognosis.
- Some symptoms of psychological disorders, as well as some disorders themselves, are **culture-general**, whereas others are **culture-bound.**

**Psychological Science: Gender Differences in Internalizing Versus Externalizing**

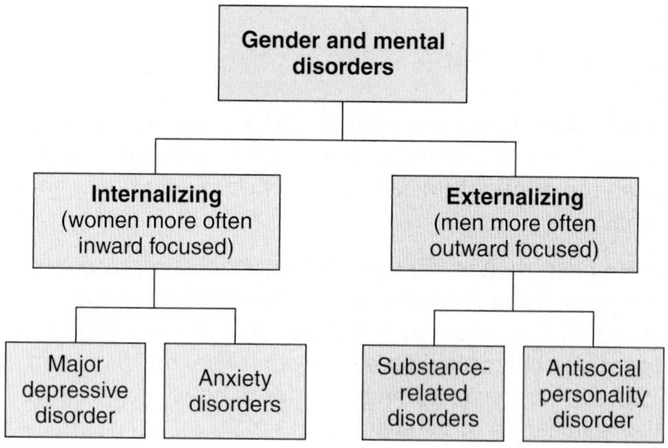


# Key Terms

**RETRIEVAL PRACTICE** Write a definition for each term before turning back to the referenced page to check your answer.

- abnormal behavior 342
- antisocial personality disorder 362
- anxiety disorder 348
- bipolar disorder 352
- borderline personality disorder (BPD) 363
- comorbidity 346
- culture-bound symptom 367
- culture-general symptom 366
- delusion 357
- *Diagnostic and Statistical Manual (DSM)* 345
- diathesis-stress model 359
- dissociative disorder 361
- dissociative identity disorder (DID) 361
- dopamine hypothesis 358
- generalized anxiety disorder (GAD) 348
- hallucination 356
- insanity 345
- learned helplessness 355
- major depressive disorder 352
- medical model 344
- mood disorder 352
- negative schizophrenia symptom 357
- neurosis 345
- obsessive-compulsive disorder (OCD) 350
- panic disorder 348
- personality disorder 362
- phobia 348
- positive schizophrenia symptom 357
- psychiatry 344
- psychosis 345
- schizophrenia 355

# Critical and Creative Thinking Questions

1. Can you think of cases where each of the four criteria for abnormal behavior might not be suitable for classifying a person's behavior as abnormal?
2. Do you think the insanity plea, as it is currently structured, should be abolished? Why or why not?
3. Why do you suppose that anxiety disorders are among the easiest to treat?
4. Have you ever felt depressed? How would you distinguish between "normal" depression and a major depressive disorder?
5. Culture clearly has strong effects on mental disorders. How does that influence how you think about what is normal or abnormal?
6. Most of the disorders discussed in this chapter have some evidence for a genetic predisposition. What would you tell a friend who has a family member with one of these disorders and fears that he or she might develop the same disorder?

# What is happening in this picture?

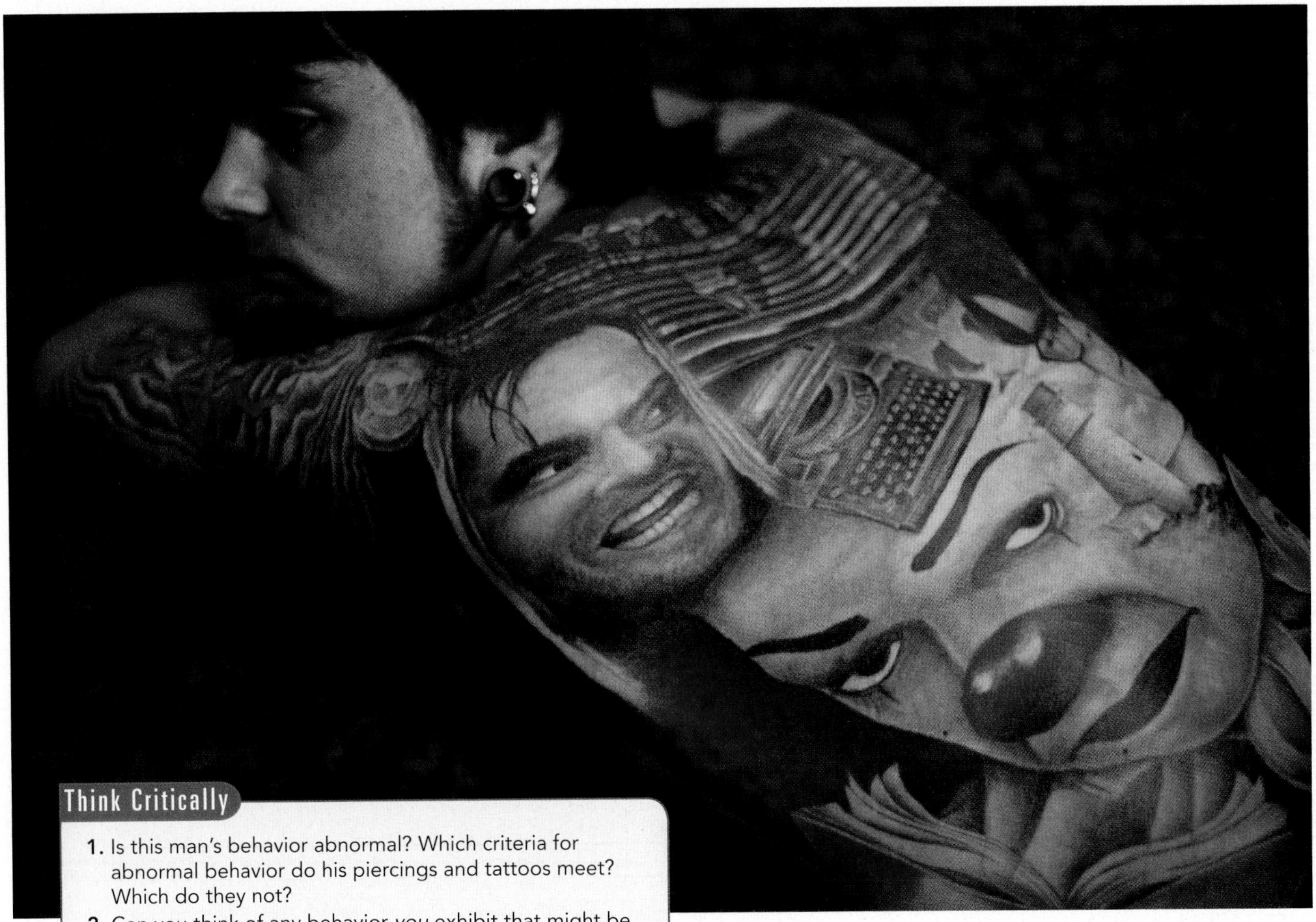

Jodi Cobb/NG Image Collection

**Think Critically**

1. Is this man's behavior abnormal? Which criteria for abnormal behavior do his piercings and tattoos meet? Which do they not?
2. Can you think of any behavior *you* exhibit that might be considered abnormal if your own cultural norms were not taken into account?

# Self-Test

**RETRIEVAL PRACTICE** Completing this self-test and comparing your answers with those in Appendix C provides immediate feedback and helpful practice for exams. Additional interactive, self-tests are available at www.wiley.com/college/carpenter.

1. Your textbook defines *abnormal behavior* as ______.
   a. statistically infrequent pattern of pathological emotion, thought, or action
   b. patterns of emotion, thought, and action that are considered pathological
   c. patterns of behaviors, thoughts, or emotions considered pathological for one or more of four reasons: deviance, dysfunction, distress, or danger.
   d. all of these options

2. ______ is the branch of medicine that deals with the diagnosis, treatment, and prevention of mental disorders.
   a. Psychology
   b. Psychiatry
   c. Psychobiology
   d. Psychodiagnostics

3. Fill in the blanks of the seven psychological perspectives on abnormal behavior in the figure.

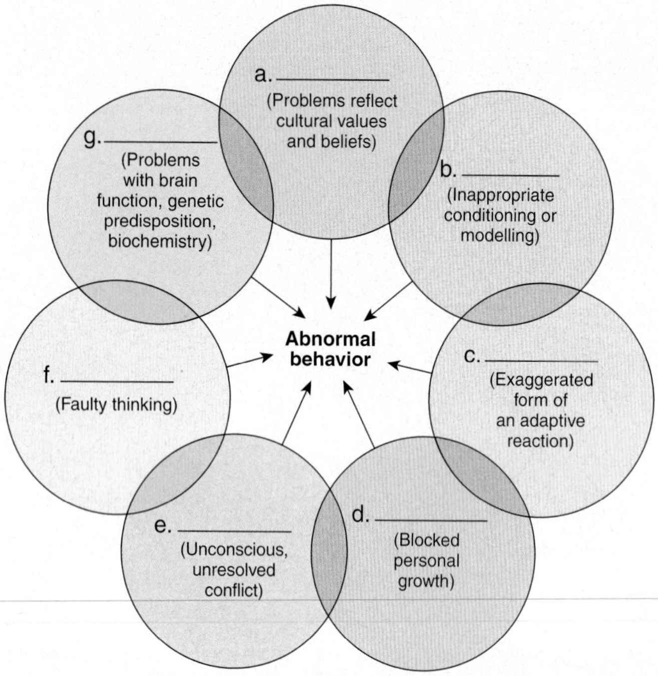


4. Anxiety disorders are ______.
   a. characterized by unrealistic, irrational fear
   b. the least frequent of the mental disorders
   c. twice as common in men as in women
   d. all of these options

5. Repetitive, ritualistic behaviors, such as hand washing, counting, or putting things in order, are called ______.
   a. obsessions
   b. compulsions
   c. ruminations
   d. phobias

6. Label the five major anxiety disorders on the figure.

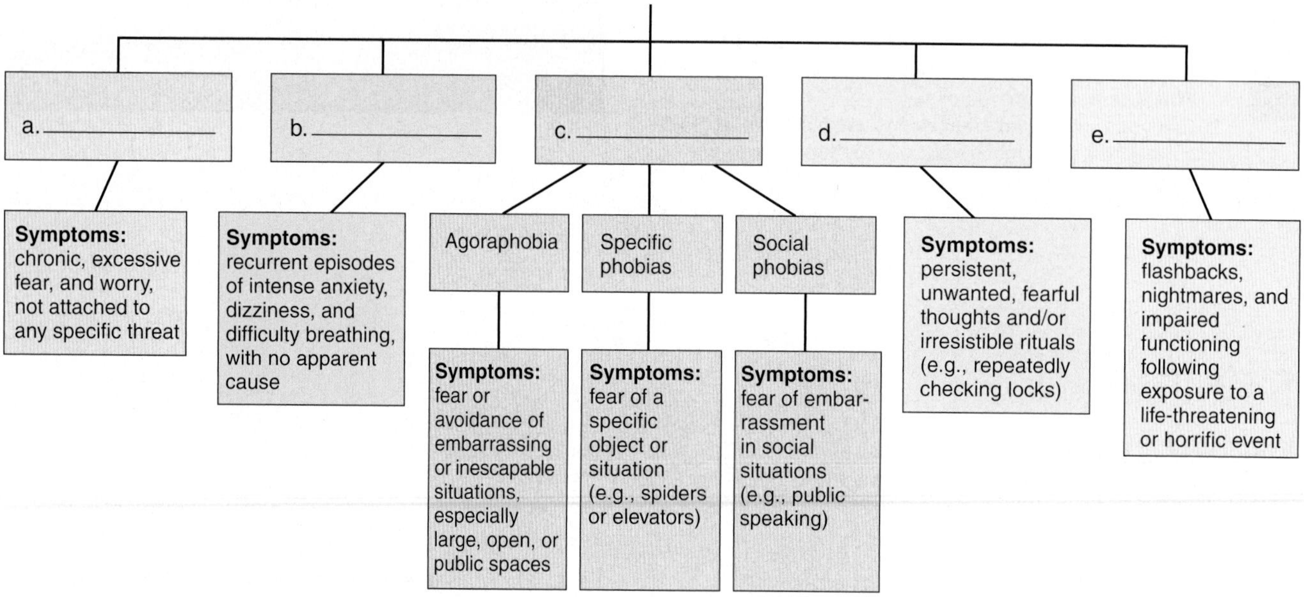

7. The two main types of mood disorders are ______.
   a. major depressive disorder and bipolar disorder
   b. mania and depression
   c. SAD and MAD
   d. learned helplessness and suicide

8. Someone who experiences repeated episodes of mania or cycles between mania and depression has a ______.
   a. disruption of circadian rhythms
   b. bipolar disorder
   c. comorbid condition
   d. diathesis-stress disorder

9. According to the theory known as ______, when faced with a painful situation from which there is no escape, animals and people enter a state of helplessness and resignation.
   a. autonomic resignation
   b. helpless resignation
   c. resigned helplessness
   d. learned helplessness

10. A psychotic disorder that is characterized by major disturbances in perception, language, thought, emotion, and behavior is ______.
   a. schizophrenia
   b. multiple personality disorder
   c. borderline psychosis
   d. neurotic psychosis

11. Perceptions for which there are no appropriate external stimuli are called ______, and the most common type among people suffering from schizophrenia is ______.
   a. hallucinations; auditory
   b. hallucinations; visual
   c. delusions; auditory
   d. delusions; visual

12. Label the five subtypes of schizophrenia on the table.

| Subtypes of schizophrenia | |
|---|---|
| a. ______________ | Dominated by delusions (persecution and grandeur) and hallucinations (hearing voices) |
| b. ______________ | Marked by motor disturbances (immobility or wild activity) and echo speech (repeating the speech of others) |
| c. ______________ | Characterized by incoherent speech, flat or exaggerated emotions, and social withdrawal |
| d. ______________ | Meets the criteria for schizophrenia but is not any of the above subtypes |
| e. ______________ | No longer meets the full criteria for schizophrenia but still shows some symptoms |

13. The disorder that is an attempt to avoid painful memories or situations and is characterized by amnesia, fugue, or multiple personalities is ______.
   a. dissociative disorder
   b. displacement disorder
   c. disoriented disorder
   d. identity disorder

14. Inflexible, maladaptive personality traits that cause significant impairment of social and occupational functioning are known as ______.
   a. nearly all mental disorders
   b. the psychotic and dissociative disorders
   c. personality disorders
   d. none of these options

15. Which of the following are examples of culture-general symptoms of mental health difficulties, useful in diagnosing disorders across cultures?
   a. trouble sleeping
   b. can't get along
   c. worry all the time
   d. all of the above

**THE PLANNER**

Review your Chapter Planner on the chapter opener and check off your completed work.

# 14 Therapy

What do you picture in your mind when you hear the words "mental illness" and "therapy"? Mental illness has been the subject of some of Hollywood's most popular and influential films. Sadly, the mentally ill characters of these films are generally depicted as either cruel, sociopathic criminals (Anthony Hopkins in *Silence of the Lambs*) or helpless, overwhelmed victims (Natalie Portman in *Black Swan*). Similarly, when Hollywood films portray therapy, it's generally in the hands of ineffectual, egomaniacs (*What About Bob*) or heartless doctors and nurses forcing dangerous medication and brutal experiments on powerless patients locked away in dark, bare-bones cells in desolate locations (*Shutter Island*).

Although these portrayals may boost movie ticket sales, they also unfortunately create and perpetuate harmful stereotypes about mental illness and therapy (Kondo, 2008; Lilienfeld et al., 2010; Rüsch et al., 2011; Thoits, 2011). To offset these myths and misconceptions, this chapter works to present a balanced, factual overview of the latest research on current treatments for mental illness. And, as you'll see, modern psychotherapy can be very affordable, humane, and effective. It also prevents much needless suffering, not only for people with psychological disorders but also for those seeking help with everyday problems in living.

© Mark Bowden/iStockphoto

## CHAPTER OUTLINE



## CHAPTER PLANNER ✓

❑ Study the picture and read the opening story.

❑ Scan the Learning Objectives in each section:
p. 376 ❑ p. 384 ❑ p. 387 ❑ p. 391 ❑

❑ Read the text and study all figures and visuals. Answer any questions.

**Analyze key features**

❑ Myth Busters, p. 376

❑ Study Organizers p. 377 ❑ p. 392 ❑

❑ What a Psychologist Sees, p. 383

❑ Process Diagram, p. 386

❑ Psychology InSight, p. 389

❑ Applying Psychology, p. 393

❑ Psychological Science, p. 395

❑ Stop: Answer the Concept Checks before you go on:
p. 384 ❑ p. 387 ❑ p. 391 ❑ p. 397 ❑

**End of chapter**

❑ Review the Summary and Key Terms.

❑ Answer the Critical and Creative Thinking Questions.

❑ Answer What is happening in this picture?

❑ Complete the Self-Test and check your answers.

# Talk Therapies

## LEARNING OBJECTIVES

**RETRIEVAL PRACTICE** While reading the upcoming sections, respond to each Learning Objective in your own words. Then compare your responses with those in Appendix B.

1. **Describe** the core treatment techniques in psychoanalysis and modern psychodynamic treatments.
2. **Summarize** the four key qualities of communication in Rogerian therapy.
3. **Explain** the principles underlying cognitive therapies.

There are many myths about therapy that aren't supported by science. Read *Myth Busters* to check whether any of your assumptions about therapy are untrue. Throughout this text, we've focused on the *scientific* study of behavior and mental processes, with an emphasis on "normal" functioning. In the previous chapter, we observed what happens when functioning goes awry, and now we'll examine how mental health workers try to help those whose behaviors, thoughts, or emotions are dysfunctional.

According to one expert (Kazdin, 1994), there may be over 400 approaches to professional **psychotherapy**. To organize our discussion, we have grouped treatments into three categories: *talk therapies, behavior therapies*, and *biomedical therapies* (**Study Organizer 14.1**). After exploring these approaches, we conclude with a discussion of topics that involve all major forms of psychotherapy.

**psychotherapy** Techniques employed to improve psychological functioning and promote adjustment to life.

## MYTH BUSTERS

THE PLANNER

### MYTHS ABOUT THERAPY

- ***Myth: There is one best therapy.***

  **Fact:** Many problems can be treated equally well with many different forms of therapy.

- ***Myth: Therapists can read your mind.***

  **Fact:** Good therapists often seem to have an uncanny ability to understand how their clients are feeling and to know when someone is trying to avoid certain topics. This is not due to any special mind-reading ability. It reflects their specialized training and daily experience working with troubled people.

- ***Myth: People who go to therapists are crazy or weak.***

  **Fact:** Most people seek counseling because of stress in their lives or because they realize that therapy can improve their level of functioning. It is difficult to be objective about our own problems. Seeking therapy is a sign not only of wisdom but also of personal strength.

- ***Myth: Only the rich can afford therapy.***

  **Fact:** Therapy can be expensive. But many clinics and therapists charge on a sliding scale based on the client's income. Some insurance plans also cover psychological services.

- ***Myth: If I am taking meds, I don't need therapy.***

  **Fact:** Medications, such as antidepressants, are only one form of therapy. They can change brain chemistry, but they can't teach us to think, feel, or behave differently. Research suggests a combination of drugs and psychotherapy may be best for some situations, whereas in other cases, psychotherapy alone is most effective.

Zigy Kaluzny/Stone/Getty Images

**Study Organizer 14.1** An overview of the three major approaches to therapy ✓ THE PLANNER

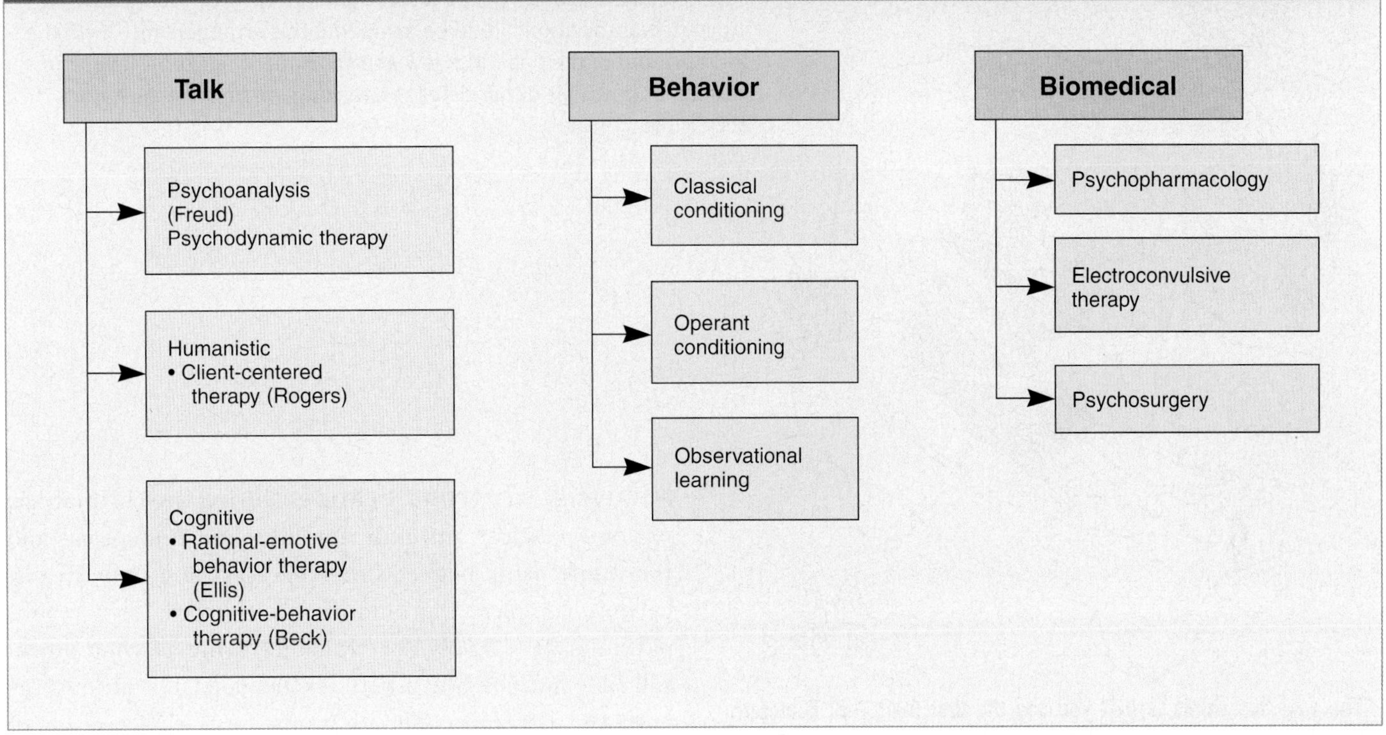


We begin our discussion of professional psychotherapy with traditional psychoanalysis and its modern counterpart, psychodynamic therapies. Then we explore humanistic and cognitive therapies. Although these therapies differ significantly, they're often grouped together as "talk therapies" because they emphasize communication between the therapist and client versus the behavioral and biomedical therapies discussed later.

## Psychoanalysis/Psychodynamic Therapies

**psychoanalysis** Freudian therapy designed to bring unconscious conflicts into conscious awareness; also Freud's theoretical school of thought.

In **psychoanalysis**, a person's *psyche* (or mind) is analyzed. Traditional psychoanalysis is based on Sigmund Freud's central belief that abnormal behavior is caused by unconscious conflicts among the three parts of the psyche—the id, ego, and superego (Chapter 12).

During psychoanalysis, these conflicts are brought to consciousness. The patient comes to understand the reasons for his or her behavior and realizes that the childhood conditions under which the conflicts developed no longer exist. Once this realization (or insight) occurs, the conflicts can be resolved and the patient can develop more adaptive behavior patterns (Cordón, 2012; Johnson, 2011).

Unfortunately, according to Freud, the ego has strong *defense mechanisms* that block unconscious thoughts from coming to light. Thus, to gain insight into the unconscious, the ego must be "tricked" into relaxing its guard. To meet that goal, psychoanalysts employ five major methods: *free association, dream analysis, analyzing resistance, analyzing transference*, and *interpretation*.

**Free association** According to Freud, when you let your mind wander and remove conscious censorship over thoughts—a process called **free association**—interesting and even bizarre connections seem to spring into awareness. Freud believed that the first thing to come to a patient's mind is often an important clue to what the person's unconscious wants to conceal. Having the patient recline

## Freud's free association • Figure 14.1

Psychoanalysis is often portrayed as a patient lying on a couch engaging in free association. Freud believed that this arrangement—with the therapist out of the patient's view and the patient relaxed—helps the patient let down his or her defenses, making the unconscious more accessible.

CartoonStock

"The way this works is that you say the first thing that comes to your mind..."

on a couch, with only the ceiling to look at, is believed to encourage free association (**Figure 14.1**).

**Dream analysis** Recall from Chapter 5 that, according to Freud, our psychological defenses are lowered during sleep. Therefore, our forbidden desires and unconscious conflicts are more freely expressed during dreams. Even while dreaming, however, these feelings and conflicts are recognized as being unacceptable and must be disguised as images that have deeper symbolic meaning. Thus, according to Freudian dream theory, a therapist might interpret a dream of riding a horse or driving a car (the surface description or **manifest content**) as a desire for, or concern about, sexual intercourse (the hidden, underlying meaning or **latent content**).

**Analysis of resistance** During free association or dream analysis, Freud found that patients often show an inability or unwillingness to discuss or reveal certain memories, thoughts, motives, or experiences. For example, suddenly "forgetting" what they were saying or completely changing the subject. It is the therapist's job to identify these cases of *resistance* and to help patients face their problems and then learn to deal with them realistically.

**Analysis of transference** During psychoanalysis, patients disclose intimate feelings and memories, and the relationship between the therapist and patient may become complex and emotionally charged. As a result, patients often apply (or transfer) some of their unresolved emotions and attitudes from past relationships onto the therapist. The therapist uses this process of **transference** to help the patient "relive" painful past relationships in a safe, therapeutic setting so that he or she can move on to healthier relationships.

**Interpretation** The core of all psychoanalytic therapy is **interpretation**. During free association, dream analysis, resistance, and transference, the analyst listens closely and tries to find patterns and hidden conflicts. At the right time, the therapist explains (or interprets) the underlying meanings to the client.

**Evaluating psychoanalysis/psychodynamic therapies** As you can see, most of psychoanalysis rests on the assumption that repressed memories and unconscious conflicts actually exist. But, as noted in Chapters 7 and 12, this assumption is the subject of a heated, ongoing debate. Critics also point to two other problems with psychoanalysis (Messer & Gurman, 2011; Miltenberger, 2011; Siegel, 2010):

- *Limited applicability* Psychoanalysis is time consuming (often lasting several years with four to five sessions a week) and expensive. In addition, critics suggest it only applies to a select group of highly motivated, articulate patients with less severe disorders versus more complex disorders, such as schizophrenia.

Interpersonal therapy (IPT) • Figure 14.2

**Interpersonal therapy (IPT)** focuses on current relationships with the goal of relieving immediate symptoms and teaching better ways to solve future interpersonal problems. Originally designed for acute depression, IPT is similarly effective for a variety of disorders, including marital conflict, eating disorders, and drug addiction (Aldenhoff, 2011; Cuijpers et al., 2011; Hardy, 2011).

Stockbyte/Superstock

- *Lack of scientific credibility* According to critics, it is difficult, if not impossible, to scientifically document the major goals of psychoanalysis. How do we prove (or disprove) the importance of unconscious conflicts or symbolic dream images?

Despite these criticisms, research shows that traditional psychoanalysis can be effective for those who have the time and money (Braaten, 2011; Herbert & Forman, 2010; Spurling, 2011).

In contrast to psychoanalysis, modern **psychodynamic therapy** treatment is briefer, the patient is treated face to face (rather than reclining on a couch), and the therapist takes a more directive approach (rather than waiting for unconscious memories and desires to slowly be uncovered).

**psychodynamic therapy** A briefer, more directive contemporary form of psychoanalysis, focusing more on conscious processes and current problems.

Also, contemporary psychodynamic therapists focus less on unconscious, early childhood roots of problems, and more on conscious processes and current problems (Hunsley & Lee, 2010; Zuckerman, 2011). Such refinements have helped make psychoanalysis shorter, more available, and more effective for an increasing number of people (**Figure 14.2**).

## Humanistic Therapies

In contrast to the rather aloof doctor-to-patient approach of the psychoanalytic and psychodynamic therapies, the humanistic approach emphasizes the *human* characteristics of a person's potential, free will, and self-awareness.

**humanistic therapy** Therapy that seeks to maximize personal growth through affective restructuring (emotional readjustment).

**Humanistic therapy** assumes that people with problems are suffering from a disruption of their normal growth potential and, hence, their self-concept. When obstacles are removed, the individual is free to become the self-accepting, genuine person everyone is capable of being.

One of the best-known humanistic therapists is Carl Rogers (Rogers, 1961, 1980), who developed an approach that encourages people to actualize their potential and to

Comstock/Superstock

## Nurturing growth • Figure 14.3

Recall how you've felt when you've been with someone who believes that you are a good person with unlimited potential, a person who believes that your "real self" is unique and valuable. These are the feelings that are nurtured in humanistic therapy.

relate to others in genuine ways. His approach is referred to as **client-centered therapy** (Rogers used the term *client* because he believed the label *patient* implied that one was sick or mentally ill, rather than responsible and competent) (**Figure 14.3**).

Client-centered therapy, like psychoanalysis and psychodynamic therapies, explores thoughts and feelings as a way to obtain insight into the causes for behaviors. For Rogerian therapists, however, the focus is on providing an accepting atmosphere and encouraging healthy emotional experiences. Clients are responsible for discovering their own maladaptive patterns.

Rogerian therapists create a therapeutic relationship by focusing on four important qualities of communication: *empathy, unconditional positive regard, genuineness*, and *active listening*.

**Empathy** Using the technique of **empathy**, a sensitive understanding and sharing of another person's inner experience, therapists pay attention to body language and listen for subtle cues to help them understand the emotional experiences of clients. To help clients explore their feelings, the therapist uses open-ended statements such as "You found that upsetting" or "You haven't been able to decide what to do about this," rather than asking questions or offering explanations.

**Unconditional positive regard** Whatever the clients' problems or behaviors, humanistic therapists offer them **unconditional positive regard**, a genuine caring and nonjudgmental attitude toward people based on their innate value as individuals. They avoid evaluative statements such as "That's good" and "You did the right thing" because such comments imply the therapist is judging the client. Rogers believed that most of us receive conditional acceptance from our parents, teachers, and others, which leads to poor self-concepts and psychological disorders (**Figure 14.4**).

## Unconditional versus conditional positive regard • Figure 14.4

According to Rogers, clients need to feel unconditionally accepted by their therapists in order to recognize and value their own emotions, thoughts, and behaviors. As this cartoon sarcastically implies, some parents withhold their love and acceptance unless the child lives up to their expectations.

Pat Byrnes/The Cartoon Bank, Inc.

*"Just remember, son, it doesn't matter whether you win or lose—unless you want Daddy's love."*

**Active listening • Figure 14.5**

Noticing a client's furrowed brow and downcast eyes while he is discussing his military experiences, a clinician might respond, "It sounds like you're angry with your situation and feeling pretty miserable right now." Can you see how this statement reflects the client's anger, paraphrases his complaint, and gives feedback to clarify the communication?

**Genuineness** Humanists believe that when therapists use **genuineness** and honestly share their thoughts and feelings with their clients, their clients will in turn develop self-trust and honest self-expression.

**Active listening** Using **active listening**, which involves reflecting, paraphrasing, and clarifying what the client is saying, the clinician communicates that he or she is genuinely interested and paying close attention (**Figure 14.5**). You can use active listening to improve your personal relationships. To *reflect* is to hold a mirror in front of the person, enabling that person to see him- or herself. To *paraphrase* is to summarize in different words what the other person is saying. To *clarify* is to check that both the speaker and listener are on the same wavelength.

**Evaluating humanistic theories** Supporters say that there is empirical evidence for the efficacy of client-centered therapy (Benjamin, 2011; Hardcastle et al., 2008; Hazler, 2011; Messer & Gurman, 2011), but critics argue that outcomes such as self-actualization and self-awareness are difficult to test scientifically. In addition, research on specific therapeutic techniques such as "empathy" and "active listening" has had mixed results (Clark, 2007; Hodges & Biswas-Diener, 2007; Norcross & Wampold, 2011).

## Cognitive Therapies

**Cognitive therapy** assumes that faulty thought processes—beliefs that are irrational, overly demanding, or that fail to match reality—create problem behaviors and emotions (Friedberg & Belsford, 2011; Miltenberger, 2011; Wright, Thase, & Beck, 2011).

**cognitive therapy** Therapy that focuses on changing faulty thought processes and beliefs to treat problem behaviors.

Like psychoanalysts, cognitive therapists believe that exploring unexamined beliefs can produce insight into the reasons for disturbed thoughts, feelings, and behaviors.

However, instead of believing that a change occurs because of insight, cognitive therapists suggest that negative **self-talk** (the unrealistic things a person tells himself or herself) is most important. Through a process called **cognitive restructuring**, this insight allows clients to challenge their thoughts, change how they interpret events, and modify maladaptive behaviors (**Figure 14.6**).

**Ellis's REBT** One of the best-known cognitive therapists, Albert Ellis, suggested that irrational beliefs are the primary culprit in problem emotions and behaviors. He proposed that most people mistakenly believe they are unhappy or upset because of external, outside events, such as receiving a bad grade on an exam. Ellis suggested that, in reality, these negative emotions result from faulty interpretations and irrational beliefs (interpreting the bad grade as a sign of your incompetence and an indication that you'll never qualify for graduate school or a good job).

To deal with these irrational beliefs, Ellis developed **rational-emotive behavior therapy (REBT)** (Ellis, 2008; Ellis & Ellis, 2011; Vernon, 2011). (See *What a Psychologist Sees.*)

**Beck's cognitive-behavior therapy** Another well-known cognitive therapist is Aaron Beck (1976, 2000; Beck & Grant, 2008). Like Ellis, Beck believes that psychological problems result from illogical thinking and destructive self-talk. But Beck seeks to directly confront and change the behaviors associated with destructive cognitions. Beck's **cognitive-behavior therapy** is designed to reduce *both* self-destructive thoughts *and* self-destructive behaviors.

One of the most successful applications of Beck's theory has been in the treatment of depression. (Beck, 1976, 2000; Hollon, 2011; Rosner, 2011; Wright, Thase, & Beck, 2011). Beck has identified several thinking patterns that he believes are associated with depression-prone people:

- *Selective perception* Focusing selectively on negative events while ignoring positive events. ("Why am I the only person alone at this party?")
- *Overgeneralization* Overgeneralizing and drawing negative conclusions about one's own self-worth. ("I'm worthless because I failed that exam.")

**Cognitive restructuring • Figure 14.6**

Cognitive therapy teaches clients to challenge and change negative self-talk in order to improve their life experience.

**a.** Negative interpretation and destructive self-talk leads to destructive and self-defeating outcomes.

**b.** Positive self-talk and beliefs lead to more positive outcomes.

## WHAT A PSYCHOLOGIST SEES

THE PLANNER

### Ellis's Rational-Emotive Behavior Therapy (REBT)

If you receive a poor performance evaluation at work, you might directly attribute your bad mood to the negative feedback. Psychologist Albert Ellis would argue that the self-talk ("I always mess up") between the event and the feeling is what actually upsets you. Furthermore, ruminating on all the other bad things in your life maintains your negative emotional state, and may even lead to anxiety disorders, depression, and other psychological disorders.

To treat these disorders Albert Ellis developed an A–B–C–D approach: **A** stands for *activating event*, **B** the person's *belief system*, **C** the emotional and behavioral *consequences*, and **D** *disputing* erroneous beliefs. During therapy, Ellis helped his clients identify the A, B, C's underlying their irrational beliefs by actively arguing, cajoling, and teasing them—sometimes in very blunt, confrontational language. Once clients recognized their self-defeating thoughts, he worked with them on how to *dispute* these beliefs and how to create and test out new, rational ones. These new beliefs then change the maladaptive emotions—thus breaking the vicious cycle.

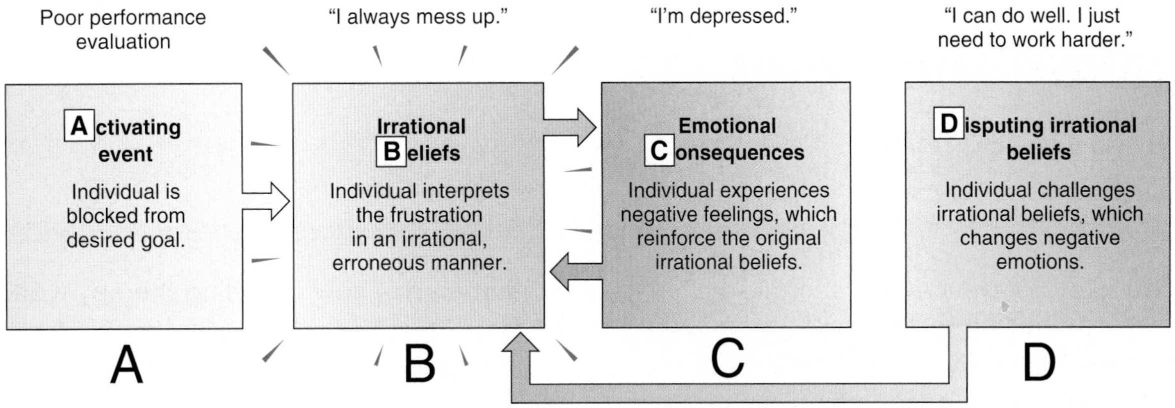


- *Magnification* Exaggerating the importance of undesirable events or personal shortcomings, and seeing them as catastrophic and unchangeable. ("She left me, and I'll never find someone like her again!")
- *All-or-nothing thinking* Seeing things as black-or-white categories—everything is either totally good or bad, right or wrong, a success or a failure. ("If I don't get straight A's, I'll never get a good job.")

In Beck's cognitive-behavior therapy, clients are first taught to recognize and keep track of their thoughts. Next, the therapist trains the client to develop ways to test these automatic thoughts against reality. This approach helps depressed people discover that negative attitudes are largely a product of faulty thought processes.

At this point, Beck introduces the second phase of therapy—persuading the client to actively pursue pleasurable activities. Depressed individuals often lose motivation, even for experiences they used to find enjoyable. Simultaneously taking an active rather than a passive role and reconnecting with enjoyable experiences help in recovering from depression.

**Evaluating cognitive therapies** Cognitive therapies are highly effective treatments for depression, anxiety disorders, bulimia nervosa, anger management, addiction, and even some symptoms of schizophrenia and insomnia (Hofmann, 2012; Kalodner, 2011; Mehta et al., 2011; Thomas et al., 2011; Young, Connor, & Feeney, 2011).

However, both Beck and Ellis have been criticized for ignoring or denying the client's unconscious dynamics, overemphasizing rationality, and minimizing the importance of the client's past (Hammack, 2003). Other critics suggest that cognitive therapies are successful because they employ behavior techniques, not because they change the underlying cognitive structure (Bandura, 1969, 1997, 2006, 2008; Messer & Gurman, 2011; Miltenberger, 2011).

Imagine that you sought treatment for depression and learned to construe events more positively and to curb your all-or-nothing thinking. Further imagine that your therapist also helped you identify activities and behaviors that would promote greater fulfillment. If you found your depression lessening, would you attribute the improvement to your changing thought patterns or to changes in your overt behavior?

**CONCEPT CHECK** STOP

1. **How** does modern psychodynamic therapy differ from traditional psychoanalysis?
2. **What** is the significance of the term *client-centered therapy*?
3. **What** are the four steps of Ellis's REBT?

# Behavior Therapies

## LEARNING OBJECTIVES

**RETRIEVAL PRACTICE** While reading the upcoming sections, respond to each Learning Objective in your own words. Then compare your responses with those in Appendix B.

1. **Identify** a key difference between these two classical conditioning techniques: systematic desensitization and aversion therapy.
2. **Explore** how operant conditioning can be used in therapy.
3. **Summarize** how modeling therapy works.
4. **Describe** two major criticisms of behavior therapies.

Sometimes having insight into a problem does not automatically solve it. In **behavior therapy**, the focus is on the problem behavior itself, rather than on any underlying causes. Although the person's feelings and interpretations are not disregarded, they are also not emphasized. The therapist diagnoses the problem by listing maladaptive behaviors that occur and adaptive behaviors that are absent. The therapist then attempts to shift the balance of the two, drawing on the learning principles of classical conditioning, operant conditioning, and observational learning (Chapter 6).

**behavior therapy** A group of techniques based on learning principles that is used to change maladaptive behaviors.

## Classical Conditioning

Behavior therapists use the principles of classical conditioning to decrease maladaptive behaviors by creating new associations to replace the faulty ones. We will explore two techniques based on these principles: *systematic desensitization* and *aversion therapy*.

Recall from Chapter 6 that classical conditioning occurs when a neutral stimulus (NS) becomes associated with an unconditioned stimulus (US) to elicit a conditioned response (CR). Sometimes a classically conditioned response, like a fear of driving on busy freeways, becomes so extreme that we call it a "phobia." To treat this phobia, behavior therapists typically use **systematic desensitization**, which begins with relaxation training, followed by imagining or directly experiencing various

What's the nature of the trouble, and when did it begin?
Let's drive it around and see what happens.
FREUDIAN
BEHAVIOR THERAPIST
S.Harris

Sidney Harris/ScienceCartoonsPlus.com

## Systematic desensitization • Figure 14.7

Together the therapist and client construct a *fear hierarchy*, or ranked listing of 10 or so anxiety-arousing images. While in a state of relaxation, the client mentally visualizes or physically experiences mildly anxiety-producing items at the lowest level of the hierarchy, and then works his or her way up to the most anxiety-producing items at the top. Each progressive step on the fear hierarchy is repeatedly paired with relaxation, until the fear response (phobia) is extinguished.

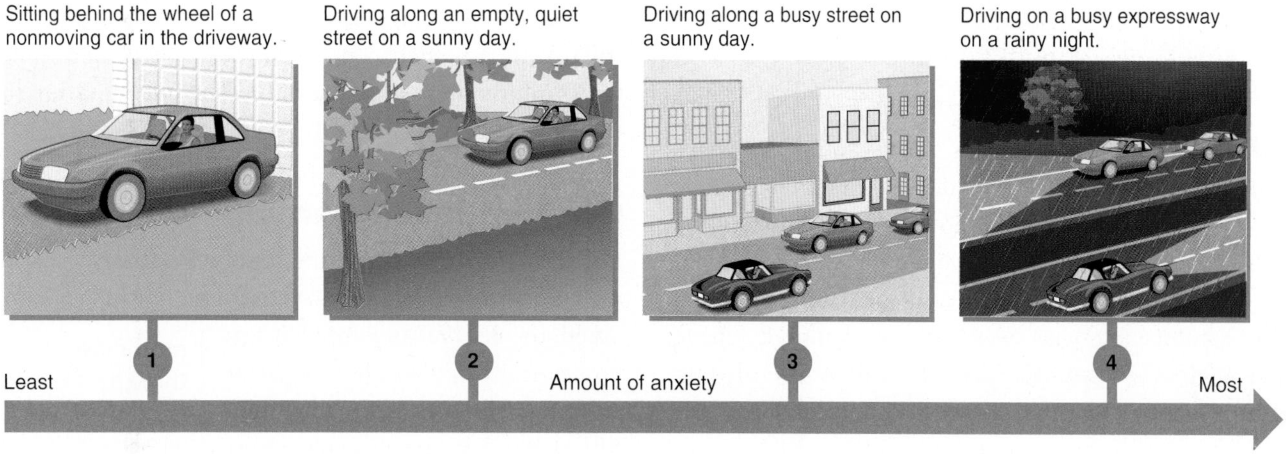

Sitting behind the wheel of a nonmoving car in the driveway.
Driving along an empty, quiet street on a sunny day.
Driving along a busy street on a sunny day.
Driving on a busy expressway on a rainy night.
1
2
3
4
Least
Amount of anxiety
Most

versions of a feared object or situation while remaining deeply relaxed (Wolpe & Plaud, 1997) (**Figure 14.7**).

How does relaxation training desensitize someone? Recall from Chapter 2 that the parasympathetic nerves control autonomic functions when we are relaxed. Because the opposing sympathetic nerves are dominant when we are anxious, it is physiologically impossible to be both relaxed and anxious at the same time. The key to success is teaching the client how to replace his or her fear response with relaxation when *exposed* to the fearful stimulus, which explains why these and related approaches are often referred to as *exposure therapies*. Modern systematic desensitization can use virtual reality technology to expose clients to feared situations right in a therapist's office (**Figure 14.8**).

## Virtual reality therapy • Figure 14.8

Rather than mental imaging or actual physical experiences of a fearful situation, virtual reality headsets and data gloves allow a client with a fear of heights, for example, to have experiences ranging from climbing a stepladder all the way to standing on the edge of a tall building.

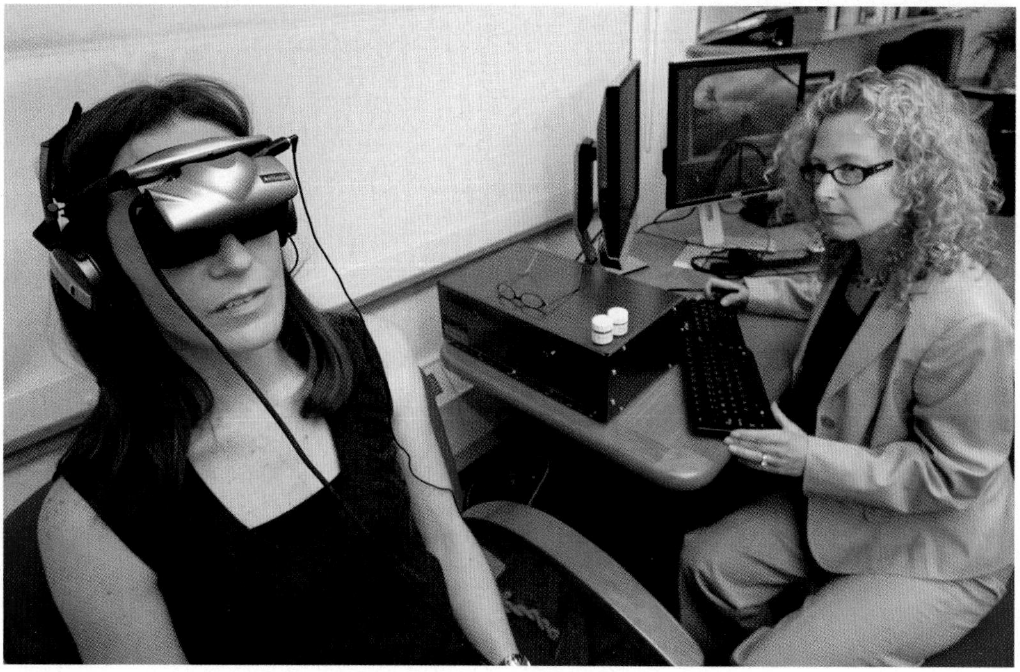

In contrast to systematic desensitization, **aversion therapy** uses classical conditioning techniques to create anxiety rather than extinguish it. People who engage in excessive drinking, for example, build up a number of pleasurable associations with alcohol. These pleasurable associations cannot always be prevented. However, aversion therapy provides *negative associations* to compete with the pleasurable ones (**Figure 14.9**).

## Operant Conditioning

One operant conditioning technique for eventually bringing about a desired (or target) behavior is **shaping**—providing rewards for successive approximations of the target behavior (Chapter 6). One of the most successful applications of shaping and reinforcement has been with developing language skills in children with autism. First, the child is rewarded for connecting pictures (or other devices) with words; later, only for using the pictures to communicate with others.

Shaping can also help people acquire social skills and greater assertiveness. If you are painfully shy, for example, a clinician might first ask you to role-play simply saying hello to someone you find attractive. Then you might practice behaviors that gradually lead you to suggest a get-together or date. During such role-playing, or behavior rehearsal, the clinician would give you feedback and reinforcement.

Adaptive behaviors can also be taught or increased with techniques that provide immediate reinforcement in the form of tokens (Miltenberger, 2011; Reed & Martens, 2011). For example, patients in an inpatient treatment facility might at first be given tokens (to be exchanged for primary rewards, such as food, treats, TV time, a private room, or outings) for merely attending group therapy sessions. Later they will be rewarded only for actually participating in the sessions. Eventually, the tokens can be discontinued when the patient receives the reinforcement of being helped by participation in the therapy sessions.

## Observational Learning

We all learn many things by observing others. Therapists use this principle in **modeling therapy**, in which clients are asked to observe and imitate appropriate models as they perform desired behaviors. For example, Albert Bandura and his colleagues (1969) asked clients with snake phobias to watch other (nonphobic) people handle snakes. After only two hours of exposure, over 92% of the phobic observers allowed a snake to crawl over their hands, arms, and necks. When the therapy combines live modeling with direct and gradual practice, it is called *participant modeling*. This type of modeling also is involved in social skills training and assertiveness training (**Figure 14.10**).

**modeling therapy** A learning technique in which the subject watches and imitates models who demonstrate desirable behaviors.

PROCESS DIAGRAM

THE PLANNER

**Aversion therapy • Figure 14.9**

The goal of aversion therapy is to create a negative (*aversive*) response to a stimulus a person would like to avoid, like alcohol.

**1 During conditioning** Someone who wants to stop drinking, for example, could take a drug called Antabuse that causes vomiting whenever alcohol enters the system.

**2 After conditioning** When the new connection between alcohol and nausea has been classically conditioned, engaging in the once desirable habit will cause an immediate negative response.

Observational learning • Figure 14.10

During modeling therapy, a client might learn how to interview for a job by first watching the therapist role-play the part of the interviewee. The client then imitates the therapist's behavior and plays the same role. Over the course of several sessions, the client becomes gradually desensitized to the anxiety of interviews.

Digital Vision/Getty Images

## Evaluating Behavior Therapies

Behavior therapy has been effective with various problems, including phobias, obsessive-compulsive disorder, eating disorders, sexual dysfunctions, autism, intellectual disabilities, and delinquency (Antony & Roemer, 2011; Haynes, O'Brien, & Kaholokula, 2011; Miltenberger, 2011; Truscott, 2010). Critics of behavior therapy, however, raise important questions that fall into two major categories:

- *Generalizability* Critics argue that in the "real world" patients are not consistently reinforced, and their newly acquired behaviors may disappear. To deal with this possibility, behavior therapists work to gradually shape clients toward real-world rewards.
- *Ethics* Critics contend that it is unethical for one person to control another's behavior. Behaviorists, however, argue that rewards and punishments already control our behaviors and that behavior therapy actually increases a person's freedom by making these controls overt and by teaching people how to change their own behavior.

**CONCEPT CHECK** STOP

1. **What** is the function of a fear hierarchy?
2. **How** can shaping be used to develop desired behaviors?
3. **How** could participant modeling help someone to overcome severe shyness?
4. **How** could behavior therapy be unethical?

# Biomedical Therapies

## LEARNING OBJECTIVES

**RETRIEVAL PRACTICE** While reading the upcoming sections, respond to each Learning Objective in your own words. Then compare your responses with those in Appendix B.

1. **Identify** the major types of drugs used to treat psychological disorders.
2. **Explain** what happens in electroconvulsive therapy and psychosurgery.
3. **Describe** the risks associated with biomedical therapies.

Some problem behaviors seem to be caused, at least in part, by chemical imbalances or disturbed nervous system functioning, and so can be treated with **biomedical therapies**. Psychiatrists, or other medical personnel, are generally the only ones who use biomedical therapies. However, in some states licensed psychologists can prescribe certain medications, and they often work with patients receiving biomedical therapies. In this section, we will discuss three aspects of biomedical therapies: *psychopharmacology*, *electroconvulsive therapy (ECT)*, and *psychosurgery*.

**biomedical therapy** Using physiological interventions (drugs, electroconvulsive therapy, and psychosurgery) to reduce or alleviate symptoms of psychological disorders.

## Psychopharmacology

Since the 1950s, drug companies have developed an amazing variety of chemicals to treat abnormal behaviors. In some cases, discoveries from **psychopharmacology** have helped correct chemical imbalances. In these instances, using a drug is similar to administering insulin to people with diabetes, whose own bodies fail to manufacture enough. In other cases, drugs have been used to relieve or suppress the symptoms of psychological disturbances even when the underlying cause was not thought to be biological. As shown in **Table 14.1**, psychiatric drugs are classified into four major categories: *antianxiety*, *antipsychotic*, *mood stabilizer*, and *antidepressant*.

**psychopharmacology** The study of drug effects on the mind and behavior.

How do drug treatments such as antidepressants actually work? Although we don't fully understand all the mechanisms at play, some studies suggest that antidepressants increase *neurogenesis* (the production of new neurons), *synaptogenesis*

**Drug treatments for psychological disorders Table 14.1**

| Description | Examples (trade names) |
|---|---|
| **Antianxiety drugs** (also known as *anxiolytics* and "minor tranquilizers") lower the sympathetic activity of the brain—the crisis mode of operation—so that anxiety is diminished and the person is calmer and less tense. However, they are potentially dangerous because they can reduce alertness, coordination, and reaction time. They also can have a synergistic (intensifying) effect with other drugs, which may lead to severe drug reactions and even death. | Ativan<br>Halcion<br>Librium<br>Restoril<br>Valium<br>Xanax |
| **Antipsychotic drugs**, or *neuroleptics,* reduce the agitated behaviors, hallucinations, delusions, and other symptoms associated with psychotic disorders, such as schizophrenia. Traditional antipsychotics work by decreasing activity at the dopamine receptors in the brain. A large majority of patients markedly improve when treated with antipsychotic drugs. | Clozaril<br>Haldol<br>Mellaril<br>Navane<br>Prolixin<br>Risperdal<br>Seroquel<br>Thorazine |
| **Mood-stabilizer drugs**, such as *lithium,* help steady the mood swings of those suffering from bipolar disorder, the condition marked by mood swings from mania to depression. Because lithium acts relatively slowly—it can take 3 or 4 weeks before it takes effect—its primary use is in preventing future episodes and helping to break the manic-depressive cycle. | Eskalith CR<br>Lithobid<br>Tegretol |
| **Antidepressant drugs** are used primarily to treat people with depression, but they are also effective for some anxiety disorders and eating disorders. There are five types of antidepressant drugs: *tricyclics, monoamine oxidase inhibitors (MAOIs), selective serotonin reuptake inhibitors (SSRIs), serotonin* and *norepinephrine reuptake inhibitors* (*SNRIs*), and *atypical antidepressants.* Each class of drugs affects neurochemical pathways in the brain in slightly different ways, increasing or decreasing the availability of certain chemicals. SSRIs (such as *Paxil* and *Prozac*) are by far the most commonly prescribed antidepressants. The atypical antidepressants are prescribed for patients who fail to respond or experience undesirable side effects to the other antidepressants. | Anafranil<br>Celexa<br>Cymbalta<br>Effexor<br>Elavil<br>Nardil<br>Norpramin<br>Parnate<br>Paxil<br>Prozac<br>Tofranil<br>Wellbutrin<br>Zoloft |

*"Before Prozac, she loathed company."*

# Psychology InSight

## How antidepressants affect the brain • Figure 14.11

Antidepressants work at the neural transmission level by increasing the availability of serotonin or norepinephrine, neurotransmitters that normally elevate mood and arousal. Shown here is the action of some of the most popular antidepressants—Prozac, Paxil, and other selective serotonin reuptake inhibitors (SSRIs).

**a. Serotonin's effect on the brain**
People with depression are known to have lower levels of serotonin. Serotonin works in the prefrontal cortex, the hippocampus, and other parts of the brain to regulate mood, sleep, and appetite, among other things.

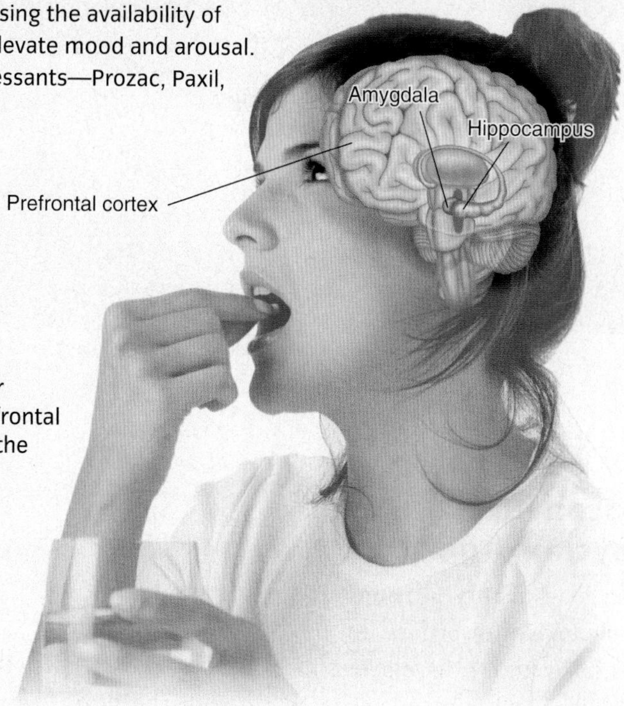

Amygdala
Hippocampus
Prefrontal cortex

MedioImages/Photodisc/Getty Images, Inc.

**b. Normal neural transmission**
The sending neuron releases an excess of neurotransmitters, including serotonin. Some of the serotonin locks into receptors on the receiving neuron, but excess serotonin is pumped back into the sending neuron (called *reuptake*) for storage and reuse. If serotonin is reabsorbed too quickly, there is less available to the brain, which may result in depression.

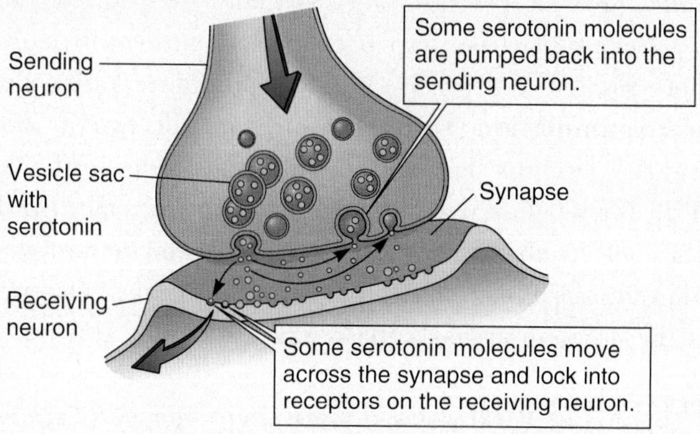

Sending neuron
Some serotonin molecules are pumped back into the sending neuron.
Vesicle sac with serotonin
Synapse
Receiving neuron
Some serotonin molecules move across the synapse and lock into receptors on the receiving neuron.

**c. Partial blockage of reuptake by SSRIs**
SSRIs, like Prozac, partially block the normal reuptake of excess serotonin, which leaves more serotonin molecules free to stimulate receptors on the receiving neuron. This increased neural transmission restores the normal balance of serotonin in the brain.

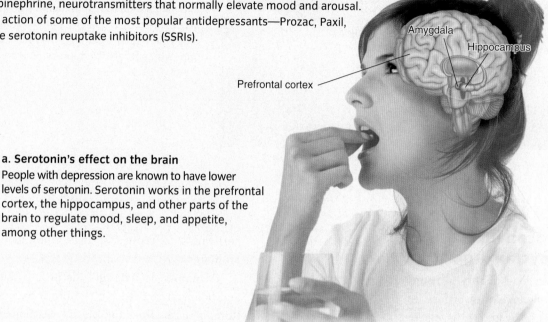

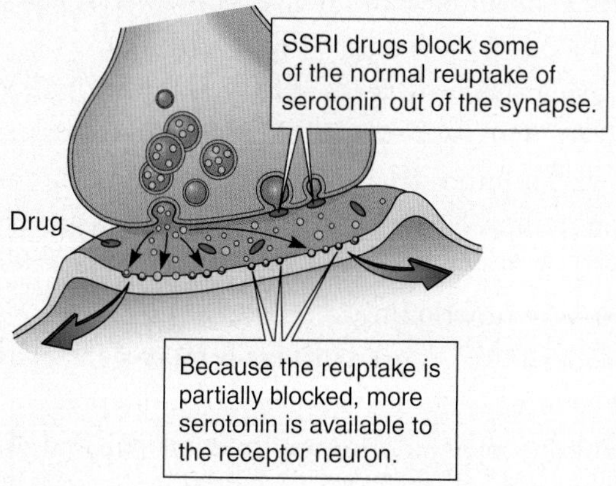

SSRI drugs block some of the normal reuptake of serotonin out of the synapse.
Drug
Because the reuptake is partially blocked, more serotonin is available to the receptor neuron.

(the production of new synapses), and/or stimulate activity in various areas of the brain (Anacker et al., 2011; Ferreira et al., 2012; Li et al., 2012; Surget et al., 2011). The best understood action of the drugs has to do with correcting an imbalance in the levels of neurotransmitters in the brain (**Figure 14.11**). This well-established association of a chemical imbalance in the brain with depression counteracts the somewhat common myth that depression is "all in your head."

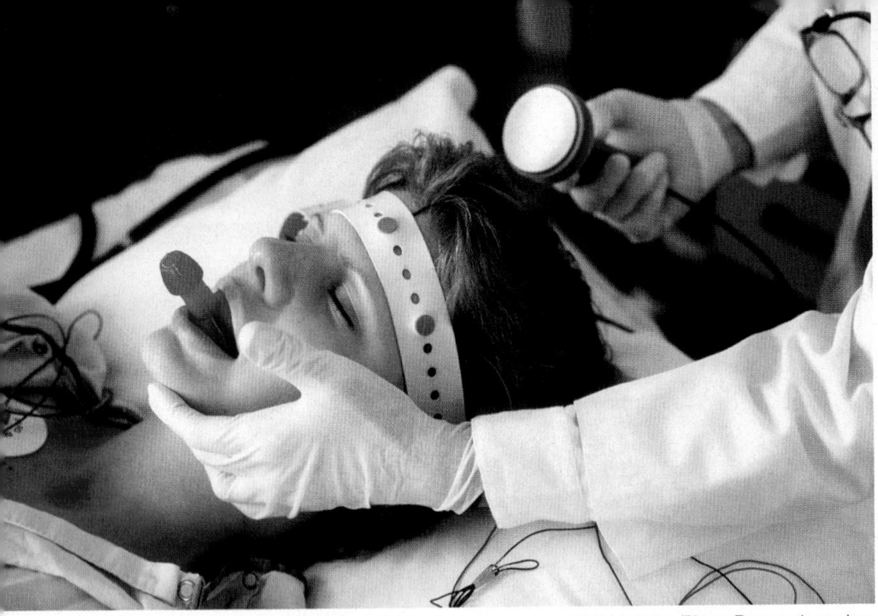

Will McIntyre/Photo Researchers, Inc.

**Electroconvulsive therapy (ECT) • Figure 14.12**

Modern ECT treatments are conducted with considerable safety precautions, including muscle-relaxant drugs that dramatically reduce muscle contractions and medication to help patients sleep through the procedure. ECT is used less often today and generally only when other treatments have failed, despite a 70% improvement rate for depression (Perugi et al., 2012).

## Electroconvulsive Therapy and Psychosurgery

There is a long history of using electrical stimulation to treat psychological disorders. In **electroconvulsive therapy (ECT)**, also known as electroshock therapy (EST), a moderate electrical current is passed through the brain between two electrodes placed on the outside of the head. The current triggers a widespread firing of neurons, or convulsions. The convulsions produce many changes in the central and peripheral nervous systems, including activation of the autonomic nervous system, increased secretion of various hormones and neurotransmitters, and changes in the blood–brain barrier (**Figure 14.12**).

**electroconvulsive therapy (ECT)** Biomedical therapy in which electrical current is passed through the brain.

During the early years of ECT, some patients received hundreds of treatments (Fink, 1999), but today most receive 12 or fewer treatments. Sometimes the electrical current is applied only to the right hemisphere, which causes less interference with verbal memories and left hemisphere functioning.

Modern ECT is used primarily on patients with severe depression who do not respond to antidepressant drugs or psychotherapy and on suicidal patients because it works faster than antidepressant drugs (Kobeissi et al., 2011; Loo et al., 2011; Pfeiffer et al., 2011).

The most extreme, and least used, biomedical therapy is **psychosurgery**—brain surgery performed to reduce serious, debilitating psychological problems.

**psychosurgery** Operative procedures on the brain designed to relieve severe mental symptoms that have not responded to other forms of treatment.

Attempts to change disturbed thoughts, feelings, and behavior by altering the brain have a long history. In Roman times, for example, it was believed that a sword wound to the head could relieve insanity. In 1936, Portuguese neurologist Egaz Moniz first treated uncontrollable psychoses by a form of psychosurgery called a **lobotomy**, in which he cut the nerve fibers between the frontal lobes (where association areas for monitoring and planning behavior are found) and the thalamus and hypothalamus.

Although these surgeries did reduce emotional outbursts and aggressiveness, many patients were left with permanent brain damage. In the mid-1950s, when antipsychotic drugs came into use, psychosurgery virtually stopped. Recently, however, psychiatrists have been experimenting with a much more limited and precise surgical procedure called **deep brain stimulation (DBS)**. The surgeon drills two tiny holes into the patient's skull and implants electrodes in the area of the brain believed to be associated with a specific disorder. These electrodes are connected to a "pacemaker" implanted in the chest or stomach that sends low-voltage electricity to the problem areas in the brain. Over time, this repeated stimulation brings significant improvement in Parkinson's disease and epilepsy (Campbell et al., 2012; Chopra et al., 2012; Sironi, 2011), and recently in major depression and other disorders (Bystritsky et al., 2011; Kennedy et al., 2011; Mayberg, 2006; Piallat et all., 2011; Schlaepfer et al., 2011).

## Evaluating Biomedical Therapies

Like all forms of therapy, biomedical therapies have both proponents and critics.

**Pitfalls of psychopharmacology** Drug therapy poses several potential problems. First, although drugs may relieve symptoms for some people, they seldom provide "cures." In addition, some patients become physically dependent on the drugs. Also, researchers are still learning about the drugs' long-term effects and potential interactions. Furthermore, psychiatric medications can cause a variety of side effects, ranging from mild fatigue to severe impairments in memory and movement.

A final potential problem with drug treatment is that its relatively low cost and generally fast results have led to its overuse in some cases. One report found that antidepressants are prescribed roughly 50% of the time a patient walks into a psychiatrist's office (Olfson et al., 1998).

Despite the problems associated with them, psychotherapeutic drugs have led to revolutionary changes in mental health. Before the use of drugs, some patients were destined to spend a lifetime in psychiatric institutions. Today, most patients improve enough to return to their homes and live successful lives if they continue to take their medications to prevent relapse.

**Challenges to ECT and psychosurgery** As we mentioned, ECT is a controversial treatment for several reasons. However, it does serve as a valuable last resort treatment for severe depression. Interestingly, similar benefits to ECT also may be available thanks to **repetitive transcranial magnetic stimulation (rTMS)**, which delivers a brief (but powerful) electrical current through a coil of wire placed on the head. Unlike ECT, which passes a strong electrical current directly through the brain, the rTMS coil creates a strong magnetic field that is applied to certain areas in the brain. When used to treat depression, the coil is usually placed over the prefrontal cortex, a region linked to deeper parts of the brain that regulate mood. Currently, the benefits of rTMS over ECT remain uncertain, but studies have shown marked improvement in depression, and, unlike with ECT, patients experience no seizures or memory loss (Hadley et al., 2011; Husain & Lisanby, 2011; Polley et al., 2011).

Because all forms of psychosurgery are potentially dangerous and have serious or even fatal side effects, some critics suggest that they should be banned altogether. Furthermore, the consequences are generally irreversible. For these reasons, psychosurgery is considered experimental and remains a highly controversial treatment.

**CONCEPT CHECK** STOP

1. **How** does Prozac affect serotonin levels in the brain?
2. **How** does modern ECT differ from the therapy's early use?
3. **What** potential problems are there with psychopharmacology?

# Psychotherapy in Perspective

## LEARNING OBJECTIVES

**RETRIEVAL PRACTICE** While reading the upcoming sections, respond to each Learning Objective in your own words. Then compare your responses with those in Appendix B.

1. **Summarize** the goals that are common to all major forms of psychotherapy.
2. **Describe** the situations in which group, marital, or family therapy would be most appropriate.
3. **Describe** some key cross-cultural similarities and differences in therapy.
4. **Explain** why therapists need to be sensitive to gender issues that pertain to mental illness.

Earlier, we mentioned that there may be more than 400 forms of therapy. How would you choose one for yourself or someone you know? In the first part of this section, we discuss five goals that are common to all psychotherapies. Then we explore specific formats for therapy as well as considerations about culture and gender. Our goal is to help you synthesize the material in this chapter and to put what you have learned about each of the major forms of therapy into a broader context.

Most therapies focus on one or more of these five goals. Can you identify which would be of most interest to psychodynamic, humanistic, cognitive, and behavioristic therapists?

**Therapy** attempts to change

**Disturbed thoughts**

Therapists work to change faulty or destructive thoughts, provide new ideas or information, and guide individuals toward finding solutions to problems.

**Disturbed emotions**

Therapists help clients understand and control their emotions and relieve their emotional discomfort.

**Disturbed behaviors**

Therapists help clients eliminate troublesome behaviors and guide them toward more effective lives.

**Interpersonal and life situation difficulties**

Therapists help clients improve their relationships with others and avoid or minimize sources of stress in their lives.

**Biomedical disturbances**

Therapists work to relieve biological disruptions that directly cause or contribute to psychological difficulties (for example, chemical imbalances that lead to depression).

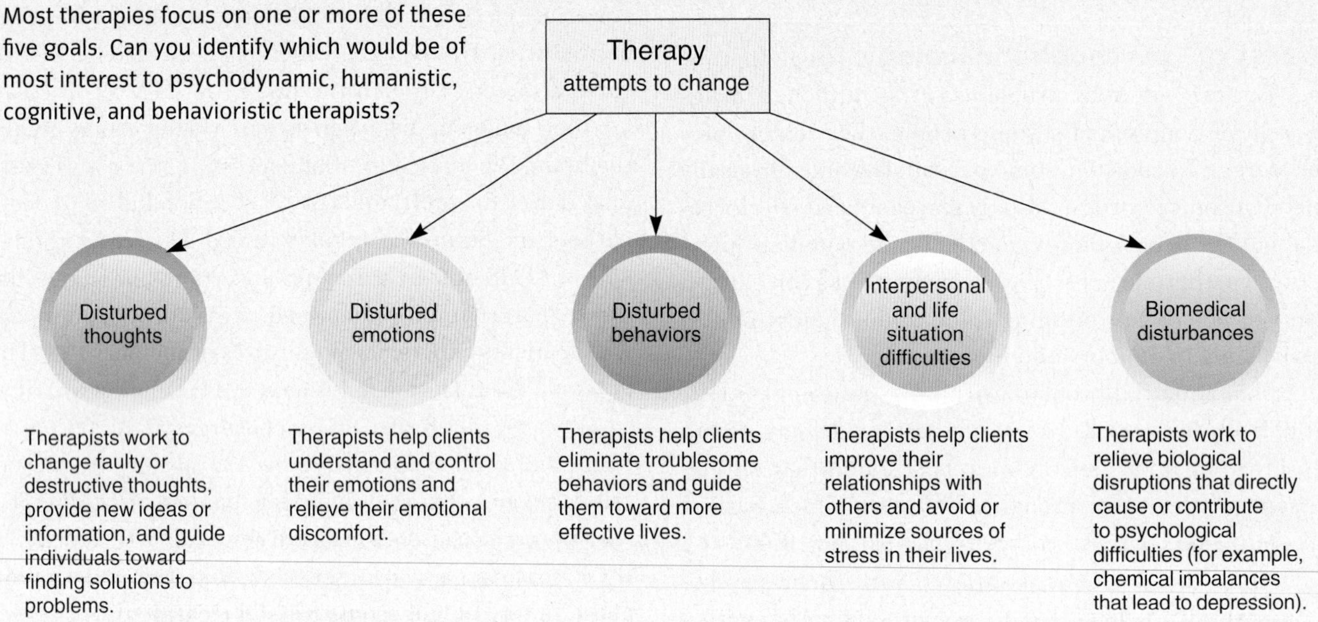

## Therapy Goals and Effectiveness

All major forms of therapy are designed to help the client in five specific areas (**Study Organizer 14.2**).

Although most therapists work with clients in several of these areas, the emphasis varies according to the therapist's training (psychodynamic, cognitive, humanistic, behaviorist, or biomedical). Clinicians who regularly borrow freely from various theories are said to take an **eclectic approach**.

Does therapy work? After years of controlled research and *meta-analysis* (a method of statistically combining and analyzing data from many studies), we have fairly clear evidence that it does. For example, one early meta-analytic review combined a total of almost 25,000 people and found that the average person who received treatment was better off than 75% of the untreated control clients (**Figure 14.13**).

Studies also show that short-term treatments can be as effective as long-term treatments, and that most therapies are equally as effective for various disorders (Dewan, Steenbarger, & Greenberg, 2011; Knekt et al., 2008; Lazar, 2010; Loewenthal & Winter, 2006; Luborsky et al., 1975, 2006; Wachtel, 2011).

## Is therapy generally effective?
### • Figure 14.13

An analysis of 375 studies of treatment effectiveness shows that the average person who gets treatment of any kind is better off after treatment than a similar person who does not get treatment. *Sources:* Smith, Glass, & Miller, 1980; Smith & Glass, 1977.

Improvement among untreated clients

Improvement among treated clients

Average untreated person
Average treated person
Number of clients
75% of untreated persons
None
Degree of Improvement
Outstanding

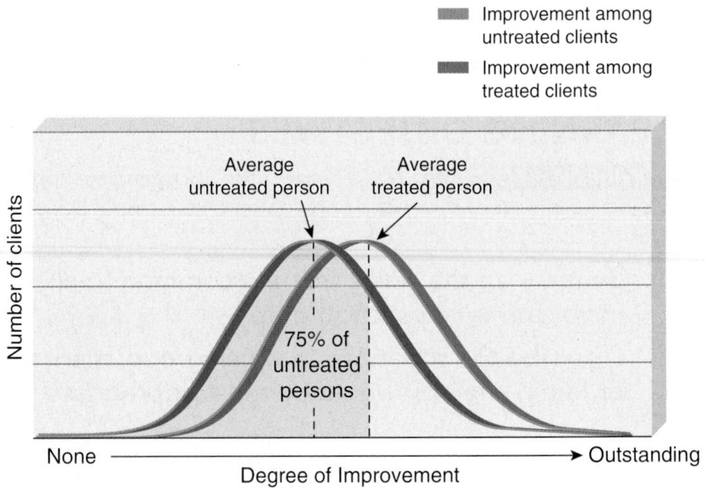

# Applying Psychology

THE PLANNER

## Choosing a Therapist

How do we find a good therapist for our specific needs? If you have the time (and money) to explore options, there are several steps you can take to find a therapist best suited to your specific goals. First, you might consult your psychology instructor, college counseling system, or family physician for specific referrals. In addition, most HMOs and health insurers provide a list of qualified professionals. Next, call the referred therapists and ask for an opportunity to ask them some questions. You could ask about their training, what approach they use, what their fees are, and whether they participate in your insurance plan.

As you can see, finding a therapist takes time and energy. If you need immediate help—if you're having suicidal thoughts or are the victim of abuse—most communities have medical hospital emergency services and telephone hotlines that provide counseling services on a 24-hour basis. And most colleges and universities have counseling centers that provide immediate, short-term therapy to students free of charge.

Finally, if you're concerned about a friend or family member who might need therapy, you can follow the previous tips to help locate a therapist and then possibly offer to go with him or her on the first appointment. If the individual refuses help and the problem affects you, it is often a good idea to seek therapy yourself. You will gain insights and skills that will help you deal with the situation more effectively.

For general help with locating a skilled therapist, identifying what types of initial questions to ask, learning how to gain the most benefits during therapy, and so on, consult the American Psychological Association (APA) website.

Jedd Cadge/The Image Bank/Getty Images

**Think Critically**

1. If you were looking for a therapist, would you want the therapist's gender to be the same as yours? Why or why not?
2. Some therapists treat clients online. What might be the advantages and disadvantages of this approach?

Interestingly, for certain specific problems, there are some therapies that are more effective than others. For example, phobias and marital problems seem to respond best to behavioral therapies, whereas depression and obsessive-compulsive disorders can be significantly relieved with cognitive-behavior therapy (CBT) accompanied by medication (Craske, 2010; Doyle & Pollack, 2004; Siev & Chambless, 2007; Wilson, 2011).

Finally, in recent years, the *empirically supported* or *evidence-based practice (EBP) movement* has been gaining momentum because it seeks to identify which therapies have received the clearest research support for particular disorders (e.g., Kennedy, 2011; Sharf, 2012; Thyer & Myers, 2011). Like all movements, it has been criticized, but this type of research and widely shared information promises to be helpful for therapists and clients alike in their treatment decisions. For specific tips on finding therapy for yourself or a loved one, see *Applying Psychology*.

### Therapy Formats

The therapies described earlier in this chapter are conducted primarily in a face-to-face, one-on-one (therapist-to-client) format. In this section, we focus on several major alternatives: group, family, and marital therapies, which treat multiple individuals simultaneously, and telehealth/electronic therapy, which treats individuals via the Internet, e-mail, and/or smartphones.

**Group therapy** In **group therapy**, multiple people meet together to work toward therapeutic goals. Typically, a group of 8 to 10 people meet with a therapist on a regular basis to talk about problems in their lives.

> **group therapy** A form of therapy in which a number of people meet together to work toward therapeutic goals.

A variation on group therapy is the **self-help group**. Unlike other group therapy approaches, a professional does not guide these groups. They are simply groups of people who share a common problem (such as alcoholism, obesity, or breast cancer) and who meet to give and receive support. Faith-based 12-step programs such as Alcoholics Anonymous, Narcotics Anonymous, and Spenders Anonymous are examples of self-help groups. Although group members don't get the same level of individual attention found in one-on-one therapies, group and self-help therapies provide their own unique advantages (Corey, 2011; Jaffe & Kelly, 2011; Kivlighan, London, & Miles, 2012; Messer & Gurman, 2011; Qualls, 2008). Compared with one-on-one therapies, they are less expensive and provide a broader base of social support. Group members can learn from each other's mistakes, share insights and coping strategies, and role-play social interactions together.

Therapists often refer their patients to group therapy and self-help groups to supplement individual therapy. Research on self-help groups for alcoholism, obesity, and other disorders suggests that they can be very effective, either alone or in addition to individual psychotherapy (Aderka et al., 2011; Jaffe & Kelly, 2011; McEvoy, 2007; Oei & Dingle, 2008; Silverman et al., 2008).

**Marital and family therapies** Because a family or marriage is a system of interdependent parts, the problem of any one individual unavoidably affects all the others, and therapy can help everyone involved (Dattilio & Nichols, 2011; Friedlander et al., 2011; Stratton et al., 2011). The line between marital (or couples) therapy and family therapy is often blurred. Here, our discussion will focus on **family therapy**, in which the primary aim is to change maladaptive family interaction patterns (**Figure 14.14**). All members of the family attend therapy sessions, though at times the therapist may see family members individually or in twos or threes.

> **family therapy** Treatment to change maladaptive interaction patterns within a family.

Family therapy is also useful in treating a number of disorders and clinical problems. As we discussed in Chapter 13, schizophrenic patients are more likely to relapse if their family members express emotions, attitudes, and behaviors that involve criticism, hostility, or emotional overinvolvement (Hooley & Hiller, 2001; Lefley, 2000; Quinn, Barowclough, & Tarrier, 2003). Family therapy can help family members modify their behavior toward the patient. Family therapy can also be the most favorable setting for the treatment of adolescent drug abuse (Dakof, Godley, & Smith, 2011; Mead, 2012; Minuchin, 2011; O'Farrell, 2011).

**Family therapy • Figure 14.14**

Many families initially come into therapy believing that one member is the cause of all their problems. However, family therapists generally find that this "identified patient" is a scapegoat for deeper disturbances. How could changing ways of interacting within the family system promote the health of individual family members and the family as a whole?

Tony Freeman/PhotoEdit

## Psychological Science

THE PLANNER

# Therapy—Is There an App for That?

Studies have long shown that therapy outcomes improve with increased client contact, and the electronic/telehealth format may be the easiest and most cost-effective way to increase this contact (DeAngelis, 2012; Harwood et al., 2011).

One of the latest technological innovations for therapy is the use of smartphones (Clay, 2012). For example, a recent research project developed a four-week cognitive-behavior therapy (CBT) intervention for patients dealing with chronic widespread pain (CWP) via smartphones (Kristjansdottir et al., 2011). After first meeting one on one with a therapist, each participant received a text message on their phone three times a day for four weeks reminding them to fill out an online diary, which asked questions about their current thoughts and pain awareness. Within 90 minutes of the diary submission, participants received online feedback from a therapist.

© Doyeol Ahn/iStockphoto

Effectiveness of the therapeutic intervention was measured by participants' responses to the Chronic Pain Acceptance Questionnaire (CPAQ) and the Pain Catastrophizing Scale (PCS), given to participants before and after treatment. The researchers plan to use this initial feasibility study to develop a full-scale intervention program that will help patients with CWP to self-manage their pain (Using CBT and Smart Phone, 2011).

### Identify the Research Method

1. What is the most likely research method used for the Kristjansdottir et al., 2011 study?
2. If you chose
   - the experimental method, label the IV, DV, experimental group, and control group.
   - the descriptive method, is this a naturalistic observation, survey, or case study?
   - the correlational method, is this a positive, negative, or zero correlation?
   - the biological method, identify the specific research tool (e.g., brain dissection, CT scan).

(Check your answers in Appendix C.)

**Telehealth/electronic therapy** Today millions of people are receiving advice and professional therapy in newer, electronic formats, such as the Internet, e-mail, virtual reality (VR), and interactive web-based conference systems (e.g., "skyping"). This latest form of electronic therapy, often referred to as **telehealth**, allows clinicians to reach more clients and provide them with greater access to information regarding their specific problems (see *Psychological Science*).

Using electronic options, such as the Internet and smartphones, does provide alternatives to traditional one-on-one therapies, but, as you might expect, these unique approaches also raise concerns (DeAngelis, 2012; Nagy, 2012). Professional therapists fear, among other things, that without interstate and international licensing, or a governing body to regulate this type of therapy, there are no checks and balances to protect the client from unethical and unsavory practices. What do you think? Would you be more likely to go to therapy if it was offered via your smartphone, e-mail, or an Internet website? Or is this too impersonal and high-tech for you?

## Cultural Issues in Therapy

The therapies described in this chapter are based on western European and North American culture. Does this mean they are unique? Or do our psychotherapists do some of the same things that, say, a native healer or shaman does? Are there similarities in therapies across cultures? Conversely, are there fundamental therapeutic differences among cultures?

When we look at therapies in all cultures, we find that they have certain key features in common (Berry et al., 2011; Brislin, 2000; Buss 2011; Markus & Kitayama, 2010; Sue, Sue, & Sue, 2010):

- *Naming the problem* People often feel better just by knowing that others experience the same problem and that the therapist has had experience with their particular problem.
- *Qualities of the therapist* Clients must feel that the therapist is caring, competent, approachable, and concerned with finding solutions to their problem.
- *Establishing credibility* Word-of-mouth testimonials and status symbols (such as diplomas on the wall) establish the therapist's credibility. Among native healers, in lieu of diplomas, credibility may be established by the healer having served as an apprentice to a revered healer.
- *Placing the problem in a familiar framework* If the client believes that evil spirits cause psychological disorders, the therapist will direct treatment toward eliminating these spirits. Similarly, if the client believes in the importance of early childhood experiences and the unconscious mind, psychoanalysis will be the likely treatment of choice.
- *Applying techniques to bring relief* In all cultures, therapy involves action. Either the client or the therapist must do something. Moreover, what therapists do must fit the client's expectations—whether it is performing a ceremony to expel demons or talking with the client about his or her thoughts and feelings.
- *A special time and place* The fact that therapy occurs outside the client's everyday experiences seems to be an important feature of all therapies.

Although there are basic similarities in therapies across cultures, there are also important differences. In the traditional western European and North American model, the emphasis is on the "self" and on independence and control over one's life—qualities that are highly valued in individualistic cultures. In collectivist cultures, however, the focus of therapy is on interdependence and accepting the realities of one's life (Sue & Sue, 2008) (**Figure 14.15**).

Not only does culture affect the types of therapy that are developed, it also influences the perceptions of the therapist. What one culture considers abnormal behavior may be quite common—and even healthy—in others. For this reason, recognizing cultural differences is very important for building trust between therapists and clients and for effecting behavioral change (Frew & Spiegier, 2012; Moodley, 2012; Smith, Rodriquez, & Bernal, 2011).

Sky Bonillo/PhotoEdit

**Emphasizing interdependence • Figure 14.15**

In Japanese Naikan therapy, patients sit quietly from 5:30 A.M. to 9:00 P.M. for seven days and are visited by an interviewer every 90 minutes. During this time, they reflect on their relationships with others in order to discover personal guilt for having been ungrateful and troublesome and developing gratitude toward those who have helped them (Moodley & Sutherland, 2010; Nakamura, 2006; Ozawa-de Silva, 2007; Shinfuku & Kitanishi, 2010).

## Gender and Therapy

Within our individualistic Western culture, men and women present different needs and problems to therapists. For example, compared with men, women are more comfortable and familiar with their emotions, have fewer negative attitudes toward therapy, and are more likely to seek psychological help (Komiya, Good, & Sherrod, 2000). Research has identified five unique concerns related to gender and psychotherapy (Bruns & Kaschak, 2011; Calogero et al., 2011; Jordan, 2011; Matlin, 2012; Russo & Tartaro, 2008).

1. *Rates of diagnosis and treatment of mental disorders* Women are diagnosed and treated for mental illness at a much higher rate than men. Is this because women are "sicker" than men as a group, or are they just more willing to admit their problems? Or perhaps the categories for illness are biased against women. More research is needed to answer this question.
2. *Stresses of poverty* Women are disproportionately likely to be poor. Poverty contributes to stress, which is directly related to many psychological disorders.
3. *Stresses of aging* Aging brings special mental health concerns for women. Elderly women, primarily those with age-related dementia, account for over 70% of the chronically mentally ill who live in nursing homes in the United States.
4. *Violence against women* Rape, incest, and sexual harassment—which are much more likely to happen to women than to men—may lead to depression, insomnia, posttraumatic stress disorders, eating disorders, and other problems.
5. *Stresses of multiple roles* Women today are mothers, wives, homemakers, wage earners, students, and so on. The conflicting demands of their multiple roles often create special stresses.

Therapists must be sensitive to possible connections between clients' problems and their gender. Rather than prescribing drugs to relieve depression in women, for example, it may be more appropriate for therapists to explore ways to relieve the stresses of multiple roles or poverty. Can you see how helping a single mother identify parenting resources, such as play groups, parent support groups, and high-quality child care, might be just as effective at relieving depression as prescribing drugs? In the case of men, how might relieving loneliness or depression help decrease their greater problems with substance abuse and aggression?

**CONCEPT CHECK** STOP

1. **What** is the evidence that therapy works?
2. **How** can family therapy help an individual cope with his or her psychological disorder?
3. **How** do individualistic and collectivistic cultures differ in their approaches to therapy?
4. **What** are two main sources of stress for women that might lead them to psychotherapy?

# Summary

THE PLANNER

## 1 Talk Therapies 376

- "Talk therapies," shown in the diagram, are forms of **psychotherapy** that seek to increase insight into clients' difficulties.

**Study Organizer 14.1 An overview of the three major approaches to therapy**

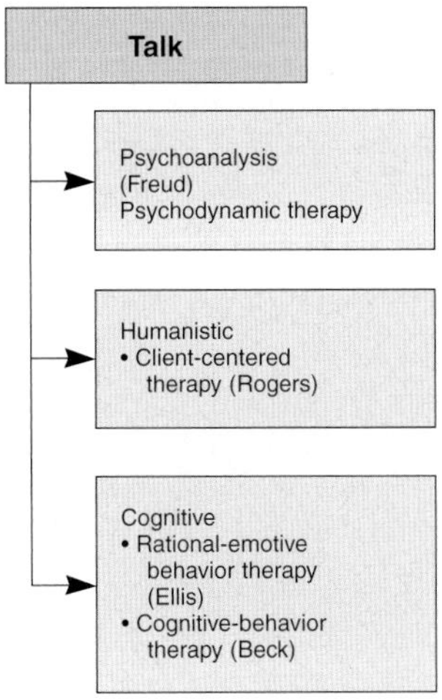


- In **psychoanalysis**, the therapist seeks to identify the patient's unconscious conflicts and to help the patient resolve them. In modern **psychodynamic therapy**, treatment is briefer and the therapist takes a more directive approach (and puts less emphasis on unconscious childhood memories) than in traditional psychoanalysis.
- **Humanistic therapy**, such as Rogers's **client-centered therapy**, seeks to maximize personal growth, encouraging people to actualize their potential and relate to others in genuine ways.
- **Cognitive therapy** seeks to help clients challenge faulty thought processes and adjust maladaptive behaviors. Ellis's **rational-emotive behavior therapy (REBT)** and Beck's **cognitive-behavior therapy** are important examples of cognitive therapy.

## 2 Behavior Therapies 384

- In **behavior therapy**, the focus is on the problem behavior itself, rather than on any underlying causes. The therapist uses learning principles to change behavior.
- Classical conditioning techniques include **systematic desensitization** and **aversion therapy**, shown in the diagram.

### Aversion therapy • Figure 14.9

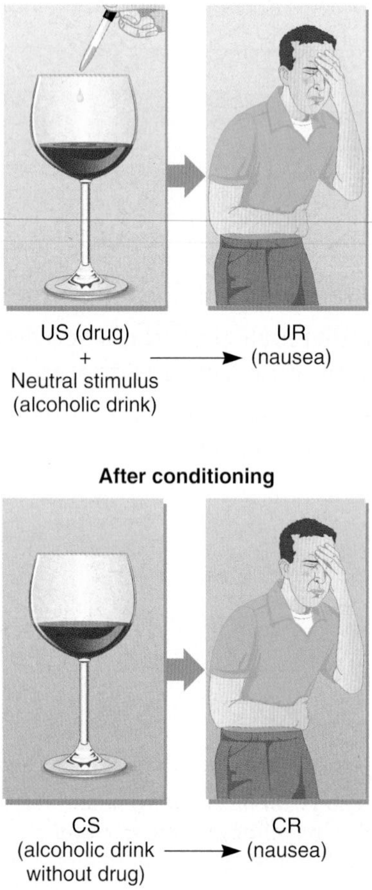

During conditioning
US (drug)
+
Neutral stimulus
(alcoholic drink)
UR
(nausea)
After conditioning
CS
(alcoholic drink
without drug)
CR
(nausea)

- Operant conditioning techniques used to increase adaptive behaviors include **shaping** and reinforcement.
- In **modeling therapy**, clients observe and imitate others who are performing the desired behaviors.

## 3 Biomedical Therapies 387

- **Biomedical therapies** are based on the premise that chemical imbalances or disturbed nervous system functioning contribute to problem behaviors.
- **Psychopharmacology** is the most common form of biomedical therapy. Major classes of drugs used to treat psychological disorders are **antianxiety drugs, antipsychotic drugs, mood stabilizer drugs**, and **antidepressant drugs** (shown in the diagram).

### How antidepressants affect the brain • Figure 14.11

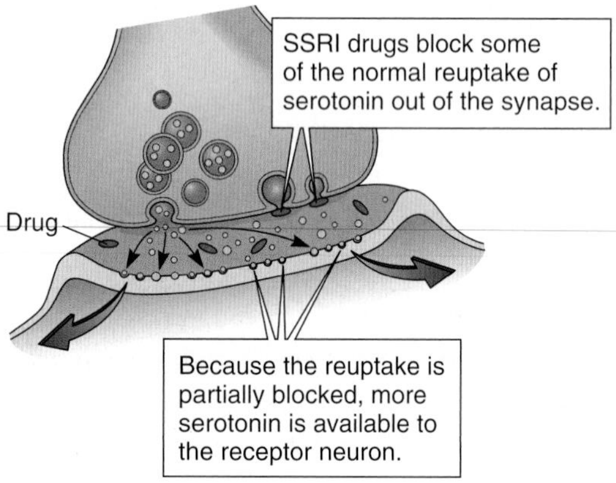

SSRI drugs block some of the normal reuptake of serotonin out of the synapse.
Drug
Because the reuptake is partially blocked, more serotonin is available to the receptor neuron.

- In **electroconvulsive therapy (ECT)**, an electrical current is passed through the brain, stimulating convulsions that produce changes in the central and peripheral nervous systems. ECT is used primarily in cases of severe depression that do not respond to other treatments.
- The most extreme biomedical therapy is **psychosurgery**. **Lobotomy**, an older form of psychosurgery, is now outmoded. Recently, psychiatrists have been experimenting with a more limited and precise surgical procedure called **deep brain stimulation (DBS).**

## 4 Psychotherapy in Perspective 391

- All major forms of therapy are designed to address disturbed thoughts, disturbed emotions, disturbed behaviors, interpersonal and life situation difficulties, and biomedical disturbances. Research indicates that, overall, therapy does work (as shown in the graph).

**Is therapy generally effective? • Figure 14.13**

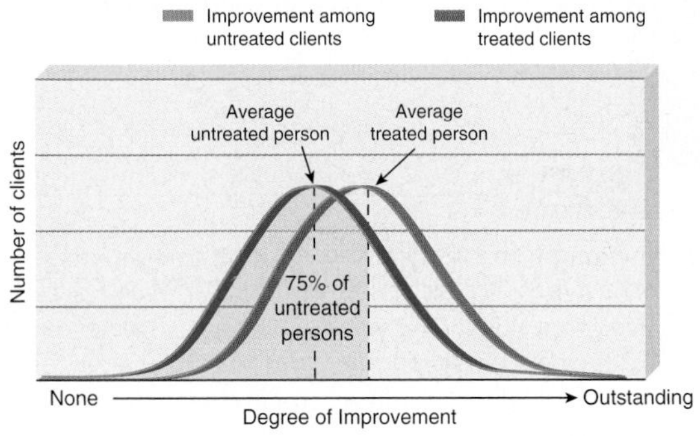


- In **group therapy**, multiple people meet together to work toward therapeutic goals. A variation is the **self-help group**, which is not guided by a professional. Therapists often refer their patients to group therapy and self-help groups to supplement individual therapy.
- In **family therapy**, the aim is to change maladaptive family interaction patterns. All members of the family attend therapy sessions, though at times the therapist may see family members individually or in twos or threes.
- Therapies in all cultures have certain key features in common; however, there are also important differences among cultures. Therapists must recognize cultural differences in order to build trust with clients and effect behavioral change. Therapists must also be sensitive to possible gender issues in therapy.

# Key Terms

**RETRIEVAL PRACTICE** Write a definition for each term before turning back to the referenced page to check your answer.

- active listening 381
- antianxiety drugs 388
- antidepressant drugs 388
- antipsychotic drugs 388
- aversion therapy 386
- behavior therapy 384
- biomedical therapy 387
- client-centered therapy 380
- cognitive restructuring 382
- cognitive therapy 381
- cognitive-behavior therapy 382
- deep brain stimulation (DBS) 390
- eclectic approach 392
- electroconvulsive therapy (ECT) 390
- empathy 380
- family therapy 394
- free association 377
- genuineness 381
- group therapy 394
- humanistic therapy 379
- interpersonal therapy (IPT) 379
- interpretation 378
- latent content 378
- lobotomy 390
- manifest content 378
- modeling therapy 386
- mood-stabilizer drugs 388
- psychoanalysis 377
- psychodynamic therapy 379
- psychopharmacology 388
- psychosurgery 390
- psychotherapy 376
- rational-emotive behavior therapy (REBT) 382
- repetitive transcranial magnetic stimulation (rTMS) 391
- self-help group 394
- self-talk 382
- shaping 386
- systematic desensitization 384
- telehealth 395
- transference 378
- unconditional positive regard 380

# Critical and Creative Thinking Questions

1. You undoubtedly had certain beliefs and ideas about therapy before reading this chapter. Has studying this chapter changed these beliefs and ideas? Explain.
2. Which form of therapy do you personally find most appealing? Why?
3. What do you consider the most important commonalities among the major forms of therapy described in this chapter? What are the most important differences?
4. Imagine that you were going to use the principles of cognitive-behavioral therapy to change some aspect of your own thinking and behavior. (Maybe you'd like to quit smoking, or be more organized, or overcome your fear of riding in elevators.) How would you identify faulty thinking? What could you do to change your thinking patterns and behavior?

# What is happening in this picture?

Photos 12/Alamy

In the 2008 movie, *The Changeling*, Angelina Jolie portrays a grief-stricken mother (Christine) who loudly and publicly challenges the police and press who try to force her to accept an impostor as her abducted son. The authorities try everything to silence her, including involuntary commitment to the county hospital's "psychopathic ward." While confined, she is forced to take mood-altering drugs and is told that her only chance for release is to admit she was mistaken about the identity of her son.

**Think Critically**

1. How might this film contribute to the negative stereotypes of psychotherapy?
2. Given that this movie was based on real-life events in 1928, what special danger does this "true story" pose?
3. Can you think of a Hollywood film that offers a positive portrayal of psychotherapy?

# Self-Test

**RETRIEVAL PRACTICE** Completing this self-test and comparing your answers with those in Appendix C provides immediate feedback and helpful practice for exams. Additional interactive, self-tests are available at www.wiley.com/college/carpenter.

1. Psychoanalysis/psychodynamic, humanistic, and cognitive therapies are often grouped together as _____.
   a. talk therapy
   b. behavior therapy
   c. humanistic and operant conditioning
   d. cognitive restructuring

2. The system of psychotherapy developed by Freud that seeks to bring unconscious conflicts into conscious awareness is known as _____.
   a. transference
   b. cognitive restructuring
   c. psychoanalysis
   d. the "hot seat" technique

3. _____ therapy emphasizes conscious processes and current problems.
   a. Self-talk
   b. Belief-behavior
   c. Psychodynamic
   d. Thought analysis

4. _____ therapy seeks to maximize personal growth through affective restructuring.
   a. Cognitive-emotive
   b. Emotive
   c. Humanistic
   d. Actualization

5. In Rogerian therapy, the _____ is responsible for discovering maladaptive patterns.
   a. therapist
   b. analyst
   c. doctor
   d. client

6. The process by which the therapist and client work to change destructive ways of thinking is called _____.
   a. problem solving
   b. self-talk
   c. cognitive restructuring
   d. rational recovery

7. Beck practices _____, which attempts to change not only destructive thoughts and beliefs, but the associated behaviors as well.
   a. psycho-behavior therapy
   b. cognitive-behavior therapy
   c. thinking-acting therapy
   d. belief-behavior therapy

8. The main focus in behavior therapy is to increase _____ and decrease _____.
   a. positive thoughts and feelings; negative thoughts and feelings
   b. adaptive behaviors; maladaptive behaviors
   c. coping resources; coping deficits
   d. all of these options

9. The three steps in systematic desensitization include all EXCEPT _____.
   a. learning to become deeply relaxed
   b. arranging anxiety-arousing stimuli into a hierarchy from least to most arousing
   c. practicing relaxation to anxiety-arousing stimuli, starting at the most arousing
   d. all of these options are included

10. The diagram shows an example of the process of _____, through which the man is learning to associate a negative response of nausea with the alcoholic drink.
   a. systematic desensitization
   b. operant conditioning
   c. observational learning
   d. aversion therapy

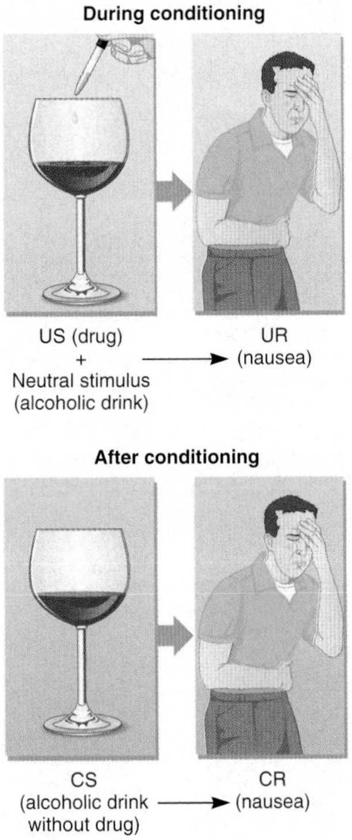


11. Label the four major categories of psychiatric drugs on the chart.

**Drug treatments for psychological disorders**

| Type of drug | Psychological disorder |
|---|---|
| a. ____________ | Anxiety disorders |
| b. ____________ | Schizophrenia |
| c. ____________ | Bipolar disorder |
| d. ____________ | Depressive disorders |

12. In electroconvulsive therapy (ECT), _____.
   a. current is never applied to the left hemisphere
   b. convulsions activate the amygdala, causing a change in maladaptive emotions
   c. electrical current passes through a coil of wire placed on the head
   d. none of the above

13. Label the five most common goals of therapies on the figure.

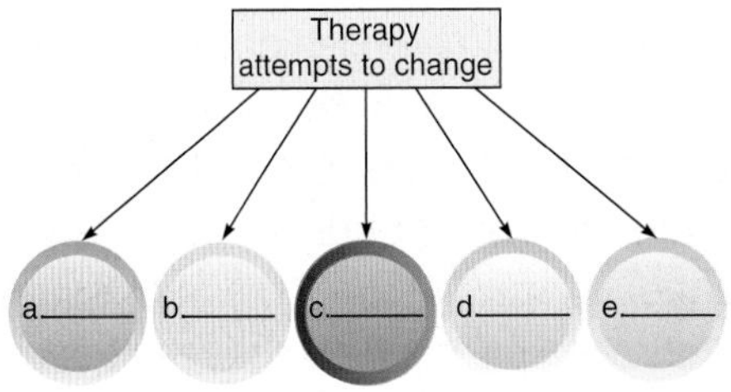


14. A(n) _____ group does not have a professional leader, and members assist each other in coping with a specific problem.
   a. self-help
   b. encounter
   c. peer
   d. behavior

15. In Japanese _____ therapy, the client is asked to discover personal guilt for having been ungrateful and troublesome to others.
   a. Kyoto
   b. Okado
   c. Naikan
   d. Obeisance

**THE PLANNER** ✓

Review your Chapter Planner on the chapter opener and check off your completed work.

# 15 Social Psychology

In the opening paragraphs for Chapter 1, we asked you to picture in your mind's eye the words "psychology" and "psychologist." Now, as we open the final chapter of this text, we hope that these initial images and thoughts have changed and that you now have a much larger and fuller appreciation for the magnitude of our field.

This text began with a brief introduction to psychology's history and research methods (Chapter 1), followed by an exploration of the smallest, micro level of behavior and mental processes—the neuron, brain, and other parts of the nervous system (Chapter 2). And now we close with a look at the largest, macro level—other people and the interconnectedness of humankind.

Unfortunately, just as a "fish doesn't know it's in water," as individuals living alone inside our own heads, we often underestimate the power of the social world that surrounds us. Beginning at the moment of conception until the moment of our death, our very existence is the result of other people. Therefore, to truly understand why and how we think, feel, or do what we do, we must look outward to the larger social factors that so dramatically determine who we are and who we will become.

© Aldo Murillo/iStockphoto

## CHAPTER OUTLINE



## CHAPTER PLANNER ✓

- ❑ Study the picture and read the opening story.
- ❑ Scan the Learning Objectives in each section: p. 404 ❑ p. 408 ❑ p. 417 ❑
- ❑ Read the text and study all figures and visuals. Answer any questions.

**Analyze key features**

- ❑ Myth Busters, p. 404
- ❑ Psychological Science, p. 406
- ❑ Process Diagram, p. 407
- ❑ Applying Psychology, p. 409
- ❑ What a Psychologist Sees p. 416 ❑ p. 426 ❑
- ❑ Psychology InSight, p. 421
- ❑ Stop: Answer the Concept Checks before you go on: p. 407 ❑ p. 416 ❑ p. 426 ❑

**End of chapter**

- ❑ Review the Summary and Key Terms.
- ❑ Answer the Critical and Creative Thinking Questions.
- ❑ Answer What is happening in this picture?
- ❑ Complete the Self-Test and check your answers.

# Social Cognition

## LEARNING OBJECTIVES

**RETRIEVAL PRACTICE** While reading the upcoming sections, respond to each Learning Objective in your own words. Then compare your responses with those in Appendix B.

1. **Explain** how attributions affect the way we perceive and judge others.
2. **Summarize** the three components of attitudes.

For many students and psychologists, your authors included, **social psychology** is the most exciting of all fields because it is about you and me, and because almost everything we do is *social*. Unlike earlier chapters that focused on individual processes, like the brain and other parts of the nervous system or personality, this field studies how large social forces, such as groups, social roles, and norms bring out the best and worst in all of us. This chapter is organized around three central themes: **social cognition** (how we think about and interpret ourselves and others), **social influence** (how situational factors and other people affect us), and **social relations** (how we develop and are affected by interpersonal relationships, including prejudice, aggression, altruism, and attraction). Before you read on, check on some misconceptions you may have about social relationships in *Myth Busters*.

**social psychology** The scientific study of how people's thoughts, feelings, and actions are affected by others.

**attributions** Explanations for behaviors or events.

## Attributions

One critical aspect of *social cognition* is the search for reasons and explanations for our own and others' behavior. It's natural to want to understand and explain why people behave as they do and why events occur as they do. Many social psychologists believe that developing logical **attributions** for people's behavior makes us feel safer and more in control (Baumeister & Vohs, 2011; Cushman & Greene, 2012; Ferrucci et al., 2011; Heider, 1958; Krueger, 2007). To do so, most people begin with the basic question of whether a given action stems mainly from the person's internal disposition or from the external situation.

**Mistaken attributions** Making the correct choice between disposition and situation is central to accurately judging why people do what they do. Unfortunately, our attributions are frequently marred by two major errors:

## MYTH BUSTERS

THE PLANNER

### OUR SOCIAL SELVES

**TRUE OR FALSE?**

___ 1. Most people judge others more harshly than they judge themselves.

___ 2. Inducing cognitive dissonance is an effective way to change attitudes.

___ 3. Groups generally make riskier or more conservative decisions than a single individual does.

___ 4. People wearing masks are more likely to engage in aggressive acts.

___ 5. Emphasizing gender differences may create and perpetuate prejudice.

___ 6. Substance abuse (particularly alcohol) is a major factor in aggression.

___ 7. When people are alone, they are more likely to help another individual than when they are in a group.

___ 8. Looks are the primary factor in our initial feelings of attraction, liking, and romantic love.

___ 9. Opposites attract.

___ 10. Romantic love generally starts to fade after 6 to 30 months.

Answers: Only one of these statements is false. You will find the answers within this chapter.

Chris Fortuna/Getty Images, Inc.

## The actor-observer effect • Figure 15.1

We tend to explain our own behavior in terms of external factors (situational attributions) and others' behavior in terms of their internal characteristics (dispositional attributions).

**Actor**

**Situational attribution**

Focuses attention on external factors

"I don't even like drinking beer, but it's the best way to meet women."

Christine Schneider/© Corbis

**Observer**

**Dispositional attribution**

Focuses on the personality of the actor

"He seems to always have a beer in his hand; he must have a drinking problem."

the fundamental attribution error and the self-serving bias. Let's explore each of these.

When we take into account *situational* influences on behavior, we generally make accurate attributions. However, given that people have enduring personality traits (Chapter 12), and our shared tendency to take cognitive shortcuts (Chapter 8), we more often choose dispositional attributions—that is, we blame the person.

For example, suppose a new student joins your class and seems distant, cold, and uninterested in interaction. It's easy to conclude that she's unfriendly, and maybe even "stuck-up"—a dispositional (personality) attribution. If you later saw her in a one-to-one interaction with close friends, you might be surprised to find that she is very warm and friendly. In other words, her behavior depends on the situation. This bias toward personal, dispositional factors rather than situational factors in our explanations for others' behavior is so common that it is called the **fundamental attribution error (FAE)** (Arkes & Kajdasz, 2011; Gebauer, Krempl, & Fleisch, 2008; Kennedy, 2010; Lennon et al., 2011).

**fundamental attribution error (FAE)** Attributing people's behavior to internal (dispositional) causes rather than external (situational) factors.

One reason that we tend to jump to internal, personal explanations is that human personalities and behaviors are more salient (or noticeable) than situational factors. This **saliency bias** helps explain why people sometimes suggest that homeless people begging for money "should just go out and get a job"—a phenomenon also called "blaming the victim."

When we explain our own behavior, we tend to favor internal attributions for our successes and external attributions for our failures. This **self-serving bias** is motivated by a desire to maintain positive self-esteem and a good public image (Alloy et al., 2011; Krusemark, Campbell, & Clementz, 2008; Martin & Carron, 2012; McClure et al., 2011). For example, students often take personal credit for doing well on an exam. If they fail the test, however, they tend to blame the instructor, the textbook, or the "tricky" questions.

How do we explain the discrepancy between the attributions we make for ourselves versus those we make for others? According to the **actor-observer effect** (Jones & Nisbett, 1971), when examining our own behaviors, we are the "actors" in the situation and therefore naturally look to the environment for explanations. "I tripped because of a crack in the sidewalk." In contrast, when explaining the behavior of others, we are "observing" the actors and therefore tend to blame the person versus the situation (a personality attribution). "She tripped because she's clumsy." (See **Figure 15.1**.)

**Culture and attributional biases** Both the fundamental attribution error and the self-serving bias may depend in part on cultural factors (Bozkurt & Aydin, 2004; Han et al., 2011; Imada & Ellsworth, 2011; Mason & Morris, 2010). In highly individualistic cultures, like that in the United States, people are defined and understood as individual selves—largely responsible for their successes and failures. But in collectivistic cultures, like that in Japan, people are primarily defined as members of their social network—responsible for doing as others expect. Accordingly, they tend to be more aware of situational constraints on behavior, making the FAE less likely (Leung et al., 2012; Matsumoto, 2000; McClure et al., 2011).

The self-serving bias is also much less common in Eastern nations. In Japan, for instance, the ideal person is someone who is aware of his or her shortcomings and continually works to overcome them—not someone who thinks highly of himself or herself (Heine & Renshaw, 2002). In the East (as well as in other collectivistic cultures), self-esteem is not related to doing better than others but to fitting in with the group (Berry et al., 2011; Markus & Kitayama, 2003; Smith, 2011).

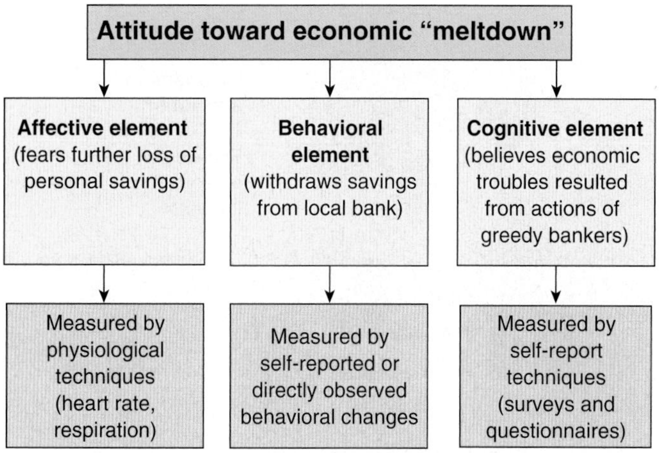


**Three components of attitudes • Figure 15.2**

When social psychologists study attitudes, they measure each of the three components: cognitive, affective, and behavioral.

## Attitudes

When we observe and respond to the world around us, we are seldom completely neutral. Rather, our responses toward subjects as diverse as pizza, AIDS, and abortion reflect our **attitudes**. We learn our attitudes both from direct experience and from watching others.

> **attitudes** Learned predisposition to respond to objects, people, and events in a particular way.

Social psychologists generally agree that most attitudes have three ABC components: *Affect* (feelings), *Behavior* (actions), and *Cognitions* (thoughts and beliefs) (**Figure 15.2**).

Although attitudes begin to form in early childhood, they are not permanent (a fact that advertisers and politicians know and exploit). One way to change attitudes is to make direct, persuasive appeals. As discussed in *Psychological Science*,

## Psychological Science

THE PLANNER

### Cognitive Dissonance

In one of the best-known tests of cognitive dissonance theory, Leon Festinger and J. Merrill Carlsmith (1959) paid participants either $1 or $20 to lie to other participants that a boring experimental task in which they had just participated was actually very enjoyable and fun. Those who were paid $1 subsequently changed their minds and actually reported more positive attitudes toward the task than those who were paid $20.

Why was there more attitude change among those who were only paid $1? All participants who lied to other participants presumably recognized the discrepancy between their initial attitude (the task was boring) and their behavior (lying to others that it was enjoyable and fun). However, the participants who were given insufficient monetary justification for lying (the $1 liars) apparently experienced greater cognitive dissonance. Therefore, they expressed more liking for the dull task than those who received sufficient monetary justification (the $20 liars). This second group had little or no motivation to change their attitude—they lied for the money! (Note that $20 in 1959 would be worth about $200 today.)

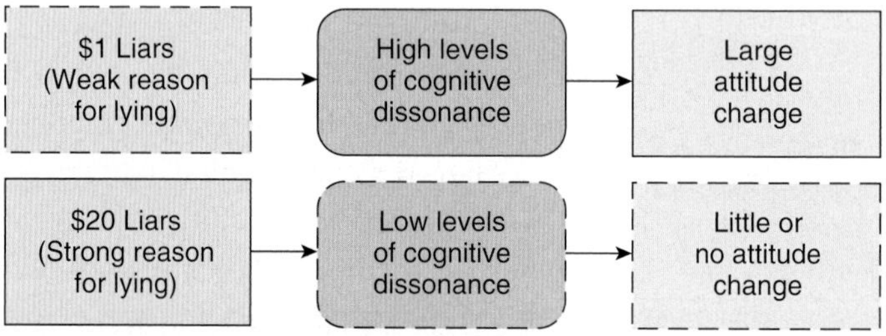


#### Identify the Research Method

1. What is the most likely research method used for Festinger and Carlsmith's cognitive dissonance study?
2. If you chose
   - the experimental method, label the IV, DV, experimental group, and control group.
   - the descriptive method, is this a naturalistic observation, survey, or case study?
   - the correlational method, is this a positive, negative, or zero correlation?
   - the biological method, identify the specific research tool (e.g., brain dissection, CT scan).

(Check your answers in Appendix C.)

PROCESS DIAGRAM

## Reducing cognitive dissonance • Figure 15.3

THE PLANNER

Why do some health professionals, who obviously know the dangers of smoking, continue to smoke?

**1** When inconsistencies or conflicts exist between our thoughts, feelings, and actions, they can lead to strong tension and discomfort (*cognitive dissonance*).

"I smoke cigarettes."
"I know smoking cigarettes leads to cancer."
REGIONAL CANCER CENTER
SMOKING INSIDE PATIO ONLY
Spencer Grant/Photo Edit


**2** To reduce this cognitive dissonance, we are motivated to change our behavior or our attitude.

**Change behavior** — **Change attitude**

**3a** Changing behavior, such as quitting smoking, can be hard to do. → "I don't smoke cigarettes any more."

**3b** If unable or unwilling to change their behavior, individuals can use one or more of the methods shown here to change their attitude. →

- **Change perceived importance of one of the conflicting cognitions:** "Experiments showing that smoking causes cancer have only been done on animals."
- **Modify one or both of the conflicting cognitions:** "I don't smoke that much."
- **Add additional cognitions:** "I only eat healthy foods, so I'm better protected from cancer."
- **Deny conflicting cognitions are related:** "There's no real evidence linking cigarettes and cancer."

an even more efficient strategy is to create **cognitive dissonance** (Baumeister & Bushman, 2011; Harmon-Jones, Amodio, & Harmon-Jones, 2010; Samnani, Boekhorst, & Harrison, 2012; Stone & Focella, 2011). Contradictions between our attitudes and behaviors can motivate us to change our attitudes to agree with our behaviors (Festinger, 1957) (**Figure 15.3**).

**cognitive dissonance** A feeling of discomfort caused by a discrepancy between two conflicting cognitions (thoughts) or between an attitude and a behavior.

The experience of cognitive dissonance may depend on a distinctly Western way of thinking about and evaluating the self. As we mentioned earlier, people in Eastern cultures tend not to define themselves in terms of their individual accomplishments. For this reason, making a bad decision may not pose the same threat to self-esteem that it would in more individualistic cultures, such as the United States (Berry et al., 2011; Choi & Nisbett, 2000; Dessalles, 2011; Imada & Kitayama, 2010).

### CONCEPT CHECK

STOP

1. **Why** do we tend to blame others for their misfortunes but deny responsibility for our own failures?
2. **How** can people resolve cognitive dissonance?

# Social Influence

## LEARNING OBJECTIVES

**RETRIEVAL PRACTICE** While reading the upcoming sections, respond to each Learning Objective in your own words. Then compare your responses with those in Appendix B.

1. **Identify** the factors that contribute to conformity.
2. **Describe** the factors that determine obedience.
3. **Explain** how groups affect behavior and decision making.

Having explored how we think about and interpret ourselves and others (social cognition), we now focus on how situational factors and other people affect us (social influence). In this section, we explore three key topics—*conformity*, *obedience*, and *group processes*.

## Conformity

Imagine that you have volunteered for a psychology experiment on visual perception. All participants are shown two cards. The first card has only a single vertical line on it, while the second card has three vertical lines of varying lengths. Your task is to determine which of the three lines on the second card is the same length as the single line on the first card. You are seated around a table with six other people and everyone is called on in order. Because you are seated to the left of the seventh participant, you are always next to last to provide your answers.

On the first two trials, everyone agrees on the correct line. However, on the third trial, your group is shown two cards with vertical lines labeled X on one or A, B, and C on the other, as in **Figure 15.4**. The first participant chooses line A as the closest in length to line X, an obvious wrong answer! When the second, third, fourth, and fifth participants also say line A, you really start to wonder: "What's going on here? Are they blind? Or am I?"

What do you think you would do at this point in the experiment? Would you stick with your convictions and say line B, regardless of what the others have answered? Or would you go along with the group? In the original version of this experiment, conducted by Solomon Asch (1951), six of the seven participants were actually *confederates* of the experimenter (that is, they were working with the experimenter and purposely gave wrong answers). Their incorrect responses were designed to test the participant's degree of **conformity**.

**conformity** The act of changing behavior as a result of real or imagined group pressure.

More than one-third of Asch's participants conformed—they agreed with the group's obviously incorrect choice. (Participants in a control group experienced no group pressure and almost always chose correctly.) Asch's study has been

**Solomon Asch's study of conformity • Figure 15.4**

Which line (A, B, or C) is most like line X? Could anyone convince you otherwise?

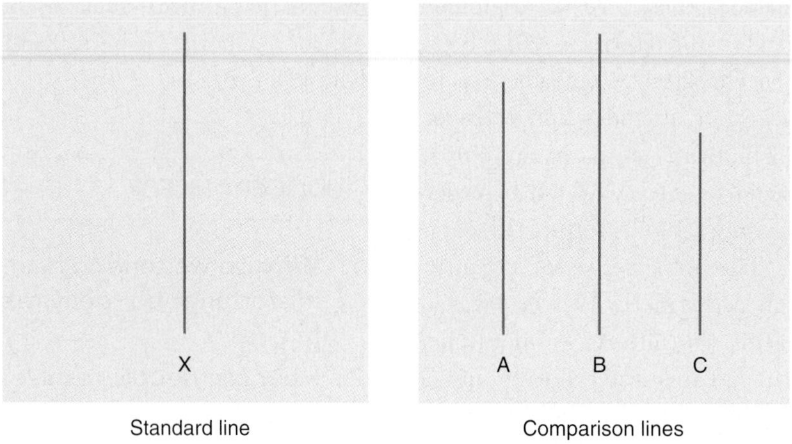

# Applying Psychology

THE PLANNER

## Cultural Norms for Personal Space

Culture and socialization have a lot to do with shaping norms for personal space. If someone invades the invisible "personal bubble" around our bodies, we generally feel very uncomfortable. People from Mediterranean and Latin American countries tend to maintain smaller interpersonal distances than do North Americans and Northern Europeans (Axtell, 2007; Steinhart, 1986). As you can see in this photo, these Middle-Eastern men are apparently comfortable with a smaller personal space and with showing male-to-male affection.

Interestingly, children in our own Western culture also tend to stand very close to others until they are socialized to recognize and maintain a greater personal distance. Furthermore, friends stand closer than strangers, women tend to stand closer than men, and violent prisoners prefer approximately three times the personal space of nonviolent prisoners (Axtell, 2007; Gilmour & Walkey, 1981; Lawrence & Andrews, 2004).

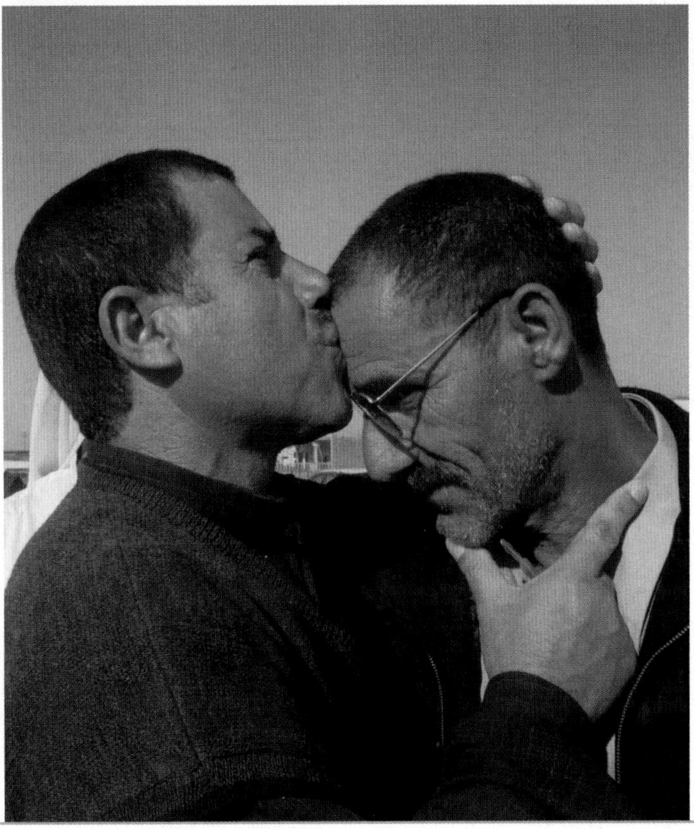
AFP/Getty Images, Inc.

### Think Critically

1. How might cultural differences in personal space help explain why Americans traveling abroad are sometimes seen as being "too loud and brassy"?
2. If men and women have different norms for personal space, what effect might this have on their relationships?

conducted dozens of times, in at least 17 countries, and always with similar results (Baumeister & Vohs, 2011; Mori & Arai, 2010; Takano & Sogon, 2008).

Why would so many people conform? To the onlooker, conformity is often difficult to understand. Even the conformer sometimes has a hard time explaining his or her behavior. Let's look at three factors that drive conformity:

- *Normative social influence* Often, people conform to group pressure out of a need for approval and acceptance by the group. **Norms** are expected behaviors that are adhered to by members of a group (see *Applying Psychology*). Most often, norms are quite subtle and implicit. Have you ever asked what others are wearing to a party, or watched your neighbor to be sure you pick up the right fork? Such behavior reflects your desire to conform and the power of normative social influence.
- *Informational social influence* Have you ever bought a specific product simply because of a friend's recommendation? You conform not to gain your friend's approval (normative social influence) but because you assume he or she has more information than you do. Given that participants in Asch's experiment observed all the other participants give unanimous decisions on the length of the lines, they also may have conformed because they believed the others had more information.
- *Reference groups* The third major factor in conformity is the power of **reference groups**—people we most admire, like, and want to resemble. Attractive actors and popular sports stars are paid millions of dollars to endorse products because advertisers know that we want to be as cool as LeBron James or as beautiful as Natalie Portman. Of course, we also have more important reference groups in our lives—parents, friends, family members, teachers, religious leaders, and so on.

Damian Dovarganes/©AP/Wide World Photos

**Advantages of conformity and obedience • Figure 15.5**

These people willingly obey the firefighters who order them to evacuate a building, and many lives are saved. What would happen to our everyday functioning if most people did not go along with the crowd or generally did not obey orders?

## Obedience

As we've seen, conformity involves going along with the group. A second form of social influence, **obedience**, involves going along with a direct command, usually from someone in a position of authority. From very early childhood, we're taught (socialized) to respect and obey our parents, teachers, and other authority figures.

**obedience** The act of following a direct command, usually from an authority figure.

Conformity and obedience aren't always bad (**Figure 15.5**). In fact, most people conform and obey most of the time because it is in their own best interest (and everyone else's) to do so. Like most North Americans, you stand in line at the movie theatre instead of pushing ahead of others. This allows an orderly purchasing of tickets. Conformity and obedience allow social life to proceed with safety, order, and predictability.

However, on some occasions it is important not to conform or obey. We don't want teenagers (or adults) engaging in risky sex or drug use just to be part of the crowd. And we don't want soldiers (or anyone else) mindlessly following orders just because they were told to do so by an authority figure. Recognizing and resisting destructive forms of obedience are particularly important to our society—and to social psychology. Let's start with an examination of a classic series of experiments on obedience by Stanley Milgram (1963, 1974).

Imagine that you have responded to a newspaper ad that is seeking volunteers for a study on memory. At the Yale University laboratory, an experimenter explains to you and another participant that he is studying the effects of punishment on learning and memory. You are selected to play the role of the "teacher." The experimenter leads you into a room where he straps the other participant—the "learner"—into a chair. He applies electrode paste to the learner's wrist "to avoid blisters and burns" and attaches an electrode that is connected to a shock generator.

You are shown into an adjacent room and told to sit in front of this same shock generator, which is wired through the wall to the chair of the learner. The shock machine consists of 30 switches representing successively higher levels of shock, from 15 volts to 450 volts. Written labels appear below each group of switches, ranging from "Slight Shock" to "Danger: Severe Shock," all the way to "XXX." The experimenter explains that it is your job to teach the learner a list of word pairs and to punish any errors by administering a shock. With each wrong answer, you are to increase the shock by one level. (The setup for the experiment is illustrated in **Figure 15.6**.)

You begin teaching the word pairs, but the learner's responses are often wrong. Before long, you are inflicting shocks that you can only assume must be extremely painful. After you administer 150 volts, the learner begins to protest: "Get me out of here. . . . I refuse to go on."

You hesitate, and the experimenter tells you to continue. He insists that even if the learner refuses to answer, you must keep increasing the shock levels. But the other person is obviously in pain. What should you do?

Actual participants in this research—the "teachers"—suffered real conflict and distress when confronted with this problem. They sweated, trembled, stuttered, laughed nervously, and repeatedly protested that they did not want to hurt the learner. But still they obeyed.

The psychologist who designed this study, Stanley Milgram, was actually investigating not punishment and learning but obedience to authority: Would participants obey the experimenter's prompts and commands to shock another human being? In Milgram's public survey, fewer than 25% thought they would go beyond 150 volts. And no respondents predicted they would go past the 300-volt level. Yet 65% of the teacher-participants in this series of studies obeyed completely—going all the way to the end of the scale, even beyond the point when the "learner" (Milgram's confederate, who actually received no shocks at all) stopped responding altogether.

Even Milgram was surprised by his results. Before the study began, he polled a group of psychiatrists, and they predicted that most people would refuse to go

## Milgram's study on obedience • Figure 15.6

Under orders from an experimenter, would you, as "teacher," use this shock generator to shock a man (the "learner") who is screaming and begging to be released? Few people believe they would. But research shows otherwise.

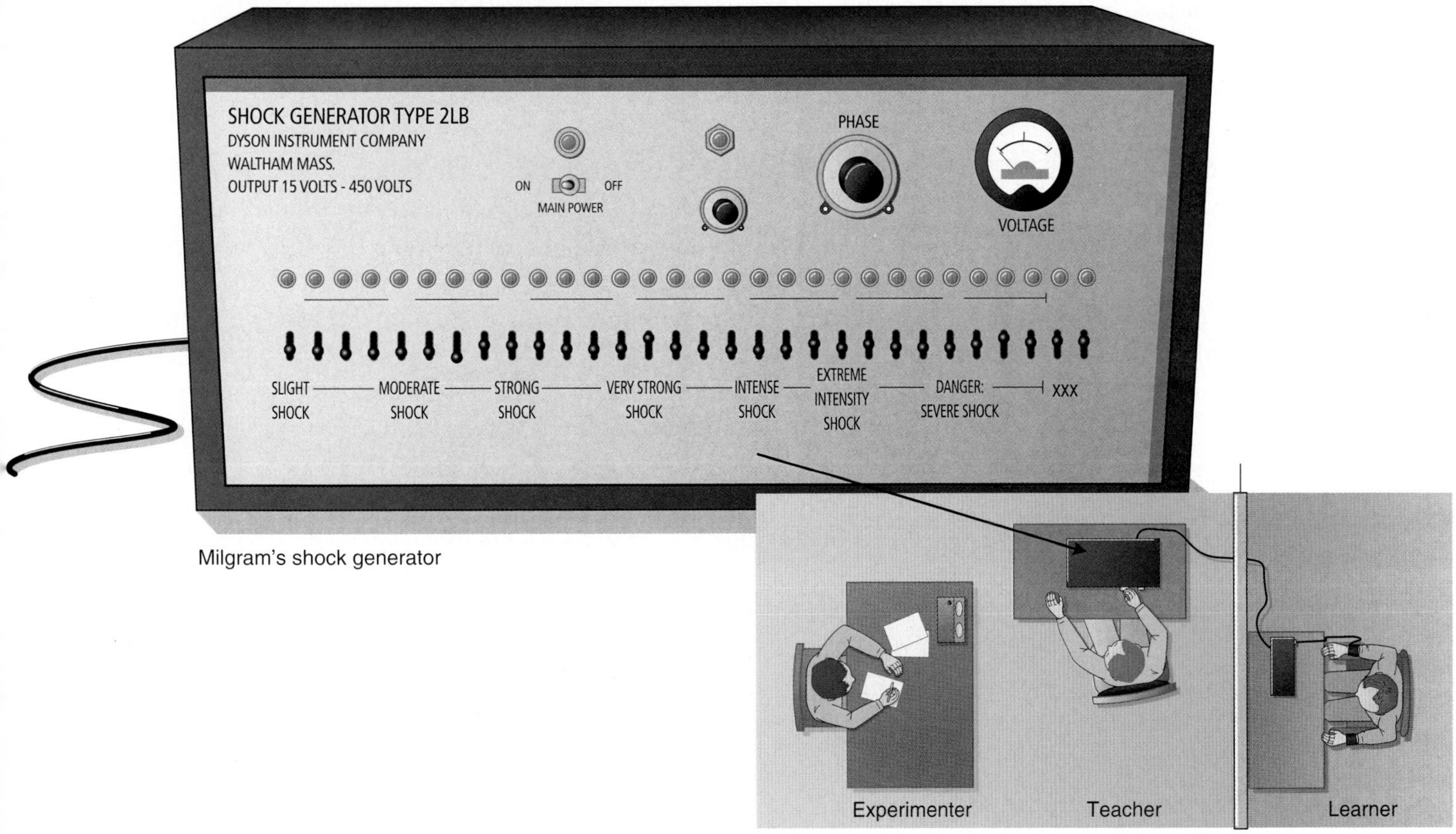


Milgram's shock generator

## Four factors in obedience • Figure 15.7

As you can see in the first bar on the graph, 65% of the participants in Milgram's early studies gave the learner the full 450-volt level of shocks. Milgram later conducted a series of studies to discover the specific conditions that either increased or decreased obedience to authority.

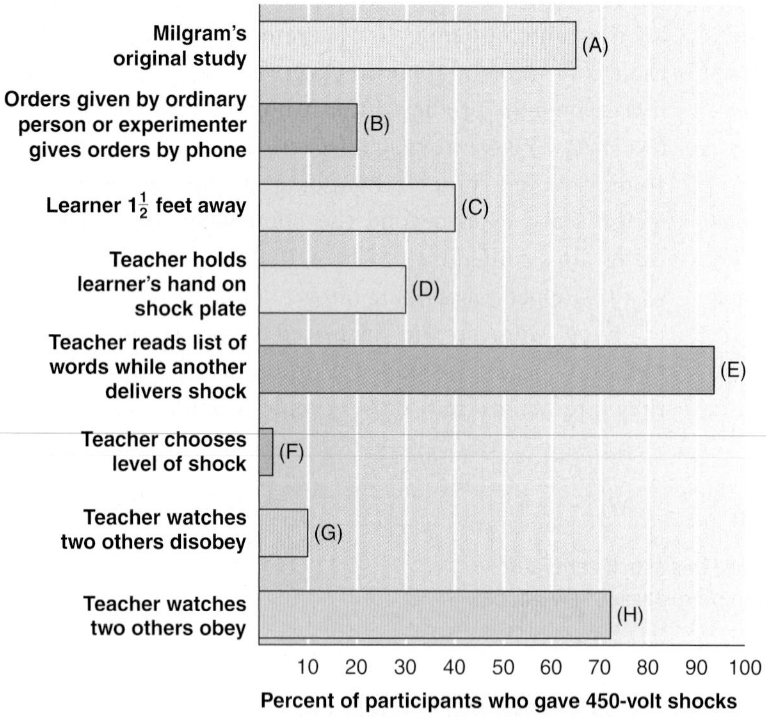


**1. Legitimacy and closeness of the authority figure**
When orders came from an ordinary person, and when the experimenter left the room and gave orders by phone, the participants' obedience dropped to 20%. (Bar B on graph.)

**2. Remoteness of the victim**
When the "learner" was only $1\frac{1}{2}$ feet away from the teacher, obedience dropped to 40%. When the teacher had to actually hold the learner's hand on the shock plate, obedience was only 30%. (Bars C and D on graph.)

**3. Assignment of responsibility**
When the teacher simply read the list of words, while another delivered the shock, obedience jumped to almost 94%. However, when the teacher was responsible for choosing the level of shock, only 3% obeyed. (Bars E and F on graph.)

**4. Modeling or imitating others**
When teachers watched two other teachers disobey, their own obedience was only 10%. But after watching other teachers obey, their own obedience jumped to over 70% (Milgram, 1963, 1974). (Bars G and H on graph.)

beyond 150 volts and that less than 1% of those tested would "go all the way." But, as Milgram discovered, most of his participants—men and women, of all ages, and from all walks of life—administered the highest voltage. The study was replicated many times and in many other countries, with similarly high levels of obedience.

Although there have been some recent partial replications of Milgram's study (e.g., Burger, 2009), Milgram's original setup could never be undertaken today. Many people are upset both by Milgram's findings and by the treatment of the participants in his research. Deception is a necessary part of some research, but the degree of deception and discomfort of participants in Milgram's research is now viewed as highly unethical. On the other hand, Milgram carefully debriefed every subject after the study and followed up with the participants for several months. Most of his "teachers" reported the experience as being personally informative and valuable.

One final, important reminder: *The "learner" was an accomplice of the experimenter and only pretended to be shocked.* Milgram provided specific scripts that they followed at every stage of the experiment. In contrast, the "teachers" were true volunteers who believed they were administering real shocks. Although they suffered and protested, in the final analysis, they still obeyed.

Why did the teachers in Milgram's study obey the orders to shock a fellow subject despite their moral objections? Are there specific circumstances that increase or decrease obedience? In a series of follow-up studies, Milgram found several important factors that influenced obedience: (1) legitimacy and closeness of the authority figure, (2) remoteness of the victim, (3) assignment of responsibility, and (4) modeling or imitating others (Blass, 1991, 2000; De Vos, 2010; Meeus & Raaijmakers, 1989; Pina e Cunha, Rego, & Clegg, 2010; Snyder, 2003) (**Figure 15.7**).

In addition to the four factors that Milgram identified, other social psychologists have emphasized several additional influences, including the following: socialization, the foot-in-the-door technique, and relaxed moral guard, among other things.

- *Socialization* This factor helps explain many instances of mindless obedience to immoral requests from people in positions of authority. For example, participants in Milgram's study came into the research lab with a lifetime of socialization toward the value of scientific research and respect for the experimenter's authority. They couldn't suddenly step outside themselves and question the morality of this particular experimenter and his orders. American soldiers accused of atrocities in Iraq, particularly those in the Abu Ghraib prison situation, shared this general socialization toward respect for authority combined with military training requiring immediate, unquestioning obedience.
- *Foot-in-the-door* The gradual nature of many obedience situations may also help explain why so many people were willing to give the maximum shocks in Milgram's study. The initial mild level of shocks may have worked as a **foot-in-the-door technique**, in which a first, small request is used as a setup for later, larger requests. Once Milgram's participants complied with the initial request, they might have felt obligated to continue.
- *Relaxed moral guard* One common intellectual illusion that hinders critical thinking about obedience is the belief that only evil people do evil things or that evil announces itself. The experimenter in Milgram's study looked and acted like a reasonable person who was simply carrying out a research project. Because he was not seen as personally corrupt and evil, the participants' normal moral guard was down, which can maximize obedience. As philosopher Hannah Arendt has suggested, the horrifying thing about the Nazis was not that they were so deviant but that they were so "terrifyingly normal."

Although the forces underlying obedience can be loud and powerful, even one, quiet, courageous, dissenting voice can make a difference. Perhaps the most beautiful and historically important example of just this type of bravery occurred in Alabama in 1955. Rosa Parks boarded a bus and, as expected in those times, obediently sat in the back section marked "Negroes." When the bus became crowded, the driver told her to give up her seat to a white man. Surprisingly for those days, Parks quietly but firmly refused and was eventually forced off the bus by police and arrested. This single act of disobedience was the catalyst for the small, but growing, civil rights movement and the later repeal of Jim Crow laws in the South. Today, Rosa Parks's courageous stand also inspires the rest of us to carefully consider when it is appropriate and good to obey authorities versus times when we must resist unethical or dangerous demands.

## Group Processes

Although we seldom recognize the power of group membership, social psychologists have identified several important ways that groups affect us.

**Group membership** How do the roles that we play within groups affect our behavior? This question fascinated social psychologist Philip Zimbardo. In his famous study at Stanford University, 24 carefully screened, well-adjusted young college men were paid $15 a day for participating in a two-week simulation of prison life (Haney, Banks, & Zimbardo, 1978; Zimbardo, 1993).

The students were randomly assigned to the role of either prisoner or guard. Prisoners were "arrested," frisked, photographed, fingerprinted, and booked at the police station. They were then blindfolded and driven to the "Stanford Prison." There, they were given ID numbers, deloused, issued prison clothing (tight nylon caps, shapeless gowns, and no underwear), and locked in cells. Participants assigned to be guards were outfitted with official-looking uniforms, billy clubs, and whistles, and they were given complete control.

Not even Zimbardo foresaw how the study would turn out. Although some guards were nicer to the prisoners than others, they all engaged in some abuse of power. The slightest disobedience was punished with degrading tasks or the loss of "privileges" (such as eating, sleeping, and washing). As demands increased and abuses began, the prisoners became passive and depressed. Only one prisoner fought back with a hunger strike, which ended with a forced feeding by the guards.

Four prisoners had to be released within the first four days because of severe psychological reactions. The study was stopped after only six days because of the alarming psychological changes in the participants.

Although this was not a true experiment in that it lacked control groups and clear measurements of the dependent variable (Chapter 1), it offers insights into the potential effects of roles on individual behavior (**Figure 15.8**). According to interviews conducted after the study, the students became so absorbed in their roles that they forgot they were participants in a university study (Zimbardo, Ebbeson, & Maslach, 1977).

Zimbardo's study also demonstrates **deindividuation**. To be deindividuated means that you feel less self-conscious, less inhibited, and less personally responsible as a member of a group than when you're alone (**Figure 15.9**). Deindividuation can be healthy and positive when you're among a happy, celebratory crowd, but it also sometimes leads to angry mobs, rioters, gang rapes, lynchings, and hate crimes (Haines & Mann, 2011; Lammers & Stapel, 2011; Rodrigues, Assmar, & Jablonski, 2005; Spears, 2011; Zimbardo, 2007). Groups sometimes actively promote deindividuation by requiring its members to wear uniforms, for example, as a way to increase allegiance and conformity.

**Group decision making** We have seen that the groups we belong to and the roles that we play within them influence how we think about ourselves. How do groups affect our decisions? Are two heads truly better than one?

Most people assume that group decisions are more conservative, cautious, and middle-of-the-road than individual decisions. But is this true? Initial investigations indicated that after discussing an issue, groups actually support riskier decisions than decisions they made as individuals before the discussion (Stoner, 1961).

Subsequent research on this **risky-shift phenomenon**, however, shows that some groups support riskier decisions while others support conservative decisions (Liu & Latané, 1998). How can we tell if a given group's decision will be risky or conservative? The final decision (risky or conservative) depends primarily on the dominant preexisting tendencies of the group. That is, as individuals interact and discuss their opinions, their initial positions become more exaggerated (or polarized).

### Power corrupts • Figure 15.8

Zimbardo's prison study showed how the demands of roles and situations could produce dramatic changes in behavior in just a few days. Can you imagine what happens to prisoners during life imprisonment, six-year sentences, or even a few nights in jail?

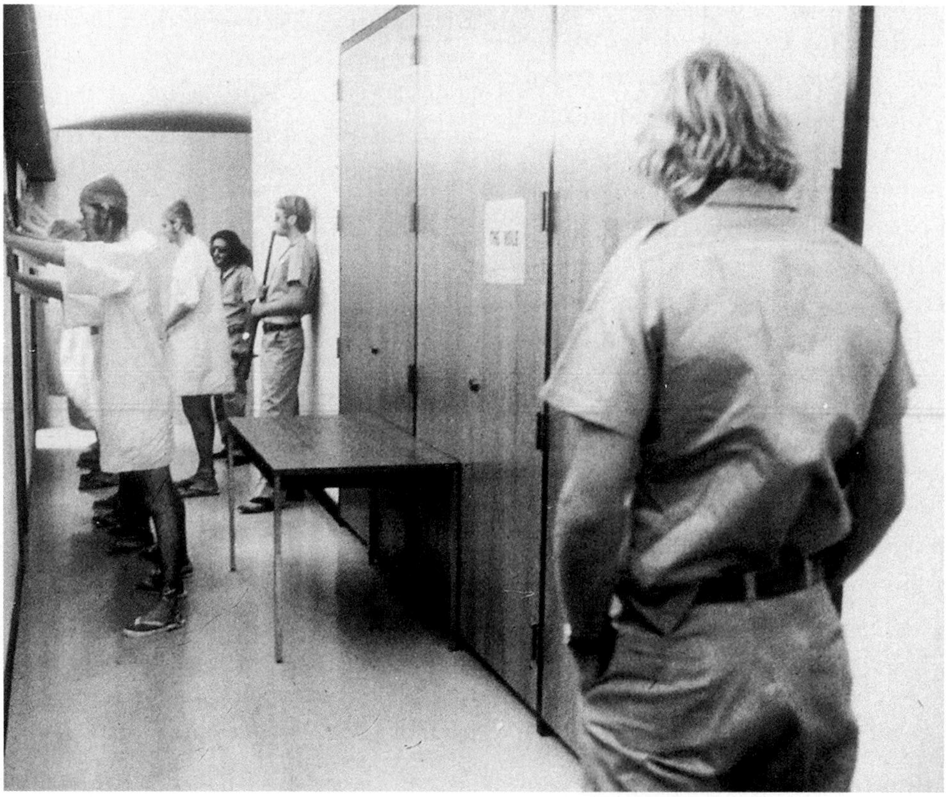

The Wisconsin State Journal, John Maniaci/©AP/Wide World Photos

**Lost in the crowd • Figure 15.9**

One of the most compelling explanations for deindividuation is the fact that the presence of others tends to increase arousal and feelings of anonymity. Anonymity is a powerful disinhibitor, helping to explain why vandalism seems to increase on Halloween (when people commonly wear masks) and why most crimes and riots occur at night—under the cover of darkness. Can you imagine your own behavior changing under such conditions?

**group polarization** A group's movement toward either riskier or more conservative behavior, depending on the members' initial dominant tendencies.

Why are group decisions more exaggerated? This tendency toward **group polarization** stems from increased exposure to persuasive arguments that reinforce the group's original opinion (Baumeister & Bushman, 2011; Tindale & Posavec, 2011; Yaniv, 2011).

A related phenomenon is **groupthink** (see *What a Psychologist Sees* on the next page). When a group is highly cohesive (a couple, a family, a panel of military advisers, an athletic team), the members' desire for agreement may lead them to ignore important information or points of view held by outsiders or critics (Hergovich & Olbrich, 2003; Vaughn, 1996).

**groupthink** Faulty decision making that occurs when a highly cohesive group strives for agreement and avoids inconsistent information.

Many presidential errors—from Franklin Roosevelt's failure to anticipate the attack on Pearl Harbor to Ronald Reagan's Iran-Contra scandal—have been blamed on groupthink. In addition, groupthink may have contributed to the losses of the space shuttles *Challenger* and *Columbia*, the terrorist attack of September 11, and the war in Iraq (Barlow, 2008; Burnette, Pollack, & Forsyth, 2011; Ehrenreich, 2004; Janis, 1972, 1989; Landay & Kuhnehenn, 2004; Post, 2011; Strozier, 2011).

# WHAT A PSYCHOLOGIST SEES

THE PLANNER

## How Groupthink Occurs

Few people realize that the decision to marry can be a form of groupthink. (Remember that a "group" can have as few as two members.) When planning for marriage, a couple may show symptoms of groupthink such as an illusion of invulnerability ("We're different—we won't ever get divorced"), collective rationalizations ("Two can live more cheaply than one"), shared stereotypes of the outgroup ("Couples with problems just don't know how to communicate"), and pressure on dissenters ("If you don't support our decision to marry, we don't want you at the wedding!").

**Antecedent conditions**
- A highly cohesive group of decision makers
- Insulation of the group from outside influences
- A directive leader
- Lack of procedures to ensure careful consideration of the pros and cons of alternative actions
- High stress from external threats with little hope of finding a better solution than that favored by the leader

↓

Strong desire for group consensus—the groupthink tendency

↓

**Symptoms of groupthink**
- Illusion of invulnerability
- Belief in the morality of the group
- Collective rationalizations
- Stereotypes of outgroups
- Self-censorship of doubts and dissenting opinions
- Illusion of unanimity
- Direct pressure on dissenters

↓

**Symptoms of poor decision making**
- An incomplete survey of alternative courses of action
- An incomplete survey of group objectives
- Failure to examine risks of the preferred choice
- Failure to reappraise rejected alternatives
- Poor search for relevant information
- Selective bias in processing information
- Failure to develop contingency plans

↓

**Low probability of successful outcome**

Joel Satore/NG Image Collection

CONCEPT CHECK STOP

1. **How** do norms, need for information, and reference groups produce conformity?
2. **What** four factors in obedience did Milgram identify?
3. **What** circumstances tend to lead to groupthink?

# Social Relations

## LEARNING OBJECTIVES

**RETRIEVAL PRACTICE** While reading the upcoming sections, respond to each Learning Objective in your own words. Then compare your responses with those in Appendix B.

1. **Explain** the difference between prejudice and discrimination.
2. **Summarize** the biological and psychosocial factors believed to be involved in aggression.
3. **Describe** the theories that attempt to explain altruism.
4. **Summarize** the factors that influence interpersonal attraction.

Kurt Lewin (1890–1947), often considered the "father of social psychology," was among the first to suggest that all behavior results from interactions between the individual and the environment. In this section, we explore how we develop and are affected by prejudice, aggression, altruism, and interpersonal attraction (all part of the general topic of *social relations*).

## Prejudice and Discrimination

**Prejudice**, which literally means prejudgment, creates enormous problems for its victims and also limits the perpetrator's ability to accurately judge others and to process information.

> **prejudice** A learned, generally negative attitude directed toward specific people solely because of their membership in an identified group.

Like all attitudes, prejudice is composed of three elements: *affective* (emotions about the group), *behavioral* (predispositions and negative actions toward members of the group), and *cognitive* (thoughts and stereotypical beliefs about group members). Although the terms *prejudice* and *discrimination* are often used interchangeably, they are not the same. Prejudice refers to an attitude, whereas **discrimination** refers to action (Fiske, 1998). The two often coincide, but not always (**Figure 15.10**).

How do prejudice and discrimination originate? Why do they persist? Five commonly cited sources of prejudice are *learning, personal experience, limited resources, displaced aggression*, and *mental shortcuts*.

**Learning** People learn prejudice the same way they learn all attitudes—primarily through *classical* and *operant conditioning* and *social learning* (Chapter 6). For example, repeated exposure to stereotypical portrayals of minorities and women on TV, in movies, and in magazines teach children that such images are correct. Similarly, hearing parents, friends, and teachers express their prejudices also reinforces prejudice (Jackson, 2011; Kassin, Fein, & Markus, 2008; Levitan, 2008; Livingston & Drwecki, 2007; Tropp & Mallett, 2011). *Ethnocentrism*, believing one's own culture represents the norm or is superior to all others, is also a form of learned prejudice.

**Personal experience** People also develop prejudice through direct experience. For example, when people make prejudicial remarks or "jokes," they often gain attention and even approval from others. Sadly, denigrating others also seems to boost one's self-esteem (Fein & Spencer, 1997; Hodson, Rush, & MacInnis, 2010; Plummer, 2001). Also, once someone has one or more negative interactions or experiences with members of a specific group, they may generalize their bad feelings and prejudice to all members of that group.

**Limited resources** Most people understand that prejudice and discrimination exact a high price on their victims, but few acknowledge the significant economic

### Prejudice and discrimination • Figure 15.10

Prejudice and discrimination are closely related, but either condition can exist without the other. The only situation in this example without prejudice or discrimination is when someone is given a job simply because he or she is the best candidate.

| Discrimination \ Prejudice | Yes | No |
|---|---|---|
| Yes | An African American is *denied* a job because the owner of a business is prejudiced. | An African American is *denied* a job because the owner of a business fears white customers won't buy from an African American salesperson. |
| No | An African American is *given* a job because the owner of a business hopes to attract African American customers. | An African American is *given* a job because he or she is the best suited for it. |

Librado Romero/The New York Times/Redux Pictures

**Prejudice and immigration • Figure 15.11**

Can you identify which of the five sources of prejudice best explains this behavior?

and political advantages they offer to the dominant group (Costello & Hodson, 2011; Kteily, Sidanius, & Levin, 2011; Perry & Sibley, 2011; Schaefer, 2008). For example, the stereotype that African Americans and Mexicans are inferior to European Americans helps justify and perpetuate a social order in the United States in which European Americans hold disproportionate power and resources.

**Displaced aggression** As we discuss in the next section, frustration sometimes leads people to attack the source of frustration. But, as history has shown, when the cause of frustration is ambiguous, or too powerful and capable of retaliation, people often redirect their aggression toward an alternative, innocent target, known as a *scapegoat*. There is strong historical evidence for the dangers of scapegoating. Jewish people were blamed for economic troubles in Germany prior to World War II. In America, beginning in the 1980s, gay men were blamed for the AIDS epidemic, and certain immigrant groups are now being blamed for the recent economic troubles (**Figure 15.11**).

**Mental shortcuts** Prejudice may stem from normal attempts to simplify a complex social world (Ambady & Adams, 2011; Dotsch, Wigboldus, & van Knippenberg, 2011; Kulik, 2005; Sternberg, 2007, 2009). Stereotypes allow people to make quick judgments about others, thereby freeing up their mental resources for other activities. In fact, stereotypes and prejudice can occur even without a person's conscious awareness or control. This process is known as "automatic" or **implicit bias** (Cattaneo et al., 2011; Columb & Plant, 2011; Dovidio, Pagotto, & Hebl, 2011). The most prominent method for measuring implicit bias, the *Implicit Association Test (IAT)*, measures how quickly people sort stimuli into particular categories.

People use stereotypes as mental shortcuts when they classify others in terms of their membership in a group. Given that people generally classify themselves as part of the preferred group, they also create ingroups and outgroups. An *ingroup* is any category that people see themselves as belonging to; an *outgroup* is any other category.

Compared with how they judge members of the outgroup, people tend to judge ingroup members as being more attractive, having better personalities, and so on—a phenomenon known as **ingroup favoritism** (Ahmed, 2007; DiDonato, Ullrich, & Krueger, 2011; Harth, Kessler, & Leach, 2008). People also tend to recognize greater diversity among members of their ingroup than they do among members of outgroups (Bendle, 2011; Cehajic, Brown, & Castano, 2008; Ryan, Casas, & Thompson, 2010; Wilson & Hugenberg, 2010). A danger of this **outgroup homogeneity effect** is that when members of minority groups are not recognized as varied and complex individuals, it is easier to treat them in discriminatory ways.

A good example is war. Viewing people on the other side as simply faceless enemies makes it easier to kill large numbers of soldiers and civilians. This dehumanization and facelessness also helps perpetuate our current high levels of fear and anxiety associated with terrorism (Dunham, 2011; Haslam et al., 2007; Hutchison & Rosenthal, 2011).

**Reducing prejudice** What can we do to combat prejudice? Four major approaches have been suggested: *cooperation and common goals, intergroup contact, cognitive retraining*, and *cognitive dissonance* (**Figure 15.12**).

**1. *Cooperation and common goals*** Research shows that one of the best ways to combat prejudice is to encourage *cooperation* rather than *competition* (Cunningham, 2002; Jackson, 2011; Sassenberg et al., 2007). Muzafer Sherif and his colleagues (1966, 1998) conducted an ingenious study to show the role of competition in promoting prejudice. The researchers artificially created strong feelings of ingroup and outgroup identification in a group of 11- and 12-year-old boys at a summer camp. They did this by physically separating the boys in different cabins and assigning different projects to each group, such as building a diving board or cooking out in the woods.

Once each group developed strong feelings of group identity and allegiance, the researchers set up a series of competitive games, including tug-of-war and touch football. They awarded desirable prizes to the winning teams. Because of this treatment, the groups began to pick fights, call each other names, and raid each other's camps. Researchers pointed to these behaviors as evidence of the experimentally produced prejudice.

After using competition to create prejudice between the two groups, the researchers created "minicrises" and tasks that required expertise, labor, and cooperation from both groups, and prizes were awarded to all. The prejudice between the groups slowly began to dissipate. By the end of the camp, the earlier hostilities and *ingroup favoritism* had vanished. Sherif's study showed not only the importance of cooperation as opposed to competition but also the importance of **superordinate goals** (the minicrises) in reducing prejudice.

**2. *Intergroup contact*** A second approach to reducing prejudice is increasing contact between groups (Butz & Plant, 2011; Cameron, Rutland, & Brown, 2007; Dovidio, Eller, & Hewstone, 2011; Gómez & Huici, 2008; Migacheva, Crocker, & Tropp, 2011; Wagner, Christ, & Pettigrew, 2008). Once people have positive experiences with a group, they tend to generalize their experiences to other groups. But as you just discovered with Sherif's study of the boys at the summer camp, contact can sometimes increase prejudice. Increasing contact only works under certain conditions that provide for *close interaction, interdependence* (superordinate goals that require cooperation), and *equal status*.

**3. *Cognitive retraining*** One of the most recent strategies in prejudice reduction requires taking another's perspective or undoing associations of negative stereotypical traits (Buswell, 2006; Müller et al., 2011; Todd et al., 2011). People can also learn to be nonprejudiced if they are taught to selectively pay attention to *similarities* rather than *differences* (Motyl et al., 2011; Phillips & Ziller, 1997). Can you see how emphasizing gender differences (*Men Are from Mars, Women Are from Venus*) may increase and perpetuate gender stereotypes?

**4. *Cognitive dissonance*** As mentioned earlier, one of the most efficient methods to change an attitude uses the principle of *cognitive dissonance*. Each time we meet someone who does not conform to our prejudiced views, we experience dissonance—"I thought all gay men were effeminate. This guy is a deep-voiced professional athlete. I'm confused." To resolve the dissonance, we can maintain our stereotypes by saying, "This gay man is an exception to the rule." However, if we continue our contact with a large variety of gay men, this "exception to the rule" defense eventually breaks down and attitude change occurs.

## How can we reduce prejudice? • Figure 15.12

Can you see how the psychological principles involved in a simple game of tug-of-war might reduce prejudice? Similarly, how might large changes in social policy, such as school busing, integrated housing, and increased civil rights legislation gradually change attitudes and eventually lead to decreased prejudice and discrimination?

Bob Daemmrich/PhotoEdit

## Aggression

Why do people act aggressively? In this section, we explore both biological and psychosocial explanations for **aggression** and how aggression can be controlled or reduced.

**aggression** Any behavior intended to harm someone.

**Biological explanations** Because aggression has such a long history and is found in all cultures, some scientists believe that humans are instinctively aggressive (Buss, 2008, 2011; Durrant & Ward, 2011; Glenn, Kurzban, & Raine, 2011; Kardong, 2008). Most social psychologists, however, reject this view, while accepting other biological factors. For example, twin studies suggest that some individuals are genetically predisposed to have hostile, irritable temperaments and to engage in aggressive acts (Haberstick et al., 2006; Hartwell, 2008; Rhee & Waldman, 2011; Weyandt, Verdi, & Swentosky, 2011). Research with brain injuries and other disorders also has identified possible aggression circuits in the brain (Denson, 2011; Hammond & Hall, 2011; Pardini et al., 2011; Siever, 2008). In addition, studies have linked the male hormone testosterone and lowered levels of some neurotransmitters with aggressive behavior (Geniole, Carré, & McCormick, 2011; Hermans, Ramsey, & van Honk, 2008; Juntii, Coats, & Shah, 2008; Popma et al., 2007; Sanchez-Martin et al., 2011; Takahashi et al., 2011).

**Psychosocial explanations** Substance abuse (particularly alcohol) is a major factor in many forms of aggression (Tremblay, Graham, & Wells, 2008; Levinthal, 2011). Similarly, aversive stimuli, such as loud noise, heat, pain, bullying, insults, and foul odors, also may increase aggression (Anderson, 2001; Anderson, Buckley, & Carnegy, 2008; Barash & Lipton, 2011; Monks, Ortega-Ruiz, & Rodriguez-Hildalgo, 2008). According to the **frustration–aggression hypothesis**, developed by John Dollard and colleagues (1939), another aversive stimulus—frustration—creates anger, which for some leads to aggression.

Research also finds that the widespread media violence on TV, the Internet, movies, and video games may contribute to aggression in both children and adults (Anderson et al., 2010; Bastian, Jetten, & Radke, 2012; Donnerstein, 2011; Montag et al., 2012). However, the research is controversial (e.g., Ferguson, 2010, 2011, 2012; Valadez & Ferguson, 2012), and the link between violent media and aggression appears to be at least a two-way street. Laboratory studies, correlational research, and cross-cultural studies have found that exposure to violence increases aggressiveness and that aggressive children tend to seek out violent programs (Barlett, Harris, & Bruey, 2008; Carnagey, Anderson, & Bartholow, 2007; Giumetti & Markey, 2007; Kalnin et al., 2011; Krahé et al., 2011).

Finally, social-learning theory (Chapter 6) suggests that people raised in a culture with aggressive models will learn aggressive responses (Cohen & Leung, 2011; Matsumoto & Juang, 2008; Rhee & Waldman, 2011). For example, the United States has a high rate of violent crimes, and American children grow up with numerous models for aggression, which they tend to imitate.

How can people control or eliminate aggression? Some therapists advise people to release their aggressive impulses by engaging in harmless forms of aggression, such as vigorous exercise, punching a pillow, or watching competitive sports. But studies suggest that this type of *catharsis* doesn't really help and may only intensify the feeling (Bushman, 2002; Kuperstok, 2008).

A more effective approach is to introduce incompatible responses. Because certain emotional responses, such as empathy and humor, are incompatible with aggression, purposely making a joke or showing some sympathy for an opposing person's point of view can reduce anger and frustration (Garrick, 2006; Kassin, Fein, & Markus, 2008; Kaukiainen et al., 1999; Oshima, 2000; Weiner, 2006). Improving social and communication skills can also help keep aggression in check.

## Altruism

After reading about prejudice, discrimination, and aggression, you will no doubt be relieved to discover that human beings also behave in positive ways. People help and support one another by donating blood, giving time and money to charities, helping stranded motorists, and so on. There are also times when people do not help. Let's consider both responses.

**Altruism** (or *prosocial behavior*) refers to actions designed to help others with no obvious benefit to the helper. There are three key approaches to explaining why we help others, as well as a five-step decision process explaining when and why we don't help (**Figure 15.13**).

**altruism** Actions designed to help others with no obvious benefit to the helper.

# Psychology InSight Helping • Figure 15.13

Why, and under what conditions, do we sometimes reach out to help others, and then at other times ignore their need for assistance?

**a. Three models for helping**

Each model focuses on a different motivation for helping. Can you see how the empathy–altrusim model might be the best explanation for why someone might give food or money to the man in this photo?

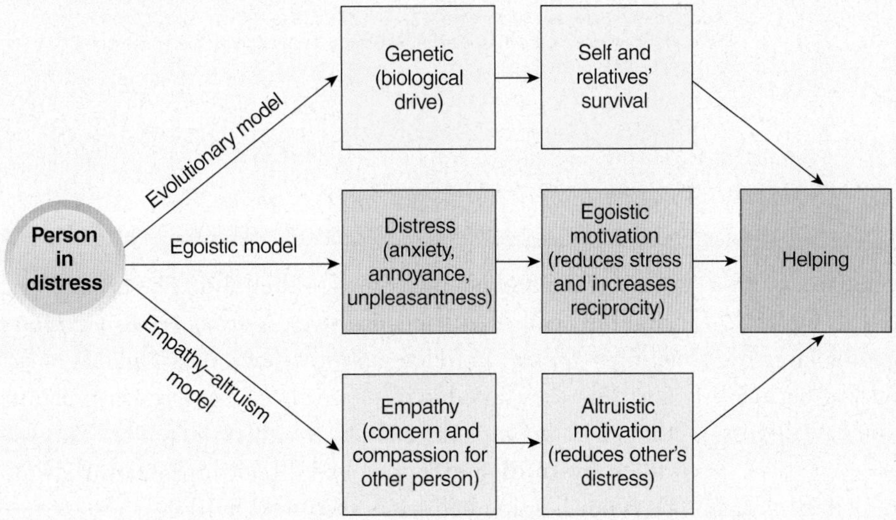

Bruce Ayers/Getty Images

**b. When and why we don't help**

According to Latané and Darley's five-step decision process (1968), if the answer at each step is yes, help is given. If the answer is no at any point, the helping process ends.

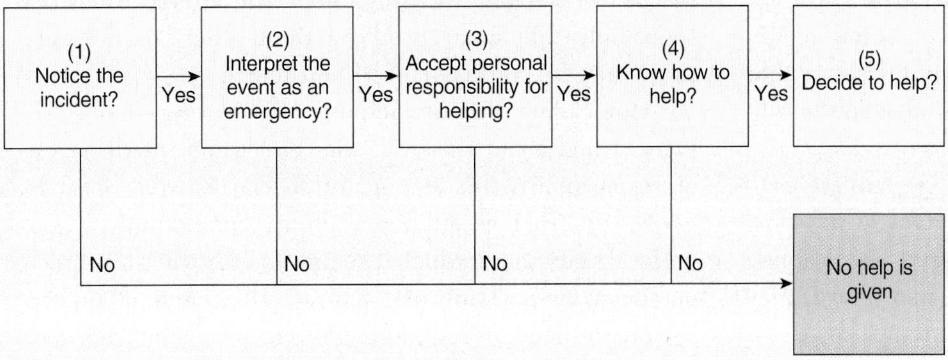


Sam Sarkis/Photodisc/Getty

**Evolutionary theory of helping** suggests that altruism is an instinctual behavior that has evolved because it favors survival of one's genes (Boyd, Richerson, & Henrich, 2011; Marshall, 2011; McNamara et al., 2008; Swami, 2011). By helping your child or other relative, for example, you increase the odds of your genes' survival.

## Bystander intervention • Figure 15.14

Social psychologists have a long history of research on the impact of others on helping behavior (e.g., Fischer et al., 2011). In one of the earliest studies, in the graph, participants were either working alone in a room or with others to complete a questionnaire, while smoke was pumped into the room through a wall vent to simulate an emergency. As you can see, about 75% of the participants quickly reported the smoke when they were alone. In contrast, fewer than 40% reported smelling smoke when others were in the room. Keep this study in mind when you're in a true emergency situation. It may save your life!

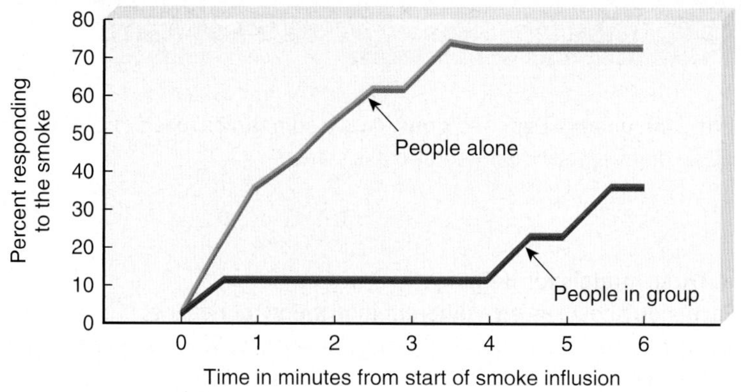


Adapted with permission from Myers, D. G. (2002). *Social Psychology* (7th ed.). New York: McGraw-Hill, Figure 12-4. © The McGraw-Hill Companies, Inc.; based on Darley, J. M., & Latané, B. 1968.

Other research suggests that helping others may actually be self-interest in disguise. According to this **egoistic model**, we help others only because we hope for later reciprocation, because it makes us feel virtuous, or because it helps us avoid feeling guilty (Cialdini, 2009; Williams et al., 1998).

Opposing the evolutionary and egoistic models is the **empathy–altruism hypothesis** (Batson, 1991, 1998, 2006, 2011; Park, Troisi, & Maner, 2011). This perspective holds that simply seeing or hearing of another person's suffering can create *empathy*—a subjective grasp of that person's feelings or experiences. And when we feel empathic toward another, we are motivated to help that person for his or her own sake. The ability to empathize may even be innate. Research with infants in the first few hours of life shows that they become distressed and cry at the sound of another infant's cries (Hay, 1994; Hoffman, 1993).

Many theories have been proposed to explain why people help, but few explain why we do not. One of the most comprehensive explanations for helping or not helping comes from the research of Bibb Latané and John Darley (1968) (see Figure 15.13b). They found that whether or not someone helps depends on a series of interconnected events and decisions: The potential helper must first notice what is happening, interpret the event as an emergency, accept personal responsibility for helping, decide how to help, and then actually initiate the helping behavior.

How does this sequence explain television news reports and "caught on tape" situations in which people are robbed or attacked and no one comes to their aid? The breakdown generally comes at the third stage—*taking personal responsibility for helping*. In follow-up interviews, most onlookers report that they failed to intervene because they were certain that someone must already have called the police. This so-called **bystander effect** is a well-known problem that affects our helping behaviors (**Figure 15.14**).

Why are we less likely to help when others are around? Researchers have labeled this reaction the **diffusion of responsibility** phenomenon—the dilution (or diffusion) of personal responsibility for acting by spreading it among all other group members. If you see someone drowning at the beach and you are the only person around, then responsibility falls squarely on you. But if others are present, there may be a diffusion of responsibility.

How can we promote helping? Considering what we've just learned about the bystander effect and the diffusion of responsibility, it's important to clarify when help is needed and then assign responsibility. For example, if you notice a situation in which it seems unclear whether someone needs help, simply ask. On the other hand, if you are the one in need of help, look directly at anyone who may be watching and give specific directions, such as "Call the police. I am being attacked!"

Highly publicized television programs, like "CNN Heroes," which honor and reward exceptional forms of altruism, also increase helping. Finally, enacting laws that protect helpers from legal liability also encourages helping.

## Interpersonal Attraction

What causes us to feel admiration, liking, friendship, intimacy, lust, or love? All of these social experiences are reflections of **interpersonal attraction**. Psychologists have found three compelling factors in interpersonal attraction: *physical attractiveness, proximity,* and *similarity*. Each influences our attraction in different ways.

**interpersonal attraction** Positive feelings toward another.

**Physical attractiveness** How we look (facial characteristics, body size and shape, manner of dress) is one of the most important factors in our initial liking or loving of others (Back et al., 2011; Bryan, Webster, & Mahaffey, 2011; Buss, 2003, 2005, 2007, 2008, 2011; Cunningham, Fink, & Kenix, 2008; Lippa, 2007; Regan, 2011). Attractive individuals are seen as more poised, interesting, cooperative, achieving, sociable, independent, intelligent, healthy, and sexually warm than unattractive people (Jaeger, 2011; Moore, Filippou, & Perrett, 2011; Tsukiura & Cabeza, 2011; Willis, Esqueda, & Schacht, 2008).

Interestingly, standards for attractiveness appear consistent across cultures. Women are valued more for looks and youth, whereas men are valued more for maturity, ambitiousness, and financial resources (Bryan, Webster, & Mahaffey, 2011; Buss, 1989, 1994, 2005, 2008, 2011; Chang et al., 2011; Li, Valentine, & Patel, 2011; Swami, 2011). Why? Evolutionary psychologists would suggest that both men and women prefer attractive people because good looks generally indicate good health, sound genes, and high fertility. Similarly, women prefer men with maturity and financial resources because the responsibility of rearing and nurturing children more often falls on women's shoulders. However, what is considered beautiful also varies from culture to culture (**Figure 15.15**).

So how do those of us who are not "superstar beautiful" manage to find mates? The good news is that people usually do not hold out for partners who are "ideally

### Standards for attractiveness • Figure 15.15

Evolutionary psychologists would suggest these women are attractive because of their youth and symmetrical features—indicators of their valuable fertility. However, you probably find the woman from your own culture more attractive, highlighting the role of cultural background in standards of beauty.

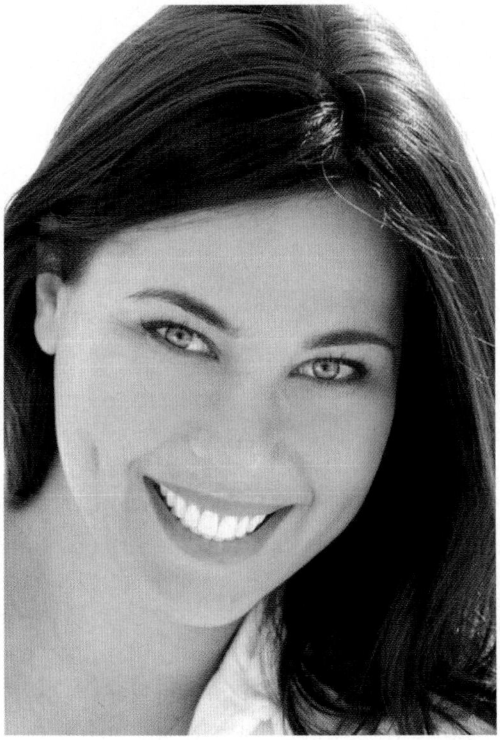

Nancy Brown/Photographer's Choice/Getty Images, Inc.

Alfredo Maiquez/Lonely Planet Images/Getty Images, Inc.

## The face in the mirror • Figure 15.16

According to the mere exposure effect, people generally prefer a reversed image of themselves (the one on the left) because it is the familiar one we are accustomed to seeing in the mirror. However, when presented with reversed and nonreversed (normal camera angle) photos, close friends prefer the nonreversed images (Mita, Dermer, & Knight, 1977).

Bernhard Kuhmsted/Retna

Bernhard Kuhmsted/Retna

attractive." Instead, both men and women tend to select partners whose physical attractiveness approximately matches their own (Regan, 1998, 2011; Sprecher & Regan, 2002; Taylor et al., 2011). Also, people use nonverbal flirting behavior to increase their attractiveness and signal interest to a potential romantic partner (Lott, 2000; Moore, 1998). Although both men and women flirt, in heterosexual couples women generally initiate courtship.

**Proximity** Attraction also depends on people being in the same place at the same time. Thus, *proximity*, or geographic nearness, is another major factor in attraction. A study of friendship in college dormitories found that the person next door was more often liked than the person two doors away, the person two doors away was liked more than someone three doors away, and so on (Priest & Sawyer, 1967).

Why is proximity so important? It's largely because of the **mere exposure effect**. Just as familiar people become more physically attractive over time, repeated exposure also increases overall liking (Monin, 2003; Rhodes, Halberstadt, & Brajkovich, 2001). This makes sense from an evolutionary point of view. Things we have seen before are less likely to pose a threat than novel stimuli. It also explains why modern advertisers tend to run highly redundant ad campaigns with familiar faces and jingles. In short, repeated exposure increases liking! (See **Figure 15.16**.)

**Similarity** The major cementing factor for long-term relationships, whether liking or loving, is *similarity*. We tend to prefer, and stay with, people (and organizations) who are most like us, those who share our ethnic background, social class, interests, and attitudes (Böhm et al., 2010; Caprara et al., 2007; Morry, Kito, & Ortiz, 2011; Sanbonmatsu, Uchino, & Birmingham, 2011). In other words, "Birds of a feather flock together."

What about the old saying "opposites attract"? The term *opposites* here probably refers to personality traits rather than to social background or values. An attraction to a seemingly opposite person is more often based on the recognition that in one or two important areas that person offers something we lack. In sum, lovers can enjoy some differences, but the more alike people are, the more both their loving and liking endure.

**Loving others** We can think of interpersonal attraction as a fundamental building block of how we feel about others. But how do we make sense of love? Many people find the subject to be alternately mysterious, exhilarating, comforting—and even maddening. In this section, we explore three perspectives on love—*Sternberg's theory, romantic love,* and *companionate love*.

Robert Sternberg, a well-known researcher on creativity and intelligence (Chapter 8), produced a **triarchic** (triangular) **theory of love** (Sternberg, 1986, 1988, 2006) (**Figure 15.17**), with three components based on feelings of:

- *Intimacy*—emotional closeness and connectedness, mutual trust, friendship, warmth, private sharing of self-disclosure, and forming of "love maps."
- *Passion*—sexual attraction and desirability, physical excitement, "a state of intense longing to be with the other."
- *Commitment*—permanence and stability, the decision to stay in the relationship for the long haul, and the feelings of security that go with this intention.

For Sternberg, a healthy degree of all three components in both partners characterizes the fullest form of love, **consummate love**, and other research agrees that these three components are all "positively correlated with successful relationships" (Martin, 2008). Trouble occurs when one of the partners has higher or lower need for one or more of the components. For example, if one partner has a much higher need for intimacy and the other partner has a stronger interest in passion, this lack of compatibility can be fatal to the relationship—unless the partners are willing to compromise and strike a mutually satisfying balance (Sternberg, 1988).

**Romantic love** When you think of **romantic love**, do you think of falling in love, a magical experience that puts you on cloud nine? Romantic love has intrigued people throughout history. Its intense joys and sorrows have also inspired countless poems, novels, movies, and songs around the world. A cross-cultural study by anthropologists William Jankowiak and Edward Fischer found romantic love in 147 of the 166 societies they studied. The researchers concluded that "romantic love constitutes a human universal or, at the least, a near universal" (1992, p. 154). Romantic love may be almost universal, but it's not perfect. And what about the best friend's type of **companionate love** that evolves over time? (See *What a Psychologist Sees* on the next page for answers.)

## The triangle of love • Figure 15.17

According to Sternberg, we all experience various forms and stages of love, six of which are seen as being on the outside of the triangle. Only true *consummate love* is inside the triangle because it involves a healthy balance of intimacy, passion, and commitment. The balance among these three components naturally shifts and changes over the course of a relationship, but relationships based on only one or two of these elements are generally less fulfilling and less likely to survive.

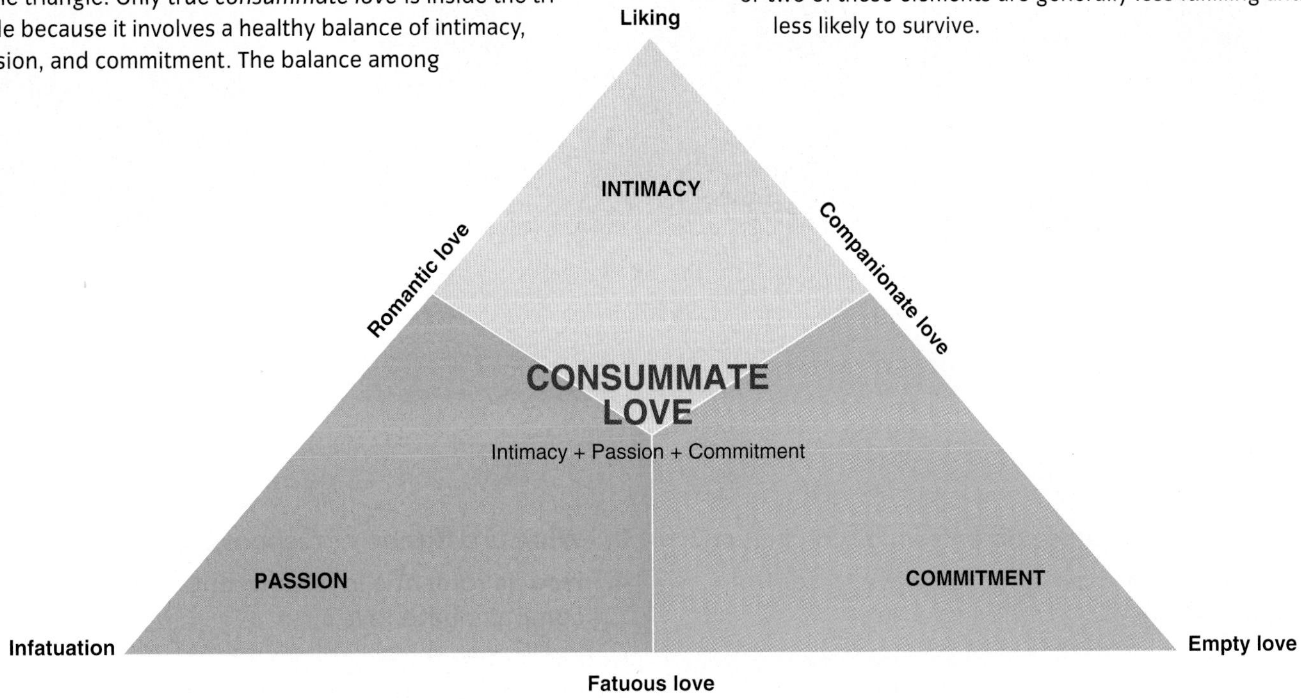

# WHAT A PSYCHOLOGIST SEES ✓ THE PLANNER

## Love over the Lifespan

Even in the most devoted couples, the intense attraction and excitement of *romantic love* generally begin to fade 6 to 30 months after the relationship begins (Hatfield & Rapson, 1996; Livingston, 1999). This is because romantic love is largely based on mystery and fantasy—people often fall in love with what they want another person to be (Fletcher & Simpson, 2000; Levine, 2001). These illusions usually fade with the realities of everyday living.

*Companionate love* is a strong, lasting attraction based on admiration, respect, trust, deep caring, and commitment. Studies of close friendships show that satisfaction grows with time as we come to recognize the value of companionship and intimacy (Gottman, 2011; Kim & Hatfield, 2004; Regan, 2011). One tip for maintaining companionate love is to overlook each other's faults. People are more satisfied with relationships when they have a somewhat idealized perception of their partner (Barelds & Dijkstra, 2011; Campbell et al., 2001; Fletcher & Simpson, 2000; Regan, 2011). This makes sense in light of research on cognitive dissonance: Idealizing our mates allows us to believe we have a good deal.

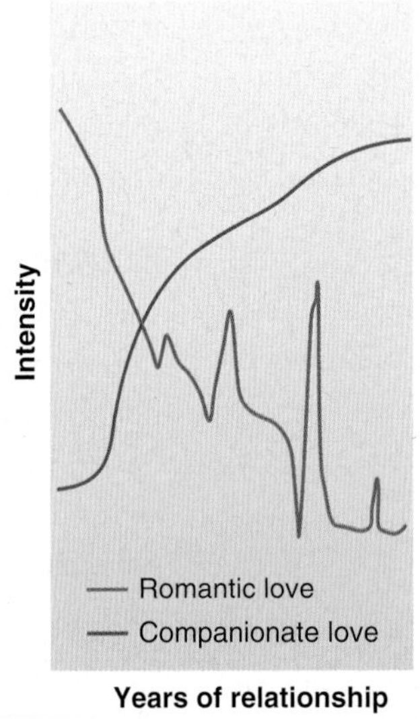

Sisse Brimberg & Cotton Coulson/NG Image Collection

Annie Griffiths Belt/NG Image Collection

Alaska Stock Images/NG Image Collection

### CONCEPT CHECK

1. **How** does prejudice differ from discrimination?
2. **What** are the most effective ways to control aggression?
3. **What** is diffusion of responsibility?
4. **How** is romantic love different from companionate love?

# Summary

THE PLANNER

## 1 Social Cognition 404

- **Social psychology** is the study of how other people influence our thoughts, feelings and actions. **Social cognition**, how we think about and interpret ourselves and others, involves **attributions**, which help us explain behaviors and events. However, these attributions are frequently marred by the **fundamental attribution error (FAE)** and the **self-serving bias**. Both biases may depend in part on cultural factors.
- **Attitudes** have three ABC components: *Affective, Behavioral,* and *Cognitive* (shown in the diagram). An efficient strategy for changing attitudes is to create **cognitive dissonance**. Like attributional biases, the experience of cognitive dissonance may depend on culture.

**Three components of attitudes • Figure 15.2**

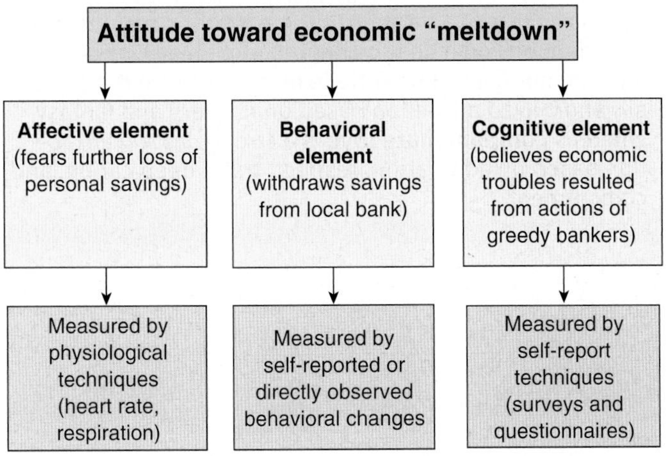


## 2 Social Influence 408

- Three factors drive **conformity**: *normative social influence, informational social influence*, and the role of *reference groups*. Conformity involves going along with the group, while **obedience** involves going along with a direct command, usually from someone in a position of authority. Milgram's research demonstrated the startling power of social situations to create obedience (see graph). The degree of deception and discomfort that Milgram's participants were subjected to raises serious ethical questions, and the same study would never be done today.

**Four factors in obedience • Figure 15.7**

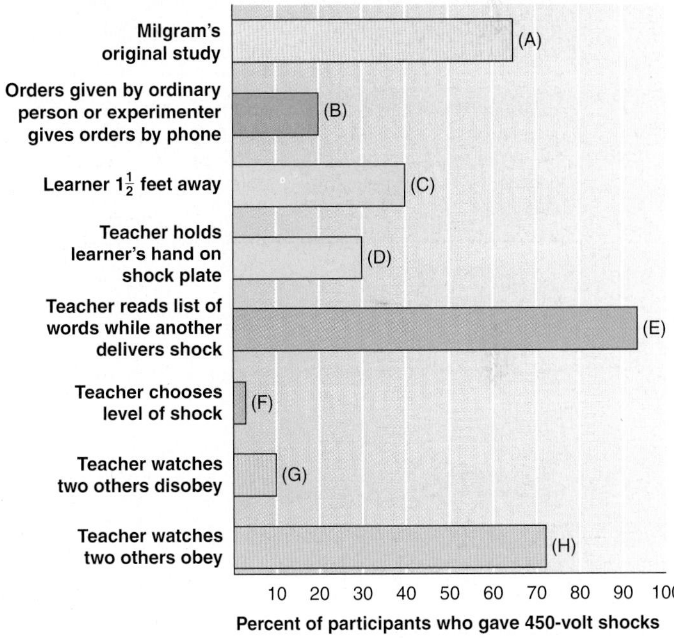


- The roles that we play within groups strongly affect our behavior, as Zimbardo's Stanford Prison experiment showed. Zimbardo's study also demonstrated **deindividuation**, In addition, groups affect our decisions. As individuals interact and discuss their opinions, **group polarization** (a group's movement toward more extreme decisions) and **groupthink** (a group's tendency to strive for agreement and to avoid inconsistent information) tend to occur. Both processes may hinder effective decision making.

## 3 Social Relations 417

- Like all attitudes, **prejudice** involves three ABC components: affective, behavioral, and cognitive. Although the terms prejudice and **discrimination** are often used interchangeably, they are not the same (see the diagram). Five commonly cited sources of prejudice are *learning, personal experience, limited resources, displaced aggression,* and *mental shortcuts*.

**Prejudice and discrimination • Figure 15.10**

| | Prejudice: Yes | Prejudice: No |
|---|---|---|
| **Discrimination: Yes** | An African American is *denied* a job because the owner of a business is prejudiced. | An African American is *denied* a job because the owner of a business fears white customers won't buy from an African American salesperson. |
| **Discrimination: No** | An African American is *given* a job because the owner of a business hopes to attract African American customers. | An African American is *given* a job because he or she is the best suited for it. |

- Several biological factors may help explain **aggression**, including genetic predisposition, aggression circuits in the brain, hormones, and neurotransmitters. Psychosocial explanations for aggression include substance abuse, aversive stimuli, media violence, and social learning.
- Evolutionary theory suggests that **altruism** is an evolved, instinctual behavior. Other research suggests that helping may actually be self-interest in disguise (**egoistic model**). The **empathy–altruism hypothesis** proposes that although altruism is occasionally based in selfish motivations, it is sometimes truly selfless and motivated by concern for others (empathy). Latané and Darley found that in order for helping to occur, the potential helper must notice what is happening, interpret the event as an emergency, take personal responsibility for helping, decide how to help, and then actually initiate the helping behavior.
- Psychologists have found three compelling factors in **interpersonal attraction:** physical attractiveness, proximity, and similarity. Sternberg suggests that **consummate love** depends on a healthy degree of intimacy, passion, and commitment. **Romantic love** is an intense but generally short-lived attraction based on mystery and fantasy, whereas **companionate love** is a strong, lasting attraction based on admiration, respect, trust, deep caring, and commitment.

# Key Terms

**RETRIEVAL PRACTICE** Write a definition for each term before turning back to the referenced page to check your answer.

- actor-observer effect 405
- aggression 420
- altruism 420
- attitude 406
- attribution 404
- bystander effect 422
- cognitive dissonance 407
- companionate love 425
- conformity 408
- consummate love 425
- deindividuation 414
- diffusion of responsibility 422
- discrimination 417
- egoistic model 422
- empathy-altruism hypothesis 422
- evolutionary theory of helping 421
- foot-in-the-door technique 413
- frustration-aggression hypothesis 420
- fundamental attribution error (FAE) 405
- group polarization 415
- groupthink 415
- implicit bias 418
- ingroup favoritism 418
- interpersonal attraction 423
- mere exposure effect 424
- norm 409
- obedience 410
- outgroup homogeneity effect 418
- prejudice 417
- reference group 409
- risky-shift phenomenon 414
- romantic love 425
- saliency bias 405
- self-serving bias 405
- social cognition 404
- social influence 404
- social psychology 404
- social relations 404
- superordinate goal 419
- triarchic theory of love 425

# Critical and Creative Thinking Questions

1. Have you ever changed a strongly held attitude? What caused the change for you?
2. Have you ever done something in a group that you would not have done if you were alone? What happened? How did you feel? What have you learned from this chapter that might help you avoid this behavior in the future?
3. How do Milgram's results—particularly the finding that the remoteness of the victim affected obedience—relate to some aspects of modern warfare?
4. What are some of the similarities between Zimbardo's prison study and the abuses at the Abu Ghraib prison in Iraq?
5. Do you believe that you are free of prejudice? After reading this chapter, which of the many factors that cause prejudice do you think is most important to change?
6. Can you think of situations when the egoistic model of altruism seems most likely correct? What about the empathy–altruism hypothesis?

# What is happening in this picture?

From a very early age, Andrew Golden was taught how to fire hunting rifles. At age 11, he and a friend killed four classmates and a teacher in a school shooting.

**Think Critically**

1. What factors might have contributed to Golden's tragically aggressive act?
2. Should Golden's parents be held legally or morally responsible for his behavior?
3. Should very young children be taught how to fire hunting rifles? Why or why not?

Najlah Feanny-Hicks/Corbis

# Self-Test

**RETRIEVAL PRACTICE** Completing this self-test and comparing your answers with those in Appendix C provides immediate feedback and helpful practice for exams. Additional interactive self-tests are available at www.wiley.com/college/carpenter.

**1.** The study of how other people influence our thoughts, feelings, and actions is called ______.

a. sociology

b. social science

c. social psychology

d. sociobehavioral psychology

**2.** The two major attribution mistakes people make are ______.

a. the fundamental attribution error and self-serving bias

b. situational attributions and dispositional attributions

c. the actor bias and the observer bias

d. stereotypes and biases

**3.** Label the three components of attitudes.

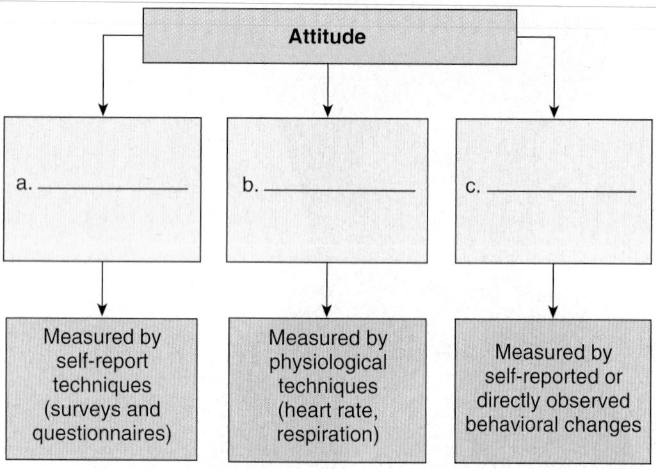


**4.** ______ theory says that contradictions between our attitudes and behavior can motivate us to change our attitudes to agree with our behavior.

a. Social learning

b. Cognitive dissonance

c. Defense mechanisms

d. Power of inconsistencies

**5.** The act of changing behavior as a result of real or imagined group pressure is called ______.

a. norm compliance

b. obedience

c. conformity

d. mob rule

**6.** Stanley Milgram was investigating ______ in his classic teacher-learner shock study.

a. the effects of punishment on learning

b. the effects of reinforcement on learning

c. obedience to authority

d. None of these options is correct.

**7.** Which of the following factors may contribute to destructive obedience?

a. remoteness of the victim

b. foot-in-the-door

c. socialization

d. All of these options are correct.

**8.** Faulty decision making resulting from a highly cohesive group striving for agreement to the point of avoiding inconsistent information is known as ______.

a. the risky-shift

b. group polarization

c. groupthink

d. destructive conformity

**9.** ______ is a learned, generally negative, attitude toward specific people solely because of their membership in an identified group.

a. Discrimination

b. Stereotyping

c. Cognitive biasing

d. Prejudice

10. As seen in this photo, people who feel anonymous within a group or crowd may experience increased arousal and decreased self-consciousness, inhibitions, and personal responsibility, which is referred to as ______.

a. groupthink

b. group polarization

c. authoritarianism

d. deindividuation

11. Research suggests that one of the best ways to decrease prejudice is to encourage ______.

a. cooperation

b. friendly competition

c. reciprocity of liking

d. conformity

12. Actions designed to help others with no obvious benefit to the helper are collectively known as ______.

a. empathy

b. sympathy

c. altruism

d. egoism

13. Label the three major explanations for helping.

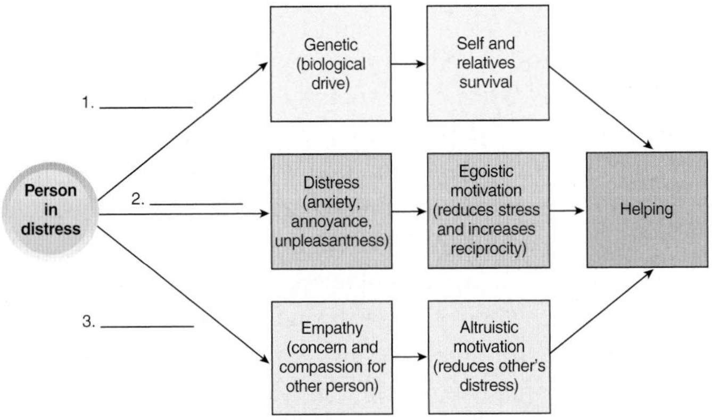


14. The degree of positive feelings you have toward others is called ______.

a. affective relations

b. interpersonal attraction

c. interpersonal attitudes

d. affective connections

15. A strong and lasting attraction characterized by admiration, respect, trust, deep caring, and commitment is called ______.

a. companionate love

b. intimate love

c. passionate love

d. all of these options

**THE PLANNER**

Review your Chapter Planner on the chapter opener and check off your completed work.

We are constantly bombarded by numbers: "On sale for 30% off," "9 out of 10 doctors recommend," "Your scores on the SAT were in the 75th percentile." Businesses and advertisers use numbers to convince us to buy their products. College admission officers use SAT percentile scores to help them decide who to admit to their programs. And, as you've seen throughout this text, psychologists use numbers to support or refute psychological theories and demonstrate that certain behaviors are indeed results of specific causal factors.

When people use numbers in these ways, they are using statistics. **Statistics** is a branch of applied mathematics that uses numbers to describe and analyze information on a subject.

Statistics make it possible for psychologists to quantify the information they obtain in their studies. They can then critically analyze and evaluate this information. Statistical analysis is imperative for researchers to describe, predict, or explain behavior. For instance, Albert Bandura (1973) proposed that watching violence on television causes aggressive behavior in children. In carefully controlled experiments, he gathered numerical information and analyzed it according to specific statistical methods. The statistical analysis helped him substantiate that the aggression of his subjects and the aggressive acts they had seen on television were related, and that the relationship was not mere coincidence.

Although statistics is a branch of applied mathematics, you don't have to be a math whiz to use statistics. Simple arithmetic is all you need for most of the calculations. For more complex statistics involving more complicated mathematics, computer programs are readily available. What is more important than learning the mathematical computations, however, is developing an understanding of when and why each type of statistic is used. The purpose of this appendix is to help you understand the significance of the statistics most commonly used.

EIGHTY-TWO PER CENT OF THE POPULATION DOESN'T GET ENOUGH FIBRE IN ITS DIET!
TWO OUT OF THREE KIDS IN AMERICA CAN'T SPELL "CHOLESTEROL"!
NINETY-NINE PER CENT OF THOSE INTERVIEWED SAID THEY HATED CARROTS!
STATISTICS ARE TOTALLY WRONG FIFTY-SIX PER CENT OF THE TIME!
THE STAT FAMILY

Drawing by M. Stevens; © 1989 the New Yorker Magazine, Inc.

## Gathering and Organizing Data

Psychologists design their studies to facilitate gathering information about the factors they want to study. The information they obtain is known as **data** (*data* is plural; its singular is *datum*). When the data are gathered, they are generally in the form of numbers; if they aren't, they are converted to numbers. After they are gathered, the data must be organized in such a way that statistical analysis is possible. In the following section, we will examine the methods used to gather and organize information.

### Variables

When studying a behavior, psychologists normally focus on one particular factor to determine whether it has an effect on the behavior. This factor is known as a **variable**, which is in effect anything that can assume more than one value (see Chapter 1). Height, weight, sex, eye color, and scores on an IQ test or a video game are all factors that can assume more than one value and are therefore variables. Some will vary between people, such as sex (you are either male *or* female but not both at the same time). Some may even vary within one person, such as scores on a video game (the same person might get 10,000 points on one try and only 800 on another). Opposed to a variable, anything that remains the same and does not vary is called a **constant**. If researchers use only females in their research, then sex is a constant, not a variable.

In nonexperimental studies, variables can be factors that are merely observed through naturalistic observation or case studies, or they can be factors about which people are questioned in a test or survey. In experimental studies, the two major types of variables are independent and dependent variables.

**Independent variables** are those that are manipulated by the experimenter. For example, suppose we were to conduct a study to determine whether the sex of the debater influences the outcome of a debate. In this study, one group of subjects watches a videotape of a debate between a male arguing the "pro" side and a female arguing the "con"; another group watches the same debate, but with the pro and con roles reversed. In such a study, the form of the presentation viewed by each group (whether "pro" is argued by a male or a female) is the independent variable because the experimenter manipulates the form of presentation seen by each group. Another example might be a study to determine whether a particular drug has any effect on a manual dexterity task. To study this question, we would administer the drug to one group and no drug to another. The independent variable would be the amount of drug given (some or none). The independent variable is particularly important when using **inferential statistics**, which we will discuss later.

The **dependent variable** is a factor that results from, or depends on, the independent variable. It is a measure of some outcome or, most commonly, a measure of the subjects' behavior. In the debate example, each subject's choice of the winner of the debate would be the dependent variable. In the drug experiment, the dependent variable would be each subject's score on the manual dexterity task.

## Frequency Distributions

After conducting a study and obtaining measures of the variable(s) being studied, psychologists need to organize the data in a meaningful way. **Table A.1** presents test scores from a Math Aptitude Test collected from 50 college students. This information is called **raw data** because there is no order to the numbers—they are simply presented as they were collected.

The lack of order in raw data makes them difficult to study. Thus, the first step in understanding the results of an experiment is to impose some order on the raw data. There are several ways to do this. One of the simplest is to create a **frequency distribution**, which shows the number of times a score or event occurs. Although frequency distributions are helpful in several ways, the major advantages are that they allow us to see the data in an organized manner and they make it easier to represent the data on a graph.

The simplest way to make a frequency distribution is to list all the possible test scores, then tally the number of people ($N$) who received those scores. **Table A.2** presents a frequency distribution using the raw data from Table A.1. As you can see, the data are now easier to read. From looking at the frequency distribution, you can see that most of the test scores lie in the middle with only a few at the very high or very low end. This was not at all evident from looking at the raw data.

**Math aptitude test scores for 50 college students**
**Table A.1**

| | | | | |
|---|---|---|---|---|
| 73 | 57 | 63 | 59 | 50 |
| 72 | 66 | 50 | 67 | 51 |
| 63 | 59 | 65 | 62 | 65 |
| 62 | 72 | 64 | 73 | 66 |
| 61 | 68 | 62 | 68 | 63 |
| 59 | 61 | 72 | 63 | 52 |
| 59 | 58 | 57 | 68 | 57 |
| 64 | 56 | 65 | 59 | 60 |
| 50 | 62 | 68 | 54 | 63 |
| 52 | 62 | 70 | 60 | 68 |

**Frequency distribution of 50 students on math aptitude test Table A.2**

| Score | Frequency |
|---|---|
| 73 | 2 |
| 72 | 3 |
| 71 | 0 |
| 70 | 1 |
| 69 | 0 |
| 68 | 5 |
| 67 | 1 |
| 66 | 2 |
| 65 | 3 |
| 64 | 2 |
| 63 | 5 |
| 62 | 5 |
| 61 | 2 |
| 60 | 2 |
| 59 | 5 |
| 58 | 1 |
| 57 | 3 |
| 56 | 1 |
| 55 | 0 |
| 54 | 1 |
| 53 | 0 |
| 52 | 2 |
| 51 | 1 |
| 50 | 3 |
| Total | 50 |

| Psychology aptitude test scores for 50 college students Table A.3 | | | | |
|---|---|---|---|---|
| 1350 | 750 | 530 | 540 | 750 |
| 1120 | 410 | 780 | 1020 | 430 |
| 720 | 1080 | 1110 | 770 | 610 |
| 1130 | 620 | 510 | 1160 | 630 |
| 640 | 1220 | 920 | 650 | 870 |
| 930 | 660 | 480 | 940 | 670 |
| 1070 | 950 | 680 | 450 | 990 |
| 690 | 1010 | 800 | 660 | 500 |
| 860 | 520 | 540 | 880 | 1090 |
| 580 | 730 | 570 | 560 | 740 |

This type of frequency distribution is practical when the number of possible scores is 20 or fewer. However, when there are more than 20 possible scores it can be even harder to make sense out of the frequency distribution than the raw data. This can be seen in **Table A.3**, which presents the hypothetical Psychology Aptitude Test scores for 50 students. Even though there are only 50 actual scores in this table, the number of possible scores ranges from a high of 1390 to a low of 400. If we included zero frequencies there would be 100 entries in a frequency distribution of this data, making the frequency distribution much more difficult to understand than the raw data. If there are more than 20 possible scores, therefore, a group frequency distribution is normally used.

In a **group frequency distribution**, individual scores are represented as members of a group of scores or as a range of scores (see **Table A.4**). These groups are called **class intervals**. Grouping these scores makes it much easier to make sense out of the distribution, as you can see from the relative ease in understanding Table A.4 as compared to Table A.3. Group frequency distributions are easier to represent on a graph.

**Group frequency distribution of psychology aptitude test scores for 50 college students Table A.4**

| Class Interval | Frequency |
|---|---|
| 1300–1390 | 1 |
| 1200–1290 | 1 |
| 1100–1190 | 4 |
| 1000–1090 | 5 |
| 900–990 | 5 |
| 800–890 | 4 |
| 700–790 | 7 |
| 600–690 | 10 |
| 500–590 | 9 |
| 400–490 | 4 |
| Total | 50 |

When graphing data from frequency distributions, the class intervals are represented along the **abscissa** (the horizontal or $x$ axis). The frequency is represented along the **ordinate** (the vertical or $y$ axis). Information can be graphed in the form of a bar graph, called a **histogram**, or in the form of a point or line graph, called a **polygon**. **Figure A.1** shows a histogram presenting the data from Table A.4. Note that the class intervals are represented along the bottom line of the graph (the $x$ axis) and the height of the bars indicates the frequency in each class interval. Now look at **Figure A.2**. The information

## Figure A.1

A histogram illustrating the information found in Table A.4.

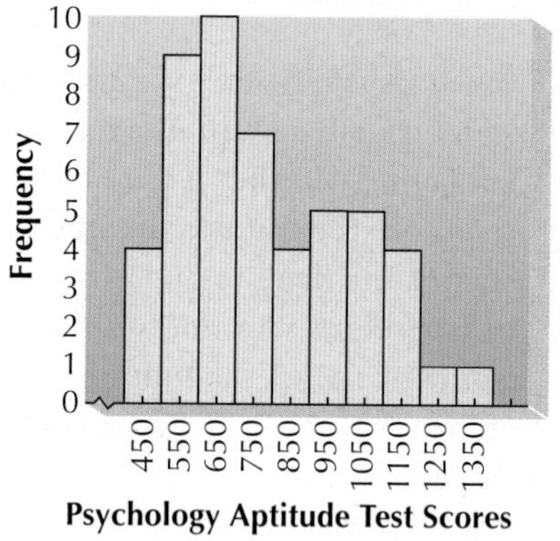


## Figure A.2

A polygon illustrating the information found in Table A.4.

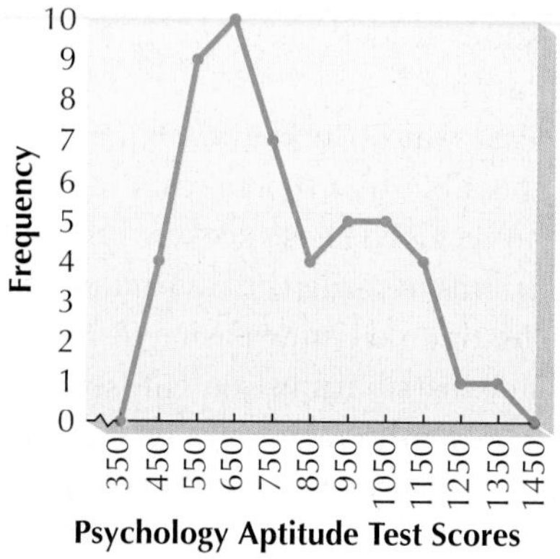

presented here is exactly the same as that in Figure A.1 but is represented in the form of a polygon rather than a histogram. Can you see how both graphs illustrate the same information? Even though graphs like these are quite common today, we have found that many students have never been formally taught how to read graphs.

### How to Read a Graph

Every graph has several major parts. The most important are the labels, the axes (the vertical and horizontal lines), and the points, lines, or bars. Find these parts in Figure A.1.

The first thing you should notice when reading a graph are the labels because they tell what data are portrayed. Usually the data consist of the descriptive statistics, or the numbers used to measure the dependent variables. For example, in Figure A.1 the horizontal axis is labeled "Psychology Aptitude Test Scores," which is the dependent variable measure; the vertical axis is labeled "Frequency," which means the number of occurrences. If a graph is not labeled, as we sometimes see in TV commercials or magazine ads, it is useless and should be ignored. Even when a graph is labeled, the labels can be misleading. For example, if graph designers want to distort the information, they can elongate one of the axes. Thus, it is important to pay careful attention to the numbers as well as the words in graph labels.

Next, you should focus your attention on the bars, points, or lines on the graph. In the case of histograms like the one in Figure A.1, each bar represents the class interval. The width of the bar stands for the width of the class interval, whereas the height of the bar stands for the frequency in that interval. Look at the third bar from the left in Figure A.1. This bar represents the interval "600 to 690 SAT Scores," which has a frequency of 10. You can see that this directly corresponds to the same class interval in Table A.4, since graphs and tables are both merely alternate ways of illustrating information.

Reading point or line graphs is the same as reading a histogram. In a point graph, each point represents two numbers, one found along the horizontal axis and the other found along the vertical axis. A polygon is identical to a point graph except that it has lines connecting the points. Figure A.2 is an example of a polygon, where each point represents a class interval and is placed at the center of the interval and at the height corresponding to the frequency of that interval. To make the graph easier to read, the points are connected by straight lines.

Displaying the data in a frequency distribution or in a graph is much more useful than merely presenting raw data and can be especially helpful when researchers are trying to spot relations between certain factors. However, as we explained earlier, if psychologists want to make precise predictions or explanations, we need to perform specific mathematical computations. How we use these statistics is the topic of our next section.

## Uses of the Various Statistics

The statistics psychologists use in a study depend on whether they are trying to describe and predict behavior or explain it. When they use statistics to describe behavior, as in reporting the average score on an aptitude test, they are using **descriptive statistics**. When they use them to explain behavior, as Bandura did in his study of children modeling aggressive behavior seen on TV, they are using **inferential statistics**.

### Descriptive Statistics

Descriptive statistics are the numbers used to describe the dependent variable. They can be used to describe characteristics of a **population** (an entire group, such as all people living in the United States) or a **sample** (a part of a group, such as a randomly selected group of 25 students from Cornell University). The major descriptive statistics include measures of central tendency (mean, median, and mode), measures of variation (variance and standard deviation), and correlation.

#### Measures of Central Tendency

Statistics indicating the center of the distribution are called **measures of central tendency** and include the mean, median, and mode. They are all scores that are typical of the center of the distribution. The **mean** is what most of us think of when we hear the word "average." The **median** is the middle score. The **mode** is the score that occurs most often.

**Mean** What is your average golf score? What is the average yearly rainfall in your part of the country? What is the average reading test score in your city? When these questions ask for the average, they are really asking for the *arithmetic mean*, which is the weighted average of all the raw scores. To find this arithmetic mean, you simply total all the raw scores and then divide that total by the number of scores added together. In statistical computation, the mean is represented by an "X" with a bar above it ($\overline{X}$, pronounced "X bar"), each individual raw score by an "$X$," and the total number of scores by an "$N$." For

**Computation of the mean for 10 IQ scores Table A.5**

| IQ Scores $X$ |
|---|
| 143 |
| 127 |
| 116 |
| 98 |
| 85 |
| 107 |
| 106 |
| 98 |
| 104 |
| 116 |
| $\Sigma X = 1{,}100$ |

$$\text{Mean} = \overline{X} = \frac{\Sigma X}{N} = \frac{1{,}100}{10} = 110$$

example, if we wanted to compute the $\overline{X}$, of the raw statistics test scores in Table A.1, we would sum all the $X$'s ($\Sigma X$, with $\Sigma$ meaning sum) and divide by $N$ (number of scores). In Table A.1, the sum of all the scores is equal to 3,100 and there are 50 scores. Therefore, the mean of these scores is

$$\overline{X} = \frac{3{,}100}{50} = 62$$

**Table A.5** illustrates how to calculate the mean for 10 IQ scores.

**Median** The **median** is the middle score in the distribution once all the scores have been arranged in rank order. If $N$ (the number of scores) is odd, then there actually is a middle score and that middle score is the median. When $N$ is even, there are two middle scores and the median is the mean of those two scores. **Table A.6** shows the computation of the median for two different sets of scores, one set with 15 scores and one with 10.

**Mode** Of all the measures of central tendency, the easiest to compute is the **mode**, which is merely the most frequent score. It is computed by finding the score that occurs most often. Whereas there is always only one mean and only one median for each distribution, there can be more than one mode. **Table A.7** shows how to find the mode in a distribution with one mode (unimodal) and in a distribution with two modes (bimodal).

**Computation of median for odd and even numbers of IQ scores Table A.6**

| IQ | IQ |
|---|---|
| 139 | 137 |
| 130 | 135 |
| 121 | 121 |
| 116 | 116 |
| 107 | 108 ← middle score |
| 101 | 106 ← middle score |
| 98 | 105 |
| 96 ← middle score | 101 |
| 84 | 98 |
| 83 | 97 |
| 82 | $N = 10$ |
| 75 | $N$ is even |
| 75 | |
| 68 | $\text{Median} = \frac{106 + 108}{2} = 107$ |
| 65 | |
| $N = 15$<br>$N$ is odd | |

There are several advantages to each of these measures of central tendency, but in psychological research the mean is used most often. If you're interested, a general statistics textbook will provide a more thorough discussion of the relative values of these measures.

**Finding the mode for two different distributions Table A.7**

| IQ | IQ |
|---|---|
| 139 | 139 |
| 138 | 138 |
| 125 | 125 |
| 116 ← | 116 ← |
| 116 ← | 116 ← |
| 116 ← | 116 ← |
| 107 | 107 |
| 100 | 98 ← |
| 98 | 98 ← |
| 98 | 98 ← |
| Mode = most frequent score | Mode = 116 and 98 |
| Mode = 116 | |

## Figure A.3

Three distributions having the same mean but a different variability.

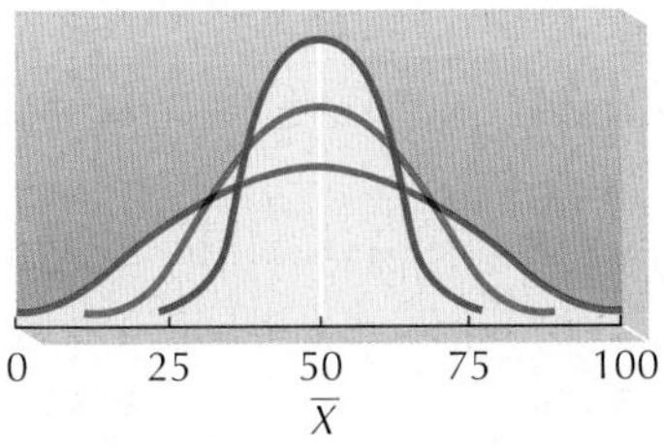


### Measures of Variation

When describing a distribution, it is not sufficient merely to give the central tendency; it is also necessary to give a **measure of variation**, which is a measure of the spread of the scores. By examining the spread, we can determine whether the scores are bunched around the middle or tend to extend away from the middle. **Figure A.3** shows three different distributions, all with the same mean but with different spreads of scores. You can see from this figure that, in order to describe these different distributions accurately, there must be some measures of the variation in their spread. The most widely used measure of variation is the standard deviation, which is represented by a lowercase "s." The **standard deviation** is a measurement of how much the scores in a distribution deviate from the mean. The formula for the standard deviation is

$$s = \sqrt{\frac{\sum (X - \overline{X})^2}{N}}$$

**Table A.8** illustrates how to compute the standard deviation.

Most distributions of psychological data are bell-shaped. That is, most of the scores are grouped around the mean, and the farther the scores are from the mean in either direction, the fewer the scores. Notice the bell shape of the distribution in **Figure A.4**. Distributions such as this are called **normal distributions**. In normal distributions, as shown in Figure A.4, approximately two-thirds of the scores fall within a range that is one standard deviation below the mean to one standard deviation above the mean. For example, the Wechsler IQ tests (see Chapter 7) have a mean of 100 and a standard deviation of 15. This means that approximately two-thirds of the people taking these tests will have scores between 85 and 115.

**Computation of the standard deviation for 10 IQ scores**
**Table A.8**

| IQ Scores $X$ | $X - \overline{X}$ | $(X - \overline{X})^2$ |
|---|---|---|
| 143 | 33 | 1089 |
| 127 | 17 | 289 |
| 116 | 6 | 36 |
| 98 | −12 | 144 |
| 85 | −25 | 625 |
| 107 | −3 | 9 |
| 106 | −4 | 16 |
| 98 | −12 | 144 |
| 104 | −6 | 36 |
| 116 | 6 | 36 |
| $\Sigma X = 1100$ | | $\Sigma(X - \overline{X})^2 = 2424$ |

Standard Deviation = $s$

$$= \sqrt{\frac{\sum (X - \overline{X})^2}{N}} = \sqrt{\frac{2424}{10}}$$

$$= \sqrt{242.4} = 15.569$$

### Correlation

Suppose for a moment that you are sitting in the student union with a friend. To pass the time, you and your friend decide to play a game in which you try to guess, ahead of time, the height of the next male who will enter the building. The winner, the one whose guess is closest to the person's actual height, gets a piece of pie paid for by the loser. When it is your turn, what do you guess? If you are like most people, you will probably try to estimate the mean of all the males in the union and use that as your guess. The mean is always your best guess if you have no other information.

## Figure A.4

The normal distribution forms a bell-shaped curve. In a normal distribution, two-thirds of the scores lie between one standard deviation above and one standard deviation below the mean.

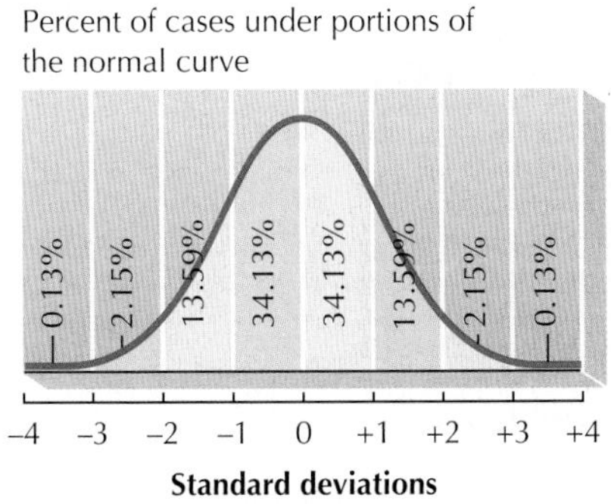

Now let's change the game a little and add a friend who stands outside the union and weighs the next male to enter the union. Before the male enters the union, your friend says "125 pounds." Given this new information, will you still guess the mean height? Probably not—you will probably predict *below* the mean. Why? Because you intuitively understand that there is a **correlation**, a relationship, between height and weight, with tall people usually weighing more than short people. Given that 125 pounds is less than the average weight for males, you will probably guess a less-than-average height. The statistic used to measure this type of relationship between two variables is called a correlation coefficient.

**Correlation Coefficient** A correlation coefficient measures the relationship between two variables, such as height and weight or IQ and SAT scores. Given any two variables, there are three possible relationships between them: positive, negative, and zero (no relationship). A positive relationship exists when the two variables vary in the same direction (e.g., as height increases, weight normally also increases). A negative relationship occurs when the two variables vary in opposite directions (e.g., as temperatures go up, hot chocolate sales go down). There is no relationship when the two variables vary totally independently of one another (e.g., there is no relationship between peoples' height and the color of their toothbrushes). **Figure A.5** illustrates these three types of correlations.

The computation and the formula for a correlation coefficient (correlation coefficient is delineated by the letter "r") are shown in **Table A.9**. The correlation coefficient ($r$) always has a value between $+1$ and $-1$ (it is never greater than $+1$ and it is never smaller than $-1$). When r is close to $+1$, it signifies a high positive relationship between the two variables (as one variable goes up, the other variable also goes up). When r is close to $-1$, it signifies a high negative relationship between the two variables (as one variable goes up, the other variable goes down). When $r$ is 0, there is no linear relationship between the two variables being measured.

Correlation coefficients can be quite helpful in making predictions. Bear in mind, however, that predictions are just that: predictions. They will have some error as long as the correlation coefficients on which they are based are not perfect ($+1$ or $-1$). Also, correlations cannot reveal any information regarding causation. Merely because two factors are correlated, it does not mean that one factor

**Computation of correlation coefficient between height and weight for 10 males Table A.9**

| Height (inches) | | Weight (pounds) | | |
|---|---|---|---|---|
| *X* | *X*² | *Y* | *Y*² | *XY* |
| 73 | 5,329 | 210 | 44,100 | 15,330 |
| 64 | 4,096 | 133 | 17,689 | 8,512 |
| 65 | 4,225 | 128 | 16,384 | 8,320 |
| 70 | 4,900 | 156 | 24,336 | 10,920 |
| 74 | 5,476 | 189 | 35,721 | 13,986 |
| 68 | 4,624 | 145 | 21,025 | 9,860 |
| 67 | 4,489 | 145 | 21,025 | 9,715 |
| 72 | 5,184 | 166 | 27,556 | 11,952 |
| 76 | 5,776 | 199 | 37,601 | 15,124 |
| 71 | 5,041 | 159 | 25,281 | 11,289 |
| 700 | 49,140 | 1,630 | 272,718 | 115,008 |

$$r = \frac{N \cdot \Sigma XY - \Sigma X \cdot \Sigma Y}{\sqrt{[N \cdot \Sigma X^2 - (\Sigma X)^2]}\sqrt{[N \cdot \Sigma Y^2 - (\Sigma Y)^2]}}$$

$$r = \frac{10 \cdot 115{,}008 - 700 \cdot 1{,}630}{\sqrt{[10 \cdot 49{,}140 - 700^2]}\sqrt{[10 \cdot 272{,}718 - 1{,}630^2]}}$$

$r = 0.92$

## Figure A.5

Three types of correlation. Positive correlation (top left): As the number of days of class attendance increases, so does the number of correct exam items. Negative correlation (top right): As the number of days of class attendance increases, the number of incorrect exam items decreases. Zero correlation (bottom): The day of the month on which one is born has no relationship to the number of exam items correct.

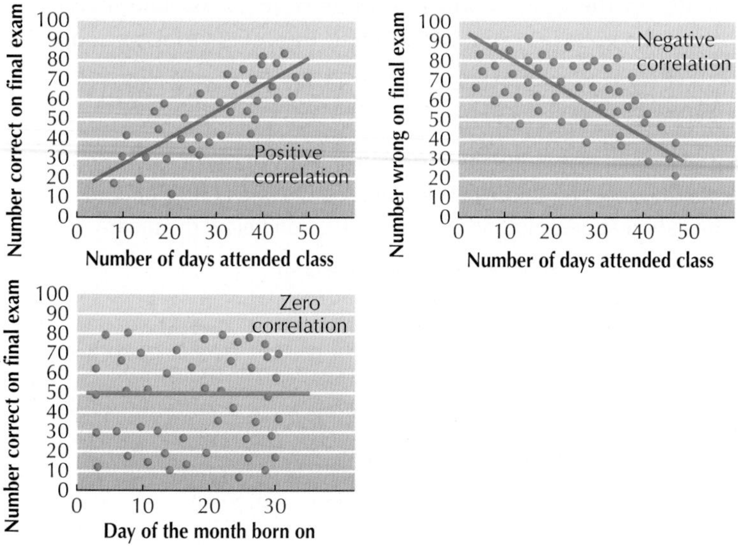

causes the other. Consider, for example, ice cream consumption and swimming pool use. These two variables are positively correlated with one another, in that as ice cream consumption increases, so does swimming pool use. But nobody would suggest that eating ice cream causes swimming, or vice versa. Similarly, just because Michael Jordan eats Wheaties and can do a slam dunk it does not mean that you will be able to do one if you eat the same breakfast. The only way to determine the cause of behavior is to conduct an experiment and analyze the results by using inferential statistics.

## Inferential Statistics

Knowing the descriptive statistics associated with different distributions, such as the mean and standard deviation, can enable us to make comparisons between various distributions. By making these comparisons, we may be able to observe whether one variable is related to another or whether one variable has a causal effect on another. When we design an experiment specifically to measure causal effects between two or more variables, we use **inferential statistics** to analyze the data collected. Although there are many inferential statistics, the one we will discuss here is the *t*-test, since it is the simplest.

**t-Test** Suppose we believe that drinking alcohol causes a person's reaction time to slow down. To test this hypothesis, we recruit 20 participants and separate them into two groups. We ask the participants in one group to drink a large glass of orange juice with one ounce of alcohol for every 100 pounds of body weight (e.g., a person weighing 150 pounds would get 1.5 ounces of alcohol). We ask the control group to drink an equivalent amount of orange juice with no alcohol added. Fifteen minutes after the drinks, we have each participant perform a reaction time test that consists of pushing a button as soon as a light is flashed. (The reaction time is the time between the onset of the light and the pressing of the button.) **Table A.10** shows the data from this hypothetical experiment. It is clear from the data that there is definitely a difference in the reaction times of the two groups: There is an obvious difference between the means. However, it is possible that this difference is due merely to chance. To determine whether the difference is real or due to chance, we can conduct a t-test. We have run a sample *t*-test in Table A.10.

**Reaction times in milliseconds (msec) for subjects in alcohol and no alcohol conditions and computation of *t* Table A.10**

| RT (msec) Alcohol $X_1$ | RT (msec) No Alcohol $X_2$ |
|---|---|
| 200 | 143 |
| 210 | 137 |
| 140 | 179 |
| 160 | 184 |
| 180 | 156 |
| 187 | 132 |
| 196 | 176 |
| 198 | 148 |
| 140 | 125 |
| 159 | 120 |
| $SX_1 = 1{,}770$ | $SX_2 = 1{,}500$ |
| $N_1 = 10$ | $N_2 = 10$ |
| $\overline{X}_1 = 177$ | $\overline{X}_2 = 150$ |
| $s_1 = 24.25$ | $s_2 = 21.86$ |
| $\Sigma_{\overline{X}_1} = \frac{s}{\sqrt{N_1 - 1}} = 8.08$ | $\Sigma_{\overline{X}_2} = \frac{s}{\sqrt{N_2 - 1}} = 7.29$ |

$$S_{\overline{X}_1 - \overline{X}_2} = \sqrt{S_{\overline{X}_1}^2 + S_{\overline{X}_2}^2} = \sqrt{8.08^2 + 7.29^2} = 10.88$$

$$t = \frac{\overline{X}_1 - \overline{X}_2}{S_{\overline{X}_1 - \overline{X}_2}} = \frac{177 - 150}{10.88} = 2.48$$

$t = 2.48,\ p < 0.5$

The logic behind a *t*-test is relatively simple. In our experiment we have two samples. If each of these samples is from the same population (e.g., the population of all people, whether drunk or sober), then any difference between the samples will be due to chance. On the other hand, if the two samples are from different populations (e.g., the population of drunk people and the population of sober people), then the difference is a significant difference and not due to chance.

If there is a significant difference between the two samples, then the independent variable must have caused that difference. In our example, there is a significant difference between the alcohol and the no alcohol groups. We can tell this because *p* (the probability that this t value will occur by chance) is less than .05. To obtain the *p*, we need only look up the t value in a statistical table, which is found in any statistics book. In our example, because there is a significant difference between the groups, we can reasonably conclude that the alcohol did cause a slower reaction time.

# Appendix C Answers to *Identify the Research Method* Questions and Self-Tests

## Answers to *Identify the Research Method* Questions

### Chapter 1

**1.** Experimental. **2.** IV = Multitasking on each of the three tests (attention to detail, memory, ability to task switch). DV = Performance scores on each of the three tests. Experimental Group = Regular/heavy multitaskers. Control Group = Nonmultitaskers. (Note that sophisticated research often has more than one level (or treatment) for the independent variable, and that in this experiment there also are multiple measures for the dependent variable.)

### Chapter 2

**1.** Descriptive. **2.** Case Study.

### Chapter 3

**1.** Correlational. **2.** Positive correlation—as stress increases incidence of ulcers also increases.

### Chapter 4

**1.** Experimental. **2.** IV = Subliminal presentation of a happy or angry face followed by a neutral face. DV = Judgments of participants' matching unconscious facial expressions. Experimental Group = Participants viewing happy or angry faces. Control Group = No description in text of control group.

### Chapter 5

**1.** Descriptive. **2.** Survey.

### Chapter 6

**1.** Experimental. **2.** IV = Live vs. televised model punching, throwing, and hitting Bobo doll. DV = Number of aggressive acts delivered by children to the Bobo doll. Experimental group = Children watching live or televised models. Control group = No description in text of control group.

### Chapter 7

**1.** Experimental (note that because Ebbinghaus had only one subject, himself, this would not qualify as a true, scientific experiment). **2.** IV = List of nonsense syllables, DV = Percentage of items correctly recalled one hour, one day, and one week later. Experimental group = Ebbinghaus was the only subject. Control group = No control group. (Again, because there was only one subject and no control group this would not qualify as a true experiment.)

### Chapter 8

**1.** Experimental. **2.** IV = Performance exam supposedly measuring intellectual abilities vs. laboratory task. DV = Exam scores. Experimental group = European-American vs. African-American students. Control group = No description in text of control group.

### Chapter 9

**1.** Descriptive. **2.** Case study.

### Chapter 10

**1.** Descriptive. **2.** Survey.

### Chapter 11

**1.** Experimental. **2.** IV(Level 1) Epinephrine vs. saline, knowledge of drug effects vs. no knowledge. IV(Level 2) = Happy vs. angry confederate. DV(Measure 1) = Self-ratings as happy or angry. DV(Measure 2) = Number of acts of euphoria vs. number of acts of anger. Experimental groups = Informed, ignorant, misinformed. Control group = Placebo. (Note that sophisticated research often has more than one level (or treatment) for the independent variable, and that in this experiment there also are multiple measures for the dependent variable.)

### Chapter 12

**1.** Descriptive. **2.** Survey.

### Chapter 13

**1.** Descriptive. **2.** Survey/interview.

## Chapter 14

**1.** Experimental. **2.** IV = Text messages sent via smartphone, DV = Scores on Chronic Pain Acceptance Questionnaire (CPAQ) and the Pain Catastrophizing Scale (PCS), Experimental Group = text message recipients, Control Group = No description in text of control group. (Note that sophisticated research often has more than one level (or treatment) for the independent variable, and that in this experiment there also are multiple measures for the dependent variable.)

## Chapter 15

**1.** Experimental. **2.** IV = Payment or no payment. DV = Rating of experiment. Experimental group = Participants paid $1 or $20 to lie. Control group = No description in text of control group. (Although in this particular experiment there is no control group, many experiments legitimately employ comparison groups, which still allow us to legitimately state that the IV was the cause of the effect.)

# Answers to Self-Tests

## Chapter 1

**1.** d (p. 4); **2.** d (p. 4); **3.** c (p. 7); **4.** a (p. 10); **5.** b (p. 12); **6.** Step 1: Question and literature review
Step 2: Testable hypothesis
Step 3: Research design
Step 4: Data collection and analysis.
Step 5: Publication.
Step 6: Theory development (p. 13);
**7.** d (p. 13); **8.** b (p. 14); **9.** d (p. 17); **10.** c (p. 16); **11.** c (p. 18); **12.** b (p. 19); **13.** b (p. 19); **14.** c (p. 20); **15.** 1. Survey, 2. Question, 3. Read, 4. Recite, 5. Review, 6. wRite (p. 24).

## Chapter 2

**1.** b (p. 34); **2.** d (p. 37); **3.** a (p. 37); **4.** a (p. 37); **5.** See Figure 2.6 (p. 38); **6.** c (p. 39); **7.** b (p. 42); **8.** c (p. 42); **9.** See Figure 2.10 (p. 43); **10.** a (p. 44); **11.** d (p. 44); **12.** c (p. 48); **13.** See Figure 2.15 (p. 51); **14.** See Figure 2.18 (p. 54); **15.** d (p. 57).

## Chapter 3

**1.** c (p. 66); **2.** a (p. 66); **3.** See Figure 3.1 (p. 66); **4.** b (p. 67); **5.** c (p. 69); **6.** c (p. 68); **7.** b (p. 69); **8.** See Figure 3.3 (p. 70); **9.** c (p. 71); **10.** c (p. 74); **11.** b (p. 75); **12.** d (p. 76); **13.** a (p. 78); **14.** b (p. 80); **15.** social support (p. 81).

## Chapter 4

**1.** c (p. 89); **2.** c (p. 89); **3.** See Figure 4.6 (p. 95); **4.** c (p. 94); **5.** See Figure 4.7 (p. 96); **6.** c (p. 97); **7.** c (p. 99); **8.** c (p. 99); **9.** a. 1, b. 2 (p. 100); **10.** c (p. 102); **11.** c (p. 105); **12.** a (p. 108); **13.** b (p. 109); **14.** b (p. 110); **15.** d (p. 112).

## Chapter 5

**1.** b (p. 120); **2.** d (p. 120); **3.** d (p. 121); **4.** See Figure 5.2 (p. 121); **5.** c (p. 123); **6.** c (p. 123); **7.** c (p. 124); **8.** c (p. 126); **9.** c (p. 129); **10.** a (p. 130); **11.** a. 1, b. 2, c. 4, d. 3 (p. 132); **12.** c (p. 132); **13.** d (p. 120); **14.** d (p. 137); **15.** c (p. 139).

## Chapter 6

**1.** c (p. 148); **2.** c (p. 148); **3.** d (p. 150); **4.** d (p. 151); **5.** d (p. 152); **6.** c (p. 153); **7.** c (p. 153); **8.** d (p. 154); **9.** c (p. 155); **10.** c (p. 160); **11.** a (p. 160); **12.** d (p. 164); **13.** c (p. 163); **14.** b (p. 168); **15.** a (p. 169).

## Chapter 7

**1.** a. encoding, b. retrieval, c. storage (p. 176); **2.** a. sensory memory, b. short-term memory (STM), c. long-term memory (LTM) (p. 177); **3.** d (p. 177); **4.** a (p. 179); **5.** a. explicit/declarative memory, b, implicit/nondeclarative memory (p. 181); **6.** b (p. 182); **7.** c (p. 184); **8.** c (p. 185); **9.** a. decay, b. interference, c. motivated forgetting, d. encoding failure, e. retrieval failure (p. 186); **10.** b (p. 188); **11.** c (p. 189); **12.** a. retrograde, b. anterograde (p. 191); **13.** c (p. 191); **14.** a (p. 193); **15.** d (p. 194).

## Chapter 8

**1.** b (p. 202); **2.** c (p. 202); **3.** evaluation (p. 204); **4.** a (p. 204); **5.** d (p. 206); **6.** d (p. 208); **7.** a. phonemes, b. morphemes, c. grammar (p. 208); **8.** b (p. 210); **9.** d (p. 213); **10.** b (p. 216); **11.** b (p. 217); **12.** a (p. 218); **13.** c (p. 219); **14.** c (p. 220); **15.** d (p. 222).

## Chapter 9

**1.** c (p. 230); **2.** c (p. 230); **3.** a. cross-sectional; b. longitudinal (p. 232); **4.** d (p. 235); **5.** d (p. 237); **6.** a (p. 239); **7.** c (p. 240); **8.** d (p. 242); **9.** c (p. 243); **10.** d (p. 244); **11.** a. sensorimotor, b. preoperational, c. concrete operational, d. formal operational (p. 245); **12.** d (p. 246); **13.** c (p. 245); **14.** a (p. 247); **15.** d (p. 250).

## Chapter 10

**1.** b (p. 258); **2.** a (p. 259); **3.** a (p. 262); **4.** c (p. 262);
**5.** Stage 1 Trust versus Mistrust,
Stage 2 Autonomy versus Shame and Doubt,
Stage 3 Initiative versus Guilt,
Stage 4 Industry versus Inferiority (p. 265);
**6.** Stage 5 Identity versus Role Confusion,
Stage 6 Intimacy versus Isolation,
Stage 7 Generativity versus Stagnation,
Stage 8 Ego Integrity versus Despair (p. 265);
**7.** b (p. 266); **8.** c (p. 269); **9.** a (p. 270); **10.** d (p. 274); **11.** a (p. 275); **12.** c (p. 275); **13.** d (p. 276); **14.** d (p. 276); **15.** a (p. 277).

## Chapter 11

**1.** a (p. 284); **2.** c (p. 286); **3.** a (p. 286); **4.** b (p. 287); **5.** d (p. 289); **6.** d (p. 290); **7.** c (p. 292); **8.** c (p. 294); **9.** a. excitement phase; b. plateau phase; c. orgasm phase; d. resolution phase (p. 295); **10.** b (p. 297); **11.** a (p. 297); **12.** a (p. 299); **13.** a (p. 300); **14.** d (p. 302); **15.** c (p. 304).

## Chapter 12

**1.** d (p. 314); **2.** a. conscious, b. preconscious, c. unconscious (p. 314); **3.** d (p. 315); **4.** c (p. 315); **5.** 1. oral, 2. anal, 3. phallic, 4. latency, 5. genital (p. 317); **6.** d (p. 318); **7.** c (p. 320); **8.** 1. openness, 2. conscientiousness, 3. extroversion, 4. agreeableness, 5. neuroticism (p. 320); **9.** b (p. 324); **10.** c (p. 324); **11.** c (p. 326); **12.** d (p. 326); **13.** c (p. 329); **14.** a (p. 331); **15.** d (p. 333).

## Chapter 13

**1.** c (p. 342); **2.** b (p. 344); **3.** a. sociocultural, b. behavioral, c. evolutionary, d. humanistic, e. psychodynamic, f. cognitive, g. biological (p. 344); **4.** a (p. 348); **5.** a (p. 350); **6.** a. generalized anxiety disorder (GAD), b. panic disorder, c., phobias, d. obsessive-compulsive disorder (OCD), e. posttraumatic stress disorder (PTSD) (p. 349); **7.** a (p. 352); **8.** b (p. 352); **9.** d (p. 355); **10.** a (p. 355); **11.** a (p. 356); **12.** a. paranoid, b. catatonic, c. disorganized, d. undifferentiated, e. residual (p. 357); **13.** a (p. 361); **14.** c (p. 362); **15.** d (p. 366).

## Chapter 14

**1.** a (p. 376); **2.** c (p. 377); **3.** c (p. 379); **4.** c (p. 379); **5.** d (p. 380); **6.** c (p. 381); **7.** b (p. 382); **8.** b (p. 384); **9.** c (p. 385); **10.** d (p. 386); **11.** a. antianxiety, b. antipsychotic, c. mood stabilizer, d. antidepressant (p. 388); **12.** d (p. 390); **13.** a. disturbed thoughts, b. disturbed emotions, c. disturbed behaviors, d. interpersonal and life situation difficulties, e. biomedical disturbances (p. 392); **14.** a (p. 394); **15.** c (p. 396).

## Chapter 15

**1.** c (p. 404); **2.** a (p. 405); **3.** a. cognitive element, b. affective element, c. behavioral element (p. 406); **4.** b (p. 407); **5.** c (p. 408); **6.** c (p. 410); **7.** d (p. 413); **8.** c (p. 415); **9.** d (p. 417); **10.** d (p. 415); **11.** a (p. 419); **12.** c (p. 420); **13.** 1. evolutionary model, 2. egoistic model, 3. empathy-altruism model (p. 421); **14.** b (p. 423); **15.** a (p. 426).

**abnormal behavior** Patterns of emotion, thought, and action that are considered pathological (diseased or disordered) for one or more of these four reasons: deviance, dysfunction, distress, or danger.
**accommodation** In Piaget's theory, the process of adjusting old schemas or developing new ones to better fit with new information.
**achievement motivation** A desire to excel, especially in competition with others.
**activity theory of aging** Theory that successful aging is fostered by a full and active commitment to life.
**aggression** Any behavior intended to harm someone.
**alternate state of consciousness (ASC)** The mental states found during sleep, dreaming, psychoactive drug use, hypnosis, and so on.
**altruism** Actions designed to help others with no obvious benefit to the helper.
**Alzheimer's disease (AD)** A progressive mental deterioration characterized by severe memory loss.
**anorexia nervosa** An eating disorder characterized by a severe loss of weight resulting from self-imposed starvation and an obsessive fear of obesity.
**anxiety disorder** A type of abnormal behavior characterized by unrealistic, irrational fear.
**applied research** Research designed to solve practical problems.
**assimilation** In Piaget's theory, the process of absorbing new information into existing schemas.
**attachment** A strong affectional bond with special others that endures over time.
**attitudes** Learned predisposition to respond to objects, people, and events in a particular way.
**attributions** Explanations for behaviors or events.
**autonomic nervous system (ANS)** Subdivision of the peripheral nervous system (PNS) that controls involuntary functions. It includes the *sympathetic* nervous system and the *parasympathetic* nervous system.
**basic research** Research conducted to advance scientific knowledge rather than for practical application.
**behavior therapy** A group of techniques based on learning principles that is used to change maladaptive behaviors.
**behavioral genetics** The study of the relative effects of genetic and environmental influences on behavior and mental processes.
**biological preparedness** The built-in (innate) readiness to form associations between certain stimuli and responses.
**biological research** The scientific study of the brain and other parts of the nervous system.
**biomedical therapy** Using physiological interventions (drugs, electroconvulsive therapy, and psychosurgery) to reduce or alleviate symptoms of psychological disorders.
**biopsychosocial model** A unifying theme of modern psychology that considers biological, psychological, and social processes.
**bipolar disorder** Repeated episodes of mania (unreasonable elation and hyperactivity) and depression.
**bulimia nervosa** An eating disorder involving the consumption of large quantities of food (bingeing), followed by vomiting, extreme exercise, or laxative use (purging).
**central nervous system (CNS)** The brain and spinal cord.
**cerebral cortex** The thin surface layer on the cerebral hemispheres that regulates most complex behavior, including receiving sensations, motor control, and higher mental processes.
**chunking** The act of grouping separate pieces of information into a single unit (or chunk).
**circadian** [ser-KAY-dee-an] **rhythms** The biological changes that occur on a 24-hour cycle (in Latin, *circa* means about, and *dies* means day).
**classical conditioning** Learning through involuntarily paired associations; it occurs when a neutral stimulus (NS) becomes paired (associated) with an unconditioned stimulus (US) to elicit a conditioned response (CR). (Also known as Pavlovian conditioning.)
**cognition** Mental activities involved in acquiring, storing, retrieving, and using knowledge.
**cognitive dissonance** A feeling of discomfort caused by a discrepancy between two conflicting cognitions (thoughts) or between an attitude and a behavior.
**cognitive therapy** Therapy that focuses on changing faulty thought processes and beliefs to treat problem behaviors.
**cognitive-social learning theory** A perspective that emphasizes the roles of thinking and social learning in behavior.
**collectivistic culture** Culture in which the needs and goals of the group are emphasized over the needs and goals of the individual.
**comorbidity** Co-occurrence of two or more disorders in the same person at the same time, as when a person suffers from both depression and alcoholism.
**concrete operational stage** Piaget's third stage (roughly ages 7 to 11), in which the child can perform mental operations on concrete objects and understand reversibility and conservation, though abstract thinking is not yet present.
**conditioned response (CR)** A learned reaction to a previously neutral (but now conditioned) stimulus (CS) (e.g., tone).
**conditioned stimulus (CS)** A learned (previously neutral) stimulus (e.g., tone) that elicits a conditioned response (CR) (e.g., salivation) as a result of repeated pairings with an unconditioned stimulus (US) (e.g., food).
**conformity** The act of changing behavior as a result of real or imagined group pressure.
**conscious** Freud's term for thoughts or motives that a person is currently aware of or is remembering.
**consciousness** An organism's awareness of its own self and surroundings (Damasio, 1999).
**conventional level** Kohlberg's second level of moral development, in which moral judgments are based on compliance with the rules and values of society.
**correlation coefficient** Number indicating strength and direction of the relationship between two variables (from −1.00 to +1.00).
**correlational research** A method by which the researcher observes or measures (without directly manipulating) two or more variables to find relationships between them.
**creativity** The ability to produce valued outcomes in a novel way.
**critical period** A period of special sensitivity to specific types of learning that shapes the capacity for future development.
**critical thinking** The process of objectively evaluating, comparing, analyzing, and synthesizing information.
**cross-sectional design** Research approach that measures individuals of various ages at one point in time and gives information about age differences.
**debriefing** Upon completion of the research, participants are informed of the study's design and purpose, and explanations are provided for any possible deception.
**defense mechanisms** In Freudian theory, the ego's protective method of reducing anxiety by distorting reality and self-deception.
**delusions** Mistaken beliefs based on misrepresentations of reality.
**dependent variable (DV)** A variable that is measured; it is affected by (or dependent on) the independent variable.

**descriptive research** Research method used to observe and describe behavior and mental processes without manipulating variables.
**developmental psychology** The study of age-related changes in behavior and mental processes from conception to death.
***Diagnostic and Statistical Manual* (*DSM*)** A reference book developed by the American Psychiatric Association that classifies mental disorders on the basis of symptoms, and provides a common language and standard criteria useful for communication and research.
**disengagement theory** Theory that successful aging is characterized by mutual withdrawal between the elderly and society.
**dissociative disorders** Amnesia, fugue, or multiple personalities resulting from avoidance of painful memories or situations.
**dyssomnia** The problems in the amount, timing, and quality of sleep, include insomnia, sleep apnea, and narcolepsy.
**elaborative rehearsal** The process of linking new information to previously stored material.
**electroconvulsive therapy (ECT)** Biomedical therapy in which electrical current is passed through the brain.
**embryonic period** The second stage of prenatal development, which begins after uterine implantation and lasts through the eighth week.
**emotion** A subjective feeling that includes arousal (heart pounding), cognitions (thoughts, values, and expectations), and expressions (frowns, smiles, and running).
**emotion-focused forms of coping** Managing one's emotional reactions to a stressor
**empirical evidence** Information acquired by direct observation and measurement using systematic scientific method.
**encoding** Processing information into the memory system.
**encoding specificity principle** Retrieval of information is improved when the conditions of recovery are similar to the conditions that existed when the information was encoded.
**evolutionary psychology** A subfield of psychology studying how natural selection and adaptation help explain behavior and mental processes.
**experimental research** A carefully controlled scientific procedure that determines whether variables manipulated by the experimenter have a causal effect on other variables.
**explicit/declarative memory** The subsystem within long-term memory that consciously stores facts, information, and personal life experiences.
**extrinsic motivation** Motivation based on obvious external rewards or threats of punishment.
**family therapy** Treatment to change maladaptive interaction patterns within a family.
**feature detectors** Specialized brain cells that respond to specific features of the stimulus, like movement, shape, and angles.
**fetal period** The third stage of prenatal development (eight weeks to birth), characterized by rapid weight gain and fine detailing of organs and systems.
**five-factor model (FFM)** The trait theory of personality composed of openness, conscientiousness, extroversion, agreeableness, and neuroticism.
**formal operational stage** Piaget's fourth stage (around age 11 and beyond), which is characterized by abstract and hypothetical thinking.
**fundamental attribution error (FAE)** Attributing people's behavior to internal (dispositional) causes rather than external (situational) factors.
**gate-control theory** The theory that pain sensations are processed and altered by mechanisms within the spinal cord.
**gender** The psychological and sociocultural meanings added to biological maleness or femaleness.
**gender roles** The societal expectations for normal and appropriate male and female behavior.
**general adaptation syndome (GAS)** Selye's three-stage (alarm, resistance, exhaustion) reaction to chronic stress.
**germinal period** The first stage of prenatal development, which begins with conception and ends with implantation in the uterus (the first two weeks).
**grammar** Rules that specify how phonemes, morphemes, words, and phrases should be combined to express meaningful thoughts. These rules include syntax and semantics.
**group polarization** A group's movement toward either riskier or more conservative behavior, depending on the members' initial dominant tendencies.
**group therapy** A form of therapy in which a number of people meet together to work toward therapeutic goals.
**groupthink** Faulty decision making that occurs when a highly cohesive group strives for agreement and avoids inconsistent information.
**habituation** The tendency of the brain to ignore environmental factors that remain constant.
**hallucinations** False, imaginary sensory perceptions that occur without an external stimulus.
**health psychology** The study of how biological, psychological, and social factors affect health and illness.
**hierarchy of needs** Maslow's theory of motivation that lower motives (such as physiological and safety needs) must be met before going on to higher needs (such as belonging and self-actualization).
**homeostasis** A body's tendency to maintain a relatively stable state, such as a constant internal temperature.
**hormones** Chemicals manufactured by endocrine glands and circulated in the bloodstream to produce bodily changes or maintain normal bodily function.
**humanistic therapy** Therapy that seeks to maximize personal growth through affective restructuring (emotional readjustment).
**hypnosis** Trance-like state of heightened suggestibility, deep relaxation, and intense focus.
**implicit/nondeclarative memory** The subsystem within long-term memory that consists of unconscious procedural skills, simple classically conditioned responses (Chapter 6), and priming.
**independent variable (IV)** A variable that is manipulated to determine its causal effect on the dependent variable.
**individualistic culture** Culture in which the needs and goals of the individual are emphasized over the needs and goals of the group.
**informed consent** A participant's agreement to take part in a study after being told what to expect.
**insight** The sudden understanding of a problem that implies the solution.
**instinctive drift** The tendency of some conditioned responses (CRs) to shift (or drift) back toward an innate response pattern.
**instincts** Behavioral patterns that are unlearned, always expressed in the same way, and universal in a species.
**intelligence quotient (IQ)** Once a formula (mental age ÷ chronological age × 100), the term IQ is now used to describe intelligence test scores.
**intelligence** The global capacity to think rationally, act purposefully, and deal effectively with the environment.
**interpersonal attraction** Positive feelings toward another.

**intrinsic motivation** Motivation resulting from personal enjoyment of a task or activity.

**language** A form of communication using sounds and symbols combined according to specified rules.

**latent learning** Hidden learning that exists without behavioral signs.

**law of effect** Thorndike's rule that the probability of an action being repeated is strengthened when followed by a pleasant or satisfying consequence.

**learning** The relatively permanent change in behavior or mental processes caused by experience.

**localization of function** Specialization of various parts of the brain for particular functions.

**long-term memory (LTM)** This third memory stage stores information for long periods. Its capacity is limitless; its duration is relatively permanent.

**long-term potentiation (LTP)** A long-lasting increase in neural excitability believed to be a biological mechanism for learning and memory.

**longitudinal design** Research approach that measures a single individual or a group of same-aged individuals over an extended period and gives information about age changes.

**maintenance rehearsal** Repeating information to maintain it in short-term memory (STM).

**major depressive disorder** Long-lasting depressed mood that interferes with the ability to function, feel pleasure, or maintain interest in life (Swartz & Margolis, 2004).

**maturation** Development governed by automatic, genetically predetermined signals.

**medical model** The perspective that diseases (including mental illness) have physical causes that can be diagnosed, treated, and possibly cured.

**meditation** A group of techniques designed to refocus attention, block out all distractions, and produce an alternate state of consciousness.

**memory** An internal record or representation of some prior event or experience.

**modeling therapy** A learning technique in which the subject watches and imitates models who demonstrate desirable behaviors.

**morpheme** The smallest meaningful unit of language, formed from a combination of phonemes.

**motivation** A set of factors that activate, direct, and maintain behavior, usually toward a goal.

**nature–nurture controversy** The ongoing dispute over the relative contributors of nature (biological and genetic factors) and nurture (environment) to the development of behavior and mental processes.

**neurogenesis** The division and differentiation of nonneuronal cells to produce neurons.

**neuron** Nerve cell that processes and transmits information; basic building block of the nervous system responsible for receiving and transmitting electrochemical information.

**neuroplasticity** The brain's ability to reorganize and change its structure and function through the lifespan.

**neurotransmitters** Chemicals that neurons release, which affect other neurons.

**neutral stimulus (NS)** An unlearned stimulus (e.g., tone) that does not elicit a response.

**obedience** The act of following a direct command, usually from an authority figure.

**objective personality tests** Standardized questionnaires that require written, self-report responses, usually to multiple-choice or true-false questions.

**observational learning** Learning a new behavior or information by watching others (also known as social learning or modeling).

**operant conditioning** Learning through voluntary responses and its consequences (also known as instrumental or Skinnerian conditioning).

**opponent-process theory** The theory that color perception is based on three systems of color receptors, each of which responds in an on-off fashion to opposite-color stimuli: blue-yellow, red-green, and black- white.

**parasomnias** The abnormal disturbances occurring during sleep, including nightmares, night terrors, sleepwalking, and sleeptalking.

**perception** The process of selecting, organizing, and interpreting sensory information into meaningful patterns.

**peripheral nervous system (PNS)** All the nerves and neurons outside the brain and spinal cord that connect the CNS to the rest of the body.

**personality disorders** Inflexible, maladaptive personality traits that cause significant impairment of social and occupational functioning.

**personality** Relatively stable and enduring patterns of thoughts, feelings, and actions.

**phoneme** The smallest basic unit of speech or sound. The English language has about 40 phonemes.

**polygraph** An instrument that measures sympathetic arousal (heart rate, respiration rate, blood pressure, and skin conductivity) to detect emotional arousal, which in turn supposedly reflects lying versus truthfulness.

**postconventional level** Kohlberg's highest level of moral development, in which individuals develop personal standards for right and wrong, and they define morality in terms of abstract principles and values that apply to all situations and societies.

**posttraumatic stress disorder (PTSD)** An anxiety disorder following exposure to a life-threatening or other extreme event that evoked great horror or helplessness. It is characterized by flashbacks, nightmares, and impaired functioning.

**preconscious** Freud's term for thoughts or motives that are just beneath the surface of awareness and can be easily brought to mind.

**preconventional level** Kohlberg's first level of moral development, in which morality is based on rewards, punishment, and the exchange of favors.

**prejudice** A learned, generally negative attitude directed toward specific people solely because of their membership in an identified group.

**preoperational stage** Piaget's second stage (roughly ages 2 to 7 years), which is characterized by the ability to employ significant language and to think symbolically, though the child lacks operations (reversible mental processes), and thinking is egocentric and animistic.

**problem-focused forms of coping** Dealing directly with a stressor to decrease or eliminate it.

**projective tests** Psychological tests that use ambiguous stimuli, such as inkblots or drawings, which allow the test taker to project his or her unconscious thoughts onto the test material.

**psychiatry** The branch of medicine that deals with the diagnosis, treatment, and prevention of mental disorders.

**psychopharmacology** The study of drug effects on the mind and behavior.

**psychoactive drugs** Chemicals that change conscious awareness, mood, or perception.

**psychoanalysis** Freudian therapy designed to bring unconscious conflicts, into conscious awareness; also Freud's theoretical school of thought.

**psychodynamic therapy** A briefer, more directive contemporary form of psychoanalysis, focusing more on conscious processes and current problems.

**psychology** The scientific study of behavior and mental processes.

**psychoneuro-immunology** [sye-koh-NEW-roh-IM-you-NOLL-oh-gee] The interdisciplinary field that studies the effects of psychological factors on the immune system.

**psychosexual stages** In Freudian theory, the five developmental periods (oral, anal, phallic, latency, and genital) during which particular kinds of pleasures must be gratified if personality development is to proceed normally.

**psychosocial stages** The eight developmental stages, each involving a crisis that must be successfully resolved, that individuals pass through in Erikson's theory of psychosocial development.

**psychosurgery** Operative procedures on the brain designed to relieve severe mental symptoms that have not responded to other forms of treatment.

**psychotherapy** Techniques employed to improve psychological functioning and promote adjustment to life.

**punishment** A consequence that weakens a response and makes it less likely to recur.

**reciprocal determinism** Bandura's belief that cognitions, behaviors, and the environment interact to produce personality.

**reinforcement** A consequence that strengthens a response and makes it more likely to recur.

**reliability** A measure of the consistency and stability of test scores when a test is readministered.

**retrieval cue** A clue or prompt that helps stimulate recall and retrieval of a stored piece of information from long-term memory.

**retrieval** Recovering information from memory storage.

**savant syndrome** A condition in which a person with generally limited mental abilities exihibits exceptional skill or brilliance in some limited field.

**schemas** Cognitive structures or patterns consisting of a number of organized ideas that grow and differentiate with experience.

**schizophrenia** A group of psychotic disorders involving major disturbances in perception, language, thought, emotion, and behavior. The individual withdraws from people and reality, often into a fantasy life of delusions and hallucinations.

**selective attention** Filtering out and attending only to important sensory messages.

**self-concept** Rogers's term for all the information and beliefs that individuals have about their own nature, qualities, and behavior.

**self-efficacy** Bandura's term for the learned belief that one is capable of producing desired results, such as mastering new skills and achieving personal goals.

**sensation** The process of detecting, converting, and transmitting raw sensory information from the external and internal environments to the brain.

**sensorimotor stage** Piaget's first stage (birth to approximately age 2), in which schemas are developed through sensory and motor activities.

**sensory adaptation** Repeated or constant stimulation decreases the number of sensory messages sent to the brain, which causes decreased sensation.

**sensory memory** This first memory stage holds sensory information. It has a relatively large capacity, but duration is only a few seconds.

**sensory reduction** Filtering and analyzing incoming sensations before sending a neural message to the cortex.

**sex** Biological maleness and femaleness, including chromosomal sex. Also, activities related to sexual behaviors, such as masturbation and intercourse.

**sexual orientation** Primary erotic attraction toward members of the same sex (homosexual, gay or lesbian), both sexes (bisexual), or the other sex (heterosexual).

**sexual prejudice** Negative attitudes toward an individual because of her or his sexual orientation.

**sexual response cycle** Masters and Johnson's description of the four-stage bodily response to sexual arousal, which consists of excitement, plateau, orgasm, and resolution.

**shaping** Reinforcement by a series of successively improved steps leading to desired response.

**short-term memory (STM)** This second memory stage temporarily stores sensory information and decides whether to send it on to long-term memory (LTM). Its capacity is limited to five to nine items, and its duration is about 30 seconds.

**social psychology** The scientific study of how people's thoughts, feelings, and actions are affected by others.

**socioemotional selectivity theory** A natural decline in social contact as older adults become more selective with their time.

**somatic nervous system (SNS)** Subdivision of the peripheral nervous system (PNS) that connects the sensory receptors and controls the skeletal muscles.

**standardization** Establishment of the norms and uniform procedures for giving and scoring a test.

**stem cells** Precursor (immature) cells that give birth to new specialized cells; a stem cell holds all the information it needs to make bone, blood, brain—any part of a human body—and can also copy itself to maintain a stock of stem cells.

**stereotype threat** Negative stereotypes about minority groups cause some members to doubt their abilities.

**storage** Retaining information over time.

**stress** The body's nonspecific response to any demand made on it; physical and mental arousal to situations or events that we perceive as threatening or challenging.

**traits** Relatively stable and consistent characteristics that can be used to describe someone.

**trichromatic theory** The theory that color perception results from mixing three distinct color systems—red, green, and blue.

**Type A personality** Behavior characteristics including intense ambition, competition, exaggerated time urgency, and a cynical, hostile outlook.

**Type B personality** Behavior characteristics consistent with a calm, patient, relaxed attitude.

**unconditional positive regard** Rogers's term for a caring, nonjudgmental attitude toward a person with no contingencies attached.

**unconditioned response (UR)** An unlearned, naturally occurring response (e.g., salivation) to an unconditioned stimulus (US) (e.g., food).

**unconditioned stimulus (US)** An unlearned stimulus (e.g., food) that naturally and automatically elicits an unconditioned response (UR) (e.g., salivation).

**unconscious** Freud's term for thoughts or motives that lie beyond a person's normal awareness, which still exert great influence.

**validity** The ability of a test to measure what it was designed to measure.

APA (2009). *Center for workforce studies*. Washington, DC: American Psychological Association.
APA (2011). *Trauma, PTSD, and Recovery*. Retrieved from http://www.apa.org/topics/trauma/ptsd/index.aspx
Aamodt, M. G. (2010). *Industrial/organizational psychology* (6th ed.). Belmont, CA: Cengage.
Aarts, H. (2007). Unconscious authorship ascription: The effects of success and effect-specific information priming on experienced authorship. *Journal of Experimental Social Psychology*, *43*, 119–126.
Abadinsky, H. (2011). *Drug use and abuse: A comprehensive introduction* (7th ed.). Belmont, CA: Cengage.
Aboa-Éboulé, C. (2008). Job strain and recurrent coronary heart disease events—Reply. *Journal of the American Medical Association*, *299*, 520–521.
Aboa-Éboulé, C., Brisson, C., Maunsell, E., Benoît, M., Bourbonnais, R., Vézina, M., Milot, A., Théroux, P., & Dagenais, G. R. (2007). Job strain and risk of acute recurrent coronary heart disease events. *Journal of the American Medical Association*, *298*, 1652–1660.
Acarturk, C., de Graaf, R., van Straten, A., ten Have, M., & Cuijpers, P. (2008). Social phobia and number of social fears, and their association with comorbidity, health-related quality of life and help seeking: A population-based study. *Social Psychiatry and Psychiatric Epidemiology*, *43*, 273–279.
Acerbi, A., & Nunn, C. L. (2011). Predation and the phasing of sleep: An evolutionary individual-based model. *Animal Behaviour*, *81*(4), 801–811.
Achenbaum, W. A., & Bengtson, V. L. (1994). Re-engaging the disengagement theory of aging: On the history and assessment of theory development in gerontology. *Gerontologist*, *34*, 756–763.
Acheson, K., Blondel-Lubrano, A., Oguey-Araymon, S., Beaumont, M., Emady-Azar, S., Ammon-Zufferey, C., Monnard, I., Pinaud, S., Nielsen-Moennoz, C., & Bovetto, L. (2011). Protein choices targeting thermogenesis and metabolism. *The American Journal of Clinical Nutrition*, *93*(3), 525–534.
Adams, M. (2011). Evolutionary genetics of personality in nonhuman primates. In M. Inoue-Murayama, S. Kawamura, & A. Weiss (Eds.), *From genes to animal behavior* (pp. 137–164). Tokyo, Japan: Springer.
Adelmann, P. K., & Zajonc, R. B. (1989). Facial efference and the experience of emotion. *Annual Review of Psychology*, *40*, 249–280.
Aderka, I., Hermesh, H., Marom, S., Weizman, A., & Gilboa-Schechtman, E. (2011). Cognitive behavior therapy for social phobia in large groups. *International Journal of Cognitive therapy*, *4*(1), 92–103.
Adler, A. (1964). The individual psychology of Alfred Adler. In H. L. Ansbacher & R. R. Ansbacher (Eds.), *The individual psychology of Alfred Adler*. New York, NY: Harper & Row.
Adler, A. (1998). *Understanding human nature*. Center City, MN: Hazelden Information Education.
Adolph, K. E., & Berger, S. E. (2012). Physical and motor development. In M. H. Bornstein & M. E. Lamb (Eds.), *Cognitive development: An advanced textbook* (pp. 257–318). New York, NY: Psychology Press.
Agartz, I., Brown, A., Rimol, L., Hartberg, C., Dale, A., Melle, I., Djurovic, S., & Andreassen, O. (2011). Common sequence variants in the major histocompatibility complex region associate with cerebral ventricular size in schizophrenia. *Biological Psychiatry*, *10*(7), 696–698.
Aggleton, J. P., O'Mara, S. M., Vann, S. D., Wright, N. F., Tsanov, M., & Erichsen, J. T. (2010). Hippocampal anterior thalamic pathways for memory: Uncovering a network of direct and indirect actions. *European Journal of Neuroscience*, *31*(12), 2292–2307.
Ahlsén, E. (2008). Embodiment in communication – Aphasia, apraxia, and the possible role of mirroring and imitation. *Clinical Linguistics and Phonetics*, *22*, 311–315.
Ahmed, A. M. (2007). Group identity, social distance, and intergroup bias. *Journal of Economic Psychology*, *28*, 324–337.
Ainsworth, M. D. S. (1967). *Infancy in Uganda: Infant care and the growth of love*. Baltimore, MD: Johns Hopkins University Press.
Ainsworth, M. D. S. (2010). Security and attachment. In R. Volpe (Ed.), *The secure child: Timeless lessons in parenting and childhood education* (pp. 43–53). Charlotte, NC: IAP Information Age Publishing.
Ainsworth, M. D. S., Blehar, M., Waters, E., & Wall, S. (1978). *Patterns of attachment: Observations in the strange situation and at home*. Hillsdale, NJ: Erlbaum.
Akers, R. M., & Denbow, D. (2008). *Anatomy and physiology of domestic animals*. Hoboken, NJ: John Wiley & Sons.
Alberts, A., Elkind, D., & Ginsberg, S. (2007). The personal fable and risk-taking in early adolescence. *Journal of Youth and Adolescence*, *36*, 71–76.
Alcock, J. (2011). Back from the future: Parapsychology and the Bem affair. *The Committee for Skeptical Inquiry* (CSI). Retrieved from http://www.csicop.org/specialarticles/show/back_from_the_future.
Alden, L. E., & Regambal, M. J. (2011). Interpersonal processes in the anxiety disorders. In L. M. Horowitz & S. Strack (Eds.), *Handbook of interpersonal psychology: Theory, research, assessment, and therapeutic interventions* (pp. 449–469). Hoboken, NJ: John Wiley & Sons.
Aldenhoff, J. (2011). Interpersonal therapy. *European Psychiatry*, *26*(1). 1782.
Al-Issa, I. (2000). Culture and mental illness in Algeria. In I. Al-Issa (Ed.), *Al-Junun: Mental illness in the Islamic world* (pp. 101–119). Madison, CT: International Universities Press.
Allam, M. D.-E., Soussignan, R., Patris, B., Marlier, L., & Schaal, B. (2010). Long-lasting memory for an odor acquired at the mother's breast. *Developmental Science*, *13*(6), 849–863.
Allan, R. F. J. (2011). Type A behavior pattern. In R. F. J. Allan (Ed.), *Heart and mind: The practice of cardiac psychology* (2nd ed.) (pp. 287–290). Washington, DC: American Psychological Association.
Allen, L. S., Hines, M., Shryne, J. E., & Gorski, R. A. (2007). Two sexually dimorphic cell groups in the human brain. In G. Einstein (Ed.), *Sex and the brain* (pp. 327–337). Cambridge, MA: MIT Press.
Alloy, L. B., Wagner, C. A., Black, S. K., Gerstein, R. K., & Abramson, L. Y. (2011). The breakdown of self-enhancing and self-protecting cognitive biases in depression. In M. D. Alicke & C. Sedikides (Eds.), *Handbook of self-enhancement and self-protection* (pp. 358–379). New York, NY: Guilford Press.
Allport, G. W. (1937). *Personality: A psychological interpretation*. New York, NY: Holt, Rinehart & Winston.
Allport, G. W., & Odbert, H. S. (1936). Trait-names: A psycho-lexical study. *Psychological Monographs: General and Applied*, *47*, 1–21.
Almela, M., Hidalgo, V., Villada, C., Espín, L., Gómez-Amor, J., & Salvador, A. (2011). The impact of cortisol reactivity to acute stress on memory: Sex differences in middle-aged people. *Stress: The International Journal on the Biology of Stress*, *14*(2), 117–127.
Amazing meeting. (2011). *The amazing one: James Randi*. Retrieved from http://www.amazingmeeting.com/speakers#randi
Ambady, N., & Adams, R. B., Jr. (2011). Us versus them: The social neuroscience of perceiving out-groups. In A. Todorov, S. T. Fiske, & D. A. Prentice (Eds.), *Oxford series in social cognition and social neuroscience. Social neuroscience: Toward understanding the underpinnings of the social mind* (pp. 135–143). New York, NY: Oxford University Press.
American Counseling Association. (2006). *Crisis fact sheet: 10 ways to recognize Post-Traumatic Stress Disorder*. Retrieved from http://www.counseling.org/PressRoom/PressReleases.aspx?AGuid69c6fad2-c05e-4be3-af4e-3a1771414044.

American Heart Association. (2012). *Heart disease and stroke statistics—2012 update*. Retrieved from http://circ.ahajournals.org/content/early/2011/12/15/CIR.0b013e31823ac046.
American Medical Association. (2012). *Alcohol and other drug abuse*. Retrieved from http://www.ama-assn.org/ama/pub/physician-resources/public-health/promoting-healthy-lifestyles/alcohol-other-drug-abuse/resources.
American Psychiatric Association. (2000). *Diagnostic and statistical manual of mental disorders* (4th ed. TR). Washington, DC: American Psychiatric Press.
American Psychiatric Association. (2002). *APA Let's talk facts about posttraumatic stress disorder*. Retrieved from http://www.psych.org/disasterpsych/fs/ptsd.cfm.
American Psychological Association (APA). (2009). *Center for workforce studies*. Washington, DC: American Psychological Association.
American Psychological Association (APA). (2011). *Trauma, PTSD, and Recovery*. Retrieved from http://www.apa.org/topics/trauma/ptsd/index.aspx.
Amundson, J. K., & Nuttgens, S. A. (2008). Strategic eclecticism in hypnotherapy: Effectiveness research considerations. *American Journal of Clinical Hypnosis*, *50*, 233–245.
Anacker, C., Zunszain, P. A., Cattaneo, A., Carvalho, L. A., Garabedian, M. J., Thuret, S., . . . Pariante, C. M. (2011). Antidepressants increase human hippocampal neurogenesis by activating the glucocorticoid receptor. *Molecular Psychiatry*, *16*(7), 738–750.
Anderman, E., & Dawson, H. (2011). Learning with motivation. In R. Mayer & P. Alexander (Eds.), *Handbook of research on learning and instruction* (pp. 219–242). New York, NY: Routledge.
Anderson, C. A. (2001). Heat and violence. *Current Directions in Psychological Science*, *10*(1), 33–38.
Anderson, C. A., Buckley, K. E., & Carnegey, N. L. (2008). Creating your own hostile environment: A laboratory examination of trait aggressiveness and the violence escalation cycle. *Personality and Social Psychology Bulletin*, *34*, 462–473.
Anderson, C. A., Shibuya, A., Ihori, N., Swing, E. L., Bushman, B. J., Sakamoto, A., . . . Saleem, M. (2010). Violent video game effects on aggression, empathy, and prosocial behavior in eastern and western countries: A meta-analytic review. *Psychological Bulletin*, *136*(2), 151–173.
Anderson, J. R., Betts, S., Ferris, J. L., & Fincham, J. M. (2011). Cognitive and metacognitive activity in mathematical problem solving: Prefrontal and parietal patterns. *Cognitive, Affective and Behavioral Neuroscience*, *11*(1), 52–67.
Anderson, K. M., & Bang, E. J. (2012). Assessing PTSD and resilience for females who during childhood were exposed to domestic violence. *Child and Family Social Work*, *17*(1), 55–65.
Anderson, M. C., Ochsner, K. N., Kuhl, B., Cooper, J., Robertson, E., Gabrieli, S. W., Glover, G. H., & Gabrieli, J. D. E. (2004). Neural systems underlying the suppression of unwanted memories. *Science*, *303*(5655), 232–235.
Anderson, N. E., Wan, L., Young, K. A., & Stanford, M. S. (2011). Psychopathic traits predict startle habituation but not modulation in an emotional faces task. *Personality and Individual Differences*, *50*(5), 712–716.
Andersson, J., & Walley, A. (2011). The contribution of heredity to clinical obesity. In R.H. Lustig (Ed.), *Obesity before birth* (pp. 25–52). New York, NY: Springer.
Andrade, T. G. C. S., & Graeff, F. G. (2001). Effect of electrolytic and neurotoxic lesions of the median raphe nucleus on anxiety and stress. *Pharmacology, Biochemistry and Behavior*, *70*(1), 1–14.
Ansell, E. B., & Grilo, C. M. (2007). Personality disorders. In M. Hersen, S. M. Turner, & D. C. Beidel (Eds.), *Adult psychopathology and diagnosis* (5th ed.) (pp. 633–678). Hoboken, NJ: John Wiley & Sons.
Antony, M. M., & Roemer, L. (2011). Summary. In M. M. Antony & L. Roemer, *Theories of psychotherapy: Behavior therapy* (pp. 123–128). Washington, DC: American Psychological Association.
Appel, M., & Richter, T. (2007). Persuasive effects of fictional narratives increase over time. *Media Psychology*, *10*, 113–134.
Appiah, K. A. (2008). *Experiments in ethics*. Cambridge, MA: Harvard University Press.
Appiah, K. A. (2010). More experiments in ethics. *Neuroethics*, *3*(3), 233–242.
Archer, T. (2011). Physical exercise alleviates debilities of normal aging and Alzheimer's disease. *Acta Neurologica Scandinavica*, *123*(4), 221–238.
Areepattamannil, S. (2010). Parenting practices, parenting style, and children's school achievement. *Psychological Studies*, *55*(4), 283–289.
Arkes, H., & Kajdasz, J. (2011). Intuitive theories of behavior. In National Research Council (Ed.), *Intelligence analysis: Behavioral and social scientific foundations* (pp. 143–168). Washington, DC: The National Academies Press.
Arnsten, A. F. T. (2011). Prefrontal cortical network connections: Key site of vulnerability in stress and schizophrenia. *International Journal of Developmental Neuroscience*, *29*(3), 215–223.
Aronson, J., Jannone, S., McGlone, M., & Johnson-Campbell, T. (2009). The Obama effect: An experimental test. *Journal of Experimental Social Psychology*, *45*(4), 957–960.
Arriagada, O., Constandil, L., Hernández, A., Barra, R., Soto-Moyano, R., & Laurido, C. (2007). Brief communication: Effects of interleukin-1B on spinal cord nociceptive transmission in intact and propentofyllinetreated rats. *International Journal of Neuroscience*, *117*, 617–625.
Arumugam, V., Lee, J-S., Nowak, J. K., Pohle, R. J., Nyrop, J. E., Leddy, J. J., & Pelkman, C. L. (2008). A high-glycemic meal pattern elicited increased subjective appetite sensations in overweight and obese women. *Appetite*, *50*, 215–222.
Asch, S. E. (1951). Effects of group pressure upon the modification and distortion of judgment. In H. Guetzkow (Ed.), *Groups, leadership, and men*. Pittsburgh, PA: Carnegie Press.
Atchley, R. C. (1997). *Social forces and aging* (8th ed.). Belmont, CA: Wadsworth.
Atkinson, R. C., & Shiffrin, R. M. (1968). Human memory: A proposed system and its control processes. In K. W. Spence & J. T. Spence (Eds.), *The psychology of learning and motivation* (Vol. 2) (pp. 90–91). New York, NY: Academic Press.
Atun-Einy, O., Berger, S. E., & Scher, A. (2012). Pulling to stand: Common trajectories and individual differences in development. *Developmental Psychobiology*, *54*(2), 187–198.
Axtell, R. E. (2007). *Essential do's and taboos: The complete guide to international business and leisure travel*. Hoboken, NJ: John Wiley & Sons.

Back, M. D., Penke, L., Schmukle, S. C., Sachse, K., Borkenau, P., & Asendorpf, J. B. (2011). Why mate choices are not as reciprocal as we assume: The role of personality, flirting and physical attractiveness. *European Journal of Personality*, *25*(2), 120–132.
Baddeley, A. D. (1992). Working memory. *Science*, *255*, 556–559.
Baddeley, A. D. (2007). Working memory, thought, and action. *Oxford psychology series*. New York, NY: Oxford University Press.
Baer, J. (1994). Divergent thinking is not a general trait: A multi-domain training experiment. *Creativity Research Journal*, *7*, 35–36.
Bagermihl, B. (1999). *Biological exuberance: Animal homosexuality and natural diversity*. New York, NY: St Martins Press.
Bagley, S. L., Weaver, T. L., & Buchanan, T. W. (2011). Sex differences in physiological and affective responses to stress in remitted depression. *Physiology and Behavior*, *104*(2), 180–186.

Bailey, J. M., Dunne, M. P., & Martin, N. G. (2000). Genetic and environmental influences on sexual orientation and its correlates in an Australian twin sample. *Journal of Personality and Social Psychology*, *78*(3), 524–536.

Bailey, R. K., Patel, T. C., Avenido, J., Patel, M., Jaleel, M., Barker, N. C., . . . Jabeen, S. (2011). Suicide: Current trends. *Journal of the National Medical Association*, *103*(7), 614–617.

Baillargeon, R. (2000). Reply to Bogartz, Shinskey, and Schilling; Schilling; and Cashon and Cohen. *Infancy*, *1*, 447–462.

Baillargeon, R. (2008). Innate ideas revisited: For a principle of persistence in infants' physical reasoning. *Perspectives on Psychological Science*, *3*, 2–13.

Baird, A. D., Scheffer, I. E., & Wilson, S. J. (2011). Mirror neuron system involvement in empathy: A critical look at the evidence. *Social Neuroscience*, *6*(4), 327–335.

Baker, D. B. (2012). *The Oxford handbook of the history of psychology: Global perspectives*. New York, NY: Oxford University Press.

Baker, D. G., Nievergelt, C. M., & O'Connor, D. T. (2011). Biomarkers of PTSD: Neuropeptides and immune signaling. *Neuropharmacology*, *62*(2), 663–673.

Baker, D., & Nieuwenhuijsen, M. J. (Eds.) (2008). *Environmental epidemiology: Study methods and application*. New York, NY: Oxford University Press.

Baldessarini, R. J., & Tondo, L. (2008). Lithium and suicidal risk. *Bipolar Disorders*, *10*, 114–115.

Baldwin, C. L., & Ash, I. K. (2011). Impact of sensory acuity on auditory working memory span in young and older adults. *Psychology and Aging*, *26*(1), 85–91.

Ball, H. A., McGuffin, P., & Farmer, A. E. (2008). Attributional style and depression. *British Journal of Psychiatry*, *192*, 275–278.

Bandura, A. (1969). *Principles of behavior modification*. New York, NY: Holt, Rinehart & Winston.

Bandura, A. (1986). *Social foundations of thought and action: A social cognitive theory*. Englewood Cliffs, NJ: Prentice Hall.

Bandura, A. (1989). Social cognitive theory. In R. Vasta (Ed.), *Annals of child development* (Vol. 6). Greenwich, CT: JAI Press.

Bandura, A. (1991). Human agency: The rhetoric and the reality. *American Psychologist*, *46*(2), 157–162.

Bandura, A. (1997). *Self-efficacy: The exercise of control*. New York, NY: Freeman.

Bandura, A. (2000). Exercise of human agency through collective efficacy. *Current Directions in Psychological Science*, *9*(3), 75–83.

Bandura, A. (2006). Going global with social cognitive theory: From prospect to paydirt. In D. E. Berger & K. Pezdek (Eds.), *The rise of applied psychology: New frontiers and rewarding careers* (pp. 53–79). Mahwah, NJ: Erlbaum.

Bandura, A. (2007). Albert Bandura. In G. L. W. M. Runyan (Ed.), *A history of psychology in autobiography, Vol. IX* (pp. 43–75). Washington, DC: American Psychological Association.

Bandura, A. (2008). Reconstrual of "free will" from the agentic perspective of social cognitive theory. In J. Baer, J. C. Kaufman, & R. F. Baumeister (Eds.), *Are we free? Psychology and free will* (pp. 86–127). New York, NY: Oxford University Press.

Bandura, A. (2011). A social cognitive perspective on positive psychology. *Revista de Psicología Social*, *26*(1), 7–20.

Bandura, A. (2011). The social and policy impact of social cognitive theory. In M. Mark, S. Donaldson, & B. Campbell (Eds.), *Social psychology and evaluation* (pp. 33–70). New York, NY: Guilford Press.

Bandura, A. (2012). On the functional properties of perceived self-efficacy revisited. *Journal of Management*, *38*(1), 9–44.

Bandura, A., & Walters, R. H. (1963). *Social learning and personality development*. New York, NY: Holt, Rinehart & Winston.

Bandura, A., Ross, D., & Ross, S. (1961). Transmission of aggression through imitation of aggressive models. *Journal of Abnormal and Social Psychology*, *63*, 575–582.

Banich, M. T., & Compton, R. J. (2011). *Cognitive neuroscience* (3rd ed.). Belmont, CA: Cengage.

Bankó, É. M., & Vidnyánszky, Z. (2010). Retention interval affects visual short-term memory encoding. *Journal of Neurophysiology*, *103*(3), 1425–1430.

Banks, M. S., & Salapatek, P. (1983). Infant visual perception. In M. M. Haith & J. J. Campos (Eds.), *Handbook of child psychology*. New York, NY: John Wiley & Sons.

Bao, A., & Swaab, D. (2011). Sexual differentiation of the human brain: Relation to gender identity, sexual orientation and neuropsychiatric disorders. *Frontiers in Neuroendocrinology*, *32*(2), 214–226.

Barash, D. P., & Lipton, J. E. (2011). *Payback: Why we retaliate, redirect aggression, and take revenge*. New York, NY: Oxford University Press.

Barbui, C., Cipriani, A., Patel, V., Ayuso-Mateos, J. L., & Ommeren, M. V. (2011). Efficacy of antidepressants and benzodiazepines in minor depression: Systematic review and meta-analysis. *The British Journal of Psychiatry*, *198*, 11–16.

Barczak, B., Miller, T. W., Veltkamp, L. J., Barczak, S., Hall, C., & Kraus, R. (2010). Transitioning the impact of divorce on children throughout the life cycle. In T. W. Miller (Ed.), *Handbook of stressful transitions across the lifespan* (pp. 185–215). New York, NY: Springer Science + Business Media.

Barelds, D. P. H., & Dijkstra, P. (2011). Positive illusions about a partner's personality and relationship quality. *Journal of Research in Personality*, *45*(1), 37–43.

Bargai, N., Ben-Shakar, G., & Shalev, A. Y. (2007). Posttraumatic stress disorder and depression in battered women: The mediating role of learned helplessness. *Journal of Family Violence*, *22*(5), 267–275.

Bar-Haim, Y., Dan, O., Eshel, Y., & Sagi-Schwartz, A. (2007). Predicting childrens' anxiety from early attachment relationships. *Journal of Anxiety Disorders*, *21*, 1061–1068.

Barlett, C. P., Harris, R. J., & Bruey, C. (2008). The effect of the amount of blood in a violent video game on aggression, hostility, and arousal. *Journal of Experimental Social Psychology*, *44*, 539–546.

Barlow, D. H. (Ed.). (2008). *Clinical handbook of psychological disorders: A step-by-step treatment manual (4th ed.)*. New York, NY: Guilford Press.

Barlow, D. H., & Durand, V. M. (2012). *Abnormal psychology: An integrative approach* (6th ed.). Belmont, CA: Cengage.

Barton, D. A., Esler, M. D., Dawood, T., Lambert, E. A., Haikerwal, D. et al. (2008). Elevated brain serotonin turnover in patients with depression: Effect of genotype and therapy. *Archives of General Psychiatry*, *65*, 38–46.

Barton, J. J. S. (2008). Prosopagnosia associated with a left occipitotemporal lesion. *Neuropsychologia*, *46*, 2214–2224.

Bastian, B., Jetten, J., & Radke, H. R. M. (2012). Cyber-dehumanization: Violent video game play diminishes our humanity. *Journal of Experimental Social Psychology*, *48*(2), 486–491.

Bastien, C. H. (2011). Insomnia: Neurophysiological and neuropsychological approaches. *Neuropsychology Review*, *21*(1), 22–40.

Bateman, C. R., & Valentine, S. R. (2010). Investigating the effects of gender on consumers' moral philosophies and ethical intentions. *Journal of Business Ethics*, *95*(3), 393–414.

Bates, A. L. (2007). How did you get in? Attributions of preferential selection in college admissions. *Dissertation Abstracts International: Section B: The Sciences and Engineering*, *68*(4-B), 2694.

Batson, C. D. (1991). *The altruism question: Toward a social-psychological answer*. Hillsdale, NJ: Erlbaum.

Batson, C. D. (1998). Altruism and prosocial behavior. In D.T. Gilbert, S.T. Fiske, & G. Lindzey (Eds.). *The handbook of social psychology, Vol. 2* (4th ed.) (pp. 282–316). Boston, MA: McGraw-Hill.

Batson, C. D. (2006). "Not all self-interest after all": Economics of empathy-induced altruism. In D. De Cremer, M. Zeelenberg, & J. K. Murnighan (Eds.), *Social psychology and economics* (pp. 281–299). Mahwah, NJ: Erlbaum.
Batson, C. D. (2011). *Altruism in humans*. New York, NY: Oxford University Press.
Baucom, B. R., Weusthoff, S., Atkins, D. C., & Hahlweg, K. (2012). Greater emotional arousal predicts poorer long-term memory of communication skills in couples. *Behaviour Research and Therapy*, *50*(6), 442–447.
Bauer, P. J., & Lukowski, A. F. (2010). The memory is in the details: Relations between memory for the specific features of events and long-term recall during infancy. *Journal of Experimental Child Psychology*, *107*(1), 1–14.
Baumeister, R. F., & Bushman, B. (2011). *Social psychology and human nature, comprehensive edition* (2nd ed.). Belmont, CA: Cengage.
Baumeister, R. F., & Tierney, J. (2011). *Willpower: Rediscovering the greatest human strength*. New York: NY: Penguin Press.
Baumeister, R., & Vohs, K. (2011). *New directions in social psychology*. Thousand Oaks, CA: Sage.
Baumrind, D. (1980). New directions in socialization research. *American Psychologist*, *35*, 639–652.
Baumrind, D. (1991). Effective parenting during the early adolescent transition. In P. A. Cowan & E. M. Hetherington (Eds.), *Family transition* (pp. 111–163). Hillsdale, NJ: Erlbaum.
Baumrind, D. (1995). *Child maltreatment and optimal caregiving in social contexts*. New York, NY: Garland.
Bava, S., & Tapert, S. F. (2010). Adolescent brain development and the risk for alcohol and other drug problems. *Neuropsychology Review*, *20*(4), 398–413.
Beacham, A., Stetson, B., Braekkan, K., Rothschild, C., Herbst, A., & Linfield, K. (2011). Causal attributions regarding personal exercise goal attainment in exerciser schematics and aschematics. *International Journal of Sport and Exercise Psychology*, *9*(1), 48–63.
Beachkofsky, A. L. (2010). Marital satisfaction: Ideal versus real mate. *Dissertation Abstracts International: Section B: The Sciences and Engineering*, *51*, 5891.
Beaton, E. A., & Simon, T. J. (2011). How might stress contribute to increased risk for schizophrenia in children with Chromosome 22q11.2 deletion syndrome? *Journal of Neurodevelopmental Disorders*, *3*(1), 68–75.
Beauchamp, M. R., Rhodes, R. E., Kreutzer, C., & Rupert, J. L. (2011). Experiential versus genetic accounts of inactivity: Implications for inactive individuals' self-efficacy beliefs and intentions to exercise. *Behavioral Medicine*, *37*(1), 8–14.
Beaver, K. M., Barnes, J. C., May, J. S., & Schwartz, J. A. (2011). Psychopathic personality traits, genetic risk, and gene-environment correlations. *Criminal Justice and Behavior*, *38*(9), 896–912.
Beck, A. T. (1976). *Cognitive therapy and the emotional disorders*. New York, NY: International Universities Press.
Beck, A. T. (2000). *Prisoners of hate*. New York, NY: Harper Perennial.
Beck, A. T., & Grant, P. M. (2008). Negative self-defeating attitudes: Factors that influence everyday impairment in individuals with schizophrenia. *American Journal of Psychiatry*, *165*, 772.
Beilin, H. (1992). Piaget's enduring contribution to developmental psychology. *Developmental Psychology*, *28*, 191–204.
Bekris, L. M., Yu, C.-E., Bird, T. D., & Tsuang, D. W. (2010). Genetics of Alzheimer disease. *Journal of Geriatric Psychiatry and Neurology*, *23*(4), 213–227.
Bellebaum, C., & Daum, I. (2011). Mechanisms of cerebellar involvement in associative learning. *Cortex: A Journal Devoted to the Study of the Nervous System and Behavior*, *47*(1), 128–136.
Bem, S. L. (1981). Gender schema theory: A cognitive account of sex typing. *Psychological Review*, *88*, 354–364.
Bem, S. L. (1993). *The lenses of gender: Transforming the debate on sexual inequality*. New Haven, CT: Yale University Press.
Bender, R. E., & Alloy, L. B. (2011). Life stress and kindling in bipolar disorder: Review of the evidence and integration with emerging biopsychosocial theories. *Clinical Psychology Review*, *31*(3), 383–398.
Bendersky, M., & Sullivan, M. W. (2007). Basic methods in infant research. In A. Slater & M. Lewis (Eds.), *Introduction to infant development* (2nd ed.). New York, NY: Oxford University Press.
Bendle, N. T. (2011). Out-group homogeneity bias and strategic market entry. *Dissertation Abstracts International Section A: Humanities and Social Sciences*, *71*(9-A), 3341.
Ben-Eliyahu, S., Page, G. G., & Schleifer, S. J. (2007). Stress, NK cells, and cancer: Still a promissory note. *Brain, Behavior, and Immunity*, *21*, 881–887.
Benjamin Neelon, S. E., Oken, E., Taveras, E. M., Rifas-Shiman, S. L., & Gillman, M. W. (2012). Age of achievement of gross motor milestones in infancy and adiposity at age 3 years. *Maternal and Child Health Journal*, *16*(5), 1015–1020.
Benjamin, E. (2011). Humanistic psychology and the mental health worker. *Journal of Humanistic Psychology*, *51*(1), 82–111.
Berger, M., Speckmann, E.-J., Pape, H. C., & Gorji, A. (2008). Spreading depression enhances human neocortical excitability in vitro. *Cephalalgia*, *28*, 558–562.
Bergstrom-Lynch, C. A. (2008). Becoming parents, remaining childfree: How same-sex couples are creating families and confronting social inequalities. *Dissertation Abstracts International Section A: Humanities and Social Sciences*, *68*(8-A), 3608.
Berman, M. E., Tracy, J. I., & Coccaro, E. F. (1997). The serotonin hypothesis of aggression revisited. *Clinical Psychology Review*, *17*(6), 651–665.
Bermeitinger, C., Goelz, R., Johr, N., Neumann, M., Ecker, U. K. H., & Doerr, R. (2009). The hidden persuaders break into the tired brain. *Journal of Experimental Social Psychology*, *45*(2), 320–326.
Bernabé, D. G., Tamae, A. C., Biasoli, É. R., & Oliveira, S. H.P. (2011). Stress hormones increase cell proliferation and regulates interleukin-6 secretion in human oral squamous cell carcinoma cells. *Brain, Behavior, and Immunity*, *25*(3), 574–583.
Bernard, L. L. (1924). *Instinct*. New York, NY: Holt.
Berreman, G. (1971). *Anthropology today*. Del Mar, CA: CRM Books.
Berry, J., Poortinga, Y., Breugelmans, S., Chasiotis, A., & Sam, D. (2011). *Cross-cultural psychology: Research and applications* (3rd ed.). Cambridge, UK: Cambridge University Press.
Berthoud, H-R., & Morrison, C. (2008). The brain, appetite, and obesity. *Annual Review of Psychology*, *59*, 55–92.
Berti, S. (2010). Arbeitsgedächtnis: Vergangenheit, Gegenwart und Zukunft eines theoretischen Konstruktes [Working memory: The past, the present, and the future of a theoretical construct]. *Psychologische Rundschau*, *61*(1), 3–9.
Berwick, R. C., Okanoya, K., Beckers, G. J. L., & Bolhuis, J. J. (2011). Songs to syntax: The linguistics of birdsong. *Trends in Cognitive Sciences*, *15*(3), 113–121.
Betancourt, T. S., Borisova, I. I., de la Soudière, M., & Williamson, J. (2011). Sierra Leone's child soldiers: War exposures and mental health problems by gender. *Journal of Adolescent Health*, *49*(1), 21–28.
Beyers, W., & Seiffge-Krenke, I. (2010). Does identity precede intimacy? Testing Erikson's theory on romantic development in emerging adults of the 21st century. *Journal of Adolescent Research*, *25*(3), 387–415.
Bhanoo, S. N. (2011). Altering a mouse gene turns up aggression, study says. *The New York Times*, 160, pp. 55, 303.
Bhattacharya, S. K., & Muruganandam, A. V. (2003). Adaptogenic activity of Withania somnifera: An experimental study using a

rat model of chronic stress. *Pharmacology, Biochemistry and Behavior*, *75*(3), 547–555.
Bianchi, M., Franchi, S., Ferrario, P., Sotgiu, M. L., & Sacerdote, P. (2008). Effects of the bisphosphonate ibandronate on hyperalgesia, substance P, and cytokine levels in a rat model of persistent inflammatory pain. *European Journal of Pain*, *12*, 284–292.
Billiard, M. (2007). Sleep disorders. In L. Candelise, R. Hughes, A. Liberati, B. M. J. Uitdehaag, & C. Warlow (Eds.), *Evidence-based neurology: Management of neurological disorders* (pp. 70–78). Malden, MA: Blackwell Publishing.
Binstock, R. H., & George, L. K. (Eds.). (2011). *The handbooks of aging consisting of three Vols. Handbook of aging and the social sciences* (7th ed.). San Diego, CA: Elsevier Academic Press.
Blashfield, R. K., Flanagan, E., & Raley, K. (2010). Themes in the evolution of the 20th-century DSMs. In T. Millon, R. F. Krueger, & E. Simonsen (Eds.), *Contemporary directions in psychopathology: Scientific foundations of the DSM-V and ICD-11* (pp. 53–71). New York, NY: Guilford Press.
Blass, T. (1991). Understanding behavior in the Milgram obedience experiment: The role of personality, situations, and their interactions. *Journal of Personality and Social Psychology*, *60*(3), 398–413.
Blass, T. (2000). Stanley Milgram. In A. E. Kazdin (Ed.), *Encyclopedia of psychology* (Vol. 5) (pp. 248–250). Washington, DC: American Psychological Association.
Blum, H. P. (2011). To what extent do you privilege dream interpretation in relation to other forms of mental representations? *The International Journal of Psychoanalysis*, *92*(2), 275–277.
Blume-Marcovici, A. (2010). Gender differences in dreams: Applications to dream work with male clients. *Dreaming*, *20*(3), 199–210.
Boag, S. (2012). *Freudian repression, the unconscious, and the dynamics of inhibition*. London, England: Karnac Books.
Boardman, J., Alexander, K., & Stallings, M. (2011). Stressful life events and depression among adolescent twin pairs. *Biodemography and Social Biology*, *57*(1), 53–66.
Bob, P. (2008). Pain, dissociation and subliminal self-representations. *Consciousness and Cognition: An International Journal*, *17*, 355–369.
Boccato, G., Capozza, D., Falvo, R., & Durante, F. (2008). Capture of the eyes by relevant and irrelevant onsets. *Social Cognition*, *26*, 224–234.
Böhm, R., Schütz, A., Rentzsch, K., Körner, A., & Funke, F. (2010). Are we looking for positivity or similarity in a partner's outlook on life? Similarity predicts perceptions of social attractiveness and relationship quality. *The Journal of Positive Psychology*, *5*(6), 431–438.
Bonk, W. J., & Healy, A. F. (2010). Learning and memory for sequences of pictures, words, and spatial locations: An exploration of serial position effects. *The American Journal of Psychology*, *123*(2), 137–168.
Bor, J., Brunelin, J., Sappey-Marinier, D., Ibarrola, D., d'Amato, T., Suaud-Chagny, M.-F., & Saoud, M. (2011). Thalamus abnormalities during working memory in schizophrenia. An fMRI study. *Schizophrenia Research*, *125*(1), 49–53.
Borchers, B. J. (2007). Workplace environment fit, commitment, and job satisfaction in a nonprofit association. *Dissertation Abstracts International: Section B: The Sciences and Engineering*, *67*(7-B), 4139.
Borges, G., Nock, M. K., Haro Abad, J. M., Hwang, I., Sampson, N. A., Alonso, J., . . . Kessler, R. C. (2010). Twelve-month prevalence of and risk factors for suicide attempts in the World Health Organization World Mental Health Surveys. *Journal of Clinical Psychiatry*, *71*(12), 1617–1628.
Borghans, L, Golsteyn, B., Heckman, J., & Humphries, J. (2011). Identification problems in personality psychology. (Report No. 16917). *National Bureau of Economic Research.* Retrieved from: http://www.nber.org/papers/w16917.
Borst, G., Kievit, R. A., Thompson, W. L., & Kosslyn, S. M. (2011). Mental rotation is not easily cognitively penetrable. *Journal of Cognitive Psychology*, *23*(1), 60–75.
Bosson, J., & Vandello, J. (2011). Precarious manhood and its links to action and aggression. *Current Directions in Psychological Science*, *20*(2), 82–86.
Bouchard, T. J., Jr. & McGue, M. (1981). Familial studies of intelligence: A review. *Science*, *212*(4498), 1055–1059.
Bouchard, T. J., Jr. (1994). Genes, environment, and personality. *Science*, *264*, 1700–1701.
Bouchard, T. J., Jr. (1997). The genetics of personality. In K. Blum & E. P. Noble (Eds.), *Handbook of psychiatric genetics* (pp. 273–296). Boca Raton, FL: CRC Press.
Bouchard, T. J., Jr. (1999). The search for intelligence. *Science*, *284*, 922–923.
Bouchard, T. J., Jr. (2004). Genetic influence on human psychological traits: A survey. *Current Directions in Psychological Science*, *13*(4), 148–151.
Bouchard, T. J., Jr., McGue, M., Hur, Y., & Horn, J. M. (1998). A genetic and environmental analysis of the California Psychological Inventory using adult twins reared apart and together. *European Journal of Personality*, *12*, 307–320.
Boucher, L., & Dienes, Z. (2003). Two ways of learning associations. *Cognitive Science*, *27*(6), 807–842.
Bourke, R. S., Anderson, V., Yang, J. S. C., Jackman, A. R., Killedar, A., Nixon, G. M., Davey, M., Walker, A., Trinder, J., & Horne, R. S. C. (2011). Neurobehavioral function is impaired in children with all severities of sleep disordered breathing. *Sleep Medicine*, *12*(3), 222–229.
Bourne, L. E., Dominowski, R. L., & Loftus, E. F. (1979). *Cognitive processes*. Englewood Cliffs, NJ: Prentice Hall.
Bouton, M. E. (1994). Context, ambiguity, and classical conditioning. *Current Directions in Psychological Science*, *2*, 49–53.
Bowers, K. S., & Woody, E. Z. (1996). Hypnotic amnesia and the paradox of intentional forgetting. *Journal of Abnormal Psychology*, *105*, 381–390.
Bowlby, J. (1969). *Attachment and loss: Vol. 1. Attachment*. New York, NY: Basic Books.
Bowlby, J. (1973). *Attachment and loss: Vol. 2. Separation and anxiety*. New York, NY: Basic Books.
Bowlby, J. (1982). Attachment and loss: Retrospect and prospect. *American Journal of Orthopsychiatry*, *52*, 664–678.
Bowlby, J. (1989). *Secure attachment*. New York, NY: Basic Books.
Bowlby, J. (2000). *Attachment*. New York, NY: Basic Books.
Bowling, A. C., & Mackenzie, B. D. (1996). The relationship between speed of information processing and cognitive ability. *Personality and Individual Differences*, *20*(6), 775–800.
Boyd, R., Richerson, P. J., & Henrich, J. (2011). Rapid cultural adaptation can facilitate the evolution of large-scale cooperation. *Behavioral Ecology and Sociobiology*, *65*(3), 431–444.
Boyer, P., & Bergstrom, B. (2011). Threat-detection in child development: An evolutionary perspective. *Neuroscience and Biobehavioral Reviews*, *35*(4), 1034–1041.
Boyle, S. H., Williams, R. B., Mark, D. B., Brummett, B. H., Siegler, I. C., Helms, M. J., & Barefoot, J. C. (2004). Hostility as a predictor of survival in patients with coronary artery disease. *Psychosomatic Medicine*, *66*(5), 629–632.
Boysen, G. A., & Vogel, D. L. (2007). Biased assimilation and attitude polarization in response to learning about biologicalexplanations of homosexuality. *Sex Roles*, *56*, 755–762.
Bozkurt, A. S., & Aydin, O. (2004). Temel yükleme hatasinin degisik yas ve iki alt kültürde incelenmesi. [A developmental investigation of fundamental attribution error in two subcultures.] *Türk Psikoloji Dergisi*, *19*, 91–104.
Braaten, E. B. (2011). Psychotherapy: Interpersonal and insight-oriented approaches. In E. B. Braaten (Ed.), *APA*

*LifeTools imprint. How to find mental health care for your child* (pp. 171–183). Washington, DC: American Psychological Association.
Breland, K., & Breland, M. (1961). The misbehavior of organisms. *American Psychologist*, *16*, 681–684.
Brent, D. A., & Melhem, N. (2008). Familial transmission of suicidal behavior. *Psychiatric Clinics of North America*, *31*, 157–177.
Brislin, R. W. (1997). *Understanding culture's influence on behavior* (2nd ed.). San Diego, CA: Harcourt Brace.
Brislin, R. W. (2000). *Understanding culture's influence on behavior* (3rd ed.). Ft. Worth, TX: Harcourt.
Brown, A., & Whiteside, S. P. (2008). Relations among perceived parental rearing behaviors, attachment style, and worry in anxious children. *Journal of Anxiety Disorders*, *22*, 263–272.
Brown, E., Deffenbacher, K., & Sturgill, K. (1977). Memory for faces and the circumstances of encounter. *Journal of Applied Psychology*, *62*, 311–318.
Brown, L. S. (2011). Guidelines for treating dissociative identity disorder in adults, third revision: A tour de force for the dissociation field. *Journal of Trauma and Dissociation*, *12*(2), 113–114.
Brown, R. P., & Josephs, R. A. (1999). A burden of proof: Stereotype relevance and gender differences in math performance. *Journal of Personality and Social Psychology*, *76*(2), 246–257.
Brown, R., & Kulik, J. (1977). Flashbulb memories. *Cognition*, *5*, 73–99.
Bruns, C. M., & Kaschak, E. (2011). Feminisms: Feminist therapies in the 21st century. *Women and Therapy*, *34*(1–2), 1–5.
Bryan, A. D., Webster, G. D., & Mahaffey, A. L. (2011). The big, the rich, and the powerful: Physical, financial, and social dimensions of dominance in mating and attraction. *Personality and Social Psychology Bulletin*, *37*(3), 365–382.
Buchsbaum, B. R., Padmanabhan, A., & Berman, K. F. (2011). The neural substrates of recognition memory for verbal information: Spanning the divide between short- and long-term memory. *Journal of Cognitive Neuroscience*, *23*(4), 978–991.
Buckle, J. L., & Fleming, S. J. (2011). *Death, dying, and bereavement. Parenting after the death of a child: A practitioner's guide*. New York, NY: Routledge/Taylor & Francis Group.
Bunde, J., & Suls, J. (2006). A quantitative analysis of the relationship between the Cook-Medley hostility scale and traditional coronary artery disease risk factors. *Health Psychology*, *25*, 493–500.
Burger, J. M. (2009). Replicating Milgram: Would people still obey today? *American Psychologist*, *64*(1), 1–11.
Burger, J. M. (2011). *Personality* (8th ed.). Belmont, CA: Cengage.
Burnette, J., Pollack, J., & Forsyth, D. (2011). Leadership in extreme contexts: A groupthink analysis of the May 1996 Mount Everest disaster. *Journal of Leadership Studies*, *4*(4), 29–40.
Burns, S. M. (2008). Unique and interactive predictors of mental health quality of life among men living with prostate cancer. *Dissertation Abstracts International: Section B: The Sciences and Engineering*, *68*(10-B), 6953.
Burr, A., Santo, J. B., & Pushkar, D. (2011). Affective well-being in retirement: The influence of values, money, and health across three years. *Journal of Happiness Studies*, *12*(1), 17–40.
Burt, K. B., & Masten, A. S. (2010). Development in the transition to adulthood: Vulnerabilities and opportunities. In J. E. Grant & M. N. Potenza (Eds.), *Young adult mental health* (pp. 5–18). New York, NY: Oxford University Press.
Burt, S. A. (2011). The importance of the phenotype in explorations of gene-environment interplay. In A. Booth, S. M. McHale, & N. S. Landale (Eds.), *National symposium on family issues. Biosocial foundations of family processes* (pp. 85–94). New York, NY: Springer.
Bushman, B. J. (2002). Does venting anger feed or extinguish the flame? Catharsis, rumination, distraction, anger and aggressive responding. *Personality and Social Psychology Bulletin*, *28*(6), 724–731.
Buss, D. M. (1989). Sex differences in human mate preferences: Evolutionary hypotheses tested in 37 cultures. *Behavioral and Brain Sciences*, *12*, 1–49.
Buss, D. M. (1994). Mate preferences in 37 countries. In W. J. Lonner & R. Malpass (Eds.), *Psychology and culture*. Boston, MA: Allyn & Bacon.
Buss, D. M. (2003). *The evolution of desire: Strategies of human mating*. New York, NY: Basic Books.
Buss, D. M. (2005). *The handbook of evolutionary psychology*. Hoboken, NJ: John Wiley & Sons.
Buss, D. M. (2007). The evolution of human mating. *Acta Psychologica Sinica*. *39*, 502–512.
Buss, D. M. (2008). *Evolutionary psychology: The new science of the mind* (3rd ed.). Boston: Allyn & Bacon.
Buss, D. M. (2011). *Evolutionary psychology: The new science of the mind* (4th ed.). Upper Saddle River, NJ: Prentice-Hall.
Buswell, B. N. (2006). The role of empathy, responsibility, and motivations to respond without prejudice in reducing prejudice. *Dissertation Abstracts International: Section B: The Sciences and Engineering*, *66*, 6968.
Butcher, J. N. (2000). Revising psychological tests: Lessons learned from the revision of the MMPI. *Psychological Assessment*, *12*(3), 263–271.
Butcher, J. N. (2011). *A beginner's guide to the MMPI-2* (3rd ed.). Washington, DC: American Psychological Association.
Butcher, J. N., & Perry, J. N. (2008). *Personality assessment in treatment planning: Use of the MMPI-2 and BTPI*. New York, NY: Oxford University Press.
Butler, R., Sati, S., & Abas, M. (2011). Assessing mental health in different cultures. In M. Abou-Saleh, C. Katona, & A. Kumar (Eds.), *Principle and practice of geriatric psychiatry* (3rd ed.) (pp. 711–716). Hoboken, NJ: John Wiley & Sons.
Butz, D. A., & Plant, E. A. (2011). Approaching versus avoiding intergroup contact: The role of expectancies and motivation. In L. R. Tropp & R. K. Mallett (Eds.), *Moving beyond prejudice reduction: Pathways to positive intergroup relations* (pp. 81–98). Washington, DC: American Psychological Association.
Buzawa, E., Buzawa, C., & Stark, E. (2012). *Responding to domestic violence: The integration of criminal justice and human services* (4th ed.). Thousand Oaks, CA: Sage.
Bystritsky, A., Korb, A. S., Douglas, P. K., Cohen, M. S., Melega, W. P., Mulgaonkar, A. P., . . . Yoo, S. S. (2011). A review of low-intensity focused ultrasound pulsation. *Brain Stimulation*, *4*(3), 125–136.

Cadoret, R. J., Leve, L. D., & Devor, E. (1997). Genetics of aggressive and violent behavior. *Psychiatric Clinics of North America*, *20*, 301–322.
Caggiano, V., Fogassi, L., Rizzolatti, G., Pomper, J. K., Their, P., Giese, M. A., & Casile, A. (2011). View-based encoding of actions in mirror neurons of area F5 in macaque premotor cortex. *Current Biology*, *21*, 144–148.
Cal, W-H., Blundell, J., Han, J., Greene, R. W., & Powell, C. M. (2006). Postreactivation glucocorticoids impair recall of established fear memory. *Journal of Neuroscience*, *26*(37), 9560–9566.
Calkins, S. D., & Keane, S. P. (2009). Developmental origins of early antisocial behavior. *Development and Psychopathology*, *21*(4), 1095–1109.
Call, J. (2011). How artificial communication affects the communication and cognition of the great apes. *Mind and Language*, *26*(1), 1–20.
Callanan, V., & Davis, M. S. (2012). Gender differences in suicide methods. *Social Psychiatry and Psychiatric Epidemiology*, *47*(6), 857–869.
Calogero, R. M., Rachel M., Tantleff-Dunn, S., & Thompson, J. K. (2011). Future directions for research and practice. In R. M. Calogero, S. Tantleff-Dunn, & J.

K. Thompson (Eds.), Self-objectification in women: Causes, consequences, and counteractions (pp. 217–237). Washington, DC: American Psychological Association.
Camarena, B., Santiago, H., Aguilar, A., Ruvinskis, E., González-Barranco, J., & Nicolini, H. (2004). Family-based association study between the monoamine oxidase A gene and obesity: Implications for psychopharmacogenetic studies. *Neuropsychobiology*, *49*(3), 126–129.
Cameron, L., Rutland, A. & Brown, R. (2007). Promoting children's positive intergroup attitudes towards stigmatized groups: Extended contact and multiple classification skills training. *International Journal of Behavioral Development*, *31*, 454–466.
Camos, V., & Barrouillet, P. (2011). Developmental change in working memory strategies: From passive maintenance to active refreshing. *Developmental Psychology*, *47*(3), 898–904.
Campbell, A., & Muncer, S. (2008). Intent to harm or injure? Gender and the expression of anger. *Aggressive Behavior*, *34*, 282–293.
Campbell, J. R., & Feng, A. X. (2011). Comparing adult productivity of American mathematics, physics, and chemistry Olympians with Terman's longitudinal study. *Roeper Review: A Journal on Gifted Education*, *33*(1), 18–25.
Campbell, L., Simpson, J. A., Kashy, D. A., & Fletcher, G. J. O. (2001). Ideal standards, the self, and flexibility of ideals in close relationships. *Personality and Social Psychology Bulletin*, 27(4), 447–462.
Campbell, L., Vasquez, M., Behnke, S., & Kinscherff, R. (2010). Therapy. In L. Campbell, M. Vasquez, S. Behnke, & R. Kinscherff (Eds.), *APA Ethics Code commentary and case illustrations, (pp. 339–372).* Washington, DC: American Psychological Association.
Campbell, M. C., Black, K. J., Weaver, P. M., Lugar, H. M., Videen, T. O., Tabbal, S. D., . . . Hershey, T. (2012). Mood response to deep brain stimulation of the subthalamic nucleus in Parkinson's Disease. *Journal of Neuropsychiatry and Clinical Neurosciences*, *24*(1), 28–36.
Capellini, I., Preston, B. T., McNamara, P., Barton, R. A., & Nunn, C. L. (2010). Ecological constraints on mammalian sleep architecture. In P. McNamara, R. A. Barton, & C. L. Nunn (Eds.), *Evolution of sleep: Phylogenetic and functional perspectives* (pp. 12–33). New York, NY: Cambridge University
Caprara, G. V., Vecchione, M., Barbaranelli, C., & Fraley, R. C. (2007). When likeness goes with liking: The case of political preference. *Political Psychology*, *28*, 609–632.
Caqueo-Urízar, A., Ferrer-García, M., Toro, J., Gutiérrez-Maldonado, J., Peñaloza, C., Cuadros-Sosa, Y., & Gálvez-Madrid, M. J. (2011). Associations between sociocultural pressures to be thin, body distress, and eating disorder symptomatology among Chilean adolescent girls. *Body Image*, *8*(1), 78–81.
Carels, R. A., Konrad, K., Young, K. M., Darby, L. A., Coit, C., Clayton, A. M., & Oemig, C. K. (2008). Taking control of your personal eating and exercise environment: A weight maintenance program. *Eating Behaviors*, *9*, 228–237.
Carey, B. (2009). A dream interpretation: Tune ups for the brain. *The New York Times*, p. 159.
Carlesimo, G. A., Costa, A., Serra, L., Bozzali, M., Fadda, L., & Caltagirone, C. (2011). Prospective memory in thalamic amnesia. *Neuropsychologia*, *49*(8), 2199–2208.
Carlson, N. (2011). *Foundations of behavioral neuroscience* (8th ed.). Upper Saddle River, NJ: Prentice Hall.
Carnagey, N. L., Anderson, C. A., & Bartholow, B. D. (2007). Media violence and social neuroscience: New questions and new opportunities. *Current Directions in Psychological Science*, *16*, 178–182.
Carré, J., McCormick, C., & Hariri, A. (2011). The social neuroendocrinology of human aggression. *Psychoneuroendocrinology*, *36*(7), 935–944.
Carroll, J. E., Low, C. A., Prather, A. A., Cohen, S., Fury, J. M., Ross, D. C., & Marsland, A. L. (2011). Negative affective responses to a speech task predict changes in interleukin (IL)-6. *Brain, Behavior, and Immunity*, *25*(2), 232–238.
Carroll, J. E., Low, C. A., Prather, A. A., Cohen, S., Fury, J. M., Ross, D. C., & Marsland, A. L. (2011). Negative affective responses to a speech task predict changes in interleukin (IL)-6. *Brain, Behavior, and Immunity*, *25*(2), 232–238.
Carstensen, L. L., Turan, B., Scheibe, S., Ram, N., Ersner-Hershfield, H., Samanez-Larkin, G. R., Brooks, K., & Nesselroade, J. R. (2011). Emotional experience improves with age: Evidence based on over 10 years of experience sampling. *Psychology and Aging*, *26*(1), 21–33.
Carvalho, C., Mazzoni, G., Kirsch, I., Meo, M., & Santandrea, M. (2008). The effect of posthypnotic suggestion, hypnotic suggestibility, and goal intentions on adherence to medical instructions. *International Journal of Clinical and Experimental Hypnosis*, *56*, 143–155.
Casarett, D. J. (2011). Rethinking hospice eligibility criteria. *JAMA: Journal of the American Medical Association*, *305*(10), 1031–1032.
Castelli, L., Corazzini, L. L., & Geminiani, G. C. (2008). Spatial navigation in large-scale virtual environments: Gender differences in survey tasks. *Computers in Human Behavior*, *24*, 1643–1667.
Cathers-Schiffman, T. A., & Thompson, M. S. (2007). Assessment of English- and Spanish-speaking students with the WISC-III and Leiter-R. *Journal of Psychoeducational Assessment*, *25*, 41–52.
Cattaneo, Z., Mattavelli, G., Platania, E., & Papagno, C. (2011). The role of the prefrontal cortex in controlling gender-stereotypical associations: A TMS investigation. *NeuroImage*, *56*(3), 1839–1846.
Cattell, R. B. (1950). *Personality: A systematic, theoretical, and factual study*. New York, NY: McGraw-Hill.
Cattell, R. B. (1963). Theory of fluid and crystallized intelligence: A critical experiment. *Journal of Educational Psychology*, *54*, 1–22.
Cattell, R. B. (1965). *The scientific analysis of personality*. Baltimore, MD: Penguin.
Cattell, R. B. (1971). *Abilities: Their structure, growth, and action*. Boston, MA: Houghton Mifflin.
Cattell, R. B. (1990). Advances in Cattellian personality theory. In L. A. Pervin (Ed.), *Handbook of personality: Theory and research*. New York, NY: Guilford Press.
Cautin, R. L. (2011). A century of psychotherapy, 1860–1960. In J. C. Norcross, G. R. VandenBos, & D. K. Freedheim (Eds.), *History of psychotherapy: Continuity and change* (2nd ed.) (pp. 3–38). Washington, DC: American Psychological Association.
Cautin, R. L. (2011). Invoking history to teach about the scientist-practitioner gap. *History of Psychology*, *14*(2), 197–203.
Cechnicki, A., Angermeyer, M. C., & Bielańska, A. (2011). Anticipated and experienced stigma among people with schizophrenia: Its nature and correlates. *Social Psychiatry and Psychiatric Epidemiology*, *46*(7), 643–650.
Cehajic, S., Brown, R., & Castano, E. (2008). Forgive and forget? Antecedents and consequences of intergroup forgiveness in Bosnia and Herzegovina. *Political Psychology*, *29*, 351–367.
Celada, T. C. (2011). Parenting styles as related to parental self-efficacy and years living in the United States among Latino immigrant mothers. *Dissertation Abstracts International: Section B: The Sciences and Engineering*, *71*(9-B), 5783.
Celes, L. A. M. (2010). Clinica psicanalitica: Aproximações histórico-conceituais e contemporâneas e perspectivas futuras [Psychoanalytic practice: Historical, conceptual and contemporary approaches and future perspectives]. *Psicologia: Teoria e Pesquisa*, *26*(*SpecIssue*), 65–80.

Cervone, D., & Pervin, L. A. (2010). *Personality theory and research* (11th ed.). Hoboken, NJ: John Wiley & Sons.

Cesaro, P., & Ollat, H. (1997). Pain and its treatments. *European Neurology*, *38*, 209–215.

Ceschi, G., & Scherer, K. R. (2001). Contrôler l'expression faciale et changer l'émotion: Une approche développementale. The role of facial expression in emotion: A developmental perspective. *Enfance*, *53*(3), 257–269.

Chadwick, R. (2011). Personal genomes: No bad news? *Bioethics*, *25*(2), 62–65.

Chafin, S., Christenfeld, N., & Gerin, W. (2008). Improving cardiovascular recovery from stress with brief poststress exercise. *Journal of Health Psychology*, *27*(1S), S64–72.

Challem, J., Berkson, B., Smith, M. D., & Berkson, B. (2000). *Syndrome X: The complete program to prevent and reverse insulin resistance*. New York: John Wiley & Sons.

Challies, D. M., Hunt, M., Garry, M., & Harper, D. N. (2011). Whatever gave you that idea? False memories following equivalence training: A behavioral account of the misinformation effect. *Journal of the Experimental Analysis of Behavior*, *96*(3), 343–362.

Chamberlin, N. L., & Saper, C. B. (2009). The agony of the ecstasy: Serotonin and obstructive sleep apnea. *Neurology*, *73*(23), 1947–1948.

Chamorro-Premuzic, T. (2011). *Personality and individual differences*. Malden, MA: Blackwell.

Chance, P. (2009). *Learning and behavior: Active learning edition* (6th ed.). Belmont, CA: Cengage.

Chandrashekar, J., Hoon, M. A., Ryba, N. J. P., & Zuker, C. S. (2006). The receptors and cells for mammalian taste. *Nature*, *444*, 288–294.

Chang, G., Orav, J., McNamara, T. K., Tong, MY., & Antin, J. H. (2005). Psychosocial function after hematopoietic stem cell transplantation. *Psychosomatics: Journal of Consultation Liaison Psychiatry*, *46*(1), 34–40.

Chang, L., Wang, Y., Shackelford, T. K., & Buss, D. M. (2011). Chinese mate preferences: Cultural evolution and continuity across a quarter of a century. *Personality and Individual Differences*, *50*(5), 678–683.

Chang, W., Tang, J., Hui, C., Chiu, C., Lam, M., Wong, G., Chung, D., Law, C., Tso, S., Chan, K., Hung, S., & Chen, E. (2011). Gender differences in patients presenting with first-episode psychosis in Hong Kong: A three-year follow up study. *Australian and New Zealand Journal of Psychiatry*, *45*(3), 199–205.

Chapman, A. L., Leung, D. W., & Lynch, T. R. (2008). Impulsivity and emotion dysregulation in borderline personality disorder. *Journal of Personality Disorders*, *22*, 148–164.

Charles, S. T., & Carstensen, L. L. (2007). Emotion regulation and aging. In J. J. Gross (Ed.), *Handbook of emotion regulation*. New York: Guilford Press.

Cheever, N. A. (2010). The cultivation of social identity in single women: The role of single female characterizations and marriage and romantic relationship portrayals on television. *Dissertation Abstracts International: Section B: The Sciences and Engineering*, *71*(6–B), 3979.

Chellappa, S. L., Frey, S., Knoblauch, V., & Cajochen, C. (2011). Cortical activation patterns herald successful dream recall after NREM and REM sleep. *Biological Psychology*, *87*(2), 251–256.

Cheung, F., van de Vijver, F., & Leong, F. (2011, January 24). Toward a new approach to the study of personality in culture. *American Psychologist, 66*(7), 596–603.

Cheung, M. S., Gilbert, P., & Irons, C. (2004). An exploration of shame, social rank, and rumination in relation to depression. *Personality and Individual Differences, 36*(5), 1143–1153.

Choi, I., & Nisbett, R. E. (2000). Cultural psychology of surprise: Holistic theories and recognition of contradiction. *Journal of Personality and Social Psychology*, *79*(6), 890–905.

Choi, N. (2004). Sex role group differences in specific, academic, and general self-efficacy. *Journal of Psychology: Interdisciplinary and Applied*, *138*(2), 149–159.

Chomsky, N. (1968). *Language and mind*. New York, NY: Harcourt, Brace, World.

Chomsky, N. (1980). *Rules and representations*. New York, NY: Columbia University Press.

Chopra, A., Tye, S. J., Lee, K. H., Sampson, S., Matsumoto, J., Adams, A., . . . Frye, M. A. (2012). Underlying neurobiology and clinical correlates of mania status after subthalamic nucleus deep brain stimulation in Parkinson's Disease: A review of the literature. *Journal of Neuropsychiatry and Clinical Neurosciences*, *24*(1), 102–110.

Chou, Y.-C., Chiao, C., & Fu, L.-Y. (2011). Health status, social support, and quality of life among family carers of adults with profound intellectual and multiple disabilities (PIMD) in Taiwan. *Journal of Intellectual and Developmental Disability*, *36*(1), 73–79.

Chrisler, J. C. (2008). The menstrual cycle in a biopsychosocial context. In F. Denmark & M. A. Paludi (Eds.), *Psychology of women: A handbook of issues and theories* (2nd ed.) (pp. 400–439). *Women's psychology*. Westport, CT: Praeger/Greenwood.

Christakou, A., Brammer, M., & Rubia, K. (2011). Maturation of limbic corticostriatal activation and connectivity associated with developmental changes in temporal discounting. *NeuroImage*, *54*(2), 1344–1354.

Christandl, F., Fetchenhauer, D., & Hoelzl, E. (2011). Price perception and confirmation bias in the context of a VAT increase. *Journal of Economic Psychology*, *32*(1), 131–141.

Christensen, D. (2000). Is snoring a diZZZease? Nighttime noises may serve as a wake-up call for future illness. *Science News*, *157*, 172–173.

Christensen, H., Anstey, K. J., Leach, L. S., & Mackinnon, A. J. (2008). Intelligence, education, and the brain reserve hypothesis. In F. I. M. Craik & T. A. Salthouse (Eds.), *The handbook of aging and cognition* (3rd ed.) (pp. 133–188). New York, NY: Psychology Press.

Christensen, K. A. (2011). PSTD symptoms expressed in direct and indirect child victims of domestic violence. *Dissertation Abstracts International: Section B: The Sciences and Engineering*, *71*(8-B), 5115.

Christopher, K., Lutz-Zois, C. J., & Reinhardt, A. R. (2007). Female sexual-offenders: Personality pathology as a mediator of the relationship between childhood sexual abuse history and sexual abuse. *Child Abuse and Neglect*, *31*, 871–883.

Chu, J. (2011). *Rebuilding shattered lives: Treating complex PTSD and Dissociative Disorders* (2nd ed.). Hoboken, NJ: John Wiley & Sons.

Chuang, D. (1998). Cited in J. Travis, Simulating clue hints how lithium works. *Science News*, *153*, 165.

Cialdini, R. B. (2009). *Influence: Science and practice* (5th ed.). Boston, MA: Allyn & Bacon.

Claes, L., Bijttebier, P., Mitchell, J. E., de Zwaan, M., & Mueller, A. (2011). The relationship between compulsive buying, eating disorder symptoms, and temperament in a sample of female students. *Comprehensive Psychiatry*, *52*(1), 50–55.

Clark, A. J. (2007). *Empathy in counseling and psychotherapy: Perspectives and practices*. Mahwah, NJ: Erlbaum.

Claxton-Oldfield, S., Wasylkiw, L., Mark, M., & Claxton-Oldfield, J. (2011). The Inventory of motivations for hospice palliative care volunteerism: A tool for recruitment and retention. *American Journal of Hospice and Palliative Medicine*, *28*(1), 35–43.

Clay, R. (2003). An empty nest can promote freedom, improved relationships. *Monitor on Psychology*, *34*(4). Retrieved from http://psycnet.apa.org/psycextra/300092003-024.pdf

Clay, R. A. (2012). Beyond psychotherapy. *Monitor on Psychology, 43*(1), 46.
Claydon, L. (2012). Are there lessons to be learned from a more scientific approach to mental condition defences? *International Journal of Law and Psychiatry, 35*(2), 88–98.
Cleeremans, A., & Sarrazin, J. C. (2007). Time, action, and consciousness. *Human Movement Science, 26*, 180–202.
Clifford, J. S., Boufal, M. M., & Kurtz, J. E. (2004). Personality traits and critical thinking: Skills in college students empirical tests of a two-factor theory. *Assessment, 11*(2), 169–176.
Cohen, D., & Leung, A. K. (2011). Violence and character: A CuPS (culture 3 person 3 situation) perspective. In P. R. Shaver & M. Mikulincer (Eds.), *Herzilya series on personality and social psychology. Human aggression and violence: Causes, manifestations, and consequences* (pp. 187–200). Washington, DC: American Psychological Association.
Cohen, S., & Lemay, E. P. (2007). Why would social networks be linked to affect and health practices? *Health Psychology, 26*, 410–417.
Cohen, S., Hamrick, N., Rodriguez, M. S., Feldman, P. J., Rabin, B. S., & Manuck, S. B. (2002). Reactivity and vulnerability to stress associated risk for upper respiratory illness. *Psychosomatic Medicine, 64*(2), 302–310.
Cole, M., & Gajdamaschko, N. (2007). Vygotsky and culture. In H. Daniels, J. Wertsch, & M. Cole (Eds.), *The Cambridge companion to Vygotsky* (pp. 193–211). New York, NY: Cambridge University Press.
Coleborne, C., & MacKinnon, D. (2011). *Exhibiting madness in museums: Remembering psychiatry through collection and display*. New York, NY: Routledge.
Collins, R. (2011). Content analysis of gender roles in media: Where are we now and where should we go? *Sex Roles, 64*(3–4), 290–298.
Colrain, I. M. (2011). Sleep and the brain. *Neuropsychology Review, 21*(1), 1–4.
Columb, C., & Plant, E. A. (2011). Revisiting the Obama Effect: Exposure to Obama reduces implicit prejudice. *Journal of Experimental Social Psychology, 47*(2), 499–501.
Combs, D. R., Basso, M. R., Wanner, J. L., & Ledet, S. N. (2008). Schizophrenia. In M. Hersen & J. Rosqvist (Eds.), *Handbook of psychological assessment, case conceptualization, and treatment, Vol 1: Adults* (pp. 352–402). Hoboken, NJ: John Wiley & Sons.
Confer, J. C., Easton, J. A., Fleischman, D. S., Goetz, C. D., Lewis, D. M. G., Perilloux, C., & Buss, D. M. (2010). Evolutionary psychology: Controversies, questions, prospects, and limitations. *American Psychologist, 65*(2), 110–126.
Congdon, E., & Canli, T. (2008). A neurogenetic approach to impulsivity. *Journal of Personality, 76*, 1447–1484.
Conger, K. (2011). Embryonic stem cell therapy for paralysis given to first patient in western United States. *Stanford School of Medicine*. Retrieved from http://med.stanford.edu/ism/2011/september/geron.html
Considering a Career. (2011). Becoming a health psychologist. *APA Division 38*. Retrieved from http://www.health-psych.org/AboutHowtoBecome.cfm
Constantino, M. J., Manber, R., Ong, J., Kuo, T. F., Huang, J. S., & Arnow, B. A. (2007). Patient expectations and therapeutic alliance as predictors of outcome in group cognitive-behavioral therapy for insomnia. *Behavioral Sleep Medicine, 5*, 210–228.
Conzen, P. (2010). Erik H. Erikson: Pionier der psychoanalytischen identitätstheorie [Erik H. Erikson: A pioneer of the psychoanalytic identity theory]. *Forum der Psychoanalyse: Zeitschrift für klinische Theorie and Praxis, 26*(4), 389–411.
Cook, M., & Mineka, S. (1989). Observational conditioning of fear to fear-relevant versus fear-irrelevant stimuli in rhesus monkeys. *Journal of Abnormal Psychology, 98*, 448–459.
Cooper, J. A., Watras, A. C., Paton, C. M., Wegner, F. H., Adams, A. K., & Schoeller, D. A. (2011). Impact of exercise and dietary fatty acid composition from a high-fat diet on markers of hunger and satiety. *Appetite, 56*(1), 171–178.
Cooper, W. E. Jr., Pérez-Mellado, V., Vitt, L. J., & Budzinsky, B. (2002). Behavioral responses to plant toxins in two omnivorous lizard species. *Physiology and Behavior, 76*(2), 297–303.
Cordón, L. (2012). *All things Freud: An encyclopedia of Freud's world* (Vols. 1–2). Santa Barbara, CA: Greenwood.
Corey, G. (2011). *Theory and practice of group counseling* (8th ed.). Belmont, CA: Cengage.
Cornum, R., Matthews, M. D., & Seligman, M. E. P. (2011). Comprehensive soldier fitness: Building resilience in a challenging institutional context. *American Psychologist, 66*(1), 4–9.
Corr, C. A., Nabe, C. M., & Corr, D. M. (2009). *Death and dying: Life and living* (6th ed.). Belmont, CA: Wadsworth.
Corr, P. J. (2011). Anxiety: Splitting the phenomenological atom. *Personality and Individual Differences, 50*(7), 889–897.
Costa, P. T., Jr., & McCrae, R. R. (2011). The five-factor model, five-factor theory, and interpersonal psychology. In L. M. Horowitz & S. Strack (Eds.), *Handbook of interpersonal psychology: Theory, research, assessment, and therapeutic interventions* (pp. 91–104). Hoboken, NJ: John Wiley & Sons.
Costa Jr., P. T., McCrae, R. R., & Martin, T. A. (2008). Incipient adult personality: The NEO-PI-3 in middle-school-aged children. *British Journal of Developmental Psychology, 26*, 71–89.
Costa, J. L., Brennen, M. B., & Hochgeschwender, U. (2002). The human genetics of eating disorders: Lessons from the leptin/melanocortin system. *Child and Adolescent Psychiatric Clinics of North America, 11*(2), 387–397.
Costello, K., & Hodson, G. (2011). Social dominance-based threat reactions to immigrants in need of assistance. *European Journal of Social Psychology, 41*(2), 220–231.
Cotter, D., Hermsen, J., & Vanneman, R. (2011). The end of the gender revolution? Gender role attitudes from 1977–2008. *American Journal of Sociology, 116*(4), 1–31.
Cougle, J. R., Bonn-Miller, M. O., Vujanovic, A. A., Zvolensky, M. J., & Hawkins, K. A. (2011). Posttraumatic stress disorder and Cannabis use in a nationally representative sample. *Psychology of Addictive Behaviors, 25*(3), 554–558.
Courage, M. L., & Adams, R. J. (1990). Visual acuity assessment from birth to three years using the acuity card procedures: Cross-sectional and longitudinal samples. *Optometry and Vision Science, 67*, 713–718.
Cox, C. L., Gotimer, K., Roy, A. K., Castellanos, F. X., Milham, M. P., & Kelly, C. (2010). Your resting brain CAREs about your risky behavior. *PLoS ONE, 5*(8).
Coyne, J. C., & Tennen, H. (2010). Positive psychology in cancer care: Bad science, exaggerated claims, and unproven medicine. *Annals of Behavioral Medicine, 39*(1), 16–26.
Craig, A. D., & Bushnell, M. C. (1994). The thermal grill illusion: Unmasking the burn of cold pain. *Science, 265*, 252–255.
Crandall, C. S., & Martinez, R. (1996). Culture, ideology, and antifat attitudes. *Personality and Social Psychology Bulletin, 22*, 1165–1176.
Craske, M. G. (2010). Evaluation. In M. G. Craske (Ed.), *Cognitive-behavioral therapy* (pp. 115–126). Washington, DC: American Psychological Associaton.
Crawford, C. S. (2008). Ghost in the machine: A genealogy of phantom-prosthetic relations (amputation, dismemberment, prosthetic). *Dissertation Abstracts International Section A: Humanities and Social Sciences, 68*(7-A), 3173.
Crespo-Facorro, B., Barbadillo, L., Pelayo-Terán, J., Rodríguez-Sánchez, J. M., & Teran, J. M. (2007). Neuropsychological functioning and brain structure in

schizophrenia. *International Review of Psychiatry*, *19*, 325–336.
Cresswell, M. (2008). Szasz and his interlocutors: Reconsidering Thomas Szasz's "Myth of Mental Illness" thesis. *Journal for the Theory of Social Behaviour*, *38*, 23–44.
Crooks, R. & Bauer, K. (2011). *Our sexuality* (11th ed.). Belmont, CA: Cengage.
Crowell, S. E., Beauchaine, T. P., & Lenzenweger, M. F. (2008). The development of borderline personality disorder and self-injurious behavior. In T. P. Beauchaine & S. P. Hinshaw (Eds.), *Child and adolescent psychopathology* (pp. 510–539). Hoboken, NJ: John Wiley & Sons.
Cuijpers, P., Geraedts, A., van Oppen, P., Andersson, G., Markowitz, J., & van Straten, A. (2011). Interpersonal psychotherapy for depression: A meta-analysis. *American Journal of Psychiatry*, *168*, 581–592.
Cullen, D., & Gotell, L. (2002). From orgasms to organizations: Maslow, women's sexuality and the gendered foundations of the needs hierarchy. *Gender, Work and Organization*, *9*(5), 537–555.
Cully, J. A., & Stanley, M. A. (2008). Assessment and treatment of anxiety in later life. In K. Laidlaw & B. Knight (Eds.), *Handbook of emotional disorders in later life: Assessment and treatment* (pp. 233–256). New York, NY: Oxford University Press.
Cummings, D. E. (2006). Ghrelin and the short-and long-term regulation of appetite and body weight. *Physiology and Behavior*, *89*, 71–84.
Cummings, E., & Henry, W. E. (1961). *Growing old: The process of disengagement*. New York, NY: Basic Books.
Cunningham, G. B. (2002). Diversity and recategorization: Examining the effects of cooperation on bias and work outcomes. *Dissertation Abstracts International Section A: Humanities and Social Sciences*, *63*, 1288.
Cunningham, G. B., Fink, J. S., & Kenix, L. J. (2008). Choosing an endorser for a women's sporting event: The interaction of attractiveness and expertise. *Sex Roles*, *58*, 371–378.
Curtiss, S. (1977). *Genie: A psycholinguistic study of a modern-day "wild child."* New York, NY: Academic Press.
Cushman, F., & Greene, J. (2012). The philosopher in the theater. In M. M. P. R. Shaver (Ed.), *The social psychology of morality: Exploring the causes of good and evil* (pp. 33–50). Washington, DC: American Psychological Association.
Cusimano, M., Biziato, D., Brambilla, E., Donegà, M., Alfaro-Cervello, C., Snider, S., . . . Pluchino, S. (2012). Transplanted neural stem/precursor cells instruct phagocytes and reduce secondary tissue damage in the injured spinal cord. *Brain: A Journal of Neurology*, *135*(2), 447–460.
Cvetkovic, D., & Cosic, I. (Eds.). (2011). *States of Consciousness: Experimental Insights into Meditation, Waking, Sleep and Dreams*. New York, NY: Springer.

Daan, S. (2011). How and why? The lab versus the field. *Sleep and Biological Rhythms*, *9*(1), 1–2.
Dackis, C. A., & O'Brien, C. P. (2001). Cocaine dependence: A disease of the brain's reward centers. *Journal of Substance Abuse Treatment*, *21*(3), 111–117.
Daffner, K. R. (2010). Promoting successful cognitive aging: A comprehensive review. *Journal of Alzheimer's Disease*, *19*(4), 1101–1122.
Dakof, G. A., Godley, S. H., & Smith, J. E. (2011). The adolescent community reinforcement approach and multidimensional family therapy: Addressing relapse during treatment. In Y. Kaminer & K. C. Winters (Eds.), *Clinical manual of adolescent substance abuse treatment* (pp. 239–268). Arlington, VA: American Psychiatric Publishing.
Dalenberg, C., Loewenstein, R., Spiegel, D., Brewin, C., Lanius, R., Frankel, S., Gold, S., Van der Kolk, B., Simeon, D., Vermetten, E., Butler, L., Koopman, C., Courtois, C., Dell, P., Nijenhuis, E., Chu, J., Sar, V., Palesh, O., Cuevas, C., & Paulson, K. (2007). Scientific study of the dissociative disorders. *Psychotherapy and Psychosomatics*, *76*, 400–401.
Dalley, J. W., Fryer, T. D., Brichard, L., Robinson, E. S. J, Theobald, D. E. H., Lääne, K., Peña, Y., Murphy, E. R., Shah, Y., Probst, K., Abakumova, I., Aigbirhio, F. I., Richards, H. K., Hong, Y., Baron, J-C., Everitt, B. J., & Robbins, T. W. (2007). Nucleus accumbens D2/3 receptors predict trait impulsivity and cocaine reinforcement. *Science*, *315*, 1267–1270.
Dana, R. H. (2005). *Multicultural assessment: Principles, applications, and examples*. Mahwah, NJ: Erlbaum.
Dandekar, M. P., Nakhate, K. T., Kokare, D. M., & Subhedar, N. K. (2011). Effect of nicotine on feeding and body weight in rats: Involvement of cocaine- and amphetamine-regulated transcript peptide. *Behavioural Brain Research*, *219*(1), 31–38.
Dantzer, R., O'Connor, J. C., Freund, G. C., Johnson, R. W., & Kelley, K. W. (2008). From inflammation to sickness and depression: When the immune system subjugates the brain. *Nature Reviews Neuroscience*, *9*, 46–57.
Darley, J. M. & Latané, B. (1968). Bystander intervention in emergencies: Diffusion of responsibility. *Journal of Personality and Social Psychology*, *8*, 377–383.
Darmani, N. A., & Crim, J. L. (2005) Delta-9- tetrahydrocannabinol prevents emesis more potently than enhanced locomotor activity produced by chemically diverse dopamine D2/D3 receptor agonists in the least shrew (Cryptotis parva). *Pharmacology Biochemistry and Behavior*, *80*, 35–44.
Dattilio, F. M., & Nichols, M. P. (2011). Reuniting estranged family members: A cognitive-behavioral-systemic perspective. *American Journal of Family Therapy*, *39*(2), 88–99.
Davies, I. (1998). A study of colour grouping in three languages: A test of the linguistic relativity hypothesis. *British Journal of Psychology*, *89*, 433–452.
Davis, C. C., & Balzano, Q. (2011). Cell phone activation and brain glucose metabolism. *Journal of the American Medical Association*, *305*(20), 2066.
Dawson, D. A., Goldstein, R. B., Moss, H. B., Li, T. K., & Grant, B. F. (2010). Gender differences in the relationship of internalizing and externalizing psychopathology to alcohol dependence: Likelihood, expression and course. *Drug and Alcohol Dependence*, *112*(1/2), 9–17.
Dawson, D., Noy, Y. I., Härmä, M., Åkerstedt, T., & Belenky, G. (2011). Modelling fatigue and the use of fatigue models in work settings. *Accident Analysis and Prevention*, *43*(2), 549–564.
de Charms, R., & Moeller, G. H. (1962). Values expressed in American children's readers: 1800–1950. *Journal of Abnormal and Social Psychology*, *64*(2), 136–142.
De Coteau, T. J., Hope, D. A., & Anderson, J. (2003). Anxiety, stress, and health in northern plains Native Americans. *Behavior Therapy*, *34*(3), 365–380.
de Oliveira-Souza, R., Moll, J., Ignácio, F. A., & Hare, R. D. (2008). Psychopathy in a civil psychiatric outpatient sample. *Criminal Justice and Behavior*, *35*, 427–437.
De Sousa, A. (2011). Freudian theory and consciousness: A conceptual analysis. *Brain, Mind and Consciousness*, *9*(1), 210–217.
De Vos, J. (2010). From Milgram to Zimbardo: The double birth of postwar psychology/psychologization. *History of the Human Sciences*, *23*(5), 156–175.
De Witte, L., Brouns, R., Kavadias, D., Engelborghs, S., De Deyn, P. P., & Mariën, P. (2011). Cognitive, affective and behavioural disturbances following vascular thalamic lesions: A review. *Cortex: A Journal Devoted to the Study of the Nervous System and Behavior*, *47*(3), 273–319.
DeAngelis, T. (2012). Practicing distance therapy, legally and ethically. *Monitor on Psychology*, *43*(3), 52.
Deary, I. J., Ferguson, K. J., Bastin, M. E., Barrow, G. W. S., Reid, L. M., Seckl, J. R., Wardlaw, J. M., & MacLullich, A. M. J. (2007). Skull size and intelligence, and King Robert Bruce's IQ. *Intelligence*, *35*, 519–528.

Dębiec, J., & LeDoux, J. E. (2006). Noradrenergic signaling in the amygdala contributes to the reconsolidation of fear memory: Treatment implications for PTSD. In R. Yehuda (Ed.), Psychobiology of posttraumatic stress disorders: A decade of progress (Vol. 1071). *Annals of the New York Academy of Sciences* (pp. 521–524). Malden: Blackwell Publishing.

DeCasper, A. J., & Fifer, W. D. (1980). Of human bonding: Newborns prefer their mother's voices. *Science*, 208, 1174–1176.

Deci, E. L. (1995). *Why we do what we do: The dynamics of personal autonomy*. New York, NY: Putnam's Sons.

Deci, E. L., & Moller, A. C. (2005). The concept of competence: A starting place for understanding intrinsic motivation and self-determined extrinsic motivation. In A. J. Elliot & C. S. Dweck (Eds.), *Handbook of competence and motivation* (pp. 579–597). New York, NY: Guilford.

DeClue, G. (2003). The polygraph and lie detection. *Journal of Psychiatry and Law*, *31*(3), 361–368.

Dement, W. C., & Wolpert, E. (1958). The relation of eye movements, bodily motility, and external stimuli to dream content. *Journal of Experimental Psychology*, *53*, 543–553.

Dempster, M., McCorry, N. K., Brennan, E., Donnelly, M., Murray, L. J., & Johnston, B. T. (2011). Do changes in illness perceptions predict changes in psychological distress among oesophageal cancer survivors? *Journal of Health Psychology*, *16*(3), 500–509.

Denmark, F. L., Rabinovitz, V. C., & Sechzer, J. A. (2005). *Engendering psychology: Women and gender revisited* (2nd ed.). Boston, MA: Allyn and Bacon.

Dennis, W., & Dennis, M. G. (1940). Cradles and cradling customs of the Pueblo Indians. *American Anthropologist*, *42*, 107–115.

Denson, T. F. (2011). A social neuroscience perspective on the neurobiological bases of aggression. In P. R. Shaver & M. Mikulincer (Eds.), *Herzilya series on personality and social psychology. Human aggression and violence: Causes, manifestations, and consequences* (pp. 105–120). Washington, DC: American Psychological Association.

Depue, B. E. (2012). A neuroanatomical model of prefrontal inhibitory modulation of memory retrieval. *Neuroscience and Biobehavioral Reviews*, *36*(5), 1382–1399.

DeSpelder, L. A., & Strickland, A. (2007). Culture, socialization, and death education. In D. Balk, C. Wogrin, G. Thornton, & D. Meagher (Eds.), *Handbook of thanatology: The essential body of knowledge for the study of death, dying, and bereavement* (pp. 303–314). New York, NY: Routledge/Taylor & Francis Group.

Dessalles, J. (2011). Sharing cognitive dissonance as a way to reach social harmony. *Social Science Information/Sur Les Sciences Sociales*, *50*, 116–127.

Deutscher, G. (2010). *Through the language glass: Why the world looks different in other languages*. New York, NY: Metropolitan Books/Henry Holt & Company.

DeValois, R. L. (1965). Behavioral and electrophysiological studies of primate vision. In W. D. Neff (Ed). *Contributions to sensory physiology* (Vol. 1). New York, NY: Academic Press.

Devlin, M. J., Yanovski, S. Z., & Wilson, G. T. (2000). Obesity: What mental health professionals need to know. *American Journal of Psychiatry*, *157*(6), 854–866.

Dewan, M. J., Steenbarger, B. N., & Greenberg, R. P. (2011). Brief psychotherapies. In R. E. Hales, S. C. Yudofsky, & G. O. Gabbard (Eds.), *Essentials of psychiatry* (3rd ed.) (pp. 525–539). Arlington, VA: American Psychiatric Publishing.

Dhikav, V., Aggarwal, N., Gupta, S., Jadhavi, R., & Singh, K. (2008). Depression in Dhat syndrome. *Journal of Sexual Medicine*, *5*, 841–844.

Diaz-Berciano, C., de Vicente, F., & Fontecha, E. (2008). Modulating effects in learned helplessness of dyadic dominance-submission relations. *Aggressive Behavior*, *34*(3), 273–281.

Dickens, W. T., & Flynn, J. R. (2001). Heritability estimates versus large environmental effects: The IQ paradox resolved. *Psychological Review*, *108*(2), 346–369.

DiDonato, T. E., Ullrich, J., & Krueger, J. I. (2011). Social perception as induction and inference: An integrative model of intergroup differentiation, ingroup favoritism, and differential accuracy. *Journal of Personality and Social Psychology*, *100*(1), 66–83.

Diego, M. A., & Jones, N. A. (2007). Neonatal antecedents for empathy. In T. Farrow & P. Woodruff (Eds.), *Empathy in mental illness* (pp. 145–167). New York, NY: Cambridge University Press.

Diem-Wille, G. (2011). *The early years of life: Psychoanalytical development theory according Freud, Klein, and Bion*. London, UK: Karnac Books.

Diener, E. (2008). Myths in the science of happiness, and directions for future research. In M. Eid & R. J. Larsen (Eds.), *The science of subjective well-being* (pp. 493–514). New York, NY: Guilford Press.

Dierdorff, E. C., Bell, S. T., & Belohlav, J. A. (2011). The power of "we": Effects of psychological collectivism on team performance over time. *Journal of Applied Psychology*, *96*(2), 247–262.

Dijksterhuis, A., Aarts, H., & Smith, P. K. (2005). The power of the subliminal: On subliminal persuasion and other potential applications. In R. R. Hassin, J. S. Uleman, & J. A. Bargh (Eds.), The new unconscious (pp. 77–106). *Oxford series in social cognition and social neuroscience*. New York, NY: Oxford University Press.

Dillard, D. A., Smith, J. J., Ferucci, E. D., & Lanier, A. P. (2012). Depression prevalence and associated factors among Alaska native people: The Alaska education and research toward health (EARTH) study. *Journal of Affective Disorders*, *136*(3), 1088–1097.

Dimberg, U., & Söderkvist, S. (2011). The voluntary facial action technique: A method to test the facial feedback hypothesis. *Journal of Nonverbal Behavior*, *35*(1), 17–33.

Dimberg, U., Thunberg, M., & Elmehed, K. (2000). Unconscious facial reactions to emotional facial expressions. *Psychological Science*, *11*(1), 86–89.

Dimitry, L. (2012). A systematic review on the mental health of children and adolescents in areas of armed conflict in the Middle East. *Child: Care, Health and Development*, *38*(2), 153–161.

Diniz, D. G., Foro, C. A. R., Rego, C. M. D., Gloria, D. A., de Oliveira, F. R. R., Paes, J. M. P., de Sousa, A., Tokuhashi, T., Trindade, L., Turiel, M., Vasconcelos, E., Torres, J., Cunnigham, C., Perry, H., da Costa Vasconcelos, P., & Diniz, C. W. P. (2010). Environmental impoverishment and aging alter object recognition, spatial learning, and dentate gyrus astrocytes. *European Journal of Neuroscience*, *32*(3), 509–519.

DiPietro, J. A. (2000). Baby and the brain: Advances in child development. *Annual Review of Public Health*, *21*, 455–71.

Distel, M. A., Middeldorp, C. M., Trul, T. J., Derom, C. A, Willemsen, G., & Boomsma, D. I. (2011). Life events and borderline personality features: The influence of gene-environment interaction and gene-environment correlation. *Psychological Medicine: A Journal of Research in Psychiatry and the Allied Sciences*, *41*(4), 849–860.

Doane, L. D., Kremen, W. S., Eaves, L. J., Eisen, S. A., Hauger, R., Hellhammer, D., Levine, S., Lupien, S., Lyons, M. J., Mendoza, S., Prom-Wormley, E., Xian, H., York, T.P., Franz, C., & Jacobson, K. C. (2010). Associations between jet lag and cortisol diurnal rhythms after domestic travel. *Health Psychology*, *29*(2), 117–123.

Doghramji, K. (2000, December). *Sleepless in America: Diagnosing and treating insomnia*. Retrieved from http://psychiatry.medscape.com/Medscape/psychiatry/ClinicalMgmt/CM.v02/public/indexCM.v02.html.

Dollard, J., Doob, L., Miller, N., Mowrer, O. H., & Sears, R. R. (1939). *Frustration and*

*aggression*. New Haven, CT: Yale University Press.
Domhoff, G. W. (2003). *The scientific study of dreams: Neural networks, cognitive development, and content analysis*. Washington, DC: American Psychological Association.
Domhoff, G. W. (2004). Why did empirical dream researchers reject Freud? A critique of historical claims by Mark Solms. *Dreaming, 14*(1), 3–17.
Domhoff, G. W. (2005). A reply to Hobson. (2005). *Dreaming, 15*(1), 30–32.
Domhoff, G. W. (2007). Realistic simulation and bizarreness in dream content: Past findings and suggestions for future research. In D. Barrett & P. McNamara (Eds.), *The new science of dreaming: Volume 2. Content, recall, and personality correlates* (pp. 1–27). Westport, CT: Praeger.
Domhoff, G. W. (2010). Dream content is continuous with waking thought, based on preoccupations, concerns, and interests. *Sleep Medicine Clinics, 5*(2), 203–215.
Domhoff, G. W., & Schneider, A. (2008). Similarities and differences in dream content at the cross-cultural, gender, and individual levels. *Consciousness and Cognition: An International Journal, 17*(4), 1257–1265.
Domjan, M. (2005). Pavlovian conditioning: A functional perspective. *Annual Review of Psychology, 56*, 179–206.
Donnerstein, E. (2011). The media and aggression: From TV to the internet. In J. P. Forgas, A. W. Kruglanski & K. D. Williams (Eds.), *The psychology of social conflict and aggression* (pp. 267–284). New York, NY: Psychology Press.
Donohue, R. (2006). Person-environment congruence in relation to career change and career persistence. *Journal of Vocational Behavior, 68*, 504–515.
Dorrian, J., Sweeney, M., & Dawson, D. (2011). Modeling fatigue-related truck accidents: Prior sleep duration, recency and continuity. *Sleep and Biological Rhythms, 9*(1), 3–11.
Dotsch, R., Wigboldus, D. H. J., & van Knippenberg, A. (2011). Biased allocation of faces to social categories. *Journal of Personality and Social Psychology, 100*(6), 999–1014.
Douglas, K., Chan, G., Gelernter, J., Arias, A., Anton, R., Poling, J., Farrer, L, & Kranzler, H. (2011). 5-HTTLPR as a potential moderator of the effects of adverse childhood experiences on risk of antisocial personality disorder. *Psychiatric Genetics, 21*(5), 240–248.
Douglass, A. B. & Pavletic, A. (2012). Eyewitness confidence malleability. In B. L. Cutler (Ed.), *Conviction of the innocent: Lessons from psychological research* (pp. 149–165). Washington DC: American Psychological Association.
Dovidio, J. F., Eller, A., & Hewstone, M. (2011). Improving intergroup relations through direct, extended and other forms of indirect contact. *Group Processes and Intergroup Relations, 14*(2), 147–160.
Dovidio, J. F., Pagotto, L., & Hebl, M. R. (2011). Implicit attitudes and discrimination against people with physical disabilities. In R. Wiener & S. L. Willborn (Eds.), *Disability and aging discrimination: Perspectives in law and psychology* (pp. 157–183). New York, NY: Springer Science 1 Business Media.
Doyle, A., & Pollack, M. H. (2004). Long-term management of panic disorder. *Journal of Clinical Psychiatry. 65*(Suppl5), 24–28.
Drew, L. J., Crabtree, G. W., Markx, S., Stark, K. L., Chaverneff, F., Xu, B., Mukai, J., Fenelon, K., Hsu, P., Gogos, J., & Karayiorgou, M. (2011). The 22q11.2 microdeletion: Fifteen years of insights into the genetic and neural complexity of psychiatric disorders. *International Journal of Developmental Neuroscience, 29*(3), 259–281.
Driscoll, A. K., Russell, S. T., & Crockett, L. J. (2008). Parenting styles and youth well-being across immigrant generations. *Journal of Family Issues, 29*, 185–209.
Drosopoulos, S., Harrer, D., & Born, J. (2011). Sleep and awareness about presence of regularity speed the transition from implicit to explicit knowledge. *Biological Psychology, 86*(3), 168–173.
Drucker, D. J. (2010). Male sexuality and Alfred Kinsey's 0–6 Scale: Toward "a sound understanding of the realities of sex." *Journal of Homosexuality, 57*(9), 1105–1123.
Dufresne, T. (2007). *Against Freud: Critics talk back*. Palo Alto, CA: Stanford University Press.
Dunham, Y. (2011). An angry 5 Outgroup effect. *Journal of Experimental Social Psychology, 47*(3), 668–671.
Duniec, E., & Raz, M. (2011). Vitamins for the soul: John Bowlby's thesis of maternal deprivation, biomedical metaphors and the deficiency model of disease. *History of Psychiatry, 22*(1), 93–107.
Durrant, R., & Ward, T. (2011). Evolutionary explanations in the social and behavioral sciences: Introduction and overview. *Aggression and Violent Behavior, 16*(5), 361–370.
Dutra, L., Callahan, K., Forman, E., Mendelsohn, M., & Herman, J. (2008). Core schemas and suicidality in a chronically traumatized population. *Journal of Nervous and Mental Disease, 196*, 71–74.
Duval, D. C. (2006). The relationship between African centeredness and psychological androgyny among African American women in middle adulthood. *Dissertation Abstracts International: Section B: The Sciences and Engineering. 67*(2-B), 1146.
Eap, S., Gobin, R. L., Ng, J., Hall, G. C., Nagayama, T. (2010). Sociocultural issues in the diagnosis and assessment of psychological disorders. In J. E. Maddux & J. P. Tangney (Eds), *Social psychological foundations of clinical psychology* (pp. 312–328). New York, NY: Guilford Press.
Easterbrooks, M. A., Chaudhuri, J. H., Bartlett, J. D., & Copeman, A. (2011). Resilience in parenting among young mothers: Family and ecological risks and opportunities. *Children and Youth Services Review, 33*(1), 42–50.
Eaton, N. R., Keyes, K. M., Krueger, R. F., Balsis, S., Skodol, A. E., Markon, K. E., . . . Hasin, D. S. (2012). An invariant dimensional liability model of gender differences in mental disorder prevalence: Evidence from a national sample. *Journal of Abnormal Psychology, 121*(1), 282–288.
Eaton, W., Chen, C., & Bromet, E. (2011). Epidemiology of schizophrenia. In M. Tsuang, M. Tohen, & P. Jones (Eds.), *Textbook in Psychiatric Epidemiology* (3rd ed.) (pp. 263–287). Hoboken, NJ: John Wiley & Sons.
Ecker, U. K. H., Lewandowsky, S., & Apai, J. (2011). Terrorists brought down the plane!—No, actually it was a technical fault: Processing corrections of emotive information. *The Quarterly Journal of Experimental Psychology, 64*(2), 283–310.
Eddy, K. T., Hennessey, M., & Thompson-Brenner, H. (2007). Eating pathology in East African women: The role of media exposure and globalization. *Journal of Nervous and Mental Disease, 195*, 196–202.
Ehrenreich, B. (2004, July 15). *All together now*. Retrieved from http://select.nytimes.com/gst/abstract.html?res5F00E16FA3C5E0C768DDDAE0894DC404482.
Eisenberger, R., & Armeli, S. (1997). Can salient reward increase creative performance without reducing intrinsic creative interest? *Journal of Personality and Social Psychology, 72*, 652–663.
Eisenberger, R., & Rhoades, L. (2002). Incremental effects of reward on creativity. *Journal of Personality and Social Psychology, 81*(4), 728–741.
Ekman, P. (1993). Facial expression and emotion. *American Psychologist, 48*, 384–392.
Ekman, P. (2004). *Emotions revealed: Recognizing faces and feelings to improve communication and emotional life*. Thousand Oaks, CA: Owl Books.
Ekman, P., & Keltner, D. (1997). Universal facial expressions of emotion: An old controversy and new findings. In U. C. Segerstrale & P. Molnar (Eds.), *Nonverbal communication: Where nature meets culture* (pp. 27–46). Mahwah, NJ: Erlbaum.

Elder, B., & Mosack, V. (2011). Genetics of depression: An overview of the current science. *Issues in Mental Health Nursing*, *32*(4), 192–202.

Elkind, D. (2000). A quixotic approach to issues in early childhood education. *Human Development, 43*(4–5), 279–283.

Elkind, D. (1981). *The hurried child*. Reading, MA: Addison-Wesley.

Elkind, D. (2007). *The hurried child: Growing up too fast too soon* (25th anniversary ed.). Cambridge, ME: Da Capo Press.

Ellis, A. (2008). Rational emotive behavior therapy. In K. Jordan (Ed.), *The quick theory reference guide: A resource for expert and novice mental health professionals*, (pp. 127–139). Hauppauge, NY: Nova Science Publishers.

Ellis, A., & Ellis, D. J. (2011a). *Rational emotive behavior therapy*. Washington, DC: American Psychological Association.

Ellis, A., & Ellis, D. J. (2011b). *Theories of psychotherapy. Rational emotive behavior therapy*. Washington, DC: American Psychological Association.

Ellis, L., Ficek, C., Burke, D., & Das, S. (2008). Eye color, hair color, blood type, and the rhesus factor: Exploring possible genetic links to sexual orientation. *Archives of Sexual Behavior*, *37*, 145–149.

Enoch, M.-A. (2011). The role of early life stress as a predictor for alcohol and drug dependence. *Psychopharmacology*, *214*(1), 17–31.

Enticott, P. G., Kennedy, H. A., Rinehart, N. J., Tonge, B. J., Bradshaw, J. L., Taffe, J. R., . . . Fitzgerald, P. B. (2012). Mirror neuron activity associated with social impairments but not age in autism spectrum disorder. *Biological Psychiatry*, *71*(5), 427–433.

Epley, N. (2008) Rebate psychology. *New York Times*. Retrieved from http://select.nytimes.com/mem/tnt.html?tntget2008/01/31/opinion/31epley.html.

Erdelyi, M. H. (2010). The ups and downs of memory. *American Psychologist*, *65*(7), 623–633.

Erdelyi, M. H., & Applebaum, A. G. (1973). Cognitive masking: The disruptive effect of an emotional stimulus upon the perception of contiguous neutral items. *Bulletin of the Psychonomic Society*, *1*, 59–61.

Erikson, E. (1950). *Childhood and society*. New York, NY: W. W. Norton & Co.

Erlacher, D., & Schredl, M. (2004). Dreams reflecting waking sport activities: A comparison of sport and psychology students. *International Journal of Sport Psychology*, *35*(4), 301–308.

Ertekin-Taner, N. (2007). Genetics of Alzheimer's disease: A centennial review. *Neurologic Clinics*, *25*, 611–667.

Eslick, A. N., Fazio, L. K., & Marsh, E. J. (2011). Ironic effects of drawing attention to story errors. *Memory*, *19*(2), 184–191.

Espelage, D. L., Aragon, S. R., Birkett, M., & Koenig, B. W. (2008). Homophobic teasing, psychological outcomes, and sexual orientation among high school students: What influence do parents and schools have? *School Psychology Review*, *37*, 202–216.

Espy, K. A., Fang, H., Johnson, C., Stopp, C., Wiebe, S. A., & Respass, J. (2011). Prenatal tobacco exposure: Developmental outcomes in the neonatal period. *Developmental Psychology*, *47*(1), 153–169.

Ethical Principles of Psychologists and Code of Conduct. (2002). Retrieved from http://www.apa.org/ethics/code2002.html.

Ethical Principles of Psychologists and Code of Conduct. (2010). *American Psychologist*, *65*(5), 493.

Evans, S., Ferrando, S., Findler, M., Stowell, C., Smart, C., & Haglin, D. (2008). Mindfulness-based cognitive therapy for generalized anxiety disorder. *Journal of Anxiety Disorders*, *22*, 716–721.

Eysenck, H. J. (1967). *The biological basis of personality*. Springfield, IL: Charles C Thomas.

Eysenck, H. J. (1982). *Personality, genetics, and behavior: Selected papers*. New York, NY: Prager.

Eysenck, H. J. (1990). Biological dimensions of personality. In L. A. Pervin (Ed.), *Handbook of personality: Theory and research* (pp. 244–276). New York, NY: Guilford Press.

Faddiman, A. (1997). *The spirit catches you and you fall down*. New York, NY: Straus & Giroux.

Fairburn, C. G., Cooper, Z., Shafran, R., & Wilson, G. T. (2008). Eating disorders: A transdiagnostic protocol. In D. H. Barlow (Ed.), *Clinical handbook of psychological disorders: A step-by-step treatment manual* (4th ed.) (pp. 578–614). New York, NY: Guilford Press.

Famous People with Mental Illness (2011). *NAMI*. Retrieved from http://www.nami.org/Content/ContentGroups/Policy/Updates/Updates_andamp__What_s_New___Healthcare_Reform__EEOC__Famous_People.htm

Famous People with Schizophrenia (2011). *HubPages*. Retrieved from http://hubpages.com/hub/famous-people-with-schizophrenia

Fang, X., & Corso, P. S. (2007). Child maltreatment, youth violence, and intimate partner violence: Developmental relationships. *American Journal of Preventive Medicine*, *33*, 281–290.

Fantz, R. L. (1956). A method for studying early visual development. *Perceptual and Motor Skills*, *6*, 13–15.

Fantz, R. L. (1963). Pattern vision in newborn infants. *Science*, *140*, 296–297.

Faraone, S. V. (2008). Statistical and molecular genetic approaches to developmental psychopathology: The pathway forward. In J. J. Hudziak (Ed.), *Developmental psychopathology and wellness: Genetic and environmental influences* (pp. 245–265). Arlington, VA: American Psychiatric Publishing.

Fein, S., & Spencer, S. J. (1997). Prejudice as selfimage maintenance: Affirming the self through derogating others. *Journal of Personality and Social Psychology*, *73*(1), 31–44.

Feldman, D. C., & Beehr, T. A. (2011). A three-phase model of retirement decision making. *American Psychologist*, *66*(3), 193–203.

Feliciano, L., Segal, D. L., & Vair, C. L. (2011). Major depressive disorder. In K. H. Sorocco & S. Lauderdale (Eds.), *Cognitive behavior therapy with older adults: Innovations across care settings* (pp. 31–64). New York, NY: Springer.

Fennis, B. M., & Stroebe, W. (2010). *The psychology of advertising*. New York, NY: Psychology Press.

Ferguson, C. J. (2010). Violent crime research: An introduction. In C. J. Ferguson (Ed.), *Violent crime: Clinical and social implications* (pp. 3–18). Thousand Oaks, CA: Sage.

Ferguson, C. J. (2011). Video games and youth violence: A prospective analysis in adolescents. *Journal of Youth and Adolescence*, *40*(4), 377–391.

Ferguson, C. J., Colwell, J., Mlačić, B., Milas, G., & Miklošić, I. (2011). Personality and media influences on violence and depression in a cross-national sample of young adults: Data from Mexican–Americans, English and Croatians. *Computers in Human Behavior*, *27*(3), 1195–1200.

Ferguson, C. J., San Miguel, C., Garza, A., & Jerabeck, J. M. (2012). A longitudinal test of video game violence influences on dating and aggression: A 3-year longitudinal study of adolescents. *Journal of Psychiatric Research*, *46*(2), 141–146.

Ferguson, C. J., Winegard, B., & Winegard, B. M. (2011). Who is the fairest one of all? How evolution guides peer and media influences on female body dissatisfaction. *Review of General Psychology*, *15*(1), 11–28.

Fernández-Ballesteros, R., Zamarrón, M. D., Díez-Nicolás, J., López-Bravo, M. D., Molina, M. Á., & Schettini, R. (2011). Productivity in old age. *Research on Aging*, *33*(2), 205–226.

Ferrari, M., Robinson, D. K., & Yasnitsky, A. (2010). Wundt, Vygotsky and Bandura: A cultural-historical science of consciousness in three acts. *History of the Human Sciences*, *23*(3), 95–118.

Ferrari, P. F., Rozzi, S., & Fogassi, L. (2005). Mirror neurons responding to observation of actions made with tools in monkey ventral premotor cortex. *Journal of Cognitive Neuroscience*, *17*, 212–226.

Ferreira, G. K., Rezin, G. T., Cardoso, M. R., Gonçalves, C. L., Borges, L. S., Vieira, J. S., . . . Streck, E. L. (2012). Brain energy metabolism is increased by chronic administration of bupropion. *Acta Neuropsychiatrica*, *24*(2), 115–121.

Ferretti, P., Copp, A., Tickle, C., & Moore, G. (Eds.) (2006). *Embryos, genes, and birth defects* (2nd ed.). Hoboken, NJ: John Wiley & Sons.

Ferrucci, L. M., Cartmel, B., Turkman, Y. E., Murphy, M. E., Smith, T., Stein, K. D., & McCorkle, R. (2011). Causal attribution among cancer survivors of the 10 most common cancers. *Journal of Psychosocial Oncology*, *29*(2), 121–140.

Festinger, L. A. (1957). *A theory of cognitive dissonance*. Palo Alto, CA: Stanford University Press.

Festinger, L. A., & Carlsmith, L. M. (1959). Cognitive consequences of forced compliance. *Journal of Abnormal and Social Psychology*, *58*, 203–210.

Field, A. P. (2006). Is conditioning a useful framework for understanding the development and treatment of phobias? *Clinical Psychology Review*, *26*, 857–875.

Field, K. M., Woodson, R., Greenberg, R., & Cohen, D. (1982). Discrimination and imitation of facial expressions by neonates. *Science*, *218*, 179–181.

Fields, A., & Cochran, S. (2011). Men and depression: Current perspectives for health care professionals. *American Journal of Lifestyle Medicine*, *5*(1), 92–100.

Fields, D. (2011). Preliminary human experiments to test safety of nerve cell transplants for spinal cord paralysis. Retrieved from http://www.scientificamerican.com/article.cfm?idpreliminary-human-experiments-test-safety-paralysis-treatment-nerve-cells.

Fife, R., & Schrager, S. (2011). *Family violence: What health care providers need to know*. Sudbury, MA: Jones & Bartlett Learning.

Figueredo, A., Jacobs, W., Burger, S., Gladden, P., & Olderbak, S. (2011). The biology of personality. In G. Terzis & R. Arp (Eds.), *Information and living systems: Philosophical and scientific perspectives* (pp. 371–406). Boston, MA: MIT Press.

Fink, G. (2011). Stress controversies: Post-traumatic stress disorder, hippocampal volume, gastroduodenal ulceration. *Journal of Neuroendocrinology*, *23*(2), 107–117.

Fink, M. (1999). *Electroshock: Restoring the mind*. London: Oxford University Press.

Fischer, P., Krueger, J. I., Greitemeyer, T., Vogrincic, C., Kastenmüller, A., Frey, D., Heene, M., Wicher, M., & Kainbacher, M. (2011). The bystander-effect: A meta-analytic review on bystander intervention in dangerous and non-dangerous emergencies. *Psychological Bulletin*, *137*(4), 517–537.

Fisher, G. L. (2011). Ponytails & death-ray looks: A review of research on relational aggression among adolescent girls. *Dissertation Abstracts International: Section B: The Sciences and Engineering*, *71*(8-B), 5182.

Fisher, J. O., & Kral, T. V. E. (2008). Super-size me: Portion size effects on young children's eating. *Physiology and Behavior*, *94*, 39–47.

Fisk, J. E., Bury, A. S., & Holden, R. (2006). Reasoning about complex probabilistic concepts in childhood. *Scandinavian Journal of Psychology*, *47*, 497–504.

Fiske, S. T. (1998). Stereotyping, prejudice, and discrimination. In D. T. Gilbert, S. T. Fiske, & G. Lindzey (Eds.), *The handbook of social psychology* (4th ed., Vol. 2, pp. 357–411). New York, NY: Oxford University Press.

Flaherty, S. C., & Sadler, L. S. (2011). A review of attachment theory in the context of adolescent parenting. *Journal of Pediatric Health Care*, *25*(2), 114–121.

Flaskerud, J. H. (2010). DSM proposed changes, part I: Criticisms and influences on changes. *Issues in Mental Health Nursing*, *31*(10), 686–688.

Flavell, J. H., Miller, P. H., & Miller, S. A. (2002). *Cognitive development* (4th ed.). Upper Saddle River, NJ: Prentice-Hall.

Fleischhaker, C., Böhme, R., Sixt, B., Brück, C., Schneider, C., & Schulz, E. (2011). Dialectical Behavioral Therapy for Adolescents (DBT-A): A clinical trial for patients with suicidal and self-injurious behavior and borderline symptoms with a one-year follow-up. *Child and Adolescent Psychiatry and Mental Health*, *5*, Article ID 3.

Fleischmann, B. K., & Welz, A. (2008). Cardiovascular regeneration and stem cell therapy. *Journal of the American Medical Association*, *299*(6), 700–701.

Fletcher, G. J. O., & Simpson, J. A. (2000). Ideal standards in close relationships: Their structure and functions. *Current Directions in Psychological Science*, *9*, 102–105.

Flynn, J. R. (1987). Massive IQ gains in 14 nations: What IQ tests really measure. *Psychological Bulletin*, *101*, 171–191.

Flynn, J. R. (2007). *What is intelligence?: Beyond the Flynn Effect*. New York, NY: Cambridge University Press.

Flynn, J. R. (2010). Problems with IQ gains: The huge vocabulary gap. *Journal of Psychoeducational Assessment*, *28*(5), 412–433.

Foerde, K., & Shohamy, D. (2011). The role of the basal ganglia in learning and memory: Insight from Parkinson's disease. *Neurobiology of Learning and Memory*, *96*(4), 624–636.

Fogarty, A., Rawstorne, P., Prestage, G., Crawford, J., Grierson, J., & Kippax, S. (2007). Marijuana as therapy for people living with HIV/AIDS: Social and health aspects. *AIDS Care*, *19*, 295–301.

Fogel, S. M., & Smith, C. T. (2011). The function of the sleep spindle: A physiological index of intelligence and a mechanism for sleep-dependent memory consolidation. *Neuroscience and Biobehavioral Reviews*, *35*(5), 1154–1165.

Fok, H. K., Hui, C. M., Bond, M. H., Matsumoto, D., & Yoo, S. H. (2008). Integrating personality, context, relationship, and emotion type into a model of display rules. *Journal of Research in Personality*, *42*, 133–150.

Font, E., & Carazo, P. (2010). Animals in translation: Why there is meaning (but probably no message) in animal communication. *Animal Behaviour*, *80*(2), e1–e6.

Forshaw, M. (2013). *Critical thinking for psychology: A student guide*. Malden, MA: Wiley-Blackwell.

Fortune, E. E., & Goodie, A. S. (2012). Cognitive distortions as a component and treatment focus of pathological gambling: A review. *Psychology of Addictive Behaviors*, *26*(2), 298–310.

Foulkes, D. (1993). Children's dreaming. In D. Foulkes & C. Cavallero (Eds.), *Dreaming as cognition* (pp. 114–132). New York, NY: Harvester Wheatsheaf.

Foulkes, D. (1999). *Children's dreaming and the development of consciousness*. Cambridge, MA: Harvard University Press.

Fraley, R. C., & Shaver, P. R. (1997). Adult attachment and the suppression of unwanted thoughts. *Journal of Personality and Social Psychology*, *73*, 1080–1091.

Freckelton, I. (2011). Review of The architecture of madness: Insane asylums in the United States. (Review of the book The architecture of madness: Insane asylums in the United States). *Psychiatry, Psychology and Law*, *18*(1), 160–162.

Freeman, J. B., & Ambady, N. (2011). A dynamic interactive theory of person construal. *Psychological Review*, *118*(2), 247–279.

Frenda, S. J., Nichols, R. M., & Loftus, E. F. (2011). Current issues and advances in misinformation research. *Current Directions in Psychological Science*, *20*(1), 20–23.

Freund, A. M., & Ritter, J. O. (2009). Midlife crisis: A debate. *Gerontology*, *55*(5), 582–591.

Frew, J., & Spiegler, M. (Eds.). (2012). *Contemporary psychotherapies for a diverse world*. New York, NY: Routledge.

Frick, W. B. (2000). Remembering Maslow: Reflections on a 1968 interview. *Journal of Humanistic Psychology*, *40*(2), 128–147.

Friedberg, R. D., & Brelsford, G. M. (2011). Core principles in cognitive therapy with youth. *Child and Adolescent Psychiatric Clinics of North America, 20*(2), 369–378.

Friedlander, M. L., Escudero, V., Heatherington, L., & Diamond, G. M. (2011). Alliance in couple and family therapy. *Psychotherapy, 48*(1), 25–33.

Fu, M., & Zuo, Y. (2011). Experience-dependent structural plasticity in the cortex. *Trends in Neurosciences, 34*(4), 177–187.

Fujimura, J. H., Rajagopalan, R., Ossorio, P. N., & Doksum, K. A. (2010). Race and ancestry: Operationalizing populations in human genetic variation studies. In I. Whitmarsh & D. S. Jones (Eds.), *What's the use of race? Modern governance and the biology of difference* (pp. 169–183). Cambridge, MA: MIT Press.

Fukase, Y., & Okamoto, Y. (2010). Psychosocial tasks of the elderly: A reconsideration of the eighth stage of Erikson's epigenetic scheme. *Japanese Journal of Developmental Psychology, 21*(3), 266–277.

Fulcher, M. (2011). Individual differences in children's occupational aspirations as a function of parental traditionality. *Sex Roles, 64*(1–2), 117–131.

Fumagalli, M., Ferrucci, R., Mameli, F., Marceglia, S., Mrakic-Sposta, S., Zago, S., Lucchiari, C., Consonni, D., Nordio, F., Pravettoni, G., Cappa, S., & Priori, A. (2010). Gender-related differences in moral judgments. *Cognitive Processing, 11*(3), 219–226.

Funder, D. C. (2000). Personality. *Annual Review of Psychology, 52*, 197–221.

Furguson, E., Chamorro-Premuzic, T., Pickering, A., & Weiss, A. (2011). Five into one does go: A critique of the general factor of personality. In T. Chamorro-Premuzic, S. von Stumm, & A. Furnam (Eds.), *Wiley-Blackwell handbook of individual differences* (pp. 162–186). Chichester, UK: Wiley-Blackwell.

Furman, E. (1990, November). Plant a potato, learn about life (and death). *Young Children, 46*(1), 15–20.

Furukawa, T., Nakano, H., Hirayama, K., Tanahashi, T., Yoshihara, K., Sudo, N., Kubo, C., & Nishima, S. (2010). Relationship between snoring sound intensity and daytime blood pressure. *Sleep and Biological Rhythms, 8*(4), 245–253.

Fusar-Poli, P., Borgwardt, S., Crescini, A., Deste, G., Kempton, M. J., Lawrie, S., Mc Guire, P., & Sacchetti, E. (2011). Neuroanatomy of vulnerability to psychosis: A voxel-based meta-analysis. *Neuroscience and Biobehavioral Reviews, 35*(5), 1175–1185.

Gabbard, G. O. (2006). Mente, cervello e disturbi di personalita. / Mind, brain, and personality disorders. *Psicoterapia e scienze umane, 40*, 9–26.

Gabry, K. E., Chrousos, G. P., Rice, K. C., Mostafa, R. M., Sternberg, E., Negrao, A. B., Webster, E. L., McCann, S. M., & Gold, P. W. (2002). Marked suppression of gastric ulcerogenesis and intestinal responses to stress by a novel class of drugs. *Molecular Psychiatry, 7*(5), 474–483.

Gacono, C. B., Evans, F. B., & Viglione, D. J. (2008). Essential issues in the forensic use of the Rorschach. In C. B. Gacono (Ed.), F. B. Evans (Ed.), N. Kaser-Boyd (Col.), & L. A. Gacono (Col.), *The handbook of forensic Rorschach assessment (pp. 3–20). The LEA series in personality and clinical psychology.* New York, NY: Routledge/Taylor & Francis Group.

Gaetz, M., Weinberg, H., Rzempoluck, E., & Jantzen, K. J. (1998). Neural network classifications and correlational analysis of EEG and MEG activity accompanying spontaneous reversals of the Necker Cube. *Cognitive Brain Research, 6*, 335–346.

Galderisi, S., Quarantelli, M., Volpe, U., Mucci, A., Cassano, G. B., Invernizzi, G., Rossi, A., Vita, A., Pini, S., Cassano, P., Daneluzzo, E., De Peri, L., Stratta, P., Brunetti, A., & Maj, M. (2008). Patterns of structural MRI abnormalities in deficit and nondeficit schizophrenia. *Schizophrenia Bulletin, 34*, 393–401.

Gansler, D. A., Lee, A. K. W., Emerton, B. C., D'Amato, C., Bhadelia, R., Jerram, M., & Fulwiler, C. (2011). Prefrontal regional correlates of self-control in male psychiatric patients: Impulsivity facets and aggression. *Psychiatry Research: Neuroimaging, 191*(1), 16–23.

Garb, H. N., Wood, J. M., Lilienfeld, S. O., & Nezworski, M. T. (2005). Roots of the Rorschach controversy. *Clinical Psychology Review, 25*, 97–118.

García, E. E., & Náñez, J. E., Sr. (2011). Education circumstances. In E. E. García & J. E. Náñez, Sr. (Eds), *Bilingualism and cognition: Informing research, pedagogy, and policy* (pp. 103–114). Washington, DC: American Psychological Association.

Garcia, G. M., & Stafford, M. E. (2000). Prediction of reading by Ga and Gc specific cognitive abilities for low-SES White and Hispanic English-speaking children. *Psychology in the Schools, 37*(3), 227–235.

Garcia, J. (2003). Psychology is not an enclave. In R. Sternberg (Ed.), *Psychologists defying the crowd: Stories of those who battled the establishment and won* (pp. 67–77). Washington, DC: American Psychological Association.

Garcia, J., & Koelling, R. S. (1966). Relation of cue to consequence in avoidance learning. *Psychonomic Science, 4*, 123–124.

Gardner, A., & Gardner, B. T. (1969). Teaching sign language to a chimpanzee. *Science, New Series, 165*, 664–672.

Gardner, H. (1983). *Frames of mind.* New York, NY: Basic Books.

Gardner, H. (1999, February). Who owns intelligence? *Atlantic Monthly*, pp. 67–76.

Gardner, H. (2002). The pursuit of excellence through education. In M. Ferrari (Ed.), *Learning from extraordinary minds* (pp. 3–35). Mahwah, NJ: Erlbaum.

Gardner, H. (2008). Who owns intelligence? In M. H. Immordino-Yang (Ed.), *Jossey-Bass Education Team. The Jossey-Bass reader on the brain and learning* (pp. 120–132). San Francisco, CA: Jossey-Bass.

Gardner, H. (2011). *Frames of mind.* New York, NY: Basic Books.

Garrick, J. (2006). The humor of trauma survivors: Its application in a therapeutic milieu. *Journal of Aggression, Maltreatment and Trauma, 12*, 169–182.

Gasser, L., & Malti, T. (2011). Relationale und physische aggression in der mittleren Kindheit: Zusammenhänge mit moralischem Wissen und moralischen Gefühlen [Relational and physical aggression in middle childhood: Relations with moral knowledge and moral emotions]. *Zeitschrift für Entwicklungspsychologie und Pädagogische Psychologie, 43*(1), 29–38.

Gazzaniga, M. S. (2009). The fictional self. Personal identity and fractured selves: Perspectives from philosophy, ethics, and neuroscience. In D. J. H. Mathews, H. Bok, & P. V. Rabins (Eds.), *Personal identity and fractured selves: Perspectives from philosophy, ethics, and neuroscience* (pp. 174–185). Baltimore, MD: Johns Hopkins University Press.

Geangu, E., Benga, O., Stahl, D., & Striano, T. (2010). Contagious crying beyond the first days of life. *Infant Behavior and Development, 33*(3), 279–288.

Gebauer, H., Krempl, R., & Fleisch, E. (2008). Exploring the effect of cognitive biases on customer support services. *Creativity and Innovation Management, 17*, 58–70.

Gelder, B. D., Meeren, H. K., Righart, R. Stock, J. V., van de Riet, W. A., & Tamietto, M. (2006). Beyond the face: Exploring rapid influences of context on face processing. *Progress in Brain Research, 155*, 37–48.

Gellerman, D. M., & Lu, F. G. (2011). Religious and spiritual issues in the outline for cultural formulation. In J. R. Peteet, F. G. Lu, & W. E Narrow (Eds.), *Religious and spiritual issues in psychiatric diagnosis: A research agenda for DSM-V* (pp. 207–220). Washington, DC: American Psychiatric Association.

Geniole, S. N., Carré, J. M., & McCormick, C. M. (2011). State, not trait, neuroendocrine function predicts costly reactive aggression in men after social exclusion and inclusion. *Biological Psychology, 87*(1), 137–145.

George, J. B. E., & Franko, D. L. (2010). Cultural issues in eating pathology and body image among children and adolescents. *Journal of Pediatric Psychology, 35*(3), 231–242.

Gerber, A. J., Posner, J., Gorman, D., Colibazzi, T., Yu, S., Wang, Z., Kangarlu, A., Zhu, H., Russell, J., & Peterson, B. S. (2008). An affective circumplex model of neural systems subserving valence, arousal, and cognitive overlay during the appraisal of emotional faces. *Neuropsychologia, 46*, 2129–2139.

Gergen, K. J. (2010). The acculturated brain. *Theory and Psychology, 20*(6), 795–816.

Gershon, E. S., & Rieder, R. O. (1993). Major disorders of mind and brain. *Mind and brain: Readings from Scientific American Magazine* (pp. 91–100). New York, NY: Freeman.

Giacobbi Jr., P., Foore, B., & Weinberg, R. S. (2004). Broken clubs and expletives: The sources of stress and coping responses of skilled and moderately skilled golfers. *Journal of Applied Sport Psychology, 16*(2), 166–182.

Giancola, P. R., & Parrott, D. J. (2008). Further evidence for the validity of the Taylor aggression paradigm. *Aggressive Behavior, 34*, 214–229.

Gibbs, N. (1995, October 2). The EQ factor. *Time*, pp. 60–68.

Gibson, E. J., & Walk, R. D. (1960). The visual cliff. *Scientific American, 202*(2), 67–71.

Gigerenzer, G. (1996). On narrow norms and vague heuristics: A reply to Kahneman and Tversky (1996). *Psychological Review, 103*(3), 592.

Gigerenzer, G., & Goldstein, D. G. (2011). The recognition heuristic: A decade of research. *Judgment and Decision Making, 6*(1), 100–121.

Gilligan, C. (1977). In a different voice: Women's conception of morality. *Harvard Educational Review, 47*(4), 481–517.

Gilligan, C. (1990). Teaching Shakespeare's sister. In C. Gilligan, N. Lyons, & T. Hanmer (Eds.), *Mapping the moral domain* (pp. 73–86). Cambridge, MA: Harvard University Press.

Gilligan, C. (1993). Adolescent development reconsidered. In A. Garrod (Ed.), *Approaches to moral development: New research and emerging themes*. New York, NY: Teachers College Press.

Gilmour, D. R., & Walkey, F. H. (1981). Identifying violent offenders using a video measure of interpersonal distance. *Journal of Consulting and Clinical Psychology, 49*, 287–291.

Gimbel, S. I., & Brewer, J. B. (2011). Reaction time, memory strength, and fMRI activity during memory retrieval: Hippocampus and default network are differentially responsive during recollection and familiarity judgments. *Cognitive Neuroscience, 2*(1), 19–26.

Gini, G., Pozzoli, T., & Hauser, M. (2011). Bullies have enhanced moral competence to judge relative to victims, but lack moral compassion. *Personality and Individual Differences, 50*(5), 603–608.

Gini, M., Oppenheim, D., & Sagi-Schwartz, A. (2007). Negotiation styles in mother-child narrative co-construction in middle childhood: Associations with early attachment. *International Journal of Behavioral Development, 31*, 149–160.

Girardi, P., Monaco, E., Prestigiacomo, C., Talamo, A., Ruberto, A., & Tatarelli, R. (2007). Personality and psychopathological profiles in individuals exposed to mobbing. *Violence and Victims, 22*, 172–188.

Giumetti, G. W., & Markey, P. M. (2007). Violent video games and anger as predictors of aggression. *Journal of Research in Personality, 41*, 1234–1243.

Glenn, A. L., Kurzban, R., & Raine, A. (2011). Evolutionary theory and psychopathy. *Aggression and Violent Behavior. Attention, Perception, and Psychophysics, 73*(3), 708–713.

Glicksohn, A., & Cohen, A. (2011). The role of Gestalt grouping principles in visual statistical learning. *Attention, Perception, and Psychophysics, 73*(3), 708–713.

Glover, V. (2011). Annual research review: Prenatal stress and the origins of psychopathology: An evolutionary perspective. *Journal of Child Psychology and Psychiatry, 52*(4), 356–367.

Golden, W. L. (2006). Hypnotherapy for anxiety, phobias and psychophysiological disorders. In R. A. Chapman (Ed.), *The clinical use of hypnosis in cognitive behavior therapy: A practitioner's casebook* (pp. 101–137). New York, NY: Springer.

Goldstein, I. (2010). Looking at sexual behavior 60 years after Kinsey. *Journal of Sexual Medicine, 7(Suppl 5)*, 246–247.

Goldstein, M., (2011). Adolescent Pregnancy. In M. Golstein (Ed.), *The MassGeneral Hospital for Children Adolescent Handbook* (pp. 111–133). New York, NY: Springer.

Goleman, D. (1980, February). 1,528 little geniuses and how they grew. *Psychology Today*, 28–53.

Goleman, D. (1995). *Emotional intelligence: Why it can matter more than IQ*. New York, NY: Bantam.

Goleman, D. (2000). *Working with emotional intelligence*. New York, NY: Bantam Doubleday.

Goleman, D. (2008). Leading resonant teams. In F. Hesselbein & A. Shrader (Eds.), *Leader to leader 2: Enduring insights on leadership from the Leader to Leader Institute's award-winning journal* (pp. 186–195). San Francisco, CA: Jossey-Bass.

Gómez, Á., & Huici, C. (2008). Vicarious intergroup contact and role of authorities in prejudice reduction. *The Spanish Journal of Psychology, 11*, 103–114.

Gómez-Laplaza, L. M., & Gerlai, R. (2010). Latent learning in zebrafish (Danio rerio). *Behavioural Brain Research, 208*(2), 509–515.

Gonzaga, G. C., Carter, S., & Buckwalter, J. (2010). Assortative mating, convergence, and satisfaction in married couples. *Personal Relationships, 17*(4), 634–644.

Goodwin, C. J. (2011). *Research in psychology: Methods and design* (6th ed.). Hoboken, NJ: John Wiley & Sons.

Goodwin, C. J. (2012). *A history of modern psychology* (4th ed.). Hoboken, NJ: John Wiley & Sons.

Gosling, S. D. (2008). Personality in non-human animals. *Social and Personality Compass, 2*, 985–1001.

Gottesman, I. I. (1991). *Schizophrenia genesis: The origins of madness*. New York, NY: Freeman.

Gottfredson, G. D., & Duffy, R. D. (2008). Using a theory of vocational personalities and work environments to explore subjective well-being. *Journal of Career Assessment, 16*, 44–59.

Gottman, J. M. (2011). *The science of trust: Emotional attunement for couples*. New York, NY: W. W. Norton & Co.

Gottman, J. M., & Levenson, R. W. (2002). A two-factor model for predicting when a couple will divorce: Exploratory analyses using 14-year longitudinal data. *Family Process, 41*(1), 83–96.

Gottman, J. M., & Notarius, C. I. (2000). Decade review: Observing marital interaction. *Journal of Marriage and the Family, 62*(4), 927–947.

Gould, R. L. (1975, August). Adult life stages: Growth toward self-tolerance. *Psychology Today*, pp. 74–78.

Gracely, R. H., Farrell, M. J., & Grant, M. A. (2002). Temperature and pain perception. In H. Pashler & S. Yantis (Eds.), *Steven's handbook of experimental psychology: Vol. 1. Sensation and perception* (3rd ed.) (pp. 619–651). Hoboken, NJ: John Wiley & Sons.

Graham, J. M. (2011). Measuring love in romantic relationships: A meta-analysis. *Journal of Social and Personal Relationships, 28*(6), 748–771.

Graham, J. R. (1991). Comments on Duckworth's review of the Minnesota Multiphasic Personality Inventory-2.

*Journal of Counseling and Development, 69*, 570–571.
Grant, J. A., Courtemanche, J., & Rainville, P. (2011). A non-elaborative mental stance and decoupling of executive and pain-related cortices predicts low pain sensitivity in Zen meditators. *Pain, 152*(1), 150–156.
Gratz, K. L., Latzman, R. D., Tull, M. T., Reynolds, E. K., & Lejuez, C.W. (2011). Exploring the association between emotional abuse and childhood borderline personality features: The moderating role of personality traits. *Behavior Therapy, 42*(3), 493–508.
Green, A. J., & De-Vries, K. (2010). Cannabis use in palliative care—An examination of the evidence and the implications for nurses. *Journal of Clinical Nursing, 19*(17–18), 2454–2462.
Green, B., Stedman, T., Chapple, B., & Griffin, C. (2011). Criminal justice outcomes of those appearing before the mental health tribunal: A record linkage study. *Psychiatry, Psychology and Law, 18*(4), 573–587.
Green, C. D. (2009). Darwinian theory, functionalism, and the first American psychological revolution. *American Psychologist, 64*(2), 75–83.
Green, J. P., & Lynn, S. J. (2011). Hypnotic responsiveness: Expectancy, attitudes, fantasy proneness, absorption, and gender. *International Journal of Clinical and Experimental Hypnosis, 59*(1), 103–121.
Greenberg, D. L. (2004). President Bush's false "flashbulb" memory of 9/11/01. *Applied Cognitive Psychology, 18*(3), 363–370.
Greenberg, J. (2002). Who stole the money, and when? Individual and situational determinants of employee theft. *Organizational Behavior and Human Decision Processes, 89*(1), 985–1003.
Gregory, R. J. (2011). *Psychological testing: History, principles, and applications* (6th ed.). Boston, MA: Allyn & Bacon.
Grondin, S., Ouellet, B., & Roussel, M. (2004). Benefits and limits of explicit counting for discriminating temporal intervals. *Canadian Journal of Experimental Psychology, 58*(1), 1–12.
Grubin, D. (2010). The polygraph and forensic psychiatry. *Journal of the American Academy of Psychiatry and the Law, 38*(4), 446–451.
Guidelines for Ethical Conduct. (2008). *Introduction: Guidelines for ethical conduct in the care and use of nonhuman animals in research*. Retrieved from http://www.apa.org/science/anguide.html.
Guidelines for the Treatment of Animals in Behavioural Research and Teaching. (2005). *Animal Behaviour, 69*(1), i–vi.
Guilford, J. P. (1967). *The nature of human intelligence*. New York, NY: McGraw-Hill.
Gunderman, R., & Kamer, A. (2011). Rewards. *Journal of the American College of Radiology, 8*(5), 341–344.
Gunderson, J. G. (2011). Borderline personality disorder. *The New England Journal of Medicine, 364*(21), 2037–2042.
Gunnar, M. R. (2010). A commentary on deprivation-specific psychological patterns: Effects of institutional deprivation. *Monographs of the Society for Research in Child Development, 75*(1), 232–247.
Gustavson, C. R., & Garcia, J. (1974, August). Pulling a gag on the wily coyote. *Psychology Today*, 68–72.
Gutiérrez-Rojas, L., Jurado, D., & Gurpegui, M. (2011). Factors associated with work, social life and family life disability in bipolar disorder patients. *Psychiatry Research, 186*(2–3), 254–260.

Haab, L., Trenado, C., Mariam, M., & Strauss, D. J. (2011). Neurofunctional model of large-scale correlates of selective attention governed by stimulus-novelty. *Cognitive Neurodynamics, 5*(1), 103–111.
Haaken, J. (2010). Transformative remembering: Feminism, psychoanalysis, and recollection of abuse. In J. Haaken & P. Reavey (Eds.), *Memory matters: Contexts for understanding sexual abuse recollections* (pp. 216–229). New York, NY: Routledge/Taylor & Francis Group.
Haberstick, B. C., Schmitz, S. Young, S. E., & Hewitt, J. K. (2006). Genes and developmental stability of aggressive behavior problems at home and school in a community sample of twins aged 7–12. *Behavior Genetics, 36*, 809–819.
Hadley, D., Anderson, B. S., Borckardt, J. J., Arana, A., Li, X., Nahas, Z., & George, M. S. (2011). Safety, tolerability, and effectiveness of high doses of adjunctive daily left prefrontal repetitive transcranial magnetic stimulation for treatment-resistant depression in a clinical setting. *The Journal of ECT, 27*(1), 18–25.
Haines, R., & Mann, J. (2011). A new perspective on de-individuation via computer-mediated communication. *European Journal of Information Systems, 20*, 156–167.
Haith, M. M., & Benson, J. B. (1998). Infant cognition. In W. Damon & R. M. Lerner (Eds.), *Handbook of child psychology (Vol. 1)*. New York: Wiley.
Hakvoort, E. M., Bos, H. M. W., Van Balen, F., & Hermanns, J. M. A. (2011). Postdivorce relationships in families and children's psychosocial adjustment. *Journal of Divorce and Remarriage, 52*(2), 125–146.
Hall, C. S., & Van de Castle, R. L. (1996). *The content analysis of dreams*. New York, NY: Appleton-Century-Crofts.
Hall, P. A., & Schaeff, C. M. (2008). Sexual orientation and fluctuating asymmetry in men and women. *Archives of Sexual Behavior, 37*, 158–165.
Hall, S., & Adams, R. (2011). Newlyweds' unexpected adjustments to marriage. *Family and Consumer Sciences Research Journal, 39*(4), 375–387.
Hall, W., & Degenhardt, L. (2010). "Adverse health effects of non-medical cannabis use"— Authors' reply. *The Lancet, 375*(9710), 197.
Halpern, D. F. (1997). Sex differences in intelligence: Implications for education. *American Psychologist, 52*, 1091–1102.
Halpern, D. F. (2000). *Sex differences in cognitive abilities*. Hillsdale, NJ: Erlbaum.
Hamilton, J. P., & Gotlib, I. H. (2008). Neural substrates of increased memory sensitivity for negative stimui in major depression. *Biological Psychiatry, 63*, 1155–1162.
Hammack, P. L. (2003). The question of cognitive therapy in a postmodern world. *Ethical Human Sciences and Services, 5*(3), 209–224.
Hammond, J. L., & Hall, S. S. (2011). Functional analysis and treatment of aggressive behavior following resection of a craniopharyngioma. *Developmental Medicine and Child Neurology, 53*(4), 369–374.
Hampton, T. (2006). Stem cells probed as diabetes treatment. *Journal of American Medical Association, 296*, 2785–2786.
Hampton, T. (2007). Stem cells ease Parkinson symptoms in monkeys. *Journal of American Medical Association, 298*, 165.
Han, S., Mao, L., Qin, J., Friederici, A. D., & Ge, J. (2011). Functional roles and cultural modulations of the medial prefrontal and parietal activity associated with causal attribution. *Neuropsychologia, 49*(1), 83–91.
Handler, M., Honts, C. R., Krapohl, D. J., Nelson, R., & Griffin, S. (2009). Integration of pre-employment polygraph screening into the police selection process. *Journal of Police and Criminal Psychology, 24*(2), 69–86
Haney, C., Banks, C., & Zimbardo, P. (1978). Interpersonal dynamics in a simulated prison. *International Journal of Criminology and Penology, 1*, 69–97.
Hansell, J. H., & Damour, L. K. (2008). *Abnormal psychology* (2nd ed.). Hoboken, NJ: John Wiley & Sons.
Harati, H., Majchrzak, M., Cosquer, B., Galani, R., Kelche, C., Cassel, J.-C., & Barbelivien, A. (2011). Attention and memory in aged rats: Impact of lifelong environmental enrichment. *Neurobiology of Aging, 32*(4), 718–736.
Hardcastle, S., Taylor, A., Bailey, M., & Castle, R. (2008). A randomized controlled trial on the effectiveness of

a primary health care based counseling intervention on physical activity, diet and CHD risk. *Patient Education and Counseling*, *70*, 31–39.

Hardy, G., Barkham, M., Shapiro, D., Guthrie, E., & Margison, F. (Eds.). (2011). *Psychodynamic-interpersonal therapy*. Thousand Oaks, CA: Sage.

Harkness, K. L., Alavi, N., Monroe, S. M., Slavich, G. M., Gotlib, I. H., & Bagby, R. M. (2010). Gender differences in life events prior to onset of major depressive disorder: The moderating effect of age. *Journal of Abnormal Psychology*, *119*(4), 791–803.

Harlow, H. F., & Harlow, M. K. (1966). Learning to love. *American Scientist*, *54*, 244–272.

Harlow, H. F., & Zimmerman, R. R. (1959). Affectional responses in the infant monkey. *Science*, *130*, 421–432.

Harlow, H. F., Harlow, M. K., & Meyer, D. R. (1950). Learning motivated by a manipulation drive. *Journal of Experimental Psychology*, *40*, 228–234.

Harlow, H. F., Harlow, M. K., & Suomi, S. J. (1971). From thought to therapy: Lessons from a primate laboratory. *American Scientist*, *59*, 538–549.

Harlow, J. (1868). Recovery from the passage of an iron bar through the head. *Publications of the Massachusetts Medical Society*, *2*, 237–246.

Harmon-Jones, E., Amodio, D. M., & Harmon-Jones, C. (2010). Action-based model of dissonance: On cognitive conflict and attitude change. In J. P. Forgas, J. Cooper & W. D. Crano (Eds.), *The psychology of attitudes and attitude change* (pp. 163–181). New York, NY: Psychology Press.

Harper, F. D., Harper, J. A., & Stills, A. B. (2003). Counseling children in crisis based on Maslow's hierarchy of basic needs. *International Journal for the Advancement of Counselling*, 25(1), 10–25.

Harrison, E. (2005). *How meditation heals: Scientific evidence and practical applications* (2nd ed.). Berkeley, CA: Ulysses Press.

Harth, N. S., Kessler, T., & Leach, C. W. (2008). Advantaged group's emotional reactions to intergroup inequality. The dynamics of pride, guilt, and sympathy. *Personality and Social Psychology Bulletin*, *34*, 115–129.

Hartwell, L. (2008). *Genetics* (3ed ed.). New York, NY: McGraw-Hill.

Harwood, T. M., Pratt, D., Beutler, L. E., Bongar, B. M., Lenore, S., & Forrester, B. T. (2011). Technology, telehealth, treatment enhancement, and selection. *Professional Psychology: Research and Practice*, *42*(6), 448–454.

Haslam, N., Loughnan, S., Reynolds, C., & Wilson, S. (2007). Dehumanization: A new perspective. *Social and Personality Psychology Compass*, *1*, 409–422.

Hatfield, E., & Rapson, R. L. (1996). *Love and Sex: Cross-cultural perspectives*. Needham Heights, MA: Allyn & Bacon.

Hatfield, E., & Rapson, R. L. (2010). Culture, attachment style, and romantic relationships. In P. E. K.-M. Ng (Ed.), *Attachment: Expanding the cultural connections* (pp. 227–242). New York, NY: Routledge/Taylor & Francis Group.

Haukkala, A., Konttinen, H., Laatikainen, T., Kawachi, I., & Uutela, A. (2010). Hostility, anger control, and anger expression as predictors of cardiovascular disease. *Psychosomatic Medicine*, *72*(6), 556–562.

Hay, D. F. (1994). Prosocial development. *Journal of Child Psychology and Psychiatry*, *35*, 29–71.

Hay, D., Nash, A., Caplan, M., Swartzentruber, J., Ishikawa, F., & Vespo, J. (2011). The emergence of gender differences in physical aggression in the context of conflict between young peers. *British Journal of Developmental Psychology*, *29*(2), 158–175.

Hayflick, L. (1977). The cellular basis for biological aging. In C. E. Finch & L. Hayflick (Eds.), *Handbook of the biology of aging* (pp. 159–186). New York, NY: Van Nostrand Reinhold.

Hayflick, L. (1996). *How and why we age*. New York, NY: Ballantine Books.

Haynes, S., O'Brien, W., & Kaholokula, J. (2011). *Behavioral assessment and case formulation*. Hoboken, NJ: John Wiley & Sons.

Hays, W. S. T. (2003). Human pheromones: Have they been demonstrated? *Behavioral Ecology and Sociobiology*, *54*(2), 89–97.

Hazan, C., & Shaver, P. (1987). Romantic love conceptualized as an attachment process. *Journal of Personality and Social Psychology*, *52*, 511–524.

Hazler, R. J. (2011). Person-centered theory. In D. Capuzzi & D. R. Gross (Eds.), *Counseling and psychotherapy* (5th ed.) (pp. 143–166). Alexandria, VA: American Counseling Association.

Heffelfinger, A. K., & Newcomer, J. W. (2001). Glucocorticoid effects on memory function over the human life span. *Development and Psychopathology*, *13*(3), 491–513.

Heider, F. (1958). *The psychology of interpersonal relations*. New York, NY: John Wiley & Sons.

Heimann, M., & Meltzoff, A. N. (1996). Deferred imitation in 9-and 14-month-old infants. *British Journal of Developmental Psychology*, *14*, 55–64.

Heine, S. J., & Renshaw, K. (2002). Interjudge agreement, self-enhancement, and liking: Cross-cultural divergences. *Personality and Social Psychology Bulletin*, *28*(5), 578–587.

Heinrich, A., & Schneider, B. A. (2011). Elucidating the effects of ageing on remembering perceptually distorted word pairs. *The Quarterly Journal of Experimental Psychology*, *64*(1), 186–205.

Helms, J. E., & Cook, D. A. (1999). *Using race and culture in counseling and psychotherapy: Theory and process*. Boston, MA: Allyn & Bacon.

Henderson, L. M. (2010). Study of perceived parenting traits and how they may contribute to the development of antisocial, negativistic, paranoid and schizotypal personality disorders based on Millon's theory. *Dissertation Abstracts International: Section B. The Sciences and Engineering*, *70*(10), 6596.

Hennessey, B. A., & Amabile, T. M. (1998). Reward, intrinsic motivation, and creativity. *American Psychologist*, *53*, 674–675.

Herbert, J. D., & Forman, E. M. (2010). *Acceptance and mindfulness in cognitive behavior therapy: Understanding and applying the new therapies*. Malden, MA: Wiley-Blackwell.

Hergovich, A., & Olbrich, A. (2003). The impact of the Northern Ireland conflict on social identity, groupthink and integrative complexity in Great Britain. *Review of Psychology*, *10*(2), 95–106.

Herholz, S. C., Halpern, A. R., & Zatorre, R. J. (2012). Neuronal correlates of perception, imagery, and memory for familiar tunes. *Journal of Cognitive Neuroscience*, *24*(6), 1382–1397.

Herman, C. P., & Polivy, J. (2008). External cues in the control of food intake in humans: The sensory-normative distinction. *Physiology and Behavior*, *94*, 722–728.

Herman, L. M., Richards, D. G., & Woltz, J. P. (1984). Comprehension of sentences by bottlenosed dolphins. *Cognition*, *16*, 129–139.

Hermans, E. J., Ramsey, N. F., & van Honk, J. (2008). Exogenous testosterone enhances responsiveness to social threat in the neural circuitry of social aggression in humans. *Biological Psychiatry*, *63*(3), 263–270.

Herrnstein, R. J., & Murray, C. (1994). *The bell curve: Intelligence and class structure in American life*. New York, NY: Free Press.

Hervé, H., Mitchell, D., Cooper, B. S., Spidel, A., & Hare, R. D. (2004). Psychopathy and unlawful confinement: An examination of perpetrator and event characteristics. *Canadian Journal of Behavioural Science*, *36*(2), 137–145.

Hess, T. M., Popham, L. E., Emery, L., & Elliott, T. (2012). Mood, motivation, and misinformation: Aging and affective state influences on memory. *Aging*,

*Neuropsychology and Cognition*, *19*(1/2), 13–34. doi: 10.1080/13825585.2011.622740.
Higley, E. R. (2008). Nighttime interactions and mother-infant attachment at one year. *Dissertation Abstracts International*: *Section B*: *The Sciences and Engineering*, *68*, 2008, 5575.
Hilgard, E. R. (1978). Hypnosis and consciousness. *Human Nature*, *1*, 42–51.
Hilgard, E. R. (1992). Divided consciousness and dissociation. *Consciousness and Cognition*, *1*, 16–31.
Hill, P. L., & Lapsley, D. K. (2011). Adaptive and maladaptive narcissism in adolescent development. In C. T. Barry, P. K. Kerig, K. K. Stellwagen, & T. D. Barry (Eds.), *Narcissism and Machiavellianism in youth*: *Implications for the development of adaptive and maladaptive behavior* (pp. 89–105). Washington, DC: American Psychological Association.
Hines, M. (2011). Gender development and the human brain. *Annual Review of Neuroscience*, *34*, *69–88*.
Hoar, S. D., Evans, M. B., & Link, C. A. (2012). How do master athletes cope with pre-competitive stress at a "Senior Games?" *Journal of Sport Behavior*, *35*(2),181–203.
Hobson, J. A. (1999). *Dreaming as delirium*: *How the brain goes out of its mind*. Cambridge, MA: MIT Press.
Hobson, J. A. (2005). In bed with Mark Solms? What a nightmare! A reply to Domhoff. *Dreaming*, *15*(1), 21–29.
Hobson, J. A., & McCarley, R. W. (1977). The brain as a dream state generator: An activation-synthesis hypothesis of the dream process. *American Journal of Psychiatry*, *134*, 1335–1348.
Hobson, J. A., Sangsanguan, S., Arantes, H., & Kahn, D. (2011). Dream logic—The inferential reasoning paradigm. *Dreaming*, *21*(1), 1–15.
Hobson, J. A., & Silvestri, L. (1999). Parasomnias. *The Harvard Mental Health Letter*, *15*(8), 3–5.
Hodges, S. D., & Biswas-Diener, R. (2007). Balancing the empathy expense account: Strategies for regulating empathic response. In T. Farrow & P. Woodruff (Eds.), *Empathy in mental illness* (pp. 389–407). New York, NY: Cambridge University Press.
Hodson, G., Rush, J., & MacInnis, C. C. (2010). A joke is just a joke (except when it isn't): Cavalier humor beliefs facilitate the expression of group dominance motives. *Journal of Personality and Social Psychology*, *99*(4), 660–682.
Hoek, H. W., van Harten, P. N., Hermans, K. M. E., Katzman, M. A., Matroos, G. E., & Susser, E. S. (2005). The incidence of anorexia nervosa on Curacao. *American Journal of Psychiatry*, *162*, 748–752.
Hofer, J., Busch, H., Bender, M., Ming, L., & Hagemeyer, B. (2010). Arousal of achievement motivation among student samples in three different cultural contexts: Self and social standards of evaluation. *Journal of Cross-Cultural Psychology*, *41*(5–6), 758–775.
Hoff, E. (2009). *Language development* (4th ed.). Belmont, CA: Wadsworth.
Hoffman, M. L. (1993). Empathy, social cognition, and moral education. In A. Garrod (Ed.), *Approaches to moral development*: *New research and emerging themes*. New York, NY: Teachers College Press.
Hofmann, S. (2012). *An introduction to modern CBT*: *Psychological solutions to mental health problems*. Malden, MA: Wiley-Blackwell.
Hogan, T. P. (2006). *Psychological testing*: *A practical Introduction* (*2nd ed.*). Hoboken, NJ: Wiley.
Holland, J. L. (1985). *Making vocational choices*: *A theory of vocational personalites and work environments* (2nd ed). Englewood Cliffs, NJ: Prentice Hall.
Holland, J. L. (1994). *Self-directed search form R*. Lutz, Fl: Psychological Assessment Resources.
Hollander, E., & Simeon, D. (2011). Anxiety disorders. In R. E. Hales, S. C. Yudofsky, & G. O. Gabbard (Eds.), *Essentials of psychiatry* (3rd ed.) (pp. 185–228). Arlington, VA: American Psychiatric Publishing.
Hollander, J., Renfrow, D., & Howard, J. (2011). *Gendered situations, gendered selves*: *A gender lens on social psychology* (2nd ed.). Lanham, MD: Rowman and Littlefield.
Hollon, S. D. (2011). Cognitive and behavior therapy in the treatment and prevention of depression. *Depression and Anxiety*, *28*(4), 263–266.
Holtzworth-Munroe, A. (2000). A typology of men who are violent toward their female partners: Making sense of the heterogeneity in husband violence. *Current Directions in Psychological Science*, *9*(4), 140–143.
Honts, C. R., & Kircher, J. C. (1994). Mental and physical countermeasures reduce the accuracy of polygraph tests. *Journal of Applied Psychology*, *79*(2), 252–259.
Hooley, J. M., & Hiller, J. B. (2001). Family relationships and major mental disorder: Risk factors and preventive strategies. In B. R. Sarason & S. Duck (Eds.), *Personal relationships*: *implications for clinical and community psychology*. New York: John Wiley & Sons.
Hopfer, C. (2011). Club drug, prescription drug, and over-the-counter medication abuse: Description, diagnosis, and intervention. In Y. Kaminer & K. C. Winters (Eds.), *Clinical manual of adolescent substance abuse treatment* (pp. 187–212). Arlington, VA: American Psychiatric Publishing.
Hoppe, A. (2011). Psychosocial working conditions and well-being among immigrant and German low-wage workers. *Journal of Occupational Health Psychology*, *16*(2), 187–201.
Hopwood, C. J., Donnellan, M. B., Blonigen, D. M., Krueger, R. F., McGue, M., Iacono, W. G., & Burt, S. A. (2011). Genetic and environmental influences on personality trait stability and growth during the transition to adulthood: A three-wave longitudinal study. *Journal of Personality and Social Psychology*, *100*(3), 545–556.
Horney, K. (1939). *New ways in psychoanalysis*. New York, NY: International Universities Press.
Horney, K. (1945). *Our inner conflicts*: *A constructive theory of neurosis*. New York, NY: W. W. W. W. Norton & Co.
Horwath, E., & Gould, F. (2011). Epidemiology of anxiety disorders. In M. Tsuang, M. Tohen, & P. Jones (Eds.), *Textbook in Psychiatric Epidemiology*, (3rd ed.) (pp. 311–328). Hoboken, NJ: John Wiley & Sons.
Houlfort, N. (2006). The impact of performance-contingent rewards on perceived autonomy and intrinsic motivation. *Dissertation Abstracts International Section A*: *Humanities and Social Sciences*, *67*(2–A), 460.
Houtz, J. C., Matos, H., Park, M-K. S., Scheinholtz, J., & Selby, E. (2007). Problem-solving style and motivational attributions. *Psychological Reports*, *101*, 823–830.
Hovland, C. I. (1937). The generalization of conditioned responses: II. The sensory generalization of conditioned responses with varying intensities of tone. *Journal of Genetic Psychology*, *51*, 279–291.
Howe, M. L., & Malone, C. (2011). Mood-congruent true and false memory: Effects of depression. *Memory*, *19*(2), 192–201.
Howell, K. K., Coles, C. D., & Kable, J. A. (2008). The medical and developmental consequences of prenatal drug exposure. In J. Brick (Ed.), *Handbook of the medical consequences of alcohol and drug abuse* (2nd ed.) (pp. 219–249). *The Haworth Press series in neuropharmacology*. New York, NY: Haworth Press/Taylor and Francis Group.
Huang, M., & Hauser, R. M. (1998). Trends in Black–White test-score differentials: II. The WORDSUM Vocabulary Test. In U. Neisser (Ed.), *The rising curve*: *Long-term gains in IQ and related measures* (pp. 303–334). Washington, DC: American Psychological Association.
Huas, C., Caille, A., Godart, N., Foulon, C., Pham, S. A., Divac, S., Dechartres, A., Lavoisy, G., Guelfi, J. D., Rouillon, F., & Falissard, B. (2011). Factors predictive of ten-year mortality in severe anorexia nervosa patients. *Acta Psychiatrica Scandinavica*, *123*(1), 62–70.

Hudziak, J. J. (Ed). (2008). *Developmental psychopathology and wellness: Genetic and environmental influences*. Arlington, VA: American Psychiatric Publishing.
Huesmann, L. R., & Kirwil, L. (2007). Why observing violence increases the risk of violent behavior by the observer. In D. J. Flannery, A. T. Vazsonyi, & I. D. Waldman (Eds.), *The Cambridge handbook of violent behavior and aggression* (pp. 545–570), New York, NY: Cambridge University Press.
Huesmann, L. R., Dubow, E. F., & Boxer, P. (2011). The transmission of aggressiveness across generations: Biological, contextual, and social learning processes. In P. R. Shaver & M. Mikulincer (Eds.), *Herzilya series on personality and social psychology. Human aggression and violence: Causes, manifestations, and consequences* (pp. 123–142). Washington, DC: American Psychological Association.
Huffman, C. J., Matthews, T. D., & Gagne, P. E. (2001). The role of part-set cuing in the recall of chess positions: Influence of chunking in memory. *North American Journal of Psychology*, *3*(3), 535–542.
Hulette, A., Freyd, J., & Fisher, P. (2011). Dissociation in middle childhood among foster children with early maltreatment experiences. *Child Abuse and Neglect*, *35*, 123–126.
Hull, C. (1952). *A behavior system*. New Haven, CT: Yale University Press.
Hull, E. M. (2011). Sex, drugs and gluttony: How the brain controls motivated behaviors. *Physiology and Behavior*, *104*(1), 173–177.
Hulette, A., Freyd, J., & Fisher, P. (2011). Dissociation in middle childhood among foster children with early maltreatment experiences. *Child Abuse and Neglect*, *35*, 123–126.
Hunsley, J., & Lee, E. M. (2010). *Introduction to clinical psychology* (2nd ed.). Toronto, ON: John Wiley & Sons.
Hunt, E. (2011). *Human intelligence*. New York, NY: Cambridge University Press.
Hurlemann, R., Matusch, A., Hawellek, B., Klingmuller, D., Kolsch, H., Maier, W., & Dolan, R. J. (2007). Emotion-induced retrograde amnesia varies as a function of noradrenergic-glucocorticord activity. *Psychopharmacology*, *194*, 261–269.
Husain, M. M., & Lisanby, S. H. (2011). Repetitive transcranial magnetic stimulation (rTMS): A noninvasive neuromodulation probe and intervention. *The Journal of ECT*, *27*(1), 2.
Hutchison, P., & Rosenthal, H. E. S. (2011). Prejudice against Muslims: Anxiety as a mediator between intergroup contact and attitudes, perceived group variability and behavioural intentions. *Ethnic and Racial Studies*, *34*(1), 40–61.
Hyde, J. S. (2005). The genetics of sexual orientation. In J. S. Hyde (Ed.), *Biological substrates of human sexuality* (pp. 9–20). Washington, DC: American Psychological Association.
Hyde, J. S. (2007). *Half the human experience: The psychology of women* (7th ed.). Boston, MA: Houghton Mifflin.
Hyde, J. S., & DeLamater, J. D. (2011). *Understanding human sexuality* (11th ed.). New York: McGraw-Hill.
Hyman, R. (1996). The evidence for psychic functioning: Claims vs. reality. *Skeptical Inquirer*, *20*, 24–26.

Iacono, W. G. (2008). Accuracy of polygraph techniques: Problems using confessions to determine ground truth. *Physiology and Behavior*, *95*(1–2), 24–26.
Ibiloglu, A. O., & Caykoylu, A. (2011). The comorbidity of anxiety disorders in bipolar I and bipolar II patients among Turkish population. *Journal of Anxiety Disorders*, *25*(5), 661–667.
ILAR, (2009). Institute for Laboratory Animal Research (ILAR). Retrieved from http://dels.nas.edu/ilar_n/ilarhome/
Imada, T., & Ellsworth, P. C. (2011). Proud Americans and lucky Japanese: Cultural differences in appraisal and corresponding emotion. *Emotion*, *11*(2), 329–345.
Imada, T., & Kitayama, S. (2010). Social eyes and choice justification: Culture and dissonance revisited. *Social Cognition*, *28*(5), 589–608.
Infurna, F. J., Gerstorf, D., Ram, N., Schupp, J., & Wagner, G. G. (2011). Long-term antecedents and outcomes of perceived control. *Psychology and Aging*, *26*(3), 559–575.
Institute for Laboratory Animal Research (ILAR). (2009). *Home Page*. Retrieved from http://dels.nas.edu/ilar_n/ilarhome/
Inui, K., Urakawa, T., Yamashiro, K., Otsuru, N., Takeshima, Y., Nishihara, M., Motomura, E., Kida, T., & Kakigi, R. (2010). Echoic memory of a single pure tone indexed by change-related brain activity. *BMC Neuroscience*, *11*, Article ID 135.
Inzlicht, M., & Al-Khindi, T. (2012). ERN and the placebo: A misattribution approach to studying the arousal properties of the error-related negativity. *Journal of Experimental Psychology: General*.
Irish, M., Lawlor, B. A., O'Mara, S. M., & Coen, R. F. (2011). Impaired capacity for autonoetic reliving during autobiographical event recall in mild Alzheimer's disease. *Cortex: A Journal Devoted to the Study of the Nervous System and Behavior*, *47*(2), 236–249.
Irvin, M. J. (2012). Role of student engagement in the resilience of African American adolescents from low-income rural communities. *Psychology in the Schools*, *49*(2), 176–193.
Irwin, M. (2008).There's no good proof the real *Medium*, Allison DuBois, has ever cracked a case, but her fans don't care. *Phoenix New Times*. Retrieved from http://www.phoenixnewtimes.com/2008-06-12/news/there-s-no-good-proof-the-real-medium-allison-dubois-has-ever-cracked-a-case-but-her-fans-don-t-care/
Irwin, M., Mascovich, A., Gillin, J. C., Willoughby, R., Pike, J., & Smith, T. L. (1994). Partial sleep deprivation reduced natural killer cell activity in humans. *Psychosomatic Medicine*, *56*(6), 493–498.
Ivanovic, D. M., Leiva, B. P., Pérez, H. T., Olivares, M. G., Díaz, N. S., Urrutia, M. S. C., Almagià, A. F., Toro, T. D., Miller, P. T., Bosch, E. O., & Larraín, C. G. (2004). Head size and intelligence, learning, nutritional status and brain development: Head, IQ, learning, nutrition and brain. *Neuropsychologia*, *42*(8), 1118–1131.
Iwata, Y., Suzuki, K., Takei, N., Toulopoulou, T., Tsuchiya, K. J., Matsumoto, K., Takagai, S., Oshiro, M., Nakamura, K., & Mori, N. (2011). Jiko-shisen-kyofu (fear of one's own glance), but not taijin-kyofusho (fear of interpersonal relations), is an east Asian culture-related specific syndrome. *Australian and New Zealand Journal of Psychiatry*, *45*(2), 148–152.

Jackendoff, R. (2003). Foundations of language, brain, meaning, grammar, evolution. *Applied Cognitive Psychology*, *17*(1), 121–122.
Jackson, L. M. (2011). Development of prejudice in children. In L. M. Jackson, (Ed.), *The psychology of prejudice: From attitudes to social action* (pp. 81–101). Washington, DC: American Psychological Association.
Jacob, P. (2008). What do mirror neurons contribute to human social cognition? *Mind and Language*, *23*, 190–223.
Jacobi, C., Hayward, C., de Zwaan, M., Kraemer, H. C., & Agras, W. S. (2004). Coming to terms with risk factors for eating disorders: Application of risk terminology and suggestions for a general taxonomy. *Psychological Bulletin*, *130*(1), 19–65.
Jacquette, D. (Ed.). (2010). *Philosophy for everyone. Cannabis: What were we just talking about?* Malden, MA: Wiley-Blackwell.
Jaeger, A., Cox, J. C., & Dobbins, I. G. (2012). Recognition confidence under violated and confirmed memory expectations. *Journal of Experimental Psychology: General*, *141*(2), 282–301.
Jaeger, M. (2011). "A thing of beauty is a joy forever"? Returns to physical attractiveness over the life course. *Social Forces*, *89*(3), 983–1003.
Jaehne, E. J., Majumder, I., Salem, A., & Irvine, R. J. (2011). Increased effects of 3, 4-methylenedioxymethamphetamine

(ecstasy) in a rat model of depression. *Addiction Biology*, *16*(1), 7–19.
Jaffe, S. L., & Kelly, J. F. (2011). Twelve-step mutual-help programs for adolescents. In Y. Kaminer & K. C. Winters (Eds.), *Clinical manual of adolescent substance abuse treatment* (pp. 269–282). Arlington, VA: American Psychiatric Publishing.
James, W. (1890). *The principles of psychology* (Vol. 2). New York, NY: Holt.
Jamieson, G. A., & Hasegawa, H. (2007). New paradigms of hypnosis research. In G. A. Jamieson (Ed.), *Hypnosis and conscious states: The cognitive neuroscience perspective* (pp. 133–144). New York, NY: Oxford University Press
Janis, I. L. (1972). *Victims of groupthink: A psychological study of foreign-policy decisions and fiascoes*. Boston, MA: Houghton Mifflin.
Janis, I. L. (1989). *Crucial decisions: Leadership in policymaking and crisis management*. New York, NY: Free Press.
Jankowiak, W., & Fischer, E. (1992). Cross-cultural perspective on romantic love. *Ethnology*, *31*, 149–155.
Jannini, E. A., Blanchard, R., Camperio-Ciani, A., & Bancroft, J. (2010). Male homosexuality: Nature or culture? *Journal of Sexual Medicine*, *7*(10), 3245–3253.
Jarvis, S. (2011). Conceptual transfer: Crosslinguistic effects in categorization and construal. *Bilingualism: Language and Cognition*, *14*(1), 1–8.
Jensen, L. A. (2011). The cultural-developmental theory of moral psychology: A new synthesis. In L. A. Jensen (Ed.), *Bridging cultural and developmental approaches to psychology: New syntheses in theory, research, and policy* (pp. 3–25). New York, NY: Oxford University Press.
Jensen, M. P., Hakimian, S., Sherlin, L. H., & Fregni, F. (2008). New insights into -neuromodulatory approaches for the treatment of pain. *The Journal of Pain*, *9*, 193–199.
Jinap, S., & Hajeb, P. (2010).Glutamate. Its applications in food and contribution to health. *Appetite*, *55*(1), 1–10.
Johnson, A. L. (2011). Psychoanalytic theory. In D. Capuzzi & D. R. Gross (Eds.), *Counseling and psychotherapy* (5th ed.) (pp. 59–76). Alexandria, VA: American Counseling Association.
Johnson, D. M., Delahanty, D. L., & Pinna, K. (2008). The cortisol awakening response as a function of PTSD severity and abuse chronicity in sheltered battered women. *Journal of Anxiety Disorders*, *22*, 793–800.
Johnson, L. (2011). The genetic epidemiology of obesity: A case study. In M. Teare (Ed.), *Genetic epidemiology: Methods in molecular biology* (pp. 227–237). New York, NY: Springer.
Johnson, N. J. (2008). Leadership styles and passive-aggressive behavior in organizations. *Dissertation Abstracts International: Section B: The Sciences and Engineering*, *68*(7-B), 4828.
Johnson, S. C., Dweck, C. S., & Chen, F. S. (2007). Evidence for infants' internal working methods of attachment. *Psychological Science*, *18*, 501–502.
Johnson, S. K., Murphy, S. E., Zewdie, S., & Reichard, R. J. (2008). The strong, sensitive type: Effects of gender stereotypes and leadership prototypes on the evaluation of male and female leaders. *Organizational Behavior and Human Decision Processes*, *106*, 39–60.
Johnson, W., & Bouchard Jr., T. J. (2007). Sex differences in mental ability: A proposed means to link them to brain structure and function. *Intelligence*, *35*, 197–209.
Johnson, W., Brett, C. E., & Deary, I. J. (2010). The pivotal role of education in the association between ability and social class attainment: A look across three generations. *Intelligence*, *38*(1), 55–65.
Jolliffe, C. D., & Nicholas, M. K. (2004). Verbally reinforcing pain reports: An experimental test of the operant conditioning of chronic pain. *Pain*, *107*, 167–175.
Jones, E. E., & Nisbett, R. E. (1971). *The actor and the observer: Divergent perceptions of the causes of behavior*. Morristown, NJ: General Learning Press.
Jopp, D. S., & Schmitt, M. (2010). Dealing with negative life events: Differential effects of personal resources, coping strategies, and control beliefs. *European Journal of Ageing*, *7*(3), 167–180.
Jordan, J. V. (2011). The Stone Center and relational–cultural theory. In J. C. Norcross, G. R. VandenBos, & D. K. Freedheim (Eds.), *History of psychotherapy: Continuity and change* (2nd ed.) (pp. 357–362). -Washington, DC: American Psychological Association.
Jost, K., Bryck, R. L., Vogel, E. K., & Mayr, U. (2011) Are old adults just like low working memory young adults? Filtering efficiency and age differences in visual working memory. *Cerebral Cortex*, *21*(5), 1147–1154.
Jung, C. G. (1946). *Psychological types*. New York, NY: Harcourt Brace.
Jung, C. G. (1959). The archetypes and the collective unconscious. In H. Read, M. Fordham, & G. Adler (Eds.), *The collected works of C. G. Jung, Vol. 9*. New York, NY: Pantheon.
Jung, C. G. (1969). The concept of collective unconscious. In *Collected Works* (Vol. 9, Part 1). Princeton, NJ: Princeton University Press (Original work published 1936).
Jung, R. E., & Haier, R. J. (2007). The Parieto-Frontal Integration Theory (P-FIT) of intelligence: Converging neuroimaging evidence. *Behavioral and Brain Sciences*, *30*, 135–154.
Juntii, S. A., Coats, J. K., & Shah, N. M. (2008). A genetic approach to dissect sexually dimorphic behaviors. *Hormones and Behavior*, *53*, 627–637.
Jurado-Berbel, P., Costa-Miserachs, D., Torras-Garcia, M., Coll-Andreu, M., & Portell-Cortés, I. (2010). Standard object recognition memory and "what" and "where" components: Improvement by post-training epinephrine in highly habituated rats. *Behavioural Brain Research*, *207*(1), 44–50.

Kagan, J. (1998). Biology and the child. In W. Damon & R. M. Lerner (Eds.), *Handbook of child psychology* (*Vol. 1*) (pp. 177–235). New York, NY: John Wiley & Sons.
Kahneman, D., & Tversky, A. (1973). On the psychology of prediction. *Psychological Review*, *80*, 237–251.
Kalat, J. W. (2013). *Biological psychology* (11th ed.). Belmont, CA: Wadsworth, Cengage Learning.
Kalat, J. W., & Shiota, M. N. (2012). *Emotion* (2nd ed.). Belmont, CA: Cengage.
Kalnin, A. J., Edwards, C. R., Wang, Y., Kronenberger, W. G., Hummer, T. A., Mosier, K. M., Dunn, D., & Mathews, V. P. (2011). The interacting role of media violence exposure and aggressive–disruptive behavior in adolescent brain activation during an emotional Stroop task. *Psychiatry Research: Neuroimaging*, *192*(1), 12–19.
Kalodner, C. R. (2011). Cognitive-behavioral theories. In D. Capuzzi & D. R. Gross (Eds.), *Counseling and psychotherapy* (5th ed.) (pp. 193–213). Alexandria, VA: American Counseling Association.
Kana, R. K., Wadsworth, H. M., & Travers, B. G. (2011). A systems level analysis of the mirror neuron hypothesis and imitation impairments in autism spectrum disorders. *Neuroscience and Biobehavioral Reviews*, *35*(3), 894–902.
Kanayama, G., & Pope, H. G., Jr. (2011). Gods, men, and muscle dysmorphia. *Harvard Review of Psychiatry*, *19*(2), 95–98.
Kandler, C., Bleidorn, W., Riemann, R., Angleitner, A., & Spinath, F. (2012). Life events as environmental states and genetic traits and the role of personality: A longitudinal twin study. *Behavior Genetics*, *42*(1), 57–72.
Kandler, C., Riemann, R., & Kämpfe, N. (2009). Genetic and environmental mediation between measures of personality and family environment in twins reared together. *Behavioral Genetics*, *39*, 24–35.
Kaplan, L. E. (2006). Moral reasoning of MSW social workers and the influence of

education. *Journal of Social Work Education*, *42*, 507–522.
Karatsoreos, I. N., Bhagat, S., Bloss, E. B., Morrison, J. H., & McEwen, B. S. (2011). Disruption of circadian clocks has ramifications for metabolism, brain, and behavior. *PNAS Proceedings of the National Academy of Sciences of the United States of America*, *108*(4), 1657–1662.
Kardong, K. (2008). *Introduction to biological evolution* (2nd ed.). New York, NY: McGraw-Hill.
Kareev, Y. (2000). Seven (indeed, plus or minus two) and the detection of correlations. *Psychological Review*, *107*(2):397–402.
Karmiloff, K., & Karmiloff-Smith, A. (2002). *Pathways to language: From fetus to adolescent*. Cambridge, MA: Harvard University Press.
Karpicke, J. D., & Bauernschmidt, A. (2011). Spaced retrieval: Absolute spacing enhances learning regardless of relative spacing. *Journal of Experimental Psychology: Learning, Memory, and Cognition*, 37(5), 1250–1257.
Karremans, J. C., Stroebe, W., & Claus, J. (2006). Beyond Vicary's fantasies: The impact of subliminal priming and brand choice. *Journal of Experimental Social Psychology*, *42*, 792–798.
Kassin, S., Fein, S., & Markus, H. R. (2008). *Social psychology* (7th ed.). Belmont, CA: Cengage.
Kastenbaum, R. J. (2007). *Death, society, and human experience* (9th ed.). Upper Saddle River, NJ: Prentice Hall.
Katz, I., Kaplan, A., & Buzukashvily, T. (2011). The role of parents' motivation in students' autonomous motivation for doing homework. *Learning and Individual Differences, 21*(4), 376–386.
Kaukiainen, A., Björkqvist, K., Lagerspetz, K., Österman, K., Salmivalli, C., Rothberg, S., et al. (1999). The relationships between social intelligence, empathy, and three types of aggression. *Aggressive Behavior*, *25*(2), 81–89.
Kaye, W. (2008). Neurobiology of anorexia and bulimia nervosa. *Physiology and Behavior*, *94*, 121–135.
Kaysen, D., Pantalone, D. W., Chawla, N., Lindgrren, K. P., Clum, G. A., Lee, C., & Resick, P. A. (2008). Posttraumatic stress disorder, alcohol use, and physical health concerns. *Journal of Behavioral Medicine*, *31*, 115–125.
Kazdin, A. E. (1994). Methodology, design, and evaluation in psychotherapy research. In A.E. Bergin & S.L. Garfield (Eds.), *Handbook of psychotheraphy and behavior change* (4th ed.) (pp. 19–71). New York, NY: John Wiley & Sons.
Kazdin, A. E. (2011). *Single-case research designs: Methods for clinical and applied settings* (2nd ed.). New York, NY: Oxford University Press.
Keats, D. M. (1982). Cultural bases of concepts of intelligence: A Chinese versus Australian comparison. In P. Sukontasarp, N. Yongsiri, P. Intasuwan, N. Jotiban, & C. Suvannathat (Eds.), *Proceedings of the second Asian workshop on child and adolescent development* (pp. 67–75). Bangkok, India: Burapasilpa Press.
Keen, E. (2011). Emotional narratives: Depression as sadness—Anxiety as fear. *The Humanistic Psychologist*, *39*(1), 66–70.
Keller, J., & Bless, H. (2008). The interplay of stereotype threat and regulatory focus. In Y. Kashima, K. Fiedler, & P. Freytag (Eds.), *Stereotype dynamics: Language-based approaches to the formation, maintenance, and transformation of stereotypes* (pp. 367–389). Mahwah, NJ: Erlbaum.
Kellogg, R. (2011). *Fundamentals of Cognitive Psychology* (2nd ed.). Thousand Oaks, CA: Sage.
Kelly, G., Brown, S., Todd, J., & Kremer, P. (2008). Challenging behaviour profiles of people with acquired brain injury living in community settings. *Brain Injury*, *22*, 457–470.
Keltner, D., Kring, A. M., & Bonanno, G. A. (1999). Fleeting signs of the course of life: Facial expression and personal adjustment. *Current Directions in Psychological Science*, *8*(1), 18–22.
Kemeny, M. E. (2007). Psychoneuroimmunology. In H. S. Friedman & R. C. Silver (Eds.), *Foundations of health psychology* (pp. 92–116). New York, NY: Oxford University Press.
Kendler, K. S., & Prescott, C. A. (2006). *Genes, environment, and psychopathology: Understanding the causes of psychiatric and substance use disorders*. New York, NY: Guilford Press.
Kendler, K. S., Gallagher, T. J., Abelson, J. M., & Kessler, R. C. (1996). Lifetime prevalence, demographic risk factors, and diagnostic validity of nonaffective psychosis as assessed in a U. S. community sample. *Archives of General Psychiatry*, *53*, 1022–1031.
Kennedy, M. M. (2010). Attribution error and the quest for teacher quality. *Educational Researcher*, *39*(8), 591–598.
Kennedy, R. P. (2011). Incorporating evidence-based practices into psychotherapy training in clinical psychology Ph.D. programs. *Dissertation Abstracts International: Section B: The Sciences and Engineering*, *72*(5-B), 3098.
Kennedy, S. H., Giacobbe, P., Rizvi, S. J., Placenza, F. M., Nishikawa, Y., Mayberg, H. S., & Lozano, A. M. (2011). Deep brain stimulation for treatment-resistant depression: Follow-up after 3 to 6 years. *American Journal of Psychiatry*, *168*(5), 502–510.
Kershaw, S. (2008). Sharing their demons on the Web. *New York Times*. Retrieved from http://www.nytimes.com/2008/11/13/fashion/13psych.html
Kessler, R. C., McGonagle, K. A., Zhao, S., Nelson, C. B., Hughes, M., Eshleman, S., Wittchen, H., & Kendler, K. S. (1994). Lifetime and 12-month prevalence of DSM-IIIR psychiatric disorders in the United States. *Archives of General Psychiatry*, *51*, 8–19.
Ketterer, H. L. (2011). Examining the measurement invariance of the MMPI-2 Restructured Clinical (RC) scales across Korean and American normative samples. *Dissertation Abstracts International: Section B: The Sciences and Engineering*, *71*(7-B), 4532.
Kieffer, K. M., Schinka, J. A., & Curtiss, G. (2004). Person-environment congruence and personality domains in the prediction of job performance and work quality. *Journal of Counseling Psychology*, *51*(2), 168–177.
Kihlstrom, J. F. (2005). Dissociative disorders. *Annual Review of Clinical Psychology*, *1*, 227–253.
Killen, M., & Hart, D. (1999). *Morality in everyday life: Developmental perspectives*. New York, NY: Cambridge University Press.
Kilpatrick, L. A., Suyenobu, B. Y., Smith, S. R., Bueller, J. A., Goodman, T., Creswell, J. D., Tillisch, K., Mayer, E., & Naliboff, B. D. (2011). Impact of mindfulness-based stress reduction training on intrinsic brain connectivity. *NeuroImage*, *56*(1), 290–298.
Kim, J., & Hatfield, E. (2004). Love types and subjective well-being: A cross cultural study. *Social Behavior and Personality*, *32*(2), 173–182.
Kim, Y. H. (2008). Rebounding from learned helplessness: A measure of academic resilience using anagrams. *Dissertation Abstracts International: Section B: The Sciences and Engineering*. *68*(10-B), 6947.
Kimmel, M. S. (2000). *The gendered society*. London: Oxford University Press.
King, B. M. (2012). *Human sexuality today* (7th ed.). Boston: Allyn & Bacon.
Kirkman, C. A. (2002). Non-incarcerated psychopaths: Why we need to know more about the psychopaths who live amongst us. *Journal of Psychiatric and Mental Health Nursing*, *9*(2), 155–160.
Kirsch, I., Mazzoni, G., & Montgomery, G. H. (2006). Remembrance of hypnosis past. *American Journal of Clinical Hypnosis*, *49*, 171–178.
Kisilevsky, B. S., Hains, S. M. J., Lee, K., Xie, X., Huang, H., Ye, H., Zhang, K., & Wang, Z. (2003). Effects of experience on fetal voice recognition. *Psychological Science*, *14*(3), 220–224.
Kivlighan Jr, D. M., London, K., & Miles, J. R. (2012). Are two heads better than one? The relationship between number of

group leaders and group members, and group climate and group member benefit from therapy. *Group Dynamics: Theory, Research, and Practice*, *16*(1), 1–13.

Klatzky, R. L. (1984). *Memory and awareness*. New York, NY: Freeman.

Kleider, H. M., Pezdek, K., Goldinger, S. D., & Kirk, A. (2008). Schema-driven source misattribution errors: Remembering the expected from a witnessed event. *Applied Cognitive Psychology*, *22*, 1–20.

Klein, S. B. (2012). *Learning: Principles and applications* (6th ed.). Thousand Oaks, CA: SAGE.

Klimes-Dougan, B., Lee, C-Y. S., Ronsaville, D., & Martinez, P. (2008). Suicidal risk in young adult offspring of mothers with bipolar or major depressive disorder: A longitudinal family risk study. *Journal of Clinical Psychology*, *64*, 531–540.

Kluger, J., & Masters, C. (2006, August 28). How to spot a liar. *Time*, 46–48.

Knekt, P., Lindfors, O., Laaksonen, M. A., Raitasalo, R., Haaramo, P., Järvikoski, A., & The Helsinki Psychotherapy Study Group, Helsinki, Finland. (2008). Effectiveness of short-term and long-term psychotherapy on work ability and functional capacity—A randomized clinical trial on depressive and anxiety disorders. *Journal of Affective Disorders*, *107*, 95–106.

Knoblauch, K., Vital-Durand, F., & Barbur, J. L. (2000). Variation of chromatic sensitivity across the life span. *Vision Research*, *41*(1), 23–36.

Kobasa, S. (1979). Stressful life events, personality, and health: An inquiry into hardiness. *Journal of Personality and Social Psychology*, *37*, 1–11.

Kobeissi, J., Aloysi, A., Tobias, K., Popeo, D., & Kellner, C. H. (2011). Resolution of severe suicidality with a single electroconvulsive therapy. *The Journal of ECT*, *27*(1), 86–88.

Köfalvi, A. (Ed). (2008). *Cannabinoids and the brain*. New York, NY: Springer Science 1 Business Media.

Kohlberg, L. (1964). Development of moral character and moral behavior. In L. W. Hoffman & M. L. Hoffman (Eds.), *Review of child development research* (*Vol. 1*)(pp. 383–431). New York, NY: Sage.

Kohlberg, L. (1969). Stage and sequence: The cognitive-developmental approach to socialization. In D. A. Goslin (Ed.), *Handbook of socialization theory and research* (pp. 347–480). Chicago, IL: Rand McNally.

Kohlberg, L. (1984). *The psychology of moral development: Essays on moral development* (*Vol. 2*). San Francisco: Harper & Row.

Köhler, W. (1925). *The mentality of apes*. New York, NY: Harcourt, Brace.

Kohn, A. (2000). *Punished by rewards: The trouble with gold stars, incentive plans, A's, and other bribes*. New York, NY: Houghton Mifflin.

Komiya, N., Good, G. E., & Sherrod, N. B. (2000). Emotional openness as a predictor of college students' attitudes toward seeking psychological help. *Journal of Counseling Psychology*, *47*(1), 138–143.

Kondo, N. (2008). Mental illness in film. *Psychiatric Rehabilitation Journal*, *31*, 250–252.

Konigsberg, R. D. (2011). *The truth about grief: The myth of its five stages and the new science of loss*. New York, NY: Simon & Schuster.

Koopmann-Holm, B., & Matsumoto, D. (2011). Values and display rules for specific emotions. *Journal of Cross-Cultural Psychology*, *42*(3), 355–371.

Koss, M. P., White, J. W., & Kazdin, A. E. (2011). Violence against women and children: Perspectives and next steps. In M. P. Koss, J. W. White, & A. E. Kazdin (Eds.), *Violence against women and children, Vol. 2. Navigating solutions* (pp. 261–305). Washington, DC: American Psychological Association.

Kracher, B., & Marble, R. P. (2008). The significance of gender in predicting the cognitive moral development of business practitioners using the Socioemotional Reflection Objective Measure. *Journal of Business Ethics*, *78*, 503–526.

Krahé, B., Möller, I., Huesmann, L. R., Kirwil, L., Felber, J., & Berger, A. (2011). Desensitization to media violence: Links with habitual media violence exposure, aggressive cognitions, and aggressive behavior. *Journal of Personality and Social Psychology, 100*(4), 630–646.

Kress, T., Aviles, C., Taylor, C., & Winchell, M. (2011). Individual/collective human needs: (Re) theorizing Maslow using critical, sociocultural, feminist, and indigenous lenses. In C. Malott & B. Porfilio (Eds.), *Critical pedagogy in the twenty-first century: A new generation of scholars* (pp. 135–157). Charlotte, NC: Information Age Publishing.

Kring, A. M., Johnson, S. L. , Davison, G. C., & Neale, J. M. (2012). *Abnormal psychology* (12th ed.). Hoboken, NJ: John Wiley & Sons.

Krishna, G. (1999). *The dawn of a new science*. Los Angeles, CA: Institute for Consciousness Research.

Kristjánsdóttir, Ó. B., Fors, E. A., Eide, E., Finset, A., van Dulmen, S., Wigers, S. H., & Eide, H. (2011). Written online situational feedback via mobile phone to support self-management of chronic widespread pain: A usability study of a web-based intervention. *BMC Musculoskeletal Disorders*, *12*(1), 51–59.

Krueger, J. I. (2007). From social projection to social behaviour. *European Review of Social Psychology*, *18*, 1–35.

Krukowski, K., Eddy, J., Kosik, K. L., Konley, T., Janusek, L. W., & Mathews, H. L. (2011). Glucocorticoid dysregulation of natural killer cell function through epigenetic modification. *Brain, Behavior, and Immunity*, *25*(2), 239–249.

Krusemark, E. A., Campbell, W. K., & Clementz, B. A. (2008). Attributions, deception, and event related potentials: An investigation of the self-serving bias. *Psychophysiology*, *45*, 511–515.

Krypel, M. N., & Henderson-King, D. (2010). Stress, coping styles, and optimism: Are they related to meaning of education in students' lives? *Social Psychology of Education*, *13*(3), 409–424.

Ksir, C. J., Hart, C. I., & Ray, O. S. (2008). *Drugs, society, and human behavior*. New York, NY: McGraw-Hill.

Kteily, N. S., Sidanius, J., & Levin, S. (2011). Social dominance orientation: Cause or 'mere effect'?: Evidence for SDO as a causal predictor of prejudice and discrimination against ethnic and racial outgroups. *Journal of Experimental Social Psychology*, *47*(1), 208–214.

Kubiak, T., Vögele, C., Siering, M., Schiel, R., & Weber, H. (2008). Daily hassles and emotional eating in obese adolescents under restricted dietary conditions—The role of ruminative thinking. *Appetite*, *51*, 206–209.

Kubicek, B., Korunka, C., Raymo, J. M., & Hoonakker, P. (2011). Psychological well-being in retirement: The effects of personal and gendered contextual resources. *Journal of Occupational Health Psychology*, *16*(2), 230–246.

Kübler-Ross, E. (1983). *On children and death*. New York, NY: Macmillan.

Kübler-Ross, E. (1997). *Death: The final stage of growth*. New York, NY: Simon & Schuster.

Kübler-Ross, E. (1999). *On death and dying*. New York, NY: Simon & Schuster.

Kühn, S., & Gallinat, J. (2011). Common biology of craving across legal and illegal drugs—A quantitative meta-analysis of cue-reactivity brain response. *European Journal of Neuroscience*, *33*(7), 1318–1326.

Kulik, L. (2005). Intrafamiliar congruence in gender role attitudes and ethnic stereotypes: The Israeli case. *Journal of Comparative Family Studies*, *36*(2), 289–303.

Kullmann, D. M., & Lamsa, K. P. (2011). LTP and LTD in cortical GABAergic interneurons: Emerging rules and roles. *Neuropharmacology*, *60*(5), 712–719.

Kumar, G., Markert, R. J., & Patel, R. (2011). Assessment of hospice patients' goals of care at the end of life. *American Journal of Hospice and Palliative Medicine*, *28*(1), 31–34.

Kuperstok, N. (2008). Effects of exposure to differentiated aggressive films, equated for levels of interest and excitation, and the vicarious hostility catharsis hypothesis. *Dissertation Abstracts International: Section B: The Sciences and Engineering 68*(7-B), 4806.

Kvavilashvili, L., Mirani, J., Schlagman, S., Erskine, J. A. K., & Kornbrot, D. E. (2010). Effects of age on phenomenology and consistency of flashbulb memories of September 11 and a staged control event. *Psychology and Aging*, *25*(2), 391–404.

Kyriacou, C. P., & Hastings, M. H. (2010). Circadian clocks: Genes, sleep, and cognition. *Trends in Cognitive Sciences*, *14*(6), 259–267.

Lachman, M. E. (2004). Development in midlife. *Annual Review of Psychology*, *55*, 305–331.

Lader, M., Cardinali, D. P., & Pandi-Perumal, S. R. (Eds.). (2006). *Sleep and sleep disorders: A neuropsychopharmacological approach*. New York, NY: Springer-Verlag.

Lahav. O., & Mioduser, D. (2008). Haptic-feedback support for cognitive mapping of unknown spaces by people who are blind. *International Journal of Human-Computer Studies*, *66*, 23–35.

Lambert, A. J., Good, K. S., & Kirk, I. J. (2010). Testing the repression hypothesis: Effects of emotional valence on memory suppression in the think—no think task. *Consciousness and Cognition*, *19*(1), 281–293.

Lammers, J., & Stapel, D. (2011). Power increases dehumanization. *Group Processes and Intergroup Relations*, *14*(1), 113–126.

Lancaster, R. S. (2007). *Stop Sylvia Browne: Sylvia Browne's best evidence*. Retrieved from http://www.stopsylvia.com/articles/ac360_brownesbestevidence.shtml.

Lanciano, T., Curci, A., & Semin, G. R. (2010). The emotional and reconstructive determinants of emotional memories: An experimental approach to flashbulb memory investigation. *Memory*, *18*(5), 473–485.

Landay, J. S., & Kuhnehenn, J. (2004, July 10). Probe blasts CIA on Iraq data. *The Philadelphia Inquirer*, p. AO1.

Landsbergis, P. A., Schnall, P. L., Belkic, K. L., Baker, D., Schwartz, J. E., & Pickering, T. G. (2011). Workplace and cardiovascular disease: Relevance and potential role for occupational health psychology. In J. C. Quick & L. E. Tetrick (Eds.), *Handbook of occupational health psychology* (2nd ed.) (pp. 243–264). Washington, DC: American Psychological Association.

Langdon, P. E., Clare, I. C. H., & Murphy, G. H. (2011). Moral reasoning theory and illegal behaviour by adults with intellectual disabilities. *Psychology, Crime and Law*, *17*(2), 101–115.

Langenecker, S. A., Bieliauskas, L. A., Rapport, L. J., Zubieta, J-K., Wilde, E. A., & Berent, S. (2005). Face emotion perception and executive functioning deficits in depression. *Journal of Clinical and Experimental Neuropsychology*, *27*(3), 320–333.

Långström, N., Rahman, Q., Carlström, E., & Lichtenstein, P. (2010). Genetic and environmental effects on same-sex sexual behavior: A population study of twins in Sweden. *Archives of Sexual Behavior*, *39*(1), 75–80.

LaPointe, L. L. (Ed.. (2005). Feral children. *Journal of Medical Speech-Language Pathology*, *13*, vii–ix.

Larkby, C. A., Goldschmidt, L., Hanusa, B. H., & Day, N. L. (2011). Prenatal alcohol exposure is associated with conduct disorder in adolescence: Findings from a birth cohort. *Journal of the American Academy of Child and Adolescent Psychiatry*, *50*(3), 262–271.

Latane, B., & Darley, J. M. (1968). Group inhibition of bystander intervention in emergencies. *Journal of Personality and Social Psychology*, *10*(3), (215–221).

Latane, B., & Darley, J. M. (1970). *The unresponsive bystander: Why doesn't he help?* New York, NY: Appleton-Century-Crofts.

Latinus, M., VanRullen, R., & Taylor, M. J. (2010). Top-down and bottom-up modulation in processing bimodal face/voice stimuli. *BMC Neuroscience*, *11*, 36.

Lawrence, C., & Andrews, K. (2004). The influence of perceived prison crowding on male inmates' perception of aggressive events. *Aggressive Behavior*, *30*, 273–283.

Lawrence, M. (2008). Review of the bifurcation of the self: The history and theory of dissociation and its disorders. *American Journal of Clinical Hypnosis*, *50*, 281–282.

Lawrence, V., Murray, J., Klugman, A., & Banerjee, S. (2011). Cross-cultural variation in the experience of depression in older people in the UK. In M. Abou-Saleh, C. Katona, & A. Kumar (Eds.), *Principle and practice of geriatric psychiatry* (3rd ed.) (pp. 711–716). Hoboken, NJ: John Wiley & Sons.

Lazar, S. (2010). *Psychotherapy is worth it: A comprehensive review of its cost-effectiveness*. Arlington, VA: American Psychiatric Publishing.

Lazar, S. W., Kerr, C. E., Wasserman, R. H., Gray, J. R., Greve, D. N., Treadway, M. T., McGarvey, M., Quinn, B. T., Dusek, J. A., Benson, H., Rauch, S. L., Moore, C. I., & Fischl, B. (2005). Meditation experience is associated with increased cortical thickness. *Neuroreport*, *16*(17), 1893–1897.

Lazarus, R. S. (1991). *Emotion and adaptation*. New York, NY: Oxford University Press.

Lazarus, R. S. (1998). *The life and work of an eminent psychologist*. New York, NY: Oxford University Press.

Leaper, C. (2000). Gender, affiliation, assertion, and the interactive context of parent-child play. *Developmental Psychology*, *36*(3), 381–393.

Leaper, C. (2011). Research in developmental psychology on gender and relationships: Reflections on the past and looking into the future. *British Journal of Developmental Psychology*, *29*(2), 347–356.

LeDoux, J. E. (1996b). Sensory systems and emotion: A model of affective processing. *Integrative Psychiatry*, *4*, 237–243.

LeDoux, J. E. (1998). *The emotional brain*. New York, NY: Simon & Schuster.

LeDoux, J. E. (2002). *Synaptic self: How our brains become who we are*. New York, NY: Viking.

LeDoux, J. E. (2003). The emotional brain, fear, and the amygdala. *Cellular and Molecular Neurobiology*, *23*, 727–738.

LeDoux, J. E. (2007). Emotional memory. *Scholarpedia*, *2*, 180.

Lee, A. M., & Messing, R. O. (2011). Protein kinase C epsilon modulates nicotine consumption and dopamine reward signals in the nucleus accumbens. *PNAS Proceedings of the National Academy of Sciences of the United States of America*, *108*(38), 16080–16085.

Lee, J., Jiang, J., Sim, K., Tay, J., Subramaniam, M., & Chong, S. (2011). Gender differences in Singaporean Chinese patients with schizophrenia. *Asian Journal of Psychiatry*, *4*(1), 60–64.

Lee, J.-H., Chung, W.-H., Kang, E.-H., Chung, D.-J., Choi, C.-B., Chang, H.-S., Lee, J.-H., Hwang, S.-H., Han, H., Choe, B.Y., & Kim, H.-Y. (2011). Schwann cell-like remyelination following transplantation of human umbilical cord blood (hUCB)-derived mesenchymal stem cells in dogs with acute spinal cord injury. *Journal of the Neurological Sciences*, *300*(1–2), 86–96.

Lee, J-H., Kwon, Y-D., Hong, S-H., Jeong, H-J., Kim, H-M., & Um, J-Y. (2008). Interleukin-1 beta gene polymorphism and traditional constitution in obese women. *International Journal of Neuroscience*, *118*, 793–805.

Lee, R., Chong, B., & Coccaro, E. (2011). Growth hormone responses to GABAB receptor challenge with baclofen and impulsivity in healthy control and personality disorder subjects. *Psychopharmacology*, *215*(1), 41–48.

Lee, S. S. (2011). Deviant peer affiliation and antisocial behavior: Interaction with monoamine oxidase A (MAOA) genotype. *Journal of Abnormal Child Psychology: An official publication of the International Society for Research in Child and Adolescent Psychopathology*, *39*(3), 321–332.

Lefley, H. P. (2000). Cultural perspectives on families, mental illness, and the law. *International Journal of Law and Psychiatry*, *23*, 229–243.

Lefrancois, G. R. (2012). *Theories of human learning: What the professor said* (6th ed.). Belmont, CA: Wadsworth/Cengage.

Leglise, A. (2008). *Progress in circadian rhythm research*. Hauppauge, NY: Nova Science.

Legrand, F. D., Gomà-i-freixanet, M., Kaltenbach, M. L., & Joly, P. M. (2007). Association between sensation seeking and alcohol consumption in French college students: Some ecological data collected in "open bar" parties. *Personality and Individual Differences, 43*, 1950–1959.

Leichsenring, F., Leibing, E., Kruse, J., New, A., & Leweke, F. (2011). Borderline personality disorder. *The Lancet, 377*(9759), 74–84.

Leichtman, M. D. (2006). Cultural and maturational influences on long-term event memory. In L. Balter & C. S. Tamis-LeMonda (Eds.), *Child psychology: A handbook of contemporary issues* (2nd ed.) (pp. 565–589). New York, NY: Psychology Press.

Lennon, A., Watson, B., Arlidge, C., & Fraine, G. (2011). 'You're a bad driver but I just made a mistake': Attribution differences between the 'victims' and 'perpetrators' of scenario-based aggressive driving incidents. *Transportation Research Part F: Traffic Psychology and Behaviour, 14*(3), 209–221.

Lenz, K. M., & McCarthy, M. M. (2010). Organized for sex—Steroid hormones and the developing hypothalamus. *European Journal of Neuroscience, 32*(12), 2096–2104.

Leonard, B. E. (2003). *Fundamentals of psychopharmacology* (3rd ed.). Hoboken, NJ: Wiley.

LePage, P., Akar, H., Temli, Y., Sen, D., Hasser, N., & Ivins, I. (2011). Comparing teachers' views on morality and moral education, a comparative study in Turkey and the United States. *Teaching and Teacher Education, 27*(2), 366–375.

Lepistö, S., Luukkaala, T., & Paavilainen, E. (2011). Witnessing and experiencing domestic violence: A descriptive study of adolescents. *Scandinavian Journal of Caring Sciences, 25*(1), 70–80.

Lepper, M. R., Greene, D., & Nisbett, R. E. (1973). Undermining children's intrinsic interest with extrinsic rewards: A test of the overjustification hypothesis. *Journal of Personality and Social Psychology, 28*, 129–137.

Leri, A., Anversa, P., & Frishman, W. H. (Eds.) (2007). *Cardiovascular regeneration and stem cell therapy*. Malden, MA: Wiley-Blackwell.

Leslie, M. (2000, July/August). The Vexing Legacy of Lewis Terman. *Stanford Magazine*. Retrieved from http://www.stanfordalumni.org/news/magazine/2000/julaug/articles/terman.html.

Leung, A. K. Y., Kim, Y. H., Zhang, Z. X., Tam, K. P., & Chiu, C. Y. (2012). Cultural construction of success and epistemic motives moderate American-Chinese differences in reward allocation biases. *Journal of Cross-Cultural Psychology, 43*(1), 46–52.

LeVay, S. (2003). Queer science: The use and abuse of research into homosexuality. *Archives of Sexual Behavior, 32*(2), 187–189.

LeVay, S. (2011). *Gay, straight, and the reason why: The science of sexual orientation*. New York, NY: Oxford University Press.

Levine, J. R. (2001). *Why do fools fall in love: Experiencing the magic, mystery, and meaning of successful relationships*. New York, NY: Jossey-Bass.

Levinson, D. J. (1977). The mid-life transition, *Psychiatry, 40*, 99–112.

Levinson, D. J. (1996). *The seasons of a woman's life*. New York, NY: Knopf.

Levinthal, C. (2011). *Drugs, behavior, and modern society* (7th ed.). Boston, MA: Prentice Hall.

Levitan, L. C. (2008). Giving prejudice an attitude adjustment: The implications of attitude strength and social network attitudinal composition for prejudice and prejudice reduction. *Dissertation Abstracts International: Section B: The Sciences and Engineering, 68*(8-B), 5634.

Levy, B., Wegman, D., Baron, S., & Sokas, R. (2011). *Occupational and environmental health: Recognizing and preventing disease and injury* (6th ed.). New York, NY: Oxford University Press.

Lew, A. R. (2011). Looking beyond the boundaries: Time to put landmarks back on the cognitive map? *Psychological Bulletin, 137*(3), 484–507.

Lewin, D., & Herron, H. (2007). Signs, symptoms, and risk factors: Health visitors' perspectives of child neglect. *Child Abuse Review, 16*, 93–107.

Li, J-C. A. (2008). Rethinking the case against divorce. *Dissertation Abstracts International Section A: Humanities and Social Sciences, 68*, 4093.

Li, N. P., Valentine, K. A., & Patel, L. (2011). Mate preferences in the US and Singapore: A cross-cultural test of the mate preference priority model. *Personality and Individual Differences, 50*(2), 291–294.

Li, Y., Li, Y., McKay, R., Riethmacher, D., & Parada, L. F. (2012). Neurofibromin modulates adult hippocampal neurogenesis and behavioral effects of antidepressants. *The Journal of Neuroscience, 32*(10), 3529–3539.

Libon, D. J., Xie, S. X., Moore, P., Farmer, J., Antani, S., McCawley, G., Cross, K., & Grossman, M. (2007). Patterns of neuropsychological impairment in frontotemporal dementia. *Neurology, 68*, 369–375.

Lilienfeld, S. O., Lynn, S. J., Ruscio, J., & Beyerstein, B. L. (2010). *50 great myths of popular psychology: Shattering widespread misconceptions about human behavior*. Malden, MA: Wiley-Blackwell.

Lim, L., Chang, W., Yu, X., Chiu, H., Chong, M., & Kua, E. (2011). Depression in Chinese elderly populations. *Asia-Pacific Psychiatry, 3*(2), 46–53.

Linden, E. (1993). Can animals think? *Time*, pp. 54–61.

Linnet, J., Møller, A., Peterson, E., Gjedde, A., & Doudet, D. (2011). Dopamine release in ventral striatum during Iowa Gambling Task performance is associated with increased excitement levels in pathological gambling. *Addiction, 106*(2), 383–390.

Lippa, R. A. (2007). The preferred traits of mates in a cross-national study of heterosexual and homosexual men and women: An examination of biological and cultural influences. *Archives of Sexual Behavior, 36*, 193–208.

Lipsanen, T., Korkeila, J., Peltola, P., Järvinen, J., Langen, K., & Lauerma, H. (2004). Dissociative disorders among psychiatric patients: Comparison with a nonclinical sample. *European Psychiatry, 19*(1), 53–55.

Liu, J. H., & Latané, B. (1998). Extremitization of attitudes: Does thought- and discussion-induced polarization cumulate? *Basic and Applied Social Psychology, 20*, 103–110.

Livingston, J. A. (1999). Something old and something new: Love, creativity, and the enduring relationship. *Bulletin of the Menninger Clinic, 63*(1), 40–52.

Livingston, M. (2011). A longitudinal analysis of alcohol outlet density and domestic violence. *Addiction, 106*(5), 919–925.

Livingston, R. W., & Drwecki, B. B. (2007). Why are some individuals not racially biased? Susceptibility to affective conditioning predicts nonprejudice toward Blacks. *Psychological Science, 18*, 816–823.

Lizarraga, M. L. S., & Ganuza, J. M. G. (2003). Improvement of mental rotation in girls and boys. *Sex Roles, 49*(5–6), 277–286.

Lodge, D. J., & Grace, A. A. (2011). Developmental pathology, dopamine, stress and schizophrenia. *International Journal of Developmental Neuroscience, 29*(3), 207–213.

Loewenthal, D., & Winter, D. (Eds.). (2006). *What is psychotherapeutic research?* London, England: Karnac Books.

Loftus, E. (1982). Memory and its distortions. In A. G. Kraut (Ed.), *The G. Stanley Hall Lecture Series Vol. 2*, (pp. 123–154). Washington, DC: American Psychological Association.

Loftus, E. F. (2000). Remembering what never happened. In E. Tulving, et al. (Eds.), *Memory, consciousness, and the brain:*

*The Tallinn Conference* (pp. 106–118). Philadelphia, PA: Psychology Press/Taylor & Francis.

Loftus, E. F. (2001). Imagining the past. *Psychologist, 14*(11), 584–587.

Loftus, E. F. (2007). Memory distortions: Problems solved and unsolved. In M. Garry, & H. Hayne (Eds.), *Do justice and let the skies fall* (pp. 1–14). Mahwah, NJ: Erlbaum.

Loftus, E. F. (2011). How I got started: From semantic memory to expert testimony. *Applied Cognitive Psychology, 25*(2), 347–348.

Loftus, E. F., & Cahill, L. (2007). Memory distortion from misattribution to rich false memory. In J. S. Nairne (Ed.), *The foundations of remembering: Essays in honor of Henry L. Roediger, III* (pp. 413–425). New York, NY: Psychology Press.

Loftus, E., & Ketcham, K. (1994). *The myth of repressed memories: False memories and allegations of sexual abuse*. New York, NY: St. Martin's Press.

Loo, C. K., Mahon, M., Katalinic, N., Lyndon, B., & Hadzi-Pavlovic, D. (2011). Predictors of response to ultrabrief right unilateral electroconvulsive therapy. *Journal of Affective Disorders, 130*(1–2), 192–197.

Lopez-Munoz, F., & Alamo, C. (2011). *Neurobiology of depression*. Boca Raton, FL: CRC Press.

Lorenz, K. Z. (1937). The companion in the bird's world. *Auk, 54*, 245–273.

Lores-Arnaiz, S., Bustamante, J., Czernizyniec, A., Galeano, P., Gervasoni, M. G., Martinez, A. R., Paglia, N., Cores, V., & Lores-Arnaiz, M. R. (2007). Exposure to enriched environments increases brain nitric oxide synthase and improves cognitive performance in prepubertal but not in young rats. *Behavioural Brain Research, 184*, 117–123.

Lott, D. A. (2000). *The new flirting game*. London, England: Sage.

Lovibond, P. (2011). Learning and anxiety: A cognitive perspective. In T. Schachtman & S. Reilly (Eds.), *Associative learning and conditioning theory: Human and non-human applications* (pp. 104–120). Oxford, UK: Oxford University Press.

Lozano, A. M. (2011). Harnessing plasticity to reset dysfunctional neurons. *The New England Journal of Medicine, 364*(14), 1367–1368.

Luborsky, L., Singer, B., & Luborsky, L. (1975). Comparative studies of psychotherapies. Is it true that "everyone has won and all must have prizes"? *Archives of General Psychiatry, 32*(8), 995–1008.

Luborsky, L. B., Barrett, M. S., Antonuccio, D. O., Shoenberger, D., & Stricker, G. (2006). What else materially influences what is represented and published as evidence? In J.C. Norcross, L.E. Beutler, & R.F. Levant (Eds.), *Evidence-based practices in mental health: Debate and dialogue on the fundamental questions*, (pp. 257–298). Washington, DC: American Psychological Association.

Lundberg, U. (2011). Neuroendocrine measures. In R. J. Contrada & A. Baum (Eds.), *The handbook of stress science: Biology, psychology, and health* (pp. 531–542). New York, NY: Springer.

Luppa, M., Sikorski, C., Luck, T., Ehreke, L., Konnopka, A., Wiese, B., . . . Riedel-Heller, S. G. (2012). Age- and gender-specific prevalence of depression in latest-life – Systematic review and meta-analysis. *Journal of Affective Disorders, 136*(3), 212–221.

Luria, A. R. (1968). *The mind of a mnemonist: A little book about a vast memory.* New York, NY: Basic Books.

Lyn, H., Greenfield, P. M., Savage-Rumbaugh, S., Gillespie-Lynch, K., & Hopkins, W. D. (2011). Nonhuman primates do declare! A comparison of declarative symbol and gesture use in two children, two bonobos, and a chimpanzee. *Language and Communication, 31*(1), 63–74.

Lynam, D. R., Miller, J. D., Miller, D. J., Bornovalova, M. A., & Lejuez, C. W. (2011). Testing the relations between impulsivity-related traits, suicidality, and nonsuicidal self-injury: A test of the incremental validity of the UPPS model. *Personality Disorders: Theory, Research, and Treatment, 2*(2), 151–160.

Lynn, S. J., Rhue, J. W. & Kirsch, I. (Eds.). (2010). *Handbook of clinical hypnosis* (2nd ed.). Washington, DC: American Psychological Association.

Lyubimov, N. N. (1992). Electrophysiological characteristics of sensory processing and mobilization of hidden brain reserves. *2nd Russian-Swedish Symposium New Research in Neurobiology*. Moscow: Russian Academy of Science Institute of Human Brain.

Maas, J., B., Wherry, M. L., Axelrod, D. J., Hogan, B. R., & Bloomin, J. (1999). *Power sleep: The revolutionary program that prepares your mind for peak performance*. New York, NY: HarperPerennial.

Macmillan, M. (2008). Phineas Gage — Unravelling the myth. *Psychologist, 21*(9), 828–831.

Macmillan, M. B. (2000). *An odd kind of fame: Stories of Phineas Gage*. Cambridge, MA: MIT Press.

Macmillan, M., & Lena, M.L. (2010). Rehabilitating Phineas Gage. *Neuropsychological Rehabilitation, 20*(5), 641–658.

Maddi, S. R., Harvey, R. H., Khoshaba, D. M., Fazel, M., & Resurreccion, N. (2012). The relationship of hardiness and some other relevant variables to college performance. *Journal of Humanistic Psychology, 52*(2), 190–205.

Madey, S. F., & Rodgers, L. (2009). The effect of attachment and Sternberg's Triangular Theory of Love on relationship satisfaction. *Individual Differences Research, 7*(2), 76–84.

Madole, K. L., Oakes, L. M., & Rakison, D. H. (2011). Information-processing approaches to infants' developing representation of dynamic features. In L. M. Oakes, C. H. Cashon, M. Casasola, & D. H. Rakison (Eds.), *Infant perception and cognition: Recent advances, emerging theories, and future directions* (pp. 153–177). New York, NY: Oxford University Press.

Maehr, M. L., & Urdan, T. C. (2000). *Advances in motivation and achievement: The role of context*. Greewich, CT: JAI Press.

Mahoney, C. R., Castellani, J., Kramer, F. M., Young, A., & Lieberman, H. R. (2007). Tyrosine supplementation mitigates working memory decrements during cold exposure. *Physiology and Behavior, 92*, 575–582.

Main, M., & Solomon, J. (1986). Discovery of an insecure-disorganized attachment pattern. In T. Brazelton & M. W. Yogman (Eds.), *Affective development in infancy* (pp. 95–124). Westport, CT: Ablex Publishing.

Main, M., & Solomon, J. (1990). Procedures for identifying infants as disorganized/disoriented during the Ainsworth Strange Situation. In M. T. Greenberg, D. Cicchetti, & E. M. Cummings (Eds.), *Attachment in the preschool years: Theory, research, and intervention,The John D and Catherine T. MacArthur Foundation series on mental health and development* (pp. 121–160). Chicago, IL: University of Chicago Press.

Maisto, S. A., Galizio, M., & Connors, G. J. (2011). *Drug use and abuse* (6th ed.). Belmont: Cengage.

Major, B., Spencer, S., Schmader, T., Wolfe, C., & Crocker, J. (1998). Coping with negative stereotypes about intellectual performance: The role of psychological disengagement. *Personality and Social Psychology Bulletin, 24*(1), 34–50.

Malik, S., McGlone, F., & Dagher, A. (2011). State of expectancy modulates the neural response to visual food stimuli in humans. *Appetite, 56*(2), 302–309.

Mancia, M., & Baggott, J. (2008). The early unrepressed unconscious in relation to Matte-Blanco's thought. *International Forum of Psychoanalysis, 17*(4), 201–212.

Manly, J. J., Byrd, D., Touradji, P., Sanchez, D., & Stern, Y. (2004). Literacy and cognitive change among ethnically diverse elders. *International Journal of Psychology, 39*(1), 47–60.

Manning, J. (2007). The use of meridian-based therapy for anxiety and phobias.

*Australian Journal of Clinical Hypnotherapy and Hypnosis*, *28*, 45–50.
Marazziti, D., Masala, I., Baroni, S., Polini, M., Massimetti, G., Giannaccini, G., Betti, L., Italiani, P., Fabbrini, L., Caglieresi, C., Moschini, C., Canale, D., Lucacchini, A., & Mauri, M. (2010). Male axillary extracts modify the affinity of the platelet serotonin transporter and impulsiveness in women. *Physiology and Behavior*, *100*(4), 364–368.
Market Opinion Research International (MORI) (2005, January). *Use of animals in medical research for coalition for medical progress*. London, England: Author.
Markham, J. A., & Koenig, J. I. (2011). Prenatal stress: Role in psychotic and depressive diseases. *Psychopharmacology*, *214*(1), 89–106.
Marks, D. F., Murray, M. D., Evans, B., & Estacio, E. V., (2011). *Health psychology: Theory, research and practice* (3rd ed.). London, UK: Sage.
Marks, L. D., Hopkins, K. C., Monroe, P. A., Nesteruk, O., & Sasser, D. D. (2008). "Together, we are strong": A qualitative study of happy, enduring African American marriages. *Family Relations*, *57*, 172–185.
Markstrom, C. A., & Marshall, S. K. (2007). The psychosocial inventory of ego strengths: Examination of theory and psychometric properties. *Journal of Adolescence*, *30*, 63–79.
Markus, H. R., & Kitayama, S. (2003). Culture, self, and the reality of the social. *Psychological Inquiry*, *14*(3–4), 277–283.
Markus, H. R., & Kitayama, S. (2010). Cultures and selves: A cycle of mutual constitution. *Perspectives on Psychological Science*, *5*(4), 420–430.
Marques, L., Robinaugh, D., LeBlanc, N., & Hinton, D. (2011). Cross-cultural variations in the prevalence and presentation of anxiety disorders. *Expert Review of Neurotherapeutics*, *11*(2), 313–322.
Marsee, M. A., Weems, C. F., Taylor, L. K. (2008). Exploring the association between aggression and anxiety in youth: A look at aggressive subtypes, gender, and social cognition. *Journal of Child and Family Studies*, *17*, 154–168.
Marshall, J. A. R. (2011). Ultimate causes and the evolution of altruism. *Behavioral Ecology and Sociobiology*, *65*(3), 503–512.
Marshall, V., & Bengston, V. (2011). Theoretical perspectives on the sociology of aging. In R. Settersten & J. Angel (Eds.), *Handbook of Sociology of Aging* (pp. 17–33). New York, NY: Springer.
Martella, D., Casagrande, M., & Lupiáñez, J. (2011). Alerting, orienting and executive control: The effects of sleep deprivation on attentional networks. *Experimental Brain Research*, *210*(1), 81–89.
Martens, U., Ansorge, U., & Kiefer, M. (2011). Controlling the unconscious: Attentional task sets modulate subliminal semantic and visuomotor processes differentially. *Psychological Science*, *22*(2), 282–291.
Marti, F., Arib, O., Morel, C., Dufresne, V., Maskos, U., Corringer, P-J., de Beaurepaire, R., & Faure, P. (2012). Smoke extracts and nicotine, but not tobacco extracts, potentiate firing and burst activity of ventral tegmental area dopaminergic neurons in mice. *Neuropsychopharmacology*, *36*(11), 2244–2257.
Martin, C. L., & Fabes, R. (2009). *Discovering child development* (2nd ed.). Belmont, CA: Cengage.
Martin, L. J., & Carron, A. V. (2012). Team attributions in sport: A meta-analysis. *Journal of Applied Sport Psychology*, *24*(2), 157–174.
Martin, S. (2008). How psychology helps you every day. *Psychology Today Monitor*. Retrieved from http://www.apa.org/monitor/2008/10/sternberg.aspx
Martinko, M. J., Harvey, P., & Dasborough, M. T. (2011). Attribution theory in the organizational sciences: A case of unrealized potential. *Journal of Organizational Behavior*, *32*(1), 144–149.
Marusich, J. A., Darna, M., Charnigo, R. J., Dwoskin, L. P., & Bardo, M. T. (2011). A multivariate assessment of individual differences in sensation seeking and impulsivity as predictors of amphetamine self-administration and prefrontal dopamine function in rats. *Experimental and Clinical Psychopharmacology*. *Experimental and Clinical Psychopharmacology*, *19*(4), 275–284.
Marx, D. M., Ko, S. J., & Friedman, R. A. (2009). The "Obama effect": How a salient role model reduces race-based performance differences. *Journal of Experimental Social Psychology*, *45*(4), 953–956.
Maslow, A. H. (1954). *Motivation and personality*. New York, NY: Harper & Row.
Maslow, A. H. (1970). *Motivation and personality (2nd ed.)*. New York, NY: Harper & Row.
Maslow, A. H. (1999). *Toward a psychology of being* (3rd ed.). New York, NY: John Wiley & Sons.
Mason, M. F., & Morris, M. W. (2010). Culture, attribution and automaticity: A social cognitive neuroscience view. *Social Cognitive and Affective Neuroscience*, *5*(2–3), 292–306.
Masten A. S., & Coatsworth, J. D. (1998). The development of competence in favorable and unfavorable environments. *American Psychologist*, *53*, 205–220.
Masten, A. S., & Narayan, A. J. (2012). Child development in the context of disaster, war, and terrorism: Pathways of risk and resilience. *Annual Review of Psychology*, *63*(1), 227–257.
Masten, A. S., & Wright, M. O. (2010). Resilience over the lifespan: Developmental perspectives on resistance, recovery, and transformation. In J. W. Reich, J. Alex, & J. S. Stuart (Eds.), *Handbook of adult resilience* (pp. 213–237). New York, NY: Guilford Press.
Masters, W. H., & Johnson, V. E. (1961). Orgasm, anatomy of the female. In A. Ellis & A. Abarbonel (Eds.), *Encyclopedia of Sexual Behavior*, Vol. 2. New York, NY: Hawthorn.
Masters, W. H., & Johnson, V. E. (1966). *Human sexual response*. Boston, MA: Little, Brown.
Masters, W. H., & Johnson, V. E. (1970). *Human sexual inadequacy*. Boston, MA: Little, Brown.
Matlin, M. W. (2012). *The psychology of women* (7th ed.). Belmont, CA: Cengage.
Matlin, M. W. (2013). *Cognition* (8th ed.). Hoboken, NJ: John Wiley & Sons.
Matlin, M. W., & Foley, H. J. (1997). *Sensation and perception* (4th ed.). Boston, MA: Allyn and Bacon.
Matsumoto, D. (1992). More evidence for the universality of a contempt expression. *Motivation and Emotion*, *16*, 363–368.
Matsumoto, D. (2000). *Culture and psychology: People around the world*. Belmont, CA: Wadsworth.
Matsumoto, D. (2010). *APA handbook of interpersonal communication*. Washington, DC: American Psychological Association.
Matsumoto, D., & Hwang, H. S. (2011). Culture, emotion, and expression. In M. J. Gelfand, C.-y. Chiu, & Y.-Y. Hong (Eds.), *Advances in culture and psychology*. *Advances in culture and psychology* (Vol. 1) (pp. 53–98). New York, NY: Oxford University Press.
Matsumoto, D., & Juang, L. (2008). *Culture and psychology* (4th ed.). Belmont, CA: Cengage.
Mayberg, H. S. (2006). Defining neurocircuits in depression: Strategies toward treatment selection based on neuroimaging phenotypes. *Psychiatric Annals*, *36*(4), 259–268.
May-Collado, L. J. (2010). Changes in whistle structure of two dolphin species during interspecific associations. *Ethology*, *116*(11), 1065–1074.
Mazzoni, G., & Vannucci, M. (2007). Hindsight bias, the misinformation effect, and false autobiographical memories. *Social Cognition*, *25*, 203–220.
McCabe, C., & Rolls, E. T. (2007). Umami: A delicious flavor formed by convergence of taste and olfactory pathways in the human brain. *European Journal of Neuroscience*, *25*, 1855–1864.
McCart, M. R., Zajac, K., Danielson, C. K., Strachan, M., Ruggiero, K. J., Smith,

D. W., Saunders, B., & Kilpatrick, D. G. (2011). Interpersonal victimization, posttraumatic stress disorder, and change in adolescent substance use prevalence over a ten-year period. *Journal of Clinical Child and Adolescent Psychology*, *40*(1), 136–143.

McClelland, D. C. (1958). Risk-taking in children with high and low need for achievement. In J. W. Atkinson (Ed.), *Motives in fantasy, action, and society* (pp. 306–321). Princeton, NJ: Van Nostrand.

McClelland, D. C. (1987). Characteristics of successful entrepreneurs. *Journal of Creative Behavior*, *3*, 219–233.

McClelland, D. C. (1993). Intelligence is not the best predictor of job performance. *Current Directions in Psychological Science*, *2*, 5–6.

McClure, J., Meyer, L. H., Garisch, J., Fischer, R., Weir, K. F., & Walkey, F. H. (2011). Students' attributions for their best and worst marks: Do they relate to achievement? *Contemporary Educational Psychology*, *36*(2), 71–81.

McCrae, R. (2011). Cross-Cultural Research on the Five-Factor Model of Personality. *Online Readings in Psychology and Culture, Unit 4*. Retrieved from http://scholarworks.gvsu.edu/orpc/vol4/iss4/1

McCrae, R. R. (2004). Human nature and culture: A trait perspective. *Journal of Research in Personality*, *38*(1), 3–14.

McCrae, R. R., Costa, P. T. Jr., Hrebícková, M., Urbánek, T., Martin, T. A., Oryol, V. E., Rukavishnikov, A. A., & Senin, I. G. (2004). Age differences in personality traits across cultures: Self-report and observer perspectives. *European Journal of Personality*, *18*(2), 143–157.

McCrae, R. R., Costa, P. T., Jr., Ostendorf, F., Angleitner, A., Hrebickova, M., Avia, M. D., Sanz, J., Sanchez-Bernardos, M. L., Kusdil, M. E., Woodfield, R., Saunders, P. R., & Smith, P. B. (2000). Nature over nurture: Temperament, personality, and life span development. *Journal of Personality and Social Psychology*, *78*(1), 173–186.

McDougall, W. (1908). *Social psychology*. New York, NY: Putnam's Sons.

McEvoy, P. M. (2007). Effectiveness of cognitive behavioural group therapy for social phobia in a community clinic: A benchmarking study. *Behaviour Research and Therapy*, *45*, 3030–3040.

McFarlane, W. R. (2011). Integrating the family in the treatment of psychotic disorders. In R. Hagen, D. Turkington, T. Berge, & R. W. Gråwe (Eds.), *International Society for the Psychological Treatments of the Schizophrenias and Other Psychoses. CBT for psychosis: A symptom-based approach* (pp. 193–209). New York, NY: Routledge/Taylor & Francis Group.

McGrath, J. (2011). Environmental risk factors for schizophrenia. In D. Weinberger, & P. Harrison (Eds.), *Schizophrenia* (3rd ed.) (pp. 226–244). Hoboken, NJ: John Wiley & Sons.

McGue, M., Bouchard, T. J., Iacono, W. G., & Lykken, D. T. (1993). Behavioral genetics of cognitive ability: A life-span perspective. In R. Plomin & G. McClearn (Eds.), *Nature, nurture, and psychology* (pp. 59–76). Washington, DC: American Psychological Association.

McGuinness, T. M., & Schneider, K. (2007). Poverty, child mistreatment, and foster care. *Journal of the American Psychiatric Nurses Association*, *13*, 296–303.

McKellar, P. (1972). Imagery from the standpoint of introspection. In P. W. Sheehan (Ed.), *The function and nature of imagery*. New York, NY: Academic Press.

McKim, W. A. (2002). *Drugs and behavior: An introduction to behavioral pharmacology* (5th ed). Englewood Cliffs, NJ: Prentice Hall.

McKinney, C., Donnelly, R., & Renk, K. (2008). Perceived parenting, positive and negative perceptions of parents, and late adolescent emotional adjustment. *Child and Adolescent Mental Health*, *13*, 66–73.

McKoon, G., & Ratcliff, R. (2012). Aging and IQ effects on associative recognition and priming in item recognition. *Journal of Memory and Language*, *66*(3), 416–437.

McLoyd, V. C. (2011). How money matters for children's socioemotional adjustment: Family processes and parental investment. In G. Carlo, L. J. Crockett, & M. A. Carranza (Eds.), *Nebraska symposium on motivation. Health disparities in youth and families: Research and applications* (pp. 33–72). New York, NY: Springer.

McNamara, J. M., Barta, Z., Fromhage, L., & Houston, A. I. (2008). The coevolution of choosiness and cooperation. *Nature*, *451*, 189–201.

Mead, E. (2012). *Family therapy education and supervision*. Hoboken, NJ: John Wiley & Sons.

Meeus, W., & Raaijmakers, Q. (1989). Autoritätsgehorsam in Experimenten des Milgram-Typs: Eine Forschungsübersicht (Obedience to authority in Milgram-type studies: A research review). *Zeitschrift für Sozialpsychologie*, *20*(2), 70–85.

Mehrabian, A. (1968). A relationship of attitude to seated posture orientation and distance. *Journal of Personality and Social Psychology*, *10*, 26–30.

Mehrabian, A. (1971). *Silent messages*. Belmont, CA: Wadsworth.

Mehrabian, A. (2007). *Nonverbal communication*. New Brunswick, NJ: Aldine Transaction.

Mehta, P. H., & Gosling, S. D. (2006). How can animal studies contribute to research on the biological bases of personality? In T. Canli (Ed.), *Biology of personality and individual differences* (pp. 427–448). New York, NY: Guilford.

Mehta, S., Orenczuk, S., Hansen, K. T., Aubut, J.-A. L., Hitzig, S. L., Legassic, M., & Teasell, R. (2011). An evidence-based review of the effectiveness of cognitive behavioral therapy for psychosocial issues post-spinal cord injury. *Rehabilitation Psychology*, *56*(1), 15–25.

Meltzoff, A. N., & Moore, M. K. (1977). Imitation of facial and manual gestures by human neonates. *Science*, *198*, 75–78.

Meltzoff, A. N., & Moore, M. K. (1985). Cognitive foundations and social functions of imitation and intermodal representation in infancy. In J. Mehler & R. Fox (Eds.), *Neonate cognition: Beyond the blooming buzzing confusion* (pp. 139–156). Hillsdale, NJ: Erlbaum.

Meltzoff, A. N., & Moore, M. K. (1994). Imitation, memory, and the representation of persons. *Infant Behavior and Development*, *17*, 83–99.

Melzack, R. (1999). Pain and stress: A new perspective. In R. J. Gatchel & D. C. Turk (Eds.), *Psychosocial factors in pain: Critical perspectives* (pp. 89–106). New York, NY: Guilford Press.

Melzack, R., & Wall, P. D. (1965). Pain mechanisms: A new theory. *Science*, *150*, 971–979.

Mendelson, T., K. Turner, A., & Tandon, S. D. (2010). Violence exposure and depressive symptoms among adolescents and young adults disconnected from school and work. *Journal of Community Psychology*, *38*(5), 607–621.

Mercadillo, R. E., Díaz, J. L., Pasaye, E. H., & Barrios, F. A. (2011). Perception of suffering and compassion experience: Brain gender disparities. *Brain and Cognition*, *76*(1), 5–14.

Merrill, D. A., & Small, G. W. (2011). Prevention in psychiatry: Effects of healthy lifestyle on cognition. *Psychiatric Clinics of North America*, *34*(1), 249–261.

Messer, S., & Gurman, A. (2011). *Essential psychotherapies: Theory and practice* (3rd ed.). New York, NY: Guilford Press.

Messinger, L.M. (2001). *Georgia O'Keeffe*. London: Thames & Hudson.

Metcalf, P., & Huntington, R. (1991). *Celebrations of death: The anthropology of mortuary ritual* (2nd ed.). Cambridge, England: Cambridge University Press.

Meyer, L. H. (2011). Making sense of differently able minds. *PsyCRITIQUES*. Retrieved from http://psycnet.apa.org/critiques/56/9/6.html.

Meyers, E. M., Xue-Lian, Q., & Constantinidis, C. (2012). Incorporation of new information into prefrontal cortical activity after learning working memory tasks. *Proceedings of the National Academy of Sciences of the United States of America*, *109*(12), 4651–4656.

Mieg, H. A. (2011). Focused cognition: Information integration and complex problem solving by top inventors. In K. L. Mosier & U. M. Fischer (Eds.), *Expertise: Research and applications. Informed by knowledge: Expert performance in complex situations* (pp. 41–54). New York, NY: Psychology Press.

Migacheva, K., Crocker, J, & Tropp, L. (2011). Focusing beyond the self: Goal orientations in intergroup relations. In Tropp, L. R., & Mallett, R. K. (Eds.): *Moving beyond prejudice reduction; Pathways to positive intergroup relations* (pp. 99–115). Washington: American Psychological Association.

Mihic, L., Wells, S., Graham, K., Tremblay, P. F., & Demers, A. (2009). Situational and respondent-level motives for drinking and alcohol-related aggression: A multilevel analysis of drinking events in a sample of Canadian university students. *Addictive Behaviors*, *34*(3), 264–269.

Milgram, S. (1963). Behavioral study of obedience. *Journal of Abnormal and Social Psychology*, 67, 371–378.

Milgram, S. (1974). *Obedience to authority: An experimental view*. New York, NY: Harper & Row.

Miller, G. A. (1956). The magical number seven, plus or minus two: Some limits on our capacity for processing information. *Psychological Review*, *63*, 81–97.

Miller, J. G., & Bersoff, D. M. (1998). The role of liking in perceptions of the moral responsibility to help: A cultural perspective. *Journal of Experimental Social Psychology*, *34*, 443–469.

Miller, P. K., Rowe, L., Cronin, C., & Bampouras, T. M. (2012). Heuristic reasoning and the observer's view: The influence of example-availability on ad-hoc frequency judgments in sport. *Journal of Applied Sport Psychology*, *24*(3), 290–302.

Millon, T. (2004). *Masters of the mind: Exploring the story of mental illness from ancient times to the new millennium*. Hoboken, NJ: John Wiley & Sons.

Miltenberger, R. G. (2011). *Behavior modification: Principles and procedures*. (5th ed.). Beverly, MA: Wadsworth.

Mineka, S., & Oehlberg, K. (2008). The relevance of recent developments in classical conditioning to understanding the etiology and maintenance of anxiety disorders. *Acta Psychologica*, *127*, 567–580.

Mingo, C., Herman, C. J., & Jasperse, M. (2000). Women's stories: Ethnic variations in women's attitudes and experiences of menopause, hysterectomy, and hormone replacement therapy. *Journal of Women's Health and Gender Based Medicine*, *9*, S27–S38.

Mingroni, M. A. (2004). The secular rise in IQ. *Intelligence*, *32*, 65–83.

Minuchin, S. (2011). *Families and family therapy* (2nd ed.). New York, NY: Routledge.

Miracle, T. S., Miracle, A. W., & Baumeister, R. F. (2006). *Human sexuality: Meeting your basic needs* (2nd ed.). Upper Saddle River, NJ: Pearson Education.

Mischel, W, Shoda, Y., & Ayduk, O. (2008). *Introduction to personality: Toward an integrative science of the person* (8th ed.). Hoboken, NJ: John Wiley & Sons.

Mischel, W. (1968). *Personality and assessment*. New York, NY: John Wiley & Sons.

Mischel, W. (1984). Convergences and challenges in the search for consistency. *American Psychologist*, *39*(4), 351–364.

Mischel, W. (2004). Toward an integrative science of the person. *Annual Review of Psychology*, *55*(1), 1–22.

Mischel, W., & Shoda, Y. (2008). Toward a unified theory of personality: Integrating dispositions and processing dynamics within the cognitive-affective processing system. In O. P. John, R. W. Robins & L. A. Pervin (Eds.), *Handbook of personality: Theory and research* (*3rd ed.*) (pp. 208–241). New York, NY: Guilford Press.

Mita, T. H., Dermer, M., & Knight, J. (1977). Reversed facial images and the mere-exposure hypothesis. *Journal of Personality and Social Psychology*, *35*(8), 597–601.

Mitchell, B. A., & Lovegreen, L. D. (2009). The empty nest syndrome in midlife families: A multimethod exploration of parental gender differences and cultural dynamics. *Journal of Family Issues*, *30*(12), 1651–1670.

Mittag, O., & Maurischat, C. (2004). Die Cook-Medley Hostility Scale (Ho-Skala) im Vergleich zu den Inhaltsskalen "Zynismus", "Ärger" sowie "Typ A" aus dem MMPI-2: Zur zukünftigen Operationalisierung von Feindseligkeit [A comparison of the Cook-Medley Hostility Scale (Ho-scale) and the content scales "cynicism," "anger," and "type A" out of the MMPI-2: On the future assessment of hostility]. *Zeitschrift für Medizinische Psychologie*, *13*(1), 7–12.

Miyake, N., Thompson, J., Skinbjerg, M., & Abi-Dargham, A. (2011). Presynaptic dopamine in schizophrenia. *CNS Neuroscience and Therapeutics*, *17*(2), 104–109.

Mograss, M. A., Guillem, F., & Godbout, R. (2008). Event-related potentials differentiates the processes involved in the effects of sleep on recognition memory. *Psychophysiology*, *45*, 420–434.

Moilanen, J., Aalto, A.-M., Hemminki, E., Aro, A. R., Raitanen, J., & Luoto, R. (2010). Prevalence of menopause symptoms and their association with lifestyle among Finnish middle-aged women. *Maturitas*, *67*(4), 368–374.

Moitra, E., Beard, C., Weisberg, R. B., & Keller, M. B. (2011). Occupational impairment and social anxiety disorder in a sample of primary care patients. *Journal of Affective Disorders*, *130*(1–2), 209–212.

Mond, J., & Arrighi, A. (2011). Gender differences in perceptions of the severity and prevalence of eating disorders. *Early Intervention in Psychiatry*, *5*(1), 41–49.

Mondimore, F., & Kelly, P. (2011). *Borderline personality disorder: New reasons for hope*. Baltimore, MD: Johns Hopkins University Press.

Moneta, G. B., & Siu, C. M. Y. (2002). Trait intrinsic and extrinsic motivations, academic performance, and creativity in Hong Kong college students. *Journal of College Student Development*, *43*(5), 664–683.

Monin, B. (2003). The warm glow heuristic: When liking leads to familiarity. *Journal of Personality and Social Psychology*, *85*(6), 1035–1048.

Monks, C. P., Ortega-Ruiz, R., & Rodríguez-Hidalgo, A. J. (2008). Peer victimization in multicultural schools in Spain and England. *European Journal of Developmental Psychology*, *5*, 507–535.

Montag, C., Weber, B., Trautner, P., Newport, B., Markett, S., Walter, N. T., . . . Reuter, M. (2012). Does excessive play of violent first-person-shooter-video-games dampen brain activity in response to emotional stimuli? *Biological Psychology*, *89*(1), 107–111.

Montoro-García, S., Shantsila, E., & Lip, G. Y. H. (2011). Platelet reactivity in prolonged stress disorders—A link with cardiovascular disease? *Psychoneuroendocrinology*, *36*(2), 159–160.

Moodley, R. (2012). *Handbook of counseling and psychotherapy in an international context*. New York, NY: Routledge.

Moodley, R., & Sutherland, P. (2010). Psychic retreats in other places: Clients who seek healing with traditional healers and psychotherapists. *Counselling Psychology Quarterly*, *23*(3), 267–282.

Moore, F., Filippou, D., & Perrett, D. (2011). Intelligence and attractiveness in the face: Beyond the attractiveness halo effect. *Journal of Evolutionary Psychology*, *9*(3), 205–217.

Moore, M. M. (1998). The science of sexual signaling. In G. C. Brannigan, E. R. Allgeier, & A. R. Allgeier (Eds.), *The sex scientists* (pp. 61–75). New York, NY: Longman.

Moran, T., & Dailey, M. (2011). Intestinal feedback signaling and satiety. *Physiology and Behavior*, *105*(1), 77–81.

Morewedge, C. K., & Norton, M. J. (2009). When dreaming is believing: The (motivated) interpretation of dreams. *Journal of Personality and Social Psychology*, *96*, 249–264.

Morgan, C. D., & Murphy, C. (2010). Differential effects of active attention and age on event-related potentials to visual

and olfactory stimuli. *International Journal of Psychophysiology*, *78*(2), 190–199.
MORI. (2005, January). Use of animals in medical research for coalition for medical progress. *Market Opinion Research International*. London, England: Author.
Mori, K., & Arai, M. (2010). No need to fake it: Reproduction of the Asch experiment without confederates. *International Journal of Psychology*, *45*(5), 390–397.
Morris, S. G. (2007). Influences on childrens' narrative coherence: Age, memory breadth, and verbal comprehension. *Dissertation Abstracts International: Section B: The Sciences and Engineering*, *68*(6-B), 4157.
Morry, M. M., Kito, M., & Ortiz, L. (2011). The attraction–similarity model and dating couples: Projection, perceived similarity, and psychological benefits. *Personal Relationships*, *18*(1), 125–143.
Moshman, D. (2011). *Adolescent rationality and development: Cognition, morality, and* identity (3rd ed.). New York, NY: Psychology Press.
Mota, N. P., Medved, M., Wang, J., Asmundson, G. J. G., Whitney, D., & Sareen, J. (2012). Stress and mental disorders in female military personnel: Comparisons between the sexes in a male dominated profession. *Journal of Psychiatric Research*, *46*(2), 159–167.
Mottus, R., Johnson, W., & Deary, I. J. (2012). Personality traits in old age: Measurement and rank-order stability and some mean-level change. *Psychology and Aging*, *27*(1), 243–249.
Motyl, M., Hart, J., Pyszczynski, T., Weise, D., Maxfield, M., & Siedel, A. (2011). Subtle priming of shared human experiences eliminates threat-induced negativity toward arabs, immigrants, and peace-making. *Journal of Experimental Social Psychology*, *47*(6), 1179–1184.
Moulin, C. (Ed.). (2011). *Human memory* (Vols. 1–4). Thousand Oaks, CA: Sage.
Moulton, S. T., & Kosslyn, S. M. (2011). Imagining predictions: Mental imagery as mental emulation. In M. Bar (Ed.), *Predictions in the brain: Using our past to generate a future* (pp. 95–106). New York, NY: Oxford University Press.
Moutinho, A., Pereira, A., & Jorge, G. (2011). Biology of homosexuality. *European Psychiatry*, *26*, 1741–1753.
Muehlenkamp, J. J., Ertelt, T. W., Miller, A. L., & Claes, L. (2011). Borderline personality symptoms differentiate non-suicidal and suicidal self-injury in ethnically diverse adolescent outpatients. *Journal of Child Psychology and Psychiatry*, *52*(2), 148–155.
Mulckhuyse, M., van Zoest, W., & Theeuwes, J. (2008). Capture of the eyes by relevant and irrelevant onsets. *Experimental Brain Research*, *186*, 225–235.
Müller, B., Kühn, S., van Baaren, R., Dotsch, R., Brass, M., & Dijksterhuis, A. (2011). Perspective taking eliminates differences in co-representation of out-group members' actions. *Experimental Brain Research*, *211*(3–4), 423–428.
Mundia, L. (2011). Social desirability, non-response bias and reliability in a long self-report measure: Illustrations from the MMPI-2 administered to Brunei student teachers. *Educational Psychology*, *31*(2), 207–224.
Munoz, L., & Anastassiou-Hadjicharalambous, X. (2011). Disinhibited behaviors in young children: Relations with impulsivity and autonomic psychophysiology, *Biological Psychology*, *86*(3), 349–359.
Murray, A. (2009). *Suicide in the Middle Ages: The violent against themselves* (Vol. 1). New York, NY: Oxford University Press.
Murray, A. D., Staff, R. T., McNeil, C. J., Salarirad, S., Starr, J. M., Deary, I. J., & Whalley, L. J. (2011). Brain lesions, hypertension and cognitive ageing in the 1921 and 1936 Aberdeen birth cohorts. *Age*, *34*(2), 451–459.
Murray, H. A. (1938). *Explorations in personality*. New York, NY: Oxford University Press.
Murty, V. P., Ritchey, M., Adcock, R. A., & LaBar, K. S. (2011). Reprint of: fMRI studies of successful emotional memory encoding: A quantitative meta-analysis. *Neuropsychologia*, *49*(4), 695–705.
Music, G. (2011). *Nurturing natures: Attachment and children's emotional, sociocultural and brain development*. New York, NY: Psychology Press.
Myerson, J., Rank, M. R., Raines, F. Q., & Schnitzler, M. A. (1998). Race and general cognitive ability: The myth of diminishing returns to education. *Psychological Science*, *9*, 139–142.

Nabar, K. K. (2011). Individualistic ideology as contained in the "Diagnostic & Statistical Manual of Mental Disorders-Fourth Edition-Text Revision" personality disorders: A relational-cultural critique. *Dissertation Abstracts International: Section B: The Sciences and Engineering*, *71*(9-B), 5800.
Nabi, R. L., Moyer-Gusé, E., & Byrne, S. (2007). All joking aside: A serious investigation into the persuasive effect of funny social issue messages. *Communication Monographs*, *74*, 29–54.
Naglieri, J. A., & Ronning, M. E. (2000). Comparison of White, African American, Hispanic, and Asian children on the Naglieri Nonverbal Ability Test. *Psychological Assessment*, *12*(3), 328–334.
Nagy, T. F. (2012). Competence. In S. J. Knapp, M. C. Gottlieb, M. M. Handelsman & L. D. VandeCreek (Eds.), *APA handbook of ethics in psychology, Vol 1: Moral foundations and common themes* (pp. 147–174). Washington, DC: American Psychological Association.
Naisch, P. L. N. (2006). Time to explain the nature of hypnosis? *Contemporary Hypnosis*, *23*, 33–46.
Nakamura, K. (2006). The history of psychotherapy in Japan. *International Medical Journal*, *13*, 13–18.
Nasehi, M., Piri, M., Jamali-Raeufy, N., & Zarrindast, M. R. (2010). Influence of intracerebral administration of NO agents in dorsal hippocampus (CA1) on cannabinoid state-dependent memory in the step-down passive avoidance test. *Physiology and Behavior*, *100*(4), 297–304.
Nash, M., & Barnier, A. (Eds.). (2008). *The Oxford handbook of hypnosis*. New York, NY: Oxford University Press.
Nathan, R., Rollinson, L., Harvey, K., & Hill, J. (2003). The Liverpool violence assessment: An investigator-based measure of serious violence. *Criminal Behaviour and Mental Health*, *13*(2), 106–120.
National Institute of Mental Health (NIMH). (2011, June 20). Obsessive-compulsive disorder, OCD. Retrieved from http://www.nimh.nih.gov/health/topics/obsessive-compulsive-disorder-ocd/index.shtml.
National Institute of Mental Health (NIMH). (2011, May 3). Anxiety disorders. Retrieved from http://www.nimh.nih.gov/health/topics/anxiety-disorders/index.shtml.
National Institute on Drug Abuse (NIDA). (2010). Research report series: Cocaine abuse and addiction. Retrieved from http://www.nida.nih.gov/PDF/RRCocaine.pdf.
National Institute on Drug Abuse (NIDA). (2012). Drug facts: Club drugs (GHB, Ketamine, and Rohypnol) Retrieved from http://www.drugabuse.gov/publications/drugfacts/club-drugs-ghb-ketamine-rohypnol.
National Organization on Fetal Alcohol Syndrome. (2008). *Facts about FAS and FASD*. Retrieved from http://www.nofas.org/family/facts.aspx.
National Sleep Foundation. (2007). *Myths and facts about sleep*. Retrieved from http://www.sleepfoundation.org/site/c.huIXKjM0IxF/b.2419251/k.2773/Myths__and_Facts__About_Sleep.
Navarro, M. (2008). Who are we? New dialogue on mixed race. *New York Times*. Retrieved from http://www.nytimes.com/2008/03/31/us/politics/31race.html.
Neal, D., & Chartrand, T. (2011). Embodied emotion perception: Amplifying and dampening facial feedback modulates emotion perception accuracy. *Social Psychological and Personality Science*. Retrieved from

http://spp.sagepub.com/content/early/2011/04/21/1948550611406138.
Neale, J. M., Oltmanns, T. F., & Winters, K. C. (1983). Recent developments in the assessment and conceptualization of schizophrenia. *Behavioral Assessment*, *5*, 33–54.
Neher, A. (1991). Maslow's theory of motivation: A critique. *Journal of Humanistic Psychology*, *31*, 89–112.
Neisser, U. (1967). *Cognitive psychology*. New York, NY: Appleton-Century-Crofts.
Nelson, R. J., & Chiavegatto, S. (2001). Molecular basis of aggression. *Trends in Neurosciences*, *24*(12), 713–719.
Nemoda, Z., Szekely, A., & Sasvari-Szekely, M. (2011). Psychopathological aspects of dopaminergic gene polymorphisms in adolescence and young adulthood. *Neuroscience and Biobehavioral Reviews*, *35*(*8*), 1665–1686.
Neri, A. L., Yassuda, M. S., Fortes-Burgos, A. C., Mantovani, E. P., Arbex, F. S., de Souza Torres, S. V., . . . Guariento, M. E. (2012). Relationships between gender, age, family conditions, physical and mental health, and social isolation of elderly caregivers. *International Psychogeriatrics*, *24*(3), 472–483.
Neria, Y., Bromet, E. J., Sievers, S., Lavelle, J., & Fochtmann, L. J. (2002). Trauma exposure and posttraumatic stress disorder in psychosis: Findings from a first-admission cohort. *Journal of Consulting and Clinical Psychology*, *70*(1), 246–251.
Nettle, D. (2011). Normality, disorder, and evolved function: The case of depression. In P. Adriaens & A. de Block (Eds.), *Maladapting minds*: *Philosophy*, *psychiatry*, *and evolutionary theory* (pp.198–215). New York, NY: Oxford University Press.
Neubauer, A. C., Grabner, R. H., Freudenthaler, H. H., Beckmann, J. F., & Guthke, J. (2004). Intelligence and individual differences in becoming neurally efficient. *Acta Psychologica*, *116*(1), 55–74.
Neugarten, B. L., Havighurst, R. J., & Tobin, S. S. (1968). The measurement of life satisfaction. *Journal of Gerontology*, *16*, 134–143.
Newby, J. M., & Moulds, M. L. (2011). Intrusive memories of negative events in depression: Is the centrality of the event important? *Journal of Behavior Therapy and Experimental Psychiatry*, *42*(3), 277–283.
Newcomb, M. E., & Mustanski, B. (2010). Internalized homophobia and internalizing mental health problems: A meta-analytic review. *Clinical Psychology Review*, *30*(8), 1019–1029.
NICHY. (2012). *Categories of disability under IDEA*. Retrieved from http://nichcy.org/disability/categories
Nickell, J. (2001). John Edward: Hustling the bereaved. *Skeptical Inquirer*, *25*, 6.
Nickerson, R. (1998). Confirmation bias: A ubiquitous phenomenon in many guises. *Review of General Psychology*, *2*, 175–220.
Nicolaidis, S. (2011). Metabolic and humoral mechanisms of feeding and genesis of the ATP/ADP/AMP concept. *Physiology and Behavior*, *104*(1), 8–14.
Nielsen, T. A., Zadra, A. L., Simard, V., Saucier, S., Saucier, S., Stenstrom, P., Smith, C., & Kuiken, D. (2003). The typical dreams of Canadian university students. *Dreaming*, *13*, 211–235.
Nikolaou, A., Schiza, S. E., Chatzi, L., Koudas, V., Fokos, S., Solidaki, E., & Bitsios, P. (2011). Evidence of dysregulated affect indicated by high alexithymia in obstructive sleep apnea. *Journal of Sleep Research*, 20(1, *Pt1*), 92–100.
Nisbet, M. (2000). The best case for ESP? Generation sXeptic. *Skeptical Inquirer*. Retrieved from http://www.csicop.org/specialarticles/show/best_case_for_esp/
Nishimoto, R. (1988). A cross-cultural analysis of psychiatric symptom expression using Langer's twenty-two item index. *Journal of Sociology and Social Welfare*, *15*, 45–62.
Nogueiras, R., & Tschöp, M. (2005). Separation of conjoined hormones yields appetite rivals. *Science*, *310*, 985–986.
Nolen-Hoeksema, S., Larson, J., & Grayson, C. (2000). Explaining the gender difference in depressive symptoms. *Journal of Personality and Social Psychology*, *77*, 1061–1072.
Nolen-Hoeksema, S., Wisco, B. E., & Lyubomirsky, S. (2008). Rethinking rumination. *Perspectives on Psychological Science*, *3*, 400–424.
Norcross, J. C., & Wampold, B. E. (2011). Evidence-based therapy relationships: Research conclusions and clinical practices. *Psychotherapy*, *48*(1), 98–102.
Nord, M., & Farde, L. (2011). Antipsychotic occupancy of dopamine receptors in schizophrenia. *CNS Neuroscience and Therapeutics*, *17*(2), 97–103.
Nosko, A., Thanh-Thanh, T., Lawford, H., & Pratt, M. W. (2011). How do I love thee? Let me count the ways: Parenting during adolescence, attachment styles, and romantic narratives in emerging adulthood. *Developmental Psychology*, *47*(3), 645–657.
Nouchi, R., & Hyodo, M. (2007). The congruence between the emotional valences of recalled episodes and mood states influences the mood congruence effect. *Japanese Journal of Psychology*, *78*, 25–32.
Nusbaum, F., Redouté, J., Le Bars, D., Volckmann, P., Simon, F., Hannoun, S., Ribes, G., Gaucher, J., Laurent, B., & Sappey-Marinier, D. (2011). Chronic low-back pain modulation is enhanced by hypnotic analgesic suggestion by recruiting an emotional network: A PET imaging study. *International Journal of Clinical and Experimental Hypnosis*, *59*(1), 27–44.
O'Brien, C. W., & Moorey, S. (2010). Outlook and adaptation in advanced cancer: A systematic review. *Psycho-oncology*, *19*(1), 1239–1249.
O'Connell, K. L. (2008). What can we learn? Adult outcomes in children of seriously mentally ill mothers. *Journal of Child and Adolescent Psychiatric Nursing*, *21*, 89–104.
O'Farrell, T. J. (2011). Family therapy. In M. Galanter & H. D. Kleber (Eds.), *Psychotherapy for the treatment of substance abuse* (pp. 329–350). Arlington, VA: American Psychiatric Publishing.
O'Neill, J. W., & Davis, K. (2011). Work stress and well-being in the hotel industry. *International Journal of Hospitality Management*, *30*(2), 385–390.
O'Rourke, M. (2010, February 1). Good grief: Is there a better way to be bereaved? *New Yorker*. Retrieved from http://www.newyorker.com/arts/critics/atlarge/2010/02/01/100201crat_atlarge_orourke.
Oei, T. P. S., & Dingle, G. (2008). The effectiveness of group cognitive behaviour therapy for unipolar depressive disorders. *Journal of Affective Disorders*, *107*, 5–21.
Ogilvie, R. D., Wilkinson, R. T., & Allison, S. (1989). The detection of sleep onset: Behavioral, physiological, and subjective convergence. *Sleep*, *12*(5), 458–474.
Olds, J., & Milner, P. M. (1954). Positive reinforcement produced by electrical stimulation of septal area and other regions of rat brains. *Journal of Comparative and Physiological Psychology*, *47*, 419–427.
Olfson, M., Marcus, S., Pincus, H. A., Zito, J. M., Thompson, J. W., & Zarin, D. A. (1998). Antidepressant prescribing practices of outpatient psychiatrists. *Archives of General Psychiatry*, *55*, 310, 316.
Olson, I. R., Berryhill, M. E., Drowos, D. B., Brown, L., & Chatterjee, A. (2010). A calendar savant with episodic memory impairments. *Neurocase*, *16*(3), 208–218.
Olszewski-Kubilius, P., & Lee, S.-Y. (2011). Gender and other group differences in performance on off-level tests: Changes in the 21st century. *Gifted Child Quarterly*, *55*(1), 54–73.
Ophir, E, Nass, C., & Wagner, A.D. (2009). Cognitive control in media multitaskers. *Proceedings of the National Academy of Sciences* (*PNSA*). Retrieved from http://www.scribd.com/doc/19081547/Cognitive-control-in-media-multitaskers.
Orne, M. T. (2006). The nature of hypnosis. artifact and essence: An experimental study. *Dissertation Abstracts International*: *Section B*: *The Sciences and Engineering*, *67*(2-B), 1207.

Ortiz, F. A., Church, A. T., Vargas-Flores, J. D. J., Ibáñez-Reyes, J., Flores-Galaz, M., Luit-Briceño, J. I., &Escamilla, J. M. (2007). Are indigenous personality dimensions culture-specific? Mexican inventories and the Five-Factor Model. *Journal of Research in Personality*, *41*, 618–649.

Orzel-Gryglewska, J. (2010). Consequences of sleep deprivation. *International Journal of Occupational Medicine and Environmental Health*, *23*(1), 95–114.

Oshima, K. (2000). Ethnic jokes and social function in Hawaii. *Humor: International Journal of Humor Research*, *13*(1), 41–57.

Osler, M., McGue, M., Lund, R., & Christensen, K. (2008). Marital status and twins' health and behavior: An analysis of middle-aged Danish twins. *Psychosomatic Medicine*, *70*, 482–487.

Overstreet, M. F., & Healy, A. F. (2011). Item and order information in semantic memory: Students' retention of the "CU fight song" lyrics. *Memory and Cognition*, *39*(2), 251–259.

Owens, J., & Massey, D. S. (2011). Stereotype threat and college academic performance: A latent variables approach. *Social Science Research*, *40*(1), 150–166.

Owens, R. (2011). *Language development: An introduction* (8th ed.). Boston, MA: Allyn & Bacon.

Owens, S., Rijsdijk, F., Picchioni, M., Stahl, D., Nenadic, I., Murray, R., & Toulopoulou, T. (2011). Genetic overlap between schizophrenia and selective components of executive function. *Schizophrenia Research*, *127*(1), 181–187.

Ozawa-de Silva, C. (2007). Demystifying Japanese therapy: An analysis of Naikan and the Ajase complex through Buddhist thought. *Ethos*, *35*, 411–446.

Pace, T. W. W., & Heim, C. M. (2011). A short review on the psychoneuroimmunology of posttraumatic stress disorder: From risk factors to medical comorbidities. *Brain, Behavior, and Immunity*, *25*(1), 6–13.

Padwa, H., & Cunningham, J. (2010). *Addiction: A reference encyclopedia*. Santa Barbara, CA: ABC-CLIO.

Palermo, R., Willis, M. L., Rivolta, D., McKone, E., Wilson, C. E., & Calder, A. J. (2011). Impaired holistic coding of facial expression and facial identity in congenital prosopagnosia. *Neuropsychologia*, *49*(5), 1225–1235.

Panksepp, J. (2005). Affective consciousness: Core emotional feelings in animals and humans. *Consciousness & Cognition: An International Journal*, *14*(1), 30–80.

Panksepp, J., & Watt, D. (2011). Why does depression hurt? Ancestral primary-process separation-distress (PANIC/GRIEF) and diminished brain reward (SEEKING) processes in the genesis of depressive affect. *Psychiatry: Interpersonal and Biological Processes*, *74*(1), 5–13.

Papageorgiou, G., Cañas, F., Zink, M., & Rossi, A. (2011). Country differences in patient characteristics and treatment in schizophrenia: Data from a physician-based survey in Europe. *European Psychiatry*, *26*(1, Suppl 1), 17–28.

Papathanassoglou, E. D. E., Giannakopoulou, M., Mpouzika, M., Bozas, E., & Karabinis, A. (2010). Potential effects of stress in critical illness through the role of stress neuropeptides. *Nursing in Critical Care*, *15*(4), 204–216.

Pardini, M., Krueger, F., Hodgkinson, C., Raymont, V., Ferrier, C., Goldman, D., Strenziok, M., Guida, S., & Grafman, J. (2011). Prefrontal cortex lesions and MAO-A modulate aggression in penetrating traumatic brain injury. *Neurology*, *76*(12), 1038–1045.

Park, A. (2011, February 22). Study: Cell phones cause changes in brain activity. *Time Healthland*. Retrieved from http://healthland.time.com/2011/02/22/study-cell-phones-cause-changes-in-brain-activity/

Park, D. C., & Bischof, G. N. (2011). Neuroplasticity, aging, and cognitive function. In K. W. Schaie & S. L. Willis (Eds.), *The handbooks of aging consisting of three Vols. Handbook of the psychology of aging* (7th ed.) (pp. 109–119). San Diego, CA: Elsevier Academic Press.

Park, L., Troisi, J., & Maner, J. (2011). Egoistic versus altruistic concerns in communal relationships. *Journal of Social and Personal Relationships*, *28*(3), 315–335.

Paterson, H. M., Kemp, R. I., & Ng, J. R. (2011). Combating co-witness contamination: Attempting to decrease the negative effects of discussion on eyewitness memory. *Applied Cognitive Psychology*, *25*(1), 43–52.

Patterson, F. (2002). *Penny's journal: Koko wants to have a baby*. Retrieved from http://www.koko.org/world/journal.phtml?offset58.

Patterson, F., & Linden, E. (1981). *The education of Koko*. New York, NY: Holt, Rinehart and Winston.

Patti, C. L., Zanin, K. A., Sanday, L., Kameda, S. R., Fernandes-Santos, L., Fernandes, H. A., Anderson, M., Tufik, S., & Frussa-Filho, R. (2010). Effects of sleep deprivation on memory in mice: Role of state-dependent learning. *Sleep: Journal of Sleep and Sleep Disorders Research*, *33*(12), 1669–1679.

Paul, D. B., & Blumenthal, A. L. (1989). On the trail of little Albert. *The Psychological Record*, *39*, 547–553.

Paul, M. A., Gray, G. W., Lieberman, H. R., Love, R. J., Miller, J. C., Trouborst, M., & Arendt, J. (2011). Phase advance with separate and combined melatonin and light treatment. *Psychopharmacology*, *214*(2), 515–523.

Pechtel, P., & Pizzagalli, D. A. (2011). Effects of early life stress on cognitive and affective function: An integrated review of human literature. *Psychopharmacology*, *214*(1), 55–70.

Pedrazzoli, M., Pontes, J. C., Peirano, P., & Tufik, S. (2007). HLA-DQB1 genotyping in a family with narcolepsy-cataplexy. *Brain Research*, *1165*, 1–4.

Peleg, G., Katzir, G., Peleg, O., Kamara, M., Brodsky, L., Hel-Or, H., et al. (2006). Hereditary family signature of facial expression. *Proceedings of the National Academy of Sciences*, *103*, 15921–15926.

Pellegrino, J. E., & Pellegrino, L. (2008). Fetal alcohol syndrome and related disorders. In P. J. Accardo (Ed.), *Capute and Accardo's neurodevelopmental disabilities in infancy and childhood: Vol 1: Neurodevelopmental diagnosis and treatment* (3rd ed.) (pp. 269–284). Baltimore, MD: Paul H Brookes.

Peng, A. C., Riolli, L. T., Schaubroeck, J., & Spain, E. S. P. (2012). A moderated mediation test of personality, coping, and health among deployed soldiers. *Journal of Organizational Behavior*, *33*(4), 512–530. doi: 10.1002/job.766.

Pérez-Mata, N., & Diges, M. (2007). False recollections and the congruence of suggested information. *Memory*, *15*, 701–717.

Perlman, A., Pothos, E. M., Edwards, D. J., & Tzelgov, J. (2010). Task-relevant chunking in sequence learning. *Journal of Experimental Psychology: Human Perception and Performance*, *36*(3), 649–661.

Perry, R., & Sibley, C. G. (2011). Social dominance orientation: Mapping a baseline individual difference component across self-categorizations. *Journal of Individual Differences*, *32*(2), 110–116.

Persson, J., Lind, J., Larsson, A., Ingvar, M., Sleegers, K., Van Broeckhoven, C., Adolfsson, R., Nilsson, L-G., & Nyberg, L. (2008). Altered deactivation in individuals with genetic risk for Alzheimer's disease. *Neuropsychologia*, *46*, 1679–1687.

Perugi, G., Medda, P., Zanello, S., Toni, C., & Cassano, G. B. (2012). Episode length and mixed features as predictors of ect nonresponse in patients with medication-resistant major depression. *Brain Stimulation*, *5*(11), 18–24.

Pessoa, L. (2010). Emotion and cognition and the amygdala: From "what is it?" to "what's to be done?" *Neuropsychologia*, *48*(12), 3416–3429.

Petry, Y. (2011). 'Many things surpass our knowledge': An early modern surgeon on magic, witchcraft and demonic possession. *Social History of Medicine*, *25*(1), 47–64.

Pezdek, K. (2012). Fallible eyewitness memory and identification. In B. Cutler (Ed.), *Conviction of the innocent: Lessons from psychological research* (pp. 105–124). Washington, DC: American Psychological Association.

Pfeiffer, P. N., Valenstein, M., Hoggatt, K. J., Ganoczy, D., Maixner, D., Miller, E. M., & Zivin, K. (2011). Electroconvulsive therapy for major depression within the Veterans Health Administration. *Journal of Affective Disorders, 130*(1–2), 21–25.

Philibert, R., Wernett, P., Plume, J., Packer, H., Brody, G., & Beach, S. (2011). Gene environment interactions with a novel variable Monoamine Oxidase A transcriptional enhancer are associated with antisocial personality disorder. *Biological Psychology, 87*(3), 366–371.

Phillips, A. C., & Hughes, B. M. (2011). Introductory paper: Cardiovascular reactivity at a crossroads: Where are we now? *Biological Psychology, 86*(2), 95–97.

Phillips, S. T., & Ziller, R. C. (1997). Toward a theory and measure of the nature of non-prejudice. *Journal of Personality and Social Psychology, 72*, 420–434.

Piaget, J. (1962). *Play, dreams and imitation in Childhood*. New York, NY: W. W. Norton & Co.

Piallat, B., Polosan, M., Fraix, V., Goetz, L., David, O., Fenoy, A., . . . Chabardes, S. (2011). Subthalamic neuronal firing in obsessive-compulsive disorder and Parkinson disease. *Annals Neurology, 69*(5), 793–802. doi: 10.1002/ana.22222

Pietschnig, J., Voracek, M., & Formann, A. K. (2011). Female Flynn effects: No sex differences in generational IQ gains. *Personality and Individual Differences, 50*(5), 759–762.

Pina e Cunha, M., Rego, A., & Clegg, S. R. (2010). Obedience and evil: From Milgram and Kampuchea to normal organizations. *Journal of Business Ethics, 97*(2), 291–309.

Ping-Delfos, W., & Soares, M. (2011). Diet induced thermogenesis, fat oxidation and food intake following sequential meals: Influence of calcium and vitamin D. *Clinical Nutrition, 30*(3), 376–383.

Plassmann, H., O'Doherty, J., Shiv, B., & Rangel, A. (2008). Marketing actions can modulate neural representations of experienced pleasantness. *Proceedings of the National Academy of Sciences, 105*, 1050–1054.

Plomin, R. (1990). The role of inheritance in behavior. *Science, 248*, 183–188.

Plomin, R. (1999). Genetics and general cognitive ability. *Nature, 402*, C25–C29.

Plomin, R., DeFries, J. C., & Fulker, D. W. (2007). *Nature and nurture during infancy and early childhood*. New York, NY: Cambridge University Press.

Plucker, J., & Esping, A. (2013). *Intelligence 101*. New York, NY: Springer.

Plummer, D. C. (2001). The quest for modern manhood: Masculine stereotypes, peer culture and the social significance of homophobia. *Journal of Adolescence, 24*(1), 15–23.

Plutchik, R. (1984). Emotions: A general psychoevolutionary theory. In K. R. Scherer, & P. Ekman (Eds.), *Approaches to emotion*. Hillsdale, NJ: Erlbaum.

Plutchik, R. (1994). *The psychology and biology of emotion*. New York, NY: HarperCollins.

Plutchik, R. (2000). *Emotions in the practice of psychotherapy: Clinical implications of affect theories*. Washington, DC: American Psychological Association.

Podgorski, K., Dunfield, D., & Haas, K. (2012). Functional clustering drives encoding improvement in a developing brain network during awake visual learning. *PLoS Biology, 10*(1), e1001236.

Poling, A. (2010). Progressive-ratio schedules and applied behavior analysis. *Journal of Applied Behavior Analysis, 43*(2), 347–349.

Polley, K. H., Navarro, R., Avery, D. H., George, M. S., & Holtzheimer, P. E. (2011). 2010 updated Avery-George-Holtzheimer Database of rTMS depression studies. *Brain Stimulation, 4*(2), 115–116.

Pomponio, A. T. (2002). *Psychological consequences of terror*. New York, NY: John Wiley & Sons.

Pope Jr., H. G., Barry, S., Bodkin, J. A., & Hudson, J. (2007). "Scientific study of the dissociative disorders": Reply. *Psychotherapy and Psychosomatics, 76*, 401–403.

Pope, K. S., & Vasquez, M. J. T. (2011). *Ethics in psychotherapy and counseling: A practical guide* (4th ed.). Hoboken, NJ: John Wiley & Sons.

Popma, A., Vermeiren, R., Geluk, C. A. M. L., Rinne, T., van den Brink, W., Knol, D. L., Jansen, L. M. C., van Engeland, H., & Doreleijers, T. A. H. (2007). Cortisol moderates the relationship between testosterone and aggression in delinquent male adolescents. *Biological Psychiatry, 61*, 405–411.

Popova, S., Stade, B., Johnston, M., MacKay, H., Lange, S., Bekmuradov, D., & Rehm, J. (2011). Evaluating the cost of fetal alcohol spectrum disorder. *Journal of Studies on Alcohol and Drugs, 72*(1), 163–164.

Porzelius, L. K., Dinsmore, B. D., & Staffelbach, D. (2001). Eating disorders. In M. Hersen & V. B. Van Hasselt (Eds.), *Advanced abnormal psychology* (2nd ed.) (pp. 261–281). Dordrecht, Netherlands: Kluwer Academic Publishers.

Post, J. M. (2011). Crimes of obedience: "Groupthink" at Abu Ghraib. *International Journal of Group Psychotherapy, 61*(1), 49–66

Posthuma, D., de Geus, E. J. C., & Boomsma, D. I. (2001). Perceptual speed and IQ are associated through common genetic factors. *Behavior Genetics, 31*(6), 593–602.

Pournaghash-Tehrani, S. (2011). Domestic violence in Iran: A literature review. *Aggression and Violent Behavior, 16*(1), 1–5.

Powell, D. H. (1998), *The nine myths of aging: Maximizing the quality of later life*. San Francisco, CA: Freeman.

Prado, J., Carp, J., & Weissman, D. H. (2011). Variations of response time in a selective attention task are linked to variations of functional connectivity in the attentional network. *NeuroImage, 54*(1), 541–549.

Preuss, U. W., Zetzsche, T., Jäger, M., Groll, C., Frodl, T., Bottlender, R., Leinsinger, G., Hegerl, U., Hahn, K., Möller, H. J., & Meisenzahl, E. M. (2005). Thalamic volume in first-episode and chronic schizophrenic subjects: A volumetric MRI study. *Schizophrenia Research, 73*(1), 91–101.

Preventing suicide. (2009). *Preventing suicide in people who have schizophrenia*. Retrieved from http://www.schizophrenia.com/suicide.html

Price, D. D., Finniss, D. G., & Benedetti, F. (2008). A comprehensive review of the placebo effect: Recent advances and current thought. *Annual Review of Psychology, 59*, 565–590.

Priest, R. F., & Sawyer, J. (1967). Proximity and peership: Bases of balance in interpersonal attraction. *American Journal of Sociology, 72*, 633–649.

Prigatano, G. P., & Gray, J. A. (2008). Predictors of performance on three developmentally sensitive neuropsychological tests in children with and without traumatic brain injury. *Brain Injury, 22*, 491–500.

Pring, L., Woolf, K., & Tadic, V. (2008). Melody and pitch processing in five musical savants with congenital blindness. *Perception, 37*, 290–307.

Prkachin, K. M., & Silverman, B. E. (2002). Hostility and facial expression in young men and women: Is social regulation more important than negative affect? *Health Psychology, 21*(1), 33–39.

Prouix, M. J. (2007). Bottom-up guidance in visual search for conjunctions. *Journal of Experimental Psychology, 33*, 48–56.

Pruchnicki, S. A., Wu, L. J., & Belenky, G. (2011). An exploration of the utility of mathematical modeling predicting fatigue from sleep/wake history and circadian phase applied in accident analysis and prevention: The crash of Comair Fight 5191. *Accident Analysis and Prevention, 43*(3), 1056–1061.

Pu, S., Yamada, T., Yokoyama, K., Matsumura, H., Kobayashi, H., Sasaki, N., Mitani, H., Adachi, A., Kaneko,

K., & Nakagome, K. (2011). A multi-channel near-infrared spectroscopy study of prefrontal cortex activation during working memory task in major depressive disorder. *Neuroscience Research, 70*(1), 91–97.
Pullum, G. K. (1991). *The great Eskimo vocabulary hoax and other irreverent essays on the study of language*. Chicago, IL: University of Chicago Press.

Qualls, S. H. (2008). Caregiver family therapy. In K. Laidlaw & B. Knight (Eds.), *Handbook of emotional disorders in later life: Assessment and treatment* (pp. 183–209). New York, NY: Oxford University Press.
Quas, J. A., Yim, I. S., Rush, E., & Sumaroka, M. (2012). Hypothalamic pituitary adrenal axis and sympathetic activation: Joint predictors of memory in children, adolescents, and adults. *Biological Psychology, 89*(2), 335–341.
Quinn, J., Barrowclough, C., & Tarrier, N. (2003). The Family Questionnaire (FQ): A scale for measuring symptom appraisal in relatives of schizophrenic patients. *Acta Psychiatrica Scandinavica, 108*(4), 290–296.
Quintanilla, Y. T. (2007). Achievement motivation strategies: An integrative achievement motivation program with first year seminar students. *Dissertation Abstracts International Section A: Humanities and Social Sciences, 68*(6-A), 2339.

Radcliff, K., & Joseph, L. (2011) Girls just being girls: Mediating relational aggression and victimization. *Preventing School Failure: Alternative Education for Children and Youth, 55*(3), 171–179.
Radvansky, G. A. (2011). *Human memory* (2nd ed.). Upper Saddle River, NJ: Pearson.
Raevuori, A., Keski-Rahkonen, A., Hoek, H. W., Sihvola, E., Rissanen, A., & Kaprio, J. (2008). Lifetime anorexia nervosa in young men in the community: Five cases and their co-twins. *International Journal of Eating Disorders, 41*, 458–463.
Raggi, A., Plazzi, G., Pennisi, G., Tasca, D., & Ferri, R. (2011). Cognitive evoked potentials in narcolepsy: A review of the literature. *Neuroscience and Biobehavioral Reviews, 35*(5), 1144–1153.
Rai, T. S., & Fiske, A. P. (2011). Moral psychology is relationship regulation: Moral motives for unity, hierarchy, equality, and proportionality. *Psychological Review, 118*(1), 57–75.
Ramsay, C. E., Stewart, T., & Compton, M. T. (2012). Unemployment among patients with newly diagnosed first-episode psychosis: Prevalence and clinical correlates in a US sample. *Social Psychiatry and Psychiatric Epidemiology, 47*(5), 797–803.
Randi, J. (1997). *An encyclopedia of claims, frauds, and hoaxes of the occult and supernatural: James Randi's decidedly skeptical definitions of alternate realities*. New York, NY: St Martin's Press.
Raphel, S. (2008). Kinship care and the situation for grandparents. *Journal of Child and Adolescent Psychiatric Nursing, 21*, 118–120.
Rattaz, C., Goubet, N., & Bullinger, A. (2005). The calming effect of a familiar odor on full-term newborns. *Journal of Developmental and Behavioral Pediatrics, 26*, 86–92.
Rauer, A. J. (2007). Identifying happy, healthy marriages for men, women, and children. *Dissertation Abstracts International: Section B: The Sciences and Engineering, 67*(10-B), 6098.
Read, D., & Grushka-Cockayne, Y. (2011). The similarity heuristic. *Journal of Behavioral Decision Making, 24*(1), 23–46.
Rechtschaffen, A., & Siegel, J.M. (2000). Sleep and Dreaming. In E. R. Kandel, J. H. Schwartz, & T. M. Jessel (Eds.), *Principles of Neuroscience* (4th ed.) (pp. 936–947). New York, NY: McGraw-Hill.
Reed, D. D., & Martens, B. K. (2011). Temporal discounting predicts student responsiveness to exchange delays in a classroom token system. *Journal of Applied Behavior Analysis, 44*(1), 1–18.
Regan, P. (1998). What if you can't get what you want? Willingness to compromise ideal mate selection standards as a function of sex, mate value, and relationship context. *Personality and Social Psychology Bulletin, 24*, 1294–1303.
Regan, P. (2011). *Close relationships*. New York, NY: Routledge/Taylor & Francis Group.
Regier, D. A., Narrow, W. E., Kuhl, E. A., & Kupfer, D. J. (Eds). (2011). *The conceptual evolution of DSM-5*. Arlington, VA: American Psychiatric Publishing.
Regier, D. A., Narrow, W. E., Rae, D. S., Manderscheid, R. W., Locke, B. Z., & Goodwin, F. K. (1993). The de facto U.S. mental and addictive disorders service system. *Archives of General Psychiatry, 50*, 85–93.
Reich, D. A. (2004). What you expect is not always what you get: The roles of extremity, optimism, and pessimism in the behavioral confirmation process. *Journal of Experimental Social Psychology, 40*(2), 199–215.
Reifman, A. (2000). Revisiting the Bell Curve. *Psychology, 11*, 21–29.
Reinhard, M.-A., & Dickhäuser, O. (2011). How affective states, task difficulty, and self-concepts influence the formation and consequences of performance expectancies. *Cognition and Emotion, 25*(2), 220–228.
Renzetti, C., Curran, D., & Kennedy-Bergen, R. (2006). *Understanding diversity*. Boston, MA: Allyn & Bacon/Longman.
Rest, J., Narvaez, D., Bebeau, M., & Thoma, S. (1999). A neo-Kohlbergian approach: The DIT and schema theory. *Educational Psychology Review, 11*(4), 291–324.
Reynolds, M. R., Keith, T. Z., Ridley, K. P., & Patel, P. G. (2008). Sex differences in latent general and broad cognitive abilities for children and youth: Evidence from higher-order MG-MACS and MIMIC models. *Intelligence, 36*, 236–260.
Rezayat, M., Niasari, H., Ahmadi, S., Parsaei, L., & Zarrindast, M. R. (2010). N-methyl-D-aspartate receptors are involved in lithium-induced state-dependent learning in mice. *Journal of Psychopharmacology, 24*(6), 915–921.
Rhee, S. H., & Waldman, I. D. (2011). Genetic and environmental influences on aggression. In P. R. Shaver & M. Mikulincer (Eds.), *Herzilya series on personality and social psychology. Human aggression and violence: Causes, manifestations, and consequences* (pp. 143–163). Washington, DC: American Psychological Association.
Rhodes, G., Halberstadt, J., & Brajkovich, G. (2001). Generalization of mere exposure effects to averaged composite faces. *Social Cognition, 19*(1), 57–70.
Rial, R. V., Akaârir, M., Gamundí, A., Nicolau, C., Garau, C., Aparicio, S., Tejada, S., Gené, L., González, J., De Vera, L. M., Coenen, Anton M., Barceló, P., & Esteban, S. (2010). Evolution of wakefulness, sleep and hibernation: From reptiles to mammals. *Neuroscience and Biobehavioral Reviews, 34(8)*, 1144–1160.
Richardson, D. S., & Hammock, G. S. (2007). Social context of human aggression: Are we paying too much attention to gender? *Aggression and Violent Behavior, 12*, 417–426.
Riebe, D., Garber, C. E., Rossi, J. S., Greaney, M. L., Nigg, C. R., Lees, F. D., Burbank, P. M., & Clark, P. G. (2005). Physical activity, physical function, and stages of change in older adults. *American Journal of Health Behavior, 29*, 70–80.
Riemann, D., & Voderholzer, U. (2003). Primary insomnia: A risk factor to develop depression? *Journal of Affective Disorders, 76*(1–3), 255–259.
Riley, B., & Kendler, K. (2011). Classical genetic studies of schizophrenia. In D. Weinberger, & P. Harrison (Eds.), *Schizophrenia* (3rd ed.) (pp. 245–268). Hoboken, NJ: John Wiley & Sons.
Ritchey, M., LaBar, K. S., & Cabeza, R. (2011). Level of processing modulates the neural correlates of emotional memory formation. *Journal of Cognitive Neuroscience, 23*(4), 757–771.
Riva, M. A., Tremolizzo, L., Spicci, M., Ferrarese, C., De Vito, G., Cesana, G. C.,

& Sironi, V. A. (2011). The disease of the moon: The linguistic and pathological evolution of the English term "lunatic." *Journal of the History of the Neurosciences*, *20*(1), 65–73.

Rivers, I. (2011). *Homophobic bullying: Research and theoretical perspectives*. New York, NY: Oxford University Press.

Rizzolatti, G., & Fabbri-Destro, M. (2010). Mirror neurons: From discovery to autism. *Experimental Brain Research*, *200*(3–4), 223–237.

Rizzolatti, G., Fadiga, L., Fogassi, L. & Gallese, V. (2002). From mirror neurons to imitation: Facts and speculations. In A. N. Meltzoff, & W. Prinz (Eds.), *The imitative mind: Development, evolution, and brain bases*. Cambridge, MA: Cambridge University Press.

Rizzolatti, G., Fogassi, L. & Gallese, V. (2006, November). In the mind. *Scientific American*, 54–61.

Rodrigues, A., Assmar, E. M., & Jablonski, B. (2005). Social-psychology and the invasion of Iraq. *Revista de Psicología Social*, *20*, 387–398.

Rogers, C. R. (1961). *On becoming a person*. Boston, MA: Houghton Mifflin.

Rogers, C. R. (1980). *A way of being*. Boston, MA: Houghton Mifflin.

Roisko, R., Wahlberg, K.-E., Hakko, H., Wynne, L., & Tienari, P. (2011). Communication deviance in parents of families with adoptees at a high or low risk of schizophrenia-spectrum disorders and its associations with attributes of the adoptee and the adoptive parents. *Psychiatry Research*, *185*(1–2), 66–71.

Romero, S. G., McFarland, D. J., Faust, R., Farrell, L., & Cacace, A. T. (2008). Electrophysiological markers of skill-related neuroplasticity. *Biological Psychology*, *78*, 221–230.

Rosch, E. (1978). Principles of organization. In E. Rosch & H. L. Lloyd (Eds.), *Cognition and categorization*. Hillsdale, NJ: Erlbaum.

Rosch, E. H. (1973). Natural categories. *Cognitive Psychology*, *4*, 328–350.

Rose, S. A., Feldman, J. F., Jankowski, J. J., & Van Rossem, R. (2011). The structure of memory in infants and toddlers: An SEM study with full-terms and preterms. *Developmental Science*, *14*(1), 83–91.

Roselli, C., Reddy, R., & Kaufman, K. (2011). The development of male-oriented behavior in rams. *Frontiers in Neuroendocrinology*, *32*(2), 164–169.

Rosenhan, D. L. (1973). On being sane in insane places. *Science*, *179*, 250–258.

Rosenström, T., Hintsanen, M., Kivimäki, M., Jokela, M., Juonala, M., Viikari, J. S., Raitakari, O., & Keltikangas-Järvinen, L. (2011). "Change in job strain and progression of atherosclerosis: The cardiovascular risk in young Finns study": Correction to Rosenström et al. (2011). *Journal of Occupational Health Psychology*, *16*(2), 201.

Rosenzweig, M. R., Bennett, E. L., & Diamond, M. C. (1972). Brain changes in response to experience. *Scientific American*, *226*, 22–29.

Rosling, A., Sparén, P, Norring, C., & von Knorring, A. (2011). Mortality of eating disorders: A follow-up study of treatment in a specialist unit 1974–2000. *International Journal of Eating Disorder*, *44*(4), 304–310.

Rosner, R. I. (2011). Aaron T. Beck's drawings and the psychoanalytic origin story of cognitive therapy. *History of Psychology*, *15*(1), 1–18.

Ross, D., Kincaid, H., Spurrett, D., & Collins, P. (Eds.). (2010). *What is addiction?* Cambridge, MA: MIT Press.

Rossi, S. L. (2011). The characterization and transplantation of human embryonic stem cell-derived motor neuron progenitors following bilateral cervical spinal cord injury. *Dissertation Abstracts International: Section B. The Sciences and Engineering*, *71*(10), 5963.

Rossignol, S., Barrière, G., Frigon, A., Barthélemy, D., Bouyer, L., Provencher, J., Leblond, H., & Bernard, G. (2008). Plasticity of locomotor sensorimotor interactions after peripheral and/or spinal lesions. *Brain Research Reviews*, *57*, 228–240.

Roth, M. L., Tripp, D. A. Harrison, M. H., Sullivan, M., & Carson, P. (2007). Demographic and psychosocial predictors of acute perioperative pain for total knee arthroplasty. *Pain Research and Management*, *12*, 185–194.

Rotter, J. B. (1954). *Social learning and clinical psychology*. Englewood Cliffs, NJ: Prentice Hall.

Rotter, J. B. (1990). Internal versus external control of reinforcement: A case history of a variable. *American Psychologist*, *45*, 489–493.

Rowe, B., & Levine, D. (2011). *Concise introduction to linguistics* (3rd ed.). Upper Saddle River, NJ: Prentice Hall.

Rozencwajg, P., Cherfi, M., Ferrandez, A. M., Lautrey, J., Lemoine, C., & Loarer, E. (2005). Age related differences in the strategies used by middle aged adults to solve a block design task. *International Journal of Aging and Human Development*, *60*(2), 159–182.

Rubinstein, E. (2008). Judicial perceptions of eyewitness testimony. *Dissertation Abstracts International: Section B: The Sciences and Engineering*, *68*(*8-B*), 5592.

Rumbaugh, D. M., von Glasersfeld, E. C., Warner, H., Pisani, P., & Gill, T. V. (1974). Lana (chimpanzee) learning language: A progress report. *Brain and Language*, *1*(2), 205–212.

Rüsch, N., Corrigan, P. W., Todd, A. R., & Bodenhausen, G. V. (2011). Automatic stereotyping against people with schizophrenia, schizoaffective and affective disorders. *Psychiatry Research*, *186*(1), 34–39.

Rushton, P., & Irwing, P. (2011). The general factor of personality: Normal and abnormal. In T. Chamorro-Premuzic, S., von Stumm, & A. Furnam (Eds.), *Wiley-Blackwell handbook of individual differences* (pp. 132–161). Chichester, UK: Wiley-Blackwell.

Russo, N. F., & Tartaro, J. (2008). Women and mental health. In F. L. Denmark & M. A. Paludi (Eds.), *Psychology of women: A handbook of issues and theories* (2nd ed.) (pp. 440–483). Westport, CT: Praeger/Greenwood.

Rutter, M. (2007). Gene-environment interdependence. *Developmental Science*, *10*, 12–18.

Ruzek, J. I., Schnurr, P. P., Vasterling, J. J., & Friedman, M. J. (2011). *Caring for veterans with deployment-related stress disorders: Iraq, Afghanistan, and beyond*. Washington, DC: American Psychological Association.

Ryan, C. S., Casas, J. F., & Thompson, B. K. (2010). Interethnic ideology, intergroup perceptions, and cultural orientation. *Journal of Social Issues*, *66*(1), 29–44.

Rymer, R. (1993). *Genie: An abused child's first flight from silence*. New York, NY: HarperCollins.

Sabet, K. A. (2007). The (often unheard) case against marijuana leniency. In M. Earleywine (Ed), *Pot politics: Marijuana and the costs of prohibition* (pp. 325–352). New York, NY: Oxford University Press.

Sacchetti, B., Sacco, T., & Strata, P. (2007). Reversible inactivation of amygdala and cerebellum but not perirhinal cortex impairs reactivated fear memories. *European Journal of Neuroscience*, *25*(9), 2875–2884.

Sachdev, P., Mondraty, N., Wen, W., & Gulliford, K. (2008). Brains of anorexia nervosa patients process self-images differently from non-self-images: An fMRI study. *Neuropsychologia*, *46*, 2161–2168.

Sack, R. L. (2010). Jet lag. *The New England Journal of Medicine*, *362*(5), 440–447.

Sack, R. L., Auckley, D., Auger, R. R., Carskadon, M. A., Wright, Jr., K. P., Vitiello, M. V., & Zhdanova, I. V. (2007). Circadian rhythm sleep disorders: Part I, basic principles, shift work and jet lag disorders: An American academy of sleep medicine review. *Sleep: Journal of Sleep and Sleep Disorders Research*, *30*, 1460–1483.

Sacks, O. (1995) *An anthropologist on Mars*. New York, NY: Vintage books.

Said, C., Haxby, J., & Todorov, A. (2011). Brain systems for assessing the affective

value of faces. *Philosophical Transactions of the Royal Society B*, *366*, 1660–1670.
Salgado-Pineda, P., Fakra, E., Delaveau, P., McKenna, P. J., Pomarol-Clotet, E., & Blin, O. (2011). Correlated structural and functional brain abnormalities in the default mode network in schizophrenia patients. *Schizophrenia Research*, *125*(2–3), 101–109.
Salovey, P., & Mayer, J. D. (1990). Emotional intelligence. *Imagination, Cognition, and Personality*, *9*, 185–211.
Salovey, P., Bedell, B. T., Detweiler, J. B., & Mayer, J. D. (2000). Current directions in emotional intelligence research. In M. Lewis & J. M. Haviland (Eds.), *Handbook of emotions* (2nd ed.) (pp. 504–520). New York, NY: Guilford Press.
Salvatore, P., Ghidini, S., Zita, G., De Panfilis, C., Lambertino, S., Maggini, C., & Baldessarini, R. J. (2008). Circadian activity rhythm abnormalities in ill and recovered bipolar I disorder patients. *Bipolar Disorders*, *10*, 256–265.
Samnani, A. K., Boekhorst, J. A., & Harrison, J. A. (2012). Acculturation strategy and individual outcomes: Cultural diversity implications for human resource management. *Human Resource Management Review*, No Pagination Specified. doi: 10.1016/j.hrmr.2012.04.001
Sampselle, C. M., Harris, V. Harlow, S. D., & Sowers, M. F. (2002). Midlife development and menopause in African American and Caucasian women. *Health Care for Women International*, *23*(4), 351–363.
Sanbonmatsu, D. M., Uchino, B. N., & Birmingham, W. (2011). On the importance of knowing your partner's views: Attitude familiarity is associated with better interpersonal functioning and lower ambulatory blood pressure in daily life. *Annals of Behavioral Medicine*, *41*(1), 131–137.
Sanchez Jr., J. (2006). Life satisfaction factors impacting the older Cuban-American population. *Dissertation Abstracts International Section A: Humanities and Social Sciences*, *67*(5-A), 1864.
Sánchez-Martín, J. R., Azurmendi, A., Pascual-Sagastizabal, E., Cardas, J., Braza, F., Braza, P., Carreras, M., & Muñoz, J. M. (2011). Androgen levels and anger and impulsivity measures as predictors of physical, verbal and indirect aggression in boys and girls. *Psychoneuroendocrinology*, *36*(5), 750–760.
Sanderson, C. A. (2013). *Health psychology* (2nd ed.). Hoboken, NJ: John Wiley & Sons.
Sangwan, S. (2001). Ecological factors as related to I.Q. of children. *Psycho-Lingua*, *31*(2), 89–92.
Sapolsky, R. (2003). Taming stress. *Scientific American*, pp. 87–95.
Sapolsky, R. M. (1992). *Stress, the aging brain, and the mechanisms of neuron death*. Cambridge, MA: MIT Press.
Sapolsky, R. M. (2007). Why Zebras don't get ulcers: Stress, metabolism, and liquidating your assets. In A. Monat, R. S. Lazarus, & G. Reevy (Eds.), *The Praeger handbook on stress and coping* (Vol. 1), (pp. 181–197). Westport, CT: Praeger Publishers/Greenwood Publishing Group.
Sara, S. J. (2010). Reactivation, retrieval, replay and reconsolidation in and out of sleep: Connecting the dots. *Frontiers in Behavioral Neuroscience*, *4*, Article ID 185.
Sarafino, E. P. (2012). *Applied behavior analysis: Principles and procedures for modifying behavior*. Hoboken, NJ: John Wiley & Sons.
Sarafino, E. P., & Smith, T. W. (2011). *Health psychology: Biopsychosocial interactions* (7th ed.). Hoboken, NJ: John Wiley & Sons.
Sass, H. (2011). Conceptual history of classification. *European Psychiatry*, *26*(1), 1778.
Sassenberg, K., Moskowitz, G. B., Jacoby, J., & Hansen, N. (2007). The carry-over effect of competition: The impact of competition on prejudice towards uninvolved outgroups. *Journal of Experimental Social Psychology*, *43*, 529–538.
Satcher, N. D. (2007). Social and moral reasoning of high school athletes and non-athletes. *Dissertation Abstracts International Section A: Humanities and Social Sciences*, *68*(3–A), 928.
Sathyaprabha, T. N., Satishchandra, P., Pradhan, C., Sinha, S., Kaveri, B., Thennarasu, K., Murthy, B. T. C., & Raju, T. R. (2008). Modulation of cardiac autonomic balance with adjuvant yoga therapy in patients with refractory epilepsy. *Epilepsy and Behavior*, *12*, 245–252.
Satler, C., Garrido, L. M., Sarmiento, E. P., Leme, S., Conde, C., & Tomaz, C. (2007). Emotional arousal enhances declarative memory in patients with Alzheimer's disease. *Acta Neurologica Scandinavica*, *116*, 355–360.
Savage-Rumbaugh, E. S. (1990). Language acquisition in a nonhuman species: Implications for the innateness debate. *Developmental Psychobiology*, *23*, 599–620.
Savic, I., Berglund, H., & Lindström, P. (2007). Brain response to putative pheromones in homosexual men. In G. Einstein (Ed.), *Sex and the brain* (pp. 731–738). Cambridge, MA: MIT Press.
Saxton, M. (2010). *Child language: Acquisition and development*. Thousand Oaks, CA: Sage.
Scarborough, P., Nnoaham, K. E., Clarke, D., Capewell, S., & Rayner, M. (2012). Modelling the impact of a healthy diet on cardiovascular disease and cancer mortality. *Journal of Epidemiology and Community Health*, *66*(5), 420–426.
Schachter, S., & Singer, J. E. (1962). Cognitive, social, and physiological determinants of emotional state. *Psychological Review*, *69*, 379–399.
Schachtman, T., & Reilly, S. (2011). *Associative learning and conditioning theory: Human and non-human applications*. Oxford, UK: Oxford University Press.
Schaefer, R. T. (2008). Power and power elite. In V. Parillo (Ed.), *Encyclopedia of social problems*. Thousand Oaks, CA: Sage.
Schaie, K. W. (1994). The life course of adult intellectual development. *American Psychologist*, *49*, 304–313.
Schaie, K. W. (2008). A lifespan developmental perspective of psychological ageing. In K. Laidlaw & B. Knight (Eds.), *Handbook of emotional disorders in later life: Assessment and treatment* (pp. 3–32). New York, NY: Oxford University Press.
Schermer, J., Vernon, P., Maio, G., & Jang, K. (2011). A behavior genetic study of the connection between social values and personality. *Twin Research and Human Genetics*, *14*(3), 233–239.
Schiffer, B., Muller, B., Scherbaum, N., Hodgins, S., Forsting, M., Wiltfang, J., Gizewski, E., & Leygraf, N. (2011). Disentangling structural brain alterations associated with violent behavior from those associated with substance use disorders. *Archives of General Psychiatry*, *68*(10), 1039–1049.
Schifferstein, H. N. J., & Hilscher, M. C. (2010). Multisensory images for everyday products: How modality importance, stimulus congruence and emotional context affect image descriptions and modality vividness ratings. *Journal of Mental Imagery*, *34*(3–4), 63–98.
Schlaepfer, T. E., Bewernick, B., Kayser, S., & Lenz, D. (2011). Modulating affect, cognition, and behavior-prospects of deep brain stimulation for treatment-resistant psychiatric disorders. *Frontiers in Integrative Neuroscience*, *5*, 29.
Schmitz, M., & Wentura, D. (2012). Evaluative priming of naming and semantic categorization responses revisited: A mutual facilitation explanation. *Journal of Experimental Psychology: Learning, Memory, and Cognition*, *38*(4), 984–1000.
Schneidman, E. S. (1969). Suicide, lethality and the psychological autopsy. *International Psychiatry Clinics*, 6, 225–250.
Schneidman, E. S. (1981). Suicide. *Suicide and Life-Threatening Behavior*, *11*(4), 198–220.
Schredl, M. (2012). Review of the nature and functions of dreaming. *Dreaming*, *22*(2), 150–155.
Schuett, S., Heywood, C. A., Kentridge, R. W., & Zihl, J. (2008). The significance of visual information processing in reading: Insights from hemianopic dyslexia. *Neuropsychologia*, *46*, 2445–2462.

Schülein, J. A. (2007). Science and psychoanalysis. *Scandinavian Psychoanalytic Review*, *30*, 13–21.
Schulze, L., Domes, G., Krüger, A., Berger, C., Fleischer, M., Prehn, K., Schmahl, C., Grossmann, A., Hauenstein, K., & Herpertz, S. C. (2011). Neuronal correlates of cognitive reappraisal in borderline patients with affective instability. *Biological Psychiatry*, *69*(6), 564–573.
Schunk, D. H. (2008). Attributions as motivators of self-regulated learning. In D. H. Schunk & B. J. Zimmerman (Eds.), *Motivation and self-regulated learning: Theory, research, and applications* (pp. 245–266). Mahwah, NJ: Lawrence Erlbaum.
Schweckendiek, J., Klucken, T., Merz, C. J., Tabbert, K., Walter, B., Ambach, W., Vaitl, D., & Stark, R. (2011). Weaving the (neuronal) web: Fear learning in spider phobia. *NeuroImage*, *54*(1), 681–688.
Scott, L. N., Levy, K. N., Adams, R. B., Jr., & Stevenson, M. T. (2011). Mental state decoding abilities in young adults with borderline personality disorder traits. *Personality Disorders: Theory, Research, and Treatment*, *2*(2), 98–112.
Scribner, S. (1977). Modes of thinking and ways of speaking: Culture and logic reconsidered. In P. N. Johnson-Laird & P. C. Wason (Eds.), *Thinking: Readings in cognitive science* (pp. 324–339). New York, NY: Cambridge University Press.
Seedat, S., Scott, K. M., Angermeyer, M. C., Berglund, P., Bromet, E. J., …Kessler, R. C. (2009). Cross-national associations between gender and mental disorders in the World Health Organization World Mental Health Surveys. *Archives of General Psychiatry*, *66*(7), 785–795.
Seeman, P. (2011). All roads to schizophrenia lead to dopamine supersensitivity and elevated dopamine D2High receptors. *CNS Neuroscience and Therapeutics*, *17*(2), 118–132.
Segal, N. L., & Bouchard, T. J. (2000). *Entwined lives: Twins and what they tell us about human behavior*. New York, NY: Plumsock.
Segerstrom, S., C., & Miller, G. E. (2004). Psychological stress and the human immune system: A meta-analytic study of 30 years of inquiry. *Psychological Bulletin*, *130*(4), 601–630.
Seligman, M. E. P. (1975) *Helplessness: On depression, development, and death*. San Francisco, CA: Freeman.
Seligman, M. E. P. (1994). *What you can change and what you can't*. New York, NY: Alfred A. Knopf.
Seligman, M. E. P. (2003). The past and future of positive psychology. In C. L. M. Keyes & J. Daidt (Eds.), *Flourishing: Positive psychology and the life well-lived* (pp. xi–xx). Washington, DC: American Psychological Association.
Seligman, M. E. P. (2007). Coaching and positive psychology. *Australian Psychologist*, *42*, 266–267.
Seligman, M. E. P. (2011). *Flourish: A visionary new understanding of happiness and well-being*. New York, NY: Free Press.
Selye, H. (1974). *Stress without distress*. New York, NY: Harper & Row.
Senko, C., Durik, A. M., & Harackiewicz, J. M. (2008). Historical perspectives and new directions in achievement goal theory: Understanding the effects of mastery and performance-approach goals. In J. Y. Shah & W. L. Gardner (Eds.), *Handbook of motivation science* (pp. 100–113). New York, NY: Guilford Press.
Shaffer, R., & Jadwiszczok, A. (2010). Psychic defective: Sylvia Browne's history of failure. *Skeptical Inquirer*, 34. 2. Retrieved from http://www.csicop.org/si/show/psychic_defective_sylvia_brownes_history_of_failure/
Shahim, S. (2008). Sex differences in relational aggression in preschool children in Iran. *Psychological Reports*, *102*, 235–238.
Shapiro, J. R. (2011). Different groups, different threats: A multi-threat approach to the experience of stereotype threats. *Personality and Social Psychology Bulletin*, *37*(4), 464–480.
Sharf, R. S. (2012). *Theories of psychotherapy and counseling: Concepts and cases* (5th ed.). Pacific Grove, CA: Brooks/Cole.
Shea, D. J. (2008). Effects of sexual abuse by Catholic priests on adults victimized as children. *Sexual Addiction and Compulsivity*, *15*, 250–268.
Sheehy, G. (1976). *Passages. Predctable crises of adult life*. New York, NY: Dutton.
Sherif, M. (1966). *In common predicament: Social psychology of intergroup conflict and cooperation*. Boston, MA: Houghton Mifflin.
Sherif, M. (1998). Experiments in group conflict. In J. M. Jenkins, K. Oatley, & N. L. Stein (Eds.), *Human emotions: A reader* (pp. 245–252). Malden, MA: Blackwell.
Shin, R.-M., Tully, K., Li, Y., Cho, J.-H., Higuchi, M., Suhara, T., & Bolshakov, V. Y. (2010). Hierarchical order of coexisting pre- and postsynaptic forms of long-term potentiation at synapses in amygdala. *PNAS Proceedings of the National Academy of Sciences of the United States of America*, *107*(44), 19073–19078.
Shinfuku, N., & Kitanishi, K. (2010). Buddhism and psychotherapy in Japan. Religion and psychiatry: Beyond boundaries. In P. J. Verhagen, H. M. van Praag, J. J., J. Jr. López-Ibor, J. L. Cox, & D. Moussaoui (Eds.), *Religion and psychiatry: Beyond boundaries* (pp. 181–191). Malden, MA: Wiley-Blackwell.
Shoji, H., & Mizoguchi, K. (2011). Aging-related changes in the effects of social isolation on social behavior in rats. *Physiology and Behavior, 102*(1), 58–62.
Shweder, R. A. (2011). Commentary: Ontogenetic cultural psychology. In L. A. Jensen (Ed.), *Bridging cultural and developmental approaches to psychology: New syntheses in theory, research, and policy* (pp. 303–310). New York, NY: Oxford University Press.
Sias, P. M., Heath, R. G., Perry, T., Silva, D., & Fix, B. (2004). Narratives of workplace friendship deterioration. *Journal of Social and Personal Relationships*, *21*(3), 321–340.
Sidhu, M., Malhi, P., & Jerath, J. (2010). Intelligence of children from economically disadvantaged families: Role of parental education. *Psychological Studies*, *55*(4), 358–364
Sieber-Blum, M. (2010). Epidermal neural crest stem cells and their use in mouse models of spinal cord injury. *Brain Research Bulletin*, *83*, 189–193.
Siegala, M., & Varley, R. (2008). If we could talk to the animals. *Behavioral and Brain Sciences*, *31*, 146–147.
Siegel, A. B. (2010). Dream interpretation in clinical practice: A century after Freud. *Sleep Medicine Clinics*, *5*(2), 299–313.
Siegel, J. M. (2000, January). Narcolepsy. *Scientific American*, pp. 76–81.
Siegel, J. M. (2008). Do all animals sleep? *Trends in Neurosciences*, *31*, 208–213.
Siev, J., & Chambless, D. L. (2007). Specificity of treatment effects: Cognitive therapy and relaxation for generalized anxiety and panic disorders. *Journal of Consulting and Clinical Psychology*, *75*(4), 513–522.
Siever, L. J. (2008). Neurobiology of aggression and violence. *American Journal of Psychiatry*, *165*, 429–442.
Sillén, A., Lilius, L., Forsell, C., Kimura, T., Winblad, B., & Graff, C. (2011). Linkage to the 8p21.1 region including the CLU gene in age at onset stratified Alzheimer's disease families. *Journal of Alzheimer's Disease*, *23*(1), 13–20.
Silverman, W. K., Pina, A. A., & Viswesvaran, C. (2008). Evidence-based psychosocial treatments for phobic and anxiety disorders in children and adolescents. *Journal of Clinical Child and Adolescent Psychology*, *37*, 105–130.
Silvestri, A. J., & Root, D. H. (2008). Effects of REM deprivation and an NMDA agonist on the extinction of conditioned fear. *Physiology and Behavior*, *93*, 274–281.
Simmons, A. L. (2012). Distributed practice and procedural memory consolidation in musicians' skill learning. *Journal of Research in Music Education*, *59*(4), 357–368.
Sinason, V. (Ed.). (2011). *Attachment, trauma, and multiplicity: Working with Dissociative*

*Identity Disorder* (2nd ed.). New York, NY: Routledge.
Singelis, T. M., Triandis, H. C., Bhawuk, D. S., & Gelfand, M. (1995). Horizontal and vertical dimensions of individualism and collectivism: A theoretical and measurement refinement. *Cross-Cultural Research, 29*, 240–275.
Singer, L. T., & Richardson, G. A. (2011). Introduction to "understanding developmental consequences of prenatal drug exposure: Biological and environmental effects and their interactions." *Neurotoxicology and Teratology, 33*(1), 5–8.
Sironi, V. A. (2011). Origin and evolution of deep brain stimulation. *Frontiers in Integrative Neuroscience, 5*, 42.
Skelhorn, J., Griksaitis, D., & Rowe, C. (2008). Colour biases are more than a question of taste. *Animal Behaviour, 75*, 827–835.
Skinner, B. F. (1953). *Science and human behavior.* New York, NY: Macmillan.
Skinner, B. F. (1961). Diagramming schedules of reinforcement. *Journal of the Experimental Analysis of Behavior, 1*, 67–68.
Skinner, R., Conlon, L., Gibbons, D., & McDonald, C. (2011). Cannabis use and non-clinical dimensions of psychosis in university students presenting to primary care. *Acta Psychiatrica Scandinavica, 123*(1), 21–27.
Slobogin, C. (2006). *Minding justice: Laws that deprive people with mental disability of life and liberty*. Cambridge, MA: Harvard University Press.
Slovácková, B., & Slovácek, L. (2007). Moral judgement competence and moral attitudes of medical students. *Nursing Ethics, 14*, 320–328.
Smith, B. (2011). Hypnosis today. *Monitor on Psychology, 42*(1), 6.
Smith, E. E. (1995). Concepts and categorization. In E. E. Smith & D. N. Osherson (Eds.), *Thinking: An invitation to cognitive science* (2nd ed.) (pp. 3–33). Cambridge, MA: MIT Press.
Smith, K. C. (2012). The Obama effect on African-American students' academic performance. *Dissertation Abstracts International: Section B: The Sciences and Engineering, 72*(11-B), 7114.
Smith, K. D. (2007). Spinning straw into gold: Dynamics of Rumpelstiltskin style of leadership. *Dissertation Abstracts International Section A: Humanities and Social Sciences, 68*(5-A), 1760.
Smith, M. T., Huang, M. I., & Manber, R. (2005). Cognitive behavior therapy for chronic insomnia occurring within the context of medical and psychiatric disorders. *Clinical Psychology Review, 25*, 559–592.
Smith, M., Wang, L., Cronenwett, W., Goldman, M., Mamah, D., Barch, D., & Csernansky, J. (2011). Alcohol use disorders contribute to hippocampal and subcortical shape differences in schizophrenia. *Schizophrenia Research, 131*(1–3), 174–183.
Smith, P. (2011). Cross-cultural perspectives on identity. In S. Schwartz, K. Luyckx, & V. Vignoles (Eds.), *Handbook of identity theory and research* (Vols. 1–2) (pp. 249–265). New York, NY: Springer.
Smith, T., Rodriguez, M., & Bernal, G. (2011). Culture. *Journal of Clinical Psychology, 67*(2), 166–175.
Snarey, J. R. (1985). Cross-cultural universality of social-moral development: A critical review of Kohlbergian research. *Psychological Bulletin, 97*, 202–233.
Snarey, J. R. (1995). In communitarian voice: The sociological expansion of Kohlbergian theory, research, and practice. In W. M. Kurtines & J. L. Gerwirtz (Eds.), *Moral development: An introduction* (pp. 109–134). Boston, MA: Allyn & Bacon.
Snijders, T. J., Ramsey, N. F., Koerselman, F., & van Gijn, J. (2010). Attentional modulation fails to attenuate the subjective pain experience in chronic, unexplained pain. *European Journal of Pain, 14*(3), e1–e10.
Snyder, C. R. (2003). "Me conform? No way": Classroom demonstrations for sensitizing students to their conformity. *Teaching of Psychology, 30*(1), 59–61.
Snyder, J. S., & Alain, C. (2007). Sequential auditory sense analysis is preserved in normal aging adults. *Cerebral Cortex, 17*, 501–512.
Solms, M. (1997). *The neuropsychology of dreams*. Mahwah, NJ: Lawrence Erlbaum.
Soto, C. J., John, O. P., Gosling, S. D., & Potter, J. (2011). Age differences in personality traits from 10 to 65: Big Five domains and facets in a large cross-sectional sample. *Journal of Personality and Social Psychology, 100*(2), 330–348.
Sowell, E. R., Mattson, S. N., Kan, E., Thompson, P. M., Riley, E. P., Edward, P., & Toga, A. W. (2008). Abnormal cortical thickness and brain-behavior correlation patterns in individuals with heavy prenatal alcohol exposure. *Cerebral Cortex, 18*, 136–144.
Spano, R., Koenig, T. L., Hudson, J. W., & Leiste, M. R. (2010). East meets west: A nonlinear model for understanding human growth and development. *Smith College Studies in Social Work, 80*(2–3), 198–214.
Spearman, C. (1923). *The nature of "intelligence" and the principles of cognition*. London, England: Macmillan.
Spears, R. (2011). Group identities: The social identity perspective. In S. Schwartz, K. Luyckx, & V. Vignoles (Eds.), *Handbook of identity theory and research* (Vols. 1–2) (pp. 201–224). New York, NY: Springer.
Sperling, G. (1960). The information available in brief visual presentations. *Psychological Monographs, 74* (Whole No. 498).
Spiegel, D., & Maldonado, J. R. (1999). Dissociative disorders. In R. E. Hales, S. C. Yudofsky, & J. C. Talbott (Eds.), *American psychiatric press textbook of psychiatry* (pp. 665–710). Washington, DC: American Psychiatric Press.
Spitz, R. A., & Wolf, K. M. (1946). The smiling response: A contribution to the ontogenesis of social relations. *Genetic Psychology Monographs, 34*, 57–123.
Sprecher, S., & Fehr, B. (2011). Dispositional attachment and relationship-specific attachment as predictors of compassionate love for a partner. *Journal of Social and Personal Relationships, 28*(4), 558–574.
Sprecher, S., & Regan, P. C. (2002). Liking some things (in some people) more than others: Partner preferences in romantic relationships and friendships. *Journal of Social and Personal Relationships, 19*(4), 463–481.
Spurling, L. (2011). Review of Off the Couch: Contemporary psychoanalytic approaches. *Psychodynamic Practice: Individuals, Groups and Organisations, 17*(1), 99–100.
Stafford, J., & Lynn, S. J. (2002). Cultural scripts, memories of childhood abuse, and multiple identities: A study of role-played enactments. *International Journal of Clinical and Experimental Hypnosis, 50*(1), 67–85.
Stalder, T., Kirschbaum, C., Heinze, K., Steudte, S., Foley, P., Tietze, A., & Dettenborn, L. (2010). Use of hair cortisol analysis to detect hypercortisolism during active drinking phases in alcohol-dependent individuals. *Biological Psychology, 85*(3), 357–360.
Steele, C. M., & Aronson, J. (1995). Stereotype threat and the intellectual test performance of African Americans. *Journal of Personality and Social Psychology, 69*, 797–811.
Stefanek, E., Strohmeier, D., Fandrem, H., & Spiel, C. (2012). Depressive symptoms in native and immigrant adolescents: The role of critical life events and daily hassles. *Anxiety, Stress and Coping: An International Journal, 25*(2), 201–217.
Steiner, L. M., Suarez, E. C., Sells, J. N., & Wykes, S. D. (2011). Effect of age, initiator status, and infidelity on women's divorce adjustment. *Journal of Divorce and Remarriage, 52*(1), 33–47.
Steinert, R., & Beglinger, C. (2011). Nutrient sensing in the gut: Interactions between chemosensory cells, visceral afferents and the secretion of satiation

peptides. *Physiology and Behavior*. Retrieved from http://www.ncbi.nlm.nih.gov/pubmed/21376067

Steinhart, P. (1986, March). Personal boundaries. *Audubon*, pp. 8–11.

Stelmack, R. M., Knott, V., & Beauchamp, C. M. (2003). Intelligence and neural transmission time: A brain stem auditory evoked potential analysis. *Personality and Individual Differences*, *34*(1), 97–107.

Sternberg, R. J. (1985). *Beyond IQ: A triarchic theory of human intelligence*. New York, NY: Cambridge University Press.

Sternberg, R. J. (1986). A triangular theory of love. *Psychological Review*, *93*(2), 119–135.

Sternberg, R. J. (1988). *The triangle of love*. New York, NY: Basic Books.

Sternberg, R. J. (1999). The theory of successful intelligence. *Review of General Psychology*, *3*, 292–316.

Sternberg, R. J. (2005). The importance of converging operations in the study of human intelligence. *Cortex*. *41*(2), 243–244.

Sternberg, R. J. (2006). A duplex theory of love. In R. J. Sternberg & K. Weis (Eds.). *The new psychology of love* (pp. 184–199). New Haven, CT: Yale University Press.

Sternberg, R. J. (2007). Developing successful intelligence in all children: A potential solution to underachievement in ethnic minority children. In M. C. Wang & R. D. Taylor (Eds.), *Closing the achievement gap*. Philadelphia, PA: Laboratory for Student Success at Temple University.

Sternberg, R. J. (2009). *Cognitive psychology* (5th ed.). Belmont, CA: Wadsworth.

Sternberg, R. J. (2010). Teaching for creativity. In R. A. Beghetto & J. C. Kaufman (Eds.), *Nurturing creativity in the classroom* (pp. 394–414). New York, NY: Cambridge University Press.

Sternberg, R. J. (2012). *Cognitive psychology* (6th ed.). Belmont, CA: Cengage.

Sternberg, R. J., & Grigorenko, E. L. (2008). Ability testing across cultures. In L. Suzuki (Ed.), *Handbook of multicultural assessment* (3rd ed.) (pp. 449–470). New York, NY: Jossey-Bass.

Sternberg, R. J., & Hedlund, J. (2002). Practical intelligence, g, and work psychology. *Human Performance*, *15*(1–2), 143–160.

Sternberg, R. J., & Kaufman, S. (Eds.). (2012). *The Cambridge handbook of intelligence*. Cambridge, UK: Cambridge University Press.

Sternberg, R. J., & Lubart, T. I. (1992). Buy low and sell high: An investment approach to creativity. *Current Directions in Psychological Science*, *1*(1), 1–5.

Sternberg, R. J., & Lubart, T. I. (1996). Investing in creativity. *American Psychologist*, *51*(7), 677–688.

Sternberg, R. J., Jarvin, L., & Grigorenko, E. L. (2011). *Explorations in giftedness*. New York, NY: Cambridge University Press.

Stollhoff, R., Jost, J., Elze, T., & Kennerknecht, I. (2011). Deficits in long-term recognition memory reveal dissociated subtypes in congenital prosopagnosia. *PLoS ONE*, *6*(1). Article ID e15702.

Stoltzfus, G., Nibbelink, B., Vredenburg, D., & Hyrum, E. (2011). Gender, gender role, and creativity. *Social Behavior and Personality: An International Journal*, *39*(3), 425–432.

Stompe, T. G., Ortwein-Swoboda, K., Ritter, K., & Schanda, H. (2003). Old wine in new bottles? Stability and plasticity of the contents of schizophrenic delusions. *Psychopathology*, *36*(1), 6–12.

Stone, J., & Focella, E. (2011). Postdecisional self-enhancement and self-protection: The role of the self in cognitive dissonance processes. In M. D. Alicke & C. Sedikides (Eds.), *Handbook of self-enhancement and self-protection* (pp. 192–210). New York, NY: Guilford Press.

Stone, K. L., & Redline, S. (2006). Sleep-related breathing disorders in the elderly. *Sleep Medicine Clinics*, *1*(2), 247–262.

Stoner, J. A. (1961). *A comparison of individual and group decisions involving risk*. Unpublished master's thesis, School of Industrial Management, MIT, Cambridge, MA.

Stoodley, C. J. (2012). The cerebellum and cognition: Evidence from functional imaging studies. *The Cerebellum*, *11*(2), 352–365.

Strack, F., Martin, L. L., & Stepper, S. (1988). Inhibiting and facilitating conditions of the human smile: A nonobstrusive test of the facial feedback hypothesis. *Journal of Personality and Social Psychology*, *54*, 768–777.

Strange, D., Garry, M., Bernstein, D. M., & Lindsay, D. S. (2011). Photographs cause false memories for the news. *Acta Psychologica*, *136*(1), 90–94.

Stratton, G. M. (1896). Some preliminary experiments on vision without inversion of the retinal image. *Psychological Review*, *3*, 611–617.

Stratton, P., Reibstein, J., Lask, J., Singh, R., & Asen, E. (2011). Competences and occupational standards for systemic family and couples therapy. *Journal of Family Therapy*, *33*(2), 123–143.

Straub, R. O. (2011). *Health psychology: A biopsychosocial approach* (3rd ed.). New York, NY: Worth.

Strauss, J. R. (2010). The baby boomers meet menopause: Attitudes and roles. *Dissertation Abstracts International Section A: Humanities and Social Sciences*, *70*(8-A), 3196.

Strozier, C. B. (2011). Torture, war, and the culture of fear after 9/11. *International Journal of Group Psychotherapy*, *61*(1), 67–72.

Subramanian, S., & Vollmer, R. R. (2002). Sympathetic activation fenfluramine depletes brown adipose tissue norepinephrine content in rats. *Pharmacology, Biochemistry and Behavior*, *73*(3), 639–646.

Sue, D., Sue, D. W., & Sue, S. (2010). *Understanding abnormal behavior* (9th ed.). Belmont, CA: Cengage.

Sue, D.W., & Sue, D. (2008). *Counseling the culturally diverse: Theory and practice* (5th ed.). Hoboken, NJ: John Wiley & Sons.

Sullivan, M. J. L. (2008). Toward a biopsychomotor conceptualization of pain: Implications for research and intervention. *Clinical Journal of Pain*, *24*, 281–290.

Sullivan, M. J. L., Tripp, D. A., & Santor, D. (1998). *Gender differences in pain and pain behavior: The role of catastrophizing*. Paper presented at the annual meeting of the American Psychological Association, San Francisco, CA.

Sullivan, T. P., & Holt, L. J. (2008). PTSD symptom clusters are differentially related to substance use among community women exposed to intimate partner violence. *Journal of Traumatic Stress*, *21*, 173–180.

Suls, J., Davidson, K., & Kaplan, R. (Eds.). (2010). *Handbook of health psychology and behavioral medicine* (pp. 370–380). New York, NY: Guilford Press.

Sundram, F., Deeley, Q., Sarkar, S., Daly, E., Latham, R., Barker, G., & Murphy, D. (2011). White matter microstructural abnormalities in antisocial personality disorder: A pilot diffusion tensor imaging study. *European Psychiatry*, *26*(1), 957.

Surbey, M. K. (2011). Adaptive significance of low levels of self-deception and cooperation in depression. *Evolution and Human Behavior*, *32*(1), 29–40.

Surget, A., Tanti, A., Leonardo, E. D., Laugeray, A., Rainer, Q., Touma, C., . . . Belzung, C. (2011). Antidepressants recruit new neurons to improve stress response regulation. *Molecular Psychiatry*, *16*(12), 1177–1188.

Surtees, P. G., Wainwright, N. W. J., Luben, R. N., Khaw, K-T, & Bingham, S. A. (2009). No evidence that social stress is associated with breast cancer incidence. *Breast Cancer Research and Treatment*, *120*(1), 169–174.

Swami, V. (2011). *Evolutionary psychology: A critical introduction*. Chichester, West Sussex, UK: Wiley-Blackwell.

Swartz, K. L. & Margolis, S. (2004). *Depression and anxiety*. Johns Hopkins White Papers. Baltimore: Johns Hopkins Medical Institutions.

Sweldens, S., Van Osselaer, S. M. J., & Janiszewski, C. (2010). Evaluative conditioning procedures and the resilience of conditioned brand attitudes. *Journal of Consumer Research*, 37(3), 473–489.

Szasz, T. (1960). The myth of mental illness. *American Psychologist*, *15*, 113–118.

Szasz, T. (2000). Second commentary on "Aristotle's function argument." *Philosophy, Psychiatry, and Psychology*, *7*, 3–16.

Szasz, T. (2004). The psychiatric protection order for the "battered mental patient." *British Medical Journal*, *327*(7429), 1449–1451.

Takahashi, A., Quadros, I. M., de Almeida, R. M. M., & Miczek, K. A. (2011). Brain serotonin receptors and transporters: Initiation vs. termination of escalated aggression. *Psychopharmacology*, *213*(2–3), 183–212.

Takahashi, T., Suzuki, G., Nibuya, M., Tanaka, T., Nozawa, H., et al. (2012). Therapeutic effect of paroxetine on stress-induced gastric lesions in mice. *Progress in Neuro-Psychopharmacology and Biological Psychiatry*, *36*(1), 39–43.

Takano, Y., & Sogon, S. (2008). Are Japanese more collectivistic than Americans? Examining conformity in in-groups and the reference-group effect. *Journal of Cross-Cultural Psychology*, *39*, 237–250.

Talarico, J. M., & Rubin, D. C. (2007). Flashbulb memories are special after all; in phenomenology, not accuracy. *Applied Cognitive Psychology*, *21*, 557–578.

Talwar, V., Harris, P., & Schleifer, M. (Eds.). (2011). *Children's understanding of death: From biological to religious conceptions*. Cambridge, UK: Cambridge University Press.

Tandon, R., Keshavan, M. S., & Nasrallah, H. A. (2008). Schizophrenia, "just the facts" what we know in 2008. 2. Epidemiology and etiology. *Schizophrenia Research*, *102*, 1–18.

Tang, Y.-P., Wang, H., Feng, R., Kyin, M., & Tsien, J. Z. (2001). Differential effects of enrichment on learning and memory function in NR2B transgenic mice. *Neuropharmacology*, *41*(6), 779–790.

Tarter, R. E., Kirisci, L., Mezzich, A., Cornelius, J. R., Pajer, K., Vanyukov, M., . . . Clark, D. (2003). Neurobehavioral disinhibition in childhood predicts early age at onset of substance use disorder. *American Journal of Psychiatry*, *160*(6), 1078.

Tarter, R. E., Vanyukov, M., Kirisci, L., Reynolds, M., & Clark, D. B. (2006). Predictors of marijuana use in adolescents before and after licit drug use: Examination of the gateway hypothesis. *American Journal of Psychiatry*, *163*, 2134–2140.

Tatkin, S. (2011). *Wired for love: How understanding your partner's brain and attachment style can help you defuse conflict and build a secure relationship*. Oakland, CA: New Harbinger Publications.

Taylor, D. J., Lichstein, K. L., & Durrence, H. H. (2003). Insomnia as a risk factor. *Behavioral Sleep Medicine*, *1*, 227–247.

Taylor, J. Y., Caldwell, C. H., Baser, R. E., Faison, N., & Jackson, J. S. (2007). Prevalence of eating disorders among Blacks in the national survey of American life. *International Journal of Eating Disorders*, *40*(Supl), S10-S14.

Taylor, L., Fiore, A., Mendelsohn, G., & Cheshire, C. (2011). "Out of my league": A real-world test of the matching hypothesis. *Personality and Social Psychology Bulletin*, *37*(7), 942–954.

Taylor, S. E. (2006). Tend and befriend: Biobehavioral bases of affiliation under stress. *Current Directions in Psychological Science*, *15*(6), 273–277.

Taylor, S. E. (2012). Tend and befriend theory. In P. A. M. Van Lange, A. W. Kruglanski, & E. T. Higgins (Eds.), *Handbook of theories of social psychology* (pp. 32–49). Thousand Oaks, CA: Sage Publications Ltd.

Taylor, S. E., & Sherman, D. K. (2008). Self-enhancement and self-affirmation: The consequences of positive self-thoughts for motivation and health. In J. Y. Shah & W. L. Gardner (Eds.), *Handbook of motivation science* (pp. 57–70). New York, NY: Guilford Press.

Tellegen, A. (1985). Structures of mood and personality and their relevance to assessing anxiety with an emphasis on self-report. In A. H. Tuma & J. D. Maser (Eds.), *Anxiety and the anxiety disorders* (pp. 681–706). Hillsdale, NJ: Erlbaum.

Terhune, D. B., Cardeña, E., & Lindgren, M. (2011). Dissociative tendencies and individual differences in high hypnotic suggestibility. *Cognitive Neuropsychiatry*, *16*(2), 113–135.

Terman, L. M. (1954). Scientists and nonscientists in a group of 800 gifted men. *Psychological Monographs*, *68*(7), 1–44.

Terrace, H. S. (1979, November). How Nim Chimpsky changed my mind. *Psychology Today*, pp. 65–76.

Terry, S. W. (2003). *Learning and memory: Basic principles, processes, and procedures* (2nd ed.). Boston, MA: Allyn & Bacon.

Thoits, P. A. (2010). Stress and health: Major findings and policy implications. *Journal of Health and Social Behavior*, *51*(1, *Suppl*), S41–S53.

Thoits, P. A. (2011). Resisting the stigma of mental illness. *Social Psychology Quarterly*, *74*(1), 6–28.

Thomas, N., Rossell, S., Farhall, J., Shawyer, F., & Castle, D. (2011). Cognitive behavioural therapy for auditory hallucinations: Effectiveness and predictors of outcome in a specialist clinic. *Behavioural and Cognitive Psychotherapy*, *39*(2), 129–138.

Thompson, R. F. (2005). In search of memory traces. *Annual Review of Clinical Psychology*, *56*, 1–23.

Thorndike, E. L. (1898). Animal intelligence. *Psychological Review Monograph*, *2*(8), 1–8.

Thorndike, E. L. (1911). *Animal intelligence*. New York, NY: Macmillan.

Thornhill, R., Gangestad, S. W., Miller, R., Scheyd, G., McCollough, J. K., & Franklin, M. (2003). Major histocompatibility complex genes, symmetry, and body scent attractiveness in men and women. *Behavioral Ecology*, *14*(5), 668–678.

Thornton, B., & Tizard, H. J. (2010). "Not in my back yard": Evidence for arousal moderating vested interest and oppositional behavior to proposed change. *Social Psychology*, *41*(4), 255–262.

Thyer, B. A., & Myers, L. L. (2011). The quest for evidence-based practice: A view from the United States. *Journal of Social Work*, *11*(1), 8–25.

Tindale, S., & Posavac, E. (2011). The social psychology of stakeholder processes: Group processes and interpersonal relations. In M. Mark, S. Donaldson, & B. Campbell (Eds.) *Social psychology and evaluation* (pp. 189–209). New York, NY: Guilford Press.

Todd, A. R., Bodenhausen, G. V., Richeson, J. A., & Galinsky, A. D. (2011). Perspective taking combats automatic expressions of racial bias. *Journal of Personality and Social Psychology*, *100*(6), 1027–1042.

Todd, P. M., & Gigerenzer, G. (2000). Precis of simple heuristics that make us smart. *Behavioral and Brain Sciences*, *23*(5), 727.

Tolman, E. C., & Honzik, C. H. (1930). Introduction and removal of reward and maze performance in rats. *University of California Publications in Psychology*, *4*, 257–275.

Tompkins, T. L., Hockett, A. R., Abraibesh, N., & Witt, J. L. (2011). A closer look at co-rumination: Gender, coping, peer functioning and internalizing/externalizing problems. *Journal of Adolescence*, *34*(5), 801–811.

Topham, G. L., Hubbs-Tait, L., Rutledge, J. M., Page, M. C., Kennedy, T. S., Shriver, L. H., & Harrist, A. W. (2011). Parenting styles, parental response to child emotion, and family emotional responsiveness are related to child emotional eating. *Appetite*, *56*(2), 261–264.

Torpy, J. M., Lynm, C., & Glass, R. M. (2007). Chronic stress and the heart. *Journal of the American Medical Association*, *298*, 1722.

Torry, Z. D., & Billick, S. B. (2011). Implications of antisocial parents. *Psychiatric Quarterly*, *82*(4), 275–285.
Trawick-Smith, J., & Dziurgot, T. (2011). 'Good-fit' teacher–child play interactions and the subsequent autonomous play of preschool children. *Early Childhood Research Quarterly*, *26*(1), 110–123.
Tremblay, P. F., Graham, K., & Wells, S. (2008). Severity of physical aggression reported by university students: A test of the interaction between trait aggression and alcohol consumption. *Personality and Individual Differences*, *45*, 3–9.
Triandis, H. C. (2007). Culture and psychology: A history of the study of their relationship. In S. Kitayama & D. Cohen (Eds), *Handbook of cultural psychology* (pp. 59–76). New York, NY: Guilford.
Troll, L. E., Miller, S. J., & Atchley, R. C. (1979). *Families in later life*. Belmont, CA: Wadsworth.
Tronci, V., & Balfour, D. J. K. (2011). The effects of the mGluR5 receptor antagonist 6-methyl-2-(phenylethynyl)-pyridine (MPEP) on the stimulation of dopamine release evoked by nicotine in the rat brain. *Behavioural Brain Research*, *219*(2), 354–357.
Tropp, L. R., & Mallett, R. K. (2011). *Moving beyond prejudice reduction: Pathways to positive intergroup relations*. Washington, DC: American Psychological Association.
Truscott, D. (2010). Behavioral. In D. Truscott (Ed.), *Becoming an effective psychotherapist: Adopting a theory of psychotherapy that's right for you and your client* (pp. 37–51). Washington, DC: American Psychological Association.
Tryon, W. W. (2008). Whatever happened to symptom substitution? *Clinical Psychology Review*, *28*, 963–968.
Tsien, J. Z. (2000, April). Building a brainier mouse. *Scientific American*, pp. 62–68.
Tsukiura, T., & Cabeza, R. (2011). Shared brain activity for aesthetic and moral judgments: implications for the Beauty-is-Good stereotype. *Social Cognitive and Affective Neuroscience*, *6*(3), 138–148.
Tulving, E., & Thompson, D. M. (1973). Encoding specificity and retrieval processes in episodic memory. *Psychological Review*, *80*, 352–373.
Tümkaya, S. (2007). Burnout and humor relationship among university lecturers. *Humor: International Journal of Humor Research*, *20*(1), 73–92.
Tversky, A., & Kahneman, D. (1974). Judgment under uncertainty: Heuristics and biases. *Science*, *185*, 1124–1131.
Tversky, A., & Kahneman, D. (1993). Probabilistic reasoning. In A. I. Goldman (Ed.), *Readings in philosophy and cognitive science*, (pp. 43–68). Cambridge, MA: The MIT Press.
Tyson, P. J., Jones, D., & Elcock, J. (2011). *Psychology in social context: Issues and debates*. Malden, MA: Wiley-Blackwell.
Using CBT and Smart Phones (2011). *Beck Institute Blog: Beck Institute for Cognitive Therapy and Research*. Retrieved from http://psycnet.apa.org/psycextra/534062012-001.pdf
Valadez, J. J., & Ferguson, C. J. (2012). Just a game after all: Violent video game exposure and time spent playing effects on hostile feelings, depression, and visuospatial cognition. *Computers in Human Behavior*, *28*(2), 608–616.
Valla, J., & Ceci, S. (2011). Can sex differences in science be tied to the long reach of prenatal hormones: Brain organization theory, digit ratio (2d/4d), and sex differences in preferences and cognition. *Perspectives on Psychological Science*, *6*(2), 134–146.
Van de Carr, F. R., & Lehrer, M. (1997). *While you are expecting: Your own prenatal classroom*. New York, NY: Humanics Publishing.
van der Laan, L. N., de Ridder, D. T. D., Viergever, M. A., & Smeets, P. A. M. (2011). The first taste is always with the eyes: A meta-analysis on the neural correlates of processing visual food cues. *NeuroImage*, *55*(1), 296–303.
van der Sluis, S., Derom, C., Thiery, E., Bartels, M., Polderman, T. J. C., Verhulst, F. C., Jacobs, N., van Gestel, S., de Geus, E. J. C., Dolan, C. V., Boomsma, D. I., & Posthuma, D. (2008). Sex differences on the WISC-R in Belgium and the Netherlands. *Intelligence*, *36*, 48–67.
van Marle, H. J. C. (2010). Violence in the family: An integrative approach to its control. *International Journal of Offender Therapy and Comparative Criminology*, *54*(4), 475–477.
van Stegeren, A. H. (2008). The role of the noradrenergic system in emotional memory. *Acta Psychologica*, *127*, 532–541.
Varga, S. (2011). Evolutionary psychiatry and depression: Testing two hypotheses. *Medicine, Health Care and Philosophy*, *15*(1), 41–52.
Vaughn, D. (1996). *The Challenger launch decision: Risky technology, culture, and deviance at NASA*. Chicago, IL: University of Chicago Press.
Vaz-Leal, F., Rodríguez-Santos, L, García-Herráiz, A., & Ramos-Fuentes, I. (2011). Neurobiological and psychopathological variables related to emotional instability: A study of their capability to discriminate patients with bulimia nervosa from healthy controls. *Neuropsychobiology*, *63*(4), 242–251.
Venables, N. C., Patrick, C. J., Hall, J. R., & Bernat, E. M. (2011). Clarifying relations between dispositional aggression and brain potential response: Overlapping and distinct contributions of impulsivity and stress reactivity. *Biological Psychology*, *86*(3), 279–288.
Verleger, R., Ludwig, J., Kolev, V., Yordanova, J., & Wagner, U. (2011). Sleep effects on slow-brain-potential reflections of associative learning. *Biological Psychology*, *86*(3), 219–229.
Vernon, A. (2011). Rational emotive behavior therapy. In D. Capuzzi & D. R. Gross (Eds.), *Counseling and psychotherapy* (5th ed.) (pp. 237–261). Alexandria, VA: American Counseling Association.
Vertosick, F. T. (2000). *Why we hurt: The natural history of pain*. New York, NY: Harcourt.
Veselka, L., Schermer, J. A., & Vernon, P. (2011). Beyond the big five: The dark triad and the Supernumerary Personality Inventory. *Twin Research and Human Genetics*, *14*(2), 158–168.
Veselka, L., Schermer, J. A., Martin, R. A., & Vernon, P. A. (2010). Laughter and resiliency: A behavioral genetic study of humor styles and mental toughness. *Twin Research and Human Genetics*, *13*(5), 442–449.
Victoroff, J., Quota, S., Adelman, J. R., Celinska, B., Stern, N., Wilcox, R., & Sapolsky, R. M. (2011). Support for religio-political aggression among teenaged boys in Gaza: Part 2: Neuroendocrinological findings. *Aggressive Behavior*, *37*(2), 121–132.
Vitoroulis, I., Schneider, B., Vasquez, C., Soteras del Toro, M., & Gonzales, Y. (2011). Perceived parental and peer support in relation to Canadian, Cuban, and Spanish adolescents' valuing of academics and intrinsic academic motivation. *Journal of Cross-Cultural Psychology*, *43*(5), 704–722.
Vogt, D. S., Rizvi, S. L., Shipherd, J. C., & Resick, P. A. (2008). Longitudinal investigation of reciprocal relationship between stress reactions and hardiness. *Personality and Social Psychology Bulletin*, *34*, 61–73.
Völker, S. (2007). Infants' vocal engagement oriented towards mother versus stranger at 3 months and avoidant attachment behavior at 12 months. *International Journal of Behavioral Development*, *31*, 88–95.
Volkow, N. D., Tomasi, D., Gene-Jack, W., Vaska, P., Fowler, J. S., Telang, F., Alexoff, D., Logan, J., & Wong, C. (2011). Effects of cell phone radiofrequency signal exposure on brain glucose metabolism. *Journal of the American Medical Association*, *305*(*8*), 808–813.

Wachlarowicz, M., Snyder, J., Low, S., Forgatch, M., & DeGarmo, D. (2012). The moderating effects of parent antisocial characteristics on the effects of Parent Management Training-Oregon (PMTOTM). *Prevention Science*, *13*(3), 229–240.

Wachtel, P. L. (2011). *Inside the session: What really happens in psychotherapy*. Washington, DC: American Psychological Association.

Wade, T., Keski-Rahkonen, A., & Hudson, J. (2011). Epidemiology of eating disorders. In M. Tsuang, M. Tohen, and P. B. Jones (Eds.), *Textbook in psychiatric epidemiology* (3rd ed.) (pp. 343–360). Hoboken, NJ: John Wiley & Sons.

Wagner, U., Christ, O., & Pettigrew, T. F. (2008). Prejudice and group-related behaviors in Germany. *Journal of Social Issues*, *64*, 403–416.

Wagstaff, G. G., Cole, J., Wheatcroft, J., Marshall, M., & Barsby, I. (2007). A componential approach to hypnotic memory facilitation: Focused meditation, context reinstatement and eye movements. *Contemporary Hypnosis*, *24*, 97–108.

Wahlgren, K., & Lester, D. (2003). The big four: Personality in dogs. *Psychological Reports*, *92*, 828.

Walker, F. R., Hinwood, M., Masters, L., Deilenberg, R. A., & Trevor, A. (2008). Individual differences predict susceptibility to conditioned fear arising from psychosocial trauma. *Journal of Psychiatric Research*, *42*, 371–383.

Wall, A. (2007). Review of integrating gender and culture in parenting. *The Family Journal*, *15*, 196–197.

Waller, N. G., & Ross, C. A. (1997). The prevalence and biometric structure of pathological dissociation in the general population: Taxometric and behavior genetics findings. *Journal of Abnormal Psychology*, *106*, 499–510.

Wallerstein, G. (2008). *The pleasure instinct: Why we crave adventure, chocolate, pheromones, and music*. Hoboken, NY: John Wiley & Sons.

Wallis, C. (2011). Performing gender: A content analysis of gender display in music videos. *Sex Roles*, *64*(3–4), 160–172.

Walther, E., Langer, T., Weil, R., & Komischke, M. (2011). Preferences surf on the currents of words: Implicit verb causality influences evaluative conditioning. *European Journal of Social Psychology*, *41*(1), 17–22.

Wamsley, E. J., & Stickgold, R. (2010). Dreaming and offline memory processing. *Current Biology*, *20*(23), 1010–1913.

Wamsley, E. J., Hirota, Y., Tucker, M. A., Smith, M. R., Doan, T., & Antrobus, J. S. (2007). Circadian and ultradian influences on dreaming: A dual rhythm model. *Brain Research Bulletin*, *71*, 347–354.

Wang, F., DesMeules, M., Luo, W., Dai, S., Lagace, C., & Morrison, H. (2011). Leisure-time physical activity and marital status in relation to depression between men and women: A prospective study. *Health Psychology*, *30*(2), 204–211.

Wang, J., Yuan, W., & Li, M. D. (2011). Genes and pathways co-associated with the exposure to multiple drugs of abuse, including alcohol, amphetamine/methamphetamine, cocaine, marijuana, morphine, and/or nicotine: A review of proteomics analyses. *Molecular Neurobiology*, *44*(3), 269–286.

Wardlaw, G. M., & Hampl, J. (2007). *Perspectives in nutrition* (7th ed.). New York, NY: McGraw-Hill.

Wason, P. C. (1968). Reasoning about a rule. *Quarterly Journal of Experimental Psychology*, *20*(3), 273–281.

Wass, T. S. (2008). Neuroanatomical and neurobehavioral effects of heavy prenatal alcohol exposure. In J. Brick (Ed.), *Handbook of the medical consequences of alcohol and drug abuse* (2nd ed.) (pp. 177–217). *The Haworth Press series in neuropharmacology*. New York, NY: Haworth Press/Taylor and Francis Group.

Waters, A. J., & Gobet, F. (2008). Mental imagery and chunks: Empirical and conventional findings. *Memory and Cognition*, *36*, 505–517.

Watson, J. B., & Rayner, R. (1920). Conditioned emotional reactions. *Journal of Experimental Psychology*, *3*, 1–14.

Waye, K. P., Bengtsson, J., Rylander, R., Hucklebridge, F., Evans, P., & Clow, A. (2002). Low frequency noise enhances cortisol among noise sensitive subjects during work performance. *Life Sciences*, *70*(7), 745–758.

Weaver, M. F., & Schnoll, S. H. (2008). Hallucinogens and club drugs. In M. Galanter & H. D. Kleber (Eds.), *The American Psychiatric Publishing textbook of substance abuse treatment* (4th ed.) (pp. 191–200). Arlington, VA: American Psychiatric Publishing.

Wechsler, D. (1944). *The measurement of adult intelligence* (3rd ed.). Baltimore, MD: Williams & Wilkins.

Wechsler, D. (1977). *Manual for the Wechsler Intelligence Scale for Children* (Rev.). New York, NY: Psychological Corporation.

Weinberger, D., & Harrison, P. (Eds.). (2011). *Schizophrenia* (3rd ed.). Hoboken, NJ: John Wiley & Sons.

Weiner, B. (1972). *Theories of motivation*. Chicago, IL: Rand-McNally.

Weiner, B. (1982). The emotional consequences of causal attributions. In M. S. Clark & S. T. Fiske (Eds.), *Affect and cognition*. Hillsdale, NJ: Erlbaum.

Weiner, B. (2006). *Social motivation, justice, and the moral emotions: An attributional approach*. Mahwah, NJ: Erlbaum.

Weiner, G. (2008). *Handbook of personality assessment*. Hoboken, NJ: John Wiley & Sons.

Weinstock, M. (2008). The long-term behavioural consequences of prenatal stress. *Neuroscience and Biobehavioral Reviews*, *32*, 1073–1086.

Weitlauf, J. C., Cervone, D., Smith, R. E., & Wright, P. M. (2001). Assessing-generalization in perceived self-efficacy: Multidomain and global assessments of the effects of self-defense training for women. *Personality and Social Psychology Bulletin*, *27*(12), 1683–1691.

Welch, E., Birgegård, A., Parling, T., & Ghaderi, A. (2011). Eating Disorder Examination Questionnaire and Clinical Impairment Assessment Questionnaire: General population and clinical norms for young adult women in Sweden. *Behaviour Research and Therapy*, *49*(2), 85–91.

Werner, J. S., & Wooten, B. R. (1979). Human infant color vision and color perception. *Infant Behavior and Development*, *2*(3), 241–273.

West, D. A., & Lichtenstein B. (2006). Andrea Yates and the criminalization of the filicidal maternal body. *Feminist Criminology*, *1*, 173–187.

Westen, D. (1998). Unconscious thought, feeling, and motivation: The end of a centurylong debate. In R. F. Bornstein & J. M. Masling (Eds.), *Empirical perspectives on the psychoanalytic unconscious* (pp. 1–43). Washington, DC: American Psychological Association.

Westen, D. (2007). *The political brain: The role of emotion in deciding the fate of the nation*. New York, NY: PublicAffairs.

Westover, A. N., McBride, S., & Haley, R. W. (2007). Stroke in young adults who abuse amphetamines or cocaine: A population-based study of hospitalized patients. *Archives of General Psychiatry*, *64*, 495–502.

Weyandt, L. L., Verdi, G., & Swentosky, A. (2011). Oppositional, conduct, and aggressive disorders. In S. Goldstein & C. R. Reynolds (Eds.), *Handbook of neurodevelopmental and genetic disorders in children* (2nd ed.) (pp. 151–170). New York, NY: Guilford Press.

Whaley, A., Smith, M., & Hancock, A. (2011). Ethnic/racial differences in the self-reported physical and mental health correlates of adolescent obesity. *Journal of Health Psychology*, *16*(7), 1048–1057.

Whiffen, V. E., & Demidenko, N. (2006). Mood disturbance across the life span. In J. W. C. D. Goodheart (Ed.), *Handbook of girls' and women's psychological health: Gender and well-being across the lifespan* (pp. 51–59). New York, NY: Oxford University Press.

Whitbourne, S. K. (2011). *Adult development and aging: Biopsychosocial perspectives* (4th ed.). Hoboken, NJ: John Wiley & Sons.

Whitbourne, S. K., & Mathews, M. J. (2009). Pathways in Adulthood: A counterpoint to the midlife crisis myth. *American Psychological Association Annual Convention*. Toronto, Ontario, CN.
White, T., & Hilgetag, C. C. (2011). Gyrification and neural connectivity in schizophrenia. *Development and Psychopathology*, *23*(1), 339–352.
White, T., Andreasen, N. C., & Nopoulos, P. (2002). Brain volumes and surface morphology in monozygotic twins. *Cerebral Cortex*, *12*(5), 486–493.
Whorf, B. L. (1956). Science and linguistics. In J. B. Carroll (Ed.), *Language, thought and reality*. Cambridge, MA: MIT Press.
Wickramasekera II, I. (2008a). Review of how can we help witnesses to remember more? It's an eyes open and shut case. *American Journal of Clinical Hypnosis*, *50*, 290–291.
Wickramasekera II, I. (2008b). Review of hypnotizability, absorption and negative cognitions as predictors of dental anxiety: Two pilot studies. *American Journal of Clinical Hypnosis*, *50*, 285–286.
Wieseler-Frank, J., Maier, S. F., & Watkins, L. R. (2005). Immune-to-brain communication dynamically modulates pain: Physiological and pathological consequences. *Brain, Behavior, and Immunity*, *19*(2), 104–111.
Willer, E. K., & Cupach, W. R. (2011). The meaning of girls' social aggression: Nasty or mastery? In W. R. Cupach & B. H. Spitzberg (Eds.), *The dark side of close relationships II* (pp. 297–326). New York, NY: Routledge/Taylor & Francis Group.
Williams, A. L., Haber, D., Weaver, G. D., & Freeman, J. L. (1998). Altruistic activity: Does it make a difference in the senior center? *Activities, Adaptation and Aging*, *22*(4), 31–39.
Williams, J. E. (2010). Anger/hostility and cardiovascular disease. In M. Potegal, G. Stemmler, & C. Spielberger (Eds.), *International handbook of anger: Constituent and concomitant biological, psychological, and social processes* (pp. 435–447). New York, NY: Springer Science 1 Business Media.
Williams, J. E., & Best, D. L. (1990). *Sex and psyche: Gender and self viewed cross-culturally*. Newbury Park, CA: Sage.
Williams, S. S. (2001). Sexual lying among college students in close and casual relationships. *Journal of Applied Social Psychology*, *31*(11), 2322–2338.
Williamson, A., Lombardi, D. A., Folkard, S., Stutts, J., Courtney, T. K., & Connor, J. L. (2011). The link between fatigue and safety. *Accident Analysis and Prevention*, *43*(2), 498–515.
Williamson, M. (1997). Circumcision anesthesia: A study of nursing implication for dorsal penile nerve block. *Pediatric Nursing*, *23*, 59–63.
Willingham, D.B. (2001). *Cognition: The thinking animal*. Upper Saddle River, NJ: Prentice Hall.
Willis, M. S., Esqueda, C. W., & Schacht, R. N. (2008). Social perceptions of individuals missing upper front teeth. *Perceptual and Motor Skills*, *106*, 423–435.
Wilner, B. (February 5, 2011). Vick wins Comeback Player award. *Associated Press*. Retrieved from http://news.yahoo.com/s/ap/20110206/ap_on_sp_fo_ne/fbn_comeback_player.
Wilson, E. O. (1975). *Sociobiology: The new synthesis*. Cambridge, MA: Harvard University Press.
Wilson, E. O. (1978). *On human nature*. Cambridge, MA: Harvard University Press.
Wilson, G. T. (2011). Behavior therapy. In R. J. Corsini & D. Wedding (Eds,), *Current psychotherapy* (9th ed.) (pp. 235–275). Florence, KY: Cengage Learning.
Wilson, J. P., & Hugenberg, K. (2010). When under threat, we all look the same: Distinctiveness threat induces ingroup homogeneity in face memory. *Journal of Experimental Social Psychology*, *46*(6), 1004–1010.
Wilson, R. S., Gilley, D. W., Bennett, D. A., Beckett L. A., & Evans, D. A. (2000). Person-specific paths of cognitive decline in Alzheimer's disease and their relation to age. *Psychology and Aging*, *15*(1), 18–28.
Wilson, S., & Nutt, D. (2008). *Sleep disorders*. London, England: Oxford University Press.
Winterich, J. A. (2003). Sex, menopause, and culture: Sexual orientation and the meaning of menopause for women's sex lives. *Gender and Society*, *17*(4), 627–642.
Wise, D., & Rosqvist, J. (2006). Explanatory style and well-being. In J. C. Thomas, D. L. Segal, & M. Hersen (Eds.), *Comprehensive handbook of personality and psychopathology, Vol. 1: Personality and everyday functioning* (pp. 285–305). Hoboken, NJ: John Wiley & Sons.
Wiste, A. K., Arango, V., Ellis, S. P., Mann, J. J., & Underwood, M. D. (2008). Norepinephrine and serotonin imbalance in the locus coeruleus in bipolar disorder. *Bipolar Disorders*, *10*, 349–359.
Witelson, S. F., Kigar, D. L., & Harvey, T. (1999). The exceptional brain of Albert Einstein. *The Lancet*, *353*, 2149–2153.
Wolf, G. (2008). Want to remember everything you'll ever learn? Surrender to this algorithm. *Wired Magazine*. Retrieved from http://www.wired.com/print/medtech/health/magazine/16-05/FF_Wozniak.
Wolpe, J. & Plaud, J. J. (1997). Pavlov's contributions to behavior therapy. *American Psychologist*, *52*(9), 966–972.
Woo, M., & Oei, T. P. S. (2006). The MMPI-2 Gender-Masculine and Gender-Feminine scales: Gender roles as predictors of psychological health in clinical patients. *International Journal of Psychology*, *41*, 413–422.
Wood, J. M., Nezworski, T., Lilienfeld, S. O., & Garb, H. N. (2011). *What's wrong with the Rorschach: Science confronts the controversial inkblot test*. Hoboken, NJ: Wiley/JosseyBass.
Wood, W., Christensen, P. N., Hebl, M. R., & Rothgerber, H. (1997). Conformity to sextyped norms, affect, and the self-concept. *Journal of Personality and Social Psychology*, *73*(3), 523–535.
Woodruff-Pak, D. S., & Disterhoft, J. F. (2008). Where is the trace in trace conditioning? *Trends in Neurosciences*, *31*, 105–112.
Woollams, A. M., Taylor, J. R., Karayanidis, F., & Henson, R. N. (2008). Event-related potentials associated with masked priming of test cues reveal multiple potential contributions to recognition memory. *Journal of Cognitive Neuroscience*, *20*, 1114–1129.
World Health Organization (WHO). (2011). *Depression*. Retrieved from http://www.who.int/topics/depression/en/
Wright, A. J. (2011). *Conducting psychological assessment: A guide for practitioners*. Hoboken, NJ: John Wiley & Sons.
Wright, J. H., Thase, M. E., & Beck, A. T. (2011). Cognitive therapy. In R. E. Hales, S. C. Yudofsky, & G. O. Gabbard (Eds.), *Essentials of psychiatry* (3rd ed.) (pp. 559–587). Arlington, VA: American Psychiatric Publishing.
Wynne, C. D. L. (2007). What the ape said. *Ethology*, *113*, 411–413.

Xu, L., & Barnes, L. (2011). Measurement invariance of scores from the inventory of school motivation across Chinese and U.S. college students. *International Journal of Testing*, *11*(2), 178–210.
Xu, Y., Schneier, F., Heimberg, R. G., Princisvalle, K., Liebowitz, M. R., Wang, S., & Blanco, C. (2012). Gender differences in social anxiety disorder: Results from the national epidemiologic sample on alcohol and related conditions. *Journal of Anxiety Disorders*, *26*(1), 12–19.

Yamada, H. (1997). *Different games, different rules: Why Americans and Japanese misunderstand each other*. London, England: Oxford University Press.
Yaniv, I. (2011). Group diversity and decision quality: Amplification and attenuation of the framing effect. *International Journal of Forecasting*, *27*(1), 41–49.
Yegneswaran, B., & Shapiro, C. (2007). Do sleep deprivation and alcohol have the same effects on psychomotor

performance? *Journal of Psychosomatic Research*, *63*, 569–572.
Yeung, D., Wong, C., & Lok, D. (2011). Emotion regulation mediates age differences in emotions. *Aging and Health*. *15*(3), 414–418.
You, S., Hong, S., & Ho, H. Z. (2011). Longitudinal effects of perceived control on academic achievement. *Journal of Educational Research*, *104*(4), 253–266.
Young, J. D., & Taylor, E. (1998). Meditation as a voluntary hypometabolic state of biological estivation. *News in Physiological Science*, *13*, 149–153.
Young, R. M., Connor, J. P., & Feeney, G. F. X. (2011). Alcohol expectancy changes over a 12-week cognitive–behavioral therapy program are predictive of treatment success. *Journal of Substance Abuse Treatment*, *40*(1), 18–25.
Young, T. (1802). On the theory of light and colours. *Philosophical Transactions of the Royal Society*, *92*, 12–48.
Young-Bruehl, E., & Schwartz, M. (2011). Warum die psychoanalyse keine geschichte hat (Why psychoanalysis has no history). *Psyche: Zeitschrift für Psychoanalyse und ihre Anwendungen*, *65*(2), 97–118.
Yu, X., Fumoto, M., Nakatani, Y., Sekiyama, T., Kikuchi, H., Seki, Y., Sato-Suzuki, I., & Arita, H. (2011). Activation of the anterior prefrontal cortex and serotonergic system is associated with improvements in mood and EEG changes induced by zen meditation practice in novices. *International Journal of Psychophysiology*, *80*(2), 103–111.
Zajonc, R. B. (1984). On the primacy of affect. *American Psychologist*, *39*, 117–123.
Zampieri, M., & Pedrosa de Souza, E. A. (2011). Locus of control, depression, and quality of life in Parkinson's Disease. *Journal of Health Psychology*, *16*(6), 980–987.
Zanetti, L., Picciotto, M. R., & Zoli, M. (2007). Differential effects of nicotinic antagonists perfused into the nucleus accumbens or the ventral tegmental area on cocaine-induced dopamine release in the nucleus accumbens of mice. *Psychopharmacology*, *190*, 189–199.
Zaragoza, M. S., Mitchell, K. J., Payment, K., & Drivdahl, S. (2011). False memories for suggestions: The impact of conceptual elaboration. *Journal of Memory and Language*, *64*(1), 18–31.
Zeanah, C. H. (2000). Disturbances of attachment in young children adopted from institutions. *Journal of Developmental and Behavioral Pediatrics*, *21*(3), 230–236.
Zelazo, P. D., Chandler, M., & Crone, E. (Eds.). (2010). *The Jean Piaget symposium series. Developmental social cognitive neuroscience*. New York, NY: Psychology Press.
Zerr, A. A., Holly, L. E., & Pina, A. A. (2011). Cultural influences on social anxiety in African American, Asian American, Hispanic and Latino, and Native American adolescents and young adults. In C. A. Alfano & D. C. Beidel (Eds.), *Social anxiety in adolescents and young adults: Translating developmental science into practice* (pp. 203–222). Washington, DC: American Psychological Association.
Zhang, L. F., & He, Y. F. (2011). Thinking styles and the Eriksonian stages. *Journal of Adult Development*, *18*(1), 8–17.
Zhaoping, L., & Guyader, N. (2007). Interference with bottom-up feature detection by higher-level object recognition. *Current Biology*, *17*, 26–31.
Zillmer, E. A., Spiers, M. V., & Culbertson, W. (2011). *Principles of neuropsychology* (3rd ed.). Belmont, CA: Cengage.
Zimbardo, P. (2007). *The Lucifer effect: Understanding how good people turn evil*. New York, NY: Random House.
Zimbardo, P. G. (1993). Stanford prison experiment: A 20-year retrospective. *Invited presentation at the meeting of the Western Psychological Association*, Phoenix, AZ.
Zimbardo, P. G., Ebbeson, E. B., & Maslach, C. (1977). *Influencing attitudes and changing behavior*. Reading, MA: Addison-Wesley.
Zuckerman, M. (1978, February). The search for high sensation. *Psychology Today*, 38–46.
Zuckerman, M. (1979). *Sensation seeking: Beyond the optimal level of arousal*. Hillsdale, NJ: Erlbaum.
Zuckerman, M. (1994). *Behavioral expressions and biosocial bases of sensation seeking*. New York, NY: Cambridge University Press.
Zuckerman, M. (2004). The shaping of personality: Genes, environments, and chance encounters. *Journal of Personality Assessment*, *82*(1), 11–22.
Zuckerman, M. (2008). Rose is a rose is a rose: Content and construct validity. *Personality and Individual Differences*, *45*, 110–112.
Zuckerman, M. (2011). Psychodynamic approaches. In M. Zuckerman (Ed.), *Personality science: Three approaches and their applications to the causes and treatment of depression* (pp. 11–45). Washington, DC: American Psychological Association.
Zweig, J. (2007). *Your money and your brain: How the new science of neuroeconomics can help make you rich*. New York, NY: Simon & Schuster.

D



E



F

## G



## H

## I



## J



## K

## L

## M



## N

## S

## T



## U



## V



## W

## X



## Y



## Z

# Subject Index

Note: Page numbers followed by a "t" indicate the entry may be found within a table.
Page numbers followed by an "f" indicate the entry may be found within a figure.

## C

## F



## G

## H



## I



## J

## N



## O



## P

## R



## S

## T

## U



## V



## W



## X



## Y



## Z